PIC MICROCONTROLLER AND EMBEDDED SYSTEMS

Using Assembly and C for PIC18

Muhammad Ali Mazidi
Rolin D. McKinlay
Danny Causey

PEARSON
Prentice
Hall

Upper Saddle River, New Jersey
Columbus, Ohio

Library of Congress Cataloging in Publication Data

Mazidi, Muhammad Ali.

   PIC microcontroller and embedded systems : using Assembly and C for PIC 18 /
Muhammad Ali Mazidi, Rolin D. McKinlay, Danny Causey.

      p. cm.
   ISBN-13: 978-0-13-119404-5
   ISBN-10: 0-13-119404-6
   1.Programmable controllers. 2. Embedded computer systems. I. McKinlay, Rolin D.
II. Causey, Danny. III. Title.

TJ223.P76M379 2008
629.8'95-dc22

                                                                    2006053262

Editor-in-Chief: Vernon Anthony
Executive Editor: Jeff Riley
Editorial Assistant: Lara Dimmick
Production Editor: Rex Davidson
Production Manager: Matt Ottenweller
Design Coordinator: Diane Ernsberger
Cover Designer: Thomas Mack
Cover Art: Getty Images
Director of Marketing: David Gesell
Marketing Manager: Ben Leonard
Marketing Assistant: Les Roberts

This book was set in Times Roman by M. Mazidi, Rolin McKinlay and Danny Causey. It was printed and bound by Courier Kendallville, Inc. The cover was printed by Coral Graphic Services, Inc.

Pearson Education Ltd.   Pearson Education Australia Pty. Limited
Pearson Education Singapore Pte. Ltd.   Pearson Education North Asia Ltd.
Pearson Education Canada, Ltd.   Pearson Educación de Mexico, S.A. de C.V.
Pearson Education—Japan   Pearson Education Malaysia Pte. Ltd.

10 9 8 7 6 5 4 3 2 1

ISBN-13: 978-0-13-119404-5
ISBN-10: 0-13-119404-6

Trademark Information and Acknowledgments

Regard man as a mine rich in gems of inestimable value. Education can, alone, cause it to reveal its treasures, and enable mankind to benefit therefrom.

Baha'u'llah

BRIEF CONTENTS

CHAPTERS

APPENDICES

CONTENTS

INTRODUCTION

Products using microprocessors generally fall into two categories. The first category uses high-performance microprocessors such as the Pentium in applications where system performance is critical. We have an entire book dedicated to this topic, *The 80x86 IBM PC and Compatible Computers, Volumes I and II*, from Prentice Hall. In the second category of applications, performance is secondary; issues of cost, space, power, and rapid development are more critical than raw processing power. The microprocessor for this category is often called a microcontroller.

This book is for the second category of applications. The PIC18 is a widely used microcontroller. There are many reasons for this, including the existence of massive support in both software and hardware by Microchip Technology. This book is intended for use in college-level courses teaching microcontrollers and embedded systems. It not only establishes a foundation of Assembly language programming, but also provides a comprehensive treatment of PIC18 interfacing for engineering students. From this background, the design and interfacing of microcontroller-based embedded systems can be explored. This book can also be used by practicing technicians, hardware engineers, computer scientists, and hobbyists. It is an ideal source for those building stand-alone projects, or projects in which data is collected and fed into a PC for distribution on a network.

Prerequisites

Readers should have had an introductory digital course. Knowledge of Assembly language would be helpful but is not necessary. Although the book is written for those with no background in Assembly language programming, students with prior Assembly language experience will be able to gain a mastery of PIC18 architecture very rapidly and start on their projects right away. For the PIC18 C programming sections of the book, a basic knowledge of C programming is required. We use the PIC18 C compiler from Microchip Technology throughout the book. The PIC18 C compiler is compatible with MPLAB and is available for free from the Microchip website (www.microchip.com). We encourage you to use the MPLAB to simulate and run the programs in this book.

Overview

A systematic, step-by-step approach is used to cover various aspects of PIC18 C and Assembly language programming and interfacing. Many examples and sample programs are given to clarify the concepts and provide students with an opportunity to learn by doing. Review questions are provided at the end of each section to reinforce the main points of the section.

Chapter 0 covers number systems (binary, decimal, and hex), and provides an introduction to basic logic gates and computer terminology. This is designed especially for students, such as mechanical engineering students who have not taken a digital logic course or those who need to refresh their memory on these topics.

Chapter 1 discusses the history of the PIC18 and features of other PIC family members such as the PIC16. It also provides a list of various members of the PIC18 family.

Chapter 2 discusses the internal architecture of the PIC18 and explains the use of a PIC18 assembler to create ready-to-run programs. It also explores the stack and the flag register.

In Chapter 3 the topics of loop, jump, and call instructions are discussed, with many programming examples.

Chapter 4 is dedicated to the discussion of I/O ports. This allows students who are working on a project to start experimenting with PIC18 I/O interfacing and start the project as soon as possible.

Chapter 5 is dedicated to arithmetic, logic instructions, and programs.

Chapter 6 covers the PIC18 addressing modes and explains how to access the data stored in the code space of the PIC18, as well as how to do bank switching.

The C programming of the PIC18 is covered in Chapter 7. We use the PIC18 C compiler from Microchip Technology for this and other C programs of the PIC18 family throughout the book. The PIC18 C compiler is compatible with MPLAB and is available for free from the www.microchip.com website.

In Chapter 8 we discuss the hardware connection of the PIC18 chip.

Chapter 9 describes the PIC18 timers and how to use them as event counters.

Chapter 10 is dedicated to serial data communication of the PIC18 and its interfacing to the RS232. It also shows PIC18 communication with COM ports of the x86 IBM PC and compatible computers.

Chapter 11 provides a detailed discussion of PIC18 interrupts with many examples on how to write interrupt handler programs.

Chapter 12 shows PIC18 interfacing with real-world devices such as LCDs and keyboards.

Chapter 13 shows PIC18 interfacing with real-world devices such as DAC chips, ADC chips, and sensors.

In Chapter 14 we cover how to use PIC18 Flash and EEPROM memories for data storage.

Chapter 15 covers the CCP and ECCP modules inside the PIC18 and shows how they are used.

Chapter 16 shows how to connect and program the DS1306 real-time clock chip using the SPI bus protocol.

Finally, Chapter 17 shows basic interfacing to relays, optoisolators, and motors.

The appendices have been designed to provide all reference material required for the topics covered in the book. Appendix A describes each PIC18 instruction in detail, with examples. Appendix B describes basics of wire wrapping. Appendix C covers IC technology and logic families, as well as PIC18 I/O port interfacing and fan-out. Make sure you study this before connecting the PIC18 to an external device. In Appendix D, the use of flowcharts and psuedocode is explored. Appendix E is for students familiar with x86 and 8051 architectures who need to make a rapid transition to PIC18 architecture. Appendix F provides the table of ASCII characters. Appendix G lists resources for assembler shareware, and electronics parts. Appendix H contains data sheets for the PIC18 chip.

Lab Manual

The lab manual covers some very basic labs and can be found at the **www.MicroDigitalEd.com** website. The more advanced and rigorous lab assignments are left up to the instructors depending on the course objectives, class level, and whether the course is graduate or undergraduate. The support materials for this and other books by the authors can be found on this website, too.

Solutions Manual/PowerPoint® Slides

The end-of-chapter problems cover some very basic concepts. The more challenging and rigorous homework assignments are left up to the instructors depending on the course objectives, class level, and whether the course is graduate or undergraduate. The solutions manual was produced with the help of Mr. Rasti and Prof. Faramarz Mortezae. The solutions manual and PowerPoint® slides for the drawings are available online for instructors only.

Online Instructor Resources

To access supplementary materials online, instructors need to request an instructor access code. Go to **www.prenhall.com**, click the **Instructor Resource Center** link, and then click **Register Today** for an instructor access code. Within 48 hours after registering you will receive a confirming e-mail including an instructor access code. Once you have received your code, go to the site and log on for full instructions on downloading the materials you wish to use.

Acknowledgments

This book is the result of the dedication and encouragement of many individuals. Our sincere and heartfelt appreciation goes to all of them.

First, we would like to thank Mr. Javad Rasti of Esfahan University. His detailed and thorough reading of the chapters resulted in finding and fixing some of the errors before the book was published. Many of the drawings and tables in this book were recreated from PIC18 data sheets by Pedram Mazidi. Numerous professors, professional engineers, and students found errors or made suggestions in improving this book. We would like to thank all of them sincerely for their enthusiasm and support. They are Javad Rasti (Esfahan University), Vahid Mokhtari (BIHE), Mohammadi Abdar (Azad University), Clyde Knight, Sam Waterman, and Faramarz Mortezaei (all from DeVry University), Frank Fortman, David Goodman, and Maryam Mohseni. Their encouragement meant a great deal to us in writing this book.

Thanks to the reviewers of this edition:

Shujen Chen, DeVry University – Tinley Park;
Lawrence Lam, DeVry University – Federal Way;
Vahid Mokhtari, BIHE University;
Faramarz Mortezaie, DeVry University – Fremont;
Sepehr Naimi,   BIHE University;
Javad Rasti, Esfahan University; and
Chao-Yin Wang, DeVry University – North Brunswick.

Finally, we would like to thank the people at Prentice Hall, in particular our editor Jeff Riley, who continues to support and encourage our writing, and our production editor Rex Davidson, who made the book a reality. We were lucky to get the best copy editors in the world, Janice Mazidi and Bret Workman. Thank you both for your fantastic job, as usual

We enjoyed writing this book, and hope you enjoy reading it and using it for your courses and projects. Please let us know if you have any suggestions or find any errors.

Assemblers/Compiler

The MPLAB and PIC18 C compilers can be downloaded from the following website:

http://www.microchip.com

ABOUT THE AUTHORS

Muhammad Ali Mazidi went to Tabriz University and holds Master's degrees from both Southern Methodist University and the University of Texas at Dallas. He is currently a.b.d. on his Ph.D. in the Electrical Engineering Department of Southern Methodist University. He is co-author of some widely used textbooks, including *The 80x86 IBM PC and Compatible Computers* and *The 8051 Microcontroller and Embedded Systems,* also available from Prentice Hall. He teaches microprocessor-based system design at DeVry University in Dallas, Texas. He is the founder of MicroDigitalEd.com.

Rolin McKinlay has a BSEET from DeVry University. He is co-author of *The 8051 Microcontroller and Embedded Systems.* He is working on his Master's degree and PE license in the state of Texas. He is currently self-employed as a senior embedded engineer and hardware designer, and is a partner in MicroDigitalEd.com.

Danny Causey is a U.S. Army veteran having served in Germany and Iraq. He graduated from the CET department of DeVry University. His areas of interest include networking, game development, and microcontroller and FPGA embedded system design. He is a partner in MicroDigitalEd.com.

The authors can be contacted at the following e-mail addresses if you have any comments or suggestions, or if you find any errors.

mdebooks@yahoo.com
mmazidi@microdigitaled.com
rmckinlay@microdigitaled.com
dcausey@microdigitaled.com

*This book is dedicated
to the memory of Mr. N. Akhtar-Khavari and Mr. Z. Mahrami
for their dedication to the cause of world peace.*
– Muhammad Ali Mazidi

To Tony and Jim for their friendship and faith in me over the years.
– Rolin D. McKinlay

*I dedicate my part to my brother John, who reached out to me even though we
lived in different homes. The experience that was given provided me the
inspiration to look for something more in life.*
– Danny Causey

CHAPTER 0

INTRODUCTION TO COMPUTING

OBJECTIVES

Upon completion of this chapter, you will be able to:

>> Convert any number from base 2, base 10, or base 16 to any of the other two bases
>> Add and subtract hex numbers
>> Add binary numbers
>> Represent any binary number in 2's complement
>> Represent an alphanumeric string in ASCII code
>> Describe the logical operations AND, OR, NOT, XOR, NAND, and NOR
>> Use logic gates to diagram simple circuits
>> Explain the difference between a bit, a nibble, a byte, and a word
>> Give precise mathematical definitions of the terms *kilobyte*, *megabyte*, *gigabyte*, and *terabyte*
>> Explain the difference between RAM and ROM and describe their use
>> Describe the purpose of the major components of a computer system
>> List the three types of buses found in computers and describe the purpose of each type of bus
>> Describe the role of the CPU in computer systems
>> List the major components of the CPU and describe the purpose of each

To understand the software and hardware of a microcontroller-based system, one must first master some very basic concepts underlying computer design. In this chapter (which in the tradition of digital computers is called Chapter 0), the fundamentals of numbering and coding systems are presented. After an introduction to logic gates, an overview of the workings inside the computer is given. Finally, in the last section we give a brief history of CPU architecture. Although some readers may have an adequate background in many of the topics of this chapter, it is recommended that the material be scanned, however briefly.

SECTION 0.1: NUMBERING AND CODING SYSTEMS

Whereas human beings use base 10 (*decimal*) arithmetic, computers use the base 2 (*binary*) system. In this section we explain how to convert from the decimal system to the binary system, and vice versa. The convenient representation of binary numbers, called *hexadecimal,* also is covered. Finally, the binary format of the alphanumeric code, called *ASCII*, is explored.

Decimal and binary number systems

Although there has been speculation that the origin of the base 10 system is the fact that human beings have 10 fingers, there is absolutely no speculation about the reason behind the use of the binary system in computers. The binary system is used in computers because 1 and 0 represent the two voltage levels of on and off. Whereas in base 10 there are 10 distinct symbols, 0, 1, 2, ..., 9, in base 2 there are only two, 0 and 1, with which to generate numbers. Base 10 contains digits 0 through 9; binary contains digits 0 and 1 only. These two binary digits, 0 and 1, are commonly referred to as *bits*.

Converting from decimal to binary

One method of converting from decimal to binary is to divide the decimal number by 2 repeatedly, keeping track of the remainders. This process continues until the quotient becomes zero. The remainders are then written in reverse order to obtain the binary number. This is demonstrated in Example 0-1.

Example 0-1

Convert 25_{10} to binary.

Solution:

	Quotient	Remainder	
25/2 =	12	1	LSB (least significant bit)
12/2 =	6	0	
6/2 =	3	0	
3/2 =	1	1	
1/2 =	0	1	MSB (most significant bit)

Therefore, $25_{10} = 11001_2$.

Converting from binary to decimal

To convert from binary to decimal, it is important to understand the concept of weight associated with each digit position. First, as an analogy, recall the weight of numbers in the base 10 system, as shown in the diagram. By the same token, each digit position of a number in base 2 has a weight associated with it:

$$740683_{10} \quad =$$

3×10^0	=		3
8×10^1	=		80
6×10^2	=		600
0×10^3	=		0000
4×10^4	=		40000
7×10^5	=		700000
			740683

$$110101_2 \quad =$$

					Decimal	Binary
1×2^0	=	1×1	=		1	1
0×2^1	=	0×2	=		0	00
1×2^2	=	1×4	=		4	100
0×2^3	=	0×8	=		0	0000
1×2^4	=	1×16	=		16	10000
1×2^5	=	1×32	=		32	100000
					53	110101

Knowing the weight of each bit in a binary number makes it simple to add them together to get its decimal equivalent, as shown in Example 0-2.

Example 0-2

Convert 11001_2 to decimal.

Solution:

Weight:	16	8	4	2	1
Digits:	1	1	0	0	1
Sum:	16 +	8 +	0 +	0 +	1 = 25_{10}

Knowing the weight associated with each binary bit position allows one to convert a decimal number to binary directly instead of going through the process of repeated division. This is shown in Example 0-3.

Example 0-3

Use the concept of weight to convert 39_{10} to binary.

Solution:

Weight:	32	16	8	4	2	1
	1	0	0	1	1	1
	32 +	0 +	0 +	4 +	2 +	1 = 39

Therefore, $39_{10} = 100111_2$.

Hexadecimal system

Base 16, or the *hexadecimal* system as it is called in computer literature, is used as a convenient representation of binary numbers. For example, it is much easier for a human being to represent a string of 0s and 1s such as 100010010110 as its hexadecimal equivalent of 896H. The binary system has 2 digits, 0 and 1. The base 10 system has 10 digits, 0 through 9. The hexadecimal (base 16) system has 16 digits. In base 16, the first 10 digits, 0 to 9, are the same as in decimal, and for the remaining six digits, the letters A, B, C, D, E, and F are used. Table 0-1 shows the equivalent binary, decimal, and hexadecimal representations for 0 to 15.

Table 0-1: Base 16 Number System

Decimal	Binary	Hex
0	0000	0
1	0001	1
2	0010	2
3	0011	3
4	0100	4
5	0101	5
6	0110	6
7	0111	7
8	1000	8
9	1001	9
10	1010	A
11	1011	B
12	1100	C
13	1101	D
14	1110	E
15	1111	F

Converting between binary and hex

To represent a binary number as its equivalent hexadecimal number, start from the right and group 4 bits at a time, replacing each 4-bit binary number with its hex equivalent shown in Table 0-1. To convert from hex to binary, each hex digit is replaced with its 4-bit binary equivalent. See Examples 0-4 and 0-5.

Example 0-4

Represent binary 100111110101 in hex.

Solution:
First the number is grouped into sets of 4 bits: 1001 1111 0101.
Then each group of 4 bits is replaced with its hex equivalent:

       1001   1111   0101
        9   F   5

Therefore, 100111110101_2 = 9F5 hexadecimal.

Example 0-5

Convert hex 29B to binary.

Solution:

        2   9   B
=   0010 1001 1011
Dropping the leading zeros gives 1010011011.

Converting from decimal to hex

Converting from decimal to hex could be approached in two ways:
1. Convert to binary first and then convert to hex. Example 0-6 shows this method of converting decimal to hex.
2. Convert directly from decimal to hex by repeated division, keeping track of the remainders. Experimenting with this method is left to the reader.

Example 0-6

(a) Convert 45_{10} to hex.

32	16	8	4	2	1
1	0	1	1	0	1

First, convert to binary.
$32 + 8 + 4 + 1 = 45$

$45_{10} = 0010\ 1101_2 = 2D$ hex

(b) Convert 629_{10} to hex.

512	256	128	64	32	16	8	4	2	1
1	0	0	1	1	1	0	1	0	1

$629_{10} = (512 + 64 + 32 + 16 + 4 + 1) = 0010\ 0111\ 0101_2 = 275$ hex

(c) Convert 1714_{10} to hex.

1024	512	256	128	64	32	16	8	4	2	1
1	1	0	1	0	1	1	0	0	1	0

$1714_{10} = (1024 + 512 + 128 + 32 + 16 + 2) = 0110\ 1011\ 0010_2 = 6B2$ hex

Converting from hex to decimal

Conversion from hex to decimal can also be approached in two ways:
1. Convert from hex to binary and then to decimal. Example 0-7 demonstrates this method of converting from hex to decimal.
2. Convert directly from hex to decimal by summing the weight of all digits.

Example 0-7

Convert the following hexadecimal numbers to decimal.

(a) $6B2_{16} = 0110\ 1011\ 0010_2$

1024	512	256	128	64	32	16	8	4	2	1
1	1	0	1	0	1	1	0	0	1	0

$1024 + 512 + 128 + 32 + 16 + 2 = 1714_{10}$

(b) $9F2D_{16} = 1001\ 1111\ 0010\ 1101_2$

32768	16384	8192	4096	2048	1024	512	256	128	64	32	16	8	4	2	1
1	0	0	1	1	1	1	1	0	0	1	0	1	1	0	1

$32768 + 4096 + 2048 + 1024 + 512 + 256 + 32 + 8 + 4 + 1 = 40,749_{10}$

Table 0-2: Counting in Bases

Decimal	Binary	Hex
0	00000	0
1	00001	1
2	00010	2
3	00011	3
4	00100	4
5	00101	5
6	00110	6
7	00111	7
8	01000	8
9	01001	9
10	01010	A
11	01011	B
12	01100	C
13	01101	D
14	01110	E
15	01111	F
16	10000	10
17	10001	11
18	10010	12
19	10011	13
20	10100	14
21	10101	15
22	10110	16
23	10111	17
24	11000	18
25	11001	19
26	11010	1A
27	11011	1B
28	11100	1C
29	11101	1D
30	11110	1E
31	11111	1F

Counting in bases 10, 2, and 16

To show the relationship between all three bases, in Table 0-2 we show the sequence of numbers from 0 to 31 in decimal, along with the equivalent binary and hex numbers. Notice in each base that when one more is added to the highest digit, that digit becomes zero and a 1 is carried to the next-highest digit position. For example, in decimal, $9 + 1 = 0$ with a carry to the next-highest position. In binary, $1 + 1 = 0$ with a carry; similarly, in hex, $F + 1 = 0$ with a carry.

Table 0-3: Binary Addition

A + B	Carry	Sum
0 + 0	0	0
0 + 1	0	1
1 + 0	0	1
1 + 1	1	0

Addition of binary and hex numbers

The addition of binary numbers is a very straightforward process. Table 0-3 shows the addition of two bits. The discussion of subtraction of binary numbers is bypassed since all computers use the addition process to implement subtraction. Although computers have adder circuitry, there is no separate circuitry for subtractors. Instead, adders are used in conjunction with *2's complement* circuitry to perform subtraction. In other words, to implement "$x - y$", the computer takes the 2's complement of y and adds it to x. The concept of 2's complement is reviewed next. Example 0-8 shows the addition of binary numbers.

Example 0-8

Add the following binary numbers. Check against their decimal equivalents.

Solution:

	Binary	*Decimal*
	1101	13
+	1001	9
	10110	22

2's complement

To get the 2's complement of a binary number, invert all the bits and then add 1 to the result. Inverting the bits is simply a matter of changing all 0s to 1s and 1s to 0s. This is called the *1's complement*. See Example 0-9.

Example 0-9

Take the 2's complement of 10011101.

Solution:

	10011101	binary number
	01100010	1's complement
+	1	
	01100011	2's complement

Addition and subtraction of hex numbers

In studying issues related to software and hardware of computers, it is often necessary to add or subtract hex numbers. Mastery of these techniques is essential. Hex addition and subtraction are discussed separately below.

Addition of hex numbers

This section describes the process of adding hex numbers. Starting with the least significant digits, the digits are added together. If the result is less than 16, write that digit as the sum for that position. If it is greater than 16, subtract 16 from it to get the digit and carry 1 to the next digit. The best way to explain this is by example, as shown in Example 0-10.

Example 0-10

Perform hex addition: 23D9 + 94BE.

Solution:

	23D9	LSD: $9 + 14 = 23$	$23 - 16 = 7$ with a carry
+	94BE	$1 + 13 + 11 = 25$	$25 - 16 = 9$ with a carry
	B897	$1 + 3 + 4 = 8$	
		MSD: $2 + 9 = B$	

Subtraction of hex numbers

In subtracting two hex numbers, if the second digit is greater than the first, borrow 16 from the preceding digit. See Example 0-11.

ASCII code

The discussion so far has revolved around the representation of number systems. Because all information in the computer must be represented by 0s and 1s, binary patterns must be assigned to letters and other characters. In the 1960s a standard representation called *ASCII* (American Standard Code for Information Interchange) was established. The ASCII (pronounced "ask-E") code assigns bina-

---

ry patterns for numbers 0 to 9, all the letters of the English alphabet, both uppercase (capital) and lowercase, and many control codes and punctuation marks. The great advantage of this system is that it is used by most computers, so that information can be shared among computers. The ASCII system uses a total of 7 bits to represent each code. For example, 100 0001 is assigned to the uppercase letter "A" and 110 0001 is for the

Hex	Symbol	Hex	Symbol
41	A	61	a
42	B	62	b
43	C	63	c
44	D	64	d
...	...	...	...
59	Y	79	y
5A	Z	7A	z

Figure 0-1. Selected ASCII Codes

lowercase "a". Often, a zero is placed in the most-significant bit position to make it an 8-bit code. Figure 0-1 shows selected ASCII codes. A complete list of ASCII codes is given in Appendix F. The use of ASCII is not only standard for keyboards used in the United States and many other countries but also provides a standard for printing and displaying characters by output devices such as printers and monitors.

Notice that the pattern of ASCII codes was designed to allow for easy manipulation of ASCII data. For example, digits 0 through 9 are represented by ASCII codes 30 through 39. This enables a program to easily convert ASCII to decimal by masking off the "3" in the upper nibble. Also notice that there is a relationship between the uppercase and lowercase letters. The uppercase letters are represented by ASCII codes 41 through 5A while lowercase letters are represented by codes 61 through 7A. Looking at the binary code, the only bit that is different between the uppercase "A" and lowercase "a" is bit 5. Therefore, conversion between uppercase and lowercase is as simple as changing bit 5 of the ASCII code.

Example 0-11

Perform hex subtraction: 59F – 2B8.

Solution:

$$\begin{array}{r} 59F \\ -\ 2B8 \\ \hline 2E7 \end{array}$$

LSD: 8 from 15 = 7
11 from 25 (9 + 16) = 14 (E)
2 from 4 (5 – 1) = 2

Review Questions

1. Why do computers use the binary number system instead of the decimal system?
2. Convert 34_{10} to binary and hex.
3. Convert 110101_2 to hex and decimal.
4. Perform binary addition: 101100 + 101.
5. Convert 101100_2 to its 2's complement representation.
6. Add 36BH + F6H.
7. Subtract 36BH – F6H.
8. Write "80x86 CPUs" in its ASCII code (in hex form).

SECTION 0.2: DIGITAL PRIMER

This section gives an overview of digital logic and design. First, we cover binary logic operations, then we show gates that perform these functions. Next, logic gates are put together to form simple digital circuits. Finally, we cover some logic devices commonly found in microcontroller interfacing.

Binary logic

As mentioned earlier, computers use the binary number system because the two voltage levels can be represented as the two digits 0 and 1. Signals in digital electronics have two distinct voltage levels. For example, a system may define 0 V as logic 0 and +5 V as logic 1. Figure 0-2 shows this system with the built-in tolerances for variations in the voltage. A valid digital signal in this example should be within either of the two shaded areas.

Figure 0-2. Binary Signals

Logic gates

Binary logic gates are simple circuits that take one or more input signals and send out one output signal. Several of these gates are defined below.

AND gate

The AND gate takes two or more inputs and performs a logic AND on them. See the truth table and diagram of the AND gate. Notice that if both inputs to the AND gate are 1, the output will be 1. Any other combination of inputs will give a 0 output. The example shows two inputs, x and y. Multiple outputs are also possible for logic gates. In the case of AND, if all inputs are 1, the output is 1. If any input is 0, the output is 0.

OR gate

The OR logic function will output a 1 if one or more inputs is 1. If all inputs are 0, then and only then will the output be 0.

Tri-state buffer

A buffer gate does not change the logic level of the input. It is used to isolate or amplify the signal.

Logical AND Function

Inputs	Output
X Y	X AND Y
0 0	0
0 1	0
1 0	0
1 1	1

X ——⊐D— X AND Y
Y ——

Logical OR Function

Inputs	Output
X Y	X OR Y
0 0	0
0 1	1
1 0	1
1 1	1

X ——⊐D— X OR Y
Y ——

Buffer

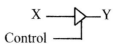

X ——▷—Y
Control ——

Inverter

The inverter, also called NOT, outputs the value opposite to that input to the gate. That is, a 1 input will give a 0 output, while a 0 input will give a 1 output.

XOR gate

The XOR gate performs an exclusive-OR operation on the inputs. Exclusive-OR produces a 1 output if one (but only one) input is 1. If both operands are 0, the output is 0. Likewise, if both operands are 1, the output is also 0. Notice from the XOR truth table, that whenever the two inputs are the same, the output is 0. This function can be used to compare two bits to see if they are the same.

NAND and NOR gates

The NAND gate functions like an AND gate with an inverter on the output. It produces a 0 output when all inputs are 1; otherwise, it produces a 1 output. The NOR gate functions like an OR gate with an inverter on the output. It produces a 1 if all inputs are 0; otherwise, it produces a 0. NAND and NOR gates are used extensively in digital design because they are easy and inexpensive to fabricate. Any circuit that can be designed with AND, OR, XOR, and INVERTER gates can be implemented using only NAND and NOR gates. A simple example of this is given below. Notice in NAND, that if any input is 0, the output is 1. Notice in NOR, that if any input is 1, the output is 0.

Logic design using gates

Next we will show a simple logic design to add two binary digits. If we add two binary digits there are four possible outcomes:

	Carry	Sum
0 + 0 =	0	0
0 + 1 =	0	1
1 + 0 =	0	1
1 + 1 =	1	0

Logical Inverter

Input	Output
X	NOT X
0	1
1	0

X —▷o— NOT X

Logical XOR Function

Inputs	Output
X Y	X XOR Y
0 0	0
0 1	1
1 0	1
1 1	0

X, Y — X XOR Y

Logical NAND Function

Inputs	Output
X Y	X NAND Y
0 0	1
0 1	1
1 0	1
1 1	0

X, Y — X NAND Y

Logical NOR Function

Inputs	Output
X Y	X NOR Y
0 0	1
0 1	0
1 0	0
1 1	0

X, Y — X NOR Y

Notice that when we add 1 + 1 we get 0 with a carry to the next higher place. We will need to determine the sum and the carry for this design. Notice that the sum column above matches the output for the XOR function, and that the carry column matches the output for the AND function. Figure 0-3(a) shows a simple adder implemented with XOR and AND gates. Figure 0-3(b) shows the same logic circuit implemented with AND and OR gates and inverters.

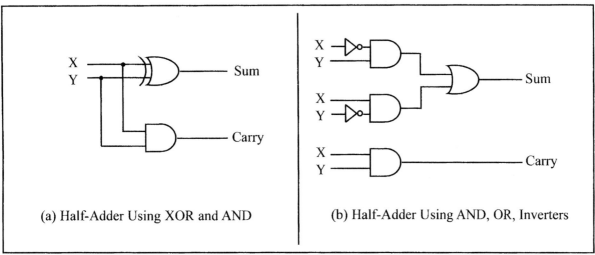

(a) Half-Adder Using XOR and AND   (b) Half-Adder Using AND, OR, Inverters

Figure 0-3. Two Implementations of a Half-Adder

Figure 0-4 shows a block diagram of a half-adder. Two half-adders can be combined to form an adder that can add three input digits. This is called a full-adder. Figure 0-5 shows the logic diagram of a full-adder, along with a block diagram that masks the details of the circuit. Figure 0-6 shows a 3-bit adder using three full-adders.

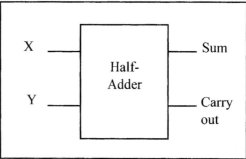

Figure 0-4. Block Diagram of a Half-Adder

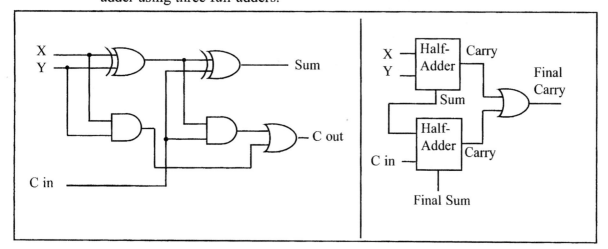

Figure 0-5. Full-Adder Built From a Half-Adder

Decoders

Another example of the application of logic gates is the decoder. Decoders are widely used for address decoding in computer design. Figure 0-7 shows decoders for 9 (1001 binary) and 5 (0101) using inverters and AND gates.

Flip-flops

A widely used component in digital systems is the flip-flop. Frequently, flip-flops are used to store data. Figure 0-8 shows the logic diagram, block diagram, and truth table for a flip-flop.

The D flip-flop is widely used to latch data. Notice from the truth table that a D-FF grabs the data at the input as the clock is activated. A D-FF holds the data as long as the power is on.

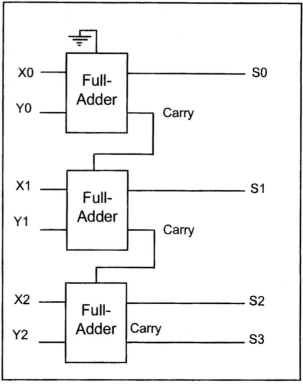

Figure 0-6. 3-Bit Adder Using Three Full-Adders

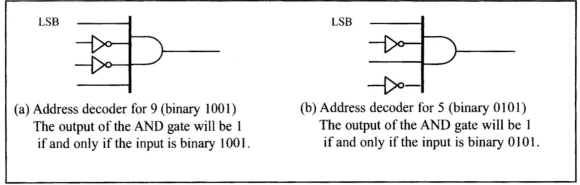

(a) Address decoder for 9 (binary 1001)
The output of the AND gate will be 1
if and only if the input is binary 1001.

(b) Address decoder for 5 (binary 0101)
The output of the AND gate will be 1
if and only if the input is binary 0101.

Figure 0-7. Address Decoders

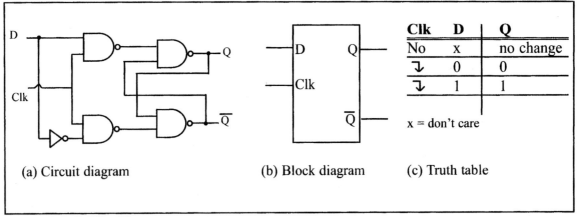

Clk	D	Q
No	x	no change
⤵	0	0
⤵	1	1

x = don't care

(a) Circuit diagram   (b) Block diagram   (c) Truth table

Figure 0-8. D Flip-Flops

Review Questions

1. The logical operation _____ gives a 1 output when all inputs are 1.
2. The logical operation _____ gives a 1 output when one or more of its inputs is 1.
3. The logical operation _____ is often used to compare two inputs to determine whether they have the same value.
4. A _____ gate does not change the logic level of the input.
5. Name a common use for flip-flops.
6. An address _____ is used to identify a predetermined binary address.

SECTION 0.3: INSIDE THE COMPUTER

In this section we provide an introduction to the organization and internal working of computers. The model used is generic, but the concepts discussed are applicable to all computers, including the IBM PC, PS/2, and compatibles. Before embarking on this subject, it will be helpful to review definitions of some of the most widely used terminology in computer literature, such as *K, mega, giga, byte, ROM, RAM,* and so on.

Some important terminology

One of the most important features of a computer is how much memory it has. Next we review terms used to describe amounts of memory in IBM PCs and compatibles. Recall from the discussion above that a *bit* is a binary digit that can have the value 0 or 1. A *byte* is defined as 8 bits. A *nibble* is half a byte, or 4 bits. A *word* is two bytes, or 16 bits. The display is intended to show the relative size of these units.

```
Bit                            0
Nibble                      0000
Byte               0000 0000
Word  0000 0000 0000 0000
```

Of course, they could all be composed of any combination of zeros and ones.

A *kilobyte* is 2^{10} bytes, which is 1024 bytes. The abbreviation K is often used to represent kilobytes. For example, some floppy disks hold 356K of data. A *megabyte*, or *meg* as some call it, is 2^{20} bytes. That is a little over 1 million bytes; it is exactly 1,048,576 bytes. Moving rapidly up the scale in size, a *gigabyte* is 2^{30} bytes (over 1 billion), and a *terabyte* is 2^{40} bytes (over 1 trillion). As an example of how some of these terms are used, suppose that a given computer has 16 megabytes of memory. That would be 16×2^{20}, or $2^4 \times 2^{20}$, which is 2^{24}. Therefore 16 megabytes is 2^{24} bytes.

Two types of memory commonly used in microcomputers are *RAM*, which stands for "random access memory" (sometimes called *read/write memory*), and *ROM*, which stands for "read-only memory." RAM is used by the computer for temporary storage of programs that it is running. That data is lost when the computer is turned off. For this reason, RAM is sometimes called *volatile memory*. ROM contains programs and information essential to operation of the computer. The information in ROM is permanent, cannot be changed by the user, and is not lost when the power is turned off. Therefore, it is called *nonvolatile memory*.

Internal organization of computers

The internal working of every computer can be broken down into three parts: CPU (central processing unit), memory, and I/O (input/output) devices (see Figure 0-9). The function of the CPU is to execute (process) information stored in memory. The function of I/O devices such as the keyboard and video monitor is to provide a means of communicating with the CPU. The CPU is connected to memory and I/O through strips of wire called a *bus*. The bus inside a computer carries information from place to place just as a street bus carries people from place to place. In every computer there are three types of buses: address bus, data bus, and control bus.

For a device (memory or I/O) to be recognized by the CPU, it must be assigned an address. The address assigned to a given device must be unique; no two devices are allowed to have the same address. The CPU puts the address (in binary, of course) on the address bus, and the decoding circuitry finds the device. Then the CPU uses the data bus either to get data from that device or to send data to it. The control buses are used to provide read or write signals to the device to indicate if the CPU is asking for information or sending information. Of the three buses, the address bus and data bus determine the capability of a given CPU.

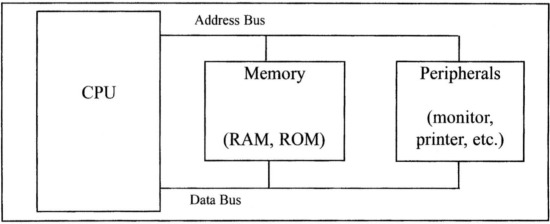

Figure 0-9. Inside the Computer

More about the data bus

Because data buses are used to carry information in and out of a CPU, the more data buses available, the better the CPU. If one thinks of data buses as highway lanes, it is clear that more lanes provide a better pathway between the CPU and its external devices (such as printers, RAM, ROM, etc.; see Figure 0-10). By the same token, that increase in the number of lanes increases the cost of construction. More data buses mean a more expensive CPU and computer. The average size of data buses in CPUs varies between 8 and 64. Early personal computers such as Apple 2 used an 8-bit data bus, while supercomputers such as Cray use a 64-bit data bus. Data buses are bidirectional, because the CPU must use them either to receive or to send data. The processing power of a computer is related to the size of its buses, because an 8-bit bus can send out 1 byte a time, but a 16-bit bus can send out 2 bytes at a time, which is twice as fast.

More about the address bus

Because the address bus is used to identify the devices and memory connected to the CPU, the more address buses available, the larger the number of devices that can be addressed. In other words, the number of address buses for a CPU determines the number of locations with which it can communicate. The number of locations is always equal to 2^x, where x is the number of address lines, regardless of the size of the data bus. For example, a CPU with 16 address lines can provide a total of 65,536 (2^{16}) or 64K of addressable memory. Each location can have a maximum of 1 byte of data. This is because all general-purpose microprocessor CPUs are what is called *byte addressable*. As another example, the IBM PC AT uses a CPU with 24 address lines and 16 data lines. Thus, the total accessible memory is 16 megabytes (2^{24} = 16 megabytes). In this example there would be 2^{24} locations, and because each location is one byte, there would be 16 megabytes of memory. The address bus is a *unidirectional* bus, which means that the CPU uses the address bus only to send out addresses. To summarize: The total number of memory locations addressable by a given CPU is always equal to 2^x where x is the number of address bits, regardless of the size of the data bus.

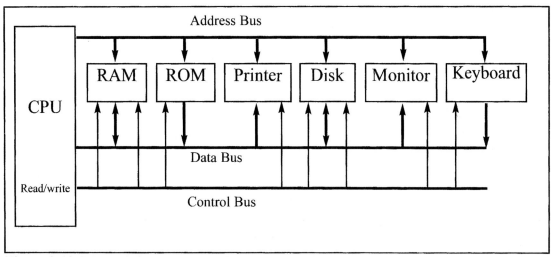

Figure 0-10. Internal Organization of a Computer

CPU and its relation to RAM and ROM

For the CPU to process information, the data must be stored in RAM or ROM. The function of ROM in computers is to provide information that is fixed and permanent. This is information such as tables for character patterns to be displayed on the video monitor, or programs that are essential to the working of the computer, such as programs for testing and finding the total amount of RAM installed on the system, or for displaying information on the video monitor. In contrast, RAM stores temporary information that can change with time, such as various versions of the operating system and application packages such as word processing or tax calculation packages. These programs are loaded from the hard drive into RAM to be processed by the CPU. The CPU cannot get the information from the disk directly because the disk is too slow. In other words, the CPU first seeks the information to be processed from RAM (or ROM). Only if the data is not there does the CPU seek it from a mass storage device such as a disk, and then it transfers the information to RAM. For this reason, RAM and ROM are sometimes referred to as *primary memory* and disks are called *secondary memory*. Figure 0-11 shows a block diagram of the internal organization of the PC.

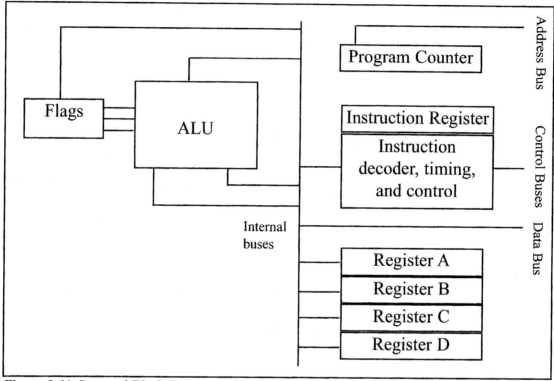

Figure 0-11. Internal Block Diagram of a CPU

Inside CPUs

A program stored in memory provides instructions to the CPU to perform an action. The action can simply be adding data such as payroll data or controlling a machine such as a robot. The function of the CPU is to fetch these instructions from memory and execute them. To perform the actions of fetch and execute, all CPUs are equipped with resources such as the following:

1. Foremost among the resources at the disposal of the CPU are a number of *registers*. The CPU uses registers to store information temporarily. The information could be two values to be processed, or the address of the value needed to be fetched from memory. Registers inside the CPU can be 8-bit, 16-bit, 32-bit, or even 64-bit registers, depending on the CPU. In general, the more and bigger the registers, the better the CPU. The disadvantage of more and bigger registers is the increased cost of such a CPU.
2. The CPU also has what is called the *ALU* (arithmetic/logic unit). The ALU section of the CPU is responsible for performing arithmetic functions such as add, subtract, multiply, and divide, and logic functions such as AND, OR, and NOT.

3. Every CPU has what is called a *program counter*. The function of the program counter is to point to the address of the next instruction to be executed. As each instruction is executed, the program counter is incremented to point to the address of the next instruction to be executed. The contents of the program counter are placed on the address bus to find and fetch the desired instruction. In the IBM PC, the program counter is a register called IP, or the instruction pointer.

4. The function of the *instruction decoder* is to interpret the instruction fetched into the CPU. One can think of the instruction decoder as a kind of dictionary, storing the meaning of each instruction and what steps the CPU should take upon receiving a given instruction. Just as a dictionary requires more pages the more words it defines, a CPU capable of understanding more instructions requires more transistors to design.

Internal working of computers

To demonstrate some of the concepts discussed above, a step-by-step analysis of the process a CPU would go through to add three numbers is given next. Assume that an imaginary CPU has registers called A, B, C, and D. It has an 8-bit data bus and a 16-bit address bus. Therefore, the CPU can access memory from addresses 0000 to FFFFH (for a total of 10000H locations). The action to be performed by the CPU is to put hexadecimal value 21 into register A, and then add to register A values 42H and 12H. Assume that the code for the CPU to move a value to register A is 1011 0000 (B0H) and the code for adding a value to register A is 0000 0100 (04H). The necessary steps and code to perform them are as follows.

Action	Code	Data
Move value 21H into register A	B0H	21H
Add value 42H to register A	04H	42H
Add value 12H to register A	04H	12H

If the program to perform the actions listed above is stored in memory locations starting at 1400H, the following would represent the contents for each memory address location:

Memory address	Contents of memory address
1400	(B0) code for moving a value to register A
1401	(21) value to be moved
1402	(04) code for adding a value to register A
1403	(42) value to be added
1404	(04) code for adding a value to register A
1405	(12) value to be added
1406	(F4) code for halt

The actions performed by the CPU to run the program above would be as follows:

1. The CPU's program counter can have a value between 0000 and FFFFH. The program counter must be set to the value 1400H, indicating the address of the

first instruction code to be executed. After the program counter has been loaded with the address of the first instruction, the CPU is ready to execute.

2. The CPU puts 1400H on the address bus and sends it out. The memory circuitry finds the location while the CPU activates the READ signal, indicating to memory that it wants the byte at location 1400H. This causes the contents of memory location 1400H, which is B0, to be put on the data bus and brought into the CPU.

3. The CPU decodes the instruction B0 with the help of its instruction decoder dictionary. When it finds the definition for that instruction it knows it must bring into register A of the CPU the byte in the next memory location. Therefore, it commands its controller circuitry to do exactly that. When it brings in value 21H from memory location 1401, it makes sure that the doors of all registers are closed except register A. Therefore, when value 21H comes into the CPU it will go directly into register A. After completing one instruction, the program counter points to the address of the next instruction to be executed, which in this case is 1402H. Address 1402 is sent out on the address bus to fetch the next instruction.

4. From memory location 1402H the CPU fetches code 04H. After decoding, the CPU knows that it must add the byte sitting at the next address (1403) to the contents of register A. After the CPU brings the value (in this case, 42H) into register A, it provides the contents of register A along with this value to the ALU to perform the addition. It then takes the result of the addition from the ALU's output and puts it in register A. Meanwhile the program counter becomes 1404, the address of the next instruction.

5. Address 1404H is put on the address bus and the code is fetched into the CPU, decoded, and executed. This code again is adding a value to register A. The program counter is updated to 1406H.

6. Finally, the contents of address 1406 are fetched in and executed. This HALT instruction tells the CPU to stop incrementing the program counter and asking for the next instruction. Without the HALT, the CPU would continue updating the program counter and fetching instructions.

Now suppose that address 1403H contained value 04 instead of 42H. How would the CPU distinguish between data 04 to be added and code 04? Remember that code 04 for this CPU means "move the next value into register A". Therefore, the CPU will not try to decode the next value. It simply moves the contents of the following memory location into register A, regardless of its value.

Review Questions

1. How many bytes is 24 kilobytes?
2. What does "RAM" stand for? How is it used in computer systems?
3. What does "ROM" stand for? How is it used in computer systems?
4. Why is RAM called volatile memory?
5. List the three major components of a computer system.
6. What does "CPU" stand for? Explain its function in a computer.
7. List the three types of buses found in computer systems and state briefly the purpose of each type of bus.

8. State which of the following is unidirectional and which is bidirectional:
    (a) data bus   (b) address bus
9. If an address bus for a given computer has 16 lines, what is the maximum amount of memory it can access?
10. What does "ALU" stand for? What is its purpose?
11. How are registers used in computer systems?
12. What is the purpose of the program counter?
13. What is the purpose of the instruction decoder?

SUMMARY

The binary number system represents all numbers with a combination of the two binary digits, 0 and 1. The use of binary systems is necessary in digital computers because only two states can be represented: on or off. Any binary number can be coded directly into its hexadecimal equivalent for the convenience of humans. Converting from binary/hex to decimal, and vice versa, is a straightforward process that becomes easy with practice. The ASCII code is a binary code used to represent alphanumeric data internally in the computer. It is frequently used in peripheral devices for input and/or output.

The logic gates AND, OR, and inverter are the basic building blocks of simple circuits. NAND, NOR, and XOR gates are also used to implement circuit design. Diagrams of half-adders and full-adders were given as examples of the use of logic gates for circuit design. Decoders are used to detect certain addresses. Flip-flops are used to latch in data until other circuits are ready for it.

The major components of any computer system are the CPU, memory, and I/O devices. "Memory" refers to temporary or permanent storage of data. In most systems, memory can be accessed as bytes or words. The terms *kilobyte*, *megabyte*, *gigabyte*, and *terabyte* are used to refer to large numbers of bytes. There are two main types of memory in computer systems: RAM and ROM. RAM (random access memory) is used for temporary storage of programs and data. ROM (read-only memory) is used for permanent storage of programs and data that the computer system must have in order to function. All components of the computer system are under the control of the CPU. Peripheral devices such as I/O (input/output) devices allow the CPU to communicate with humans or other computer systems. There are three types of buses in computers: address, control, and data. Control buses are used by the CPU to direct other devices. The address bus is used by the CPU to locate a device or a memory location. Data buses are used to send information back and forth between the CPU and other devices.

Finally, this chapter gave an overview of digital logic.

PROBLEMS

SECTION 0.1: NUMBERING AND CODING SYSTEMS

1. Convert the following decimal numbers to binary:
   (a) 12   (b) 123   (c) 63   (d) 128   (e) 1000
2. Convert the following binary numbers to decimal:
   (a) 100100   (b) 1000001   (c) 11101   (d) 1010   (e) 00100010
3. Convert the values in Problem 2 to hexadecimal.
4. Convert the following hex numbers to binary and decimal:
   (a) 2B9H   (b) F44H   (c) 912H   (d) 2BH   (e) FFFFH
5. Convert the values in Problem 1 to hex.
6. Find the 2's complement of the following binary numbers:
   (a) 1001010   (b) 111001   (c) 10000010   (d) 111110001
7. Add the following hex values:
   (a) 2CH + 3FH   (b) F34H + 5D6H   (c) 20000H + 12FFH
   (d) FFFFH + 2222H
8. Perform hex subtraction for the following:
   (a) 24FH – 129H   (b) FE9H – 5CCH   (c) 2FFFFH – FFFFFH
   (d) 9FF25H – 4DD99H
9. Show the ASCII codes for numbers 0, 1, 2, 3, ..., 9 in both hex and binary.
10. Show the ASCII code (in hex) for the following string:
    "U.S.A. is a country" CR,LF
    "in North America" CR,LF
    (CR is carriage return, LF is line feed)

SECTION 0.2: DIGITAL PRIMER

11. Draw a 3-input OR gate using a 2-input OR gate.
12. Show the truth table for a 3-input OR gate.
13. Draw a 3-input AND gate using a 2-input AND gate.
14. Show the truth table for a 3-input AND gate.
15. Design a 3-input XOR gate with a 2-input XOR gate. Show the truth table for a 3-input XOR.
16. List the truth table for a 3-input NAND.
17. List the truth table for a 3-input NOR.
18. Show the decoder for binary 1100.
19. Show the decoder for binary 11011.
20. List the truth table for a D-FF.

SECTION 0.3: INSIDE THE COMPUTER

21. Answer the following:
    (a) How many nibbles are 16 bits?
    (b) How many bytes are 32 bits?
    (c) If a word is defined as 16 bits, how many words is a 64-bit data item?
    (d) What is the exact value (in decimal) of 1 meg?

(e) How many K is 1 meg?

(f) What is the exact value (in decimal) of 1 gigabyte?

(g) How many K is 1 gigabyte?

(h) How many meg is 1 gigabyte?

(i) If a given computer has a total of 8 megabytes of memory, how many bytes (in decimal) is this? How many kilobytes is this?

22. A given mass storage device such as a hard disk can store 2 gigabytes of information. Assuming that each page of text has 25 rows and each row has 80 columns of ASCII characters (each character = 1 byte), approximately how many pages of information can this disk store?

23. In a given byte-addressable computer, memory locations 10000H to 9FFFFH are available for user programs. The first location is 10000H and the last location is 9FFFFH. Calculate the following:

(a) The total number of bytes available (in decimal)

(b) The total number of kilobytes (in decimal)

24. A given computer has a 32-bit data bus. What is the largest number that can be carried into the CPU at a time?

25. Below are listed several computers with their data bus widths. For each computer, list the maximum value that can be brought into the CPU at a time (in both hex and decimal).

(a) Apple 2 with an 8-bit data bus

(b) IBM PS/2 with a 16-bit data bus

(c) IBM PS/2 model 80 with a 32-bit data bus

(d) Cray supercomputer with a 64-bit data bus

26. Find the total amount of memory, in the units requested, for each of the following CPUs, given the size of the address buses:

(a) 16-bit address bus (in K)

(b) 24-bit address bus (in megs)

(c) 32-bit address bus (in megabytes and gigabytes)

(d) 48-bit address bus (in megabytes, gigabytes, and terabytes)

27. Regarding the data bus and address bus, which is unidirectional and which is bidirectional?

28. Which register of the CPU holds the address of the instruction to be fetched?

29. Which section of the CPU is responsible for performing addition?

30. List the three bus types present in every CPU.

ANSWERS TO REVIEW QUESTIONS

SECTION 0.1: NUMBERING AND CODING SYSTEMS

1. Computers use the binary system because each bit can have one of two voltage levels: on and off.
2. $34_{10} = 100010_2 = 22_{16}$
3. $110101_2 = 35_{16} = 53_{10}$
4. 1110001
5. 010100
6. 461
7. 275
8. 38 30 78 38 36 20 43 50 55 73

SECTION 0.2: DIGITAL PRIMER

1. AND
2. OR
3. XOR
4. Buffer
5. Storing data
6. Decoder

SECTION 0.3: INSIDE THE COMPUTER

1. 24,576
2. Random access memory; it is used for temporary storage of programs that the CPU is running, such as the operating system, word processing programs, etc.
3. Read-only memory; it is used for permanent programs such as those that control the keyboard, etc.
4. The contents of RAM are lost when the computer is powered off.
5. The CPU, memory, and I/O devices
6. Central processing unit; it can be considered the "brain" of the computer; it executes the programs and controls all other devices in the computer.
7. The address bus carries the location (address) needed by the CPU; the data bus carries information in and out of the CPU; the control bus is used by the CPU to send signals controlling I/O devices.
8. (a) bidirectional   (b) unidirectional
9. 64K, or 65,536 bytes
10. Arithmetic/logic unit; it performs all arithmetic and logic operations.
11. They are used for temporary storage of information.
12. It holds the address of the next instruction to be executed.
13. It tells the CPU what steps to perform for each instruction.

CHAPTER 1

THE PIC MICROCONTROLLERS: HISTORY AND FEATURES

OBJECTIVES

Upon completion of this chapter, you will be able to:

>>   **Compare and contrast microprocessors and microcontrollers**
>>   **Describe the advantages of microcontrollers for some applications**
>>   **Explain the concept of embedded systems**
>>   **Discuss criteria for considering a microcontroller**
>>   **Explain the variations of speed, packaging, memory, and cost per unit and how these affect choosing a microcontroller**
>>   **Compare and contrast the various members of the PIC family**
>>   **Compare the PIC with microcontrollers offered by other manufacturers**

This chapter begins with a discussion of the role and importance of microcontrollers in everyday life. In Section 1.1 we also discuss criteria to consider in choosing a microcontroller, as well as the use of microcontrollers in the embedded market. Section 1.2 covers various members of the PIC18 family and their features. In addition, we provide a brief discussion of alternatives to the PIC chip such as the 8051, AVR, and 68HC11 microcontrollers.

SECTION 1.1: MICROCONTROLLERS AND EMBEDDED PROCESSORS

In this section we discuss the need for microcontrollers and contrast them with general-purpose microprocessors such as the Pentium and other x86 microprocessors. We also look at the role of microcontrollers in the embedded market. In addition, we provide some criteria on how to choose a microcontroller.

Microcontroller versus general-purpose microprocessor

What is the difference between a microprocessor and microcontroller? By microprocessor is meant the general-purpose microprocessors such as Intel's x86 family (8086, 80286, 80386, 80486, and the Pentium) or Motorola's PowerPC family. These microprocessors contain no RAM, no ROM, and no I/O ports on the chip itself. For this reason, they are commonly referred to as *general-purpose microprocessors*. See Figure 1-1.

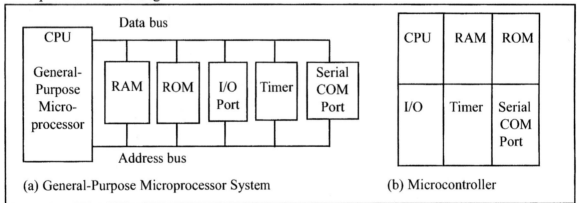

(a) General-Purpose Microprocessor System   (b) Microcontroller

Figure 1-1. Microprocessor System Contrasted With Microcontroller System

A system designer using a general-purpose microprocessor such as the Pentium or the PowerPC must add RAM, ROM, I/O ports, and timers externally to make them functional. Although the addition of external RAM, ROM, and I/O ports makes these systems bulkier and much more expensive, they have the advantage of versatility, enabling the designer to decide on the amount of RAM, ROM, and I/O ports needed to fit the task at hand. This is not the case with microcontrollers. A microcontroller has a CPU (a microprocessor) in addition to a fixed amount of RAM, ROM, I/O ports, and a timer all on a single chip. In other words, the processor, RAM, ROM, I/O ports, and timer are all embedded together on one chip; therefore, the designer cannot add any external memory, I/O, or timer to it. The fixed amount of on-chip ROM, RAM, and number of I/O ports in microcontrollers makes them ideal for many applications in which cost and space are critical. In many applications, for example a TV remote control, there is no need for

Home
Appliances
Intercom
Telephones
Security systems
Garage door openers
Answering machines
Fax machines
Home computers
TVs
Cable TV tuner
VCR
Camcorder
Remote controls
Video games
Cellular phones
Musical instruments
Sewing machines
Lighting control
Paging
Camera
Pinball machines
Toys
Exercise equipment
Office
Telephones
Computers
Security systems
Fax machine
Microwave
Copier
Laser printer
Color printer
Paging
Auto
Trip computer
Engine control
Air bag
ABS
Instrumentation
Security system
Transmission control
Entertainment
Climate control
Cellular phone
Keyless entry

Table 1-1: Some Embedded Products Using Microcontrollers

the computing power of a 486 or even an 8086 microprocessor. In many applications, the space used, the power consumed, and the price per unit are much more critical considerations than the computing power. These applications most often require some I/O operations to read signals and turn on and off certain bits. For this reason some call these processors IBP, "itty-bitty processors." (See "Good Things in Small Packages Are Generating Big Product Opportunities" by Rick Grehan, BYTE magazine, September 1994 (http://www.byte.com) for an excellent discussion of microcontrollers.)

It is interesting to note that some microcontroller manufacturers have gone as far as integrating an ADC (analog-to-digital converter) and other peripherals into the microcontroller.

Microcontrollers for embedded systems

In the literature discussing microprocessors, we often see the term *embedded system*. Microprocessors and microcontrollers are widely used in embedded system products. An embedded product is controlled by its own internal microprocessor (or microcontroller) as opposed to an external controller. Typically, in an embedded system, the microcontroller's ROM is burned with a purpose for specific functions needed for the system. A printer is an example of an embedded system because the processor inside it performs one task only; namely, getting the data and printing it. Contrast this with a Pentium-based PC (or any x86 IBM-compatible PC), which can be used for any number of applications such as word processor, print-server, bank teller terminal, video game player, network server, or Internet terminal. A PC can also load and run software for a variety of applications. Of course, the reason a PC can perform myriad tasks is that it has RAM memory and an operating system that loads the application software into RAM and lets the CPU run it. In an embedded system, typically only one application software is burned into ROM. An x86 PC contains or is connected to various embedded products such as the keyboard, printer, modem, disk controller, sound card, CD-ROM driver, mouse, and so on. Each one of these peripherals has a microcontroller inside it that performs only one task. For example, inside every mouse a microcontroller performs the task of finding the mouse's position and sending it to the PC. Table 1-1 lists some embedded products.

x86 PC embedded applications

Although microcontrollers are the preferred choice for many embedded systems, sometimes a microcontroller is inadequate for the task. For this reason, in recent years many manufacturers of general-purpose microprocessors such as Intel, Freescale Semiconductor (formerly Motorola), and AMD (Advanced Micro

Devices, Inc.) have targeted their microprocessor for the high end of the embedded market. Intel and AMD push their x86 processors for both the embedded and desktop PC markets. In the early 1990s, Apple computer began using the PowerPC microprocessors (604, 603, 620, etc.) in place of the 680x0 for the Macintosh. The PowerPC microprocessor is a joint venture between IBM and Motorola, and is targeted for the high end of the embedded market as well as the PC market. It must be noted that when a company targets a general-purpose microprocessor for the embedded market it optimizes the processor used for embedded systems. For this reason these processors are often called *high-end embedded processors*. Another chip widely used in the high end of the embedded system design is the ARM microprocessor. Very often the terms *embedded processor* and *microcontroller* are used interchangeably.

One of the most critical needs of an embedded system is to decrease power consumption and space. This can be achieved by integrating more functions into the CPU chip. All the embedded processors based on the x86 and PowerPC 6xx have low power consumption in addition to some forms of I/O, COM port, and ROM, all on a single chip. In high-performance embedded processors, the trend is to integrate more and more functions on the CPU chip and let the designer decide which features to use. This trend is invading PC system design as well. Normally, in designing the PC motherboard we need a CPU plus a chipset containing I/O, a cache controller, a flash ROM containing BIOS, and finally a secondary cache memory. New designs are emerging in industry. For example, many companies have a chip that contains the entire CPU and all the supporting logic and memory, except for DRAM. In other words, we have the entire computer on a single chip.

Currently, because of Linux, MS-DOS, and Windows standardization, many embedded systems use x86 PCs. In many cases, using x86 PCs for the high-end embedded applications not only saves money but also shortens development time because a vast library of software already exists for the Linux, DOS, and Windows platforms. The fact that Windows and Linux are widely used and well-understood platforms means that developing a Windows-based or Linux-based embedded product reduces the cost and shortens the development time considerably.

Choosing a microcontroller

There are five major 8-bit microcontrollers. They are: Freescale Semiconductor's (formerly Motorola) 68HC08/68HC11, Intel's 8051, Atmel's AVR, Zilog's Z8, and PIC from Microchip Technology. Each of the above microcontrollers has a unique instruction set and register set; therefore, they are not compatible with each other. Programs written for one will not run on the others. There are also 16-bit and 32-bit microcontrollers made by various chip makers. With all these different microcontrollers, what criteria do designers consider in choosing one? Three criteria in choosing microcontrollers are as follows: (1) meeting the computing needs of the task at hand efficiently and cost effectively; (2) availability of software and hardware development tools such as compilers, assemblers, debuggers, and emulators; and (3) wide availability and reliable sources of the microcontroller. Next, we elaborate on each of the above criteria.

Criteria for choosing a microcontroller

1. The first and foremost criterion in choosing a microcontroller is that it must meet the task at hand efficiently and cost effectively. In analyzing the needs of a microcontroller-based project, we must first see whether an 8-bit, 16-bit, or 32-bit microcontroller can best handle the computing needs of the task most effectively. Among other considerations in this category are:

    (a) Speed. What is the highest speed that the microcontroller supports?

    (b) Packaging. Does it come in a 40-pin DIP (dual inline package) or a QFP (quad flat package), or some other packaging format? This is important in terms of space, assembling, and prototyping the end product.

    (c) Power consumption. This is especially critical for battery-powered products.

    (d) The amount of RAM and ROM on the chip.

    (e) The number of I/O pins and the timer on the chip.

    (f) Ease of upgrade to higher-performance or lower-power-consumption versions.

    (g) Cost per unit. This is important in terms of the final cost of the product in which a microcontroller is used. For example, some microcontrollers cost 50 cents per unit when purchased 100,000 units at a time.

2. The second criterion in choosing a microcontroller is how easy it is to develop products around it. Key considerations include the availability of an assembler, debugger, a code-efficient C language compiler, emulator, technical support, and both in-house and outside expertise. In many cases, third-party vendor (i.e., a supplier other than the chip manufacturer) support for the chip is as good as, if not better than, support from the chip manufacturer.

3. The third criterion in choosing a microcontroller is its ready availability in needed quantities both now and in the future. For some designers this is even more important than the first two criteria. Currently, of the leading 8-bit microcontrollers, the 8051 family has the largest number of diversified (multiple source) suppliers. (Supplier means a producer besides the originator of the microcontroller.) In the case of the 8051, which was originated by Intel, several companies also currently produce (or have produced in the past) the 8051.

    Note that Freescale Semiconductor (Motorola), Atmel, Zilog, and Microchip Technology have all dedicated massive resources to ensure wide and timely availability of their products because their products are stable, mature, and single sourced. In recent years, companies have begun to sell *Field-Programmable Gate Array* (FPGA) and *Application-Specific Integrated Circuit* (ASIC) libraries for the different microcontrollers.

Mechatronics and microcontrollers

The microcontroller is playing a major role in an emerging field called *mechatronics*. Here is an excellent summary of what the field of mechatronics is all about, taken from the web site of Newcastle University (http://mechatronics2004.newcastle.edu.au/mech2004), which holds a major conference every year on this subject:

"Many technical processes and products in the area of mechanical and electrical engineering show an increasing integration of mechanics with electronics and information processing. This integration is between the components (hardware) and the information-driven functions (software), resulting in integrated systems called mechatronic systems.

The development of mechatronic systems involves finding an optimal balance between the basic mechanical structure, sensor and actuator implementation, automatic digital information processing and overall control, and this synergy results in innovative solutions. The practice of mechatronics requires multidisciplinary expertise across a range of disciplines, such as: mechanical engineering, electronics, information technology, and decision making theories."

Review Questions

1. True or false. Microcontrollers are normally less expensive than microprocessors.
2. When comparing a system board based on a microcontroller and a general-purpose microprocessor, which one is cheaper?
3. A microcontroller normally has which of the following devices on-chip?
    (a) RAM   (b) ROM   (c) I/O   (d) all of the above
4. A general-purpose microprocessor normally needs which of the following devices to be attached to it?
    (a) RAM   (b) ROM   (c) I/O   (d) all of the above
5. An embedded system is also called a dedicated system. Why?
6. What does the term *embedded system* mean?
7. Why does having multiple sources of a given product matter?

SECTION 1.2: OVERVIEW OF THE PIC18 FAMILY

In this section, we first look at the PIC family of microcontrollers and then examine the PIC18 family in more detail.

A brief history of the PIC microcontroller

In 1989, Microchip Technology Corporation introduced an 8-bit microcontroller called the PIC, which stands for Peripheral Interface Controller. This microcontroller had small amounts of data RAM, a few hundred bytes of on-chip ROM for the program, one timer, and a few pins for I/O ports, all on a single chip with only 8 pins. (See Figure 1-2.) It is amazing that a company that began with such a humble product became one of the leading suppliers of 8-bit microcontrollers in less than a decade. At the time of this writing, Microchip is the number-one supplier of 8-bit microcontrollers in the world. Since the introduction of the PIC16xxx, they have introduced an array of 8-bit microcontrollers too numerous to list here. They include the PIC families of 10xxx, 12xxx, 14xxx, 16xxx, 17xxx, and 18xxx. They are all 8-bit processors, meaning that the CPU can work on only 8 bits of data at a time. Data larger than 8 bits has to be broken into 8-bit pieces to be processed by the CPU. One of the problems with the PIC family is that they are not all 100% upwardly compatible in terms of software when going from one family to another family. For example, while the 12xxx/16xxx have 12-bit and 14-bit wide instructions, the PIC18xxx instruction is 16 bits wide with many new instructions. To run programs written for the PIC12xxx on a PIC18, we must recompile the program and possibly change some register locations before loading it into the PIC18. At the time of this writing, the PIC18xxx family has the highest performance of all the families of 8-bit PIC microcontrollers. The fact that PIC18xxx is available in 18- to 80-pin packages makes it an ideal choice for new designs because it allows an easy migration to more powerful versions of the chip without losing software compatibility. At this time, no 8-pin version of the PIC18xxx exists, and that is the main reason to choose other family members of the 10xxx–16xxx if your design calls for a small package. Because this book is about the PIC18 family, we describe some of the main features of this family and refer the reader to the Microchip web site for other families of PIC10xxx–16xxx. For those who have mastered the PIC18 family, understanding the other families is very easy and straightforward. The following is a brief description of the PIC18 series.

PIC18 features

The PIC18 has a RISC architecture that comes with some standard features such as on-chip program (code) ROM, data RAM, data EEPROM, timers, ADC, and USART and I/O ports. See Figure 1-2. Although the size of the program ROM, data RAM, data EEPROM, and I/O ports varies among the family members, they all have peripherals such as timers, ADC, and USART. See Figures 1-3 and 1-4. Due to the importance of these peripherals, we have dedicated an entire chapter to each one of them. The details of the RAM/ROM memory and I/O features of the PIC18 are given in the next few chapters.

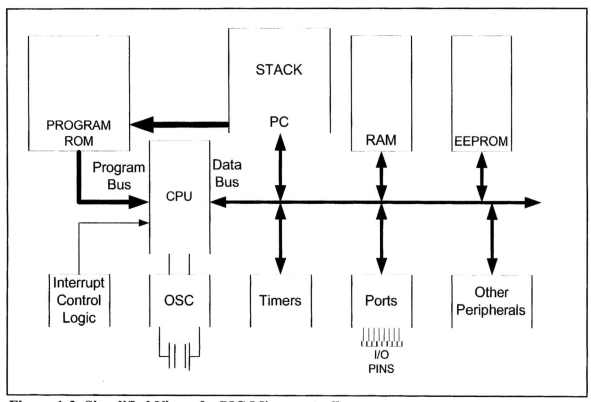

Figure 1-2. Simplified View of a PIC Microcontroller

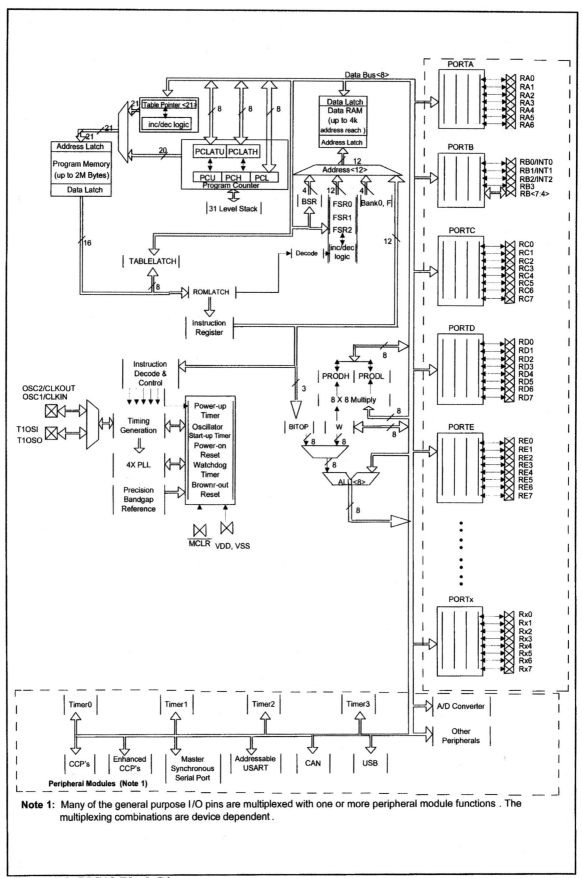

Figure 1-3. PIC18 Block Diagram

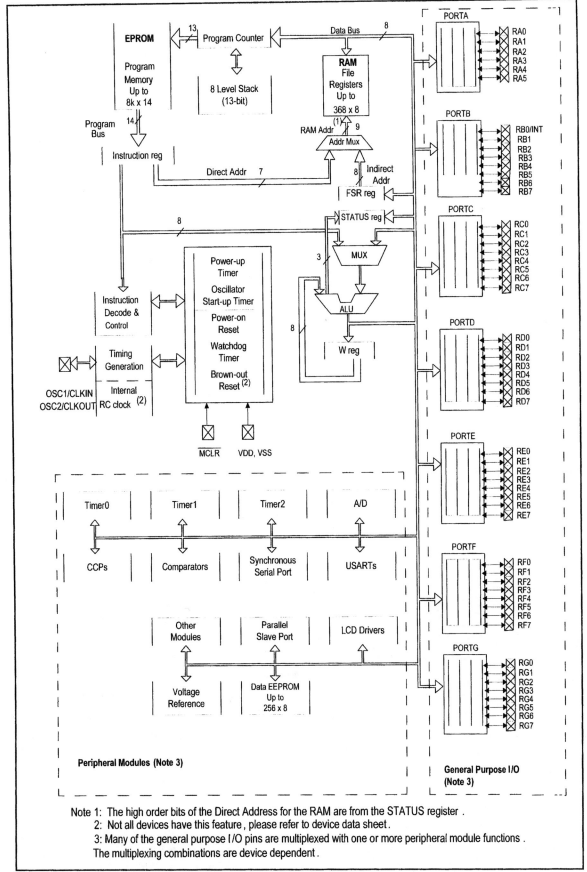

Figure 1-4. PIC16 Block Diagram

Note 1: The high order bits of the Direct Address for the RAM are from the STATUS register .
2: Not all devices have this feature , please refer to device data sheet .
3: Many of the general purpose I /O pins are multiplexed with one or more peripheral module functions .
The multiplexing combinations are device dependent .

PIC microcontroller program ROM

In microcontrollers, the ROM is used to store programs and for that reason it is called program or code ROM. Although the PIC18 has 2M (megabytes) of program (code) ROM space, not all family members come with that much ROM installed. The program ROM size can vary from 4K to 128K at the time of this writing, depending on the family member. The PIC18 program ROM is available in different memory types, such as flash, OTP, and masked, all of which have different part numbers. A discussion of the various types of ROM is given in Chapter 14, if you need to refresh your memory on these important memory technologies. Note that although different flavors of the PIC18 exist in terms of speed and amount of on-chip RAM/ROM, they are all compatible with each other as far as the instructions are concerned. This means that if you write your program for one, it will run on any of them regardless of the chip number. Next, we discuss briefly the program ROM type for the PIC18 family.

PIC microcontroller with UV-EPROM

Some of the PIC microcontrollers use UV-EPROM, for on-chip program ROM. To use these kinds of chips for development requires access to a PROM burner, as well as a UV-EPROM eraser to erase the contents of ROM. The window on the UV-EPROM chip allows the UV light to erase the ROM. The problem with the UV-EPROM is that it takes around 20 minutes to erase the chip before it can be programmed again. This has led Microchip to introduce a flash version of the PIC family. At this time flash is replacing the UV-EPROM altogether. Table 1-2 shows some members of the PIC18 family.

PIC18Fxxx with flash

Many PIC18 chips have on-chip program ROM in the form of flash memory. The flash version uses the letter F in the part number to indicate that the on-chip ROM is flash. PIC18F458 is an example of PIC18 with flash ROM. The flash version is ideal for fast development because flash memory can be erased in seconds compared to the 20 minutes or more needed for the UV-EPROM version. For this reason, the PIC18F has been used in place of the UV-EPROM to eliminate the waiting time needed to erase the chip, thereby speeding up the development time. To use the PIC18F to develop a microcontroller-based system requires a ROM burner that supports flash memory; however, a ROM eraser is not needed, because flash is an EEPROM (electrically erasable PROM). Notice that in flash memory, you must erase the entire contents of ROM in order to program it again. This erasing of flash is done by the ROM programmer itself, and so a separate eraser is not needed. We can also program the PIC18F via the PICkit 2 from MicroChip using the USB port of an IBM PC.

OTP version of the PIC

OTP (one-time-programmable) versions of the PIC are also available from Microchip. PIC16C432 chip uses OTP for program ROM. Contrast the PIC16C432 and PIC18F252. The letter C indicates the OTP ROM, while the letter F is for the flash. The flash version is typically used for product development. When a product is designed and absolutely finalized, the OTP version of the PIC is used for mass production because it is cheaper than flash in terms of price per unit. The problem with the OTP is that you cannot reprogram it if you want to modify your program.

Masked version of PIC

Microchip Corporation provides a service in which you can send in your program and they will burn the program into the PIC chip during the fabrication process of the chip. This chip is commonly referred to as masked PIC, which is one of the stages of IC fabrication. Masked PIC is the cheapest of all types, if the unit numbers are high enough. This is because there is a minimum order for the masked version of the PIC microcontrollers.

PIC microcontroller data RAM and EEPROM

While ROM is used to store program (code), the RAM space is for data storage. The PIC18 has a maximum of 4096 bytes (4K) of data RAM space. Not all of the family members come with that much RAM. The data RAM size for the PIC18 varies from 256 bytes to 4096 bytes. As we will see in the next chapter, the data RAM space has two components: General-Purpose RAM (GPR) and Special Function Registers (SFRs). Because the SFRs are fixed and every microcontroller must have them, it is the GPR's size that varies from chip to chip. For this reason, the Microchip web site gives only the GPR size. The RAM GPR space is used for read/write scratch pad and data manipulation and is divided into banks of 256 bytes each, as we will see in Chapter 6. The GPR size given for the PIC18 is always a multiple of 256 bytes. In some of the PIC18 family members, we also have a small amount of EEPROM to store critical data that does not need to be changed very often. While every PIC18 must have some data RAM for scratch pad, the EEPROM is optional, so not all versions of the PIC18 come with EEP-ROM. EEPROM is used mainly for storage of critical data, as we will see in Chapter 14.

Table 1-2: Some Members of the PIC18 Family (http://www.microchip.com)

Part Num	Code ROM	Data RAM	Data EEPROM	I/O pins	ADC	Timers	Pin numbers & Package
PIC18F1220	4K (flash)	256	256	16	10-bit	4	18 DIP
PIC18F2420	16K (flash)	768	0	25	10-bit	4	28 DIP
PIC18F2220	4K (flash)	512	256	25	10-bit	4	28 DIP
PIC18F452	32K (flash)	1536	256	34	10-bit	4	40 DIP
PIC18F4520	32K (flash)	1536	256	36	10-bit	4	40 DIP
PIC18F458	32K (flash)	1536	256	34	10-bit	4	40 DIP
PIC18F4580	32K (flash)	1536	256	36	10-bit	4	40 DIP
PIC18F8722	128K (flash)	3936	1024	70	10-bit	5	80 TQFP

Notes:
1. All ROM, RAM, and EEPROM memories are in bytes.
2. Data RAM (General-Purpose RAM) is the amount of RAM available for data manipulation (scratch pad) in addition to the Special Function Registers (SFRs) space.
3. All the above chips have USART for serial data transfer.

PIC microcontroller I/O pins

The PIC18 can have from 16 to 72 pins dedicated for I/O. The number of I/O pins depends on the number of pins in the package itself. The number of pins for the PIC18 package goes from 18 to 80 at this time. In the case of the 18-pin PIC18F1220, we have 16 pins for I/O, while in the case of the 80-pin PIC18F8722, we can use up to 72 pins for I/O. We will study I/O pins and programming in Chapter 4.

PIC microcontroller peripherals

All the members of the PIC18 family come with ADC (analog-to-digital converter), timers, and USART (Universal Synchronous Asynchronous Receiver Transmitter) as standard peripherals. As we will see in Chapter 13, the ADC is 10-bit and the number of ADC channels in each PIC chip varies from 5 to 16, depending on the number of pins in the package. The PIC18 can have up to 4 timers besides the watchdog timer. We will examine timers in Chapter 9. The USART peripheral allows us to connect the PIC18-based system to serial ports such as the COM port of the IBM PC, as we will see in Chapter 10. Many of the PIC18 family members come with the I²C and CAN bus as well.

PIC trainer

In Chapter 8, we discuss the design of the PIC18F458 trainer extensively. This trainer is programmed using the PICkit 2. The MDEPIC trainer is also compatible with other 40-pin devices from Microchip.

Other Microcontrollers

There are many other popular 8-bit microcontrollers besides the PIC chip. Among them are the 8051, 68HC11, AVR, and Z8. Besides Intel, a number of other companies make the 8051 family, as seen in Table 1-4. The AVR is made by Atmel Corp. Freescale (Motorola) makes the 68HC11 and many of its variations. Zilog produces the Z8 microcontroller. To contrast the PIC18 family with the 8051/52 chip, examine Table 1-3. For a comprehensive treatment of the 8051 microcontroller, see "The 8051 Microcontroller and Embedded Systems" by Mazidi, et. al.

Table 1-3: Comparison of 8051 and PIC18 Family (40-pin package)

Feature	8051/52	PIC18xxx
Program ROM (maximum space)	64K	2M
Data RAM (maximum space)	256 bytes	4K
Timers	3	4
I/O pins	32	33
Serial port	1	1

Table 1-4: Some of the Companies that Produce Widely Used 8-bit Microcontrollers

Company	Web Site	Architecture
Microchip	http://www.microchip.com	PIC16xxx/18xxx
Intel	http://www.intel.com/design/mcs51	8051
Atmel	http://www.atmel.com	AVR and 8051
Philips/Signetics	http://www.semiconductors.philips.com	8051
Zilog	http://www.zilog.com	Z8 and Z80
Dallas Semi/Maxim	http://www.maxim-ic.com	8051
Freescale Semi	http://www.freescale.com	68HC11/68HC08

See http://www.microcontroller.com for a complete list.

See the following web sites for PIC microcontrollers and PIC Trainer:

http://www.microchip.com

http://www.MicroDigitalEd.com

Review Questions

1. Name three features of the PIC18xxx.
2. What is the main difference between the PIC18Fxxx and PIC18Cxxx microcontrollers?
3. Give the size of RAM in each of the following:
   (a) PIC18F2420   (b) PIC18F4520
4. Give the size of the on-chip program ROM in each of the following:
   (a) PIC18F2420   (b) PIC18F4520
5. The PIC18 is a(n) _____-bit microprocessor.

SUMMARY

This chapter discussed the role and importance of microcontrollers in everyday life. Microprocessors and microcontrollers were contrasted and compared. We discussed the use of microcontrollers in the embedded market. We also discussed criteria to consider in choosing a microcontroller such as speed, memory, I/O, packaging, and cost per unit. The second section of this chapter described various families of the PIC, such as the PIC18 and PIC16, and their features. In addition, we discussed various members of the PIC18 family such as the PIC18F252 and PIC18F458.

PROBLEMS

SECTION 1.1: MICROCONTROLLERS AND EMBEDDED PROCESSORS

1. True or False. A general-purpose microprocessor has on-chip ROM.
2. True or False. Generally, a microcontroller has on-chip ROM.
3. True or False. A microcontroller has on-chip I/O ports.
4. True or False. A microcontroller has a fixed amount of RAM on the chip.
5. What components are usually put together with the microcontroller onto a single chip?
6. Intel's Pentium chips used in Windows PCs need external _____ and _____ chips to store data and code.
7. List three embedded products attached to a PC.
8. Why would someone want to use an x86 as an embedded processor?
9. Give the name and the manufacturer of some of the most widely used 8-bit microcontrollers.
10. In Question 9, which one has the most manufacture sources?
11. In a battery-based embedded product, what is the most important factor in choosing a microcontroller?
12. In an embedded controller with on-chip ROM, why does the size of the ROM matter?
13. In choosing a microcontroller, how important is it to have multiple sources for that chip?
14. What does the term "third-party support" mean?
15. Suppose that a microcontroller architecture has both 8-bit and 16-bit versions. Which of the following statements is true?
    (a) The 8-bit software will run on the 16-bit system.
    (b) The 16-bit software will run on the 8-bit system.

SECTION 1.2: OVERVIEW OF THE PIC18 FAMILY

16. The PIC18F458 has _____ bytes of on-chip program ROM.
17. The PIC18F2420 has _____ bytes of on-chip data RAM.
18. The PIC18F452 has _____ on-chip timer(s).
19. The PIC18F458 has _____ bytes of on-chip data RAM.
20. Check the Microchip web site to see if we have a ROMless version of the PIC18. Give the part number if there is one.
21. The PIC18F458 has _____ pins for I/O.
22. The PIC18Fxxx has circuitry to support _____ serial ports.
23. The PIC18F458 on-chip program ROM is of type _____.
24. The PIC16C432 on-chip program ROM is of type _____.
25. The PIC18F452 on-chip program ROM is of type _____.
26. The PIC18F8772 on-chip program ROM is of type _____.
27. Give the amount of program ROM and data RAM for the following chips:
    (a) PIC18F2420   (b) PIC18F458   (c) PIC18F8772
28. Of the PIC18 family, which memory type is the most cost effective if you are using a million of them in an embedded product?

29. What are the main differences between the PIC18F2420 and PIC18F2220?
30. The PIC18F458/4580 has _____ bytes of data EEPROM.

ANSWERS TO REVIEW QUESTIONS

SECTION 1.1: MICROCONTROLLERS AND EMBEDDED PROCESSORS

1. True
2. A microcontroller-based system
3. (d)
4. (d)
5. It is dedicated because it is dedicated to doing one type of job.
6. Embedded system means that the application and processor are combined into a single system.
7. Having multiple sources for a given part means you are not hostage to one supplier. More importantly, competition among suppliers brings about lower cost for that product.

SECTION 1.2: OVERVIEW OF THE PIC18 FAMILY

1. 4K of RAM space, 2M of on-chip ROM space, and a large number of I/O pins.
2. C is OTP while F is the flash ROM.
3. The PIC18F2420 has 768 bytes of RAM and the PIC18F4520 has 1536 bytes.
4. (a) 16K
    (b) 32K
5. 8

CHAPTER 2

PIC ARCHITECTURE & ASSEMBLY LANGUAGE PROGRAMMING

OBJECTIVES

Upon completion of this chapter, you will be able to:

>> Examine the data RAM file register of the PIC microcontroller
>> Manipulate data using the WREG and MOVE instructions
>> Perform simple operations such as ADD and MOVE using the file register and access bank in the PIC microcontroller
>> Explain the purpose of the status register
>> Discuss data RAM memory space allocation in the PIC microcontroller
>> List SFRs (special function registers) of the PIC microcontroller
>> Code simple PIC Assembly language instructions
>> Describe PIC data types and directives
>> Assemble and run a PIC program using MPLAB
>> Describe the sequence of events that occur upon PIC power-up
>> Examine programs in PIC ROM code
>> Explain the PIC ROM memory map
>> Detail the execution of PIC Assembly language instructions
>> Understand the RISC and Harvard architectures of the PIC microcontroller
>> Examine the PIC's registers and data RAM using the MPLAB simulator

CPUs use many registers to store data temporarily. To program in Assembly language, we must understand the registers and architecture of a given CPU and the role they play in processing data. In Section 2.1 we look at the WREG register of the PIC. We demonstrate the use of one of the most widely used registers of the PIC with simple instructions such as MOVE and ADD. Allocation of RAM memory inside the PIC and the access bank of the PIC18 are discussed in Section 2.2. Programming the access bank is examined in Section 2.3. In Section 2.4 we discuss the status register's flag bits and how they are affected by arithmetic instructions. In Section 2.5 we look at some widely used Assembly language directives, pseudocode, and data types related to the PIC. In Section 2.6 we examine Assembly language and machine language programming and define terms such as mnemonics, opcode, operand, and so on. The process of assembling and creating a ready-to-run program for the PIC is discussed in Section 2.7. Step-by-step execution of a PIC program and the role of the program counter are examined in Section 2.8. The merits of RISC architecture are examined in Section 2.9. Assembling and running of the PIC programs with MPLAB are discussed in Section 2.10. In that section we also examine the registers and memory of the PIC using the MPLAB simulator.

SECTION 2.1: THE WREG REGISTER IN THE PIC

PIC microcontrollers have many registers for arithmetic and logic operations. Among them is the WREG register. Because there are a large number of registers inside the PIC, we will concentrate on the widely used register WREG in this section. General-purpose registers are covered in the next section, as well as special function registers. In this section we examine the WREG register of the PIC and show its use with the simple instructions MOVE and ADD.

WREG register

In the CPU, registers are used to store information temporarily. That information could be a byte of data to be processed, or an address pointing to the data to be fetched. The vast majority of PIC registers are 8-bit registers. In the PIC there is only one data type: 8-bit. The 8 bits of a register are shown in the diagram below. These range from the MSB (most-significant bit) D7 to the LSB (least-significant bit) D0. With an 8-bit data type, any data larger than 8 bits must be broken into 8-bit chunks before it is processed.

| D7 | D6 | D5 | D4 | D3 | D2 | D1 | D0 |

The 8-bit WREG register is the most widely used register in the PIC microcontroller. WREG stands for working register, as there is only one. The WREG register is the same as the accumulator in other microprocessors. The WREG register is used for all arithmetic and logic instructions. To understand the use of the WREG register, we will show it in the context of two simple instructions: MOVE and ADD.

MOVLW instruction

Simply stated, the MOVLW instruction moves 8-bit data into the WREG register. It has the following format:

```
MOVLW K      ;move literal value K into WREG
```

K is an 8-bit value that can range from 0–255 in decimal, or 00–FF in hex. The L stands for literal, which means, literally, a number must be used. In other words, if we see the word *literal* in any instruction, we are dealing with an actual value that must be provided right there with the instruction. This is similar to the immediate value we see in other microprocessors. Notice that in MOVLW, the letter L (literal) comes first and then the letter W (WREG), which means "move a literal value to WREG," the destination. The following instruction loads the WREG register with a literal value of 25H (i.e., 25 in hex).

```
MOVLW 25H   ;move value 25H into WREG (WREG = 25H)
```

The following instruction loads the WREG register with value 87H (87 in hex).

```
MOVLW 87H   ;load 87H into WREG (WREG = 87H)
```

The following instruction loads the WREG register with value 15H (15 in hex and 21 in decimal).

```
MOVLW 15H   ;load 15H into WREG (WREG = 15H)
```

ADDLW instruction

The ADDLW instruction has the following format:

```
ADDLW K      ;ADD literal value K to WREG
```

The ADD instruction tells the CPU to add the literal value K to register WREG and put the result back in the WREG register. Notice that in ADDLW, first comes the letter L (literal) and then the letter W (WREG), which means "add a literal value to WREG," the destination. To add two numbers such as 25H and 34H, one can do the following:

```
MOVLW 25H   ;load 25H into WREG
ADDLW 34H   ;add value 34 to W(W = W + 34H)
```

Executing the above lines results in WREG = 59H (25H + 34H = 59H)

Figure 2-1 shows the literal value and WREG being fed to the PIC ALU.

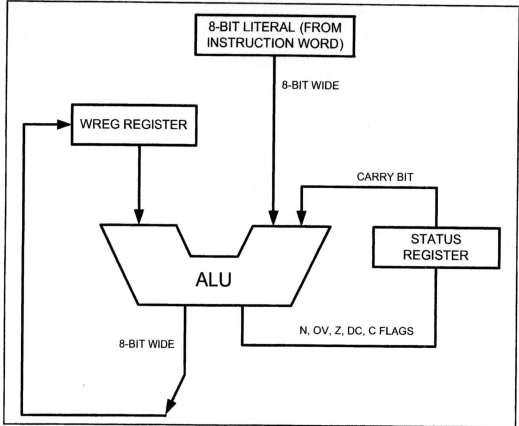

Figure 2-1. PIC WREG and ALU Using Literal Value

The following program will add values 12H, 16H, 31H, and 43H:

```
MOVLW 12H    ;load value 12H into WREG (WREG = 12H)
ADDLW 16H    ;add 16 to WREG (WREG = 28H)
ADDLW 11H    ;add 11 to WREG (WREG = 39H)
ADDLW 43H    ;add 43 to WREG (WREG = 7CH)
```

When programming the WREG register of the PIC microcontroller with a literal value, the following points should be noted:

1. Values can be loaded directly into the WREG register. There is no need for a preceding pound sign or dollar sign to indicate that a value is an immediate value as is the case with some other microcontrollers.
2. If values 0 to F are moved into an 8-bit register such as WREG, the rest of the bits are assumed to be all zeros. For example, in "MOVLW 5H" the result will be WREG = 05H; that is, WREG = 00000101 in binary.
3. Moving a value larger than 255 (FF in hex) into the WREG register will truncate the upper byte and cause a warning in the .err file.

```
MOVLW 7F2H   ;ILLEGAL 7F2H > 8 bits (FFH), becomes F2H
MOVLW 456H   ;ILLEGAL 456H > FFH, becomes 56H
MOVLW 60A5H  ;ILLEGAL but becomes A5H
```

Review Questions

1. Write instructions to move value 34H into the WREG register.
2. Write instructions to add the values 16H and CDH. Place the result in the WREG register.
3. True or false. No value can be moved directly into the WREG register.
4. What is the largest hex value that can be moved into an 8-bit register? What is the decimal equivalent of that hex value?
5. The vast majority of registers in the PIC are _____-bit.

SECTION 2.2: THE PIC FILE REGISTER

The PIC microcontroller has many other registers in addition to the WREG register. They are called data memory space to distinguish them from program (code) memory space. The data memory space in PIC is a read/write (static RAM) memory. In the PIC microcontroller literature, the data memory is also called the *file register*. In this section, we examine the various locations of file register data RAM in the PIC family and discuss their usage with simple instructions such as ADD and MOVE.

File register (data RAM) space allocation in PIC

The file register is read/write memory used by the CPU for data storage, scratch pad, and registers for internal use and functions. As with WREG, we can perform arithmetic and logic operations on many locations of the file register data RAM. The PIC microcontrollers' file register size ranges from 32 bytes to several thousand bytes depending on the chip. Even within the same family, the size of the file register data RAM varies from chip to chip. Notice that the file register data RAM has a byte-size width, just like WREG. The file register data RAM in PIC is divided into two sections: (a) Special Function Registers (SFR), and (b) General-Purpose Registers (GPR). The general-purpose register section is also referred to as General-Purpose RAM (GP RAM). We examine each section separately.

SFRs (Special Function Registers)

The Special Function Registers (SFRs) are dedicated to specific functions such as ALU status, timers, serial communication, I/O ports, ADC, and so on. The function of each SFR is fixed by the CPU designer at the time of design because it is used for control of the microcontroller or peripheral. The PIC SFRs are 8-bit registers. The number of locations in the file register set aside for SFR depends on the pin numbers and peripheral functions supported by that chip. That number can vary from chip to chip even among members of the same family. Some have as few as 7 (8-pin PIC12C508 with no on-chip analog-to-digital converter) and some have over a hundred (40-pin PIC18F458 with on-chip analog-to digital converter). For example, the more timers we have in a PIC chip, the more SFR registers we will have. We will study and use many SFRs in future chapters.

GPR (General-Purpose Registers or RAM)

The general-purpose registers are a group of RAM locations in the file register that are used for data storage and scratch pad. Each location is 8 bits wide and can be used to store any data we want as long as it is 8-bit. Again, the number of RAM locations in the file register that are set aside for general-purpose registers can vary from chip to chip, even among members of the same family. In the PIC controllers, the space that is not allocated to the SFRs typically is used for general-purpose registers. That means in a PIC chip with a thousand-byte file register, no more than 100 bytes are used for SFRs and the rest are used for general-purpose registers. A larger GPR size means more difficulties in managing these registers if you use Assembly language programming. In today's high-performance microcontroller, however, with over a thousand bytes of GPR, the job of managing them is handled by the C compilers. Indeed, the C compilers are the very reason we need a large GPR since it makes it easier for C compilers to store parameters and perform their jobs much faster. See Table 2-1 for a comparison of file registers among various PIC chips. Also see Figure 2-2.

Table 2-1: File Register Size for PIC Chips

	File Register (Bytes)	=	SFR (Bytes)	+	Available space for GPR (Bytes)
PIC12F508	32		7		25
PIC16F84	80		12		68
PIC18F1220	512		256		256
PIC18F452	1792		256		1536
PIC18F2220	768		256		512
PIC18F458	1792		256		1536
PIC18F8722	4096		158		3938

Extracted from http://www.microchip.com

GP RAM vs. EEPROM in PIC chips

Note that there are two RAM columns in the chip information section of the Microchip web site. One refers to the general-purpose registers' (GP RAM) size, and the other is the EEPROM size. GP RAM (which constitutes most of the file register) must not be confused with the EEPROM data memory. The GPRs are used by the CPU for internal data storage, whereas the EEPROMs are considered as an add-on memory that one can also add externally to the chip. In other words, while many PIC chips have zero bytes of EEPROM data memory, it is impossible for a microcontroller to have zero size for the file register. The EEPROM memory of PIC chips is covered in Chapter 14.

The Microchip website provides the data RAM size, which is the same as GPR size.

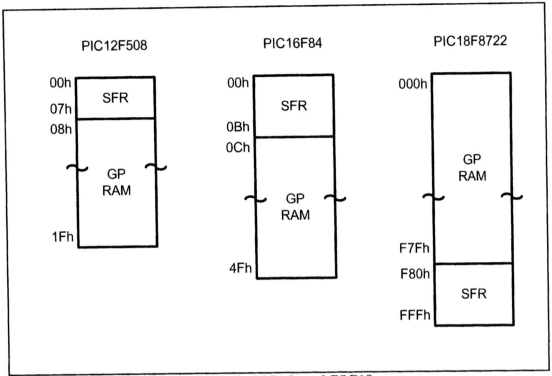

Figure 2-2. File Registers of PIC12, PIC16, and PIC18

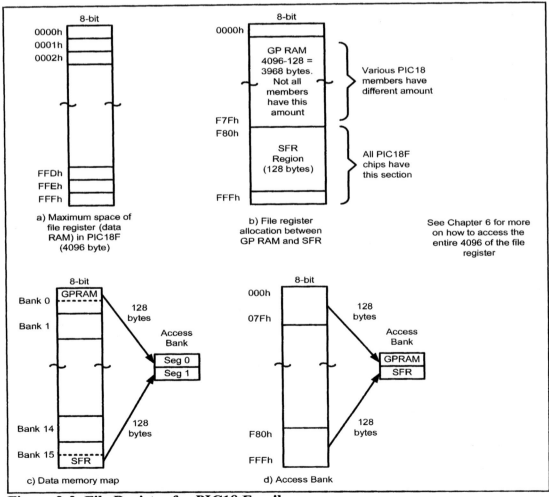

Figure 2-3. File Register for PIC18 Family

File register and access bank in the PIC18

The file register of the PIC18 family can have a maximum of 4096 (4K) bytes. With 4096 bytes, the file register has addresses of 000–FFFH. The file register in the PIC18 is divided into 256-byte banks. Therefore, we can have up to a maximum of 16 banks ($16 \times 256 = 4096$). Although not all members of the PIC18 family have that many banks, every PIC18 family member has at least one bank for the file register. This bank is called the *access bank* and is the default bank when we power up the PIC18 chip. To simplify the discussion of how to use the file register in the PIC family, we focus on this single bank that is found in every member of the PIC18 family. You can examine the file registers in other PIC families such as PIC12 and PIC16 at the Microchip website. In this book we concentrate on the PIC18 series with their large file register, although the insight gained in the process can be applied to the PIC16 and PIC12 series.

Examine the access bank for the PIC18 in Figure 2-3. The 256-byte access bank is divided into two equal sections of 128 bytes. These 128-byte sections are given to the general-purpose registers and special function registers. The 128 bytes from locations 00H to 7FH are set aside for general-purpose registers and are used for read/write storage, or what is normally called a *scratch pad*. These 128 locations of RAM are widely used for storing data and parameters by PIC18 programmers and C compilers. Each location of this 128-byte RAM of general-purpose registers can be accessed directly by its address. We will use these locations in future chapters to store data brought into the CPU via I/O and serial ports. We will also use them to define counters for time delay in Chapter 3. The other 128 bytes of the access bank is used for SFRs. It has addresses of F80H to FFFH, as shown in Figure 2-4. One might wonder why the memory space of the SFRs and GPRs in the access bank is not contiguous. The reason is to allow the RAM space between 080H and F7FH to be used for the general-purpose registers by various members of the PIC18 if they implement a larger data RAM size for the file register. A file register of more than 256 bytes will necessitate bank switching. *Bank switching* is a method used to access all the banks of the file register for PIC18 family members that have more than the minimum access bank. PIC18 members with a file register of more than 256 bytes will be discussed in more detail in Chapter 6 when we discuss bank switching.

Notice that the I/O port SFRs, PORTA, PORTB, PORTC, PORTD, and associated registers are among the most widely used SFRs in PIC. See Chapter 4 for additional information on the special function registers.

F80h	PORTA	FA0h	PIE2	FC0h	----	FE0h	BSR	
F81h	PORTB	FA1h	PIR2	FC1h	ADCON1	FE1h	FSR1L	
F82h	PORTC	FA2h	IPR2	FC2h	ADCON0	FE2h	FSR1H	
F83h	PORTD	FA3h	----	FC3h	ADRESL	FE3h	PLUSW1	*
F84h	PORTE	FA4h	----	FC4h	ADRESH	FE4h	PREINC1	*
F85h	----	FA5h	----	FC5h	SSPCON2	FE5h	POSTDEC1	*
F86h	----	FA6h	----	FC6h	SSPCON1	FE6h	POSTINC1	*
F87h	----	FA7h	----	FC7h	SSPSTAT	FE7h	INDF1	*
F88h	----	FA8h	----	FC8h	SSPADD	FE8h	WREG	
F89h	LATA	FA9h	----	FC9h	SSPBUF	FE9h	FSR0L	
F8Ah	LATB	FAAh	----	FCAh	T2CON	FEAh	FSR0H	
F8Bh	LATC	FABh	RCSTA	FCBh	PR2	FEBh	PLUSW0	*
F8Ch	LATD	FACh	TXSTA	FCCh	TMR2	FECh	PREINC0	*
F8Dh	LATE	FADh	TXREG	FCDh	T1CON	FEDh	POSTDEC0	*
F8Eh	----	FAEh	RCREG	FCEh	TMR1L	FEEh	POSTINC0	*
F8Fh	----	FAFh	SPBRG	FCFh	TMR1H	FEFh	INDF0	*
F90h	----	FB0h	----	FD0h	RCON	FF0h	INTCON3	
F91h	----	FB1h	T3CON	FD1h	WDTCON	FF1h	INTCON2	
F92h	TRISA	FB2h	TMR3L	FD2h	LVDCON	FF2h	INTCON	
F93h	TRISB	FB3h	TMR3H	FD3h	OSCCON	FF3h	PRODL	
F94h	TRISC	FB4h	----	FD4h	----	FF4h	PRODH	
F95h	TRISD	FB5h	----	FD5h	T0CON	FF5h	TABLAT	
F96h	TRISE	FB6h	----	FD6h	TMR0L	FF6h	TBLPTRL	
F97h	----	FB7h	----	FD7h	TMR0H	FF7h	TBLPTRH	
F98h	----	FB8h	----	FD8h	STATUS	FF8h	TBLPTRU	
F99h	----	FB9h	----	FD9h	FSR2L	FF9h	PCL	
F9Ah	----	FBAh	CCP2CON	FDAh	FSR2H	FFAh	PCLATH	
F9Bh	----	FBBh	CCPR2L	FDBh	PLUSW2	*	FFBh	PCLATU	
F9Ch	----	FBCh	CCPR2H	FDCh	PREINC2	*	FFCh	STKPTR	
F9Dh	PIE1	FBDh	CCP1CON	FDDh	POSTDEC2	*	FFDh	TOSL	
F9Eh	PIR1	FBEh	CCPR1L	FDEh	POSTINC2	*	FFEh	.TOSH	
F9Fh	IPR1	FBFh	CCPR1H	FDFh	INDF2	*	FFFh	TOSU	

* - These are not physical registers.

Figure 2-4. Special Function Registers of the PIC18 Family.

Review Questions

1. True or false. Data space in PIC is SRAM memory, whereas program (code) space is of the ROM type.
2. The general-purpose RAM and SFRs together are called_____.
3. True or false. The larger the file register, the more difficult it is to manage.
4. True or false. The more file register space that is assigned to the SFRs, the less is available for the GP RAM.
5. The SFR registers in PIC are _____-bit.
6. The file register space in PIC18 is divided into _____-byte banks.
7. The file register space in PIC18 can be a maximum of _____ bytes.

SECTION 2.3: USING INSTRUCTIONS WITH THE DEFAULT ACCESS BANK

The instructions we have used so far are the literal (constant) value of K and the WREG register. They also used the WREG register as their destination. We saw simple examples of using MOVLW and ADDLW earlier in Section 2.1. The PIC allows direct access to other locations in the file register for ALU and other operations. In this section we show the instructions using various locations of the file register. This is one of the most important sections in the book for mastering the topic of PIC Assembly language programming.

MOVWF instruction

As we discussed in the last section, the access bank of the file register is the default bank upon powering up the PIC18. The term *file register* must be emphasized because the instructions have the letter F in their mnemonics. In instructions such as MOVWF, the F stands for a location in the file register, while W means WREG. The MOVWF instruction tells the CPU to move (in reality, copy) the source register of WREG to a destination in the file register (F). After this instruction is executed, the location in the file register will have the same value as register WREG. The location in the file register can be one of the SFRs or a location in the general purpose registers region. For example, the "MOVWF PORTA" instruction will move the contents of WREG into the SFR register called PORTA. The following program first loads the WREG register with value 55H, then moves this value around to various SFRs of ports B, C, and D:

```
MOVLW 55H          ;WREG = 55H
MOVWF PORTB        ;copy WREG to Port B (Port B = 55H)
MOVWF PORTC        ;copy WREG to Port C (Port C = 55H)
MOVWF PORTD        ;copy WREG to Port D (Port D = 55H)
```

PORTB, PORTC, and PORTD are part of the special function registers in the file register, as was shown in Figure 2-4. They can be connected to the I/O pins of the PIC microcontroller as we will see in Chapter 4. We can also move (copy) the contents of WREG into any location in the general-purpose registers (RAM) region of the file registers. The following program will put 99H into locations 0–4 of the GPR region in the file register:

```
MOVLW 99H          ;WREG = 99H
MOVWF 0H           ;move (copy) WREG contents to location 0h
MOVWF 1H           ;move (copy) WREG contents to location 1h
MOVWF 2H
MOVWF 3H
MOVWF 4H
```

Address	Data
000	99
001	99
002	99
003	99
004	99

The chart indicates the contents of addresses 0–4 after execution of the code.

Example 2-1

State the contents of file register RAM locations after the following program:

```
        MOVLW  99H       ;load WREG with value 99H
        MOVWF  12H
        MOVLW  85H       ;load WREG with value 85H
        MOVWF  13H
        MOVLW  3FH       ;load WREG with value 3FH
        MOVWF  14H
        MOVLW  63H       ;load WREG with value 63H
        MOVWF  15H
        MOVLW  12H       ;load WREG with value 12H
        MOVWF  16H
```

Solution:

After the execution of MOVWF 12H fileReg RAM location 12H has value 99H;
After the execution of MOVWF 13H fileReg RAM location 13H has value 85H;
After the execution of MOVWF 14H fileReg RAM location 14H has value 3FH;
After the execution of MOVWF 15H fileReg RAM location 15H has value 63H;
And so on, as shown in the chart.

Address	Data
012	99
013	85
014	3F
015	63
016	12

Notice that you cannot move literal (immediate) values directly into the general-purpose RAM locations in the PIC18. They must be moved there via WREG.

More instructions involving the WREG and the access bank

There is a group of logic and arithmetic instructions that involve both the WREG and a location in the file register. The ADDWF instruction is one of them. The ADDWF instruction adds together the contents of WREG and a file register location. The file register location can be one of the SFRs or a general-purpose register. The destination for the result can be the WREG or the file register. The following format indicates the destination:

```
ADDWF  fileReg, D
```

where fileReg is the file register location and D indicates the destination bit. The D bit can be 0 or 1. If D = 0, it means that the destination is WREG. If D = 1, then the result will be placed in the file register.

The following will first put value 22H into GP RAM locations 5, 6, and 7, then add them together and put the result in WREG:

```
MOVLW 22H    ;WREG = 22H
MOVWF 5H     ;move(copy) WREG contents to location 5H
MOVWF 6H     ;move(copy) WREG contents to location 6H
MOVWF 7H     ;move(copy) WREG contents to location 7H
ADDWF 5H, 0 ;add W and loc 5, result in WREG (W = 44H)
ADDWF 6H, 0 ;add W and loc 6, result in WREG (W = 66H)
ADDWF 7H, 0 ;add W and loc 7, result in WREG (W = 88H)
```

Address	Data
005	22
006	22
007	22

Address	Data
005	22
006	22
007	22

GPR after the execution up to
"MOVWF 7H"
WREG = 22H

GPR after the execution up to
"ADDWF 7H, 0"
WREG = 88H

Now look at the same program where the result is put into file register location 7:

```
MOVLW 22H    ;WREG = 22H
MOVWF 5H     ;move (copy) WREG contents to location 5H
MOVWF 6H     ;move (copy) WREG contents to location 6H
MOVWF 7H     ;move (copy) WREG contents to location 7H
ADDWF 5, 0  ;add W and loc 5, result in WREG (W = 44H)
ADDWF 6, 0  ;add W and loc 6, result in WREG (W = 66H)
ADDWF 7, 1  ;add W and loc 7, result in location 7H
             ;now location 7 has 88H and WREG = 66H
```

Address	Data
005	22
006	22
007	22

Address	Data
005	22
006	22
007	88

GP RAM after the execution up to
"MOVWF 7H"
WREG = 22H

GP RAM after the execution up to
"ADDWF 7H, 1"
WREG = 66H

To make things less confusing as far as the D bit is concerned, the PIC assembler allows us to use the letters W or F instead of 0 or 1 to indicate the destination. Look at the following two formats:

```
ADDWF fileReg, w  ;add WREG and fileReg. WREG = the result
ADDWF fileReg, f  ;add WREG and fileReg
                  ;fileReg = the result
```

This format is much easier and will help us to avoid confusion about the destination. Look at the rewrite of the last program with the new format for the ADDWF instruction:

```
MOVLW 22H    ;WREG = 22H
MOVWF 5H     ;move (copy) WREG contents to location 5H
MOVWF 6H     ;move (copy) WREG contents to location 6H
MOVWF 7H     ;move (copy) WREG contents to location 7H
ADDWF 5H,W   ;add W and loc 5, result in WREG (W = 44H)
ADDWF 6H,W   ;add W and loc 6, result in WREG (W = 66H)
ADDWF 7H,F   ;add W and loc 7, result in location 7
             ;now location 7 has 88H and WREG = 66H
```

The above concept is important and must be understood since there are a large number of instructions with this format. Compare Examples 2-2 and 2-3.

Example 2-2

State the contents of file register RAM locations 12H and WREG after the following program:

```
MOVLW 0          ;move 0 WREG to clear it (WREG = 0)
MOVWF 12H        ;move WREG to location 12 to clear it
MOVLW 22H        ;load WREG with value 22H
ADDWF 12H, F     ;add WREG to loc 12H, loc 12 = sum
ADDWF 12H, F     ;add WREG to loc 12H, loc 12 = sum
ADDWF 12H, F     ;add WREG to loc 12H, loc 12 = sum
ADDWF 12H, F     ;add WREG to loc 12H, loc 12 = sum
```

Solution:

The program clears both the WREG and RAM location 12H in the file register. Then it loads WREG with value 22H. From then on, it adds the WREG register and location 12 together and saves the result in location 12H. It does that four times. At the end, location 12H of GP RAM has the value of 88H (4 × 22H = 88H) and WREG = 22H.

After each "ADDWF 12, F" instruction

Address	Data	Address	Data	Address	Data	Address	Data
011		011		011		011	
012	22	012	44	012	66	012	88
013		013		013		013	

WREG = 22H   WREG = 22H   WREG = 22H   WREG = 22H

Example 2-3

Rewrite the last example to place the sum in WREG as you add the file register locations and the WREG register.

```
MOVLW  0              ;move 0 WREG to clear it (WREG = 0)
MOVWF  12H            ;move WREG to location 12 to clear it
MOVLW  22H            ;load WREG with value 22H
ADDWF  12H, W         ;add WREG and loc 12H, WREG = sum
ADDWF  12H, W         ;add WREG and loc 12H, WREG = sum
ADDWF  12H, W         ;add WREG and loc 12H, WREG = sum
ADDWF  12H, W         ;add WREG and loc 12H, WREG = sum
```

Solution:

The program adds WREG and location 12H together and saves the result in WREG each time. At the end, location 12H has a value of 22H and WREG = 88H
(4 × 22H = 88H).

After each "ADDWF 12, W" instruction:

Address	Data	Address	Data	Address	Data	Address	Data
011		011		011		011	
012	22	012	22	012	22	012	22
013		013		013		013	

WREG = 22H	WREG = 44H	WREG = 66H	WREG = 88H

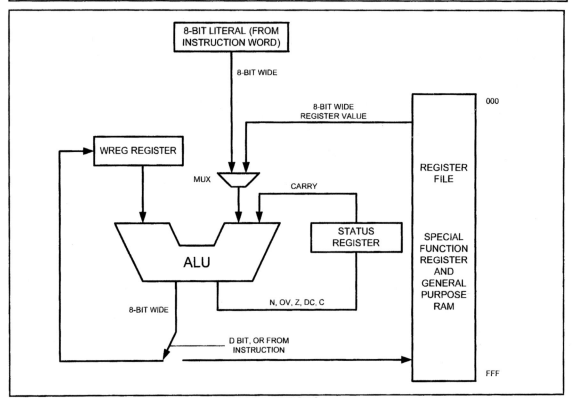

Figure 2-5. WREG, fileReg, and ALU in PIC18

Now examine the instructions in Tables 2-2 and 2-3. The instructions in Table 2-2 operate on both WREG and a file register location and then give you the option of placing the result in WREG or a file register location. The instructions in Table 2-3, however, operate on the file register only and then give you the option of placing the result in WREG or a file register location.

Table 2-2: ALU Instructions Using Both WREG and fileReg

Instruction		
ADDWF	fileReg, d	ADD WREG and fileReg
ADDWFC	fileReg, d	ADD WREG and fileReg with Carry
ANDWF	fileReg, d	AND WREG with fileReg
IORWF	fileReg, d	OR WREG with fileReg
SUBFWB	fileReg, d	Subtract fileReg from WREG with borrow
SUBWF	fileReg, d	Subtract WREG from fileReg
SUBWFB	fileReg, d	Subtract WREG from fileReg with borrow
XORWF	fileReg, d	Exclusive-OR WREG with fileReg

Note: The d bit selects the destination for the operation. If d = w; the result is stored in WREG (d = 0). If d = F; the result is stored in the fileReg (d = 1). The default is F. That means "ADDWF myfile" is the same as "ADDWF myfile,F."

See Chapter 5 for examples of the instructions in Table 2-2.

Table 2-3: File Register Instructions Using fileReg or WREG as Destination

Instruction		
COMF	fileReg, d	Complement fileReg
DECF	fileReg, d	Decrement fileReg
DECFSZ	fileReg, d	Decrement fileReg and skip if zero
DECFSNZ	fileReg, d	Decrement fileReg and skip if not zero
INCF	fileReg, d	Increment fileReg
INCFSZ	fileReg, d	Increment fileReg and skip if zero
INCSNZ	fileReg, d	Increment fileReg and skip if not zero
MOVF	fileReg, d	Move fileReg
NEGF	fileReg, d	Negative fileReg
RLCF	fileReg, d	Rotate left fileReg through carry
RLNCF	fileReg, d	Rotate left fileReg (No carry)
RRCF	fileReg, d	Rotate right fileReg through carry
RRNCF	fileReg, d	Rotate right fileReg (No carry)
SWAPF	fileReg, d	Swap nibbles in fileReg
BTG	fileReg, d	Bit Toggle fileReg

Note: The d bit selects the destination for the operation. If d = w; the result is stored in the WREG (d = 0). If d = F; the result is stored in the fileReg (d = 1). The default is F. That means "DECF myfile" is the same as "DECF myfile,F."

Chapters 3 through 6 will show how to use the instructions in Table 2-3.

COMF instruction

The "COMF fileReg, d" instruction complements (inverts) the contents of fileReg and places the result in WREG or fileReg. This is an example of what is called Read - Modify - Write and we will see more of this in future chapters. In the following program, we put 55H into WREG and then send it to SFR location of Port B. Then the content of Port B is complemented, which becomes AA in hex. The 01010101 (55H) is inverted and becomes 10101010 (AAH).

```
MOVLW 55H          ;WREG = 55h
MOVWF PORTB        ;Move WREG to Port B SFR (PB = 55h)
COMF  PORTB, F     ;complement Port B (PB = AAh)
```

Examine Example 2-4.

Example 2-4

Write a simple program to toggle the SFR of PORT B continuously forever.

Solution:

```
    MOVLW 55H            ;WREG = 55h
    MOVWF PORTB          ;move WREG to Port B SFR (PB = 55h)
B1  COMF  PORTB, F       ;complement Port B and place it in Port B
    GOTO  B1             ;repeat forever (See Chapter 3 for GOTO)
```

DECF instruction

The "DECF fileReg, d" instruction decrements (subtracts one from) the contents of fileReg and places the result in WREG or fileReg. In the following program, we put the value 3 into fileReg location 0x20. Then the value in location 0x20 is decremented and placed in fileReg.

```
MOVLW 3            ;WREG = 3
MOVWF 20H          ;move WREG to loc 20H (loc 20H = 3)
DECF  0x20, F      ;loc 20H has 2
DECF  0x20, F      ;loc 20H has 1
DECF  0x20, F      ;loc 20H has 0 and WREG = 3
```

Now, contrast the above code with the following:

```
MOVLW 3            ;WREG = 3
MOVWF 20H          ;move WREG to loc 20H (loc 20H = 3)
DECF  0x20, W      ;loc 20H has 3 (WREG = 2)
DECF  0x20, W      ;loc 20H has 3 (WREG = 2)
DECF  0x20, W      ;loc 20H has 3 (WREG = 2)
```

The above concept will be used in loops in the next chapter.

MOVF instruction

The MOVF mnemonic is intended to perform MOVFW. It has the following format:

```
MOVF        fileReg, D
```

If D = 0, it copies the content of fileReg to WREG. If D = 1, the content of fileReg is copied to itself. While typically we use the MOVF instruction to bring data into WREG from I/O pins, we sometimes use it to copy fileReg to itself for the purpose of testing fileReg contents. Examine the difference between the MOVWF and MOVF instructions. We used the MOVWF instruction earlier to move data to SFRs such as Port B. We also saw how it is widely used to load fixed (literal) data into the RAM locations of the file register because there is no way we can load data into the file register directly. In contrast, the MOVF instruction is widely used to bring data from I/O ports such as Port B into the CPU. We also use the MOVF instruction to bring data into WREG from any SFRs or from any location in the GP RAM in order to perform arithmetic and operations on them. Examine Examples 2-5 and 2-6. Note that the only time we use the "MOVF fileReg, F" instruction to copy data from fileReg to itself is when we want to affect the flag bits of the status register. The status register bits are discussed in the next section, and Chapter 3 shows how to use them.

Example 2-5

Write a program to get data from the SFRs of Port B and send it to the SFRs of PORT C continuously.

Solution:

```
AGAIN MOVF   PORTB, W    ;bring data from PortB into WREG
      MOVWF  PORTC       ;send it to Port C
      GOTO   AGAIN       ;keep doing it forever
```

In Example 2-5 we use GOTO to repeat an action indefinitely. We will study looping in Chapter 3. The details of I/O ports are discussed in Chapter 4.

Example 2-6

Write a program to get data from the SFRs of Port B. Add the value 5 to it and send it to the SFRs of Port C.

Solution:

```
      MOVF   PORTB,W     ;bring data from Port B into WREG
      ADDLW  05H         ;add 5 to WREG
      MOVWF  PORTC       ;copy WREG to Port C
```

MOVFF instruction

The MOVFF instruction copies data from one location in fileReg to another location in fileReg. The fileReg location for source and destination can be any of the 4096 locations in the data RAM space of the PIC18. The MOVFF instruction allows us to move data within the 4K space of the data RAM without going through the WREG register. (See Figure 2-6.) Compare Examples 2-5 and 2-7.

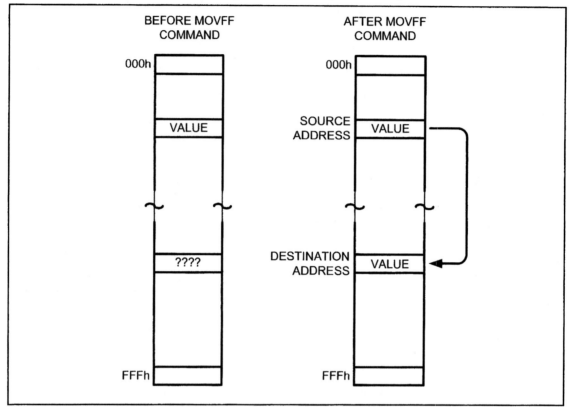

Figure 2-6. Moving Data Directly Among the fileReg Locations

Example 2-7

Write a program to get data from the SFRs of Port B and send it to the SFRs of PORT C continuously using MOVFF. Compare this to Example 2-5 and explain the difference.
Solution:

```
AGAIN MOVFF PORTB, PORTC      ;copy data from Port B to Port C
      GOTO AGAIN              ;keep doing it forever
```

In Example 2-5 we have:

```
AGAIN MOVF  PORTB, W          ;bring data from Port B into WREG
      MOVWF PORTC             ;send it to Port C
      GOTO  AGAIN             ;keep doing it forever
```

Using MOBVFF we simply copy data from one location to another location. But when we use WREG we can perform arithmetic and logic operations on data before it is moved.

Review Questions

1. True or false. The access bank is 256 bytes divided evenly between GPRs and SFRs.
2. Write instructions to add the values 16H and CDH. Place the result in location 0 of the file register.
3. True or false. No value can be moved directly into general-purpose RAM.
4. What is the largest hex value that can be moved into a location in the file register? What is the decimal equivalent of the hex value?
5. "ADDWF PORTB, W" puts the result in _____ .

SECTION 2.4: PIC STATUS REGISTER

Like all other microprocessors, the PIC has a flag register to indicate arithmetic conditions such as the carry bit. The flag register in the PIC is called the *status register*. In this section, we discuss various bits of this register and provide some examples of how it is altered. Chapters 3 and 5 show how the flag bits of the status register are used.

PIC18 status register

The status register is an 8-bit register. It is also referred to as the *flag register*. Although the status register is 8 bits wide, only 5 bits of it are used by the PIC18. The three unused bits are unimplemented and read as 0. The five flags are called *conditional flags*, meaning that they indicate some conditions that result after an instruction is executed. These five flags are C (carry), DC (digital carry), Z (zero), OV (overflow), and N (negative). See Figure 2-7 for the bits of the status register. Each of the conditional flags can be used to perform a conditional branch (jump), as we will see in Chapter 3.

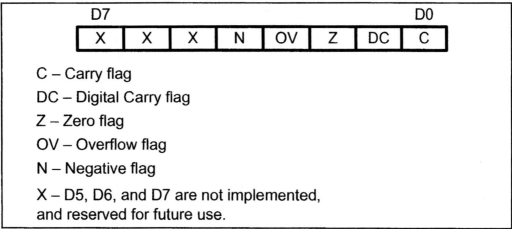

Figure 2-7. Bits of Status Register

The following is a brief explanation of the flag bits of the status register. The impact of instructions on this register is then discussed.

C, the carry flag

This flag is set whenever there is a carry out from the D7 bit. This flag bit is affected after an 8-bit addition or subtraction. Chapter 5 shows how the carry flag is used.

DC, the digital carry flag

If there is a carry from D3 to D4 during an ADD or SUB operation, this bit is set; otherwise, it is cleared. This flag bit is used by instructions that perform BCD (binary coded decimal) arithmetic. In some microprocessors this is called the AC flag (Auxiliary Carry flag). See Chapter 5 for more information.

Z, the zero flag

The zero flag reflects the result of an arithmetic or logic operation. If the result is zero, then $Z = 1$. Therefore, $Z = 0$ if the result is not zero. See Chapter 3 to see how we use the Z flag for looping.

OV, the overflow flag

This flag is set whenever the result of a signed number operation is too large, causing the high-order bit to overflow into the sign bit. In general, the carry flag is used to detect errors in unsigned arithmetic operations while the overflow flag is used to detect errors in signed arithmetic operations. The OV and N flag bits are used for the signed number arithmetic operations and are discussed in Chapter 5.

N, the negative flag

Binary representation of signed numbers uses D7 as the sign bit. The negative flag reflects the result of an arithmetic operation. If the D7 bit of the result is zero, then $N = 0$ and the result is positive. If the D7 bit is one, then $N = 1$ and the result is negative. The negative and OV flag bits are used for the signed number arithmetic operations and are discussed in Chapter 5.

ADDLW instruction and the status register

Next we examine the impact of the ADDLW instruction on the flag bits C, DC, and Z of the status register. Some examples should clarify their meanings. Although all the flag bits C, Z, DC, OV, and N are affected by the ADDLW instruction, we will focus on flags C, DC, and Z for now. The other flag bits are discussed in Chapter 5, because they relate only to signed number operations. Examine Examples 2-8 through 2-10 to see the impact of the ADD instruction on selected flag bits.

Not all instructions affect the flags

Some instructions affect all the five flag bits C, DC, Z, OV, and N (e.g., ADDWL). But some instructions affect no flag bits at all. The move instructions are in this category (except MOVF). And some instructions affect only the Z or N flag bits, or both. The logic instructions are in this category (e.g., ANDWL).

Example 2-8

Show the status of the C, DC, and Z flags after the addition of 38H and 2FH in the following instructions:

```
MOVLW 38H
ADDLW 2FH          ;add 2FH to WREG
```

Solution:

$$
\begin{array}{rl}
38H & 0011\ 1000 \\
+\ 2FH & 0010\ 1111 \\
\hline
67H & 0110\ 0111 \qquad WREG = 67H
\end{array}
$$

C = 0 because there is no carry beyond the D7 bit.
DC = 1 because there is a carry from the D3 to the D4 bit.
Z = 0 because the WREG has a value other than 0 after the addition.

Example 2-9

Show the status of the C, DC, and Z flags after the addition of 9CH and 64H in the following instructions:

```
MOVLW 9CH
ADDLW 64H          ;add 64H to WREG
```

Solution:

$$
\begin{array}{rl}
9CH & 1001\ 1100 \\
+\ 64H & 0110\ 0100 \\
\hline
100H & 0000\ 0000 \qquad WREG = 00
\end{array}
$$

C = 1 because there is a carry beyond the D7 bit.
DC = 1 because there is a carry from the D3 to the D4 bit.
Z = 1 because the WREG has a value 0 in it after the addition.

Example 2-10

Show the status of the C, DC, and Z flags after the addition of 88H and 93H in the following instructions:

```
MOVLW 88H
ADDLW 93H          ;add 93H to WREG
```

Solution:

$$
\begin{array}{rl}
88H & 1000\ 1000 \\
+\ 93H & 1001\ 0011 \\
\hline
11BH & 0001\ 1011 \qquad WREG = 1BH
\end{array}
$$

C = 1 because there is a carry beyond the D7 bit.
DC = 0 because there is no carry from the D3 to the D4 bit.
Z = 0 because the WREG has a value other than 0 after the addition.

Table 2-4 shows the instructions and the flag bits affected by them. Appendix A provides a complete list of all the instructions and their associated flag bits.

Table 2-4: Instructions That Affect Flag Bits

Instruction	C	DC	Z	OV	N
ADDLW	X	X	X	X	X
ADDWF	X	X	X	X	X
ADDWFC	X	X	X	X	X
ANDLW			X		X
ANDWF			X		X
CLRF			X		
COMF			X		X
DAW	X				
DECF	X	X	X	X	X
INCF	X	X	X	X	X
IORLW			X		X
IORWF			X		X
MOVF			X		
NEGF	X	X	X	X	X
RLCF	X		X		X
RLNCF			X		X
RRCF	X		X		X
RRNCF			X		X
SUBFWB	X	X	X	X	X
SUBLW	X	X	X	X	X
SUBWF	X	X	X	X	X
SUBWFB	X	X	X	X	X
XORLW			X		X
XORWF			X		X

Note: X can be 0 or 1.

See Chapter 5 for how to use these instructions.

Flag bits and decision making

Because status flags are also called conditional flags, there are instructions that will make a conditional jump (branch) based on the status of the flag bits. Table 2-5 provides the list. Chapter 3 will discuss the conditional branch instructions and how they are used.

Table 2-5: PIC18 Branch (Jump) Instructions Using Flag Bits

Instruction	Action
BC	Branch if C = 1
BNC	Branch if C ≠ 0
BZ	Branch if Z = 1
BNZ	Branch if Z ≠ 0
BN	Branch if N = 1
BNC	Branch if N ≠ 0
BOV	Branch if OV = 1
BNOV	Branch if OV ≠ 0

Review Questions

1. The flag register in the PIC is called the _____.
2. What is the size of the flag register in the PIC?
3. Which bits of the status register are unused?
4. Find the C, Z, and DC flag bits for the following code:
   ```
   MOVLW  9FH
   ADDLW  61H
   ```
5. Find the C, Z, and DC flag bits for the following code:
   ```
   MOVLW  82H
   ADDLW  22H
   ```
6. Find the C, Z, and DC flag bits for the following code:
   ```
   MOVLW  67H
   ADDLW  99H
   ```

SECTION 2.5: PIC DATA FORMAT AND DIRECTIVES

In this section we look at some widely used data formats and directives supported by the PIC assembler.

PIC data type

The PIC microcontroller has only one data type. It is 8 bits, and the size of each register is also 8 bits. It is the job of the programmer to break down data larger than 8 bits (00 to FFH, or 0 to 255 in decimal) to be processed by the CPU. For examples of how to process data larger than 8 bits, see Chapter 5. The data types used by the PIC can be positive or negative. A discussion of signed numbers is given in Chapter 5 also. The bit-addressable data is discussed in Chapters 4 and 6.

Data format representation

There are four ways to represent a byte of data in the PIC assembler. The numbers can be in hex, binary, decimal, or ASCII formats. The following are examples of how each work.

Hex numbers

There are four ways to show hex numbers:

1. We can use h (or H) right after the number like this: MOVLW 99H
2. Put 0x (or 0X) in front of the number like this: MOVLW 0x99
3. Put nothing in front or back of the number like this: MOVLW 99
4. Put h in front of the number, but with single quotes around the number like this: MOVLW h'99'

We use all four of these methods in this book, because many application notes out there use one of them and we need to get used to them. Notice that some PIC assemblers might give you a warning (but no error) when you use 99H because the assembler already knows that data is in hex and there is no need to remind it. We do that simply to remind ourselves (and it is a good reminder) when we do coding in Assembly.

Here are a few lines of code that use the hex format:

```
MOVLW 25    ;WREG = 25H
ADDLW 0x11  ;WREG = 25H + 11H = 36H
ADDLW 12H   ;WREG = 36H + 12H = 48H
ADDLW H'2A' ;WREG = 48H + 2AH = 72H
ADDLW 2CH   ;WREG = 72H + 2CH = 9EH
```

The following are invalid:

```
MOVLW E5H   ;invalid, it must be MOVLW 0E5H
ADDLW C6    ;invalid, it must be ADDLW 0C6
```

Notice in the last two instructions that if the value starts with the hex digits A–F, then it must be preceded with a zero. However, the following is valid:

```
MOVLW 0F    ;valid, WREG = 0FH (or 00001111 in binary)
```

Binary numbers

There is only one way to represent binary numbers in a PIC assembler. It is as follows:

```
MOVLW B'10011001' ;WREG = 10011001 or 99 in hex
```

The lowercase b will also work. Note that ' is the single quote key, which is on the same key as the double quote ". This is different from other assemblers such as the 8051 and x86. Here are some examples of how to use it:

```
MOVLW B'00100101' ;WREG = 25H
ADDLW B'00010001' ;WREG = 25H + 11H = 36H
```

Decimal numbers

There are two ways to represent decimal numbers in a PIC assembler. One way is as follows:

```
MOVLW D'12'      ;WREG = 00001100 or 0C in hex
```

The lowercase d will work also. This is different from other assemblers such as the 8051 and x86. In those assemblers, to indicate decimal numbers we simply use the decimal (e.g., 12) and nothing before or after it, while in the PIC assembler, 12 is the default for hex numbers. Here are some examples of how to use it:

```
MOVLW D'37' ;WREG = 25H (37 in decimal is 25 in hex)
ADDLW D'17' ;WREG = 37 + 17 = 54 where 54 in dec is 36H
```

The other way to represent decimal numbers is to use ".value" as seen in some application notes for PIC microcontrollers. This is shown as follows:

```
MOVLW .12        ;WREG = 00001100 = 0CH = 12
```

ASCII character

To represent ASCII data in a PIC assembler we use the letter A as follows:

```
MOVLW A'2'   ;WREG = 00110010 or 32 in hex (See Appendix F)
```

Lowercase 'a' will work as well. Again, this is different from other assemblers such as the 8051 and x86. In those assemblers, single quotes are used for single ASCII characters and double quotes are used for a string. Here are some more examples:

```
MOVLW A'9'   ;WREG = 39H, which is hex number for ASCII '9'
ADDLW A'1'   ;WREG = 39H + 31H = 70H
             ;(31 hex is for ASCII '1')
MOVLW '9'    ;WREG = 39H another way for ASCII
```

To define ASCII strings (more than one character), we use the DB (define byte) directive. We will see DB usage in Chapter 6.

Assembler directives

While instructions tell the CPU what to do, directives (also called pseudo-instructions) give directions to the assembler. For example, the MOVLW and ADDLW instructions are commands to the CPU, but EQU, ORG, and END are directives to the assembler. The following sections present some more widely used directives of the PIC and how they are used.

EQU (equate)

This is used to define a constant value or a fixed address. The EQU directive does not set aside storage for a data item, but associates a constant number with a data or an address label so that when the label appears in the program, its constant will be substituted for the label. The following uses EQU for the counter constant, and then the constant is used to load the WREG register:

```
COUNT EQU    0x25
           . . .          . . . .
      MOVLW        COUNT ;WREG = 25H
```

When executing the above instruction "MOVLW   COUNT", the register WREG will be loaded with the value 25H. What is the advantage of using EQU? Assume that a constant (a fixed value) is used throughout the program, and the programmer wants to change its value everywhere. By the use of EQU, the programmer can change it once and the assembler will change all of its occurrences throughout the program, rather than search the entire program trying to find every occurrence.

SET

This directive is used to define a constant value or a fixed address. In this regard, the SET and EQU directives are identical. The only difference is the value assigned by the SET directive may be reassigned later.

Using EQU for fixed data assignment

To get more practice using EQU to assign fixed data, examine the following:

```
                ;in hexadecimal
DATA1  EQU   39           ;hex data is the default
DATA2  EQU   0x39         ;another way for hex
DATA3  EQU   39H          ;another way for hex (redundant)
DATA4  EQU   H'39'        ;another way for hex
DATA5  EQU   h'39'        ;another way for hex

                ;in binary
DATA6  EQU   b'00110101'  ;binary (35 in hex)
DATA7  EQU   B'00110101'  ;binary (35 in hex)

                ;in decimal
DATA8  EQU   D'28'        ;decimal numbers (1C in hex)
DATA9  EQU   d'28'        ;second way for decimal

                ;in ASCII
DATA10 EQU   A'2'         ;ASCII characters
DATA11 EQU   a'2'         ;another way for ASCII char
DATA12 EQU   '2'          ;another way for ASCII char
```

We use DB to allocate code ROM memory locations for fixed data such as ASCII strings. See Chapter 6 for more examples.

Using EQU for SFR address assignment

EQU is also widely used to assign SFR addresses. Examine the following code:

```
COUNTER EQU 0x00    ;counter value 00
PORTB EQU 0xFF6     ;SFR Port B address
MOVLW COUNTER       ;WREG = 00H
MOVWF PORTB         ;Port B now has 00 too
INCF  PORTB, F      ;Port B has 01
INCF  PORTB, F      ;increment Port B (Port B = 02)
INCF  PORTB, F      ;increment Port B (Port B = 03)
```

The above is for the PIC18 family. If you use a different PIC controller such as PIC16F, where Port B is a different address, then change the EQU address for Port B and re-assemble the program and run it.

```
COUNTER EQU 0x00    ;counter value 00
PORTB EQU 0x07      ;Port B addr in PIC16F
MOVLW COUNTER       ;WREG = 00H
MOVWF PORTB         ;Port B now has 00 too
INCF  PORTB, F      ;Port B has 01
INCF  PORTB, F      ;Port B has 02
INCF  PORTB, F      ;Port B has 03
        and so on
```

Using EQU for RAM address assignment

Another common usage of EQU is for the address assignment of the general-purpose region of the file register. Examine the following rewrite of an earlier example using EQU:

```
MYREG EQU 0x12      ;assign RAM loc to MYREG

MOVLW 0             ;clear WREG (WREG = 0)
MOVWF MYREG         ;clear MYREG (loc 12H has 0)
MOVLW 22H           ;WREG = 22H
ADDWF MYREG, F      ;MYREG = WREG + MYREG
ADDWF MYREG, F      ;MYREG = WREG + MYREG
ADDWF MYREG, F      ;MYREG = WREG + MYREG
ADDWW MYREG, F      ;MYREG = WREG + MYREG
```

This is especially helpful when the address needs to be changed in order to use a different PIC chip for a given project. It is much easier to refer to a name than a number when accessing RAM address locations.

The following program will move value 9 into RAM locations 0–4, then add them together and place the sum in location 10H:

```
MYVAL EQU    9      ;MYVAL = 9
R0    EQU    0      ;assign RAM addresses to R0
R1    EQU    1      ;to R1
R2    EQU    2
R3    EQU    3
R4    EQU    4
SUM   EQU    10H

MOVLW MYVAL         ;WREG = 9
MOVWF R0            ;RAM loc 0 has 9
MOVWF R1            ;RAM loc 1 has 9
MOVWF R2            ;RAM loc 2 has 9
MOVWF R3            ;RAM loc 3 has 9
MOVWF R4            ;RAM loc 4 has 9
MOVLW 0             ;WREG = 0
ADDWF R0, W         ;WREG = R0 + WREG
ADDWF R1, W         ;WREG = R1 + WREG
ADDWF R2, W         ;WREG = R2 + WREG
ADDWF R3, W         ;WREG = R3 + WREG
ADDWF R4, W         ;WREG = R4 + WREG
MOVWF SUM
```

ORG (origin)

The ORG directive is used to indicate the beginning of the address. It can be used for both code and data. The number that comes after ORG must be in hex.

END directive

Another important pseudocode is the END directive. This indicates to the assembler the end of the source (asm) file. The END directive is the last line of the PIC program, meaning that anything after the END directive in the source code is ignored by the assembler.

LIST directive

Unlike ORG and END, which are used by all assemblers, the LIST directive is unique to the PIC assembler. It indicates to the assembler the specific PIC chip for which the program should be assembled. It is used as follows:

```
LIST P=18F458
```

The above tells the PIC assembler to assemble the program specifically for the PIC18F458 microcontroller. We use LIST to state the target chip.

#include directive

The #include directive tells the PIC assembler to use the libraries associated with the specific chip for which we are compiling the program.

_config directive

The _config directive tells the assembler the configuration bits for the targeted PIC chip. It is important to use the correct _config directive, because incorrect use may make the chip unusable. The configuration bits are read during power-up of the PIC device and are stored at location 300000H. Microchip has defined the _config directive symbols to ease the configuration. These symbols are located in the .INC file for the device that is being used. See Chapter 8.

radix directive

We can use the radix directive to indicate whether the numbering system is hexadecimal or decimal. The default is hex if we do not use the radix directive. If we use "radix dec", the default representation will change to decimal and any unformatted number will be interpreted as decimal rather than hex, as seen before.

Rules for labels in Assembly language

By choosing label names that are meaningful, a programmer can make a program much easier to read and maintain. There are several rules that names must follow. First, each label name must be unique. The names used for labels in Assembly language programming consist of alphabetic letters in both upper and lower case, the digits 0 through 9, and the special characters question mark (?), period (.), at (@), underline (_), and dollar sign ($). The first character of the label must be an alphabetic character. In other words, it cannot be a number. Every assembler has some reserved words that must not be used as labels in the program. Foremost among the reserved words are the mnemonics for the instructions. For example, "MOVWL" and "ADDWL" are reserved because they are instruction mnemonics. In addition to the mnemonics there are some other reserved words. Check your assembler for the list of reserved words.

Review Questions

1. Give three ways for hex data representation in the PIC assembler.
2. Show how to represent decimal 99 in formats of (a) hex, (b) decimal, and (c) binary in the PIC assembler.
3. What is the advantage in using the EQU directive to define a constant value?
4. Show the hex number value used by the following directives:
   (a) `ASC_DATA EQU A'4'` (b) `MY_DATA EQU B'00011111'`
5. Give the value in WREG for the following:

```
MYCOUNT    EQU    15
MOVLW      MYCOUNT
```

6. Give the value in fileReg 0x20 for the following:

```
MYCOUNT    EQU    0x95
MYREG      EQU    0x20
MOVLW      MYCOUNT
MOVWF      MYREG
```

7. Give the value in fileReg 0x63 for the following:

```
MYDATA     EQU    D'12'
MYREG      EQU    0x63
FACTOR     EQU    0x10
MOVLW      MYDATA
ADDLW      FACTOR
MOVWF      MYREG
```

SECTION 2.6: INTRODUCTION TO PIC ASSEMBLY PROGRAMMING

In this section we discuss Assembly language format and define some widely used terminology associated with Assembly language programming.

While the CPU can work only in binary, it can do so at a very high speed. It is quite tedious and slow for humans, however, to deal with 0s and 1s in order to program the computer. A program that consists of 0s and 1s is called *machine language*. In the early days of the computer, programmers coded programs in machine language. Although the hexadecimal system was used as a more efficient way to represent binary numbers, the process of working in machine code was still cumbersome for humans. Eventually, Assembly languages were developed, which provided mnemonics for the machine code instructions, plus other features that made programming faster and less prone to error. The term *mnemonic* is frequently used in computer science and engineering literature to refer to codes and abbreviations that are relatively easy to remember. Assembly language programs must be translated into machine code by a program called an *assembler*. Assembly language is referred to as a *low-level language* because it deals directly with the internal structure of the CPU. To program in Assembly language, the programmer must know all the registers of the CPU and the size of each, as well as other details.

Today, one can use many different programming languages, such as BASIC, Pascal, C, C++, Java, and numerous others. These languages are called *high-level languages* because the programmer does not have to be concerned with the internal details of the CPU. Whereas an *assembler* is used to translate an

Assembly language program into machine code (sometimes also called *object code* or opcode for operation code), high-level languages are translated into machine code by a program called a *compiler*. For instance, to write a program in C, one must use a C compiler to translate the program into machine language. Next we look at PIC Assembly language format.

Structure of Assembly language

An Assembly language program consists of, among other things, a series of lines of Assembly language instructions. An Assembly language instruction consists of a mnemonic, optionally followed by one or two operands. The operands are the data items being manipulated, and the mnemonics are the commands to the CPU, telling it what to do with those items.

An Assembly language program (see Program 2-1) is a series of statements, or lines, which are either Assembly language instructions such as ADDLW and MOVWF, or statements called directives. While instructions tell the CPU what to do, directives (also called pseudo-instructions) give directions to the assembler. For example, in Program 2-1 while the MOVWF and ADDLW instructions are commands to the CPU, ORG and END are directives to the assembler. The directive ORG tells the assembler to place the opcode at memory location 0 while END indicates the end of the source code to the assembler. In other words, one directive is for the start of the program and the other for the end of the program.

An Assembly language instruction consists of four fields:

```
[label]     mnemonic    [operands]    [;comment]
```

Brackets indicate that a field is optional and not all lines have them. Brackets should not be typed in. Regarding the above format, the following points should be noted:

1. The label field allows the program to refer to a line of code by name. The label field cannot exceed a certain number of characters. Check your assembler for the rule.
2. The Assembly language mnemonic (instruction) and operand(s) fields together perform the real work of the program and accomplish the tasks for which the program was written. In Assembly language statements such as

```
MOVLW 55H
ADDLW 67H
```

ADDLW and MOVLW are the mnemonics that produce opcodes; the "55H" and "67H" are the operands. Instead of a mnemonic and an operand, these two fields could contain assembler pseudo-instructions, or directives. Remember that directives do not generate any machine code (opcode) and are used only by the assembler, as opposed to instructions that are translated into machine code (opcode) for the CPU to execute. In Program 2-1 the commands ORG (origin) and END are examples of directives. More of these pseudo-instructions are discussed in future chapters.

3. The comment field begins with a semicolon comment indicator ";". Comments may be at the end of a line or on a line by themselves. The assembler ignores comments, but they are indispensable to programmers. Although comments are optional, it is recommended that they be used to describe the program in a way that makes it easier for someone else to read and understand.

4. Notice the label "HERE" in the label field in Program 2-1. In the GOTO the PIC is told to stay in this loop indefinitely. If your system has a monitor program you do not need this line and it should be deleted from your program. In Section 2.7 we will see how to create a ready-to-run program.

```
        ;PIC Assembly Language Program To Add Some Data.
        ;store SUM in fileReg location 10H.

SUM     EQU     10H             ;RAM loc 10H for SUM

        ORG     0H              ;start at address 0
        MOVLW   25H             ;WREG = 25
        ADDLW   0x34            ;add 34H to WREG
        ADDLW   11H             ;add 11H to WREG
        ADDLW   D'18'           ;W = W + 12H = 7CH
        ADDLW   1CH             ;W = W + 1CH = 98H
        ADDLW   B'00000110'     ;W = W + 6 = 9EH
        MOVWF   SUM             ;save the SUM in loc 10H
HERE    GOTO    HERE            ;stay here forever
        END                     ;end of asm source file
```

Program 2-1: Sample of an Assembly Language Program

Review Questions

1. What is the purpose of pseudo-instructions?
2. _____ are translated by the assembler into machine code, whereas _____ are not.
3. True or false. Assembly language is a high-level language.
4. Which of the following instructions produces opcode? List all that do.
   (a) MOVLW 25H (b) ADDLW 12 (c) ORG 2000H (d) GOTO HERE
5. Pseudo-instructions are also called _____.
6. True or false. Assembler directives are not used by the CPU itself. They are simply a guide to the assembler.
7. In Question 4, which one is an assembler directive?

SECTION 2.7: ASSEMBLING AND LINKING A PIC PROGRAM

Now that the basic form of an Assembly language program has been given, the next question is: How it is created, assembled, and made ready to run? The steps to create an executable Assembly language program (Figure 2-8) are outlined as follows:

1. First we use a text editor to type in a program similar to Program 2-1. In the case of the PIC microcontrollers, we use the MPLAB IDE, which has a text editor, assembler, linker, simulator, and much more all in one software package. It is an excellent development software that supports all the PIC chips and is free. Many editors or word processors are also available that can be used to create or edit the program. Some widely used editors are the MS-DOS EDIT, and Notepad in Windows, which comes with all Microsoft operating systems. Notice that the editor must be able to produce an ASCII file. For assemblers, the file names follow the usual DOS conventions, but the source file has the extension "asm". The "asm" extension for the source file is used by an assembler in the next step.

2. The "asm" source file containing the program code created in step 1 is fed to the PIC assembler. The assembler converts the instructions into machine code. The assembler will produce an object file and an error file. The extension for

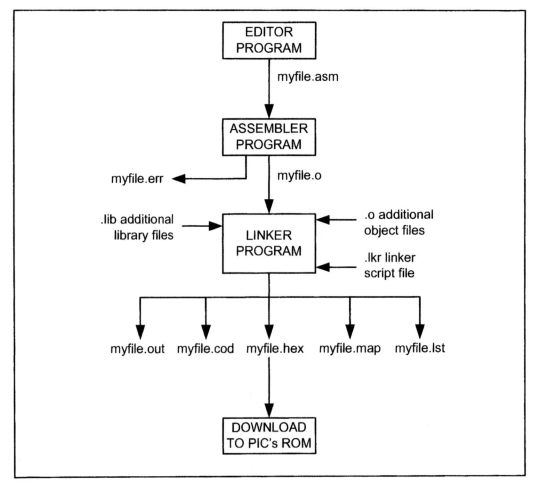

Figure 2-8. Steps to Create a Program

the object file is "o". The extension for the error file, which contains any syntax errors and their line numbers, is "err". The error file can be viewed with any text editor.

3. Assemblers require a third step called *linking*. The link program takes one or more object files and produces a hex file, a list file, a map file, an intermediate object file, and a debug file. The hex file has the extension "hex", the list file extension is "lst", the map file extension is "map", the intermediate object file extension is "out", and the debug file extension is "cod". After a successful link, the hex file is ready to be burned into the PIC's program ROM and is downloaded into the PIC Trainers. See Chapter 8 for more details.

The MPLAB IDE, a Windows-based program, combines steps 2 and 3 into one step after the program has been typed.

More about asm, err, and object files

The asm file is also called the *source* file and must have the "asm" extension. As mentioned earlier, this file is created with a text editor such as MS-DOS EDIT or Windows Notepad. Many assemblers come with a text editor. The assembler converts the asm file's Assembly language instructions into machine language and provides the o (object) file. The PIC assembler produces the object and error files. The object file as mentioned earlier, has an "o" as it extension. In modular programming, we use the linker to link many object files together to create a ready-to-burn hex file as we will see in Chapter 6. But before we can link a program to create a ready-to-run program, we must make sure that it is error free. The PIC assembler provides us the error file with the extension of "err" and this is the file we examine to see the nature of syntax errors. The linker will not link the program until all the syntax errors are fixed. We can print the error file or use Notepad to examine the nature of the errors. Then we go back to the asm file and correct all the errors before we assemble it again. A sample of an error file is shown on the next page.

```
;PIC Assembly Language Program To Add Some Data.
;store sum in fileReg location 10H.

SUM     EQU    10H              ;RAM loc 10H for sum

        ORG    0H               ;start at address 0
        MOVLW  25H              ;WREG = 25
        ADDLW  0x34             ;add 34H to WREG
        ADDLW  11H              ;add 11H to WREG
        ADDLW  D'18'            ;W = W + 12H = 7CH
        ADDLW  1CH              ;W = W + 1CH = 98H
        ADDLW  B'00000110'      ;W = W + 6 = 9EH
        MOVWF  SUM              ;save the sum in loc 10H
HERE    GOTO   HERE             ;stay here forever
        END                     ;end of asm source file
```

Program 2-1: Sample of a PIC Assembly Source Code (asm file)

```
Warning[207]  C:\MDEPIC\EXAMPLE 2-1.ASM 6  : Found label after column 1. (R4)
Warning[207]  C:\MDEPIC\EXAMPLE 2-1.ASM 13 : Found label after column 1. (movle)
Error[122]    C:\MDEPIC\EXAMPLE 2-1.ASM 13 : Illegal opcode (d)
Warning[207]  C:\MDEPIC\EXAMPLE 2-1.ASM 17 : Found label after column 1. (DEC)
Error[122]    C:\MDEPIC\EXAMPLE 2-1.ASM 17 : Illegal opcode (COUNT)
Warning[203]  C:\MDEPIC\EXAMPLE 2-1.ASM 20 : Found opcode in column 1. (movwf)
Warning[207]  C:\MDEPIC\EXAMPLE 2-1.ASM 21 : Found label after column 1. (addl)
Error[108]    C:\MDEPIC\EXAMPLE 2-1.ASM 21 : Illegal character (0)
Error[116]    C:\MDEPIC\EXAMPLE 2-1.ASM 29 : Address label duplicated or differ-
ent in second pass (AGAIN)
```

Program 2-1: Sample of a PIC Error (err file)

"lst" and "map" files

The lst (list) and map files are very useful to the programmer. The list shows the binary and source code. The map file shows the memory layout of used and unused memory locations. These files can be accessed by an editor such as Notepad and displayed on the monitor, or sent to the printer to get a hard copy. The programmer uses the list and map files to ensure correct system design.

```
LOC      OBJECT CODE LINE SOURCE TEXT            VALUE

                     00001
                     00002 ;PIC Asm Language Program To Add Some Data
                     00003 ;store SUM in fileReg location 10H
   00000010          00004 SUM EQU 10H           ;RAM loc 10H for Sum
                     00005
000000               00006 ORG 0H                ;start at address 0
000000 0E25          00007 MOVLW 25H             ;WREG = 25
000002 0F34          00008 ADDLW 0x34            ;add 34H to WREG
000004 0F11          00009 ADDLW 11H             ;add 11H to WREG
000006 0F12          00010 ADDLW D'18'           ;W = W + 12H = 7CH
000008 0F1C          00011 ADDLW 1CH             ;W = W + 1CH = 98H
00000A 0F06          00012 ADDLW B'00000110'     ;W = W + 6 = 9EH
00000C 6E10          00013 MOVWF SUM             ;save the SUM in loc 10H
00000E EF07 F000     00014 HERE GOTO HERE        ;stay here forever
                     00015 END                   ;end of asm source file
```

Program 2-1: List File

Review Questions

1. True or false. The MPLAB, MS-DOS Edit, and Windows Notepad text editors all produce an ASCII file.
2. True or false. The extension for the source file is "asm".
3. Which of the following files can be produced by the text editor?
   (a) myprog.asm (b) myprog.o (c) myprog.hex (d) myprog.lst (e) myprog.err
4. Which of the following files is produced by an assembler?
   (a) myprog.asm (b) myprog.o (c) myprog.hex (d) myprog.lst (e) myprog.err
5. Which of the following files lists syntax errors?
   (a) myprog.asm (b) myprog.o (c) myprog.hex (d) myprog.lst (e) myprog.err

SECTION 2.8: THE PROGRAM COUNTER AND PROGRAM ROM SPACE IN THE PIC

In this section we discuss the role of the program counter (PC) in executing a program and show how the code is fetched from ROM and executed. We will also discuss the program (code) ROM space for various PIC family members. Finally, we examine the Harvard architecture of the PIC18.

Program counter in the PIC

Another important register in the PIC microcontroller is the PC (program counter). The program counter is used by the CPU to point to the address of the next instruction to be executed. As the CPU fetches the opcode from the program ROM, the program counter is incremented automatically to point to the next instruction. The wider the program counter, more the memory locations a CPU can access. That means that a 14-bit program counter can access a maximum of 16K (2^{14} = 16K) of code from addresses 0000–3FFFH. The PIC family 16F has 14-bit program counters, while the program counter in PIC12F is 12-bit. In the case of a 16-bit program counter, the code space is 64K (2^{16} = 64K), which occupies the 0000–FFFFH address range. The 8051 microcontrollers have a 16-bit program counter. The program counter in the PIC18 family is 21-bit. This means that the PIC18 family can access program addresses 000000 to 1FFFFFH, a total of 2M of code. However, not all members of the PIC18 family have the entire 2M (2^{21} = 2M) of on-chip ROM installed. See Table 2-6. The 14-bit program counter in the PIC16C family had imposed the maximum code size of 16K. To overcome this major limitation, PIC designers had to introduce the tedious job of page switching in the later members of the PIC16 family. They learned their lessons and solved the problem for the PIC18 family by expanding the program counter to 21-bit for that family. See Figure 2-9. The 2M code space is plenty of space for many years to come. The data in Table 2-6 is from the Microchip web site.

ROM memory map in the PIC18 family

As we just saw, some family members have only a few K of on-chip ROM (PIC18F2220) and some, such as the PIC18F6680, have 64K of ROM. PIC18F458 has 32K of on-chip ROM. The point to remember is that no member of the PIC family can access more than 2M of opcode because the program counter in the PIC is 21 bits wide (000000 to 1FFFFF address range). It must be noted that while the first location of program ROM inside the PIC has the address of 000000, the last location can be different depending on the size of the ROM on the chip. (See Figure 2-10.) Among the PIC18 family members, the PIC18F2220 has 4K of on-chip ROM. This 4K ROM memory has memory addresses of 00000 to 000FFFH. Therefore, the first location of on-chip

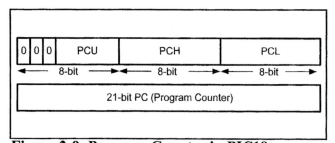

Figure 2-9. Program Counter in PIC18

ROM of this PIC has an address of 000000 and the last location has the address of 0FFFH. Look at Example 2-11 to see how this is computed.

Table 2-6: PIC18 On-chip ROM Size and Address Space

	On-Chip Code ROM (Bytes)	Code Address Range (Hex)
PIC18F2220	4K	00000–00FFF
PIC18F2410	16K	00000–03FFF
PIC18F458	32K	00000–07FFF
PIC18F6680	64K	00000–0FFFF
PIC18F8722	128K	00000–1FFFF

Example 2-11

Find the ROM memory address of each of the following PIC chips:
(a) PIC18F2220 with 4 KB
(b) PIC18F2410 with 16 KB
(c) PIC18F458 with 32 KB

Solution:

(a) With 4K of on-chip ROM memory space, we have 4096 bytes (4 × 1024 = 4096). This maps to address locations of 0000 to 0FFFH. Notice that 0 is always the first location.
(b) With 16K of on-chip ROM memory space, we have 16,384 bytes (16 × 1024 = 16,384),which gives 0000–3FFFH.
(c) With 32K we have 32,768 bytes (32 × 1024 = 32,768). Converting 32,768 to hex, we get 8000H; therefore, the memory space is 0000 to 7FFFH.

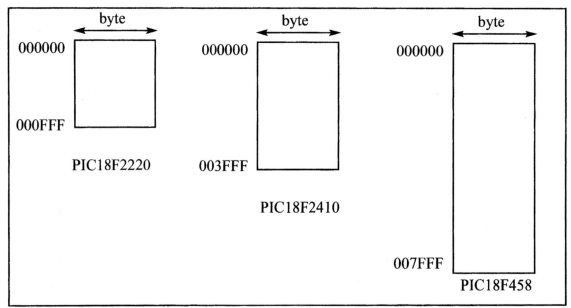

Figure 2-10. PIC18 On-Chip Program (code) ROM Address Range

Where the PIC wakes up when it is powered up

One question that we must ask about any microcontroller (or microprocessor) is: At what address does the CPU wake up when power is applied? Each microprocessor is different. In the case of the PIC family, that is, all members regardless of the family and variation, the microcontroller wakes up at memory address 0000 when it is powered up. By powering up we mean applying V_{CC} to the RESET pin as discussed in Chapter 8. In other words, when the PIC is powered up, the PC (program counter) has the value of 00000 in it. This means that it expects the first opcode to be stored at ROM address 00000H. For this reason, in the PIC system, the first opcode must be burned into memory location 00000H of program ROM because this is where it looks for the first instruction when it is booted. We achieve this by using the ORG statement in

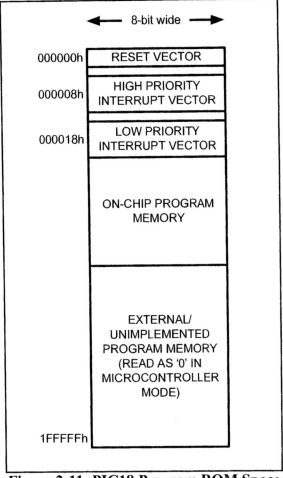

Figure 2-11. PIC18 Program ROM Space

the source program as shown earlier. Next we discuss the step-by-step action of the program counter in fetching and executing a sample program.

Placing code in program ROM

To get a better understanding of the role of the program counter in fetching and executing a program, we examine the action of the program counter as each instruction is fetched and executed. First, we examine once more the list file of the sample program and show how the code is placed in the ROM of PIC chip. As we can see, the opcode and operand for each instruction are listed on the left side of the list file.

After the program is burned into ROM of a PIC family member such as PIC18F452 or PIC18F458, the opcode and operand are placed in ROM memory locations starting at 0000 as shown in the Program 2-1 list file.

The list shows that address 0000 contains 0E, which is the opcode for moving a value into WREG, and address 0001 contains the operand (in this case 25H) to be moved to WREG. Therefore, the instruction "MOVLW 25H" has a machine code of "0E25", where 0E is the opcode and 25 is the operand. Similarly, the machine code "0F34" is located in memory locations 0002 and 0003 and represents the opcode and the operand for the instruction "ADDLW 34H". In the same way, machine code "0F11" is located in memory locations 0004 and 0005 and represents the opcode and the operand for the instruction "ADDLW 11". The memory location 0006 has the opcode of 0F, which is the opcode for the instruction "MOVLW" and memory location 0007 has the content 12, which is the operand for the decimal 18 in the ADDLW D'18' instruction. The opcode for instruction "ADDLW 1CH" is located at address 0008 and the operand 1CH at address 0009. The memory locations 000A and 0000B have the opcode and operand for the ADDLW B'00000110' instruction. The opcode for instruction "MOVWF SUM" is located at address 0000C and its address of 10H at address 0000D. The opcode for "GOTO HERE" and its target address are located in locations 0000E, F, 10, and 11. While all the instructions in this program are 2-byte instructions, the GOTO instruction is a 4-byte instruction. The reasons are explained at the end of this section.

```
LOC     OBJECT CODE LINE SOURCE TEXT          VALUE

                    00001
                    00002 ;PIC Asm Language Program To Add Some Data
                    00003 ;store sum in fileReg location 10H
   00000010         00004 SUM EQU 10H          ;RAM loc 10H for sum
                    00005
000000              00006 ORG 0H               ;start at address 0
000000  0E25        00007 MOVLW 25H            ;WREG = 25
000002  0F34        00008 ADDLW 0x34           ;add 34H to WREG
000004  0F11        00009 ADDLW 11H            ;add 11H to WREG
000006  0F12        00010 ADDLW D'18'          ;W = W + 12H = 7CH
000008  0F1C        00011 ADDLW 1CH            ;W = W + 1CH = 98H
00000A  0F06        00012 ADDLW B'00000110'    ;W = W + 6 = 9EH
00000C  6E10        00013 MOVWF SUM            ;save the sum in loc 10H
00000E  EF07 F000   00014 HERE GOTO HERE       ;stay here forever
                    00015 END                  ;end of asm source file
```

Program 2-1: List File

ROM Address	Machine Language	Assembly Language
00000	0E25	MOVLW 25H
00002	0F34	ADDLW 34
00004	0F11	ADDLW 11H
00006	0F12	ADDLW D'18'
00008	0F1C	ADDLW 1CH
0000A	0F06	ADDLW B'00000110'
0000C	6E10	MOVWF SUM
0000E	EF07 F000	HERE GOTO HERE

Executing a program byte by byte

Assuming that the above program is burned into the ROM of a PIC18 chip, the following is a step-by-step description of the action of the PIC upon applying power to it:

1. When the PIC is powered up, the PC (program counter) has 00000 and starts to fetch the first opcode from location 00000 of the program ROM. In the case of the above program the first opcode is 0E, which is the code for moving an operand to WREG. Upon executing the opcode, the CPU places the value of 25 in WREG. Now one instruction is finished. Then the program counter is incremented to point to 00002 (PC = 00002), which contains opcode 0F, the opcode for the instruction "ADDLW 34H".

2. Upon executing the opcode 0F, the value 34H is added to WREG. Then the program counter is incremented to 0004.

3. ROM location 0004 has the opcode for instruction "ADDLW 11H". This instruction is executed and now PC = 0006.

4. This process goes on until all the instructions up to "MOVWF SUM" are fetched and executed. Notice that all the above instructions are 2-byte instructions; that is, each one takes two memory locations.

5. Now PC = 000E points to the next instruction, which is "GOTO HERE". This is a 4-byte instruction. It takes ROM addresses of 0E, 0F, 10, and 11. After the execution of this instruction, PC = 0000E. This keeps the program in an infinite loop. If your PIC Trainer

Program 2-1: ROM Contents

Address	Code
000000	0E
000001	25
000002	0F
000003	34
000004	0F
000005	11
000006	0F
000007	12
000008	0F
000009	1C
00000A	0F
00000B	06
00000C	6E
00000D	10
00000E	07
00000F	EF
000010	00
000011	F0
000012	

has a monitor program you do not have to use the GOTO instruction, and the program will go back to the monitor program. The fact that the program counter points at the next instruction to be executed explains why some microprocessors (notably the x86) call the program counter the *instruction pointer*.

ROM width in the PIC18

A microprocessor's memory that holds code is byte-addressable and under the control of the program counter, as we have seen so far in this section. That means that each location of the address space holds only one byte. If we have 16 address lines, it will give us 2^{16} locations, which is 64K of memory space with an address map of 0000–FFFFH. CPUs with 8-bit data will fetch one byte at a time. This was the case in the first IBM PC and Apple computers. To bring in more information (code or data) into the CPU we can increase the width of the data bus to 16 bits. That is what IBM did to the PC AT in 1984. To increase performance even further, Intel Corporation increased the data bus width to 32 bits for the 386 and 64 bits for the Pentium. In a sense, the data bus is like traffic lanes on the highway where each lane is 8 bits wide. The more lanes, the more information we can bring into the CPU for processing. For the PIC18, the internal data bus between the code ROM and the CPU is 16 bits, as shown in Figure 2-12. Therefore, the 64K ROM space is shown as 32K × 16 using a 16-bit word data size. The same rule applies to the entire program address space of PIC18, which is 2M, organized as 1M × 16. The widening of the data path between the program ROM and the CPU is another way in which the PIC designers increased the processing power of the PIC18 family. Another reason to make the code ROM 16 bits wide is to match it with the instruction width of the PIC18 because the vast majority of the instructions are 2-byte instructions. This way, the CPU brings in an instruction from ROM everytime it makes a trip to the program ROM. That will make instruction fetch a single cycle, as we will see in the next chapter when instruction timing is discussed.

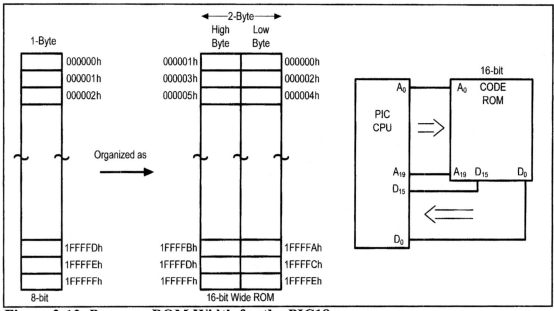

Figure 2-12. Program ROM Width for the PIC18

The PIC18 designers have made all instructions either 2-byte or 4-byte; there are no 1-byte or 3-byte instructions, as is the case with the x86 and 8051 chips. This is part of the RISC architectural philosophy, which we will study in the next section. It must be noted that not all the program ROMs in the PIC microcontrollers have 16-bit width. The PIC16 ROM width is 14-bit, while the PIC12 has a 12-bit width. It must also be noted that the data memory SRAM for the file register in the PIC microcontroller is 8-bit, and just like program ROM, it is byte addressable.

WORD ADDRESS	HIGH BYTE	LOW BYTE
000000h	0Eh	25h
000002h	0Fh	34h
000004h	0Fh	11h
000006h	0Fh	12h
000008h	0Fh	1Ch
00000Ah	0Fh	06h
00000Ch	6Eh	10h
00000Eh	EFh	07h
000010h	0Fh	00h

Figure 2-13. PIC18 Program ROM Contents for Program 2-1 List File

Little endian vs. big endian war

Examine the placing of the code in the PIC18 ROM, shown in Figure 2-13. The low byte goes to the low memory location and the high byte goes to the high memory address. This convention is called little endian to contrast it with big endian. The origin of the terms *big endian* and *little endian* is from a Gulliver's Travels story about how an egg should be opened: from the big end or the little end. In the big endian method, the high byte goes to the low address, whereas in the little endian method, the high byte goes to the high address and the low byte to the low address. All Intel microprocessors and many microcomputers, notably the Digital VAX, use the little endian convention. Freescale (Motorola) microprocessors (used in the Macintosh), along with some mainframes, use big endian. The difference might seem as trivial as whether to break an egg from the big end or the little end, but it is a nuisance in converting software from one camp to be run on a computer of the other camp. Some microprocessors, such as the PowerPC from IBM/Freescale (Motorola), let the software designer choose little endian or big endian convention.

Harvard architecture in the PIC

Every microprocessor must have memory space to store program (code) and data. As we have seen so far, the PIC is no exception with its code ROM space and data RAM (file register) space. While code provides instructions to the CPU, the data provides the information to be processed. The CPU uses buses (wire traces) to access the code ROM and data RAM memory spaces. The early computers used the same bus for accessing both the code and data. Such an architecture is commonly referred to as von Neumann (Princeton) architecture. That means for von Neumann computers, the process of accessing the code or data could cause them to get in each other's way and slow down the processing speed of the CPU, because each had to wait for the other to finish fetching. To speed up the process of program execution, some CPUs use what is called Harvard architecture. In Harvard architecture, we have separate buses for the code and data memory. That means that we need four sets of buses: (1) a set of data buses for carrying data into and out of the CPU, (2) a set of address buses for accessing the data, (3) a set of data buses for carrying code into the CPU, and (4) an address bus for accessing the code. See Figure 2-14. This is easy to implement inside an IC chip such as a microcontroller where both ROM code and data RAM are internal (on-chip) and distances are on the micron and millimeter scale. But to implement Harvard architecture for systems such as x86 IBM PC-type computers is very expensive because the RAM and ROM that hold code and data are external to the CPU. Separate wire traces for data and code on the motherboard will make the board large and expensive. For example, for a Pentium microprocessor with a 64-bit data bus and a 32-bit address bus we will need about 100 wire traces on the mother board if it is von Neumann architecture (96 for address and data, plus a few others for control signals of read and write and so on). But the number of wire traces will double to 200 if we use Harvard architecture. Harvard architecture will also necessitate a large number of pins coming out of the microprocessor itself. For this reason you do not see Harvard architecture implemented in the world of PCs and workstations. This is also the reason that microcontrollers such as PIC use Harvard architecture internally, but they still use von Neumann architecture if they need external memory for code and data space. The von Neumann architecture was developed at Princeton University, while the Harvard architecture was the work of Harvard University.

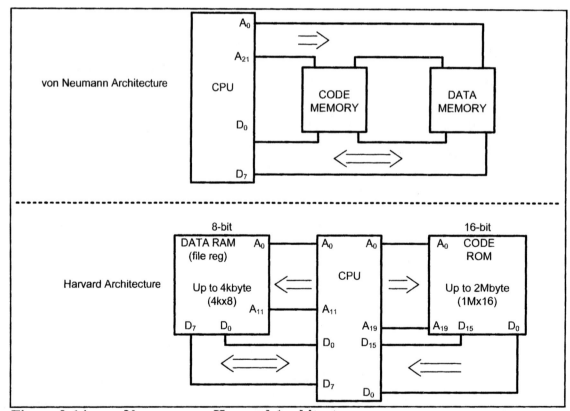

Figure 2-14. von Neumann vs. Harvard Architecture

Instruction size of the PIC18

Recall that PIC18 program memory is byte-addressable, and the instructions are either 2-byte or 4-byte. Almost all the instructions in the PIC18 are 2-byte instructions. The exceptions are MOVFF, GOTO, and a few others. Next we explore the instruction size and formation for a few of the instructions we have used in this chapter. This should give you some insights into the instructions of the PIC18.

MOVLW instruction formation

The MOVLW is a 2-byte (16-bit) instruction. Of the 16 bits, the first 8 bits are set aside for the opcode and the other 8 bits are used for the literal value of 00 to FFH. This is shown below.

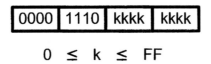

$$0 \le k \le FF$$

ADDLW instruction formation

The ADDLW is a 2-byte (16-bit) instruction. Of the 16 bits, the first 8 bits are set aside for the opcode and the other 8 bits are used for the literal value of 00 to FFH. This is shown below.

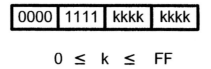

$$0 \le k \le FF$$

MOVWF instruction formation

The MOVWF is a 2-byte (16-bit) instruction. Of the 16 bits, the first 8 bits are set aside for the opcode and the other 8 bits are used for the location of the file register in the data RAM. The LSB bit of the opcode is designated by the letter a to signify the access from the access bank or the other bank in the 4096 location. If a = 0, the fileReg is in the access bank. If a = 1, then we have to use bank switching, which is covered in Chapter 6. This is shown below.

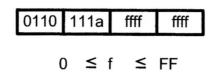

$$0 \le f \le FF$$

a = 0 : access bank is used.
a = 1 : access bank is specified by the BSR register.

See Chapter 6.

MOVFF instruction formation

The MOVFF is a 4-byte (32-bit) instruction. Of the 32 bits, the first 16 bits are set aside for the opcode and the address of the source fileReg and the other 16 bits are used for the opcode and the address of the destination. This is shown below.

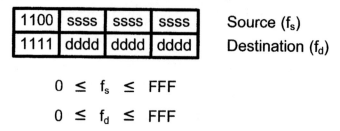

$$0 \leq f_s \leq FFF$$

$$0 \leq f_d \leq FFF$$

Notice that for both the source and destination parts of the instruction, 12 bits are used for the file register address of the PIC18. The 12 bits cover the entire range of the addresses 000–FFFH for the file register, which has 4096 bytes (4K) of data RAM space. That means that MOVFF can move the contents of any location in the file register to any other location directly. This is done without going through the WREG, as we saw in Section 2.3.

GOTO instruction formation

The GOTO is a 4-byte (32-bit) instruction. Of the 32 bits, only 12 bits are set aside for the opcode and the rest (20 bits) are used for the target address of the GOTO. This is shown below.

1110	1111	$k_7 kkk$	$kkkk_0$
1111	$k_{19}kkk$	kkkk	$kkkk_8$

$$0 \leq k \leq FFFFF$$

However, the 20-bit address gives us only 1M of address space and the PIC18 has 2M of ROM space. This is solved by making the least-significant bit (LSB) of the GOTO instruction 0, as shown below.

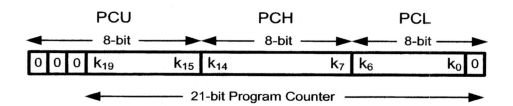

Setting the LSB of the target address to zero will make sure that the target address is an even address. As we saw in the last section, that is exactly what we want because all the instructions are either 2-byte or 4-byte. This should also avoid landing at the middle of an instruction.

Coming from other microprocessors to the PIC18

If you have a background in programming other microprocessors/microcontrollers, making the transition from these devices to the PIC18 can be easier if we remember some facts about the PIC18. They are:

1. The access bank section of the register of the PIC18 with its address range of 00–7FH can be viewed as large a number of registers, except they do not have names like other processors. We can assign any register names we want, however, as long as we are not using any of the reserved names used by SFRs, WREG, and so on. Here is an example if we are used to the 8051 or some other RISC processor:

R0	EQU	0
R1	EQU	1
R2	EQU	2
R3	EQU	3
...	....	..

Or look at the following for the x86:

BL	EQU	0
BH	EQU	1
CL	EQU	2
CH	EQU	4
DL	EQU	5
DH	EQU	6

In both of the above we can use any fileReg address of 00–7FH.

2. WREG is exactly like the accumulator in other microprocessors. It must be involved in all the arithmetic and logic operations.
3. To move data to locations in the file register or SFR, we must first move it to WREG. As was shown, we use the MOVLW instruction first to load the value into the WREG and then use MOVWF to move it to a desired location in the fileReg. In other words, no value can be moved directly to SFR or fileReg.

Review Questions

1. In the PIC18, the program counter is _____ bits wide.
2. True or false. Every member of the PIC18 family, regardless of the program ROM size, wakes up at memory 0000H when it is powered up.
3. At what ROM location do we store the first opcode of a PIC18 program?
4. The instruction "MOVLW 44H" is a ____-byte instruction.
5. What is the ROM address space for the PIC18F458?
6. The instruction "GOTO label" is a ____-byte instruction.
7. True or false. All the instructions in the PIC18 are 2- or 4-byte instructions.

SECTION 2.9: RISC ARCHITECTURE IN THE PIC

There are three ways available to microprocessor designers to increase the processing power of the CPU:

1. Increase the clock frequency of the chip. One drawback of this method is that the higher the frequency, the more power and heat dissipation. Power and heat dissipation is especially a problem for hand-held devices.
2. Use Harvard architecture by increasing the number of buses to bring more information (code and data) into the CPU to be processed. While in the case of x86 and other general purpose microprocessors this architecture is very expensive and unrealistic, in today's microcontrollers this is not a problem. As we saw in the last section, the PIC18 has Harvard architecture.
3. Change the internal architecture of the CPU and use what is called RISC architecture.

Microchip used all three methods to increase the processing power of the PIC18 microcontrollers. In this section we discuss the merits of RISC architecture and examine how the PIC18 microcontrollers have adapted it.

RISC architecture

In the early 1980s, a controversy broke out in the computer design community, but unlike most controversies, it did not go away. Since the 1960s, in all mainframes and minicomputers, designers put as many instructions as they could think of into the CPU. Some of these instructions performed complex tasks. An example is adding data memory locations and storing the sum into memory. Naturally, microprocessor designers followed the lead of minicomputer and mainframe designers. Because these microprocessors used such a large number of instructions and many of them performed highly complex activities, they came to be known as CISC (complex instruction set computer). According to several studies in the 1970s, many of these complex instructions etched into the brain of the CPU were never used by programmers and compilers. The huge cost of implementing a large number of instructions (some of them complex) into the microprocessor, plus the fact that a good portion of the transistors on the chip are used by the instruction decoder, made some designers think of simplifying and reducing the number of instructions. As this concept developed, the resulting processors came to be known as RISC (reduced instruction set computer).

Features of RISC

The following are some of the features of RISC as implemented by the PIC18 microcontroller.

Feature 1

RISC processors have a fixed instruction size. In a CISC microcontroller such as the 8051, instructions can be 1, 2, or even 3 bytes. For example, look at the following instructions in the 8051:

```
CLR C      ;Clear Carry flag          ,a 1-byte instruction
ADD Accumulator, #mybyte               ;a 2-byte instruction
LJMP  target_address                   ;a 3-byte instruction
```

This variable instruction size makes the task of the instruction decoder very difficult because the size of the incoming instruction is never known. In a RISC architecture, the size of all instructions is fixed. Therefore, the CPU can decode the instructions quickly. This is like a bricklayer working with bricks of the same size as opposed to using bricks of variable sizes. Of course, it is much more efficient to use bricks of the same size. In the last section we saw how the PIC18 uses 2-byte instructions with very few 4-byte instructions.

Feature 2

One of the major characteristics of RISC architecture is a large number of registers. All RISC architectures have at least 32 registers. Of these 32 registers, only a few are assigned to a dedicated function. One advantage of a large number of registers is that it avoids the need for a large stack to store parameters. Although a stack can be implemented on a RISC processor, it is not as essential as in CISC because so many registers are available. In the PIC microcontrollers the use of a 256-byte bank for the file register satisfies this RISC feature. The stack for the PIC18 is covered in the next chapter.

Feature 3

RISC processors have a small instruction set. RISC processors have only the basic instructions such as ADD, SUB, MUL, LOAD, STORE, AND, OR, EXOR, CALL, JUMP, and so on. The limited number of instructions is one of the criticisms leveled at the RISC processor because it makes the job of Assembly language programmers much more tedious and difficult compared to CISC Assembly language programming. This is one reason that RISC is used more commonly in high-level language environments such as the C programming language rather than Assembly language environments. It is interesting to note that some defenders of CISC have called it "complete instruction set computer" instead of "complex instruction set computer" because it has a complete set of every kind of instruction. How many of these instructions are used and how often is another matter. The limited number of instructions in RISC leads to programs that are large. Although these programs can use more memory, this is not a problem because memory is cheap. Before the advent of semiconductor memory in the 1960s, however, CISC designers had to pack as much action as possible into a single instruction to get the maximum bang for their buck. In the PIC16 we have around 35 instructions, while the PIC18 has 75 instructions. We will examine more of the instruction set for the PIC18 in future chapters.

Feature 4

At this point, one might ask, with all the difficulties associated with RISC programming, what is the gain? The most important characteristic of the RISC processor is that more than 95% of instructions are executed with only one clock cycle, in contrast to CISC instructions. Even some of the 5% of the RISC instructions that are executed with two clock cycles can be executed with one clock cycle by juggling instructions around (code scheduling). Code scheduling is most often the job of the compiler. We will examine the instruction cycle time and pipelining of the PIC18 in Chapter 3.

Feature 5

RISC processors have separate buses for data and code. In all the x86 processors, like all other CISC computers, there is one set of buses for the address (e.g., A0–A24 in the 80286) and another set of buses for data (e.g., D0–D15 in the 80286) carrying opcodes and operands in and out of the CPU. To access any section of memory, regardless of whether it contains code or data operands, the same address bus and data bus are used. In RISC processors, there are four sets of buses: (1) a set of data buses for carrying data (operands) in and out of the CPU, (2) a set of address buses for accessing the data, (3) a set of buses to carry the opcodes, and (4) a set of address buses to access the opcodes. The use of separate buses for code and data operands is commonly referred to as Harvard architecture. In the last section we examined the Harvard architecture of the PIC18.

Feature 6

Because CISC has such a large number of instructions, each with so many different addressing modes, microinstructions (microcode) are used to implement them. The implementation of microinstructions inside the CPU takes more than 40–60% of transistors in many CISC processors. In the case of RISC, however, due to the small set of instructions, they are implemented using the hardwire method. Hardwiring of RISC instructions takes no more than 10% of the transistors.

Feature 7

RISC uses load/store architecture. In CISC microprocessors, data can be manipulated while it is still in memory. For example, in instructions such as "ADD Reg, Memory", the microprocessor must bring the contents of the external memory location into the CPU, add it to the contents of the register, then move the result back to the external memory location. The problem is there might be a delay in accessing the data from external memory. Then the whole process would be stalled, preventing other instructions from proceeding in the pipeline. In RISC, designers did away with these kinds of instructions. In RISC, instructions can only load from external memory into registers or store registers into external memory locations. There is no direct way of doing arithmetic and logic operations between a register and the contents of external memory locations. All these instructions must be performed by first bringing both operands into the registers inside the CPU, then performing the arithmetic or logic operation, and then sending the result back to memory. This idea was first implemented by the Cray 1 supercomputer in 1976 and is commonly referred to as load/store architecture. In the last section, we saw that the arithmetic and logic operations are between the fileReg (an internal memory) and WREG, but none involve a ROM location and a fileReg location. For example, there is no "ADDW ROM-Loc" instruction in PIC18.

In concluding this discussion of RISC processors, it is interesting to note that RISC technology was explored by the scientists in IBM in the mid-1970s, but it was David Patterson of the University of California at Berkeley who in 1980 brought the merits of RISC concepts to the attention of computer scientists. It must also be noted that in recent years CISC processors such as the Pentium have used some of the RISC features in their design. This was the only way they could enhance the processing power of the x86 processors and stay competitive. Of course, they had to use lots of transistors to do the job, because they had to deal with all the CISC instructions of the 8086/286/386 processors and the legacy software of DOS.

Review Questions

1. What do RISC and CISC stand for?
2. True or false. The CISC architecture executes the vast majority of its instructions in 2, 3, or more clock cycles, while RISC executes them in one clock.
3. RISC processors normally have a _____ (large, small) number of general-purpose registers.
4. True or false. Instructions such as "ADD WREG, ROMmemory" do not exist in RISC microcontrollers such as the PIC18.
5. How many instructions does the PIC18 have? Does it qualify as RISC?
6. True or false. While CISC instructions are of variable sizes, RISC instructions are all the same size.
7. Which of the following operations do not exist for the ADD instruction in RISC?
   (a) register to register (b) immediate to register (c) memory to memory
8. True or false. Harvard architecture uses the same address and data buses to fetch both code and data.

SECTION 2.10: VIEWING REGISTER AND MEMORY WITH MPLAB SIMULATOR

The PIC microcontroller has one of the best tools and support systems, many of them free or inexpensive. MPLAB is an assembler, linker, and simulator provided for free by Microchip Corporation and can be downloaded from the www.microchip.com web site. See http://www.MicroDigitalEd.com for tutorials on how to use the MPLAB assembler and simulators.

Many assemblers and C compilers come with a simulator. Simulators allow us to view the contents of registers and memory after executing each instruction (single-stepping). It is strongly recommended to use a simulator to single-step some of the programs in this chapter and future chapters. Single-stepping a program with a simulator gives us a deeper understanding of microcontroller architecture, in addition to the fact that we can use it to find the nature of error in our programs. Figures 2-15 through 2-17 show screen-shots for PIC simulators from MPLAB.

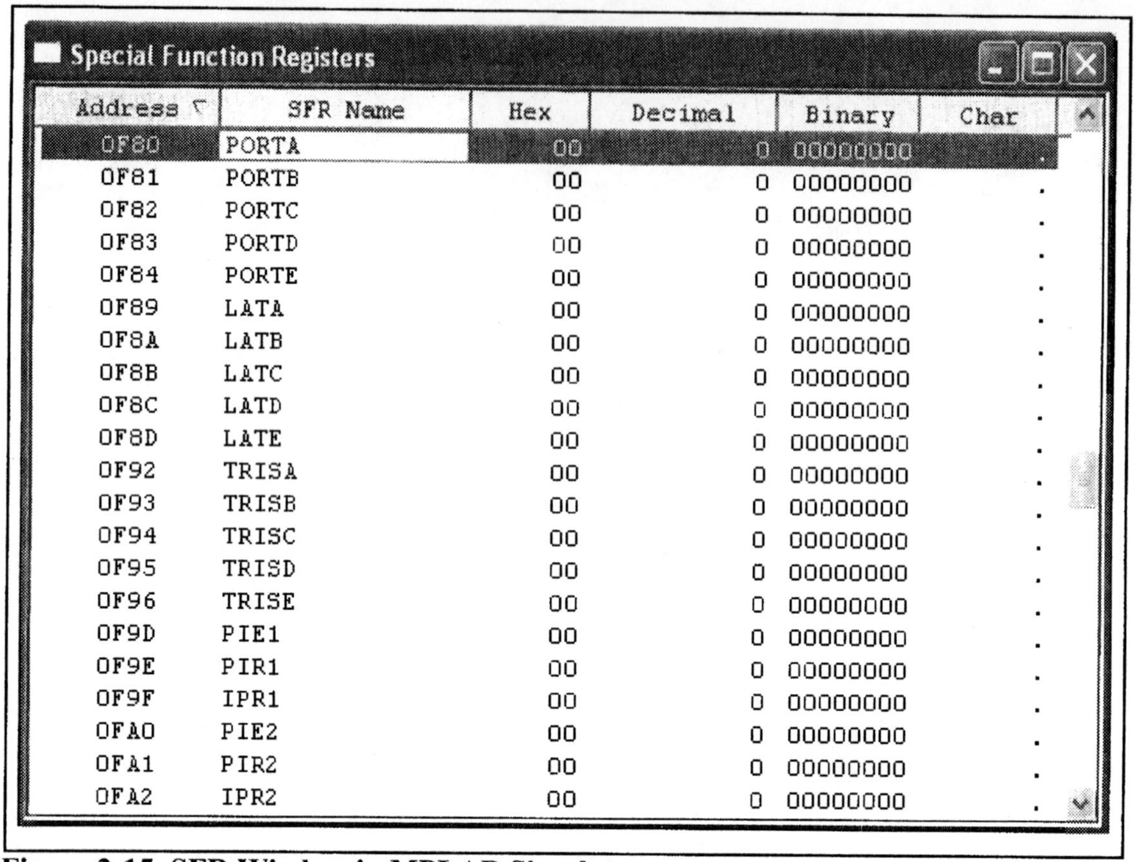

Special Function Registers

Address	SFR Name	Hex	Decimal	Binary	Char
0F80	PORTA	00	0	00000000	.
0F81	PORTB	00	0	00000000	.
0F82	PORTC	00	0	00000000	.
0F83	PORTD	00	0	00000000	.
0F84	PORTE	00	0	00000000	.
0F89	LATA	00	0	00000000	.
0F8A	LATB	00	0	00000000	.
0F8B	LATC	00	0	00000000	.
0F8C	LATD	00	0	00000000	.
0F8D	LATE	00	0	00000000	.
0F92	TRISA	00	0	00000000	.
0F93	TRISB	00	0	00000000	.
0F94	TRISC	00	0	00000000	.
0F95	TRISD	00	0	00000000	.
0F96	TRISE	00	0	00000000	.
0F9D	PIE1	00	0	00000000	.
0F9E	PIR1	00	0	00000000	.
0F9F	IPR1	00	0	00000000	.
0FA0	PIE2	00	0	00000000	.
0FA1	PIR2	00	0	00000000	.
0FA2	IPR2	00	0	00000000	.

Figure 2-15. SFR Window in MPLAB Simulator

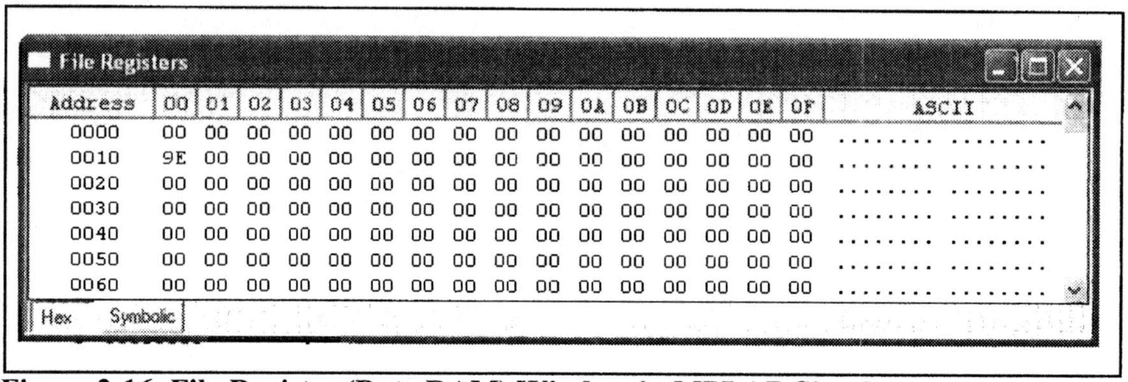

File Registers

Address	00	01	02	03	04	05	06	07	08	09	0A	0B	0C	0D	0E	0F	ASCII
0000	00	00	00	00	00	00	00	00	00	00	00	00	00	00	00	00	....
0010	9E	00	00	00	00	00	00	00	00	00	00	00	00	00	00	00	....
0020	00	00	00	00	00	00	00	00	00	00	00	00	00	00	00	00	....
0030	00	00	00	00	00	00	00	00	00	00	00	00	00	00	00	00	....
0040	00	00	00	00	00	00	00	00	00	00	00	00	00	00	00	00	....
0050	00	00	00	00	00	00	00	00	00	00	00	00	00	00	00	00	....
0060	00	00	00	00	00	00	00	00	00	00	00	00	00	00	00	00	....

Hex | Symbolic

Figure 2-16. File Register (Data RAM) Window in MPLAB Simulator

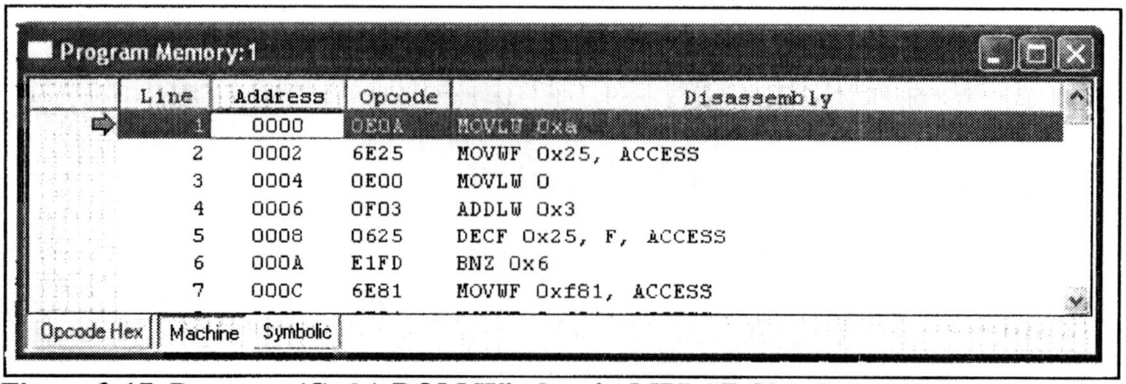

Program Memory:1

Line	Address	Opcode	Disassembly
1	0000	0E0A	MOVLW 0xa
2	0002	6E25	MOVWF 0x25, ACCESS
3	0004	0E00	MOVLW 0
4	0006	0F03	ADDLW 0x3
5	0008	0625	DECF 0x25, F, ACCESS
6	000A	E1FD	BNZ 0x6
7	000C	6E81	MOVWF 0xf81, ACCESS

Opcode Hex | Machine | Symbolic

Figure 2-17. Program (Code) ROM Window in MPLAB Simulator

See the following web site for a tutorial on using MPLAB:

http://www.MicroDigitalEd.com

SUMMARY

This chapter began with an exploration of the major registers of the PIC, including WREG, SFRs, and general-purpose data RAM, and the program counter. The use of these registers was demonstrated in the context of programming examples. The process of creating an Assembly language program was described from writing the source file, to assembling it, linking, and executing the program. The PC (program counter) register always points to the next instruction to be executed. The way the PIC uses program ROM space was explored because PIC Assembly language programmers must be aware of where programs are placed in ROM, and how much memory is available.

An Assembly language program is composed of a series of statements that are either instructions or pseudo-instructions, also called *directives*. Instructions are translated by the assembler into machine code. Pseudo-instructions are not translated into machine code: They direct the assembler in how to translate instructions into machine code. Some pseudo-instructions, called *data directives*, are used to define data. Data is allocated in byte-size increments. The data can be in binary, hex, decimal, or ASCII formats.

Flags are useful to programmers because they indicate certain conditions, such as carry or zero, that result from execution of instructions. The concepts of the RISC and Harvard architectures were also explored.

The RISC architecture allows the design of much more powerful microcontrollers. It has a simple instruction set and uses of a large number of registers. Harvard architecture allows us to bring more code and data to the CPU faster. The use of a wider data bus in the PIC18 allows us to fetch an instruction every cycle because the PIC instructions are typically 2 bytes.

PROBLEMS

SECTION 2.1: THE WREG REGISTER IN THE PIC

1. PIC18 is a(n) _____-bit microcontroller.
2. Register WREG is _____ bits wide.
3. The literal value in MOVLW is _____ bits wide.
4. The largest number that can be loaded into WREG is _____ in hex.
5. To load WREG with the value 65H, the pound sign is _____ (not necessary, optional, necessary) in the instruction "MOVLW #65H".
6. What is the result of the following code and where is it kept?

```
        MOVLW       15H
        ADDLW       13H
```

7. Which of the following is (are) illegal?
   (a) MOVLW 500   (b) MOVLW 50   (c) MOVLW 00
   (d) MOVLW 255H   (e) MOVLW 25H   (f) MOVLW F5H
   (g) MOVLW mybyte,50H
8. Which of the following is (are) illegal?
   (a) ADDLW 300H   (b) ADDLW 50H   (c) ADDLW $500
   (d) ADDLW 255H   (e) ADDLW 12H   (f) ADDLW 0F5H
   (g) ADDWL 25H
9. What is the result of the following code and where is it kept?

```
        MOVLW 25H
        ADDLW 1FH
```

10. What is the result of the following code and where is it kept?

```
        MOVLW 15H
        ADDLW 0EAH
```

11. The largest number that K can take for the instruction "ADDWL K" is _____ in hex.
12. True or false. We have many WREG registers in the PIC18.

SECTION 2.2: THE PIC FILE REGISTER

13. PIC data RAM consists of _____ (EEPROM, SRAM).
14. True or false. Data RAM in PIC is also called the file register.
15. True or false. The SFRs are part of the file register memory space.
16. True or false. The general-purpose RAM is not part of the file register memory space.
17. True or false. All members of PIC18 family have the same size file register.
18. If we add the SFR and general-purpose RAM sizes together we should get the total space for the _____.
19. Find the file register size for the following PIC chips:
    (a) PIC12508   (b) PIC16F84   (c) PIC18F8772
20. What is the difference between the EEPROM and data RAM space in the PIC18?
21. Can we have a PIC chip with no EEPROM?
22. Can we have a PIC chip with no file register?

23. The access bank has _____ bytes space.
24. Give the address map of the SFR and GP RAM section of the access bank.
25. What is the maximum number of banks that the PIC18 can have?
26. What is the maximum number of bytes that the PIC18 can have for the file register?

SECTION 2.3: USING INSTRUCTIONS WITH THE DEFAULT ACCES BANK

27. What is the address range for the scratch pad section of the access bank?
28. Show a simple code to load values 30H and 97H into locations 5 and 6 respectively.
29. Show a simple code to load value 55H into locations 0–8.
30. Show a simple code to load value 5FH into Port B SFR.
31. True or false. We can not load literal values into the scratch pad area directly.
32. True or false. The "ADDWF fileReg, D" instruction involves a fileReg and WREG.
33. In Question 32, to place the result in WREG, the D bit must be _____.
34. In Question 32, to place the result in fileReg, the D bit must be _____.
35. Show a simple code to (a) load value 11H into locations 0–5, and (b) add the values together and place the result in WREG as they are added.
36. Repeat Problem 35, except place the result in location 5 after the addition is done.
37. Show a simple code to (a) load value 15H into location 7, and (b) add it to WREG five times and place the result in WREG as the values are added. WREG should be zero before the addition starts.
38. Repeat Problem 37, except place the result in location 7 as numbers are being added together.
39. What is the difference between the MOVWF and MOVF instructions?
40. Write a simple code to complement the contents of location 8 and place the result in WREG.
41. True or false. We can use MOVFF to copy data from any location to any location in the file register.
42. Write a simple code to copy data from location 8 to PORTC (a) using WREG and (b) without using WREG.

SECTION 2.4: PIC STATUS REGISTER

43. The status register is a(n) _____ -bit register.
44. Which bits of the status register are used for the C and DC flag bits, respectively?
45. Which bits of the status register are used for the OV and N flag bits, respectively?
46. In the ADDLW instruction, when is C raised?
47. In the ADDLW instruction, when is DC raised?
48. What is the status of the C and Z flags after the following code?
```
MOVLW     FFH
ADDLW     1
```

49. Find the C flag value after each of the following codes:
   (a) MOVLW 54H   (b) MOVLW 00   (c) MOVLW FFH
        ADDLW 0C4H   ADDLW FFH   ADDLW 05H
50. Write a simple program in which the value 55H is added 5 times.

SECTION 2.5: PIC DATA FORMAT AND DIRECTIVES

51. State the value (in hex) used for each of the following data:

```
MYDAT_1    EQU    55
MYDAT_2    EQU    D'98'
MYDAT_3    EQU    A'G'
MYDAT_4    EQU    0x50
MYDAT_5    EQU    D'200'
MYDAT_6    EQU    A'A'
MYDAT_7    EQU    AAH
MYDAT_8    EQU    D'255'
MYDAT_9    EQU    B'10010000'
MYDAT_10   EQU    B'01111110'
MYDAT_11   EQU    D'10'
MYDAT_12   EQU    D'15'
```

52. State the value (in hex) for each of the following data:

```
DAT_1    EQU    22
DAT_2    EQU    56H
DAT_3    EQU    B'10011001'
DAT_4    EQU    D'32'
DAT_5    EQU    0xF6
DAT_6    EQU    B'11111011'
```

53. Show a simple code to (a) load value 11H into locations 0–5, and (b) add them together and place the result in WREG as the values are added. Use EQU to assign the names R0–R5 to locations 0–5.

SECTION 2.6: INTRODUCTION TO PIC ASSEMBLY PROGRAMMING
and
SECTION 2.7: ASSEMBLING AND LINKING A PIC PROGRAM

54. Assembly language is a_____ (low, high)-level language while C is a _____ (low, high)-level language.
55. Of C and Assembly language, which is more efficient in terms of code generation (i.e., the amount of ROM space it uses)?
56. Which program produces the o file?
57. True or false. The source file has the extension "asm".
58. Which file provides the listing of error messages?
59. True or false. The source code file can be a non-ASCII file.
60. True or false. Every source file must have ORG and END directives.
61. Do the ORG and END directives produce opcodes?
62. Why are the ORG and END directives also called pseudocode?
63. True or false. The ORG and END directives appear in the ".lst" file.

64. True or false. The linker produces the file with the extension "asm".
65. True or false. The linker produces the file with the extension "hex".
66. The file with the _____ extension is downloaded into PIC ROM.
67. Give three file extensions produced by MPLAB.

SECTION 2.8: THE PROGRAM COUNTER AND PROGRAM ROM SPACE IN THE PIC

68. Every PIC18 family member wakes up at address _____ when it is powered up.
69. A programmer puts the first opcode at address 100H. What happens when the microcontroller is powered up?
70. Find the number of bytes each of the following instructions takes:
    (a) MOVLW 5H   (b) MOVLW 9FH   (c) ADDLW 50H
    (d) ADDLW 0   (e) MOVLW 0x41   (f) MOVLW 20
    (g) ADDLW d'200'   (h) GOTO
71. Write a program to (a) place each of your 5-digit ID numbers into a RAM locations starting at address 0, (b) add each digit to WREG and store the sum in RAM location 6, and (c) use the program listing and show the ROM memory addresses and their contents.
72. Use the program listing of your choice and show the ROM memory addresses and their contents.
73. Find the address of the last location of on-chip program ROM for each of the following:
    (a) PIC with 48 KB   (b) PIC with 96 KB
    (c) PIC with 64 KB   (d) PIC with 16 KB
    (f) PIC with 128 KB
74. Show the lowest and highest values (in hex) that the PIC18 program counter can take.
75. A given PIC has 7FFFH as the address of the last location of its on-chip ROM. What is the size of on-chip ROM for this PIC?
76. Repeat Question 75 for 3FFH.
77. Find the on-chip program ROM size in K for the PIC18 with the following address ranges:
    (a) 0000–1FFF   (b) 0000–3FFF
    (c) 0000–5FFF   (d) 0000–BFFF
    (e) 0000–FFFF   (f) 00000–1FFFF
    (g) 00000–2FFFF   (h) 00000–3FFFF

78. Find the on-chip program ROM size in K for the PIC18 with the following address ranges:
    (a) 00000–4FFFF   (b) 00000–3FFFF
    (c) 00000–5FFFF   (d) 00000–7FFFF
    (e) 00000–BFFFF   (f) 00000–FFFFF
    (g) 00000–17FFFF   (h) 00000–1FFFFF
Some of the above might not be in production yet.

79. How wide is the ROM in the PIC18 chip?

80. How wide is the data bus between the CPU and the program ROM in PIC18?

81. Show the even and odd address designation for 4K × 16. See Figure 2-11.

82. In Question 81, what is the ROM size in K?

83. Show the even and odd address designation for 16K × 16. See Figure 2-11.

84. In Question 83, what is the ROM size in K?

85. Explain Harvard architecture and how it makes processing of code and data faster.

86. What is the drawback of using Harvard architecture for memories external to the CPU?

87. In instruction "MOVLW K" explain why the K value cannot be larger than 255 decimal.

88. In instruction "ADDLW K" explain why the K value cannot be larger than 255 decimal.

89. In "MOVWF fileReg", explain what the size of the instruction is and how it allows one to cover the entire range of the file register in PIC18.

90. In "MOVFF source, dest", explain what the size of the instruction is and how it allows one to cover the entire range of the file register in PIC18.

91. In the instruction "GOTO target-addr" explain why the lowest bit of the program counter is 0.

92. Explain why the instruction "GOTO target-addr" will not land at an odd address.

93. In Question 92, explain why it should not.

94. Explain how the instruction "GOTO target-addr" is able to cover the entire 2M address space of the PIC18.

SECTION 2.9. RISC ARCHITECTURE IN THE PIC

95. What do RISC and CISC stand for?

96. In _____ (RISC, CISC) architecture we can have 1-, 2-, 3-, or 4-byte instructions.

97. In _____ (RISC, CISC) architecture instructions are fixed in size.

98. In _____ (RISC, CISC) architecture instructions are mostly executed in one or two cycles.

99. In _____ (RISC, CISC) architecture we can have an instruction to ADD a register to external memory.

100. True or false. Most instructions in CISC are executed in one or two cycles.

ANSWERS TO REVIEW QUESTIONS

SECTION 2.1: THE WREG REGISTER IN THE PIC

1. MOVLW 0x34
2. MOVLW 0x16
    ADDLW 0xCD
3. False
4. FF hex and 255 in decimal
5. 8

SECTION 2.2: THE PIC FILE REGISTER

1. True
2. File register
3. True
4. True
5. 8
6. 256
7. 4096

SECTION 2.3: USING INSTRUCTIONS WITH THE DEFAULT ACCESS BANK

1. True
2.  ```
    MOVLW  0x16
    MOVWF  0
    MOVLW  0xFD
    ADDWF  0,  F
    ```
3. True
4. FF, 255
5. WREG

SECTION 2.4: PIC STATUS REGISTER

1. Status register
2. 8 bits
3. D5, D6, and D7
4.

Hex	binary	
9F	1001 1111	
+ 61	+ 0110 0001	
100	10000 0000	This leads to C = 1, DC = 1, and Z = 1.

5.

Hex	binary	
82	1000 0010	
+ 22	+ 0010 0010	
A4	1010 0100	This leads to C = 0, DC = 0, and Z = 0.

6.

Hex	binary	
67	0110 0111	
+ 99	+ 1001 1001	
100	10000 0000	This leads to C = 1, DC = 1, and Z = 1.

SECTION 2.5: PIC DATA FORMAT AND DIRECTIVES

1.  ```
    DATA1  EQU  9FH
    DATA2  EQU  0x9F
    DATA3  EQU  H'9F'
    ```
2.  ```
    DATA1  EQU  99H
    DATA2  EQU  D'99'
    DATA3  EQU  B'10011001'
    ```
3. If the value is to be changed later, it can be done once in one place instead of at every occurrence.
4. (a) 34H   (b) 1FH
5. WREG = 15H
6. value of location 0x20 = (0x95)
7. 0CH + 10H = 1CH will be in fileReg location 63H

    ```

SECTION 2.6: INTRODUCTION TO PIC ASSEMBLY PROGRAMMING

1. The real work is performed by instructions such as MOV and ADD. Pseudo-instructions, also called assembly directives, instruct the assembler in doing its job.
2. The instruction mnemonics, pseudo-instructions
3. False
4. All except (c)
5. Assembler directives
6. True
7. (c)

SECTION 2.7: ASSEMBLING AND LINKING A PIC PROGRAM

1. True
2. True
3. (a)
4. (b) through (e)
5. (d) and (e)

SECTION 2.8: THE PROGRAM COUNTER AND PROGRAM ROM SPACE IN THE PIC

1. 21
2. True
3. 0000H
4. 2
5. With 32K, we have 32768 ($32 \times 1024 = 32768$) bytes, and the ROM space is 0000 to 7FFFH.
6. 4
7. True

SECTION 2.9: RISC ARCHITECTURE IN THE PIC

1. CISC stands for complex instruction set computer; RISC is reduced instruction set computer.
2. True
3. Small
4. True
5. 75, yes
6. True
7. (c)
8. False

CHAPTER 3

BRANCH, CALL, AND TIME DELAY LOOP

OBJECTIVES

Upon completion of this chapter, you will be able to:

>> Code PIC Assembly language instructions to create loops
>> Code PIC Assembly language conditional branch instructions
>> Explain conditions that determine each conditional branch instruction
>> Code GOTO (long jump) instructions for unconditional jumps
>> Calculate target addresses for conditional branch instructions
>> Code PIC subroutines
>> Describe the stack and its use in subroutines
>> Discuss pipelining in the PIC
>> Discuss crystal frequency versus instruction cycle time in the PIC
>> Code PIC programs to generate a time delay

In the sequence of instructions to be executed, it is often necessary to transfer program control to a different location. There are many instructions in PIC to achieve this. This chapter covers the control transfer instructions available in PIC Assembly language. In Section 3.1, we discuss instructions used for looping, as well as instructions for conditional and unconditional branches (jumps). In the second section, we examine the stack and the CALL instruction. In Section 3.3, instruction pipelining of the PIC18 is examined. Instruction timing and time delay subroutines are also discussed in Section 3.3.

SECTION 3.1: BRANCH INSTRUCTIONS AND LOOPING

In this section we first discuss how to perform a looping action in PIC and then the branch (jump) instructions, both conditional and unconditional.

Looping in PIC

Repeating a sequence of instructions or an operation a certain number of times is called a *loop*. The loop is one of most widely used programming techniques. In the PIC, there are several ways to repeat an operation many times. One way is to repeat the operation over and over until it is finished, as shown below:

```
MOVLW  0      ;WREG = 0
ADDLW  3      ;add value 3 to WREG
ADDLW  3      ;add value 3 to WREG(W = 6)
ADDLW  3      ;add value 3 to WREG(W = 9)
ADDLW  3      ;add value 3 to WREG(W = 0Ch)
ADDLW  3      ;add value 3 to WREG(W = 0Fh
```

In the above program, we add 3 to WREG 5 times. That makes $5 \times 3 = 15$ = 0Fh. One problem with the above program is that too much code space would be needed to increase the number of repetitions to 50 or 100. A much better way is to use a loop. There are two ways to do a loop in PIC. Next, we describe each method.

DECFSZ instruction and looping

The DECFSZ (decrement fileReg skip zero) instruction is a widely used instruction supported across all PIC families of microcontrollers from PIC12 to PIC18. It has the following format:

DECFSZ   fileReg , d   ;decrement fileReg and skip next instruction if 0

In this instruction, the fileReg is decremented, and if its content is zero, it skips the next instruction. By placing the "GOTO target" instruction right below it we can create a loop. The target address of the "GOTO target" instruction is the beginning of the loop, as shown in Examples 3-1 and 3-2. Figure 3-1 shows the flowchart for the DECFSZ instruction. Study the flowchart structure in Appendix D to get familiar with the symbols. The flowchart is a widely used method to represent a sequence of actions pictorially. Its usage for program design is recommended very strongly.

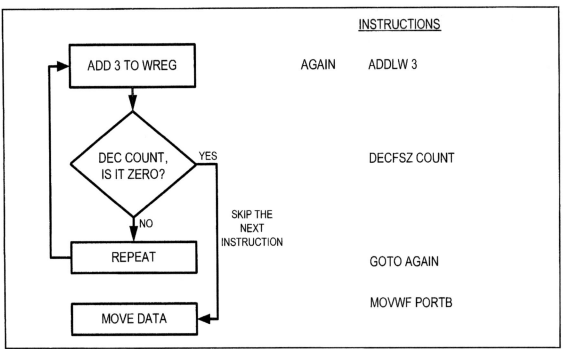

Figure 3-1. Flowchart for the DECFSZ Instruction

Example 3-1

Write a program to (a) clear WREG, and (b) add 3 to WREG ten times and place the result in SFR of PORTB. Use the DECFSZ instruction to perform looping.

Solution:

```
;this program adds value 3 to WREG ten times

COUNT   EQU 0x25            ;use loc 25H for counter

        MOVLW   d'10'       ;WREG = 10 (decimal) for counter
        MOVWF   COUNT       ;load the counter
        MOVLW   0           ;WREG = 0
AGAIN   ADDLW   3           ;add 03 to WREG (WREG = sum)
        DECFSZ  COUNT,F     ;decrement counter, skip if count = 0
        GOTO    AGAIN       ;repeat until count becomes 0
        MOVWF   PORTB       ;send sum to PORTB SFR
```

Notice that the DECFSZ instruction will decrement the counter (fileReg loc 0x25), which has 10 in it. It becomes 9. Because it is not zero, it will execute the "GOTO AGAIN" instruction. The "GOTO AGAIN" goes back to the start of the loop. Next, it decrements, our counter becomes 8, and, because it is not zero, it executes the GOTO. It goes on like that until the counter becomes zero. Upon the counter becoming zero, it skips the GOTO, which gets it out of the loop, and executes the "MOVWF PORTB" instruction. Notice that we use "DECFSZ COUNT, F" and not "DECFSZ COUNT, W" because we want the count value to change for the next iteration. We will never get out of the loop if we use "DECFSZ COUNT, W" because COUNT = 9 and the decrement value is placed in WREG.

CHAPTER 3: BRANCH, CALL, AND TIME DELAY LOOP   99

Using instruction BNZ for looping

The BNZ (branch if not zero) instruction is supported by the PIC18 family and not earlier families such as PIC16 or PIC12. It uses the zero flag in the status register. The BNZ instruction is used as follows:

```
BACK    ........    ;start of the loop
        ........    ;body of the loop
        ........    ;body of the loop
        DECF        ;decrement fileReg, Z = 1 if fileReg = 0
        BNZ BACK    ;branch to BACK if Z = 0
```

In the last two instructions, the fileReg is decremented; if it is not zero, it branches (jumps) back to the target address referred to by the label. Prior to the start of the loop, the fileReg is loaded with the counter value for the number of repetitions. Notice that the BNZ instruction refers to the Z flag of the status register affected by the previous instruction, DECF. This is shown in Example 3-2.

In the program in Example 3-2, fileReg location 0x25 is used as a counter. The counter is first set to 10. In each iteration, the DEC instruction decrements the fileReg and sets the flag bits accordingly. If fileReg is not zero ($Z \neq 0$), it jumps to the target address associated with the label "AGAIN". This looping action continues until fileReg COUNT becomes zero. After fileReg COUNT becomes zero ($Z = 0$), it falls through the loop and executes the instruction immediately below it, in this case "MOVWF PORTB". See Figure 3-2.

Notice, in the "DECF COUNT, F" instruction, that fileReg 25H is used as a register to hold the count as it decrements instead of WREG. If WREG is used as the destination of the DECF instruction, then you have an infinite loop because COUNT remains at its original value of 10.

Example 3-2

Write a program to (a) clear WREG, then (b) add 3 to WREG ten times.

Use the zero flag and BNZ.

Solution:

```
;this program adds value 3 to the WREG ten times

        COUNT EQU 0x25    ;use loc 25H for counter

        MOVLW d'10'       ;WREG = 10 (decimal) for counter
        MOVWF COUNT       ;load the counter
        MOVLW 0           ;WREG = 0
AGAIN   ADDLW 3           ;add 03 to WREG (WREG = sum)
        DECF COUNT, F     ;decrement counter
        BNZ AGAIN         ;repeat until COUNT = 0
        MOVWF PORTB       ;send sum to PORTB SFR
```

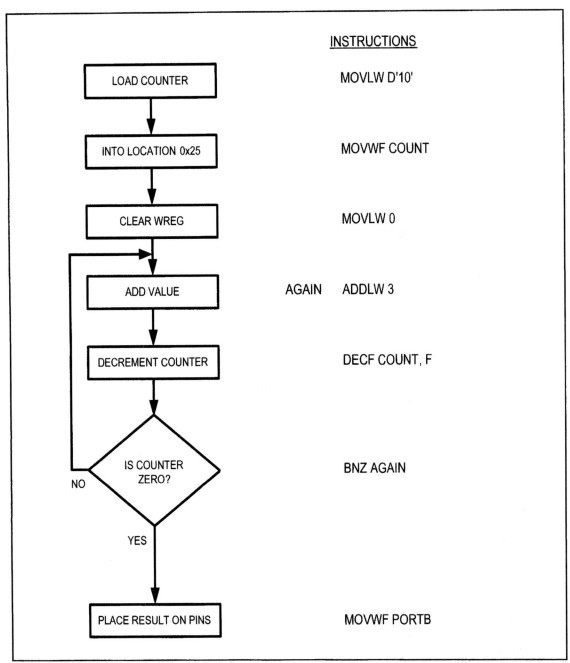

INSTRUCTIONS

LOAD COUNTER   MOVLW D'10'

INTO LOCATION 0x25   MOVWF COUNT

CLEAR WREG   MOVLW 0

ADD VALUE   AGAIN   ADDLW 3

DECREMENT COUNTER   DECF COUNT, F

IS COUNTER ZERO?   BNZ AGAIN

NO

YES

PLACE RESULT ON PINS   MOVWF PORTB

Figure 3-2. Flowchart for Example 3-2

Example 3-3

What is the maximum number of times that the loop in Example 3-2 can be repeated?

Solution:

Because location COUNT in fileReg is an 8-bit register, it can hold a maximum of FFH (255 decimal); therefore, the loop can be repeated a maximum of 255 times. See Example 3-4 to bypass this limitation.

CHAPTER 3: BRANCH, CALL, AND TIME DELAY LOOP

Loop inside a loop

As shown in Example 3-2, the maximum count is 255. What happens if we want to repeat an action more times than 255? To do that, we use a loop inside a loop, which is called a *nested loop*. In a nested loop, we use two registers to hold the count. See Example 3-4.

Example 3-4

Write a program to (a) load the PORTB SFR register with the value 55H, and (b) complement Port B 700 times.

Solution:

Because 700 is larger than 255 (the maximum capacity of any register), we use two registers to hold the count. The following code shows how to use fileReg locations 25H and 26H as a register for counters.

```
        R1 EQU 0x25
        R2 EQU 0x26
        COUNT_1 EQU d'10'
        COUNT_2 EQU d'70'
        MOVLW 0x55          ;WREG = 55h
        MOVWF PORTB         ;PORTB = 55h
        MOVLW COUNT_1       ;WREG = 10, outer loop count value
        MOVWF R1            ;load 10 into loc 25H (outer loop count)
LOP_1 MOVLW COUNT_2        ;WREG = 70, inner loop count value
        MOVWF R2            ;load 70 into loc 26H
LOP_2 COMPF PORTB, F       ;complement Port B SFR
        DECF R2, F          ;dec fileReg loc 26 (inner loop)
        BNZ LOP_2           ;repeat it 70 times
        DECF R1, F          ;dec fileReg loc 25 (outer loop)
        BNZ LOP_1           ;repeat it 10 times
```

In this program, fileReg location 0x26 is used to keep the inner loop count. In the instruction "BNZ LOP_2", whenever location 26H becomes 0 it falls through and "DECF R1, F" is executed. This instruction forces the CPU to load the inner count with 70 if it is not zero, and the inner loop starts again. This process will continue until location 25 becomes zero and the outer loop is finished.

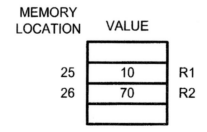

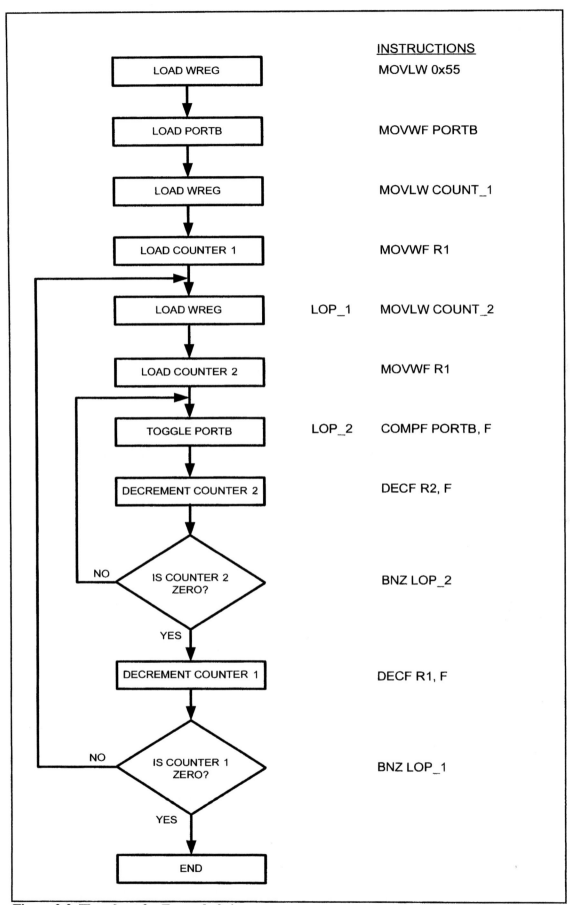

LOAD WREG		MOVLW 0x55
LOAD PORTB		MOVWF PORTB
LOAD WREG		MOVLW COUNT_1
LOAD COUNTER 1		MOVWF R1
LOAD WREG	LOP_1	MOVLW COUNT_2
LOAD COUNTER 2		MOVWF R1
TOGGLE PORTB	LOP_2	COMPF PORTB, F
DECREMENT COUNTER 2		DECF R2, F
IS COUNTER 2 ZERO?	NO	BNZ LOP_2
	YES	
DECREMENT COUNTER 1		DECF R1, F
IS COUNTER 1 ZERO?	NO	BNZ LOP_1
	YES	
END		

Figure 3-3. Flowchart for Example 3-4

CHAPTER 3: BRANCH, CALL, AND TIME DELAY LOOP 103

Looping 100,000 times

Because two registers give us a maximum value of 65025 ($255 \times 255 = 65025$), we can use three registers to get up to more than 16 million (2^{24}) iterations. The following code repeats an action 100,000 times:

```
R1 EQU 0x1                    ;assign RAM loc for the R1-R2
R2 EQU 0x2
R2 EQU 0x3
COUNT_1 EQU D'100'            ;fixed value for 100,000 times
COUNT_2 EQU D'100'
COUNT_3 EQU D'10'

        MOVLW 0x55
        MOVWF PORTB
        MOVLW COUNT_3
        MOVWF R3
LOP_3   MOVLW COUNT_2
        MOVWF R2
LOP_2   MOVLW COUNT_1
        MOVWF R1
LOP_1   COMPF PORTB, F
        DECF  R1 ,F
        BNZ   LOP_1
        DECF  R2, F
        BNZ   LOP_2
        DECF  R3, F
        BNZ   LOP_3
```

Other conditional jumps

Conditional branches for the PIC are summarized in Table 3-1. More details of each instruction are provided in Appendix A. In Table 3-1, notice that some of the instructions, such as BZ (Branch if Z = 1) and BC (Branch if C = 1), jump only if a certain condition is met. Next, we examine some conditional branch instructions with examples.

BZ (Branch if Z = 1)

In this instruction, the Z flag is checked. If it is high, it jumps to the target address. For example, look at the following code.

Table 3-1: PIC Conditional Branch (Jump) Instructions

Instruction	Action
BC	Branch if C = 1
BNC	Branch if C ≠ 0
BZ	Branch if Z = 1
BNZ	Branch if Z ≠ 0
BN	Branch if N = 1
BNN	Branch if N ≠ 0
BOV	Branch if OV = 1
BNOV	Branch if OV ≠ 0

```
OVER  MOVF PORTB,W     ;read Port B and put it in WREG
      JZ OVER          ;jump if WREG is zero
```

In this program, if PORTB is zero, it jumps to the label OVER. It stays there until PORTB has a value other than zero. Notice that the BZ instruction can be used to see whether any fileReg or WREG is zero. More importantly, you don't have to perform an arithmetic instruction such as decrement to use the BZ instruction. See Example 3-5.

Example 3-5

Write a program to determine if fileReg location 0x30 contains the value 0. If so, put 55H in it.

Solution:
```
      MYLOC EQU 0x30
      MOVF  MYLOC,F          ;copy MYLOC to itself
      BNZ   NEXT             ;branch if MYLOC is not zero
      MOVLW 0x55
      MOVWF MYLOC            ;put 0x55 if MYLOC has zero value
NEXT  ...
```

BNC (branch if no carry, branch if CY = 0)

In this instruction, the carry flag bit in the Status register is used to make the decision whether to jump. In executing "BNC label", the processor looks at the carry flag to see if it is raised (C = 1). If it is not, the CPU starts to fetch and execute instructions from the address of the label. If C = 1, it will not branch but will execute the next instruction below BNC. Study Example 3-6 to see how BNC is used to add numbers together when the sum is higher than FFH. Note that there

---

Example 3-6

Find the sum of the values 79H, F5H, and E2H. Put the sum in fileReg locations 5 (low byte) and 6 (high byte).

Solution:
```
L_Byte EQU 0x5            ;assign RAM loc 5 to L_byte of sum
H_Byte EQU 0x6            ;assign RAM loc 6 to H_byte of sum

       ORG 0h
       MOVLW 0x0          ;clear WREG (WREG = 0)
       MOVWF H_Byte       ;H_Byte = 0
       ADDLW 0x79         ;WREG = 0 + 79H = 79H, C = 0
       BNC   N_1          ;if C = 0, add next number
       INCF  H_Byte,F     ;C = 1, increment (now H_Byte = 0)
N_1    ADDLW 0xF5         ;WREG = 79 + F5 = 6E and C = 1
       BNC   N_2          ;branch if CY = 0
       INCF  H_Byte,F     ;C = 1, increment (now H_Byte = 1)
N_2    ADDLW 0xE2         ;WREG = 6E + E2 = 50 and C = 1
       BNC   OVER         ;branch if C = 0
       INCF  H_Byte,F     ;C = 1, increment (now H_Byte = 2)
OVER   MOVWF L_Byte       ;now L_Byte = 50H, and H_Byte = 02
       END
```

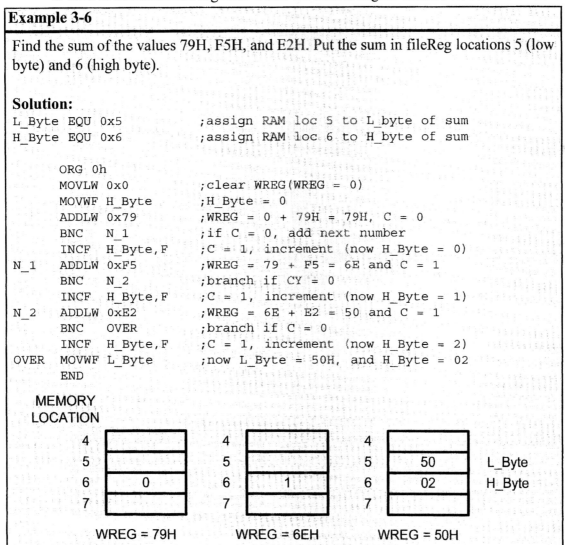

---

is also a "BC label" instruction. In the BC instruction, if C = 1 it jumps to the target address. We will give more examples of these instructions in the context of some applications in Chapter 5.

The other conditional branch instructions in Table 3-1 are discussed in Chapter 5 when arithmetic operations with signed numbers are discussed.

All conditional branches are short jumps

It must be noted that all conditional jumps are short jumps, meaning that the address of the target must be within 256 bytes of the contents of the program counter (PC). This concept is discussed next.

Example 3-7

Using the following list file of Example 3-6, verify the jump forward address calculation.

```
Line  PC              Opcode                  Mnemonic Operand

LOC       OBJECT LINE   SOURCE TEXT
          CODE
    VALUE

    00000005    00001   L_Byte EQU 0x5      ;assign RAM Loc 5 to L_byte of sum
    00000006    00002   H_Byte EQU 0x6      ;assign RAM Loc 6 to H_byte of sum
                00003
000000          00004          ORG 0h
000000 0E00     00005          MOVLW  0x0     ;clear WREG(WREG=0)
000002 6E06     00006          MOVWF  H_Byte  ;H_Byte = 0
000004 0F79     00007          ADDLW  0x79    ;WREG = 0 + 79H = 79H, C = 0
000006 E301     00008          BNC    N_1     ;if C = 0, add next number
000008 2A06     00009          INCF   H_Byte,F ;C = 1, increment (now H_Byte = 0)
00000A 0FF5     00010   N_1    ADDLW  0xF5    ;WREG = 79 + F5 = 6E and C = 1
00000C E301     00011          BNC    N_2     ;branch if CY = 0
00000E 2A06     00012          INCF   H_Byte,F ;C = 1, increment (now H_Byte = 1)
000010 0FE2     00013   N_2    ADDLW  0xE2    ;WREG = 6E + E2 = 50 and C = 1
000012 E301     00014          BNC    OVER    ;branch if C = 0
000014 2A06     00015          INCF   H_Byte,F ;C = 1, increment (now H_Byte = 2)
000016 6E05     00016   OVER   MOVWF  L_Byte  ;now L_Byte = 50H, and H_Byte = 02
                00017          END
```

Solution:

First notice that the BNC instruction jumps forward. The target address for a forward jump is calculated by adding the PC of the following instruction to the second byte of the branch instruction times 2. Recall that each instruction takes 2 bytes. In line 6 the instruction "BNC N_1" has an opcode of E3 and an operand of 01 at the addresses of 000006 and 000007. The 01 × 02 = 02 is the relative address, relative to the address of the next instruction INCF, which is 000008. By adding 000002 to 000008, the target address of the label N_1, which is 00000A, is generated. In the same way for line 000011, the "BNC N_2" instruction, and line 000014, the "BNC OVER" instruction jumps forward because the relative value is positive.

Calculating the short branch address

All conditional branches such as BNC, BZ, and BNZ are short branches due to the fact that they are all 2-byte instructions. In these instructions the first byte is the opcode and the second byte is the relative address. The target address is relative to the value of the program counter. If the second byte is positive, the the jump is forward. If the second byte is negative, then the jump is backwards. The second byte can be a value from -127 to $+128$. To calculate the target address, we add the second byte of the instruction times 2 to the PC of the next instruction [target address = (2nd byte of instruction $\times$ 2) + PC]. See Example 3-7. We do the same thing for the backward branch, although the second byte is negative. That is, we multiply the negative number by two and add it to the PC value of the next instruction. See Example 3-8.

Example 3-8

Verify the calculation of backward jumps for the listing of Example 3-2, shown below.

Solution:

```
LOC     OBJECT LINE    SOURCE TEXT
        CODE
  VALUE

  00000025  00001  COUNT  EQU 0x25        ;use loc 25H for counter
000000      00002         ORG 0h
000000 0E0A  00003         MOVLW d'10'     ;WREG = 10 (decimal) for counter
000002 6E25  00004         MOVWF COUNT     ;load the counter
000004 0E00  00005         MOVLW 0         ;WREG = 0
000006 0F03  00006  AGAIN  ADDLW 3         ;add 03 to WREG (WREG = sum)
000008 0625  00007         DECF COUNT, F   ;decrement counter
00000A E1FD  00008         BNZ AGAIN       ;repeat until COUNT = 0
00000C 6E81  00009         MOVWF PORTB     ;send sum to PORTB SFR
             00010         END
```

In the program list, "BNZ AGAIN" has opcode E1 and relative address FDH. The FDH gives us -3, which means the displacement is $-3 \times 2 = -6$. When the relative address of -6 is added to 00000CH, the address of the instruction below the byte, we have $-6 + 0CH = 06H$ (the carry is dropped). Notice that 000006 is the address of the label AGAIN. FDH is a negative number and that means it will branch backward. For further discussion of the addition of negative numbers, see Chapter 5.

Although we can use BNZ along with DECF to perform a loop, it is better to use an instruction such as DCFSNZ, because it combines the decrement and jump into a single instruction.

Unconditional branch instruction

The unconditional branch is a jump in which control is transferred unconditionally to the target location. In the PIC18 there are two unconditional branches: GOTO (go to) and BRA (branch). Deciding which one to use depends on the target address. Each instruction is explained next.

GOTO (GOTO is a long jump)

GOTO is an unconditional jump that can go to any memory location in the 2M address space of the PIC18. It is a 4-byte (32-bit) instruction in which 12 bits are used for the opcode, and the other 20 bits represent the 20-bit address of the target location. The 20-bit target address allows a jump to 1M of memory locations from 00000 to FFFFFH, instead of 2M. This problem is solved by making the lowest bit of the program counter $A0 = 0$, and the 20-bit target address of the GOTO becomes address bits A21–A1. In this way, the GOTO can cover the entire 2M address space of 00000–1FFFFH and also makes sure that the target address lands on an even address location. Because all the PIC18 instructions are 2-byte or 4-byte instructions, the GOTO will not land at the middle of an instruction. See Figure 3-4.

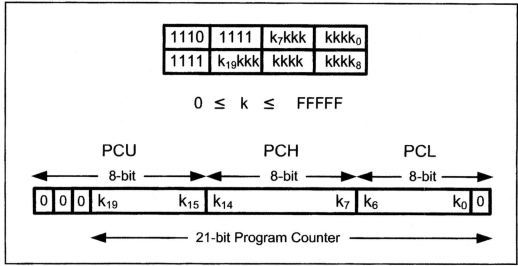

Figure 3-4. GOTO Instruction

Remember that although the program counter in the PIC18 is 21-bit (thereby giving a ROM address space of 2M), not all PIC18 family members have that much on-chip program ROM. Some of the PIC18 family members have only 4K–32K of on-chip ROM for program space; consequently, every byte is precious. For this reason there is also a BRA (branch) instruction, which is a 2-byte instruction as opposed to the 4-byte GOTO instruction. This can save some bytes of memory in many applications where ROM memory space is in short supply. BRA is discussed next.

BRA (branch)

In this 2-byte (16-bit) instruction, the first 5 bits are the opcode and the rest (lower 11 bits) is the relative address of the target location. The relative address range of 000–FFFH is divided into forward and backward jumps; that is, within

–1024 to +1023 bytes of memory relative to the address of the current PC (program counter). If the jump is forward, then the target address is positive. If the jump is backward, then the target address is negative. In this regard, BRA is like the conditional branch instructions except that 11 bits are used for the offset address instead of 8. This is shown in detail in Figure 3-5.

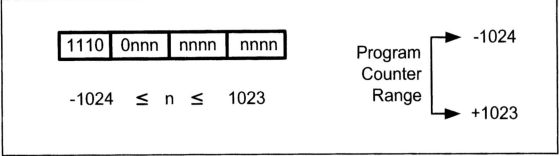

Figure 3-5. BRA (Branch Unconditionally) Instruction Address Range

Notice that this is a 2-byte instruction, and is preferred over the GOTO because it takes less ROM space. Chapter 5 examines signed numbers.

GOTO to itself using $ sign

In cases where there is no monitor program, we use the GOTO (jump) to itself in order to keep the microcontroller busy. A simple way of doing that is to use the $ sign. That means in place of this:

    HERE   GOTO   HERE

we can use the following:

    GOTO $

This will also work for the BRA instruction, as shown below:

    OVER   BRA   OVER

which is the same as:

    BRA $;$ means same line

Review Questions

1. The mnemonic BNZ stands for _____.
2. True or false. "BNZ BACK" makes its decision based on the last instruction affecting the Z flag.
3. "BNZ HERE" is a ___ -byte instruction.
4. In "JZ NEXT", which register's content is checked to see if it is zero?
5. GOTO is a(n) ___ -byte instruction.

SECTION 3.2: CALL INSTRUCTIONS AND STACK

Another control transfer instruction is the CALL instruction, which is used to call a subroutine. Subroutines are often used to perform tasks that need to be performed frequently. This makes a program more structured in addition to saving memory space. In the PIC18 there are two instructions for call: CALL (long call) and RCALL (relative call). Deciding which one to use depends on the target address. Each instruction is explained next.

CALL

In this 4-byte (32-bit) instruction, the 12 bits are used for the opcode and the other 20 bits, A21–A1, are used for the address of the target subroutine. Just as in the GOTO instruction, the lowest bit of the program counter is 0 automatically to ensure it lands on an even address. Therefore, CALL can be used to call subroutines located anywhere within the 2M address space of 00000–1FFFFH for the PIC18, as shown in Figure 3-6.

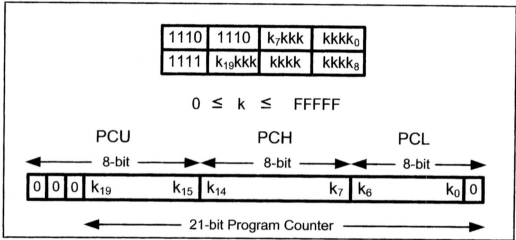

Figure 3-6. CALL Instruction

To make sure that the PIC knows where to come back to after execution of the called subroutine, the microcontroller automatically saves on the stack the address of the instruction immediately below the CALL. When a subroutine is called, control is transferred to that subroutine, and the processor saves the PC (program counter) of the next on the stack and begins to fetch instructions from the new location. After finishing execution of the subroutine, the instruction RETURN transfers control back to the caller. Every subroutine needs RETURN as the last instruction.

Stack and stack pointer in the PIC18

The stack is read/write memory (RAM) used by the CPU to store some very critical information temporarily. This information usually is an address, but it could be data as well. The CPU needs this storage area because there are only a limited number of registers. The stack in the PIC18 is 21-bit because the program counter is 21-bit. This means that it is used for the CALL instruction to make sure that the PIC knows where to come back to after execution of the called subroutine.

A 21-bit stack can take values of 00000 to 1FFFFFH, just like the program counter. If the stack is RAM, there must be a register inside the CPU to point to it. The register used to access the stack is called the SP (stack pointer) register. The PIC18 has a 5-bit stack pointer, which can take values of 00 to 1FH. That gives us a total of 32 locations where each location is 21 bits wide. This is shown in Figure 3-7. When the PIC18 is powered up, the SP register contains value 0. This means that stack location 1 is the first location used for the stack because the SP points to the last-used location. That means that location 0 of the stack is not available and we have only 31 stack locations in the PIC18.

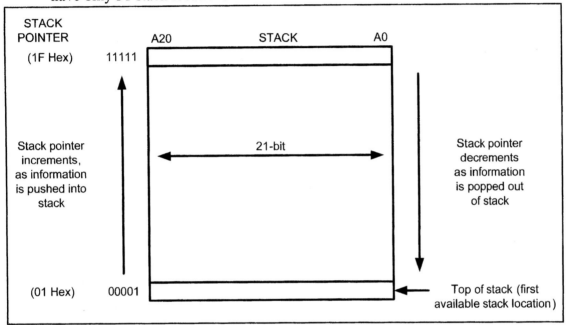

Figure 3-7. PIC Stack 31 × 21

How stacks are accessed in the PIC18

The storing of CPU information such as the program counter on the stack is called a PUSH, and loading the contents of the stack back into a CPU register is called a POP. In other words, a register is pushed onto the stack to save it and popped off the stack to retrieve it. The following describes each process.

Pushing onto the stack

In the PIC, the stack pointer (SP) is pointing to the last used location of the stack. The last-used location of the stack is referred to as the top of the stack (TOS). As data is pushed onto the stack, the stack pointer is incremented. Notice that this is different from many other microprocessors, notably x86 processors, in which the SP is decremented when data is pushed onto the stack. Examining Example 3-9, we see that as each CALL is executed, the contents of the program counter are saved on the stack and SP is incremented. Notice that for every program counter saved on the stack, SP is incremented only once.

Popping from the stack

Popping the contents of the stack back into a given register such as the program counter is the opposite process of pushing. When the RETURN instruction at the end of the subroutine is executed, the top location of the stack is copied back to the program counter and the stack pointer is decremented once. That means the stack is LIFO (Last-In-First-Out) memory.

CALL instruction and the role of the stack

In the PIC, the CPU uses the stack to save the address of the instruction just below the CALL instruction. This is how the CPU knows where to resume when it returns from the called subroutine. To understand the importance of the stack in microcontrollers, examine the contents of the stack and stack pointer for Example 3-9. This is shown in Example 3-10.

The following points should be noted for the program in Example 3-9:

1. Notice the DELAY subroutine. Upon executing the first "CALL DELAY", the address of the instruction right below it, "MOVLW 0xAA", is pushed onto the stack, and the PIC starts to execute instructions at address 000300H.
2. In the DELAY subroutine, first the counter MYREG is set to 255 (MYREG = FFH); therefore, the loop is repeated 256 times. When MYREG becomes 0, control falls to the RETURN instruction, which pops the address from the top of the stack into the program counter and resumes executing the instructions after the CALL.

The amount of time delay in Example 3-9 depends on the frequency of the PIC. How to calculate the exact time will be explained in the last section of this chapter.

Example 3-9

Toggle all the bits of the SFR register of Port B by sending to it the values 55H and AAH continuously. Put a time delay in between each issuing of data to Port B.

Solution:

```
MYREG EQU    0x08                ;use location 08 as counter
        ORG     0
BACK    MOVLW   0x55            ;load WREG with 55H
        MOVWF   PORTB           ;send 55H to port B
        CALL    DELAY           ;time delay
        MOVLW   0xAA            ;load WREG with AA (in hex)
        MOVWF   PORTB           ;send AAH to port B
        CALL    DELAY
        GOTO    BACK            ;keep doing this indefinitely
;------- this is the delay subroutine
        ORG     300H            ;put time delay at address 300H
DELAY   MOVLW   0xFF            ;WREG = 255,the counter
        MOVWF   MYREG
AGAIN   NOP                     ;no operation wastes clock cycles
        NOP
        DECF    MYREG, F
        BNZ     AGAIN           ;repeat until MYREG becomes 0
        RETURN                  ;return to caller
        END                     ;end of asm file
```

Example 3-10

Analyze the stack for the CALL instructions in the following program.

Solution:

When the first CALL is executed, the address of the instruction "MOVLW 0xAA" is saved (pushed) on the stack. The last instruction of the called subroutine must be a RETURN instruction, which directs the CPU to pop the contents of the top location of the stack into the PC and resume executing at address 000007. The diagrams show the stack frame after the CALL and RETURN instructions.

```
LOC   OBJECT CODE      LINE SOURCE TEXT
  VALUE

                       00001 #DEFINE PORTB 0xF81
   00000008            00002 MYREG EQU 0x08      ;use location 08 as counter
                       00003
                       00004
   000000              00005        ORG    0
000000 0E55            00006 BACK   MOVLW  0x55   ;load WREG with 55H
000002 6E81            00007        MOVWF  PORTB  ;send 55H to port B
000004 EC80 F001       00008        CALL   DELAY  ;time delay
000008 0EAA            00009        MOVLW  0xAA   ;load WREG with AA (in hex)
00000A 6E81            00010        MOVWF  PORTB  ;send AAH to port B
00000C EC80 F001       00011        CALL   DELAY
000010 EF00 F000       00012        GOTO   BACK   ;keep doing this indefinitely
                       00013
                       00014 ;——— this is the delay subroutine
                       00015
   000300              00016        ORG    300H   ;put delay at address 300H
000300 0EFF            00017 DELAY  MOVLW  0xFF   ;WREG = 255,the counter
000302 6E08            00018        MOVWF  MYREG
000304 0000            00019 AGAIN  NOP           ;no op wastes clock cycles
000306 0000            00020        NOP
000308 0608            00021        DECF   MYREG, F
00030A E1FC            00022        BNZ    AGAIN  ;repeat until MYREG becomes 0
00030C 0012            00023        RETURN        ;return to caller
                       00024        END           ;end of asm file
```

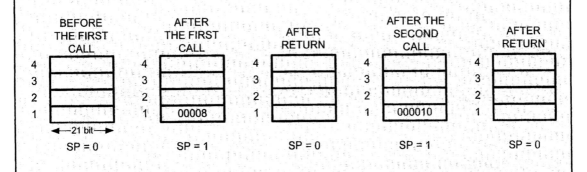

CHAPTER 3: BRANCH, CALL, AND TIME DELAY LOOP

The upper limit of the stack

As mentioned earlier, in the PIC18 there are only 31 RAM locations for the stack and they are mapped to addresses 01 to 1FH. That limits the number of calls inside of calls for a given program to 31 because stack location 00 is not available. In PIC, the stack is used for calls and interrupts. We must remember that upon calling a subroutine, the stack keeps track of where the CPU should return after completing the subroutine. For this reason, we must be very careful not to manipulate the stack contents. See Chapter 6 for more on this.

Calling many subroutines from the main program

In Assembly language programming, it is common to have one main program and many subroutines that are called from the main program. (See Figure 3-8.)This allows you to make each subroutine into a separate module. Each module can be tested separately and then brought together with the main program. More importantly, in a large program the modules can be assigned to different programmers in order to shorten development time. See Chapter 6 for discussion of modules.

```
;MAIN program calling subroutines
            ORG    0
MAIN        CALL   SUBR_1
            CALL   SUBR_2
            CALL   SUBR_3

HERE        BRA    HERE          ;stay here
;————————end of MAIN
;
SUBR_1      ....
            ....
            RETURN
;————————end of subroutine 1
;
SUBR_2      ....
            ....
            RETURN
;————————end of subroutine 2

SUBR_3      ....
            ....
            RETURN
;————————end of subroutine 3
            END                 ;end of the asm file
```

Figure 3-8. PIC Assembly Main Program That Calls Subroutines

It needs to be emphasized that in using CALL, the target address of the subroutine can be anywhere within the 2M memory space of the PIC18. (See Example 3-11.) This is not the case for the other call instruction, RCALL, which is explained next.

RCALL (relative call)

RCALL is a 2-byte instruction in contrast to CALL, which is 4 bytes.

114

Example 3-11

Write a program to count up from 00 to FFH and send the count to SFR of Port B. Use one CALL subroutine for sending the data to Port B and another one for time delay. Put a time delay in between each issuing of data to Port B.

Solution:

```
LOC    OBJECT CODE    LINE    SOURCE TEXT
  VALUE
                      00001          list  P=PIC18F458
                      00002   #include P18F458.INC
                      00003
  00000007            00004   COUNT  EQU    0x07   ;use location 07 for count-up
  00000008            00005   MYREG  EQU    0x08   ;use location 08 for delay
                      00006
000000                00007   ORG    0
000000 0E00           00008          MOVLW  0                ;WREG = 0
000002 6E07           00009          MOVWF  COUNT            ;count = 0
000004 EC06 F000      00010   BACK   CALL   DISPLAY
000008 EF02 F000      00011          GOTO   BACK
                      00012
                      00013   ;------ increment and put it in PORTB
00000C 2A07           00014   DISPLAY INCF  COUNT,F          ;increment count
00000E C007 FF81      00015          MOVFF  COUNT,PORTB  ;send it to PORTB
000012 EC80 F001      00016          CALL   DELAY
000016 0012           00017          RETURN               ;return to caller
                      00018
                      00019   ;------ this is the delay subroutine
000300                00020   ORG    300H   ;put time delay at address 300H
000300 0EFF           00021   DELAY  MOVLW  0xFF           ;WREG = 255, the counter
000302 6E08           00022          MOVWF  MYREG
000304 0000           00023   AGAIN  NOP    ;no operation wastes clock cycles
000306 0000           00024          NOP
000308 0000           00025          NOP
00030A 0608           00026          DECF   MYREG,F
00030C E1FB           00027          BNZ    AGAIN  ;repeat until MYREG becomes 0
00030E 0012           00028          RETURN           ;return to caller
                      00029          END              ;end of asm file
```

	BEFORE ANY CALL	AFTER CALL DISPLAY	AFTER CALL DELAY	AFTER DELAY RETURN	AFTER DISPLAY RETURN
4					
3					
2			00016		
1		00008	00008	00008	
	◄—21 bit—►				
	SP = 0	SP = 1	SP = 2	SP = 1	SP = 0

Because RCALL is a 2-byte instruction, the target address of the subroutine must be within 2K because only 11 bits of the 2 bytes are used for the address. There is no difference between RCALL and CALL in terms of saving the program counter on the stack or the function of the RETURN instruction. The only difference is that the target address for CALL can be anywhere within the 2M address space of the PIC18 while the target address of RCALL must be within a 2K range. In many variations of the PIC18 marketed by Microchip Corporation, on-chip ROM is as low as 4K. In such cases, the use of RCALL instead of CALL can save a number of bytes of program ROM space.

Of course, in addition to using compact instructions, we can program efficiently by having a detailed knowledge of all the instructions supported by a given microprocessor, and using them wisely. Look at Example 3-12.

Example 3-12

Rewrite the main part of Example 3-9 as efficiently as you can.

Solution:

```
        MYREG EQU 0x08
        ORG   0
        MOVLW 0x55          ;load WREG with 55H
BACK    MOVWF PORTB             ;issue value in PORTB SFR
        RCALL DELAY         ;time delay
        COMPF PORTB,F           ;complement Port B SFR
        BRA   BACK          ;keep doing this indefinitely
;———————this is the delay subroutine
DELAY   MOVLW 0xFF          ;WREG = 255, the counter
        MOVWF MYREG
AGAIN   NOP                 ;no operation wastes clock cycles
        NOP
        DECF  MYREG,F
        BNZ   AGAIN         ;repeat until MYREG becomes 0
        RETURN              ;return to caller (MYREG = 0)
        END                 ;end of asm file
```

Example 3-13

A developer is using the PIC18 microcontroller chip for a product. This chip has only 4K of on-chip flash ROM. Which of the instructions, CALL or RCALL, is more useful in programming this chip?

Solution:

The RCALL instruction is more useful because it is a 2-byte instruction. It saves two bytes each time the call instruction is used. However, we must use CALL if the target address is beyond the 2K boundary.

Review Questions

1. How wide is the size of the stack in the PIC18?
2. True or false. In the PIC18, control can be transferred anywhere within the 2M of code space by using the CALL instruction.
3. The CALL instruction is a(n) ____ -byte instruction.
4. True or false. In the PIC18, control can be transferred anywhere within the 2M of code space by using the RCALL instruction.
5. With each CALL instruction, the stack pointer register, SP, is _____ (incremented, decremented).
6. With each RETURN instruction, the SP is _____ (incremented, decremented).
7. On power-up, the PIC uses location _____ as the first location of the stack.
8. How deep is the size of the stack in the PIC18?
9. The RCALL instruction is a(n) ____ -byte instruction.
10. _____ (RCALL, CALL) takes more ROM space.

SECTION 3.3: PIC18 TIME DELAY AND INSTRUCTION PIPELINE

In the last section we used the DELAY subroutine. In this section we discuss how to generate various time delays and calculate exact delays for the PIC18. We will also discuss instruction pipelining and its impact on execution time.

Delay calculation for the PIC18

In creating a time delay using Assembly language instructions, one must be mindful of two factors that can affect the accuracy of the delay:

1. The crystal frequency: The frequency of the crystal oscillator connected to the OSC1 and OSC2 input pins is one factor in the time delay calculation. The duration of the clock period for the instruction cycle is a function of this crystal frequency.

2. The PIC design: Since the 1970s, both the field of IC technology and the architectural design of microprocessors have seen great advancements. Due to the limitations of IC technology and limited CPU design experience for many years, the instruction cycle duration was longer. Advances in both IC technology and CPU design in the 1980s and 1990s have made the single instruction cycle a common feature of many microcontrollers. Indeed, one way to increase performance without losing code compatibility with the older generation of a given family is to reduce the number of instruction cycles it takes to execute an instruction. One might wonder how microprocessors such as PIC are able to execute an instruction in one cycle. There are three ways to do that: (a) Use Harvard architecture to get the maximum amount of code and data into the CPU, (b) use RISC architecture features such as fixed-size instructions, and finally (c) use pipelining to overlap fetching and execution of instructions. We have examined the Harvard and RISC architectures in Chapter 2. Next, we discuss pipelining.

Pipelining

In early microprocessors such as the 8085, the CPU could either fetch or execute at a given time. In other words, the CPU had to fetch an instruction from memory, then execute it, and then fetch the next instruction, execute it, and so on. The idea of pipelining in its simplest form is to allow the CPU to fetch and execute at the same time, as shown in Figure 3-9.

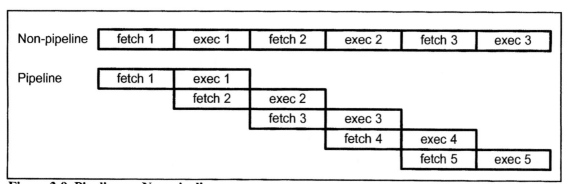

Figure 3-9. Pipeline vs. Non-pipeline

CHAPTER 3: BRANCH, CALL, AND TIME DELAY LOOP   **117**

Instruction cycle time for the PIC

It takes a certain amount of time for the CPU to execute an instruction. In the PIC, this time is referred to as *instruction cycles* (referred to as machine cycles in some other CPUs). Because all the instructions in the PIC18 are either 2-byte or 4-byte, most instructions take no more than one or two instruction cycles to execute. (Notice, however, that some instructions such as BTFSS could take up to three instruction cycles.) Appendix A provides a list of PIC18 instructions and their cycles. In the PIC family, the length of the instruction cycle depends on the frequency of the oscillator connected to the PIC system. The crystal oscillator, along with on-chip circuitry, provide the clock source for the PIC CPU (see Chapter 8). In the PIC18, one instruction cycle consists of four oscillator periods. Therefore, to calculate the instruction cycle for the PIC, we take 1/4 of the crystal frequency, then take its inverse, as shown in Example 3-14.

Example 3-14

The following shows the crystal frequency for three different PIC-based systems. Find the period of the instruction cycle in each case.
(a) 4 MHz (b) 16 MHz (c) 20 MHz

Solution:
(a) 4/4 = 1 MHz; instruction cycle is 1/1 MHz = 1 μs (microsecond)
(b) 16 MHz/4 = 4 MHz; instruction cycle = 1/4 MHz = 0.25 μs = 250 ns (nanosecond)
(c) 20 MHz/4 = 5 MHz; instruction cycle = 1/5 MHz = 0.2 μs = 200 ns

---

Branch penalty

The overlapping of fetch and execution of the instruction is widely used in today's microcontrollers such as PIC. For the concept of pipelining to work, we need a buffer or queue in which an instruction is prefetched and ready to be executed. In some circumstances, the CPU must flush out the queue. For example, when a branch instruction is executed, the CPU starts to fetch codes from the new memory location and the code in the queue that was fetched previously is discarded. In this case, the execution unit must wait until the fetch unit fetches the new instruction. This is called a branch penalty. The penalty is an extra instruction cycle to fetch the instruction from the target location instead of executing the instruction right below the branch. Remember that the instruction below the branch has already been fetched and is next in line to be executed when the CPU branches to a different address. This means that while the vast majority of PIC instructions take only one instruction cycle, some instructions take two or three instruction cycles. These are GOTO, BRA, CALL, and all the conditional branch instructions such as BNZ, BC, and so on. The conditional branch instruction can take only one instruction cycle if it does not jump. For example, the BNZ will jump if Z = 0 and that takes two instruction cycles. If Z = 1, then it falls through and it takes only one instruction cycle. See Examples 3-15 and 3-16.

Example 3-15

For a PIC18 system of 4 MHz, find how long it takes to execute each of the following instructions:

(a) MOVLW   (b) DECF   (c) MOVWF
(d) ADDLW   (e) NOP   (f) GOTO
(g) CALL   (h) BNZ

Solution:

The machine cycle for a system of 4 MHz is 1 μs, as shown in Example 3-14. Appendix A shows instruction cycles for each of the above instructions. Therefore, we have:

Instruction	Instruction cycles	Time to execute
(a) MOVLW 0x55	1	1×1 μs = 1 μs
(b) DECF MYREG	1	1×1 μs = 1 μs
(c) MOVWF	1	1×1 μs = 1 μs
(d) ADDLW	1	1×1 μs = 1 μs
(e) NOP	1	1×1 μs = 1 μs
(f) GOTO	2	2×1 μs = 2 μs
(g) CALL	2	2×1 μs = 2 μs
(h) BNZ	2/1	(2 μs taken, 1 μs if it falls through)

Example 3-16

Find the size of the delay of the code snippet below if the crystal frequency is 4 MHz:

Solution:

From Appendix A, we have the following machine cycles for each instruction of the DELAY subroutine:

```
                              Instruction Cycle
MYREG EQU   0x08         ;use location 08 as counter

DELAY     MOVLW    0xFF           1
          MOVWF    MYREG          1

AGAIN     NOP                     1
          NOP                     1
          DECF     MYREG,F        1
          BNZ      AGAIN          2

          RETURN                  1
```

Therefore, we have a time delay of $[(255 \times 5) + 1 + 1 + 1] \times 1$ μs = 1278 μs.
Notice that BNZ takes two instruction cycles if it jumps back, and takes only one when falling through the loop. That means the above number should be 1277 μs.

Delay calculation for PIC18

As seen in the last section, a delay subroutine consists of two parts: (1) setting a counter, and (2) a loop. Most of the time delay is performed by the body of the loop, as shown in Examples 3-17 and 3-18.

Example 3-18

Find the size of the delay in the following program if the crystal frequency is 4 MHz:

```
MYREG EQU    0x08                ;use location 08 as counter

        ORG     0
BACK    MOVLW   0x55             ;load WREG with 55H
        MOVWF   PORTB            ;send 55H to port B
        CALL    DELAY            ;time delay
        MOVLW   0xAA             ;load WREG with AA (in hex)
        MOVWF   PORTB            ;send AAH to port B
        CALL    DELAY
        GOTO    BACK             ;keep doing this indefinitely

;——— this is the delay subroutine
        ORG     300H             ;put time delay at address 300H
DELAY   MOVLW   0xFA             ;WREG = 250, the counter
        MOVWF   MYREG
AGAIN   NOP                      ;no operation wastes clock cycles
        NOP
        NOP
        DECF    MYREG, F
        BNZ     AGAIN            ;repeat until MYREG becomes 0
        RETURN                   ;return to caller
        END                      ;end of asm file
```

Solution:

From Appendix A, we have the following machine cycles for each instruction of the DELAY subroutine:

			Instruction Cycle
DELAY	MOVLW	0xFA	1
	MOVWF	MYREG	1
AGAIN	NOP		1
	NOP		1
	NOP		1
	DECF	MYREG, F	1
	BNZ	AGAIN	2
	RETURN		1

Therefore, we have a time delay of $[(250 \times 6) + 1 + 1 + 1] \times 1 \, \mu s = 1503 \, \mu s$.

Very often we calculate the time delay based on the instructions inside the loop and ignore the clock cycles associated with the instructions outside the loop.

In Example 3-16, the largest value the MYREG register can take is 255; therefore, one way to increase the delay is to use NOP instructions in the loop. NOP, which stands for "no operation," simply wastes time, but takes 2 bytes of program ROM space and that is too heavy a price to pay for just one instruction cycle. A better way is to use a nested loop

Loop inside a loop delay

Another way to get a large delay is to use a loop inside a loop, which is also called a *nested loop*. See Example 3-18. Compare that with Example 3-19 to see the disadvantage of using many NOPs.

Example 3-18

For a instruction cycle of 1 µs, find the time delay in the following subroutine:

```
R2      EQU     0x7
R3      EQU     0x8
DELAY                           Instruction Cycle
        MOVLW       D'200'       1
        MOVWF       R2           1
AGAIN   MOVLW       D'250'       1
        MOVWF       R3           1
HERE    NOP                      1
        NOP                      1
        DECF        R3, F        1
        BNZ         HERE         2
        DECF        R2, F        1
        BNZ         AGAIN        2
        RETURN                   1
```

Solution:
For the HERE loop, we have (5 × 250) 1 µs = 1250 µs. The AGAIN loop repeats the HERE loop 200 times; therefore, we have 200 × 1250 µs = 250000 µs, if we do not include the overhead. However, the following instructions of the outer loop add to the delay:

```
AGAIN       MOVLW       D'250'       1
            MOVWF       R3           1
            . . . . .
            DECF        R2, F        1
            BNZ         AGAIN        2
```

The above instructions at the beginning and end of the AGAIN loop add 5 × 200 × 1 µs = 1000 µs to the time delay. We should also subtract 200 µs for the times BNZ HERE falls through. As a result we have 250000 + 1000 – 200 = 250800 µs = 250.8 milliseconds for the total time delay associated with the above DELAY subroutine. Notice that in the case of a nested loop, as in all other time delay loops, the time is approximate because we have ignored the first few instructions and the last instruction, RETURN, in the subroutine. NOP is a 2-byte instruction. There are 11 instructions in the above DELAY program, and all the instructions are 2-byte instructions. That means that the loop delay takes 22 bytes of ROM code space.

Example 3-19

Find the time delay for the following subroutine, assuming a crystal frequency of 4 MHz. Discuss the disadvantage of this over Example 3-18.

```
MYREG EQU    0x8

                                Machine Cycle

DELAY    MOVLW   D'200'              1
         MOVWF   MYREG               1

AGAIN    NOP                         1
         NOP                         1
         NOP                         1
         NOP                         1
         NOP                         1
         NOP                         1
         NOP                         1
         NOP                         1
         NOP                         1
         NOP                         1
         NOP                         1
         NOP                         1
         DECF    MYREG, F            1
         BNZ     AGAIN               2

         RETURN                      1
```

Solution:

The time delay inside the AGAIN loop is $[200(13 + 2)] \times 1$ μs = 3000 μs. NOP is a 2-byte instruction, even though it does not do anything except to waste cycle time. There are 17 instructions in the above DELAY program, and all the instructions are 2-byte instructions. This means the loop delay takes 34 bytes of ROM code space, and gives us only a 3000 μs delay. That is the reason we use a nested loop instead of NOP instructions to create a time delay. Chapter 9 shows how to use PIC timers to create delays much more efficiently.

From these discussions we conclude that the use of instructions in generating time delay is not the most reliable method. To get more accurate time delay we use timers, as described in Chapter 9. We can use MPLAB's simulator to verify delay time and number of cycles used. Meanwhile, to get an accurate time delay for a given PIC microcontroller, we must use an oscilloscope to measure the exact time delay.

Example 3-20

Write a program to toggle all the bits of SFR PORTB every 1 s. Assume that the crystal frequency is 10 MHz and the system is using a PIC18F458.

Solution:

```
;tested using MPLAB with PIC18F458 operating at 10 MHz
R2      EQU    0x2
R3      EQU    0x3
R4      EQU    0x4


            MOVLW    0x55             ;load WREG with 55H
            MOVWF    PORTB            ;send 55H to PORTB B
BACK        CALL     DELAY_500MS ;time delay
            COMF     PORTB            ;complement PORTB
            GOTO     BACK             ;keep doing this indefinitely

;——— this is the delay subroutine
DELAY_500MSEC
            MOVLW      D'20'
            MOVWF      R4
BACK        MOVLW      D'100'
            MOVWF      R3
AGAIN       MOVLW      D'250'
            MOVWF      R2
HERE        NOP
            NOP
            DECF       R2, F
            BNZ        HERE
            DECF       R3, F
            BNZ        AGAIN
            DECF       R4, F
            BNZ        BACK
            RETURN
```

Delay $20 \times 100 \times 250 \times 5 \times 400$ ns $= 1,000,000,000$ ns $= 1,000,000$ μs $= 1$ s.

In this calculation, we have not included the overhead associated with the two outer loops. Use the MPLAB simulator to verify the delay.

PIC multistage execution pipeline

We can use a superpipeline to speed up execution of instructions. In superpipelining, the process of executing instructions is split into many small steps that are all executed in parallel. In this way, the execution of many instructions is overlapped. One limitation of superpipelining is that the speed of execution is limited to the slowest stage of the pipeline. Compare this to making pizza. You can split the process of making pizza into many stages, such as flattening the dough, putting on the toppings, and baking, but the process is limited to the slowest stage, baking, no matter how fast the rest of the stages are performed. What happens if we use two or three ovens for baking pizzas to speed up the process? This may work for making pizza but not for executing programs, because in the execution of instructions we must make sure that the sequence of instructions is kept intact and that there is no out-of-step execution. In the PIC18, the execution unit takes 4 clock periods of the oscillator, as shown in Figure 3-10.

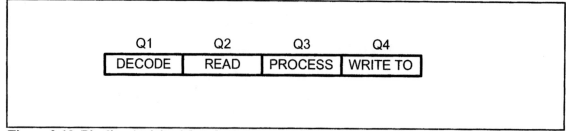

Figure 3-10. Pipeline Activity After the Instruction Has Been Fetched

Figure 3-10 explains why we divide the oscillator by 4 to get the instruction cycle. In Q1, we decode the instruction that is already fetched and sitting in the queue. In Q2, the operand is fetched from the file register. In Q3, the operation is performed: The adding of the two numbers is done. In Q4, the result is written into the destination register. In reality, one can construct the PIC18 superpipeline for four instructions, and is shown in Figure 3-11.

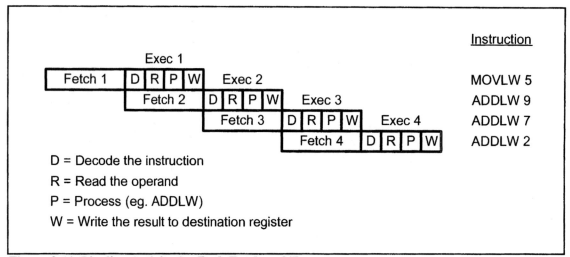

Figure 3-11. Pipeline Activity for Both Fetch and Execute

Notice, in many computer architecture books the process stage is referred to as execute and write to is called write back.

Review Questions
1. True or false. In the PIC18, the instruction cycle lasts 4 clock periods of the crystal frequency.
2. The minimum number of instruction cycles needed to execute a PIC18 instruction is _____.
3. For Question 2, what is the maximum number of cycles needed, and for which instructions?
4. Find the instruction cycle for a crystal frequency of 12 MHz.

5. Assuming a crystal frequency of 4 MHz, find the time delay associated with the loop section of the following DELAY subroutine:

```
DELAY
            MOVLW       D'100'
            MOVWF       MYREG
HERE        NOP
            NOP
            NOP
            NOP
            NOP
            DECF        MYREG,  F
            BNZ         HERE
            RETURN
```

6. True or false. In the PIC18, the instruction cycle lasts 6 clock periods of the crystal frequency.
7. Find the machine cycle for a PIC18 if the crystal frequency is 8 MHz.
8. True or false. In the PIC, the instruction fetching and execution are done at the same time.
9. True or false. BRA and CALL will always take 2 instruction cycles.
10. True or false. The BNZ instruction will always take 2 instruction cycles.

SUMMARY

The flow of a program proceeds sequentially, from instruction to instruction, unless a control transfer instruction is executed. The various types of control transfer instructions in Assembly language include conditional and unconditional branches, and call instructions.

Looping in PIC Assembly language is performed using an instruction to decrement a counter and to jump to the top of the loop if the counter is not zero. This is accomplished with the BNZ instruction. Other branch instructions jump conditionally, based on the value of the carry flag, the Z flag, or other bits of the status register. Unconditional branches can be long or short, depending on the location of the target address. Special attention must be given to the effect of CALL and RCALL instructions on the stack.

PROBLEMS

SECTION 3.1: BRANCH INSTRUCTIONS AND LOOPING

1. In the PIC, looping action with the instruction "BNZ target" is limited to _____ iterations.
2. If a conditional branch is not taken, what is the next instruction to be executed?
3. In calculating the target address for a branch, a displacement is added to the contents of register _____.
4. The mnemonic BRA stands for _____ and it is a(n) ___-byte instruction.
5. The GOTO instruction is a(n) ___-byte instruction.
6. What is the advantage of using BRA over GOTO?

7. True or false. The target of a BNZ can be anywhere in the 2M address space.
8. True or false. All PIC branch instructions can branch to anywhere in the 2M address space.
9. Which of the following instructions are 2-byte instructions.
   (a) BZ   (b) BNC   (c) GOTO   (d) BRA
10. Dissect the BRA instruction, indicating how many bits are used for the operand and the opcode, and indicate how far it can branch.
11. True or false. All conditional branches are 2-byte instructions.
12. Show code for a nested loop to perform an action 1,000 times.
13. Show code for a nested loop to perform an action 100,000 times.
14. Find the number of times the following loop is performed:

```
           MOVLW        D'200'
           MOVWF        REGA
BACK       MOVLW        D'100'
           MOVWF        REGB
HERE       DECF         REGB,F
           BNZ          HERE
           DECF         REGA,F
           BNZ          BACK
```

15. The target address of a BNZ is backward if the second byte of opcode is _____ (negative, positive).
16. The target address of a BNZ is forward if the second byte of opcode is _____ (negative, positive).

SECTION 3.2: CALL INSTRUCTIONS AND STACK

17. CALL is a(n) ___-byte instruction.
18. RCALL is a(n) ___-byte instruction.
19. True or false. The RCALL target address can be anywhere in the 2M address space.
20. True or false. The CALL target address can be anywhere in the 2M address space.
21. When CALL is executed, how many locations of the stack are used?
22. When RCALL is executed, how many locations of the stack are used?
23. Upon reset, the first available location of the stack is _____.
24. Describe the action associated with the RETURN instruction.
25. Give the size of the stack in PIC18.
26. In PIC18, which address is pushed into the stack and the stack pointer when a call instruction is executed.

SECTION 3.3: PIC18 TIME DELAY AND INSTRUCTION PIPELINE

27. Find the oscillator frequency if the instruction cycle = 1.25 μs.
28. Find the instruction cycle if the crystal frequency is 20 MHz.
29. Find the instruction cycle if the crystal frequency is 10 MHz.
30. Find the instruction cycle if the crystal frequency is 16 MHz.

31. True or false. The CALL and RCALL instructions take the same amount of time to execute even though one is a 4-byte instruction and the other is a 2-byte instruction.

32. Find the time delay for the delay subroutine shown below if the system has a PIC18 with a frequency of 4 MHz:

```
            MOVLW         D'200'
            MOVWF         REGA
BACK        MOVLW         D'100'
            MOVWF         REGB
HERE        NOP
            DECF          REGB,F
            BNZ           HERE
            DECF          REGA,F
            BNZ           BACK
```

33. Find the time delay for the delay subroutine shown below if the system has a PIC18 with a frequency of 16 MHz:

```
            MOVLW         D'200'
            MOVWF         REGA
BACK        MOVLW         D'100'
            MOVWF         REGB
HERE        NOP
            NOP
            DECF          REGB,F
            BNZ           HERE
            DECF          REGA,F
            BNZ           BACK
```

34. Find the time delay for the delay subroutine shown below if the system has a PIC18 with a frequency of 4 MHz:

```
            MOVLW D'200'
            MOVWF REGA
BACK        MOVLW D'250'
            MOVWF REGB
HERE        NOP
            DECF          REGB
            BNZ           HERE
            DECF          REGA
            BNZ           BACK
```

35. Find the time delay for the delay subroutine shown below if the system has a PIC18 with a frequency of 10 MHz:

```
            MOVLW         D'200'
            MOVWF         REGA
BACK        MOVLW         D'100'
            MOVWF         REGB
            NOP
            NOP
            NOP
HERE        DECF          REGB,F
            BNZ           HERE
            DECF          REGA,F
            BNZ           BACK
```

ANSWERS TO REVIEW QUESTIONS

SECTION 3.1: BRANCH INSTRUCTIONS AND LOOPING

1. Branch if not zero
2. True
3. 2
4. Z flag of status register
5. 4

SECTION 3.2: CALL INSTRUCTIONS AND STACK

1. 21-bit
2. True
3. 4
4. False
5. Incremented
6. Decremented
7. 1
8. 31 locations (21 × 31)
9. 2
10. CALL

SECTION 3.3: PIC18 TIME DELAY AND INSTRUCTION PIPELINE

1. True
2. 1
3. 2 and CALL. Also, the DECFSZ instruction can take up to 3 cycles
4. 12 MHz / 4 = 3 MHz, and MC = 1/3 MHz = 0.333 μs
5. [100 (1 + 1 + 1 + 1 + 1 + 1 + 2)] × 1 μs = 800 μs = 0.8 milliseconds
6. False. It takes 4 clocks.
7. 8 MHz / 4 = 2 MHz; machine cycle is 1 / 2 MHz = 500 ns
8. True
9. True
10. False. Only if it branches to the target address.

CHAPTER 4

PIC I/O PORT PROGRAMMING

This chapter describes I/O port programming of the PIC18 with many examples. In Section 4-1, we describe I/O access using byte-size data and in Section 4-2, bit manipulation of the I/O ports is discussed in detail.

SECTION 4.1: I/O PORT PROGRAMMING IN PIC18

In the PIC18 family, there are many ports for I/O operations, depending on which family member you choose. Examine Figure 4-1 for the PIC18F458 40-pin chip. A total of 33 pins are set aside for the five ports PORTA, PORTB, PORTC, PORTD, and PORTE. The rest of the pins are designated as V_{dd} (V_{cc}), V_{ss} (GND), OSC1, OSC2, MCLR (reset), and another set of V_{dd} and V_{ss}. They are discussed in Chapter 8.

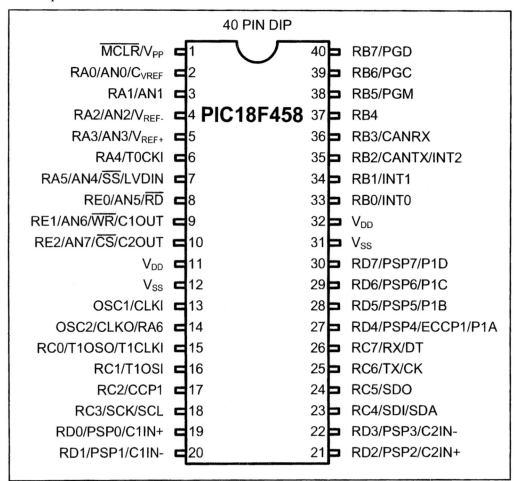

Figure 4-1. PICF458 Pin Diagram

I/O port pins and their functions

The number of ports in the PIC18 family varies depending on the number of pins on the chip. The 18-pin PIC18 has ports A and B only, while the 64-pin version has ports A through F, and the 80-pin PIC18 has ports A through L, as shown in Table 4-1. The 40-pin PIC18F458 has five ports. They are PORTA, PORTB, PORTC, PORTD, and PORTE. To use any of these ports as an input or output port, it must be programmed, as we will explain throughout this section. In addition to

Table 4-1: Number of Ports in PIC18 Family Members

Pins	18-pin	28-pin	40-pin	64-pin	80-pin
Chip	PIC18F1220	PIC18F2220	PIC18F458	PIC18F6525	PIC18F8525
Port A	X	X	X	X	X
Port B	X	X	X	X	X
Port C		X	X	X	X
Port D			X	X	X
Port E			X	X	X
Port F				X	X
Port G				X	X
Port H				X	X
Port J				X	X
Port K					X
Port L					X

Note: X indicates that the port is available.

being used for simple I/O, each port has some other functions such as ADC, timers, interrupts, and serial communication pins. Figure 4-1 also shows alternate functions for the PIC18F458 pins. We will study all these alternate functions in future chapters. In this chapter we focus on the simple I/O function of the PIC18 family. Not all ports have 8 pins. For example, in the PIC18F458, Port A has 7 pins; Ports B, C, and D each have 8 pins; and Port E has only 3 pins. Each port has three SFRs associated with it, as shown in Table 4-2. They are designated as PORTx, TRISx, and LATx. For example, for Port B we have PORTB, TRISB, and LATB. Note that TRIS stands for TRIState and LAT stands for LATch. Next, we describe how to access the SFRs associated with the ports.

Table 4-2: Ports' SFR Addresses for PIC18F458

Port	Address
PORTA	F80H
PORTB	F81H
PORTC	F82H
PORTD	F83H
PORTE	F84H
LATA	F89H
LATB	F8AH
LATC	F8BH
LATD	F8CH
LATE	F8DH
TRISA	F92H
TRISB	F93H
TRISC	F94H
TRISD	F95H
TRISE	F96H

TRIS register role in outputting data

Each of the Ports A–E in the PIC18F458 can be used for input or output. The TRISx SFR is used solely for the purpose of making a given port an input or output port. For example, to make a port an output, we write 0s to the TRISx register. In other words, to output data to any of the pins of the Port B, we must first put 0s into the TRISB register to make it an output port, and then send the data to the Port B SFR itself.

The following code will toggle all 8 bits of Port B forever with some time delay in between "on" and "off" states:

```
         MOVLW 0x0          ;WREG = 00
         MOVWF TRISB        ;make Port B an output port 0000 0000
L1       MOVLW 0x55         ;WREG = 55h
         MOVWF PORTB        ;put 55h on port B pins
         CALL  DELAY
         MOVLW 0xAA         ;WREG = AAh
         MOVWF PORTB        ;put AAh on port B pins
         CALL  DELAY
         GOTO  L1
```

It must be noted that unless we activate the TRIS bit (set it to zero), the data will not go from the port register to the pins of the PIC. This means that if we remove the first two lines of the above code, the 55H and AAH values will not get to the pins. They will be sitting in the SFR of Port B inside the CPU.

To see the role of the TRISx register in allowing the data to go from Portx to the pins, examine Figure 4-3 and Figure 4-4. If you are unfamiliar with the internals of logic gates, see Appendix C for an overview. Notice that the CMOS "on" and "off" states in Figure 4-2 are taken from Appendix C.

Note that upon reset, all ports have value FFH in their TRIS registers. This means all ports are configured as input as we will see next.

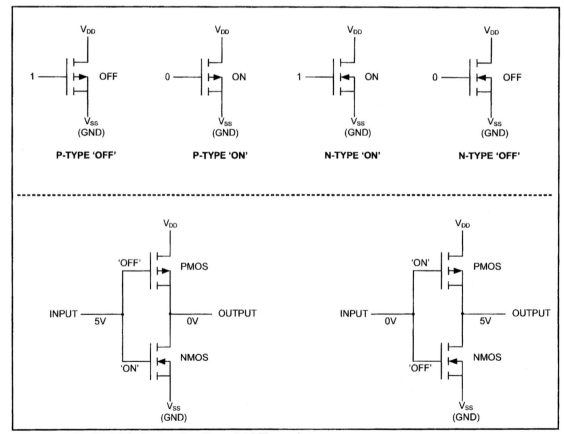

Figure 4-2. CMOS States for P and N Transistors

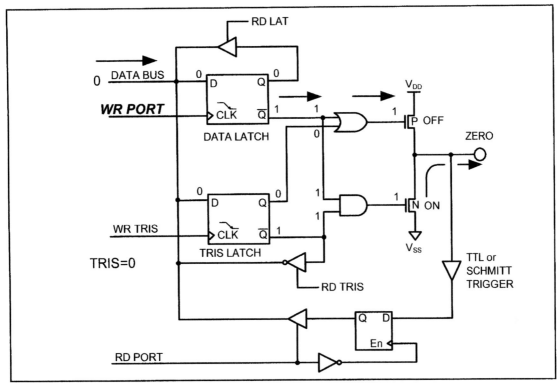

Figure 4-3. Outputting (Writing) 0 to a Pin in the PIC18

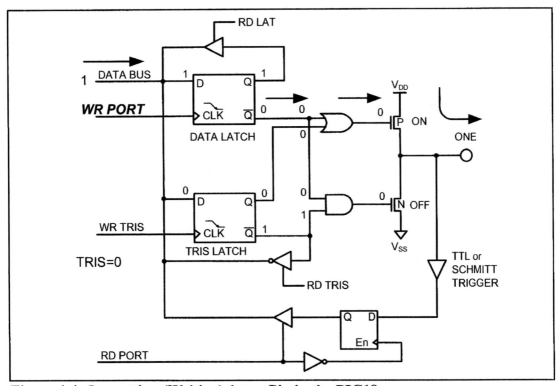

Figure 4-4. Outputting (Writing) 1 to a Pin in the PIC18

TRIS register role in inputting data

To make a port an input port, we must first put 1s into the TRISx register for that port, and then bring in (read) the data present at the pins. Notice that 0 stands for out and 1 for in. This is easy to remember because O and 0 look alike the same way that I looks like 1. The following code will get the data present at the pins of port C and send it to port B indefinitely, after adding the value 5 to it:

```
        MOVLW B'00000000'  ;WREG = 0000000 (binary)
        MOVWF TRISB        ;Port B an output port(0 for O)
        MOVLW B'11111111'  ;WREG = 11111111 (binary)
        MOVWF TRISC        ;Port C an input port (1 for I)
L2      MOVF  PORTC,W       ;move data from Port C to WREG
        ADDLW 5            ;add some value to it
        MOVWF PORTB        ;send it to Port B
        GOTO  L2           ;continue forever
```

Another, more efficient, version of the program is as follows:

```
        CLRF   TRISB       ;clear TRISB (Port B an output port)
        SETF   TRISC       ;set TRISC (Port C an input port)
L2      MOVF   PORTC,W        ;get data from port C
        ADDLW 5            ;add some value
        MOVWF PORTB        ;send it to port B
        BRA    L2
```

Again, it must be noted that unless we activate the TRIS bits (by putting 1s there), the data will not be brought into the WREG register from the pins of Port C. To see the role of the TRISx register in allowing the data to come into the CPU from the pins, examine Figures 4-5 and 4-6.

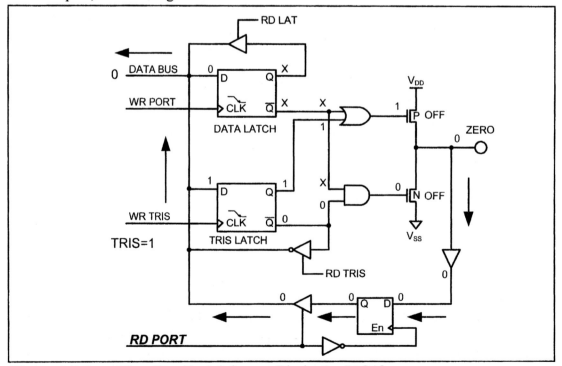

Figure 4-5. Inputting (Reading) 0 from a Pin in the PIC18

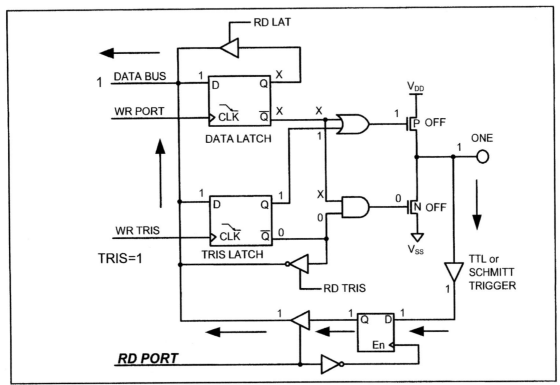

Figure 4-6. Inputting (Reading) 1 from a Pin in the PIC18

Port A

Port A occupies a total of 7 pins (RA0–RA6), but for the PIC18F458, pin A6 is used for the OSC2 pin. A6 is not available if we use a crystal oscillator to provide frequency to the PIC18 chip, as we will see in Chapter 8.

To use the pins of Port A as both input and output ports, each bit must be connected externally to the pin by enabling the bits of the TRISA register. For example, the following code will continuously send out to Port A the alternating values of 55H and AAH:

```
;toggle all bits of PORTA

        MOVLW B'00000000'  ;WREG = 00000000 (binary)
        MOVWF TRISA        ;make Port A an output port (0 for Out)
L1      MOVLW 0x55         ;WREG = 55h
        MOVWF PORTA        ;put 55h on Port A pins
        CALL  DELAY
        MOVLW 0xAA         ;WREG = AAh
        MOVWF PORTA        ;put AAh on Port A pins
        CALL  DELAY
        GOTO  L1
```

It must be noted that 55H (01010101) when complemented becomes AAH (10101010). Although by sending 55H and AAH to Port A continuously, we toggle all 8 bits of the Port A register, only 6 pins (RA0–RA5) will show the toggling data.

Port A as input

In order to make all the bits of Port A an input, TRISA must be programmed by writing 1 to all the bits. In the following code, Port A is configured first as an input port by writing all 1s to register TRISA, and then data is received from Port A and saved in some RAM location of the fileReg:

```
MYREG EQU    0X20   ;save it here

MOVLW B'11111111' ;WREG = 11111111 (binary)
MOVWF TRISA       ;make Port A an input port (1 for In)
MOVF  PORTA,W     ;move from fileReg of Port A to WREG
MOVWF MYREG       ;save it in fileReg of MYREG
```

Port B

Port B occupies a total of 8 pins (RB0–RB7). To use the pins of Port B as both input and output ports, each bit must be connected externally to the pin by enabling the bits of register TRISB.

For example, the following code will continuously send out the alternating values of 55H and AAH to Port B:

```
   ;toggle all bits of PORTB

   MOVLW B'00000000' ;WREG = 00
   MOVWF TRISB       ;make Port B an output port
L1 MOVLW 0x55        ;WREG = 55h
   MOVWF PORTB       ;put 55h on port B pins
   CALL  DELAY
   MOVLW 0xAA        ;WREG = AAh
   MOVWF PORTB       ;put AAh on port B pins
   CALL  DELAY
   GOTO  L1
```

Port B as input

In order to make all the bits of Port B an input, TRISB must be programmed by writing 1 to all the bits. In the following code, Port B is configured first as an input port by writing all 1s to register TRISB, and then data is received from Port B and saved in some RAM location of the fileReg:

```
MYREG EQU    0X25   ;save it here

MOVLW B'11111111' ;WREG = 11111111 (binary)
MOVWF TRISB       ;make Port B an input port (1 for In)
MOVF  PORTB,W     ;move from fileReg of Port B to WREG
MOVWF MYREG       ;save it in fileReg
```

Dual role of Ports A and B

The PIC18 multiplexes an analog-to-digital converter through Port A to save I/O pins. The alternate functions of the pins for Port A are shown in Table 4-3. We will show how to use Port A's ADC in Chapter 13. Because many projects use an ADC, we do not use Port A for simple I/O functions. The PIC18 multiplexes some other functions through Port B to save pins. The alternate functions of the pins for Port B are shown in Table 4-4. We will show how to use the alternate functions of Port B in future chapters.

Table 4-3: Port A Alternate Functions

Bit	Function
RA0	AN0/CVREF
RA1	AN1
RA2	AN2/VREF−
RA3	AN3/VREF+
RA4	T0CKI
RA5	AN4/SS/LVDIN
RA6	OSC2/CLKO

Table 4-4: Port B Alternate Functions

Bit	Function
RB0	INT0
RB1	INT1
RB2	INT2/CANTX
RB3	CANRX
RB4	
RB5	PGM
RB6	PGC
RB7	PGD

Port C

Port C occupies a total of 8 pins (RC0–RC7). To use the pins of Port C as both input and output ports, each bit must be connected externally to the pin by enabling the bits of register TRISC. For example, the following code will continuously send out the alternating values of 55H and AAH to Port C:

```
      ;toggle all bits of PORTB

      MOVLW B'00000000'  ;WREG = 00
      MOVWF TRISC        ;make Port C an output port
L1    MOVLW 0x55         ;WREG = 55h
      MOVWF PORTC        ;put 55h on Port C pins
      CALL  DELAY
      MOVLW 0xAA         ;WREG = AAh
      MOVWF PORTC        ;put AAh on Port C pins
      CALL  DELAY
      GOTO  L1
```

Port C as input

In order to make all the bits of Port C an input, TRISC must be programmed by writing 1 to all the bits. In the following code, Port C is configured first as an input port by writing all 1s to register TRISC, and then data is received from Port C and saved in some RAM location of the fileReg:

```
MYREG EQU 0x20      ;save it here
      MOVLW B'11111111' ;WREG = 11111111 (binary)
      MOVWF TRISC     ;make Port C an input port (1 for In)
      MOVF  PORTC,W   ;move from fileReg of Port C to WREG
      MOVWF MYREG     ;save it in fileReg
```

Port D

Port D occupies a total of 8 pins (RD0–RD7). To use the pins of Port D as both input and output ports, each bit must be connected externally to the pin by enabling the bits of register TRISD. For example, the following code will continuously send out to Port D the alternating values of 55H and AAH:

```
      ;toggle all bits of PORTD

      CLRF  TRISD     ;make Port D an output port
L1    MOVLW 0x55      ;WREG = 55h
      MOVWF PORTD     ;put 55h on Port D pins
      CALL  DELAY
      MOVLW 0xAA      ;WREG = AAh
      MOVWF PORTD     ;put AAh on Port D pins
      CALL  DELAY
      BRA   L1        ;we can use GOTO
```

Port D as input

In order to make all the bits of Port D an input, TRISD must be programmed by writing 1 to all the bits. In the following code, Port D is configured first as an input port by writing all 1s to register TRISD, and then data is received from Port D and saved in some RAM location of the fileReg:

```
      MYREG EQU 0x20  ;save it here
      SETF  TRISD     ;TRISD = 11111111 (binary) = PORTD = Input
      MOVF  PORTD,W   ;move from Port D to WREG
      MOVWF MYREG     ;save it in fileReg
```

Dual role of Ports C and D

The alternate functions of the pins for Port C are shown in Table 4-5. We will show how to use Port C's alternate functions in future chapters. The alternate functions of the pins for Port D are shown in Table 4-6. We will show how to use Port D's alternate functions in future chapters.

Table 4-5: Port C Alternate Functions

Bit	Function
RC0	T1OSO/T1CKI
RC1	T1OSI
RC2	CCP1
RC3	SCK/SCL
RC4	SDI/SDA
RC5	SDO
RC6	TX/CK
RC7	RX/DT

Table 4-6: Port D Alternate Functions

Bit	Function
RD0	PSP0/C1IN+
RD1	PSP1/C1IN-
RD2	PSP2/C2IN+
RD3	PSP3/C2IN-
RD4	PSP4/ECCP1/P1A
RD5	PSP5/P1B
RD6	PSP6/P1C
RD7	PSP7/P1D

Port E

Port E occupies a total of 3 pins (RE0–RE2) in the PIC18F458/4580. Port E is used for 3 additional analog inputs or simple I/O: AN5, AN6, and AN7. Just like other ports, Port E has alternate functions. We will show how to use them in future chapters.

Different ways of accessing the entire 8 bits

In the following code, as in many previous I/O examples, the entire 8 bits of Port B are accessed:

```
     ;toggle all bits of PORTB

     MOVLW 0x0         ;WREG = 00
     MOVWF TRISB       ;make Port B an output port
L1   MOVLW 0x55        ;WREG = 55h
     MOVWF PORTB       ;put 55h on Port B pins
     CALL  DELAY
     MOVLW 0xAA        ;WREG = AAh
     MOVWF PORTB       ;put AAh on Port B pins
     CALL  DELAY
     GOTO  L1
```

The previous code toggles all the bits of Port B continuously. Another variation of the above code follows:

```
       CLRF   TRISB           ;make Port B an output port
L1     MOVLW  0x55            ;WREG = 55h
       MOVWF  PORTB           ;put 55h on Port B pins
       CALL   DELAY
       MOVLW  0xAA            ;WREG = AAh
       MOVWF  PORTB           ;put AAh on Port B pins
       CALL   DELAY
       GOTO   L1
```

The following is another way of doing the same thing:

```
       CLRF   TRISB           ;make Port B an output port
       MOVLW  0x55            ;WREG = 55h
       MOVWF  PORTB           ;put 55h on Port B pins
L2     COMF   PORTB, F        ;toggle bits of Port B
       CALL   DELAY
       BRA    L2
```

The above code uses a technique called *read-modify-write*.

Read followed by write I/O operation

Due to the timing issue, we must be careful not to have two I/O operations one right after the other. Examine the following rewrite of an earlier code fragment in which data was read from PORTC and sent to PORTB:

```
       CLRF   TRISB ;clear TRISB to make PORTB an output port
       SETF   TRISC           ;set TRISC all 1s (Port C as Input)
L4     MOVF   PORTC,W         ;get data from Port C into WREG
       NOP                    ;NEED some NOP to ensure data is in WREG
       MOVWF  PORTB           ;before it is sent to Port B
       BRA    L4              ;keep doing it
```

We need a NOP (or some other instruction) to make sure that the data is written into WREG before it is read for outputting to Port B. This is called *data dependency* in CPU design. This type of data dependency is commonly referred to as RAW (Read-After-Write). The NOP will introduce a bubble into the pipeline to remove data dependency due to RAW. See Figure 4-7. One way to avoid this problem is to use the MOVFF instruction, which is a 4-byte instruction. This is coded as follows:

```
       CLRF   TRISB           ;make Port B an output port
       SETF   TRISC           ;TRISC = FFh (Port C Input)
L5     MOVFF  PORTC,PORTB ;get from Port C and send to PORTB
       BRA    L5              ;keep doing it
```

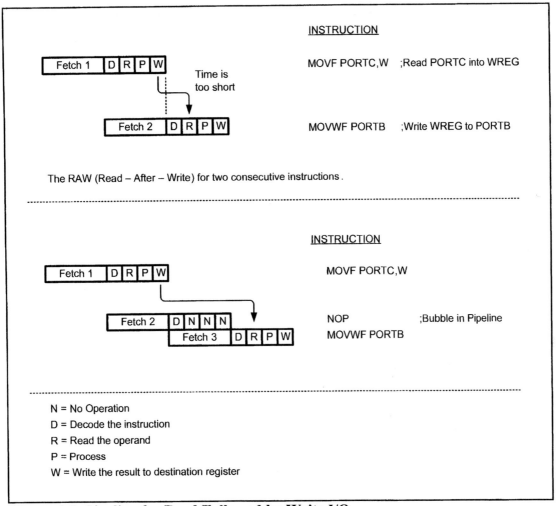

INSTRUCTION

Fetch 1 | D | R | P | W   Time is too short   MOVF PORTC,W   ;Read PORTC into WREG

Fetch 2 | D | R | P | W   MOVWF PORTB   ;Write WREG to PORTB

The RAW (Read – After – Write) for two consecutive instructions.

INSTRUCTION

Fetch 1 | D | R | P | W   MOVF PORTC,W

Fetch 2 | D | N | N | N   NOP   ;Bubble in Pipeline
Fetch 3 | D | R | P | W   MOVWF PORTB

N = No Operation
D = Decode the instruction
R = Read the operand
P = Process
W = Write the result to destination register

Figure 4-7. Pipeline for Read Followed by Write I/O

Ports status upon reset

Upon reset, all ports have value FFH on their TRIS register, as shown in Table 4-7. This makes them input ports upon reset.

Table 4-7: Reset Values of TRIS Registers for PIC18

Register	Reset Value (Binary)
TRISA	11111111
TRISB	11111111
TRISC	11111111
TRISD	11111111

Note: All ports are input ports upon reset.

Review Questions

1. There are total of _____ ports in the PIC18F458.
2. True or false. All of the PIC18F458 ports have 8 pins.
3. List all PIC18F458 ports that have 8 pins.
4. True or false. Upon power-up, the I/O pins are configured as output ports.
5. Code a simple program to send 99H to Port B and Port C.
6. To make Port B an output port, we must place _____ in register _____.
7. To make Port B an input port, we must place _____ in register _____.

Example 4-1

Write a test program for the PIC18 chip to toggle all the bits of PORTB, PORTC, and PORTD every 1/4 of a second. Assume a crystal frequency of 4 MHz.

Solution:

```
;tested with MPLAB for the PIC18F458 and XTAL = 4 MHz

        list P=PIC18F458
#include P18F458.INC

R1 equ 0x07
R2 equ 0x08

ORG 0
        CLRF   TRISB        ;make Port B an output port
        CLRF   TRISC        ;make Port C an output port
        CLRF   TRISD        ;make Port D an output port
        MOVLW 0x55          ;WREG = 55h
        MOVWF PORTB         ;put 55h on Port B pins
        MOVWF PORTC         ;put 55h on Port C pins
        MOVWF PORTD         ;put 55h on Port D pins
L3      COMF   PORTB,F      ;toggle bits of Port B
        COMF   PORTC,F      ;toggle bits of Port C
        COMF   PORTD,F      ;toggle bits of Port D
        CALL   QDELAY       ;quarter of a second delay
        BRA    L3


;-----------1/4 SECOND DELAY
QDELAY
        MOVLW D'200'
        MOVWF R1
D1      MOVLW D'250'
        MOVWF R2
D2      NOP
        NOP
        DECF R2, F
        BNZ D2
        DECF R1, F
        BNZ D1
        RETURN
        END
```

Calculations:

4 MHz / 4 = 1 MHz
1 / 1 MHz = 1 μs
Delay = 250 × 200 × 5 MC × 1 μs = 250,000 μs (if we include the overhead, we will have 250,800. See Example 3-17 in the previous chapter.)

Use the MPLAB simulator to verify the delay size.

SECTION 4.2: I/O BIT MANIPULATION PROGRAMMING

In this section we further examine the PIC18 I/O instructions. We pay special attention to I/O bit manipulation because it is a powerful and widely used feature of the PIC family.

I/O ports and bit-addressability

Sometimes we need to access only 1 or 2 bits of the port instead of the entire 8 bits. A powerful feature of PIC I/O ports is their capability to access individual bits of the port without altering the rest of the bits in that port. For all PIC ports, we can access either all 8 bits or any single bit without altering the rest. Table 4-8 lists the single-bit instructions for the PIC18. Although the instructions in Table 4-9 can be used for any registers in the data RAM file register, I/O port operations use them most often. We will see the use of these instructions throughout future chapters.

Table 4-8: Single-Bit (Bit-Oriented) Instructions for PIC18

Instruction	Function
BSF fileReg,bit	Bit Set fileReg (set the bit: bit = 1)
BCF fileReg,bit	Bit Clear fileReg (clear the bit: bit = 0)
BTG fileReg,bit	Bit Toggle fileReg (complement the bit)
BTFSC fileReg,bit	Bit test fileReg, skip if clear (skip next instruction if bit = 0)
BTFSS fileReg,bit	Bit test fileReg, skip if set (skip next instruction if bit = 1)

Table 4-9: Single-Bit Addressability of Ports for PIC18F458/4580

PORT	PORTB	PORTC	PORTD	PORTE	Port Bit
RA0	RB0	RC0	RD0	RE0	D0
RA1	RB1	RC1	RD1	RE1	D1
RA2	RB2	RC2	RD2	RE2	D2
RA3	RB3	RC3	RD3		D3
RA4	RB4	RC4	RD4		D4
RA5	RB5	RC5	RD5		D5
	RB6	RC6	RD6		D6
	RB7	RC7	RD7		D7

Next we describe all these instructions and examine their usage.

BSF (bit set fileReg)

To set HIGH a single bit of a given fileReg, we use the syntax "BSF fileReg, bit_num" where fileReg can be any location in the file register and bit_num is the desired bit number from 0 to 7. Although the bit-oriented instructions can be used for manipulation of bits D0–D7 of any file register, they are mostly used for I/O ports in embedded systems. For example, "BSF PORTB, 5" sets HIGH bit 5 of Port B.

Example 4-2

An LED is connected to each pin of Port D. Write a program to turn on each LED from pin D0 to pin D7. Call a delay module before turning on the next LED.

Solution:

```
CLRF   TRISD      ;make PORTD an output port
BSF    PORTD,0    ;bit set turns on RD0
CALL   DELAY      ;delay before next one
BSF    PORTD,1    ;turn on RD1
CALL   DELAY      ;delay before next one
BSF    PORTD,2
CALL   DELAY
BSF    PORTD,3
CALL   DELAY
BSF    PORTD,4
CALL   DELAY
BSF    PORTD,5
CALL   DELAY
BSF    PORTD,6
CALL   DELAY
BSF    PORTD,7
CALL   DELAY
```

BCF (bit clear fileReg)

To clear a single bit of a given fileReg, we use the syntax "BCF fileReg, bit_number". Remember that for I/O ports, we must activate the appropriate bit in the TRISx register if we want the pin to reflect the changes. For example, the following code toggles pin RB2 continuously:

```
      BCF    TRISB, 2    ;bit = 0, make RB2 an output pin
AGAIN BSF    PORTB, 2    ;bit set (RB2 = high)
      CALL   DELAY
      BCF    PORTB, 2    ;bit clear(RB2 = low)
      CALL   DELAY
      BRA    AGAIN
```

Example 4-3

Write the following programs:
(a) Create a square wave of 50% duty cycle on bit 0 of Port C.
(b) Create a square wave of 66% duty cycle on bit 3 of Port C.

Solution:

(a) The 50% duty cycle means that the "on" and "off" states (or the high and low portions of the pulse) have the same length. Therefore, we toggle RC0 with a time delay between each state.

```
        BCF   TRISC,0    ;clear TRIS bit for RC0 = out
HERE    BSF   PORTC,0    ;set to HIGH RC0 (RC0 = 1)
        CALL  DELAY      ;call the delay subroutine
        BCF   PORTC,0    ;RC0 = 0
        CALL  DELAY
        BRA   HERE       ;keep doing it
```

Another way to write the above program is:

```
        BCF   TRISC,0    ;make RC0 = out
HERE    BTG   PORTC,0    ;complement bit 0 of PORTC
        CALL  DELAY      ;call the delay subroutine
        BRA   HERE       ;keep doing it
```

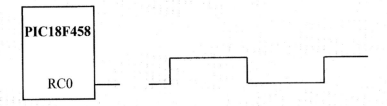

(b) A 66% duty cycle means that the "on" state is twice the "off" state.

```
        BCF   TRISC,3    ;clear TRISC3 bit for output
BACK    BSF   PORTC,3    ;RC3 = 1
        CALL  DELAY      ;call the delay subroutine
        CALL  DELAY      ;twice for 66%
        BCF   PORTC,3    ;RC3 = 0
        CALL  DELAY      ;call delay once for 33%
        BRA   BACK       ;keep doing it
```

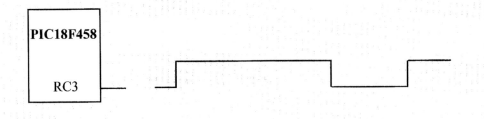

BTG (bit toggle fileReg)

To toggle a single bit of a given fileReg, we use the syntax "BTG fileReg, bit_number".

```
        BCF     TRISB, 2        ;make RB2 an output pin
BACK    BTG     PORTB, 2        ;toggle pin RB2 only
        CALL    DELAY
        BRA     BACK
```

Notice that RB2 is the third bit of Port B (the first bit is RB0, the second bit is RB1, etc.). This is shown in Table 4-9. See Example 4-2 for an example of bit manipulation of I/O bits.

Notice in Example 4-2 that unused portions of Port C are undisturbed. This single-bit addressability of I/O ports is one of most powerful features of the PIC microcontroller and is one of the reasons that many designers choose the PIC over other microcontrollers. We will see the use of the bit-addressability of I/O ports in future chapters.

Checking an input pin

To make decisions based on the status of a given bit in the file register, we use the instructions BTFSC (bit test fileReg skip if clear) and BTFSS (bit test fileReg skip if set). These single-bit instructions are widely used for I/O operations. They allow you to monitor a single pin and make a decision depending on whether it is 0 or 1. Again it must be noted that the instructions BTFSC and BTFSS can be used for any bits of the file register, including the I/O ports A, B, C, D, and so on.

BTFSS (bit test fileReg, skip if set)

To monitor the status of a single bit for HIGH, we use the BTFSS instruction. This instruction tests the bit and skips the next instruction if it is HIGH. Example 4-4 shows how it is used.

| 1010 | bbba | ffff | ffff |

$$0 \leq f \leq FF$$
$$0 \leq b \leq 7$$

BTFSC (bit test fileReg, skip if clear)

To monitor the status of a single bit for LOW, we use the BTFSC instruction. This instruction tests the bit and skips the instruction right below it if the bit is LOW. Example 4-5 shows how it is used.

| 1011 | bbba | ffff | ffff |

$$0 \leq f \leq FF$$
$$0 \leq b \leq 7$$

Example 4-4

Write a program to perform the following:
(a) Keep monitoring the RB2 bit until it becomes HIGH;
(b) When RB2 becomes HIGH, write value 45H to Port C, and also send a HIGH-to-LOW pulse to RD3.

Solution:

```
        BSF    TRISB,2     ;make RB2 an input
        CLRF   TRISC       ;make PORTC an output port
        BCF    PORTD,3     ;make RD3 an output
        MOVLW  0x45        ;WREG = 45h
AGAIN   BTFSS  PORTB,2     ;bit test RB2 for HIGH
        BRA    AGAIN       ;keep checking if LOW
        MOVWF  PORTC       ;issue WREG to Port C
        BSF    PORTD,3     ;bit set fileReg RD3 (H-to-L)
        BCF    PORTD,3     ;bit clear fileReg RD3 (L)
```

In this program, instruction "BTFSS PORTB,2" stays in the loop as long as RB2 is LOW. When RB2 becomes HIGH, it skips the branch instruction to get out of the loop, and writes the value 45H to Port C. It also sends a HIGH-to-LOW pulse to RD3.

Example 4-5

Assume that bit RB3 is an input and represents the condition of a door alarm. If it goes LOW, it means that the door is open. Monitor the bit continuously. Whenever it goes LOW, send a HIGH-to-LOW pulse to port RC5 to turn on a buzzer.

Solution:

```
        BSF    TRISB,3     ;make RB3 an input
        BCF    TRISC,5     ;make RC5 an output
HERE    BTFSC  PORTB,3     ;keep monitoring RB3 for HIGH
        BRA    HERE        ;stay in the loop
        BSF    PORTC,5     ;make RC5 HIGH
        BCF    PORTC,5     ;make RC5 LOW for H-to-L
        BRA    HERE
```

Monitoring a single bit

We can also use the bit test instructions to monitor the status of a single bit and make a decision to perform an action. See Examples 4-6 and 4-7.

Example 4-6

A switch is connected to pin RB2. Write a program to check the status of SW and perform the following:
(a) If SW = 0, send the letter 'N' to PORTD.
(b) If SW = 1, send the letter 'Y' to PORTD.

Solution:

```
        BSF    TRISB,2      ;make RB2 an input
        CLRF   TRISD        ;make PORTD an output port
AGAIN   BTFSS  PORTB,2      ;bit test RB2 for HIGH
        BRA    OVER         ;it must be LOW
        MOVLW  A'Y'         ;WREG = 'Y' ASCII letter Y
        MOVWF  PORTD        ;issue WREG to PORTD
        GOTO   AGAIN        ;we can use BRA too
OVER    MOVLW  A'N'         ;WREG = 'N' ASCII  letter N
        MOVWF  PORTD        ;issue WREG to PORTD
        GOTO   AGAIN        ;we can use BRA too
```

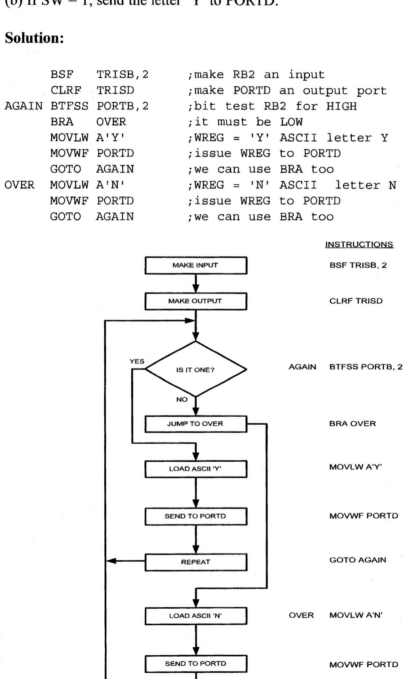

Example 4-7

A switch is connected to pin RB2. Write a program to check the status of SW and perform the following:
(a) If SW = 0, send letter 'N' to PORTD.
(b) If SW = 1, send letter 'Y' to PORTD.

Use the BTFSC instruction to check the SW status. This is another version of Example 4-6 using the BTFSC instruction instead of BTFSS.

Solution:

```
        BSF    TRISB,2    ;make RB2 an input
        CLRF   TRISD      ;make PORTD an output port
AGAIN   BTFSC  PORTB,2    ;bit test RB2 for LOW
        BRA    OVER       ;it must be HIGH
        MOVLW  A'N'       ;WREG = 'N' ASCII letter N
        MOVWF  PORTD      ;issue WREG to PORTD
        BRA    AGAIN      ;we can use GOTO
OVER    MOVLW  A'Y'       ;WREG = 'Y' ASCII letter Y
        MOVWF  PORTD      ;issue WREG to PORTD
        BRA    AGAIN      ;we can use GOTO
```

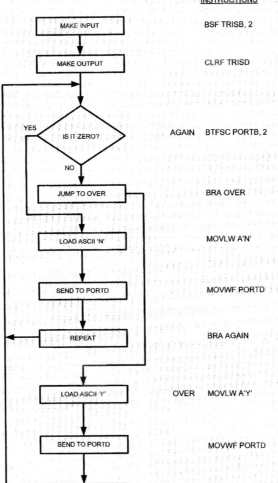

Reading a single bit

We can also use the bit test instructions to read the status of a single bit and send it to another bit or save it. This is shown in Examples 4-8 and 4-9.

Example 4-8

A switch is connected to pin RB0 and an LED to pin RB7. Write a program to get the status of SW and send it to the LED.

Solution:

```
        BSF    TRISB,0      ;make RB0 an input
        BCF    TRISB,7      ;make RB7 an output
AGAIN   BTFSS  PORTB,0      ;bit test RB0 for HIGH
        GOTO   OVER         ;it must be LOW (BRA is OK too)
        BSF    PORTB,7
        GOTO   AGAIN        ;we can use BRA too
OVER    BCF    PORTB,7
        GOTO   AGAIN        ;we can use BRA too
```

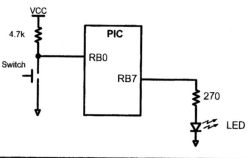

Example 4-9

A switch is connected to pin RB0. Write a program to get the status of SW and save it in D0 of fileReg location 0x20.

Solution:

```
MYBITREG EQU 0x20 ;set aside loc 0x20 reg

        BSF    TRISB,0      ;make RB0 an input
AGAIN   BTFSS  PORTB,0      ;bit test RB0 for HIGH
        GOTO   OVER         ;it must be LOW (BRA is OK too)
        BSF    MYBITREG,0   ;set bit 0 of fileReg
        GOTO   AGAIN        ;we can use BRA too
OVER    BCF    MYBITREG,0   ;clear bit 0 of fileReg
        GOTO   AGAIN        ;we can use BRA too
```

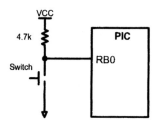

Reading input pins vs. LATx port

In reading a port, some instructions read the status of the port pins while others read the status of an internal port latch called LATx. Therefore, when reading ports there are two possibilities:

1. Read the status of the input pin.
2. Read the internal latch of the LAT register.

We must make a distinction between these two categories of instructions because confusion between them is a major source of errors in PIC programming, especially where external hardware is concerned. We will discuss these instructions shortly. However, readers must study and understand the material on this topic and on the internal working of ports that is given in Appendix C.2. Examine the structure of the ports in Figure 4-8 once again. In addition to the PORTx and TRISx register, the LATx register is the third important register associated with the PIC18 ports.

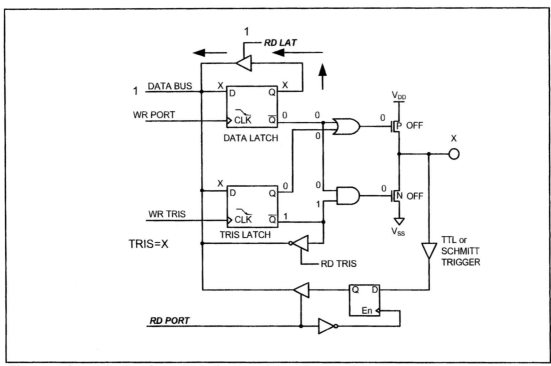

Figure 4-8. LATx Register Role in Reading a Port or Latch

Reading LATx for ports

Some instructions read the contents of an internal port latch instead of reading the status of an external pin. Table 4-10 provides a list of these instructions. For example, consider the "COMF PORTB" instruction. The sequence of actions taken when such an instruction is executed is as follows:

1. The instruction reads the internal latch of the LATB and brings that data into the CPU.

2. This data is complemented.
3. The result is rewritten back to the LATB latch.
4. The data on the pins are changed only if the TRISB bits are cleared to 0s.

It is very rare that we use an instruction to read the latch register, such as "COMF LATB, F", although it is a valid instruction.

From the above discussion, we conclude that the instructions that read the port latch normally read a latch value, perform an operation, then rewrite it back to the port latch. This is called *read-modify-write*. To use the read-modify-write, the port must be configured as output.

Table 4-10: Some of the Read-Modify-Write Instructions

Instruction		Function
ADDWF	fileReg,d	Add WREG to f
BSF	fileReg,bit	Bit Set fileReg (set the bit: bit = 1)
BCF	fileReg,bit	Bit Clear fileReg (clear the bit: bit = 0)
COMF	fileReg,d	Complement f
INCF	fileReg,d	Increment f
SUBWF	fileReg,d	Subtract WREG from f
XORWF	fileReg,d	Exclusive-OR WREG with f

Review Questions

1. True or false. The instruction "BSF PORTB, 1" makes pin RB1 HIGH while leaving other pins of PORTB unchanged, if bit 1 of the TRISB bits is configured for output.
2. Show one way to toggle the pin RB7 continuously using PIC instructions.
3. Using the instruction "BTFSS PORTC, 5" assumes that bit RC5 is an _____ (input, output) pin.
4. Write instructions to get the status of RB2 and put it on RB0.
5. Write instructions to toggle both bits of RD7 and RD0 continuously.

CAUTION

We strongly recommend that you study Section C.2 (Appendix C) before connecting any external hardware to your PIC system. Failure to use the right instruction or the right connection to port pins can damage the ports of your PIC chip.

SUMMARY

This chapter focused on the I/O ports of the PIC. The five ports of the PIC18F458, PORTA, PORTB, PORTC, PORTD, and PORTE, were explored. These ports can be used for input or output. All the ports have alternate functions. The three registers associated with each port are PORTx, TRISx, and LATx. Their role in I/O manipulation was examined. Then, I/O instructions of the PIC were explained, and numerous examples were given. We also showed the bit-address-ability of PIC ports.

PROBLEMS

SECTION 4.1: I/O PORT PROGRAMMING IN PIC18

1. The PIC18F458 has a DIP package of ____ pins.
2. In PIC18F458, how many pins are assigned to V_{CC} and GND?
3. In the PIC18F458, how many pins are designated as I/O port pins?
4. How many pins are designated as PORTA in the 40-pin DIP package and what are their numbers?
5. How many pins are designated as PORTB in the 40-pin DIP package and what are their numbers?
6. How many pins are designated as PORTC in the 40-pin DIP package and what are their numbers?
7. How many pins are designated as PORTD in the 40-pin DIP package and what are their numbers?
8. Upon reset, all the bits of ports are configured as _____ (input, output).
9. For the PIC18, which register must be programmed in order to be used as simple I/O?
10. Explain the role of TRISx and PORTx in I/O operations.
11. Write a program to get 8-bit data from PORTC and send it to ports PORTB and PORTD.
12. Write a program to get 8-bit data from PORTD and send it to ports PORTB and PORTC.
13. Which pins are for RxD and TxD?
14. Give RAM data location in the file register assigned to Ports A–C and their TRIS registers for the PIC18F458.
15. Write a program to toggle all the bits of PORTB and PORTC continuously (a) using AAH and 55H (b) using the COMF instruction.

SECTION 4.2: I/O BIT MANIPULATION PROGRAMMING

16. Which ports of the PIC18 are bit-addressable?
17. What is the advantage of bit-addressability for PIC ports?
18. When RB2 is accessed as a single-bit port, it is designated as _____.
19. Is the instruction "COMF PORTB" a valid instruction?

20. Write a program to toggle RB2 and RB5 continuously without disturbing the rest of the bits.
21. Write a program to toggle RD3, RD7, and RC5 continuously without disturbing the rest of the bits.
22. Write a program to monitor bit RC3. When it is HIGH, send 55H to PORTD.
23. Write a program to monitor the RB7 bit. When it is LOW, send 55H and AAH to PORTC continuously.
24. Write a program to monitor the RE0 bit. When it is HIGH, send 99H to PORTB. If it is LOW, send 66H to PORTC.
25. Write a program to monitor the RB5 bit. When it is HIGH, make a LOW-to-HIGH-to-LOW pulse on RB3.
26. Write a program to get the status of RC3 and put it on RC4.
27. The RB4 refers to which bit of PORTB?
28. Create a flowchart and write a program to get the statuses of RD7 and RD6 and put them on RC0 and RC7, respectively.

ANSWERS TO REVIEW QUESTIONS

SECTION 4.1: I/O PORT PROGRAMMING IN PIC18

1. 5
2. False
3. PORTB, PORTC, and PORTD
4. False
5. MOVLW 0x99
    MOVWF PORTB
    MOVWF PORTC
6. 00, TRISB
7. FFH, TRISB

SECTION 4.2: I/O BIT MANIPULATION PROGRAMMING

1. True
2.  ```
        BCF  TRISB,7
    H1  BTG  PORTB,7
        BRA  H1
    ```
3. Input
4.  ```
                BSF    TRISB,2
                BCF    TRISB,0
        AGAIN   BTFSS  PORTB,2
                BRA    OVER
                BSF    PORTB,0
                BRA    AGAIN
        OVER    BCF    PORTB,0
                BRA    AGAIN
    ```

5.  ```
                BCF  TRISD,0
                BCF  TRISD,7
        H2      BTG  PORTD,0
                BTG  PORTD,7
                BRA  H2
    ```
    ```

CHAPTER 5

ARITHMETIC, LOGIC INSTRUCTIONS, AND PROGRAMS

OBJECTIVES

Upon completion of this chapter, you will be able to:

>> Define the range of numbers possible in PIC unsigned data
>> Code addition and subtraction instructions for unsigned data
>> Perform addition of BCD data
>> Code PIC unsigned data multiplication instructions
>> Code PIC programs for division
>> Code PIC Assembly language logic instructions AND, OR, and EX-OR
>> Use PIC logic instructions for bit manipulation
>> Use compare and skip instructions for program control
>> Code PIC rotate instructions and data serialization
>> Explain the BCD (binary coded decimal) system of data representation
>> Contrast and compare packed and unpacked BCD data
>> Code PIC programs for ASCII and BCD data conversion

This chapter describes all PIC arithmetic and logic instructions. Program examples are given to illustrate the application of these instructions. In Section 5.1 we discuss instructions and programs related to addition, subtraction, multiplication, and division of unsigned numbers. Signed numbers are discussed in Section 5.2. In Section 5.3, we discuss the logic instructions AND, OR, and XOR, as well as the COMPARE instruction. The ROTATE instruction and data serialization are discussed in Section 5.4. In Section 5.5 we provide some real-world applications such as BCD and ASCII conversion.

SECTION 5.1: ARITHMETIC INSTRUCTIONS

Unsigned numbers are defined as data in which all the bits are used to represent data, and no bits are set aside for the positive or negative sign. This means that the operand can be between 00 and FFH (0 to 255 decimal) for 8-bit data.

Addition of unsigned numbers

In order to add numbers together in the PIC, the WREG register must be involved. One form of the ADD instruction is

```
ADDLW   K   ;WREG = WREG + K
```

The sum is stored in the WREG register. The instruction could change any of the C, DC, Z, N, or OV bits of the status register, depending on the operands involved. The effect of the ADDLW instruction on N and OV is discussed in Section 5.3 because these bits are relevant mainly in signed number operations. Look at Example 5-1.

Example 5-1

Show how the flag register is affected by the following instructions.

```
    MOVLW 0xF5          ;WREG = F5 hex
    ADDLW 0xB           ;WREG = F5 + 0B = 00 and C = 1
```

Solution:
```
    F5H            1111 0101
+   0BH          + 0000 1011
   100H            0000 0000
```

After the addition, register WREG contains 00 and the flags are as follows:
C = 1 because there is a carry out from D7.
Z = 1 because the result in WREG is zero.
DC = 1 because there is a carry from D3 to D4.

ADDWF and addition of individual bytes

Instruction "ADDWF fileReg, d" allows the addition of WREG and individual bytes residing in RAM locations of the file register. Notice that WREG must be involved because memory-to-memory arithmetic operations are never allowed in PIC Assembly language. To calculate the sum of any number of operands, the carry flag should be checked after the addition of each operand. Example 5-2 uses location 7 of the file register to accumulate carries as the operands are added to WREG. In Chapter 6, the loop version of this program will be shown for any number of bytes.

Example 5-2

Assume that file register RAM locations 40–43H have the following hex values. Write a program to find the sum of the values. At the end of the program, location 6 of the file register should contain the low byte and location 7 the high byte of the sum.

```
40 = (7D)
41 = (EB)
42 = (C5)
43 = (5B)
```

Solution:

```
L_Byte EQU  0x6          ;assign RAM location 6 to L_byte of sum
H_Byte EQU  0x7          ;assign RAM location 7 to H_byte of sum

       MOVLW 0           ;clear WREG (WREG = 0)
       MOVWF H_Byte      ;H_Byte = 0
       ADDWF 0x40,W      ;WREG = 0 + 7DH = 7DH , C = 0
       BNC   N_1         ;branch if C = 0
       INCF  H_Byte,F    ;increment (now H_Byte = 0)
N_1    ADDWF 0x41,W      ;WREG = 7D + EB = 68H and C = 1
       BNC   N_2         ;
       INCF  H_Byte,F    ;C = 1, increment (now H_Byte = 1)
N_2    ADDWF 0x42,W      ;WREG = 68 + C5 = 2D and C = 1
       BNC   N_3         ;
       INCF  H_Byte      ;C = 1, increment (now H_Byte = 2)
N_3    ADDWF 0x43,W      ;WREG = 2D + 5B = 88H and C = 0
       BNC   N_4         ;
       INCF  H_Byte,F    ;(H_Byte = 2)
N_4    MOVWF L_Byte      ;now L_Byte = 88h
```

At the end the fileReg location 6 = (8B), and location 7 = (02) because 7D + EB + C5 + 5B + 30 = 28BH. We can use the register indirect addressing mode to do this program much more efficiently. Chapter 6 shows how to do that.

ADDWFC and addition of 16-bit numbers

When adding two 16-bit data operands, we need to be concerned with the propagation of a carry from the lower byte to the higher byte. This is called *multi-byte addition* to distinguish it from the addition of individual bytes. The instruction ADDWFC (ADDW and fileReg with carry) is used on such occasions.

For example, look at the addition of 3CE7H + 3B8DH, as shown next.

```
         1
        3C E7
    +   3B 8D
        78 74
```

When the first byte is added, there is a carry (E7 + 8D = 74, CY = 1). The carry is propagated to the higher byte, which results in 3C + 3B + 1 = 78 (all in hex). Example 5-3 shows the above steps in a PIC program.

Example 5-3

Write a program to add two 16-bit numbers. The numbers are 3CE7H and 3B8DH. Assume that fileReg location 6 = (8D) and location 7 = (3B). Place the sum in fileReg locations 6 and 7; location 6 should have the lower byte.

Solution:
```
    ;location 6 = (8D)
    ;location 7 = (3B)

    MOVLW 0xE7      ;load the low byte now (WREG = E7H)
    ADDWF 0x6,F     ;F = W + F = E7 + 8D = 74 and CY = 1
    MOVLW 0x3C      ;load the high byte (WREG = 3CH)
    ADDWFC 0x7,F    ;F = W + F + carry, adding the upper byte
                    ;with Carry from lower byte
                    ;F = 3C + 3B + 1 = 78H (all in hex)
```

Notice the use of ADDWF for the lower byte and ADDWFC for the higher byte.

BCD (binary coded decimal) number system

BCD stands for *binary coded decimal*. BCD is needed because in everyday life we use the digits 0 to 9 for numbers, not binary or hex numbers. Binary representation of 0 to 9 is called BCD (see Figure 5-1). In computer literature, one encounters two terms for BCD numbers: (1) unpacked BCD, and (2) packed BCD. We describe each one next.

Digit	BCD
0	0000
1	0001
2	0010
3	0011
4	0100
5	0101
6	0110
7	0111
8	1000
9	1001

Figure 5-1. BCD Code

Unpacked BCD

In unpacked BCD, the lower 4 bits of the number represent the BCD number, and the rest of the bits are 0. Example: "0000 1001" and "0000 0101" are unpacked BCD for 9 and 5, respectively. Unpacked BCD requires 1 byte of memory, or an 8-bit register, to contain it.

Packed BCD

In packed BCD, a single byte has two BCD numbers in it: one in the lower 4 bits, and one in the upper 4 bits. For example, "0101 1001" is packed BCD for 59H. Only 1 byte of memory is needed to store the packed BCD operands. Thus one reason to use packed BCD is that it is twice as efficient in storing data.

There is a problem with adding BCD numbers, which must be corrected. The problem is that after adding packed BCD numbers, the result is no longer BCD. Look at the following.

```
MOVLW 0x17
ADDLW 0x28
```

Adding these two numbers gives 0011 1111B (3FH), which is not BCD! A BCD number can only have digits from 0000 to 1001 (or 0 to 9). In other words, adding two BCD numbers must give a BCD result. The result above should have been 17 + 28 = 45 (0100 0101). To correct this problem, the programmer must add 6 (0110) to the low digit: 3F + 06 = 45H. The same problem could have happened in the upper digit (for example, in 52H + 87H = D9H). Again, 6 must be added to the upper digit (D9H + 60H = 139H) to ensure that the result is BCD (52 + 87 = 139). This problem is so pervasive that most microprocessors such as the PIC18 have an instruction to deal with it. In the PIC18 instruction "DAW" is designed to correct the BCD addition problem. This is discussed next.

DAW instruction

The DAW (decimal adjust WREG) instruction in the PIC18 is provided to correct the aforementioned problem associated with BCD addition. The mnemonic "DAW" works only with an operand in the WREG register. The DAW instruction will add 6 to the lower nibble or higher nibble if needed; otherwise, it will leave the result alone. The following example will clarify these points.

```
MOVLW 0x47    ;WREG = 47H first BCD operand
ADDLW 0x25    ;hex(binary) addition (WREG = 6CH)
DAW           ;adjust for BCD addition (WREG = 72H)
```

After the program is executed, register WREG will contain 72H (47 + 25 = 72). Note that the "DAW" instruction works only on WREG.

Summary of DAW action

After any instruction,
1. If the lower nibble (4 bits) is greater than 9, or if DC = 1, add 0110 to the lower 4 bits.
2. If the upper nibble is greater than 9, or if C = 1, add 0110 to the upper 4 bits.

In reality there is no use for the DC (auxiliary carry) flag bit other than for BCD addition and correction.

```
MOVLW 0x00    ;WREG = 0
ADDLW 0x09    ;WREG = 0x09
ADDLW 0x08    ;WREG = 0x11, DC = 1
DAW           ;WREG = 0x17 (9 + 8 = 17)
```

As another example, examine the case of adding 55H and 77H. This will result in CCH, which is incorrect as far as BCD is concerned.

CHAPTER 5: ARITHMETIC, LOGIC INSTRUCTIONS, AND PROGRAMS   **159**

Hex	BCD
57	0101 0111
+ 77	+ 0111 0111
CE	1100 1110
+ 66	+ 0110 0110
134	1 0011 0100 Note C = 1

Note that unlike other processors, the PIC does not require the use of arithmetic instructions prior to execution of the "DAW" instruction. Look at the following case where no arithmetic instruction is used.

```
MOVLW 0x0C   ;WREG = 00001100
DAW          ;WREG = 00001100 + 00000110 = 00010010 = 0x12
```

Examine Example 5-4.

Example 5-4

Assume that 5 BCD data items are stored in RAM locations starting at 40H, as shown below. Write a program to find the sum of all the numbers. The result must be in BCD.

```
40 = (71)
41 = (88)
42 = (69)
43 = (97)
```

Solution:

```
L_Byte     EQU   0x6          ;assign RAM loc 6 to L_Byte of sum
H_Byte     EQU   0x7          ;assign RAM loc 7 to H_Byte of sum

      MOVLW      0            ;clear WREG (WREG = 0)
      MOVWF      H_Byte       ;H_Byte = 0
      ADDWF      0x40,W       ;WREG = 0 + 71H = 71H, C = 0
      DAW                     ;WREG = 71H
      BNC        N_1          ;branch if C = 0
      INCF       H_Byte,F     ;
N_1   ADDWF      0x41,W       ;WREG = 71 + 88 = F9H
      DAW                     ;WREG = 59H AND C = 1
      BNC        N_2          ;
      INCF       H_Byte,F     ;C = 1, increment (now H_Byte = 1)
N_2   ADDWF      0x42,W       ;WREG = 59 + 69 = C2 and Carry = 0
      DAW                     ;WREG = 28 and C = 1
      BNC        N_3          ;
      INCF       H_Byte       ;C = 1, increment (now H_Byte = 2)
N_3   ADDWF      0x43,W       ;WREG = 28 + 97 = BFH and C = 0
      DAW                     ;WREG = 25 and C = 1
      BNC        N_4          ;
      INCF       H_Byte,F     ;(now H_Byte = 3)
N_4   MOVWF      L_Byte       ;Now L_Byte = 25H
```

After this code executes, fileReg location 6 = (03), and WREG = 25 because 71 + 88 + 69 + 97 = 325H. We can use the register indirect addressing mode and looping to do this program much more efficiently. Chapter 6 shows how to do that.

Subtraction of unsigned numbers

In many microprocessors, there are two different instructions for subtraction: SUB and SUBB (subtract with borrow). In the PIC18 we have four instructions for subtraction: SUBLW, SUBWF, SUBWFB, and SUBFWB. The last two are subtract with borrow. Notice that we use the C (carry) flag for the borrow. We now will examine each of these commands.

SUBLW K (WREG = K – WREG)

In subtraction, the PIC microcontrollers (indeed, all modern CPUs) use the 2's complement method. Although every CPU contains adder circuitry, it would be too cumbersome (and take too many transistors) to design separate subtracter circuitry. For this reason, the PIC uses adder circuitry to perform the subtraction command. Assuming that the PIC is executing a simple subtract instruction and that C = 0 prior to the execution of the instruction, one can summarize the steps of the hardware of the CPU in executing the SUBLW instruction for unsigned numbers as follows:

1. Take the 2's complement of the subtrahend (WREG operand).
2. Add it to the minuend (K operand).

These two steps are performed for every SUB instruction by the internal hardware of the CPU, regardless of the source of the operands, provided that the addressing mode is supported. It is after these two steps that the result is obtained and the flags are set. Example 5-5 illustrates the two steps.

Example 5-5

Show the steps involved in the following.

```
    MOVLW 0x23          ;load 23H into WREG (WREG = 23H)
    SUBLW 0x3F          ;WREG = 3F - WREG
```

Solution:

```
     K    = 3F   0011 1111        0011 1111
   - WREG = 23   0010 0011    +   1101 1101 (2's complement)
           1C                   1 0001 1100
                                C = 1, D7 = N = 0 (result is positive)
```

The flags would be set as follows: C = 1, N = 0 (notice that D7 is the negative flag). The programmer must look at the N (or C) flag to determine if the result is positive or negative.

After the execution of SUB, if N = 0 (or C = 1), the result is positive; if N = 1 (or C = 0), the result is negative and the destination has the 2's complement of the result. Normally, the result is left in 2's complement, but the NEGF (negate, which is 2's complement) instruction can be used to change it. Another SUB instruction in PIC is SUBWF (Destination = fileReg – WREG). This is shown in Example 5-6 along with the NEGF instruction.

Example 5-6

Write a program to subtract 4C − 6E.
Solution:

```
MYREG EQU 0x20
      MOVLW 0x4C          ;load WREG (WREG = 4CH)
      MOVWF MYREG         ;MYREG = 4CH
      MOVLW 0x6E          ;WREG = 6EH
      SUBWF MYREG,W       ;WREG = MYREG - WREG. 4C - 6E = DE, N = 1
      BNN   NEXT          ;if N = 0 (C = 1), jump to NEXT target
      NEGF  WREG          ;take 2's complement of WREG
NEXT  MOVWF MYREG         ;save the result in MYREG
```

The following are the steps after the SUBWF instruction:

```
    4C      0100 1100                    0100 1100
   -6E      0110 1110    2's comp =     1001 0010
   -22                                  1101 1110
```

After SUBWF, we have N = 1 (or C = 0), and the result is negative, in 2's complement. Then it falls through and NEGF will be executed. The NEGF instruction will take the 2's complement, and we have MYREG = 22H.

SUBWFB (dest = fileReg − W − $\overline{Borrow}$) subtract with borrow

This instruction is used for multibyte numbers and will take care of the borrow of the lower byte. If C = 0 prior to executing the SUBWFB instruction, it also subtracts 1 from the result. See Example 5-7.

SUBFWB (dest = WREG − fileReg − $\overline{Borrow}$)

This instruction is also used for multibyte numbers and will take care of the borrow of the lower byte. Notice the difference between SUBWFB and SUBFWB. See Appendix A for the description of these two instructions.

Example 5-7

Write a program to subtract two 16-bit numbers. The numbers are 2762H − 1296H. Assume fileReg location 6 = (62) and location 7 = (27). Place the difference in fileReg locations 6 and 7; loc 6 should have the lower byte.

Solution:

```
      loc 6 = (62)
      loc 7 = (27)

      MOVLW 0x96          ;load the low byte (WREG = 96H)
      SUBWF 0x6,F         ;F = F - W = 62 - 96 = CCH, C = borrow = 0, N = 1
      MOVLW 0x12          ;load the high byte (WREG = 12H)
      SUBWFB 0x7,F        ;F = F - W - b̄, sub byte with the borrow
                          ;F = 27 - 12 - 1 = 14H
```

After the SUBWF, loc 6 has = 62H − 96H = CCH and the carry flag is set to 0, indicating there is a borrow (notice, N = 1). Because C = 0, when SUBWFB is executed the fileReg location 7 has = 27H − 12H − 1 = 14H. Therefore, we have 2762H − 1296H = 14CCH.

The C flag in subtraction for PIC

Notice that the PIC18 is different from other CPU's such as the x86 and the 8051 when it comes to the carry flag in subtract operations. In those CPUs, the carry is inverted by the CPU itself and we examine the C flag to see if the result is positive or negative. In the PIC18, if C = 0, the result is negative. That is the reason in subtract with borrow we have $F = F - W - \overline{b}$. Use the MPLAB simulator to gain additional insight into this important issue.

Multiplication of unsigned numbers

The PIC supports byte-by-byte multiplication only. The bytes are assumed to be unsigned data. The syntax is as follows:

```
MULLW   K     ;W × K and 16-bit is result is in PRODH:PRODL
```

In byte-by-byte multiplication, one of the operands must be in the WREG register, and the second operand must be a literal K value. After multiplication, the result is in the special function registers PRODH and PRODL; the lower byte is in PRODL, and the upper byte is in PRODH. See Table 5-1. The following example multiplies 25H by 65H.

```
MOVLW 0x25       ;load 25H to WREG (WREG = 25H)
MULLW 0x65       ;25H * 65H = E99 where
                 ;PRODH = 0EH and PRODL = 99H
```

Table 5-1: Unsigned Multiplication Summary (MULLW K)

Multiplication	Byte 1	Byte2	Result
Byte × Byte	WREG	K	PRODH = high byte, PRODL = low byte

Note: Multiplication of operands larger than 8-bit takes some manipulation.

Division of unsigned numbers

There is no single instruction for the division of byte/byte numbers in the PIC18. We can write a program to perform division by repeated subtraction. In dividing a byte by a byte, the numerator is placed in a fileReg and the denominator is subtracted from it repeatedly. The quotient is the number of times we subtracted and the remainder is in fileReg upon completion. See the following example.

```
NUM    EQU   0x19         ;set aside fileReg
MYQ    EQU   0x20
MYNMB  EQU   D'95'
MYDEN  EQU   D'10'
       CLRF  MYQ          ;quotient = 0
       MOVLW MYNMB        ;WREG = 95
       MOVWF NUM          ;numerator = 95
       MOVLW MYDEN        ;WREG = denominator = 10
B1     INCF  MYQ,F        ;increment quotient for every 10 subtr
       SUBWF NUM,F        ;subtract 10 (F = F - W)
       BC    B1           ;keep doing it until C = 0
       DECF  MYQ,F        ;once too many
       ADDWF NUM,F        ;add 10 back to get remainder
```

An application for division

Sometimes an ADC (analog-to-digital converter) is connected to a port and the ADC represents some quantity such as temperature or pressure. The 8-bit ADC provides data in hex in the range of 00–FFH. This hex data must be converted to decimal. We do that by dividing it by 10 repeatedly, saving the remainders, as shown in Example 5-8.

Example 5-8

Assume that file register location 0x15 has value FD (hex). Write a program to convert it to decimal. Save the digits in locations 0x22, 0x23, and 0x24, where the least-significant digit is in 0x22

Solution:

```
#include <P18F458.INC>
;PIC Assembly Language Program for division (by repeated subtraction)
;(Byte/Byte)

NUME        EQU   0x15          ;RAM location for NUME
QU          EQU   0x20          ;RAM location for quotient
RMND_L      EQU   0x22
RMND_M      EQU   0x23
RMND_H      EQU   0x24
MYNUM       EQU   0xFD          ;FDH = 253 in decimal
MYDEN       EQU   D'10'         ;253/10
            ORG   0H            ;start at address 0
            MOVLW MYNUM         ;WREG = 253, the numerator
            MOVWF NUME          ;load numerator
            MOVLW MYDEN         ;WREG = 10, the denominator
            CLRF  QU,F          ;clear quotient
D_1         INCF  QU,F          ;increment quotient for every sub
            SUBWF NUME          ;sub WREG from NUME value
            BC    D_1           ;if positive go back (C = 1 for positive)
            ADDWF NUME          ;once too many, this is our first digit
            DECF  QU,F          ;once too many for quotient
            MOVFF NUME,RMND_L   ;save the first digit
            MOVFF QU,NUME       ;repeat the process one more time
            CLRF  QU            ;clear QU
D_2         INCF  QU,F
            SUBWF NUME          ;sub WREG from NUME value
            BC    D_2           ;(C = 1 for positive)
            ADDWF NUME          ;once too many
            DECF  QU,F
            MOVFF NUME,RMND_M   ;2nd digit
            MOVFF QU,RMND_H     ;3rd digit
HERE        GOTO  HERE          ;stay here forever
            END                 ;end of asm source file
```

To convert a single decimal digit to ASCII format, we OR it with 30H, as shown in Sections 6.4 and 6.5.

Example 5-9

Analyze the program in Example 5-8 for a numerator of 253.

Solution:

To convert a binary (hex) value to decimal, we divide it by 10 repeatedly until the quotient is less than 10. After each division the remainder is saved. In the case of an 8-bit binary, such as FDH, we have 253 decimal, as shown below.

```
                Quotient    Remainder
253/10  =       25          3 (low digit)
25/10   =       2           5 (middle digit)
                            2 (high digit)
```

Therefore, we have FDH = 253. In order to display this data, it must be converted to ASCII, which is described in a later section in this chapter.

Review Questions

1. In multiplication of two bytes in the PIC18, we can place one byte in register _____ and for the other one we can use value K.
2. In unsigned byte-by-byte multiplication, the product will be placed in register(s) _____ .
3. Is "MULLW F" a valid PIC18 instruction? Explain your answer.
4. In PIC18, the largest two numbers that can be multiplied are _____ and _____.
5. True or false. The DAW instruction works on WREG only.
6. Is "DAW fileReg, d" a valid PIC18 instruction? Explain your answer.
7. The instruction "ADDLW K" places the sum in _____.
8. Why is the following ADD instruction illegal? "ADDLW fileReg"
9. Rewrite the instruction above to add WREG to the fileReg.
10. The instruction "ADDWFC fileReg, W" places the sum in _____.
11. Find the value of the DC and C flags in each of the following.
    (a) MOVLW 0x4F   (b)   MOVLW 0x9C
        ADDLW 0xB1   ADDLW 0x63
12. Show how the CPU would subtract 05H from 43H.
13. If C = 1, WREG = 95H, and fileReg = 4FH prior to the execution of "SUBFWB fileReg, F", what will be the contents of WREG and fileReg after the subtraction?

SECTION 5.2: SIGNED NUMBER CONCEPTS AND ARITHMETIC OPERATIONS

All data items used so far have been unsigned numbers, meaning that the entire 8-bit operand was used for the magnitude. Many applications require signed data. In this section the concept of signed numbers is discussed along with related instructions. If your applications do not involve signed numbers, you can bypass this section.

Concept of signed numbers in computers

In everyday life, numbers are used that could be positive or negative. For example, a temperature of 5 degrees below zero can be represented as −5, and 20 degrees above zero as +20. Computers must be able to accommodate such numbers. To do that, computer scientists have devised the following arrangement for the representation of signed positive and negative numbers: The most significant bit (MSB) is set aside for the sign (+ or –), while the rest of the bits are used for the magnitude. The sign is represented by 0 for positive (+) numbers and 1 for negative (–) numbers. Signed byte representation is discussed below.

Signed 8-bit operands

In signed byte operands, D7 (MSB) is the sign and D0 to D6 are set aside for the magnitude of the number. If D7 = 0, the operand is positive, and if D7 = 1, it is negative. The N flag in the status register is the D7 bit.

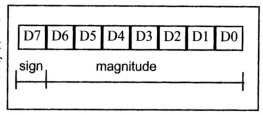

Figure 5-2. 8-Bit Signed Operand

Positive numbers

The range of positive numbers that can be represented by the format shown in Figure 5-2 is 0 to +127. If a positive number is larger than +127, a 16-bit operand must be used. Because the PIC18 does not support 16-bit data, we will not discuss it.

```
   0      0000 0000
  +1      0000 0001
  . . .      . . .
  +5      0000 0101
  . . .      . . .
+127      0111 1111
```

Negative numbers

For negative numbers, D7 is 1; however, the magnitude is represented in its 2's complement. Although the assembler does the conversion, it is still important to understand how the conversion works. To convert to negative number representation (2's complement), follow these steps:

1. Write the magnitude of the number in 8-bit binary (no sign).
2. Invert each bit.
3. Add 1 to it.

Examples 5-10, 5-11, and 5-12 demonstrate these three steps.

Example 5-10

Show how the PIC would represent –5.

Solution:

Observe the following steps.

```
1.    0000 0101        5 in 8-bit binary
2.    1111 1010        invert each bit
3     1111 1011        add 1 (which becomes FB in hex)
```

Therefore, –5 = FBH, the signed number representation in 2's complement for –5. The D7 = N = 1 indicates that the number is negative.

Example 5-11

Show how the PIC would represent –34H.

Solution:

Observe the following steps.

```
1.    0011 0100        34H given in binary
2.    1100 1011        invert each bit
3     1100 1100        add 1 (which is CC in hex)
```

Therefore, –34 = CCH, the signed number representation in 2's complement for 34H. The D7 = N = 1 indicates that the number is negative.

Example 5-12

Show how the PIC would represent –128.

Solution:

Observe the following steps.

```
1.    1000 0000        128 in 8-bit binary
2.    0111 1111        invert each bit
3     1000 0000        add 1 (which becomes 80 in hex)
```

Therefore, –128 = 80H, the signed number representation in 2's complement for –128. The D7 = N = 1 indicates that the number is negative. Notice that 128 (binary 10000000) in unsigned representation is the same as signed –128 (binary 10000000).

From the examples above, it is clear that the range of byte-sized negative numbers is –1 to –128. The following lists byte-sized signed number ranges:

Decimal	Binary	Hex
-128	1000 0000	80
-127	1000 0001	81
-126	1000 0010	82
...	....	..
-2	1111 1110	FE
-1	1111 1111	FF
0	0000 0000	00
+1	0000 0001	01
+2	0000 0010	02
..	....	..
+127	0111 1111	7F

The above explains the mystery behind the relative address of –128 to +127 in the BNZ and other conditional branch instructions discussed in Chapter 3.

Overflow problem in signed number operations

When using signed numbers, a serious problem arises that must be dealt with. This is the overflow problem. The PIC indicates the existence of an error by raising the OV (overflow) flag, but it is up to the programmer to take care of the erroneous result. The CPU understands only 0s and 1s and ignores the human convention of positive and negative numbers. What is an overflow? If the result of an operation on signed numbers is too large for the register, an overflow has occurred and the programmer must be notified. Look at Example 5-13.

Example 5-13

Examine the following code and analyze the result, including the N and OV flags.

```
    MOVLW  +D'96'       ;WREG = 0110 0000
    ADDLW  +D'70'       ;WREG = (+96) + (+70) = 1010 0110
                        ;WREG = A6H = -90 decimal, INVALID!!
```

Solution:

```
    +96    0110 0000
+   +70    0100 0110
+   166    1010 0110   N = 1 (negative) and OV = 1. Sum = -90
```

According to the CPU, the result is negative (N = 1), which is wrong. The CPU sets OV = 1 to indicate the overflow error. Remember that the N flag is the D7 bit. If N = 0, the sum is positive, but if N = 1, the sum is negative.

In Example 5-13, +96 is added to +70 and the result, according to the CPU, was –90. Why? The reason is that the result was larger than what WREG could contain. Like all other 8-bit registers, WREG could only contain up to +127. The designers of the CPU created the overflow flag specifically for the purpose of informing the programmer that the result of the signed number operation is erroneous. The N flag is D7 of the result. If N = 0, the sum is positive (+) and if N = 1, then the sum is negative.

When is the OV flag set?

In 8-bit signed number operations, OV is set to 1 if either of the following two conditions occurs:

1. There is a carry from D6 to D7 but no carry out of D7 (C = 0).
2. There is a carry from D7 out (C = 1) but no carry from D6 to D7.

In other words, the overflow flag is set to 1 if there is a carry from D6 to D7 or from D7 out, but not both. This means that if there is a carry both from D6 to D7 and from D7 out, OV = 0. In Example 5-13, because there is only a carry from D6 to D7 and no carry from D7 out, OV = 1. Study Examples 5-14, 5-15, and 5-16 to understand the overflow flag in signed arithmetic.

Example 5-14

Observe the following, noting the role of the OV and N flags:

```
    MOVLW  -D'128'      ;WREG = 1000 0000 (WREG = 80H)
    ADDLW  -D'2'        ;W = (-128) + (-2)
                        ;W = 1000000 + 11111110 = 0111 1110,
                        ;N = 0, W = 7EH = +126, invalid
```

Solution:

```
    -128          1000 0000
 +  - 2           1111 1110
    - 130         0111 1110   N = 0 (positive) and OV = 1
```

According to the CPU, the result is +126, which is wrong, and OV = 1 indicates that.

Example 5-15

Observe the following, noting the OV and N flags:

```
    MOVLW  -D'2'        ;WREG = 1111 1110 (WREG = FEH)
    ADDLW  -D'5'        ;WREG = (-2) + (-5) = -7 or F9H
                        ;correct, since OV = 0
```

Solution:

```
    -2          1111 1110
 +  -5          1111 1011
    - 7         1111 1001   and OV = 0 and N = 1. Sum is negative
```

According to the CPU, the result is –7, which is correct, and the OV flag indicates that. (OV = 0).

Example 5-16

Examine the following, noting the role of the OV and N flags:

```
        MOVLW +D'7'          ;WREG = 0000 0111
        ADDLW +D'18'         ;W = (+7) + (+18)
                             ;W = 00000111 + 00010010 = 0001 1001
                             ;W = (+7) + (+18) = +25, N = 0, positive and
                             ;correct, OV = 0
```

Solution:

```
    +  7 0000 0111
  + +18 0001 0010
    +25 0001 1001   N = 0 (positive 25) and OV = 0
```

According to the CPU, this is +25, which is correct and OV = 0 indicates that.

From the above examples, we conclude that in any signed number addition, OV indicates whether the result is valid or not. If OV = 1, the result is erroneous; if OV = 0, the result is valid. We can state emphatically that in unsigned number addition, we must monitor the status of C (carry flag), and in signed number addition, the OV (overflow) flag must be monitored by the programmer. In the PIC, instructions such as BNC and BC allow the program to branch right after the addition of unsigned numbers, as we saw in Section 5.1. There are also the BOV and the BNOV instructions for the OV flag that allow us to correct the signed number error. We also have two branch instructions for the N flag (negative), BN and BNN.

Instructions to create 2's complement

The PIC18 does have a special instruction to make the 2's complement of a number. It is called NEG fileReg (negate fileReg) and is covered in the next section.

Review Questions

1. In an 8-bit operand, bit _____ is used for the sign bit.
2. Convert –16H to its 2's complement representation.
3. The range of byte-sized signed operands is – _____ to + _____.
4. Show +9 and –9 in binary.
5. Explain the difference between a carry and an overflow.

SECTION 5.3: LOGIC AND COMPARE INSTRUCTIONS

Apart from I/O and arithmetic instructions, logic instructions are some of most widely used instructions. In this section we cover Boolean logic instructions such as AND, OR, Exclusive-OR (XOR), and complement. We will also study the compare instruction.

AND

```
ANDLW K   ;WREG = WREG AND K
```

This instruction will perform a logical AND on the two operands and place the result in WREG. There is also the "ANDWF fileReg, d" instruction where the destination can be WREG or fileReg. The fileReg operand can be any register in the data RAM file register. See Appendix A. The AND instruction will affect the Z and N flags. N is D7 of the result, and Z = 1 if the result is zero. The AND instruction is often used to mask (set to 0) certain bits of an operand. See Example 5-17.

Logical AND Function

Inputs		Output
X	Y	X AND Y
0	0	0
0	1	0
1	0	0
1	1	1

X ──┐
    ⊐D── X AND Y
Y ──┘

Example 5-17

Show the results of the following.

```
    MOVLW  0x35      ;WREG = 35H
    ANDLW  0x0F      ;W = W AND 0FH (now W = 05)
```
Solution:

```
    35H  0 0 1 1 0 1 0 1
    0FH  0 0 0 0 1 1 1 1
    05H  0 0 0 0 0 1 0 1    ;35H AND 0FH = 05H, Z = 0, N = 0
```

OR

```
IORLW K     ;WREG = WREG Inclusive-OR K
```

This instruction will perform a logical OR on the two operands and place the result in WREG. There is also an "IORWF fileReg, d" instruction where the destination can be WREG or fileReg. The fileReg operand can be any register in the data RAM file register. See Appendix A. The OR instruction will affect the Z and N flags. N is D7 of the result and Z = 1 if the result is zero. The OR instruction can be used to set certain bits of an operand to 1. See Example 5-18.

Logical OR Function

Inputs		Output
X	Y	X OR Y
0	0	0
0	1	1
1	0	1
1	1	1

X ──┐
    ⊐D── X OR Y
Y ──┘

Example 5-18

(a) Show the results of the following:
```
        MOVLW  0x04                ;WREG = 04
        IORLW  0x30                ;now WREG = 34H
```
(b) Assume that Port B bit RB2 is used to control an outdoor light, and bit RB5 to control a light inside a building. Show how to turn "on" the outdoor light and turn "off" the inside one.

Solution:

(a)
```
        04H       0000 0100
        30H       0011 0000
        34H       0011 0100        04 OR 30 = 34H,  Z = 0 and N = 0
```

(b)
```
        BCF    TRISB,2             ;make RB2 an output
        BCF    TRISB,5             ;make RB5 an output
        MOVLW  B'00000100'         ;D2 = 1
        IORWF  PORTB,F             ;make RB2 = 1 only
        MOVLW  B'11011111'         ;D5 = 0
        ANDWF  PORTB,F             ;mask RB5 = 0 only
```

Of course, the above method is unnecessary in PIC, since we can manipulate individual bits using bit-oriented operations. This is shown in Section 6.4.

EX-OR

```
        XORLW  K        ;WREG = WREG XOR K
```

This instruction will perform a logical EX-OR on the two operands and place the result in WREG. There is also an "XORWF fileReg, d" instruction where the destination can be WREG or fileReg. The fileReg operand can be any register in the data RAM file register. See Appendix A. The EX-OR instruction will affect the Z and N flags. N is D7 of the result and Z = 1 if result is zero. See Examples 5-19 and 5-20.

EX-OR can also be used to see if two registers have the same value. "XORWF fileReg, W" will EX-OR the WREG register and a fileReg location, and put the result in WREG. If both registers have the same value, 00 is placed in WREG. Then we can use the BZ instruction to make a decision based on the result. See Examples 5-20 and 5-21.

Logical XOR Function

Inputs		Output
A	**B**	**A XOR B**
0	0	0
0	1	1
1	0	1
1	1	0

A
B ⊐D⊃— A XOR B

Another widely used application of EX-OR is to toggle the bits of an operand.

```
MOVLW 0xFF        ;WREG = FFH
XORWF PORTC,F     ;EX-OR PORTC with 1111 1111 will
                  ;change all the bits of Port C to
                  ;opposite
```

Example 5-19

Show the results of the following:
```
MOVLW 0x54
XORLW 0x78
```

Solution:

54H	0 1 0 1 0 1 0 0
78H	0 1 1 1 1 0 0 0
2CH	0 0 1 0 1 1 0 0 54H XOR 78H = 2CH, Z = 0, N = 0

Example 5-20

The EX-OR instruction can be used to test the contents of a register by EX-ORing it with a known value. In the following code, we show how EX-ORing value 45H with itself will raise the Z flag:

```
OVER   MOVF   PORTB,W    ;get a byte from PORTB into WREG
       XORLW  0x45
       BNZ    OVER       ;branch if not zero
```

Solution:

45H	01000101
45H	01000101
00	00000000

EX-ORing a number with itself sets it to zero with Z = 1. We can use the BNZ instruction to make the decision. EX-ORing with any other number will result in a non-zero value.

Example 5-21

Read and test PORTB to see whether it has value 45H. If it does, send 99H to PORTC; otherwise, it stays cleared.

Solution:

```
        CLRF   TRISC       ;Port C = output
        CLRF   PORTC       ;Port C = 00
        SETF   TRISB       ;Port B = input
        MOVLW  0x45
        XORWF  PORTB,W     ;EX-OR with 0x45, Z = 1 if yes
        BNZ    EXIT        ;branch if PORTB has value other than 0
        MOVLW  0x99
        MOVWF  PORTC       ;Port C = 99h
EXIT:...
```

COMF (complement fileReg)

This instruction complements the contents of a file register. The complement action changes the 0s to 1s and the 1s to 0s. This is also called *1's complement*.

Logical Inverter	
Input	**Output**
X	**NOT X**
0	1
1	0

X ─▷o─ NOT X

```
CLRF   TRISB      ;Port B = Output
MOVLW  0x55
MOVWF  PORTB
COMF   PORTB,F    ;now PORTB = AAH
```

NEGF (negate fileReg)

This instruction takes the 2's complement of a file register. See Example 5-22.

Example 5-22

Find the 2's complement of the value 85H. Note that 85H is –123.
Solution:

```
MYREG EQU 0x10
      MOVLW 0x85              85H = 1000 0101
      MOVWF MYREG             1'S = 0111 1010
      NEGF  MYREG                   ____+_1_
                    2's comp       0111 1011 = 7BH
```

Compare instructions

The PIC18 has three instructions for the compare operation, as shown in Table 5-2. These instructions compare a value in the file register with the contents of the WREG register, and make decisions based on whether fileReg is greater than, equal to, or less than WREG. The compare instruction is really a subtraction, except that the values of the operands do not change. In PIC18, flags are not changed either after the compare instruction. It must be emphasized again that in compare instructions, the operands are not affected, regardless of the result of the comparison. We describe each of the instructions in Table 5-2 with an example.

Table 5-2: PIC18 Compare Instructions

CPFSGT	Compare FileReg with WREG, skip if greater than	FileReg > WREG
CPFSEQ	Compare FileReg with WREG, skip if equal	FileReg = WREG
CPFSLT	Compare fileReg with WREG, skip if less than	FileReg < WREG

Note: These instructions have no effect on the flag bits of the status register. Also the values in fileReg and WREG remain unchanged.

CPFSGT instruction

The CPFSGT compares a fileReg with WREG and skips the next instruction if fileReg is greater than WREG (F > W). See Figure 5-3 and Example 5-23.

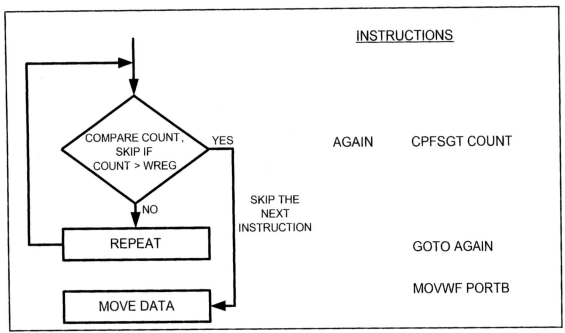

Figure 5-3. Flowchart for CPFSGT

Example 5-23

Write a program to find the greater of the two values 27 and 54, and place it in file register location 0x20.

Solution:

```
        VAL_1 EQU    D'27'
        VAL_2 EQU    D'54'
        GREG  EQU    0x20

        MOVLW   VAL_1      ;WREG = 27
        MOVWF   GREG       ;GREG = 27
        MOVLW   VAL_2      ;WREG = 54
        CPFSGT  GREG       ;skip if GREG > WREG
        MOVWF   GREG       ;place the greater in GREG
```

CPFSEQ instruction

The CPFSEQ compares a fileReg with WREG and skips the next instruction if they are equal (F = W). See Example 5-24 and Figure 5-4.

Example 5-24

Write a program to monitor PORTD continuously for the value 63H. It should stop monitoring only if PORTD = 63H.

Solution:

```
        SETF    TRISD    ;PORTD = input
        MOVLW   0x63     ;WREG = 63H
BACK    CPFSEQ  PORTD    ;skip BRA instruction if PORTD = 63H
        BRA     BACK
        . . . . .
```

CHAPTER 5: ARITHMETIC, LOGIC INSTRUCTIONS, AND PROGRAMS   175

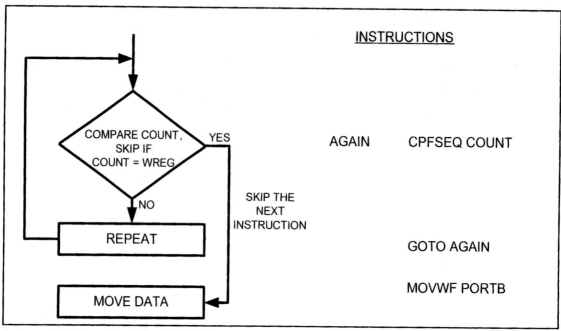

NaNINSTRUCTIONS

AGAIN   CPFSEQ COUNT

COMPARE COUNT,   YES

SKIP IF

COUNT = WREG

SKIP THE
NEXT
INSTRUCTION

NO

REPEAT

GOTO AGAIN

MOVWF PORTB

MOVE DATA

Figure 5-4. Flowchart for CPFSEQ

CPFSLT instruction

The CPFSLT compares a fileReg with WREG and skips the next instruction if fileReg is less than WREG (F < W). See Example 5-25 and Figure 5-5.

Example 5-25

Write a program to find the smaller of the two values 27 and 54, and place it in file register location 0x20.

Solution:

```
VAL_1 EQU    D'27'
VAL_2 EQU    D'54'
LREG  EQU    0x20   ;location for smaller of two

MOVLW   VAL_1        ;WREG = 27
MOVWF   LREG         ;LREG = 27
MOVLW   VAL_2        ;WREG = 54
CPFSLT  LREG         ;skip if LREG < WREG
MOVWF   LREG         ;place the smaller value in LREG
```

176

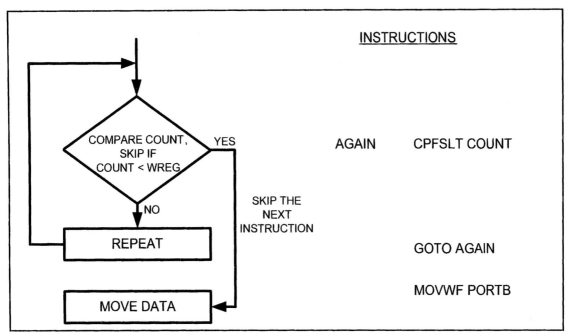

Figure 5-5. Flowchart for CPFSLT

Example 5-26

Assume that Port D is an input port connected to a temperature sensor. Write a program to read the temperature and test it for the value 75. According to the test results, place the temperature value into the registers indicated by the following.

    If T = 75   then WREG = 75
    If T > 75   then GREG = T
    If T < 75   then LREG = T

Solution:

```
LREG EQU 0x20
GREG EQU 0x21
     SETF    TRISD            ;PORTD = input
     MOVLW   D'75'            ;WREG = 75 decimal
     CPFSGT  PORTD            ;skip BRA instruction if PORTD > 75
     BRA     LEQ
     MOVFF   PORTD, GREG
     BRA     OVER
LEQ  CPFSLT  PORTD            ;skip if PORTD < 75
     BRA     OVER
     MOVFF   PORTD, LREG
OVER .....                    ;it must be equal, WREG = 75
```

CHAPTER 5: ARITHMETIC, LOGIC INSTRUCTIONS, AND PROGRAMS   177

Example 5-27

Write code to determine if data on PORTB contains the value 99H. If so, write letter 'Y' to PORTC; otherwise, make PORTC = 'N'.

Solution:

```
        CLRF    TRISC       ;PORTC = output
        MOVLW   A'N'        ;WREG = 'N' (ASCII)
        MOVWF   PORTC       ;PORTC = 'N'
        SETF    TRISB       ;PORTB = input
        MOVLW   0x99        ;WREG = 99H
        CPFSEQ  PORTB       ;skip BRA instruction if PORTB = WREG
        BRA     OVER
        MOVLW   'Y'
        MOVWF   PORTC       ;PORTC = 'Y'
OVER    .....
```

Review Questions

1. Find the content of register WREG after the following code in each case:

   (a) MOVLW 0x37 (b) MOVLW 0x37   (c) MOVLW 0x37
       ANDLW 0xCA   IORLW 0xCA   XORLW 0xCA

2. To mask certain bits of the WREG, we must AND it with _____.
3. To set certain bits of the WREG to 1, we must OR it with _____.
4. EX-ORing an operand with itself results in _____.
5. True or false. The CPFSLT instruction alters the contents of its operands.
6. What value must MYREG have in order for the following code to skip the BRA instruction?

   ```
           MOVLW 0x99
   BACK    CPFSLT MYREG
           BRA BACK
   ```

7. Find the contents of register WREG after execution of the following code:

   ```
   MOVLW 0
   IORLW 0x99
   XORLW 0xFF
   ```

SECTION 5.4: ROTATE INSTRUCTION AND DATA SERIALIZATION

In many applications there is a need to perform a bitwise rotation of an operand. In the PIC18 the rotation instructions RRCF, RRNCF, RLCF, and RLNCF are designed specifically for that purpose. They allow a program to rotate the file register right or left. We explore the rotate instructions next because they are widely used in many different applications. There are two types of rotations. One is a simple rotation of the bits of the file register, and the other is a rotation through the carry. Each is explained below.

Rotating the bits of fileReg right or left

```
RRNCF fileReg,d    ;rotate fileReg right (no carry)
```

In rotate right, the 8 bits of the fileReg are rotated right one bit, and bit D0 exits from the least-significant bit and enters into D7 (most-significant bit). After the rotation the result can be in fileReg or WREG, depending on the d bit. See the code and diagram.

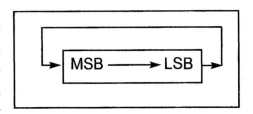

```
MREG EQU 0x20
      MOVLW 0x36          ;WREG = 0011 0110
      MOVWF MYREG
      RRNCF MYREG,F       ;MYREG = 0001 1011
      RRNCF MYREG,F       ;MYREG = 1000 1101
      RRNCF MYREG,F       ;MYREG = 1100 0110
      RRNCF MYREG,F       ;MYREG = 0110 0011
```

```
RLNCF fileReg,d    ;rotate fileReg left (no carry)
```

In rotate left, the 8 bits of the fileReg are rotated left one bit, and bit D7 exits from the MSB (most-significant bit) and enters into D0 (least-significant bit). After the rotation the result can be in fileReg or WREG, depending on the d bit. See the code and diagram.

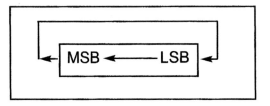

```
MREG EQU 0x20
      MOVLW 0x72          ;WREG = 0111 0010
      MOVWF MYREG
      RLNCF MYREG,F       ;MYREG = 1110 0100
      RLNCF MYREG,F       ;MYREG = 1100 1001
```

Notice in the RRNCF and RLNCF instructions that both the Z and N flags are affected.

Rotating through the carry

There are two more rotate instructions in the PIC18. They involve the carry flag. Each is shown next.

```
RRCF fileReg, d    ;rotate fileReg right through carry
```

In RRCF, as bits are rotated from left to right, the carry flag enters the MSB and the LSB exits to the carry flag. In other words, in RRCF the C is moved to the MSB, and the LSB is moved to the C. In reality, the carry flag acts as if it is part of the register, making it a 9-bit register.

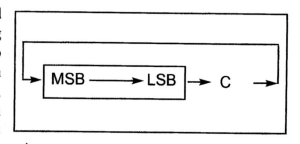

```
MREG EQU 0x20
       BCF     STATUS,C    ;make C = 0 (carry is D0 of status)
       MOVLW   0x26        ;WREG = 0010 0110
       MOVWF   MYREG
       RRCF    MYREG,F     ;MYREG = 0001 0011 C = 0
       RRCF    MYREG,F     ;MYREG = 0000 1001 C = 1
       RRCF    MYREG,F     ;MYREG = 1000 0100 C = 1
```

```
RLCF fileReg, d           ;rotate fileReg left through carry
```

In RLCF, as bits are shifted from right to left, the carry flag enters the LSB and the MSB exits to the carry flag. In other words, in RLCF the C is moved to the LSB, and the MSB is moved to the C. See the following code and diagram. Again the carry flag acts as if it is part of the register, making it a 9-bit register.

```
MREG EQU 0x20
       BSF     STATUS,C    ;make C = 1 (carry is D0 of status)
       MOVLW   0x15        ;WREG = 0001 0101
       MOVWF   MYREG
       RLCF    MYREG,F     ;MYREG = 0010 1011 C = 0
       RLCF    MYREG,F     ;MYREG = 0101 0110 C = 0
       RLCF    MYREG,F     ;MYREG = 1010 1100 C = 0
       RLCF    MYREG,F     ;MYREG = 0101 1000 C = 1
```

Serializing data

Serializing data is a way of sending a byte of data one bit at a time through a single pin of the microcontroller. There are two ways to transfer a byte of data serially:

1. Using the serial port. In using the serial port, programmers have very limited control over the sequence of data transfer. The details of serial port data transfer are discussed in Chapter 10.
2. The second method of serializing data is to transfer data one bit at a time and control the sequence of data and spaces between them. In many new generations of devices such as LCD, ADC, and ROM, the serial versions are becoming popular because they take less space on a printed circuit board. Next, we discuss how to use rotate instructions in serializing data.

Serializing a byte of data

Serializing data is one of the most widely used applications of the rotate instruction. We can use the rotate instruction to transfer a byte of data serially (one bit at a time). Example 5-28 shows how to transfer an entire byte of data serially via any PIC pin.

Example 5-28

Write a program to transfer value 41H serially (one bit at a time) via pin RB1. Put one high at the start and end of the data. Send the LSB first.

Solution:

```
        RCNT   EQU   0x20        ;fileReg loc for counter
        MYREG  EQU   0x21        ;fileReg loc for rotate

        BCF    TRISB,1           ;make RB1 an output bit
        MOVLW  0x41              ;WREG = 41
        MOVWF  MYREG             ;value to be serialized
        BCF    STATUS,C          ;C = 0
        MOVLW  0x8               ;counter
        MOVWF  RCNT              ;load the counter
        BSF    PORTB,1           ;RB1 = high
AGAIN   RRCF   MYREG,F           ;rotate right via carry
        BNC    OVER
        BSF    PORTB,1           ;set the carry bit to PB1
        BRA    NEXT
OVER    BCF    PORTB,1
NEXT    DECF   RCNT,F
        BNZ    AGAIN
        BSF    PORTB,1           ;RB1 = high
```

Example 5-29 shows how to bring in a byte of data serially (one bit at a time). We will see how to use these concepts for a serial RTC (real-time clock) chip in Chapter 16. Example 5-30 shows how to scan the bits in a byte.

Example 5-29

Write a program to bring in a byte of data serially (one bit at a time) via pin RC7 and save it in file register location 0x21. The byte comes in with the LSB first.

Solution:

```
RCNT  EQU   0x20            ;fileReg loc for counter
MYREG EQU   0x21            ;fileReg loc for incoming byte

      BSF   TRISC,7         ;make RC7 an input bit
      MOVLW 0x8             ;counter
      MOVWF RCNT            ;load the counter
AGAIN BTFSC PORTC,7         ;skip if RC7 = 0
      BSF   STATUS,C        ;carry = 1
      BTFSS PORTC,7         ;skip if RC7 = 1
      BCF   STATUS,C        ;otherwise carry = 0
      RRCF  MYREG,F         ;rotate right carry into MYREG
      DECF  RCNT,F          ;decrement the counter
      BNZ   AGAIN           ;repeat until RCNT = 0
                            ;now loc 21H has the byte
```

Example 5-30

Write a program that finds the number of 1s in a given byte.

Solution:

```
R1     EQU  0x20 ;fileReg loc for number of 1s
COUNT  EQU  0x21 ;fileReg loc for counter
VALREG EQU  0x22 ;fileReg loc for the byte

       BCF   STATUS,C    ;C = 0
       CLRF  R1          ;R1 keeps the number of 1s
       MOVLW 0x8         ;counter = 08 to rotate 8 times
       MOVWF COUNT
       MOVLW 0x97        ;find the number of 1s in 97H
       MOVWF VALREG
AGAIN  RLCF  VALREG,F    ;rotate it through the C once
       BNC   NEXT        ;check for C
       INCF  R1,F        ;if C = 1 then add one to R1 reg
NEXT   DECF  COUNT,F
       BNZ AGAIN         ;go through this 8 times
                         ;now loc 0x20 has the number of 1s
```

SWAPF fileReg, d

Another useful instruction is the SWAPF instruction. It works on the file register. It swaps the lower nibble and the higher nibble. In other words, the lower 4 bits are put into the higher 4 bits, and the higher 4 bits are put into the lower 4 bits. See the diagrams below and Example 5-31.

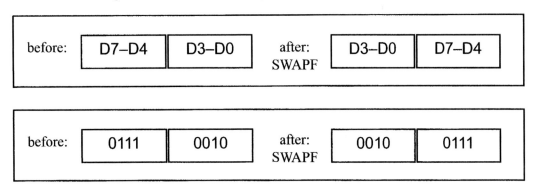

Example 5-31

(a) Find the contents of the MYREG register in the following code.
(b) In the absence of a SWAPF instruction, how would you exchange the nibbles? Write a simple program to show the process.

Solution:

(a)
```
MYREG EQU 0x20
      MOVLW 0x72        ;WREG = 72H
      MOVWF MYREG       ;MYREG = 72H
      SWAPF MYREG,F     ;MYREG = 27H
```

(b)
```
MYREG EQU 0x20
      MOVLW 0x72        ;WREG = 0111 0010
      MOVWF MYREG       ;MYREG = 0111 0010
      RLNCF MYREG,F     ;MYREG = 1110 0100
      RLNCF MYREG,F     ;MYREG = 1100 1001
      RLNCF MYREG,F     ;MYREG = 1001 0011
      RLNCF MYREG,F     ;MYREG = 0010 0111
```

Review Questions

1. What is the value of MYREG in the file register after the following code is executed?

```
MYREG EQU 0x40
MOVLW 0x25
MOVWF MYREG
RRNCF MYREG,F
RRNCF MYREG,F
RRNCF MYREG,F
RRNCF MYREG,F
```

---

2. What is the value of MYREG in the file register after the following code is executed?

```
MYREG EQU  0x40
MOVLW  0x25
MOVWF  MYREG
RLNCF  MYREG,F
RLNCF  MYREG,F
RLNCF  MYREG,F
RLNCF  MYREG,F
```

3. What is the value of MYREG after the following code is executed?

```
MYREG EQU  0x40
CLRF   MYREG
BSF    STATUS,C      ;C = 1
RRCF   MYREG,F
BSF    STATUS,C      ;C = 1
RRCF   MYREG,F
```

4. Does "RLCF W" give an error in the PIC?
5. What is in MYREG after the execution of the following code?

```
MYREG EQU  0x40
MOVLW  0x85
MOVWF  MYREG
SWAPF  MYREG,F
```

SECTION 5.5: BCD AND ASCII CONVERSION

In this section we provide some real-world examples of how to use arithmetic and logic instructions. We will cover their applications in real-world devices in future chapters. For example, many newer microcontrollers have a real-time clock (RTC), where the time and date are kept even when the power is off. These microcontrollers provide the time and date in BCD. To display them, however, they must convert BCD values to ASCII. Next, we show the application of logic and rotate instructions in the conversion of BCD and ASCII.

ASCII numbers

On ASCII keyboards, when the key "0" is activated, "011 0000" (30H) is provided to the computer. Similarly, 31H (011 0001) is provided for key "1", and so on, as shown in Table 5-3.

It must be noted that BCD numbers are universal although ASCII is standard in the United States (and many other countries). Because the keyboard, printers, and monitors all use ASCII, how does data get converted from ASCII to BCD, and vice versa? These are the subjects covered next.

Table 5-3: ASCII and BCD Codes for Digits 0-9

Key	ASCII (hex)	Binary	BCD (unpacked)
0	30	011 0000	0000 0000
1	31	011 0001	0000 0001
2	32	011 0010	0000 0010
3	33	011 0011	0000 0011
4	34	011 0100	0000 0100
5	35	011 0101	0000 0101
6	36	011 0110	0000 0110
7	37	011 0111	0000 0111
8	38	011 1000	0000 1000
9	39	011 1001	0000 1001

Packed BCD to ASCII conversion

In many systems we have what is called a *real-time clock* (RTC). The RTC provides the time of day (hour, minute, second) and the date (year, month, day) continuously, regardless of whether the power is on or off (see Chapter 16). This data, however, is provided in packed BCD. For this data to be displayed on a device such as an LCD, or to be printed by the printer, it must be in ASCII format.

To convert packed BCD to ASCII, you must first convert it to unpacked BCD. Then the unpacked BCD is tagged with 011 0000 (30H). The following demonstrates converting packed BCD to ASCII. See also Example 5-32.

```
Packed BCD   Unpacked BCD    ASCII
29H          02H  & 09H      32H  & 39H
0010 1001    0000 0010 &     0011 0010 &
             0000 1001       0011 1001
```

Example 5-32

Assume that register WREG has packed BCD. Write a program to convert packed BCD to two ASCII numbers and place them in file register locations 6 and 7.

Solution:

```
BCD_VAL EQU 0x29
L_ASC   EQU 0x06   ;set aside file register location
H_ASC   EQU 0x07   ;set aside file register location

        MOVLW BCD_VAL      ;WREG = 29H, packed BCD
        ANDLW 0x0F         ;mask the upper nibble (W = 09)
        IORLW 0x30         ;make it an ASCII, W = 39H ('9')
        MOVWF L_ASC        ;save it (L_ASC = 39H ASCII char)
        MOVLW BCD_VAL      ;W = 29H get BCD data once more
        ANDLW 0xF0         ;mask the lower nibble (W = 20H)
        SWAPF WREG,W       ;swap nibbles (WREG = 02H)
        IORLW 0x30         ;make it an ASCII, W = 32H ('2')
        MOVWF H_ASC        ;save it (H_ASC = 32H ASCII char)
```

ASCII to packed BCD conversion

To convert ASCII to packed BCD, you first convert it to unpacked BCD (to get rid of the 3), and then combine it to make packed BCD. For example, for 4 and 7 the keyboard gives 34 and 37, respectively. The goal is to produce 47H or "0100 0111", which is packed BCD. This process is illustrated next.

Key	ASCII	Unpacked BCD	Packed BCD
4	34	00000100	
7	37	00000111	01000111 which is 47H

```
MYBCD EQU 0x20      ;set aside location in file register

        MOVLW A'4'          ;WREG = 34H, hex for ASCII char 4
        ANDLW 0x0F          ;mask upper nibble (WREG = 04)
        MOVWF MYBCD         ;save it in MYBCD loc
        SWAPF MYBCD,F       ;MYBCD = 40H
        MOVLW A'7'          ;WREG = 37H, hex for ASCII char 7
        ANDLW 0x0F          ;mask upper nibble (WREG = 07)
        IORWF MYBCD,F       ;MYBCD = 47H, a packed BCD
```

After this conversion, the packed BCD numbers are processed and the result will be in packed BCD format. As we saw earlier in this chapter, a special instruction, "DAW", requires that the data be in packed BCD format.

Review Questions

1. For the following decimal numbers, give the packed BCD and unpacked BCD representations.
   (a) 15   (b) 99
2. Show the binary and hex formats for "76" and its BCD version.
3. Does the WREG register have BCD data after the following instruction is executed?
   ```
   MOVLW D'54'
   ```
4. 67H in BCD when converted to ASCII is ____H and ____H.
5. Does the following convert unpacked BCD in the WREG register to ASCII?
   ```
   MOVLW 0x09
   ADDLW 0x30
   ```

SUMMARY

This chapter discussed arithmetic instructions for both signed and unsigned data in the PIC. Unsigned data uses all 8 bits of the byte for data, making a range of 0 to 255 decimal. Signed data uses 7 bits for data and 1 for the sign bit, making a range of –128 to +127 decimal.

Binary coded decimal (BCD) data represents the digits 0 through 9. Both packed and unpacked BCD formats were discussed. The PIC contains special instructions for arithmetic operations on BCD data.

In coding arithmetic instructions for the PIC, special attention has to be given to the possibility of a carry or overflow condition.

This chapter defined the logic instructions AND, OR, XOR, and complement. In addition, PIC Assembly language instructions for these functions were described. Compare and skip instructions were described as well. These functions are often used for bit manipulation purposes.

The rotate and swap instructions of the PIC are used in many applications such as serial devices. This chapter also described BCD and ASCII formats and conversions.

PROBLEMS

SECTION 5.1: ARITHMETIC INSTRUCTIONS

1. Find the C, Z, and DC flags for each of the following:

(a)	MOVLW	0x3F
	ADDLW	0x45

(b)	MOVLW	0x99
	ADDLW	0x58

(c)	MOVLW	0xFF
	MOVWF	MYREG
	BSF	STATUS,C
	MOVLW	0
	ADDWFC	MYREG,F

(d)	MOVLW	0xFF
	ADDLW	0x1

(e)	MOVLW	0xFE
	MOVWF	MYREG
	BSF	STATUS,C
	MOVLW	0
	ADDWFC	MYREG,F

(f)	BCF	STATUS,C
	MOVLW	0xFF
	MOVWF	MYREG
	MOVLW	0
	ADDWFC	MYREG,F

2. Write a program to add all the digits of your ID number and save the result in a file register location. The result must be in BCD.
3. Write a program to add the following numbers and save the result in a file register location.
    0x25, 0x59, 0x65
4. Modify Problem 3 to make the result in BCD.
5. Write a program to (a) write the value 25H to file register RAM locations 20H–23H, and (b) add all these RAM locations contents together, and save the result in RAM location 60H.
6. State the steps that the SUB instruction will go through for each of the following.
    (a) 23H – 12H   (b) 43H – 53H   (c) 99 – 99
7. For Problem 6, write a program to perform each operation.
8. True or false. The "DAW" instruction works only on the WREG register.
9. Write a program to add 7F9AH to BC48H and save the result in RAM memory locations starting at 40H.
10. Write a program to subtract 7F9AH from BC48H, and save the result in RAM memory locations starting at 40H.
11. Write a program to add BCD 7795H to 9548H and save the BCD result in RAM memory locations starting at 40H.
12. Show how to perform 77×34 in the PIC18.

CHAPTER 5: ARITHMETIC, LOGIC INSTRUCTIONS, AND PROGRAMS   187

13. Show how to perform 77/3 in the PIC18.
14. True or false. The MULLW instruction works on any register of the PIC18.
15. The MULLW instruction places the result in registers _____ and_____.

SECTION 5.2: SIGNED NUMBER CONCEPTS AND ARITHMETIC
OPERATIONS

16. Show how the following are represented by the assembler:
    (a) –23   (b) +12   (c) –28
    (d) +6FH   (e) –128   (f) +127
17. The memory addresses in computers are _____ (signed, unsigned) numbers.
18. Write a program for each of the following and indicate the status of the OV flag for each:
    (a) (+15) + (–12)   (b) (–123) + (–127)
    (c) (+25H) + (+34H)   (d) (–127) + (+127)

19. Explain the difference between the C and OV flags and where each one is used.
20. When is the OV flag raised? Explain.
21. Which register holds the OV flag?
22. How do you detect the OV flag in the PIC18? How do you detect the C flag?

SECTION 5.3: LOGIC AND COMPARE INSTRUCTIONS

23. Assume that WREG = F0H. Perform the following operations. Indicate the result and the register where it is stored.
    Note: The operations are independent of each other.
    (a) ANDLW 0x45   (b) IORLW 0x90
    (c) XORLW 0x76   (d) ANDLW 0x90
    (e) XORLW 0x90   (f) IORLW 0x90
    (g) ANDLW 0xFF   (h) IORLW 0x99
    (i) XORLW 0xEE   (j) XORLW 0xAA
24. Find the contents of register WREG after each of the following instructions:
    (a) MOVLW 0x65   (b)   MOVLW 0x70
          ANDLW 0x76   IORL   0x6B
    (c) MOVLW 0x95   (d)   MOVLW 0x5D
          XORLW 0xAA   ANDLW 0x78
    (e) MOVLW 0x0C5 (f)   MOVLW 0x6A
          IORLW 0x12   XORLW 0x6E
    (g) MOVLW 0x37
          IORLW 0x26
25. True or false. In using the CPFSEQ instruction, we must use WREG as one of the registers.
26. Explain how the CPFSGT instruction works.
27. Does the compare instruction affect the flag bits of the status register?

188

28. Assume that MYREG = 85H. Indicate if it skips after compare is executed in each of the following cases:

(a)
```
MOVLW   0x90
CPFSGT  MYREG
INCF    MYREG,F
ADDLW   0x2
```

(b)
```
MOVLW   0x70
CPFSGT  MYREG
INCF    MYREG,F
ADDLW   0x2
```

(c)
```
MOVLW   0x85
CPFSEQ  MYREG
INCF    MYREG,F
ADDLW   0x2
```

(d)
```
MOVLW   0x5D
CPFSLT  MYREG
INCF    MYREG,F
ADDLW   0x2
```

29. In Problem 28, indicate the value in MYREG.

SECTION 5.4: ROTATE INSTRUCTION AND DATA SERIALIZATION

30. Find register WREG contents after each of the following is executed:

(a)
```
MOVLW 0x56
MOVWF MYREG
SWAPF MYREG,F
RRCF  MYREG,F
RRCF  MYREG,F
```

(b)
```
MOVLW 0x39
BCF   STATUS,C
MOVWF MYREG,F
RLCF  MYREG,F
RLCF  MYREG,F
```

(c)
```
BCF   STATUS,C
MOVLW 0x4D
MOVWF MYREG
SWAPF MYREG,F
RRCF  MYREG,F
RRCF  MYREG,F
RRCF  MYREG,F
```

(d)
```
BCF   STATUS,C
MOVLW 0x7A
MOVWF MYREG
SWAPF MYREG,F
RLCF  MYREG,F
RLCF  MYREG,F
```

31. Show the code to replace the SWAPF code:
    (a) using the rotate right instructions
    (b) using the rotate left instructions

32. Write a program that finds the number of zeros in an 8-bit data item.

33. Write a program that finds the position of the first high in an 8-bit data item. The data is scanned from D0 to D7. Give the result for 68H.

34. Write a program that finds the position of the first high in an 8-bit data item. The data is scanned from D7 to D0. Give the result for 68H.

35. A stepper motor uses the following sequence of binary numbers to move the motor. How would you generate them?
    ```
    1100, 0110, 0011, 1001
    ```

SECTION 5.5: BCD AND ASCII CONVERSION

36. Write a program to convert the following packed BCD numbers to ASCII. Place the ASCII codes in data RAM locations starting at 40H.
    ```
    MYBCD_1     EQU     0x76
    MYBCD_2     EQU     0x87
    ```

37. Write a program to convert the following ASCII numbers to packed BCD. Place the BCD data in RAM locations starting at 60H.

```
MYASC_1    EQU    A'8'
MYASC_2    EQU    A'7'
MYASC_3    EQU    A'9'
MYASC_4    EQU    A'2'
```

ANSWERS TO REVIEW QUESTIONS

SECTION 5.1: ARITHMETIC INSTRUCTIONS

1. WREG
2. PRODH and PRODL
3. No. It should be "MULWF fileReg,F"
4. 255 and 255.
5. True.
6. No. DAW works on WREG only.
7. WREG.
8. We cannot mix the literal value, WREG, and fileReg.
9. "ADDWF fileReg,F"
10. WREG
11. (a) WREG = 00, C = 1, and DC = 1
    (b) WREG = FF, C = 0, and DC = 0
12.
```
    43H  0100 0011                          0100 0011
  - 05H  0000 0101  2's complement    +     1111 1011
    3EH                                     0011 1110
```
13. fileReg = 95H – 4FH – 0 = 46H, WREG = 95H

SECTION 5.2: SIGNED NUMBER CONCEPTS AND ARITHMETIC OPERATIONS

1. D7
2. 16H is 00010110 in binary and its 2's complement is 1110 1010 or
   −16H = EA in hex.
3. −128 to +127
4. +9 = 00001001 and −9 = 11110111 or F7 in hex.
5. An overflow is a carry into the sign bit (D7) but the carry is a carry out of register.

SECTION 5.3: LOGIC AND COMPARE INSTRUCTIONS

1. (a) 02
   (b) FFH
   (c) FDH
2. Zero
3. One
4. All zeros
5. False
6. any value less than 0x99
7. 66H

190

SECTION 5.4: ROTATE INSTRUCTION AND DATA SERIALIZATION

1. 52H
2. 52H
3. C0H
4. No, because WREG is a SFR
5. 58H

SECTION 5.5: BCD AND ASCII CONVERSION

1. (a) 15H = 0001 0101 packed BCD, 0000 0001,0000 0101 unpacked BCD
    (b) 99H = 1001 1001 packed BCD, 0000 1001,0000 1001 unpacked BCD
2. 3736H = 00110111 00110110B
    and in BCD we have 76H = 0111 0110B
3. No. We need to write it as 54H (with the H) or 01010100B to make it BCD. The value 54 without the "H" is interpreted as 36H by the assembler.
4. 36H, 37H
5. Yes, because WREG = 39H

CHAPTER 6

BANK SWITCHING, TABLE PROCESSING, MACROS, AND MODULES

OBJECTIVES

Upon completion of this chapter, you will be able to:

>>   List all the addressing modes of the PIC18 microcontroller
>>   Contrast and compare the addressing modes
>>   Code PIC Assembly language instructions using each addressing mode
>>   Access the data RAM file register using various addressing modes
>>   Code PIC18 instructions to manipulate a look-up table
>>   Access fixed data residing in the program ROM space
>>   Discuss how to create macros and modules
>>   Discuss how to access the entire 4K of RAM space in the PIC18
>>   List the addresses for all 16 banks of the PIC18
>>   Discuss how to access all banks of the PIC18
>>   Discuss bank switching for the PIC18
>>   Code PIC18 programs for ASCII and BCD data conversion
>>   Code PIC18 programs to create and test the checksum byte
>>   List the advantages of macros and modules in programming

The CPU can access data in various ways. The data could be in a register, or in memory, or provided as an immediate value. These various ways of accessing data are called *addressing modes*. In this chapter we discuss PIC18 addressing modes in the context of some examples.

The various addressing modes of a microprocessor are determined when it is designed, and therefore cannot be changed by the programmer. The PIC18 provides a total of four distinct addressing modes. They are as follows:

1. Immediate
2. Direct
3. Register indirect
4. Indexed-ROM

In Section 6.1 we look at immediate and direct addressing modes. In Section 6.2 we cover accessing RAM data memory using the register indirect mode. Section 6.3 explains how to access fixed data and look-up tables stored in program ROM. Section 6.4 discusses the bit-addressability of the file register data RAM space. In Section 6.5 we discuss bank switching and show how to access banks other than the access bank. Checksum generation and BCD-ASCII conversions are discussed in Section 6.6. In Section 6.7, macros and modules are examined and modular programming is described.

SECTION 6.1: IMMEDIATE AND DIRECT ADDRESSING MODES

In this section, we examine first the immediate addressing mode and then the direct addressing mode.

Immediate addressing mode

In this addressing mode, the operand is a literal constant. In immediate addressing mode, as the name implies, the operand comes immediately after the opcode when the instruction is assembled. Notice that immediate data is called a *literal* in the PIC. This addressing mode can be used to load information into WREG and selected registers, but not to any file register. The immediate addressing mode is also used for arithmetic and logic instructions. Examine the following examples.

```
MOVLW 0x25          ;load 25H into WREG
SUBLW D'62'         ;subtract WREG from 62
ANDLW B'01000000'   ;AND WREG with 40H
```

We can use the EQU directive to access immediate data as shown below.

```
COUNT EQU 0x30
...      ..
MOVLW COUNT         ;WREG = 30h
```

Notice that we can also use immediate addressing mode to perform arithmetic and logic operations on WREG only. For example, "ADDLW 0x25" adds value 25 to WREG.

Direct addressing mode

As mentioned in Chapter 2, the 256-byte access bank file register is split into two sections: The lower addresses, 00 to 7FH, are assigned to the general purpose registers, and the upper addresses, F80–FFFH, to the SFR. The access bank is the default bank when the PIC18 is powered up. It is the minimum bank that all PIC18 processors have. The MOVFF instruction also plays a role in choosing the access bank. We will discuss that issue in Section 6.5 when we discuss bank switching.

The entire data RAM file register can be accessed using either direct or register indirect addressing modes. The register indirect addressing mode will be discussed in the next section. In direct addressing mode, the operand data is in a RAM memory location whose address is known, and this address is given as a part of the instruction. Contrast this with immediate addressing mode in which the operand data itself is provided with the instruction. While the letter "L" in the instruction means literal (immediate), the letter "F" in the instruction signifies the address of the file register location. See the example below, and note the letter F in the instructions.

```
MOVLW 0x56        ;WREG = 56H (immediate addressing mode)
MOVWF 0x40        ;copy WREG into fileReg RAM location 40H
MOVFF 0x40,0x50   ;copy data from loc 40H to 50H.
```

The last two instructions use direct addressing mode. If we dissect the opcode we see that the addresses are embedded in the instruction, as shown in Figure 6-1a.

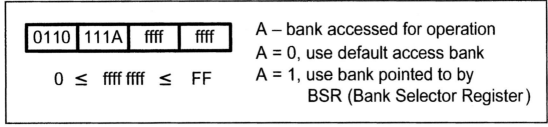

Figure 6-1a. MOVFF Direct Addressing Opcode

As shown in Figure 6-1b, the address field is an 8-bit address and can take values from 00–FFH. The A bit for bank switching is discussed in Section 6-4. Of course, it is much easier to use names instead of addresses in the program, and we have seen many examples of them in the last few chapters. It must be noted that file register data RAM does not support immediate addressing mode. In other words, to move data into any file register, we must first move it to WREG, and then move it from WREG to the file register using the MOVWF instruction.

```
┌──────┬──────┬──────┬──────┐
│ 0110 │ 111A │ ffff │ ffff │
└──────┴──────┴──────┴──────┘

   0  ≤  ffff ffff  ≤  FF
```

A – bank accessed for operation
A = 0, use default access bank
A = 1, use bank pointed to by
BSR (Bank Selector Register)

Figure 6-1b. MOVWF Direct Addressing Opcode

The difference between "INCF fileReg, W" and "INCF fileReg, F"

In direct addressing mode, when an operation is performed on a file register, we have the option of saving the result in the file register itself or in WREG. This option is a major source of errors in PIC programming and its correct use must be emphasized. The following code increments the contents of register file location 20H using the direct addressing mode, but the destination for the increment result is decided by the W or F parameter:

```
MOVLW  0          ;WREG = 0
MOVWF  0x20       ;loc 0x20 = (0), WREG = 0
INCF   0x20,W     ;loc 0x20 = (0), WREG = 1
INCF   0x20,W     ;loc 0x20 = (0), WREG = 1
INCF   0x20,W     ;loc 0x20 = (0), WREG = 1
INCF   0x20,F     ;loc 0x20 = (1), WREG = 1
INCF   0x20,F     ;loc 0x20 = (2), WREG = 1
INCF   0x20       ;loc 0x20 = (3), WREG = 1
INCF   0x20       ;loc 0x20 = (4), WREG = 1
INCF   0x20,W     ;loc 0x20 = (4), WREG = 5
```

Notice in the above code when the second parameter is not stated, it is assumed to be fileReg (F).

DECFSZ and DECF

Other instructions that need to be examined are DECFSZ and DECF. We can use either one for looping. In "DECFSZ fileReg, d" the fileReg is decremented, and the next instruction is skipped if the fileReg is zero. DECF does not skip the next instruction. Contrast the two codes complementing PORTB 5 times. The following code uses the DECFSZ instruction for looping:

```
      CLRF   TRISB      ;Port B as output
      MOVLW  5          ;WREG = 5
      MOVWF  MYREG      ;counter = 5
      CLRF   PORTB      ;clear Port B
B1    COMF   PORTB      ;complement Port B
      DECFSZ MYREG,F    ;decrement and skip if MYREG = 0
      GOTO   B1         ;go back since it is not zero
      SETF   PORTB      ;make PB = FFH
```

while the code below uses the BNZ (branch not zero) instruction:

```
      CLRF   TRISB      ;Port B as output
      MOVLW  5          ;WREG = 5
      MOVWF  MYREG      ;counter = 5
      CLRF   PORTB      ;clear Port B
B2    COMF   PORTB      ;complement Port B
      DECF   MYREG,F    ;decrement counter
      BNZ    B2         ;go back if MYREG is not zero
      SETF   PORTB      ;make PB = FFH
```

Notice that if we use "DECF MYREG, W" instead of "DECF MYREG, F" in the BNZ program, we will never get out of the loop because the values of MYREG

Table 6-1: Selected PIC18 Special Function Register (SFR) Addresses

Symbol	Name	Address
WREG	Working register	FE8H
PORTA	Port A	F80H
PORTB	Port B	F81H
PORTC	Port C	F82H
LATA	Output latch, Port A	F89H
LATB	Output latch, Port B	F8AH
LATC	Output latch, Port C	F8BH
TRISA	Data direction, Port A	F92H
TRISB	Data direction, Port B	F93H
TRISC	Data direction, Port C	F94H
INDF0	Indirect addressing register 0	FEFH
INDF1	Indirect addressing register 1	FE7H
FSR0L	Indirect data memory address pointer 0 low	FE9H
FSR0H	Indirect data memory address pointer 0 high	FEAH
FSR1L	Indirect data memory address pointer 1 low	FE1H
FSR1H	Indirect data memory address pointer 1 high	FE2H
PLUSW0	Indirect indexed address register	FEBH
PREINC0	Preincrement register 0	FECH
POSTDEC0	Post-decrement register 0	FEDH
POSTINC0	Post-increment register 0	FEEH
TBLPTRL	Table pointer, low byte	FF6H
TBLPTRH	Table pointer, high byte	FF7H
TBLPTRU	Table pointer, upper byte	FF8H
TABLAT	Program memory table latch	FF5H
STATUS	Status flag byte	FD8H

and WREG will remain 5 and 4 respectively forever.

SFR registers and their addresses

PIC18 registers for Ports A, B, and so on are part of the group of registers commonly referred to as SFRs (special function registers). There are many special function registers and they are widely used, as we will discuss in future chapters. The SFRs can be accessed by their names (which is much easier) or by their addresses. For example, Port B has address F81H, and Port C the address F82H, as shown in Table 6-1. Notice how the following pairs of instructions mean the same thing:

```
MOVWF 0xF81      ;is the same as
MOVWF PORTB      ;which means copy WREG into Port B

CLRF 0xF82       ;is the same as
CLRF PORTC       ;which means clear Port C

BSF 0xFD8,0      ;is the same as
BSF STATUS,C     ;which make C = 1
```

Table 6-1 lists selected PIC18 special function registers (SFRs) and their addresses. The following two points should be noted about SFR addresses:

1. The special function registers have addresses between F80H and FFFH. These addresses are below FFFH, because the PIC18 starts assigning SFR addresses at FFFH and goes down until all SFRs supported by that chip are assigned. Not all the members of the PIC18 family have the same peripherals; therefore, the number of locations used for SFRs varies among the PIC18 family.
2. Not all the address space of F80H to FFFH is used by the SFR. The unused locations F80H to FFFH are reserved and must not be used by the PIC18 programmer.

Example 6-1

Write code to send 55H to Port B. Include

(a) The register names.
(b) Their addresses.

Solution:

(a)
```
CLRF   TRISB        ;Port B output
MOVLW  0x55         ;WREG = 55H
MOVWF  PORTB        ;Port B = 55H
```

(b)   From Table 6-1, TRISB address = F93H and PORTB address = F81H

```
CLRF   0xF93        ;Port B output
MOVLW  0x55H        ;WREG = 55H
MOVWF  0xF81        ;Port B = 55H
```

Regarding direct addressing mode in the PIC18, notice the following points:

1. If you examine the .lst file for an Assembly language program, you will see that the SFR register names are replaced with their addresses as listed in Table 6-1.
2. The WREG register is one of the SFR registers and has address FE8H.
3. The direct addressing mode is also called *register direct* to contrast it with the register indirect addressing mode discussed in the next section.

Review Problems

1. Can the programmer of a microcontroller make up new addressing modes?
2. Show the instruction to load 1000 0000 (binary) into register WREG.
3. Why is "MOVLF myvalue, fileReg" invalid?
4. True or false. In PIC18, the PC (program counter) is part of the SFR.
5. True or false. In PIC18, the WREG register is not part of the SFR.

SECTION 6.2: REGISTER INDIRECT ADDRESSING MODE

We can use register direct or register indirect addressing modes to access data stored in the general purpose RAM section of the file register. In the last section we showed how to use direct addressing mode, which is also called register direct. The register indirect addressing mode is a very important addressing mode in the PIC18. This topic will be discussed thoroughly in this section.

Register indirect addressing mode

In the register indirect addressing mode, a register is used as a pointer to the data RAM location. In the PIC18, three registers are used for this purpose: FSR0, FSR1, and FSR2. FSR stands for *file select register* and must not be confused with SFR (special function register). The FSR is a 12-bit register allowing access to the entire 4096 bytes of data RAM space in the PIC18. We use LFSR (load FSR) to load the RAM address. In other words, when FSRx are used as pointers, they must be loaded first with the RAM addresses as shown below.

```
LFSR 0, 0x30       ;load FSR0 with 0x30
LFSR 1, 0x40       ;load FSR1 with 0x40
LFSR 2, 0x6F       ;load FSR2 with 0x6F
```

Because FSR0, FSR1, and FSR2 are 12-bit registers they cannot fit into the SFR address space unless they are split into pieces of an 8-bit size. That is exactly what PIC18 has done. The FSR registers have the low-byte and high-byte parts called FSRxL and FSRxH, as shown in the SFR table of Table 6-1. In Table 6-1 we see FSR0L and FSR0H, representing the low and high parts of the 12-bit FSR0 register. Note that the FSRxH is only 4-bit and the upper 4 bits are not used. Another register associated with the register indirect addressing mode is the INDF (indirect register). Each of the FSR0, FSR1, and FSR2 registers has an INDF register associated with it, and these are called INDF0, INDF1, and INDF2. When we move data into INDFx we are moving data into a RAM location pointed to by the FSR. In the same way, when we read data from the INDF register, we are reading data from a RAM location pointed to by the FSR. This is shown below.

```
LFSR  0, 0x30      ;FSR0 = 30H RAM location pointer
MOVWF INDF0        ;copy contents of WREG into RAM
                   ;location whose address is held by
                   ;12-bit FSR0 register
```

Advantages of register indirect addressing mode

One of the advantages of register indirect addressing mode is that it makes accessing data dynamic rather than static, as with direct addressing mode. Example 6-2 shows three cases of copying 55H into RAM locations 40H to 45H. Notice in solution (b) that two instructions are repeated numerous times. We can create a loop with those two instructions as shown in solution (c). Solution (c) is the most efficient and is possible only because of the register indirect addressing mode. In Example 6-2, we must use "INCF FSR0L, F" to increment the pointer

---

CHAPTER 6: BANK SWITCHING, TABLES, MACROS, AND MODULES   **199**

because there is no such instruction as "INCF FSR0, F". Looping is not possible in direct addressing mode, and that is the main difference between the direct and register indirect addressing modes. For example, trying to send a string of data located in consecutive locations of data RAM is much more efficient and dynamic using register indirect addressing mode than using direct addressing mode. See Example 6-3.

Example 6-2

Write a program to copy the value 55H into RAM memory locations 40H to 45H using
(a) Direct addressing mode.
(b) Register indirect addressing mode without a loop.
(c) A loop.

Solution:

(a)
```
        MOVLW   0x55        ;load WREG with value 55H
        MOVWF   0x40        ;copy WREG to RAM location 40H
        MOVWF   0x41        ;copy WREG to RAM location 41H
        MOVWF   0x42        ;copy WREG to RAM location 42H
        MOVWF   0x43        ;copy WREG to RAM location 43H
        MOVWF   0x44        ;copy WREG to RAM location 44H
```

(b)
```
        MOVLW   55H         ;load with value 55H
        LFSR    0,0x40      ;load the pointer. FSR0 = 40H
        MOVWF   INDF0       ;copy W to RAM loc FSR0 points to
        INCF    FSR0L,F     ;increment pointer. Now FSR0 = 41H
        MOVWF   INDF0       ;copy W to RAM loc FSR0 points to
        INCF    FSR0L,F     ;increment pointer. Now FSR0 = 42H
        MOVWF   INDF0       ;copy W to RAM loc FSR0 points to
        INCF    FSR0L,F     ;increment pointer. Now FSR0 = 43H
        MOVWF   INDF0       ;copy W to RAM loc FSR0 points to
        INCF    FSR0L,F     ;increment pointer. Now FSR0 = 44H
        MOVWF   INDF0       ;copy W to RAM loc FSR0 points to
```

(c)
```
        COUNT   EQU 0x10    ;location 10H for counter
        MOVLW   0x5         ;WREG = 5
        MOVWF   COUNT       ;load the counter, Count = 5
        LFSR    0,0x40      ;load pointer. FSR0 = 40H, RAM address
        MOVLW   0x55        ;WREG = 55h value to be copied
B1      MOVWF   INDF0       ;copy WREG to RAM loc SFR0 points to
        INCF    FSR0L,F     ;increment FSR0 pointer
        DECF    COUNT,F     ;decrement the counter
        BNZ     B1          ;loop until counter = zero
```

Use the MPLAB simulator to examine RAM contents after the above program is run.
```
        40 = (55)
        41 = (55)
        42 = (55)
        43 = (55)
        44 = (55)
```

Example 6-3

Assume that RAM locations 30–34H have a string of ASCII data, as shown below. Write a program to get each character and send it to Port B one byte at a time. Show the program using:

(a) Direct addressing mode.
(b) Register indirect addressing mode.

```
30 = ('H')
31 = ('E')
32 = ('L')
33 = ('L')
34 = ('O')
```

Solution:

(a) Using direct addressing mode

```
        CLRF   TRISB          ;make Port B an output
        MOVFF 0x30, PORTB      ;copy contents of loc 0x30 to PB
        MOVFF 0x31, PORTB
        MOVFF 0x32, PORTB
        MOVFF 0x33, PORTB
        MOVFF 0x34, PORTB
```

(b) Using register indirect mode

```
        COUNTREG EQU 0x20      ;fileReg loc for counter
        CNTVAL EQU 5           ;counter value
        CLRF   TRISB           ;make Port B an output (TRSI = 0 = out)
        MOVLW CNTVAL           ;WREG = 5
        MOVWF COUNTREG         ;load the counter, Count = 5
        LFSR   2,0x30          ;load pointer. FSR2 = 30H, RAM address
B3      MOVF   INDF2,W         ;copy RAM loc FSR2 points at to WREG
        MOVWF PORTB            ;copy WREG to PORTB
        INCF   FSR2L           ;increment FSR2 to point at next loc
        DECF   COUNTREG,F      ;decrement counter
        BNZ    B3              ;loop until counter = zero
```

When simulating the above program on the MPLAB, make sure that RAM locations 30H–34H have the message "HELLO". Notice that "MOVF INDF2, W" moves data from INDF2 into WREG.

When using the MPLAB simulator with examples in this chapter, you may have noticed that you cannot view INDF0, POSTDEC0, or PLUSW0 in the watch window. This is because these registers are not physically implemented memory locations. Accessing these registers indicates indirect addressing. See Figure 2-4 in Chapter 2 for the registers that are not physically implemented.

Auto-increment option for FSR

Because the FSR is a 12-bit register, it can go from 000 to FFFH, which covers the entire 4K RAM space of the PIC18. Using the "INCF FSR0L, F" instruction to increment the pointer can cause a problem when an address such as 5FFH is incremented. The instruction "INCF FSR0L, F" will not propagate the carry into the FSR1H register. The PIC18 gives us the options of auto-increment and auto-decrement for FSRn to overcome this problem. The syntax used for such cases for the CLRF instruction is shown in Table 6-2.

Table 6-2: PIC18 Auto-Increment/Decrement of FSRn for CLRF Instruction

Instruction	Function
CLRF INDFn	After clearing fileReg pointed to by FSRn, the FSRn stays the same.
CLRF POSTINCn	After clearing fileReg pointed to by FSRn, the FSRn is incremented.
CLRF PREINCn	The FSRn is incremented, then fileReg pointed to by FSRn is cleared.
CLRF POSTDECn	After clearing fileReg pointed to by FSRn, the FSRn is decremented.
CLRF PLUSWn	Clears fileReg pointed to by (FSRn +WREG), FSRn & W unchanged.

Note: This table shows the syntax for the CLRF instruction, it works for all such instructions. The auto-decrement or auto-increment affects the entire 12 bits of the FSRn and has no effect on status register. This means that FSR0 going from FFF to 000 will not raise any flag. The option of PLUSWn is widely used for a RAM-based look-up table. See Section 6.4.

Example 6-4

Write a program to clear 16 RAM locations starting at RAM address 60H.
Use the following:
(a) INCF FSRnL
(b) Auto-increment

Solution:

```
(a)
COUNTREG EQU 0x10        ;fileReg loc for counter
CNTVAL EQU D'16'         ;counter value
      MOVLW   CNTVAL     ;WREG = 16
      MOVWF   COUNTREG   ;load the counter, Count = 16
      LFSR    1,0x60     ;load pointer. FSR1 = 40H, RAM address
B2    CLRF    INDF1      ;clear RAM loc FSR1 points to
      INCF    FSR1L,F    ;increment FSR1L, point to next loc
      DECF    COUNTREG,F ;decrement counter
      BNZ     B2         ;loop until counter = zero


(b)
COUNTREG EQU 0x10        ;fileReg loc for counter
CNTVAL EQU D'16'         ;counter value
      MOVLW   CNTVAL     ;WREG = 16
      MOVWF   COUNTREG   ;load the counter, Count = 16
      LFSR    1,0x60     ;load pointer. SFR0 = 40H, RAM address
B3    CLRF    POSTINC1   ;clear RAM, increment FSR1 pointer
      DECF    COUNTREG,F ;decrement counter
      BNZ     B3         ;loop until counter = zero
```

Example 6-5

Write a program to copy a block of 5 bytes of data from RAM locations starting at 30H to RAM locations starting at 60H.

Solution:

```
        COUNTREG EQU 0x10  ;fileReg loc for counter
        CNTVAL EQU D'5'    ;counter value
        MOVLW  CNTVAL      ;WREG = 10
        MOVWF  COUNTREG    ;load the counter, count = 10
        LFSR   0,0x30      ;load pointer. FSR0 = 30H, RAM address
        LFSR   1,0x60      ;load pointer. FSR1 = 60H, RAM address
B3      MOVF   POSTINC0,W  ;copy RAM to WREG and increment FSR0
        MOVWF  POSTINC1    ;copy WREG to RAM and increment FSR1
        DECF   COUNTREG,F  ;decrement counter
        BNZ    B3          ;loop until counter = zero
```

Before we run the above program.

```
    30 = ('H')   31 = ('E')   32 = ('L')   33 = ('L')   34 = ('O')
```

After the program is run, the addresses 60–64H have the same data as 30–34H.

```
    30 = ('H')   31 = ('E')   32 = ('L')   33 = ('L')   34 = ('O')
    60 = ('H')   61 = ('E')   62 = ('L')   63 = ('L')   64 = ('O')
```

Example 6-6

Assume that RAM locations 40–43H have the following hex data. Write a program to add them together and place the result in locations 0x06 and 0x07.

$$40 = (7D) \qquad 41 = (EB) \qquad 42 = (C5) \qquad 43 = (5B)$$

Solution:

```
        COUNTREG EQU 0x20         ;fileReg loc for counter
        L_BYTE EQU 0x06           ;fileReg loc for L_Byte
        H_BYTE EQU 0x07           ;fileReg loc for L_Byte
        CNTVAL EQU 4              ;counter value
        MOVLW  CNTVAL             ;WREG = 4
        MOVWF  COUNTREG           ;load the counter
        LFSR   0,0x40             ;load pointer. FSR0 = 40H, RAM address
        CLRF   WREG               ;clear WREG
        CLRF   H_BYTE             ;clear H_BYTE
B5      ADDWF  POSTINC0, W        ;add RAM to WREG and increment FSR0
        BNC    OVER               ;if C = 0, go to next
        INCF   H_BYTE,F           ;C = 1, add 1 to high byte
OVER    DECF   COUNTREG,F         ;decrement counter
        BNZ    B5                 ;loop until counter = zero
        MOVWF  L_BYTE
```

The above is a register indirect version of Example 5-2 in Chapter 5 with a loop. Contrast them to see the difference.

To see an example of how to use all three FSRn registers, study and simulate Example 6-7.

Example 6-7

Write a program to add the following multibyte BCD numbers and save the result.

      12896577H
+   23647839H

Solution:

```
        COUNTREG EQU 0x20 ;fileReg loc for counter
        CNTVAL EQU D'4'    ;counter value
        MOVLW CNTVAL       ;WREG = 4
        MOVWF COUNTREG     ;load the counter. Count = 4
        LFSR   0,0x30      ;load pointer. FSR0 = 30H, RAM address
        LFSR   1,0x50      ;load pointer. FSR1 = 50H, RAM address
        LFSR   2,0x60      ;load pointer. FSR2 = 60H, RAM address
        BCF    STATUS,C    ;clear carry flag for the LSB
B3      MOVF   POSTINC0,W  ;copy RAM to WREG and INC FSR0
        ADDWFC POSTINC1,W  ;add RAM to WREG and INC FSR1
        DAW                ;decimal adjust WREG
        MOVWF  POSTINC2    ;copy WREG to RAM and INC FSR2
        DECF   COUNTREG,F  ;decrement counter
        BNZ    B3          ;loop until counter = zero
```

Before the addition we have:

MSByte			**LSByte**
33 = (12)	32 = (89)	31 = (65)	30 = (77)
53 = (23)	52 = (64)	51 = (78)	50 = (39)

After the addition we have:

        63 = (36)   62 = (54)   61 = (44)   60 = (16)

Notice that we are using the little endian convention of storing a low byte to a low address and a high byte to a high address. Single-step the program in MPLAB and examine the FSRx and memory contents to gain an insight into register indirect addressing mode.

Review Questions

1. The instruction "MOVWF 0x40" uses _____ addressing mode. Why?
2. What address is assigned to register FSR0L?
3. What address is assigned to register FSR0H?
4. The FSRn is a(n) _____-bit register.
5. Which registers are allowed to be used for register indirect addressing mode if the data is in the data RAM file register?

SECTION 6.3: LOOK-UP TABLE AND TABLE PROCESSING

So far, we have seen that the PIC18 has a maximum of 2M of code (program) space and 4K of data RAM space. While we never use any of the data RAM space for storing code, we can use the code space to store fixed data. In this section we discuss how to access fixed data residing in the program ROM space of the PIC18. First we examine how to store fixed data in the program ROM space using the DB (define byte) directive.

DB (define byte) and fixed data in program ROM

The DB data directive is widely used to allocate ROM program (code) memory in byte-sized chunks. In other words, DB is used to define an 8-bit fixed data. When DB is used to define fixed data, the numbers can be in decimal, binary, hex, or ASCII formats. The DB directive is used to define ASCII strings. See Example 6-8. In Example 6-8, notice that we must use single quotes (') for a single character or double quotes (") for a string.

Example 6-8

Assume that we have burned the following fixed data into program ROM of a PIC chip. Give the contents of each ROM location starting at 500H. See Appendix F for the hex values of the ASCII characters.

```
;MY DATA IN ROM
        ORG 500H            ;notice it must be an even address
DATA1 DB D'28'              ;DECIMAL(1C in hex)
DATA2 DB B'00110101'        ;BINARY (35 in hex)
DATA3 DB 0x39               ;HEX

        ORG 510H            ;notice it must be an even address
DATA4 DB 'Y'                ;single ASCII char
DATA5 DB '2','0','0','5'    ;ASCII numbers

        ORG 518H            ;notice it must be an even address
DATA6 DB "Hello ALI"        ;ASCII string
        END
```

Solution:

```
DATA1        DATA2        DATA3
500 = (1C)   501 = (35)   502 = (39)


DATA4        DATA5
510 = (59)  511 = (32)  512 = (30)  513 = (30)  514 = (35)
       Y           2           0           0           5

DATA6
518 = (48)  519 = (65)  51A = (6C)  51B = (6C)  51C = (6F)
       H           e           l           l           o
51D = (20)  51E = (41)  51F = (4C)  520 = (49)
     SPACE         A           L           I
```

MPASM also allows the use of DATA in place of DB to define values greater than 255 (0xFF) but not larger than 65535 (0xFFFF).

Reading table elements in the PIC18

Example 6-8 showed how to place fixed data into program ROM. The only problem is that the 2M of program (code) space is under the direct control of the program counter register. This means that we need to have a special function register to point to the data to be fetched from the code space. For this reason we can call it register indirect ROM addressing mode. This is an addressing mode widely used to access data elements located in the program ROM space of the PIC18. This is often called *table processing*.

There is a group of instructions in the PIC18 designed for table processing. These can be used for both table read and table write. We discuss table read first because it is the most widely used. Table 6-3 shows the instructions for table read of the PIC18. To read the fixed data byte, we need an address pointer that points to the data and a register to store the data when it is brought into the CPU. TBLP-TR is a 21-bit register and is used to point to the byte to be fetched. With the 21-bit register TBLPTR, we can cover the entire 2M program (code) space for the PIC18. The only problem is that we do not have an instruction to load the 21-bit address into TBLPTR.

TBLPTR is divided into three 8-bit parts. These are called TBLPTRL (low), TBLPTRH (high), and TBLPTRU (upper), and all are part of the SFRs. Notice that the last 2 bits of TBLPTRU (upper) are not used and are cleared to 0s.

The other SFR register used for the table processing is TABLAT. The TABLAT (TABle LATch) register is used for keeping the byte once it is fetched into the CPU. See Example 6-9. In the next few examples, we load only the TBLPTRL and TBLPTRH registers because the table elements are stored in the first 64K of the PIC18 address space (0000–FFFFH). You must load TBLPTRU as well if the data is residing on ROM addresses of 10000H and beyond.

Auto-increment option for TBLPTR

Because the TBLPTR is 21-bit register it can cover from 000000 to 1FFFFFH, which is the 2M ROM space of the PIC18. Using the "INCF TBLPTRL, F" instruction to increment the pointer can cause a problem when an address such as 5FFH is incremented. The carry will not propagate into TBLPTRH. The PIC18 gives us the options of TBLRD*+ (table read and increment), TBLRD*- (table read and decrement), and so on, as shown in Table 6-3. See Examples 6-10, 6-11, and 6-12.

Table 6-3: PIC18 Table Read Instructions

Instruction	Function	Description
TBLRD*	Table read	After read, TBLPTR stays the same
TBLRD*+	Table read with post-inc.	Reads and increments TBLPTR
TBLRD*-	Table read with post-dec.	Reads and decrements TBLPTR
TBLRD+*	Table read with pre-inc.	Increments TBLPTR and then reads

Note: The byte of data is read into the TABLATch register from code space pointed to by TBLPTR.

Example 6-9

In this program, assume that the word "USA" is burned into ROM locations starting at 500H, and that the program is burned into ROM locations starting at 0. Analyze how the program works and state where "USA" is stored after this program is run.

Solution:

```
            ORG    0000H           ;burn into ROM starting at 0
            CLRF   TRISB           ;make PB an output
            MOVLW  0x0             ;WREG = 0 look-up table low-byte addr
            MOVWF  TBLPTRL         ;look-up table low-byte addr
            MOVLW  0x05            ;WREG = 5 look-up table high-byte addr
            MOVWF  TBLPTRH         ;look-up table high-byte addr
            TBLRD*                 ;TABLAT = 'U' char pointed to by TABPTR
            MOVFF  TABLAT,PORTB    ;send it to Port B
            INCF   TBLPTRL,F       ;TBLPTRL = 01 pointing to next (501)
            TBLRD*                 ;TABLAT = 'S' char pointed to by TBLPTR
            MOVFF  TABLAT,PORTB    ;send it to Port B
            INCF   TBLPTRL,F       ;TBLPTRL = 02 pointing to next (502)
            TBLRD*                 ;TABLAT = 'A' char pointed to by TBLPTR
            MOVFF  TABLAT,PORTB    ;send it to Port B
HERE    GOTO HERE                  ;stay here forever

;data is burned into code(program) space starting at 500H
            ORG    500H
MYDATA DB   "USA"
            END                    ;end of program
```

In the above program ROM locations 500H–502H have the following contents.

```
500 = ('U') 501 = ('S') 502 = ('A')
```

We start with TBLPTR = 500H (TBLPTRH = 05 and TBLPTRL = 0). The instruction "TBLRD*" moves the contents of ROM location 500H to TABLAT. Register TABLAT contains 55H, the ASCII value for 'U'. This is moved to Port B. Next, TBLPTRL is incremented to make TBLPTR = 501H. The TBLRD instruction will get the contents of the next ROM location 501H, which holds character 'S'. After this program is run, we send the ASCII values for the characters 'U', 'S', and 'A' to Port B one character at a time. The loop version of this program is given in the next example.

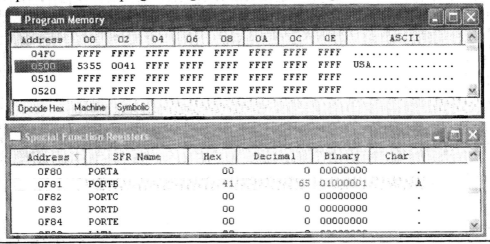

Example 6-10

Assuming that program ROM space starting at 250H contains "USA", write a program to send all the characters to Port B one byte at a time.

Solution:

(a) This method uses a counter

```
RCOUNT EQU 0x20                 ;counter loc in fileReg
CNTVAL EQU 0x3                  ;counter value
       ORG   0000H              ;burn into ROM starting at 0
       MOVLW 0x50               ;WREG = 50, low-byte addr
       MOVWF TBLPTRL            ;look-up table low-byte addr
       MOVLW 0x02               ;WREG = 2, high-byte addr
       MOVWF TBLPTRH            ;look-up table high-byte addr
       MOVLW CNTVAL             ;WREG = 03, counter value
       MOVWF RCOUNT             ;load counter
       CLRF  TRISB              ;TRSIB = 00 (Port B as output)
B6     TBLRD*                   ;read table byte pointed to by TBLPTR
       MOVFF TABLAT,PORTB       ;send it to Port B
       INCF  TBLPTRL,F          ;increment to point to next char
       DECF  RCOUNT,F           ;dec the counter
       BNZ   B6                 ;repeat if counter not zero
HERE   GOTO  HERE               ;stay here

;data is burned into code(program) space starting at 250H
       ORG   0x250
MYDATA DB    "USA"
       END
```

(b) This method uses null char for end of string

```
       ORG   0000H              ;burn into ROM starting at 0
       MOVLW 0x50               ;WREG = 50, low-byte addr
       MOVWF TBLPTRL            ;look-up table low-byte addr
       MOVLW 0x02               ;WREG = 2, high-byte addr
       MOVWF TBLPTRH            ;look-up table high-byte addr
       CLRF  TRISB              ;TRSIB = 00 (Port B as output)
B7     TBLRD*                   ;bring in next byte
       MOVF  TABLAT,W           ;copy to WREG (Z = 1, if null)
       BZ    EXIT               ;is it null char? exit if yes
       MOVWF PORTB              ;send it to Port B
       INCF  TBLPTRL,F          ;increment pointing to next
       BRA   B7                 ;continue
EXIT   GOTO  EXIT

       ORG   0x250
MYDATA DB    "USA",0            ;notice null
       END
```

Example 6-11

Repeat Example 6-10, using auto-increment.

Solution:

```
        ORG   0000H      ;burn into ROM starting at 0
        MOVLW 0x50       ;WREG = 50 low-byte addr
        MOVWF TBLPTRL    ;look-up table low-byte addr
        MOVLW 0x02       ;WREG = 2, high-byte addr
        MOVWF TBLPTRH    ;look-up table high-byte addr
        CLRF  TRISB      ;TRSIB = 00 (Port B as output)
B7      TBLRD*+          ;bring in next byte and inc TBLPTR
        MOVF  TABLAT,W   ;copy to WREG (Z = 1, if null)
        BZ    EXIT       ;is it null char? exit if yes
        MOVWF PORTB      ;send it to Port B
        BRA   B7         ;continue
EXIT    GOTO  EXIT


        ORG   0x250
MYDATA  DB    "USA",0    ;notice null
        END
```

Example 6-12

Assume that ROM space starting at 500H contains the message "The Promise of World Peace". Write a program to bring it into CPU one byte at a time and place the bytes in RAM locations starting at 40H.

Solution:

```
        ORG   0000H      ;burn into ROM starting at 0
        MOVLW 0x00       ;WREG = 00 low-byte addr
        MOVWF TBLPTRL    ;look-up table low-byte addr
        MOVLW 0x05       ;WREG = 05, high-byte addr
        MOVWF TBLPTRH    ;look-up table high-byte addr
        LFSR  2,0x40     ;load pointer. FSR2 = 40H, RAM address
B8      TBLRD*+          ;read the table, then increment TBLPTR
        MOVF  TABLAT,W   ;copy to WREG (Z = 1 if null)
        BZ    EXIT       ;exit if end of string
        MOVWF POSTINC2   ;copy WREG to RAM and INC FSR2
        BRA   B8
EXIT    GOTO  EXIT

;--------------------message
        ORG   0x500      ;data burned starting at 0x500
MYDATA  DB "The Promise of World Peace",0
        END
```

Look-up table and RETLW instruction

The look-up table is a widely used concept in microcontroller programming. It allows access to elements of a frequently used table with minimum operations. As an example, assume that for a certain application we need x^2 values in the range of 0 to 9. We can use a look-up table instead of calculating the values, which takes some time. In the PIC, to get the table element we first call the look-up table, then we add a fixed value to the PCL (low-byte portion of the program counter) to index into the look-up table. Upon return from the table, the RETLW instruction will provide the desired look-up table element in the WREG register. This is shown in Examples 6-13 and 6-14.

Example 6-13

Assume that the lower three bits of Port C are connected to three switches. Write a program to send the following ASCII characters to Port D based on the status of the switches.

000	'0'
001	'1'
010	'2'
011	'3'
100	'4'
101	'5'
110	'6'
111	'7'

Solution:

```
        ORG     0
        SETF    TRISC           ;TRISC = FFh (Port C as input)
        CLRF    TRISD           ;TRISD = 00 (Port D as output)
B1      MOVF    PORTC,W         ;read x from Port C into WREG
        ANDLW   B'00000111'     ;mask upper 5 bits
        CALL    ASCI_TABLE      ;get ASCII from look-up table
        MOVWF   PORTD           ;copy it to Port D
        BRA     B1              ;continue

;look-up table for ASCII numbers 0-7
ASCI_TABLE
        MULLW   0x2     ;align it for even address for 2-byte RETLW opcode
        MOVFF   PRODL, WREG ;put it into WREG for indexing
        ADDWF   PCL             ;PCL = PCL + WREG
        RETLW   '0'             ;ASCII for 0
        RETLW   '1'             ;ASCII for 1
        RETLW   '2'             ;ASCII for 2
        RETLW   '3'             ;notice that each ASCII value is placed
        RETLW   '4'             ;in the ROM at an even address
        RETLW   '5'
        RETLW   '6'
        RETLW   '7'
        END
```

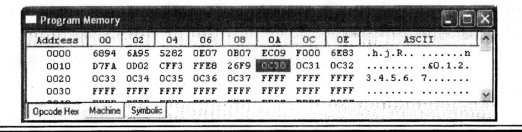

Example 6-14

Write a program to get the *x* value from Port B and send x^2 to Port C. Assume that RB3–RB0 has the *x* value of 0–9. Use a look-up table instead of a multiply instruction.

What is the value of Port C if we have 9 at Port B?

Solution:

```
        ORG   0
        SETF  TRISB        ;TRISB = FFh (Port B as input)
        CLRF  TRISC        ;TRISC = 00 (Port C as output)
B1      MOVF  PORTB,W      ;read x from Port B into WREG
        ANDLW 0x0F         ;mask upper bits
        CALL  XSQR_TABLE   ;get x² from the look-up table
        MOVWF PORTC        ;copy it to Port C
        BRA   B1           ;continue

;look-up table for square of numbers 0-9
XSQR_TABLE
        MULLW 0x2          ;align it for even address
        MOVFF PRODL, WREG  ;put it into WREG for indexing
        ADDWF PCL          ;PCL = PCL + WREG
        RETLW D'0'         ;square of 0
        RETLW D'1'         ;square of 1
        RETLW D'4'         ;square of 2
        RETLW D'9'         ;square of 3
        RETLW D'16'        ;square of 4  (10 hex)
        RETLW D'25'        ;square of 5  (19 hex)
        RETLW D'36'        ;square of 6  (24 hex)
        RETLW D'49'        ;square of 7  (31 hex)
        RETLW D'64'        ;square of 8  (40 hex)
        RETLW D'81'        ;square of 9  (51 hex)
        END
```

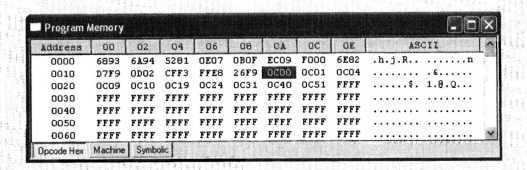

From the screen shot above, notice that location 001A has the "RETLW D'0'" opcode and operand, the square of 0. Location 001C has 01, the square of 1. Location 001E has 04, the square of 2. Location 0020 has 09, the square of 3. Location 0022 has 10, the square of 4 ($4 \times 4 = 16 = 10H$) and so on. Notice that the odd addresses have the opcode for RETLW, which is 0C hex. If we have 9 at Port B, then Port C will have 51H, which is the hex value of decimal 81 ($9^2 = 81$).

Accessing a look-up table in RAM

The look-up table elements can also be in RAM instead of ROM. Sometimes we need to bring in the elements of the look-up table from RAM because the elements are dynamic and can change. The PIC18 allows us to do that using the FSR as pointer. For example, the instruction "MOVFF PLUSW2, PORTD" will bring elements of the look-up table pointed to by the address location formed by the addition of FSR2 + WREG. In this case, WREG is used as an index into the look-up table. See Examples 6-15 and 6-16.

Example 6-15

Repeat Example 6-13 assuming that the look-up table elements are in data RAM locations starting at address 20H as shown below.

$$20 = (\text{'0'})$$
$$21 = (\text{'1'})$$
$$22 = (\text{'2'})$$
$$23 = (\text{'3'})$$
$$24 = (\text{'4'})$$
$$25 = (\text{'5'})$$
$$26 = (\text{'6'})$$
$$27 = (\text{'7'})$$

Solution:

```
        ORG    0
        SETF   TRISC           ;TRISC = FFh (Port C as input)
        CLRF   TRISD           ;TRISD = 00 (Port D as output)
        LFSR   2,0x20          ;load pointer. FSR2 = 20H, RAM address
B1      MOVF   PORTC,W         ;read x from Port C into WREG
        ANDLW  B'00000111'     ;mask upper 5 bits
        MOVFF  PLUSW2,PORTD    ;get data pointed to by FSR2 + WREG
        BRA    B1
        END
```

When simulating this program on your MPLAB, make sure the RAM locations 20–27H have the elements of the look-up table. Notice that the "MOVFF PLUSW2, PORTD" instruction will bring the value from its RAM location and put it on PORTD.

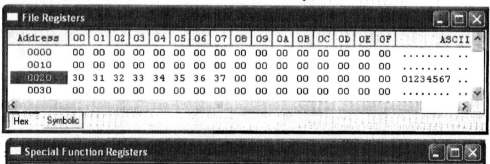

Example 6-16

Write a program to get the x value from Port B and send $x^2 + 2x + 3$ to Port C. Assume PB3–PB0 has the x value of 0–9. Use a look-up table instead of a multiply instruction.

Solution:

```
        ORG   0
        SETF  TRISB          ;TRISB = FFh (Port B as input)
        CLRF  TRISC          ;TRISC = 00 (Port C as output)
B1      MOVF  PORTB,W        ;read x from Port B into WREG
        ANDLW 0x0F           ;mask upper bits
        CALL  XSQR_TABLE     ;get x² from the look-up table
        MOVWF PORTC          ;copy it to Port C
        BRA   B1             ;continue
XSQR_TABLE
        MULLW 0x2            ;align it for even address
        MOVFF PRODL, WREG    ;put it into WREG for indexing
        ADDWF PCL            ;PCL = PCL + WREG
        RETLW D'3'           ;(0)² + 2(0) + 3 = 3
        RETLW D'6'           ;(1)² + 2(1) + 3 = 6
        RETLW D'11'          ;(2)² + 2(2) + 3 = 11
        RETLW D'18'          ;(3)² + 2(3) + 3 = 18
        RETLW D'27'          ;(4)² + 2(4) + 3 = 27
        RETLW D'38'          ;(5)² + 2(5) + 3 = 38
        RETLW D'51'          ;(6)² + 2(6) + 3 = 51
        RETLW D'66'          ;(7)² + 2(7) + 3 = 66
        RETLW D'83'          ;(8)² + 2(8) + 3 = 83
        RETLW D'102'         ;(9)² + 2(9) + 3 = 102
        END
```

Writing table elements in PIC18

In PIC18 we also have the TBLWRT instruction, which allows us to write (store) data into program ROM space. While the TBLRD instruction can be used with any family member of the PIC18 regardless of the type of ROM it has, the TBLWRT can be used only with PIC18 chips that have flash ROM for the program ROM space. The TBLWRT instruction will not work for PIC18 chips with OTP (one-time programmable) or mask ROM. In mask ROM, the information (code and data) is burned into the ROM during the chip fabrication by the Microchip Corp. Because writing to flash ROM involves manipulating the configuration bits it is discussed in Chapter 14.

Review Questions

1. The instruction "TBLRD*" uses register _____ as address pointer.
2. What register is incremented upon execution of the TBLRD* instruction?
3. What register is holding data, once it is read by the TBLRD* instruction?
4. What is the size of TBLPTR? How much ROM space does it cover?
5. What register is incremented upon execution of the TBLRD*+ instruction?
6. What is the difference between the TBLRD*+ and TBLRD+* instructions?
7. True or false. The TBLWT instruction works with all ROM versions of the PIC18 family.

SECTION 6.4: BIT-ADDRESSABILITY OF DATA RAM

Many microprocessors such as the 386 or Pentium allow programs to access registers and I/O ports in byte size only. In other words, if you need to check a single bit of an I/O port, you must read the entire byte first and then manipulate the whole byte with some logic instructions to get hold of the desired single bit. This is not the case with the PIC as we saw in Chapter 4. Indeed, one of the most important features of the PIC is its ability to access the file register's RAM location in bits as well as bytes. This means that all I/O ports, SFRs, and general purpose RAM areas for the PIC18 are bit-addressable because they are part of the file register data RAM. WREG is also bit-addressable because it is part of the SFRs. This is a very powerful feature of the PIC18 family. In this section, we provide more programming examples of the bit-addressable option of the PIC18 family.

Bit-addressable file register data RAM

The entire 4096 bytes of file register data RAM of the PIC18 are bit-addressable. This means that while ROM program space is only byte-addressable, the 4K of data RAM is both byte- and bit-addressable. To distinguish between the byte-addressable and bit-addressable options of data RAM, the PIC18 provides two categories of instructions: bit-oriented and byte-oriented instructions. Bit-oriented instructions are called bit-addressable, while byte-oriented instructions are referred to byte-addressable.

The bit-oriented instructions are given in Table 6-4. Notice that the bit-oriented instructions use only one addressing mode, the direct addressing mode. In the first three sections of this chapter we showed various addressing modes of byte-addressable space of the PIC18, among them register indirect addressing mode for both data RAM and program (code) ROM. Note that there is no register indirect addressing mode for bit-oriented instructions in the PIC18.

File register bit-addressability

As we discussed in Chapter 2, the PIC18 can have up to 4096 bytes of file register data RAM depending on the family member. We can access either the entire 8 bits or any single bit without altering the rest. When accessing a file register in a single-bit manner, we use the syntax "Bit-Oriented-instr fileReg, x" where fileReg is any register in the file register and X is the desired bit number from 0 to 7 for data bits D0 to D7. For example, the instruction "BTG 0x20, 7" will toggle D7 of RAM location 20H. As we mentioned earlier in this chapter, every register, including WREG, is assigned a byte address in the file register and

Table 6-4: Single-Bit (Bit-Oriented) Instructions for PIC18

Instruction	Function
BSF fileReg,bit	Bit Set fileReg (set the bit: bit = 1)
BCF fileReg,bit	Bit Clear fileReg (clear the bit: bit = 0)
BTG fileReg,bit	Bit Toggle fileReg (complement the bit)
BTFSC fileReg,bit	Bit test fileReg, skip if clear (skip next instruction if bit = 0)
BTFSS fileReg,bit	Bit test fileReg, skip if set (skip next instruction if bit = 1)

Note: fileReg can be any location of file register data RAM.

ports PORTA–PORTE are part of the file register. For example, "BSF PORTB, 5" sets high bit RB5 of Port B. Notice that when code such as "BSF PORTB, 5" is assembled, it becomes "8A81" because Port B has the RAM address of 81H in the access bank. Examine the next few examples in this section to gain insight into the bit addressibility of the file register in the PIC18.

Example 6-17

A switch is connected to pin RC7. Write a program to check the status of the switch and perform the following:

(a) If switch = 0, send letter 'N' to Port D.
(b) If switch = 1, send letter 'Y' to Port D.

Solution:

```
        BSF   TRISC,7     ;make RC7 an input
        CLRF  TRISD       ;make Port D an output port
AGAIN BTFSS PORTC,7       ;bit test RC7 for HIGH
        BRA   OVER        ;it must be LOW
        MOVLW A'Y'        ;WREG = 'Y' ASCII letter Y
        MOVWF PORTD       ;issue WREG to PD
        GOTO  AGAIN       ;we could use BRA instead
OVER  MOVLW A'N'          ;WREG = 'N' ASCII letter N
        MOVWF PORTD       ;issue WREG to PORTD
        GOTO  AGAIN       ;we can use BRA too
```

Example 6-18

Write a program to toggle RB1 a total of 200 times. Use file register RAM location 32H to hold your counter value.

Solution:

```
        MYREG EQU 0x32    ;set aside loc 0x20 reg
        CNTVAL EQU D'200'
        MOVLW CNTVAL      ;load counter into WREG
        MOVWF MYREG       ;load the count into MYREG location
        BCF   TRISB,1     ;TRISB bit = 0, make RB1 an output
AGAIN BTG   PORTB,1       ;toggle bit RB1
        DECF  MYREG,F     ;decrement MYREG
        BNZ   AGAIN       ;continue until counter is zero
```

Example 6-19

A switch is connected to pin RC7. Write a program to get the status of the switch and perform the following.

(a) If RC7 = 0, increment Port B.
(b) If RC7 = 1, decrement Port B.

Solution:

```
        BSF   TRISC,7      ;make RC7 an input
        CLRF  TRISB        ;make Port B an output port
AGAIN   BTFSS PORTC,7      ;bit test RC7 for HIGH
        BRA   OVER         ;it must be LOW
        INCF  PORTB,F      ;increment
        GOTO  AGAIN        ;we can use BRA too
OVER    DECF  PORTB,F      ;decrement
        GOTO  AGAIN        ;we can use BRA too
```

Example 6-20

A switch is connected to pin RB0. Write a program to get the status of the switch and save it in D0 of fileReg location 0x20.

Solution:

```
MYBITREG EQU 0x20         ;set aside loc 0x20 reg
        BSF   TRISB,0      ;make RB0 an input
AGAIN   BTFSS PORTB,0      ;bit test RB0, skip if set
        GOTO  OVER         ;it must be LOW (BRA is OK too)
        BSF   MYBITREG,0   ;set bit D0 = 1
        GOTO  AGAIN        ;we can use BRA too
OVER    BCF   MYBITREG,0   ;clear D0 (D0 = 0)
        GOTO  AGAIN        ;we can use BRA too
```

Example 6-21

Write a program to see if the RAM location 37H contains an even value. If so, send it to Port B. If not, make it even and then send it to Port B.

Solution:

```
MYREG EQU 0x37            ;set aside loc 0x37 reg
        CLRF  TRISB        ;make Port B an output port
AGAIN   BTFSS MYREG,0      ;bit test D0, skip if set
        GOTO  OVER         ;it must be LOW
        BCF   MYREG,0      ;clear bit D0 = 0
OVER    MOVFF MYREG,PORTB  ;copy it to Port B
        GOTO  AGAIN        ;we can use BRA too
```

Example 6-22

Write a program for Port B to count up from 0000 to 1111 (binary).

Solution:

```
      CLRF   TRISB      ;TRISB = 0, make PB output
      CLRF   PORTB      ;Port B = 0
AGAIN INCF   PORTB,F    ;increment Port B
      BTFSS  PORTB,4    ;test D4 bit of Port B
      BRA    AGAIN
      GOTO   $
```

Notice how it counts up from 0000 to 1111 and, when it becomes 10000, skips the loop.

Example 6-23

Write a program to check the status of the D7 bit of file register location 0x20 and make the following decisions:

(a) If D7 = 0, send "NO" to Port B.
(b) If D7 = 1, send "YES" to Port B.

Solution:

```
MYREG EQU 0x20
      CLRF   TRISB      ;make Port B an output port
AGAIN BTFSS  MYREG,7    ;bit test for HIGH
      BRA    OVER       ;it must be LOW
      MOVLW  A'Y'       ;WREG = 'Y' ASCII letter Y
      MOVWF  PORTB      ;issue WREG to Port B
      MOVLW  A'E'       ;WREG = 'E' ASCII letter Y
      MOVWF  PORTB      ;issue WREG to Port B
      MOVLW  A'S'       ;WREG = 'S' ASCII letter Y
      MOVWF  PORTB      ;issue WREG to Port B
      GOTO   AGAIN      ;we can use BRA too
OVER  MOVLW  A'N'       ;WREG = 'N' ASCII letter N
      MOVWF  PORTB      ;issue WREG to Port B
      MOVLW  A'O'       ;WREG = 'O' ASCII letter Y
      MOVWF  PORTB      ;issue WREG to Port B
      GOTO   AGAIN      ;we can use BRA too
```

Status register bit-addressability

Of the bit-addressable registers, we will concentrate on the familiar status register. The rest will be discussed in future chapters.

Now let's see how we can use bit-addressability of the status register. As we discussed in Chapter 2, 5 bits in the status register are set aside for the flags C, DC, Z, N, and OV. See Figure 6-2 and Example 6-24.

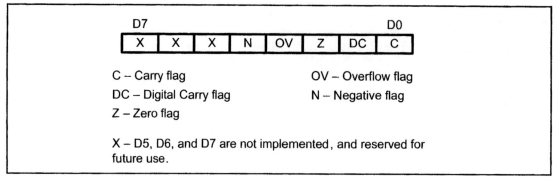

D7   D0

| X | X | X | N | OV | Z | DC | C |

C – Carry flag   OV – Overflow flag

DC – Digital Carry flag   N – Negative flag

Z – Zero flag

X – D5, D6, and D7 are not implemented, and reserved for future use.

Figure 6-2. Bits of the Status Register

Example 6-24

While we have instructions such as BC (branch carry) and BZ (branch zero) to check the carry and zero flag bits, show how would you use the status register flag to check the (a) C and (b) Z flags.

Solution:

(a) The C flag is D0 of the status register; therefore, we can use the following instruction to check the C flag:

```
BTFSS STATUS,C      ;bit test C, skip if C = 1
```

(b) The Z flag is D3 of the status register; therefore, we can use the following instruction to check the Z flag:

```
BTFSS STATUS,Z      ;bit test Z, skip if Z = 1
```

Review Questions

1. True or false. All I/O ports of the PIC18 are bit-addressable.
2. True or false. The status register of the PIC18 is bit-addressable.
3. True or false. All file register RAM locations of the PIC are bit-addressable.
4. Indicate which of the following registers are bit-addressable.
   (a) Port A (b) Port B   (c) WREG   (d) status register (e) 21-bit PC
5. Of the 4096 bytes of RAM in the PIC18, how many bytes are bit-addressable?
6. How would you check to see whether bit D1 of RAM location 3 is HIGH or LOW?
7. State what each instruction does.
   (a) `BSF 0x20,1` (b) `BCF 0x32,7`   (c) `BSF 0x12,2`
   (d) `BSF PORTB,4` (e) `BSF STATUS,1`
8. Show how to clear the carry flag.

SECTION 6.5: BANK SWITCHING IN THE PIC18

The PIC18 microcontroller has a maximum of 4K of data RAM space. Although not all members of the family have the entire RAM installed, every member of the family has at least the access bank for the file register. The file register RAM is divided into banks of 256 bytes each, which gives us a total of 16 banks in the PIC18. The minimum bank that every PIC18 has is called the access bank, as we discussed in Chapter 2. The access bank is made of 128 bytes of lower addresses and 128 bytes of higher addresses. While the lower 128 bytes of address space 000–07FH are used for general-purpose RAM, the higher 128 bytes are dedicated to the SFRs (special function registers) residing in address space F80–FFFH. The vast majority of the PIC18 chips we see on the Microchip web site have more than just the access bank. In this section we show how to use bank switching to take advantage of the entire data RAM space of the PIC18.

The A bit and bank switching

All the instructions we have used so far assumed the access bank as the default bank. This was achieved by ignoring the letter A in instructions such as "MOVWF fileReg, A". In other words, the instruction "MOVWF fileReg" is really "MOVWF fileReg, A" where the A bit can be 0 or 1. If A = 0, then the access bank is the default bank. If A = 1, however, then the instruction will use the bank selector register (BSR) to select the bank instead of using the access bank. If A is not stated in a given instruction, it means A = 0 and the access bank is the default bank. That is what we have done so far for the simple reason of making the PIC18 Assembly language easier to understand and master. Next, we examine the role of the BSR register in bank switching.

The BSR register and bank switching

To use banks other than the access bank, we need to set bit A = 1 in the coding of the instruction. With A = 1, we use the BSR (bank select register) to choose the desired bank. The BSR is an 8-bit register and is part of the SFRs. Of the 8 bits of the BSR, only 4 least-significant bits are used in the PIC18. The upper 4 bits are set to zero and are ignored by the PIC18. The 4-bit BSR gives us 16 banks, and because each bank is 256 bytes, we cover the entire 4096 ($16 \times 256 = 4096$) bytes of the data RAM file register using bank switching. The 4K (4096) bytes of the data RAM are organized as banks 0 to F, where the lowest bank, 0, has an address of 00–FFH, and the highest bank is bank F with the addresses of F00–FFFH. In the PIC18, the last 128 bytes of bank F are always set aside for the SFRs, while general purpose registers always start at address 0 of bank 0. Upon power-on reset, BSR = 0 (0000 binary), which indicates that only the lowest addresses of data RAM, from 000 to 0FFH, can be used for the general-purpose register in addition to the SFRs, which always reside in the last half of bank F. Similarly, if we make BSR = 1 (0001 binary), then PIC18 selects bank 1 using the 100–1FFH addresses in addition to the SFRs, which use only the last half of the bank with addresses of F80–FFFH. To select bank 2, we load BSR with the value 02 (0010 binary), which allows access to the bank addresses 200–2FF in addition to the SFR addresses of F80–FFFH. As we can see, no matter how much data RAM we have in the PIC18, the GP register always starts at address 000 and goes up, while the SFRs start at the other end of the 4 KB, at address FFF, and come down. At the present time PIC is using only the highest 128 bytes of bank F (F80H–FFFH address) for the SFRs. In the future they might start to use the rest of bank F and may even use bank E for SFRs, if the special functions embedded into the PIC18 keep increasing. Although the number of bytes in bank F used for SFRs in the PIC18 chip varies depending on the functions embedded into the chip, the SFRs always start at address FFF and go down. This point must be emphasized. For example, in the case of the access bank, the last half of bank F is set aside for SFRs, even though in some family members not all 128 bytes are needed due to the limited number of functions supported by that chip. See Table 6-5 for data RAM file registers of some PIC18 chips. Note that although we can use any of the addressing modes, such as immediate, direct, or register indirect to access the GP register regions, we use only direct addressing mode in accessing the SFR registers. To gain a better understanding of the bank switching, we use the PIC18F458 chip to show some examples.

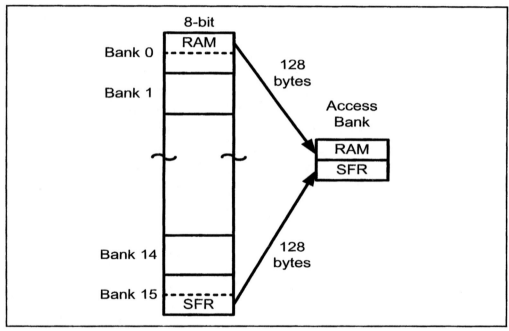

Figure 6-3. Data RAM Registers

Table 6-5: Data RAM Bank for Selected PIC18 Chips

	File Register (Bytes)	=	SFR (Bytes)	+	Available space for GPR (Bytes)
PIC18F1220	512		256		256
PIC18F452	1792		256		1536
PIC18F2220	768		256		512
PIC18F458	1792		256		1536
PIC18F8722	4096		158		3938

Note: The newer versions of the PIC18F458/452 are the PIC18F4580/4520.
Extracted from http://www.microchip.com

Bank switching and "INCF F, D, A" instruction

The PIC18F458 has a total of 1792 bytes for the data RAM file register. The bank organization for the PIC18F458 is shown in Figure 6-3. All the examples we have seen so far ignored the A bit in the instruction, which means that A = 0. With A = 0, the access bank is the default bank. Now to use banks other than the access bank, two things must be done:

1. Load the BSR with the desired bank number, and
2. Make A = 1 in the instruction itself.

Therefore, instruction "INCF MYREG, F, 1" has a totally different meaning from "INCF MYREG, F, 0". The A = 1 means to use the bank pointed to by BSR. In the following code, we first load the bank number into the BSR register using the MOVLB instruction, and then manipulate the contents of RAM location 0x240 (location 40 of bank 2):

```
MYREG EQU    0x40

MOVLB 0x2           ;load 2 into BSR (use bank 2)
MOVLW 0             ;WREG = 0
MOVWF MYREG, 1      ;loc 0x240 = (0), WREG = 0, Notice A = 1
INCF  MYREG, F, 1   ;loc 0x240 = (1), WREG = 0, Notice A = 1
INCF  MYREG, F, 1   ;loc 0x240 = (2), WREG = 0, Notice A = 1
INCF  MYREG, F, 1   ;loc 0x240 = (3), WREG = 0
```

Contrast the above program with the one below:

```
MOVLB 0x2           ;load 2 into BSR (use bank 2)
MOVLW 0             ;WREG = 0
MOVWF MYREG         ;loc 0x40 = (0), WREG = 0
INCF  MYREG, F      ;loc 0x40 = (1), WREG = 0, Notice A = 0
INCF  MYREG, F      ;loc 0x40 = (2), WREG = 0, Notice A = 0
INCF  MYREG, F      ;loc 0x40 = (3), WREG = 0
```

Although we loaded BSR, because the A bit was not indicated, MPASM defaults it to zero, which means to use location 0x40 of the access bank. The A bit in the instruction field is given in Figure 6-4.

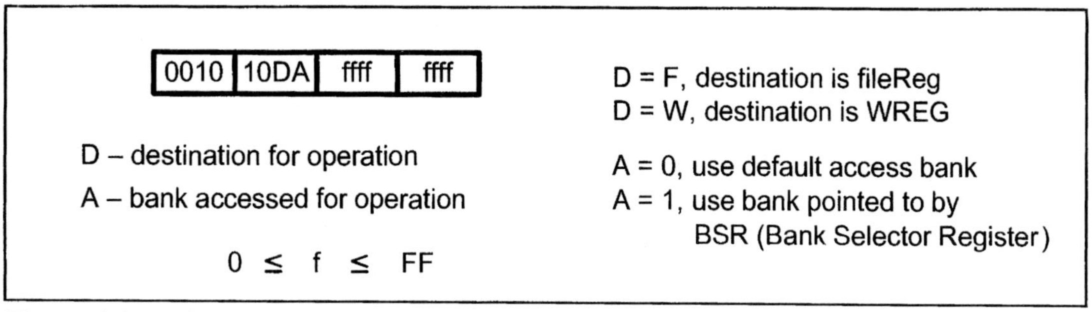

```
0010 | 10DA | ffff | ffff
```

D = F, destination is fileReg
D = W, destination is WREG

D – destination for operation
A – bank accessed for operation

A = 0, use default access bank
A = 1, use bank pointed to by
        BSR (Bank Selector Register)

$0 \leq f \leq FF$

Figure 6-4. A Bit in the Instruction Field for INCF

Examine the following code to see the role of the D and A bits:

```
MOVLB 0x2          ;load 2 into BSR (use bank 2)
MOVLW 0            ;WREG = 0
MOVWF 0x20,1       ;loc 0x220 = (0), WREG = 0, D = W, A = 1 means Bank 2
INCF  0x20,W,1     ;loc 0x220 = (0), WREG = 1, D = W, A = 1
INCF  0x20,W,1     ;loc 0x220 = (0), WREG = 1, D = W, A = 1
INCF  0x20,W,1     ;loc 0x220 = (0), WREG = 1, D = W, A = 1
INCF  0x20,F,1     ;loc 0x220 = (1), WREG = 1, D = F, A = 1
INCF  0x20,F,1     ;loc 0x220 = (2), WREG = 1, D = F, A = 1
INCF  0x20,F,1     ;loc 0x220 = (3), WREG = 1, D = F, A = 1
INCF  0x20,F,1     ;loc 0x220 = (4), WREG = 1, D = F, A = 1
```

Simulate the next few examples with MPLAB and examine the data RAM memory to see how bank switching works.

Example 6-25

Write a program to copy the value 55H into RAM memory locations 340H to 345H using:

(a) direct addressing mode.
(b) a loop.

Solution:
(a)
```
        MOVLB 0x3          ;BANK 3
        MOVLW 0x55         ;load WREG with value 55H
        MOVWF 0x40, 1      ;copy WREG to RAM location 340H
        MOVWF 0x41, 1      ;copy WREG to RAM location 341H
        MOVWF 0x42, 1      ;copy WREG to RAM location 342H
        MOVWF 0x43, 1      ;copy WREG to RAM location 343H
        MOVWF 0x44, 1      ;copy WREG to RAM location 344H
```
(b)
```
COUNT   EQU 0x10           ;loc 10h
        MOVLB 0x3          ;BANK 3
        MOVLW 0x5          ;WREG = 5
        MOVWF COUNT        ;load the counter, count = 5
        LFSR  0,0x340      ;load pointer. FSR0 = 40H, RAM address
        MOVLW 0x55         ;WREG = 55h value to be copied
B1      MOVWF INDF0,0      ;copy WREG to RAM loc FSR0 points to
        INCF  FSR0L        ;increment FSR0L pointer
        DECF  COUNT,F,0    ;decrement the counter
        BNZ   B1           ;loop until counter = zero
```

The following shows RAM contents after the above program is run:

        340 = (55)
        341 = (55)
        342 = (55)
        343 = (55)
        344 = (55)

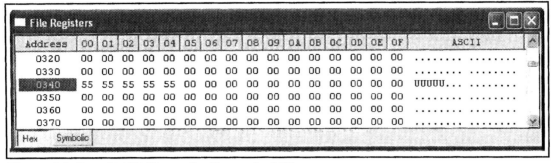

Figure 6-5. Data RAM Shown for Example 6-25

Table 6-6 shows the banks for various sizes of the data RAM in the PIC18 chip.

Table 6-6: PIC18 Data Memory Range

Data Memory	Banks
64	0, 15
128	0, 15
256	0, 15
512	0–1, 15
640	0–2, 15
768	0–2, 15
1024	0–3, 15
1280	0–4, 15
1536	0–5, 15
1792	0–6, 15
2048	0–7, 15
2304	0–8, 15
2560	0–9, 15
2816	0–10, 15
3072	0–11, 15
3328	0–12, 15
3584	0–13, 15
3840	0–14, 15
3968	0–15

MOVFF and banks

The great thing about the MOVFF instruction is that there is no need to worry about bank switching because it can move data anywhere within the 4K of RAM space. See Figure 6-4. Also see Example 6-26.

Examining Data RAM space using MPLAB simulator

The MPLAB simulator is a great tool to examine data RAM contents. We encourage its use to examine and verify the results of programs using data RAM.

Example 6-26

Assume RAM locations 330–334H of the PIC18F458 have the string of ASCII data shown below. Write a program to get each character and send it to Port B one byte at a time. Show the program using

(a) direct addressing mode.
(b) register indirect addressing mode.

$$330 = (\text{'H'})$$
$$331 = (\text{'E'})$$
$$332 = (\text{'L'})$$
$$333 = (\text{'L'})$$
$$334 = (\text{'O'})$$

Solution:

(a) Using direct addressing mode

```
CLRF  TRISB                 ;make Port B an output
MOVFF 0x330, PORTB          ;copy contents of loc 0x330 to PB
MOVFF 0x331, PORTB
MOVFF 0x332, PORTB
MOVFF 0x333, PORTB
MOVFF 0x334, PORTB
```

(b) Using register indirect addressing mode

```
COUNTREG EQU 0x20           ;fileReg loc 20 for counter
CNTVAL EQU 5                ;counter value
     CLRF   TRISB           ;make Port B an output (TRSIB = out)
     MOVLW  CNTVAL          ;WREG = 5
     MOVWF  COUNTREG        ;load the counter, count = 5
     LFSR   2,0x330         ;load pointer. FSR2 = 330H, RAM address
B3   MOVF   INDF2,W         ;copy RAM loc FSR2 points at to WREG
     MOVWF  PORTB           ;copy WREG to PORTB
     INCF   FSR2L           ;increment FSR2 to point at next loc
     DECF   COUNTREG,F      ;decrement counter
     BNZ    B3              ;loop until counter = zero
```

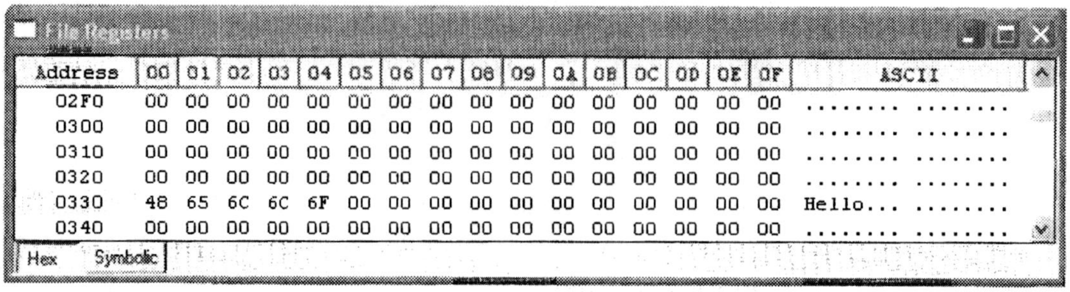

Address	00	01	02	03	04	05	06	07	08	09	0A	0B	0C	0D	0E	0F	ASCII
02F0	00	00	00	00	00	00	00	00	00	00	00	00	00	00	00	00	....
0300	00	00	00	00	00	00	00	00	00	00	00	00	00	00	00	00	....
0310	00	00	00	00	00	00	00	00	00	00	00	00	00	00	00	00	....
0320	00	00	00	00	00	00	00	00	00	00	00	00	00	00	00	00	....
0330	48	65	6C	6C	6F	00	00	00	00	00	00	00	00	00	00	00	Hello... ....
0340	00	00	00	00	00	00	00	00	00	00	00	00	00	00	00	00	....

Hex   Symbolic

Figure 6-6. Example 6-26 Data RAM Dump in MPLAB

Example 6-27

Write a program for the PIC18F452 chip to put FFH into 16 RAM locations starting at
RAM address 160H. Use:

(a) INCF FSRnL.
(b) auto-increment.

Solution:

```
(a)
COUNTREG EQU 0x10       ;fileReg loc for counter
CNTVAL EQU D'16'        ;counter value
      MOVLW CNTVAL      ;WREG = 16
      MOVWF COUNTREG    ;load the counter, count = 16
      LFSR  1,0x160     ;load pointer. FSR1 = 60H, RAM address
      MOVLW 0xFF        ;load 0xFF
B2    MOVWF INDF1,0     ;move W to RAM loc FSR1 points to
      INCF  FSR1L       ;increment FSR1L, point to next loc
      DECF  COUNTREG,F  ;decrement counter
      BNZ   B2          ;loop until counter = zero

(b)
COUNTREG EQU 0x10       ;fileReg loc for counter
CNTVAL EQU D'16'        ;counter value
      MOVLW CNTVAL      ;WREG = 16
      MOVWF COUNTREG    ;load the counter, count = 16
      LFSR  1,0x160     ;load pointer. FSR1 = 160H, RAM address
      MOVLW 0xFF        ;load 0xFF
B3    MOVWF POSTINC1, 0
      DECF  COUNTREG,F  ;decrement counter
      BNZ   B3          ;loop until counter = zero
```

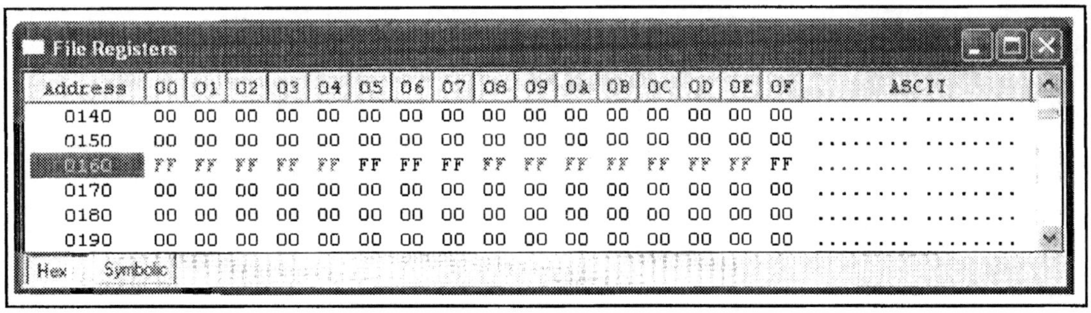

File Registers																	
Address	00	01	02	03	04	05	06	07	08	09	0A	0B	0C	0D	0E	0F	ASCII
0140	00	00	00	00	00	00	00	00	00	00	00	00	00	00	00	00	....
0150	00	00	00	00	00	00	00	00	00	00	00	00	00	00	00	00	....
0160	FF	FF	FF	FF	FF	FF	FF	FF	FF	FF	FF	FF	FF	FF	FF	FF	....
0170	00	00	00	00	00	00	00	00	00	00	00	00	00	00	00	00	....
0180	00	00	00	00	00	00	00	00	00	00	00	00	00	00	00	00	....
0190	00	00	00	00	00	00	00	00	00	00	00	00	00	00	00	00	....

Hex   Symbolic

Figure 6-7. Data Example 6-27 Data RAM Dump

Example 6-28

Write a program to copy a block of data from RAM locations 330H–33FH to 360H–36FH.

Solution:
```
COUNTREG EQU 0x20            ;fileReg loc 20 for counter
CNTVAL EQU 0x0F              ;counter value = 15
        CLRF   TRISB         ;make Port B an output (TRSIB = FFH)
        MOVLW  CNTVAL        ;WREG = 15
        MOVWF  COUNTREG,1    ;load the counter, count = 15
        LFSR   1,0x330       ;load pointer. FSR1 = 330H, RAM address
        LFSR   2,0x360       ;load pointer. FSR2 = 360H, RAM address
B3      MOVFF  INDF1, INDF2
        INCF   FSR1L         ;increment FSR1 to point at next loc
        INCF   FSR2L         ;increment FSR2 to point at next loc
        DECF   COUNTREG,F    ;decrement counter
        BNZ    B3            ;loop until counter = zero
```

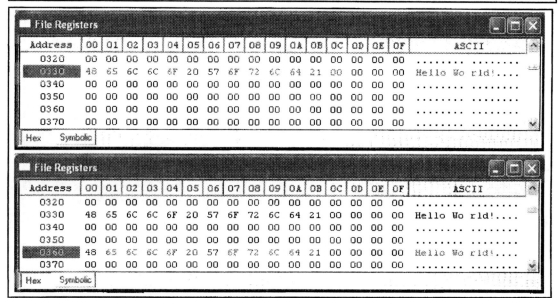

Figure 6-8. Before and After Example 6-28 is Run

Review Questions

1. True or false. The PIC18 uses the last bank for the SFRs.
2. True or false. The PIC18 uses a total of 256 bytes for each bank.
3. True or false. The first 128 bytes of RAM in the PIC18 are used for the access bank.
4. Give the address for the upper RAM used for SFR.
5. Show how to put value 99H into RAM location 202H in the PIC18F458.
6. Show how to put value 55H into RAM location 408H in a PIC18 with 4K of RAM.
7. True or false. The MOVFF instruction can copy a byte from any RAM location to any other RAM location in the PIC18.
8. The BSR register is a(n) _____-bit register, but only ___ bits of it are used for bank selection.

SECTION 6.6: CHECKSUM AND ASCII SUBROUTINES

In this section we look at some widely used subroutines such as checksum byte, BCD, and ASCII conversion. We will also examine the use of a stack in the PIC18.

Checksum byte in ROM

To ensure the integrity of ROM contents, every system must perform a checksum calculation. The checksum will detect any corruption of the contents of ROM. One cause of ROM corruption is current surge, either when the system is turned on, or during operation. To ensure data integrity in ROM, the checksum process uses what is called a *checksum byte*. The checksum byte is an extra byte that is tagged to the end of a series of bytes of data. To calculate the checksum byte of a series of bytes of data, the following steps can be taken:

1. Add the bytes together and drop the carries.
2. Take the 2's complement of the total sum, and that is the checksum byte, which becomes the last byte of the series.

To perform a checksum operation, add all the bytes, including the checksum byte. The result must be zero. If it is not zero, one or more bytes of data have been changed (corrupted). To clarify these important concepts, see Example 6-29.

Checksum program

The checksum generation and testing program is given in subroutine form. We have divided Program 6-1 into three subroutines (or subprograms). These three subroutines perform the following operations:

1. Retrieve the data from code ROM.
2. Calculate the checksum byte.
3. Test the checksum byte for any data error.

Each of these subroutines can be used in other applications. Example 6-29 shows how to manually calculate the checksum for a list of values. Also see Program 6-1.

Example 6-29

Assume that we have 4 bytes of hexadecimal data: 25H, 62H, 3FH, and 52H.

(a) Find the checksum byte.
(b) Perform the checksum operation to ensure data integrity.
(c) If the second byte, 62H, has been changed to 22H, show how the checksum method detects the error.

Solution:

(a)   Find the checksum byte.

```
        25H
     +  62H
     +  3FH
     +  52H
        118H      (Dropping the carry of 1, we have 18H. Its 2's complement is E8H. Therefore
                  checksum byte is E8H.)
```

(b)   Perform the checksum operation to ensure data integrity.

```
        25H
     +  62H
     +  3FH
     +  52H
     +  E8H
        200H      (Dropping the carries, we see 00, indicating that data is not corrupted.)
```

(c)   If the second byte 62H has been changed to 22H, show how the checksum method detects the error.

```
        25H
     +  22H
     +  3FH
     +  52H
     +  E8H
        1C0H      (Dropping the carry, we get C0H, which is not 00. This means that data
                  is corrupted.)
```

```
;PROG 6-1: CALCULATING AND TESTING CHECKSUM BYTE
       #include P18F458.inc

RAM_ADDR     EQU 40H      ;RAM space to place the bytes
COUNTREG     EQU 0x20     ;fileReg loc for counter
CNTVAL       EQU 4        ;counter value = 4 for adding 4 bytes
CNTVAL1      EQU 5        ;counter value = 5 for adding 5 bytes
                         ;including checksum byte

;-----------main program
       ORG   0
       CALL  COPY_DATA
       CALL  CAL_CHKSUM
       CALL  TEST_CHKSUM
       BRA   $
```

```
;--------copying data from code ROM address 500H to data RAM loc
COPY_DATA
        MOVLW  low(MYBYTE)        ;WREG = 00 LOW-byte addr
        MOVWF  TBLPTRL            ;ROM data LOW-byte addr
        MOVLW  hi(MYBYTE)         ;WREG = 5, HIGH-byte addr
        MOVWF  TBLPTRH            ;ROM data HIGH-byte addr
        MOVLW  upper(MYBYTE)      ;WREG = 00 upper-byte addr
        MOVWF  TBLPRTRU           ;ROM data upper-byte addr
        LFSR   0,RAM_ADDR         ;FSR0 = RAM_ADDR, place to save
C1      TBLRD*+                   ;bring in next byte and inc TBLPTR
        MOVF   TABLAT,W           ;copy to WREG (Z = 1, if null)
        BZ     EXIT               ;is it null char? exit if yes
        MOVWF  POSTINC0           ;copy WREG to RAM and inc pointer
        BRA    C1
EXIT    RETURN


;-----calculating checksum byte
CAL_CHKSUM
        MOVLW  CNTVAL             ;WREG = 4
        MOVWF  COUNTREG           ;load the counter, count = 4
        LFSR   0,RAM_ADDR         ;load pointer. FSR0 = 40H
        CLRF   WREG
C2      ADDWF  POSTINC0,W         ;add RAM to WREG and increment FSR0
        DECF   COUNTREG,F         ;decrement counter
        BNZ    C2                 ;loop until counter = zero
        XORLW  0xFF               ;1's comp
        ADDLW  1                  ;2'compl
        MOVWF  POSTINC0
        RETURN


;----------testing checksum byte
TEST_CHKSUM
        MOVLW  CNTVAL1            ;WREG = 5
        MOVWF  COUNTREG           ;load the counter, count = 5
        CLRF   TRISB              ;PORTB = output
        LFSR   0,RAM_ADDR         ;load pointer. FSR0 = 40H
        CLRF   WREG
C3      ADDWF  POSTINC0,W         ;add RAM and increment FSR0
        DECF   COUNTREG,F         ;decrement counter
        BNZ    C3                 ;loop until counter = zero
        XORLW  0x0                ;EX-OR to see if WREG = zero
        BZ     G_1                ;is result zero? then good
        MOVLW  'B'
        MOVWF  PORTB              ;if not, data is bad
        RETURN
G_1     MOVLW  'G'
        MOVWF  PORTB              ;data is not corrupted
        RETURN


;----------my data in program ROM
        ORG 0x500
MYBYTE DB 0x25, 0x62, 0x3F, 0x52, 0x00
        END
```

Note the usage of the keywords low, hi, and upper, to indicate the 21-bit address of the program ROM.

BCD to ASCII conversion program

Many RTCs (real-time clocks) provide time and date in BCD format. To display the BCD data on an LCD or a PC screen, we need to convert it to ASCII. Program 6-2 (a) transfers packed BCD data from program ROM to data RAM, (b) converts packed BCD to ASCII, and (c) sends the ASCII to port B for display. We will use a portion of this program in Chapter 16. The displaying of data on LCD will be shown in Chapter 12. See Chapter 5 for the BCD to ASCII conversion algorithm.

```
;PROG 6-2: CONVERTING PACKED BCD TO ASCII
        #include P18F458.inc

RAM_ADDR      EQU 0x40
ASC_RAM       EQU 0x50
COUNTREG      EQU 0x20            ;fileReg loc for counter
CNTVAL        EQU D'4'            ;counter value of BCD bytes
CNTVAL1       EQU D'8'            ;counter value of ASCII bytes

;------------main program
        ORG    0
        CALL   COPY_DATA
        CALL   BCD_ASC_CONV
        CALL   DISPLAY
        BRA    $

;--------copying data from code ROM to data RAM
COPY_DATA
        MOVLW  low(MYBYTE)        ;WREG = 00 LOW-byte addr
        MOVWF  TBLPTRL            ;ROM data LOW-byte addr
        MOVLW  hi(MYBYTE)         ;WREG = 5, HIGH-byte addr
        MOVWF  TBLPTRH            ;ROM data HIGH-byte addr
        MOVLW  upper(MYBYTE)      ;WREG = 00 upper-byte addr
        MOVWF  TBLPRTRU           ;ROM data upper-byte addr
        LFSR   0,RAM_ADDR         ;FSR0 = RAM_ADDR, place to save
C1      TBLRD*+                   ;bring in next byte and inc TBLPTR
        MOVF   TABLAT,W           ;copy to WREG (Z = 1, if null)
        BZ     EXIT               ;is it null char? exit if yes
        MOVWF  POSTINC0           ;copy WREG to RAM and inc pointer
        BRA    C1
EXIT    RETURN

;-----convert packed BCD to ASCII
BCD_ASC_CONV
        MOVLW  CNTVAL             ;get the counter value
        MOVWF  COUNTREG           ;load the counter
        LFSR   0,RAM_ADDR         ;FSR0 = RAM_ADR BCD byte pointer
        LFSR   1,ASC_RAM          ;FSR1 = ASC_RAM ASCII byte pointer
B2      MOVF   INDF0,W            ;copy BCD to WREG
        ANDLW  0x0F               ;mask the upper nibble (W = 09)
        IORLW  0x30               ;make it an ASCII
        MOVWF  POSTINC1           ;copy to RAM and increment FSR1
        MOVF   POSTINC0,W         ;note the use of instruction
        ANDLW  0xF0               ;mask the lower nibble (W = 20H)
        SWAPF  WREG
        IORLW  0x30               ;make it an ASCII
        MOVWF  POSTINC1           ;copy to RAM and increment FSR1
        DECF   COUNTREG,F         ;decrement counter
        BNZ    B2                 ;loop until counter = zero
```

```
                  RETURN

        ;-----send ASCII data to port B
        DISPLAY
                CLRF    TRISB           ;make PORTB output (TRSIB = FFH)
                MOVLW   CNTVAL1         ;WREG = 8, send 8 bytes of data
                MOVWF   COUNTREG        ;load the counter, count = 8
                LFSR    2,ASC_RAM       ;load pointer. FSR2 = 50H
        B3      MOVF    POSTINC2,W      ;copy RAM to WREG and inc pointer
                MOVWF   PORTB           ;copy WREG to PORTB
                DECF    COUNTREG,F      ;decrement counter
                BNZ     B3              ;loop until counter = zero
                RETURN

        ;----------my BCD data in program ROM
                ORG 0x500
        MYBYTE DB 0x25, 0x67, 0x39, 0x52, 0x00
                END
```

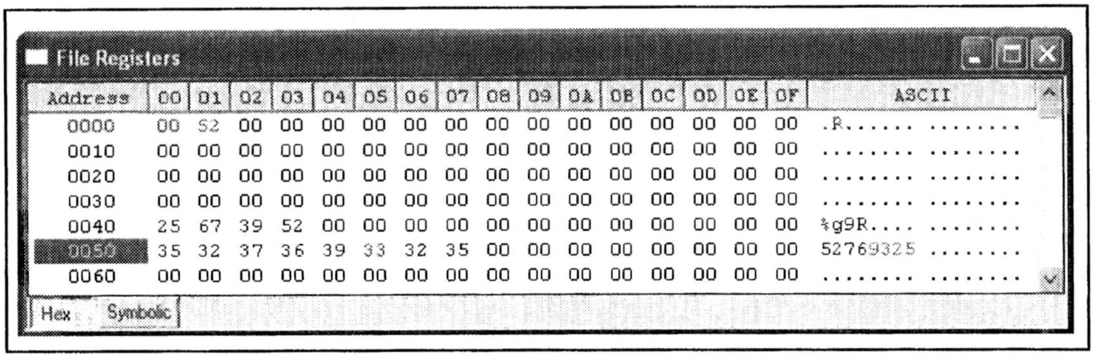

Figure 6-9. Results of Program 6-2 After it Ran.

Binary (hex) to ASCII conversion program

Many ADC (analog-to-digital converter) chips provide output data in binary (hex). To display the data on an LCD or PC screen, we need to convert it to ASCII. The code for the binary-to-ASCII conversion is shown in Program 6-3. Notice that the subroutine gets a byte of 8-bit binary (hex) data from Port B and converts it to decimal digits, and the second subroutine converts the decimal digits to ASCII digits and saves them. We are saving the low digit in the lower address location and the high digit in higher address location. This is referred to as the little-endian convention (i.e., low-byte to low-location and high-byte to high-location). All PIC18 products use the little-endian convention. For the binary-to-ASCII conversion algorithm see Chapter 5.

```
;PROG 6-3: CONVERTING BIN(HEX) TO ASCII
#include P18F458.INC

NUME            EQU     0x00            ;RAM loc for NUME
QU              EQU     0x20            ;RAM loc for quotient
RMND_L          EQU     0x30            ;the least significant digit loc
RMND_M          EQU     0x31            ;the middle significant digit loc
RMND_H          EQU     0x32            ;the most significant digit loc
```

```
MYDEN          EQU   D'10'            ;value for divide by 10

COUNTREG       EQU  0x10              ;fileReg loc for counter
CNTVAL         EQU  d'3'              ;counter value
UNPBCD_ADDR EQU 0x30
ASCII_RESULT EQU 0x40

;----------main program
       ORG    0
       SETF   TRISB                  ;make PORTB input
       CALL   BIN_DEC_CON
       CALL   DEC_ASCII_CON
       BRA    $

;-----converting BIN(HEX) TO DEC (00-FF TO 000-255)
BIN_DEC_CON
       MOVFF  PORTB,WREG              ;get the binary data from PORTB
       MOVWF  NUME                    ;load numerator
       MOVLW  MYDEN                   ;WREG = 10, the denominator
       CLRF   QU                      ;clear quotient
D_1    INCF   QU                      ;inc quotient for every subtraction
       SUBWF  NUME                    ;subtract WREG from NUME value
       BC     D_1                     ;if positive go back
       ADDWF  NUME                    ;once too many, first digit
       DECF   QU                      ;once too many for quotient
       MOVFF  NUME,RMND_L             ;save the first digit
       MOVFF  QU,NUME                 ;repeat the process one more time
       CLRF   QU                      ;clear QU
D_2    INCF   QU
       SUBWF  NUME                    ;subtract WREG from NUME value
       BC     D_2
       ADDWF  NUME                    ;once too many
       DECF   QU
       MOVFF  NUME,RMND_M             ;2nd digit
       MOVFF  QU,RMND_H               ;3rd digit
       RETURN

;----converting unpacked BCD digits to displayable ASCII digits
DEC_ASCII_CON
       MOVLW  CNTVAL                  ;WREG = 10
       MOVWF  COUNTREG                ;load the counter, count = 10
       LFSR   0,UNPBCD_ADDR           ;load pointer FSR0
       LFSR   1,ASCII_RESULT          ;load pointer FSR1
B3     MOVF   POSTINC0, W             ;copy RAM to WREG, increment FSR0
       ADDLW  0x30                    ;make it an ASCII
       MOVWF  POSTINC1                ;copy WREG and increment FSR1
       DECF   COUNTREG,F              ;decrement counter
       BNZ    B3                      ;loop until counter = zero
       RETURN
       END                           ;end of the program
```

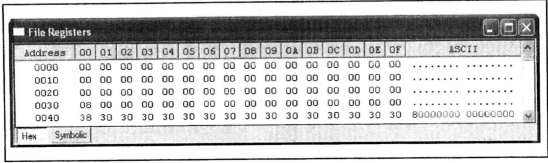

Figure 6-10. Results of Program 6-3 After it Ran. (Note the contents of locations 0030H, 0040H, and 0041H.)

Using memory banks for a stack

The stack in the PIC18 is 31 bytes deep and 21 bits wide. Because the program counter is 21 bits wide, the stack must be 21 bits wide also, as we discussed in Chapter 2. This stack is used mainly to save addresses for call and interrupt subroutines. Unlike other microprocessors, the stack in the PIC18 is not part of the data RAM space. Having access to such a large number of banks, however, makes a traditional stack unnecessary. In traditional CPUs (e.g., the x86) a limited number of registers forced us to push main registers into the stack at the beginning of a called subroutine, before we could use the main registers for data manipulation. In the case of the PIC18, all we have to do is to change the default bank to a new bank when we go into the subroutine. In other words, if we want to store a program's data on a stack in a given subroutine, we can use one of the banks of data RAM space instead of the 31 × 21-bit stack. Besides the fact that the 21-bit wide stack does not lend itself to storing 8-bit data, we must reserve all the 31 × 21-bit stack for calls and interrupts. Notice that we can still use the access bank for the storage of global variables. Global variables are discussed in the next section.

Review Questions

1. For the following ASCII numbers, give the ASCII and packed BCD representations.
   (a) '5', '7'   (b) '9', '4'
2. Show the hex format for "2005" and its BCD version.
3. Does the WREG register have BCD data after the following instruction is executed? Assume that the default the radix for the MPASM is decimal.
   ```
   MOVLW 95
   ```
4. 33H in BCD when converted to ASCII is ____H and ____H.
5. Find the ASCII value for the binary 11110010 if we want to display it on a computer screen as a 3-digit decimal number.
6. The checksum byte method is used to test data integrity in ____(RAM, ROM).
7. Find the checksum byte for the following hex values: 88H, 99H, AAH, BBH, CCH, DDH
8. True or false. If we add all the bytes, including the checksum byte, and the result is FFH, there is no error in the data.

SECTION 6.7: MACROS AND MODULES

In this section we explore macros and modules and their use in Assembly language programming. The format and usage of macros are defined, and many examples of their applications are explored. In addition, this section demonstrates modular programming along with rules for writing modules and linking them together. Some very useful modules will be given, along with methods of passing parameters among various modules. Dividing a program into several modules (in C programming these are called functions) allows us to use modules in other applications. It is common practice to divide a program into several modules, test each module, and put them into a library.

What is macro and how is it issued?

There are applications in Assembly language programming where a group of instructions performs a task that is used repeatedly. For example, moving data into a RAM location is done repeatedly in the same program. It does not make sense to rewrite this code every time it is needed. Therefore, to reduce the time that it takes to write code and reduce the possibility of errors, the concept of macros was born. Macros allow the programmer to write the task (code to perform a specific job) once only, and to invoke it whenever it is needed.

MACRO definition

Every macro definition must have three parts, as follows:

```
name         MACRO       dummy1, dummy2, ... , dummyN
             ......
             ......
             ENDM
```

The MACRO directive indicates the beginning of the macro definition and the ENDM directive signals the end. What goes in between the MACRO and ENDM directives is called the *body* of the macro. The name must be unique and must follow Assembly language naming conventions. The dummies are names, or parameters, or even registers that are mentioned in the body of the macro. After the macro has been written, it can be invoked (or called) by its name, and appropriate values are substituted for dummy parameters. Moving literal data into file register data RAM is a widely used service, but there is no instruction for that. We can use a macro to do the job as shown in the following code:

```
MOVLF MACRO K, MYREG
      MOVLW K
      MOVWF MYREG
      ENDM
```

The above is the macro definition. Note that dummy arguments of K and MYREG are mentioned in the body of macro.

The following are three examples of how to use the above macro:

```
1. MOVLF      0x55, 0x20        ;send value 55H to loc 20H

2. VAL_1      EQU 0x55
   RAM_LOC    EQU 0x20
   MOVLF      VAL_1, RAM_LOC

3. MOVLF      0x55, PORTB       ;send value 55H to Port B
```

The instruction "MOVLF 0x5, 0x20" invokes the macro. The assembler expands the macro by providing the following code in the .lst file:

```
M          MOVLW 5
M          MOVWF 0x20
```

The M indicates that the code is from the macro.

LOCAL directive

In the discussion of macros so far, the examples chosen do not have a label or name in the body of the macro. This is because if a macro is expanded more than once in a program and there are labels in the label field of the body of the macro, these labels must be declared as LOCAL. Otherwise, an assembler error would be generated when the same label was encountered in two or more places. The following rules must be observed in the body of the macro:

1. All labels in the label field must be declared LOCAL.
2. The LOCAL directive must be right after the MACRO directive. In other words, it must be placed even before comments and the body of the macro; otherwise, the assembler gives an error.
3. The LOCAL directive can be used to declare all names and labels at once as follows:
              LOCAL   name1, name2, name3
    or one at a time as:
              LOCAL   name1
              LOCAL   name2
              LOCAL   name3

To clarify these points, look at the following macro for time delay:

```
DELAY_1 MACRO V1, TREG
        LOCAL BACK
        MOVLW V1
        MOVWF TREG
BACK    NOP
        NOP
        NOP
        NOP
        DECF  TREG,F
        BNZ   BACK
        ENDM
```

Notice that the "BACK" label is defined as LOCAL right after the MACRO directive. Defining this label anywhere else causes an error. The use of a LOCAL directive allows the assembler to define the labels separately each time it encounters them. Examining the list file shows that when the macro is expanded for the first time, the list file has "??0000", for the second time it has "??0001", and for the third time it has "??0002" in place of the "BACK" label, indicating that the "BACK" label is local. To clarify this concept, see Program 6-4 without the LOCAL directive to see how the assembler will give an error. The following code is another macro for a time delay with a nested loop:

```
DELAY_2 MACRO V1, V2, R1, R2
        LOCAL BACK
        LOCAL AGAIN
        MOVLW V2
        MOVWF R2
AGAIN   MOVLW V1
        MOVWF R1
BACK    NOP
        NOP
        NOP
        NOP
        DECF   R1,F
        BNZ    BACK
        DECF   R2,F
        BNZ    AGAIN
        ENDM
```

Now examine Program 6-4 to see how to use a macro in a program.

```
;------------------------------------------
;Program 6-4: toggling Port B using macros
        #include P18F458.INC

;---------------sending data to fileReg macro
MOVLF MACRO K, MYREG
      MOVLW K
      MOVWF MYREG
      ENDM

;---------------------------time delay macro
DELAY_1 MACRO V1, TREG
        LOCAL BACK
        MOVLW V1
        MOVWF TREG
BACK    NOP
        NOP
        NOP
        NOP
        DECF   TREG,F
        BNZ    BACK
        ENDM

;--------------------------program starts
      ORG    0
      CLRF   TRISB            ;Port B as an output
```

236

```
OVER   MOVLF    0x55,PORTB
       DELAY_1 0x200,0x10
       MOVLF    0xAA,PORTB
       DELAY_1 0x200,0x10
       BRA      OVER
       END
;-------------------end of file
```

INCLUDE directive

Assume that several macros are used in every program. Must they be rewritten every time? The answer is no, if the concept of the INCLUDE directive is known. The INCLUDE directive allows a programmer to write macros and save them in a file, and later bring them into any program file. For example, assume that the following widely used macros were written and then saved under the filename "MYMACRO1.MAC".

Assuming that these macros are saved on a disk under the filename "MYMACRO1.MAC", the INCLUDE directive can be used to bring this file into any ".asm" file and then the program can call upon any of the macros as many times as needed. When a file includes all macros, the macros are listed at the beginning of the ".lst" file and, as they are expanded, will be part of the program. To understand this, see Program 6-5.

```
;Program 6-5: toggling Port B using macros
       #include P18F458.INC
       #include "MYMACRO1.MAC" ;get macros from macro file

;--------------------------program starts
       ORG   0
       CLRF  TRISB             ;Port B as an output
OVER   MOVLF 0x55,PORTB
       DELAY_1     0x200,0x10
       MOVLF 0xAA,PORTB
       DELAY_1     0x200,0x10
       BRA   OVER
       END
;------------------end of file
```

NOEXPAND/EXPAND directive

When viewing the .lst file with macros, we see them fully displayed. The expand directive is set by default, and it shows the macro at every location it is called. This is fine for two or three iterations, but when there are more, it can become cumbersome. Using the noexpand directive, we can turn off the display of macros in the list file.

```
        00001  ;Program 6-4:toggling Port B using macros
        00002          #include P18F458.INC
        00003          NOEXPAND
        00004  ;--------------sending data to fileReg macro
        00005  MOVLF MACRO K, MYREG
        00006          MOVLW     K
        00007          MOVWF     MYREG
        00008          ENDM
        00009
        00010  ;--------------------------time delay macro
        00011  DELAY_1 MACRO V1, TREG
        00012          LOCAL     BACK
        00013          MOVLW     V1
        00014          MOVWF     TREG
        00015  BACK    NOP
        00016          NOP
        00017          NOP
        00018          NOP
        00019          DECF      TREG,F
        00020          BNZ       BACK
        00021          ENDM
        00022
        00023  ;--------------------------program starts
000000  00024          ORG       0
000000 6A93 00025      CLRF      TRISB        ;Port B as an output
        00026  OVER    MOVLF     0x55,PORTB
        00027          DELAY_1   0x200,0x10
        00028          MOVLF     0xAA,PORTB
        00029          DELAY_1   0x200,0x10
00002A D7EB 00030      BRA       OVER
        00031          END
```

Figure 6-11. List File with NOEXPAND Option for Program 6-4

```
            00001  ;Program 6-4:toggling Port B using macros
            00002         #include P18F458.INC
            00003         EXPAND
            00004  ;--------------sending data to fileReg macro
            00005  MOVLF MACRO K, MYREG
            00006         MOVLW K
            00007         MOVWF MYREG
            00008         ENDM
            00009
            00010  ;-------------------------time delay macro
            00011  DELAY_1 MACRO V1, TREG
            00012         LOCAL BACK
            00013         MOVLW V1
            00014         MOVWF TREG
            00015  BACK  NOP
            00016         NOP
            00017         NOP
            00018         NOP
            00019         DECF   TREG,F
            00020         BNZ    BACK
            00021         ENDM
            00022
            00023  ;-------------------------program starts
000000      00024         ORG    0
000000 6A93 00025         CLRF   TRISB ;Port B as an output
            00026  OVER   MOVLF 0x55,PORTB
000002 0E55     M         MOVLW 0x55
000004 6E81     M         MOVWF PORTB
            00027  DELAY_1 0x200,0x10
  0000          M         LOCAL BACK
000006 0E00     M         MOVLW 0x200
000008 6E10     M         MOVWF 0x10
00000A 0000     M  BACK   NOP
00000C 0000     M         NOP
00000E 0000     M         NOP
000010 0000     M         NOP
000012 0610     M         DECF   0x10,F
000014 E1FA     M         BNZ    BACK
            00028         MOVLF 0xAA,PORTB
000016 0EAA     M         MOVLW 0xAA
000018 6E81     M         MOVWF PORTB
            00029  DELAY_1 0x200,0x10
  0000          M         LOCAL BACK
00001A 0E00     M         MOVLW 0x200
00001C 6E10     M         MOVWF 0x10
00001E 0000     M  BACK   NOP
000020 0000     M         NOP
000022 0000     M         NOP
000024 0000     M         NOP
000026 0610     M         DECF   0x10,F
000028 E1FA     M         BNZ    BACK
00002A D7EB 00030         BRA    OVER
            00031         END
```

Figure 6-12. List File with EXPAND Option for Program 6-4

Macros vs. subroutines

Macros and subroutines are useful in writing assembly programs, but each have limitations. Macros increase code size everytime they are invoked. For example, if you call a 10-instruction macro 10 times, the code size is increased by 100 instructions. Whereas, if you call the same subroutine 10 times, the code size is only that of the subroutine instructions. The only problem with subroutines is that they use stack space when called, and this can cause problems when there are nested calls (a subroutine calling another subroutine). The nested call can lead to a stack overflow and cause the program to crash. The PIC18 has provisions for stack overflow, discussed in the PIC18 Reference Manual.

Modules

It is common practice in writing software packages to break down the project into small modules and distribute the task of writing those modules among several programmers. This not only makes the project more manageable but also has other advantages, such as:

1. Each module can be written, debugged, and tested individually.
2. The failure of one module does not stop the entire project.
3. The task of locating and isolating any problem is easier and less time consuming.
4. One can use the modules to link with high-level languages such as C.
5. Parallel development shortens considerably the time required to complete a project.

Next we explain how to write and link modules to create a single executable program.

Writing modules

In programs given in the last section, a main procedure was written that called many other subroutines. In those examples, if one subroutine did not work properly, the entire program would have to be rewritten and reassembled. A more efficient way to develop software is to treat each subroutine as a separate program (or module) with a separate filename. Then each one can be assembled and tested. After testing each program and making sure that each works, they can all be brought together (linked) to make a single program. To enable these modules to be linked together, certain Assembly language directives must be used. Among these directives, the two most widely used are EXTERN (external) and GLOBAL. The GLOBAL directive is the same as PUBLIC in other Assembly language programs. Each is discussed below.

EXTERN directive

The EXTERN directive is used to notify the assembler and linker that certain names and variables that are not defined in the present module are defined externally somewhere else. In the absence of the EXTERN directive, the assembler would show an error because it cannot find where the names are defined. The EXTERN directive has the following format:

```
EXTERN name1          ;each name can be in a separate EXTERN
EXTERN name2

EXTERN name1, name2   ;or many can be listed in the same EXTERN
```

GLOBAL directive

Names or parameters defined as EXTERN (indicating that they are defined outside the present module) must be defined as GLOBAL in the module where they are defined. Defining a name as GLOBAL (PUBLIC) allows the assembler and linker to match it with its EXTERN counterpart(s). The following is the format for the GLOBAL directive:

```
GLOBAL name1          ;each name can be in a separate directive
GLOBAL name2

GLOBAL name1, name2   ;or many can be listed in the same GLOBAL
```

Program 6-6 should help to clarify these concepts. It demonstrates that for every EXTERN definition there is a GLOBAL directive defined in another module. Notice the entry and exit points of the program. Modules that are called by the main module have their own END directives. See Program 6-6.

```
;-------------------------------------------------------
;PROG 6-6: MAIN.ASM - CALCULATING AND TESTING CHECKSUM BYTE
        #include P18F458.INC

RAM_ADDR     EQU 40H
COUNTREG     EQU 0x20          ;fileReg loc for counter
CNTVAL       EQU 4             ;counter value
CNTVAL1      EQU 5             ;counter value

        EXTERN CAL_CHKSUM
        EXTERN TEST_CHKSUM

PGM CODE
;------------main program
        ORG   0
        CALL  COPY_DATA        ;this subroutine is in this file
        CALL  CAL_CHKSUM       ;this sub is in external file
        CALL  TEST_CHKSUM      ;this sub is in external file
        BRA   $
```

```
;--------copying data from code ROM to data RAM
COPY_DATA
        MOVLW  low(MYBYTE)        ;WREG = 00 LOW-byte addr.
        MOVWF  TBLPTRL            ;ROM data LOW-byte addr.
        MOVLW  hi(MYBYTE)         ;WREG = 5, HIGH-byte addr.
        MOVWF  TBLPTRH            ;ROM data HIGH-byte addr.
        MOVLW  upper(MYBYTE)      ;WREG = 00 upper-byte addr.
        MOVWF  TBLPRTRU           ;ROM data upper-byte addr.
        LFSR   0,RAM_ADDR         ;FSR0 = RAM_ADDR, place to save
C1      TBLRD*+                   ;bring in next byte and inc TBLPTR
        MOVF   TABLAT,W           ;copy to WREG (Z = 1, if null)
        BZ     EXIT               ;is it null char? exit if yes
        MOVWF  POSTINC0           ;copy WREG to RAM and inc pointer
        BRA    C1
EXIT    RETURN


;----------my data in program ROM
        ORG 0x500
MYBYTE DB 0x25, 0x62, 0x3F, 0x52, 0x00
        END



;-----------------------------------------------------
;PROG 6-6: CALCCSB.ASM - CALCULATING CHECKSUM BYTE
        #include P18F458.inc

RAM_ADDR        EQU 40H
COUNTREG        EQU 0x20           ;fileReg loc for counter
CNTVAL          EQU 4              ;counter value
CNTVAL1         EQU 5              ;counter value


        GLOBAL CAL_CHKSUM

PGM CODE                          ;we use this to inform the linker that
                                  ;the code segment has the name PGM
CAL_CHKSUM
        MOVLW CNTVAL              ;WREG = 4
        MOVWF COUNTREG            ;load the counter
        LFSR  0,RAM_ADDR          ;load pointer. FSR0 = 40H
        CLRF  WREG
C2      ADDWF POSTINC0,W          ;add RAM to WREG and increment FSR0
        DECF  COUNTREG,F          ;decrement counter
        BNZ   C2                  ;loop until counter = zero
        XORLW 0xFF                ;1's comp
        ADDLW 1                   ;2'compl
        MOVWF POSTINC0
        RETURN
        END



;-----------------------------------------------------
;PROG 6-6: TESTCSB.ASM - TESTING CHECKSUM BYTE
        #include P18F458.inc

RAM_ADDR        EQU 40H
COUNTREG        EQU 0x20           ;fileReg loc for counter
CNTVAL          EQU 4              ;counter value
CNTVAL1         EQU 5              ;counter value
```

```
        GLOBAL TEST_CHKSUM
        PGM CODE
TEST_CHKSUM
        MOVLW CNTVAL1          ;WREG = 5
        MOVWF COUNTREG         ;load the counter
        CLRF  TRISB
        LFSR  0,0x40           ;load pointer. FSR0 = 40H
        CLRF  WREG
C3      ADDWF POSTINC0,W       ;add RAM and increment FSR0
        DECF  COUNTREG,F       ;decrement counter
        BNZ   C3               ;loop until counter = zero
        XORLW 0x0              ;EX-OR to see if zero
        BZ    G_1              ;is result zero? then good
        MOVLW 'B'
        MOVWF PORTB            ;if not, data is bad
        RETURN
G_1     MOVLW 'G'
        MOVWF PORTB            ;data is not corrupted
        RETURN
        END
```

Linking modules together

Assuming that each program module in Program 6-6 is assembled separately and saved under the filenames MAIN.O, CALCCSB.O, and TESTCSB.O, the following shows how to link them together with MPLINK in order to generate a single executable file:

> MPLink.exe" "18f458.lkr" "MAIN.O" "CALCCSB.O" "TESTCSB.O" /o"PRG6-6.COF"

Program 6-6 shows how the EXTERN and GLOBAL directives can also be applied to data variables. The linker program resolves external references by matching GLOBAL and EXTERN names. The linker program will search through the files specified in the MPLINK command for the external subroutines.

The MPLAB IDE handles the compiling and linking in one step. This aids in program development by reducing time and errors in typing the command line call.

Review Questions

1. Discuss the benefits of macro programming.
2. List the three parts of a macro.
3. Explain and contrast the macro definition, invoking the macro, and expanding the macro.
4. True or false. A label defined within a macro is automatically understood by the assembler to be local.
5. The_____directive is used within a module to indicate that the named variable or subroutine can be used by another module.
6. The_____directive is used within a module to indicate that the named variable or subroutine was defined in another module.

SUMMARY

This chapter described the addressing modes of the PIC18. Immediate addressing mode uses a constant for the operand. Direct or register indirect addressing modes can be used to access data stored in data RAM file registers of the PIC18. Register indirect addressing mode uses a register as a pointer to the data. The advantage of this is that it makes addressing dynamic rather than static. Indexed ROM addressing mode is widely used in accessing data elements of look-up table entries located in the program ROM space of the PIC18. The PIC18 allows the reading of fixed data stored in program ROM space, in addition to writing to ROM if the PIC18 is of the flash type.

A group of registers called SFRs (special function registers) can be accessed by their names or their addresses. We also discussed the bit-addressable data RAM locations and ports and showed how to use single-bit instructions to access them directly. The topic of bank switching was discussed, and we showed how to use BSR registers to access all 16 banks of RAM in the PIC18.

We discussed how to break up a program into several or many subroutines and write and test each one separately. Macros and modules were also explored and their advantages were discussed.

PROBLEMS

SECTION 6.1: IMMEDIATE AND DIRECT ADDRESSING MODES

1. Which of the following are invalid uses of immediate addressing mode?
   (a) MOVLW 0x24   (b) MOVLW MYREG, 0x30   (c) MOVLW 0x60
2. Identify the addressing mode for each of the following:
   (a) MOVWF PORTB (b) MOVLW 0x50   (c) MOVWF MYREG
   (d) MOVLW 0   (e) MOVFF MYBREG, YOUREG
   (f) MOVWF YOURREG
3. Indicate the address assigned to each of the following:
   (a) PORTB   (b) WREG   (c) PORTC
   (d) PORTD   (e) PCL   (f) PCH
   (g) PCU   (h) TRISC   (i) TRISB
   (j) STATUS   (k) FSR0L
4. Which bank is used for SFRs?
5. In accessing the SFRs, we must use _____ addressing mode.
6. What does the following instruction do? "MOVLW 0xF0"
7. What does the following instruction do? "MOVWF PORTC"
8. What does the following instruction do? "MOVF PORTC,W"
9. "CLRF MYREG" is a(n) _____ (valid, invalid) instruction.
10. The byte addresses assigned to the 128 bytes of lower data RAM are _____ to _____.
11. The byte addresses assigned to the SFRs are _____ to _____.
12. Indicate the byte addresses assigned to both of the following. Is there a gap

between them?

(a) RAM locations of access bank   (b) SFR of access bank

13. Write a program to add the following data and place the result in RAM location 20H: The data values are 6, 9, 2, 5, 7

SECTION 6.2: REGISTER INDIRECT ADDRESSING MODE

14. Which registers are allowed to be used as a pointer for register indirect addressing mode when accessing data RAM? Give their names and show how they are loaded.
15. Write a program to copy FFH into RAM locations 50H to 6FH.
16. Write a program to copy 10 bytes of data starting at RAM address 40H to RAM locations starting at 70H.
17. What is the size of the FSRx register?
18. Give the SFR registers related to the FSR0 and FSR1.
19. Write a program to clear RAM locations 0 to 7FH.
20. Write a program to toggle RAM locations 50H to 5FH.
21. Explain the role of the INDFx register.
22. How much RAM space does the FSRx register cover?

SECTION 6.3: LOOK-UP TABLE AND TABLE PROCESSING

23. Compile and state the contents of each ROM location for the following data:
```
             ORG    200H
MYDAT_1:     DB     "Earth"
MYDAT_2:     DB     "987-65"
MYDAT_3:     DB     "GABEH 98"
```
24. Compile and state the contents of each ROM location for the following data:
```
             ORG    340H
DAT_1: DB 0x22,0x56, B'10011001', D'32', 0xF6, B'11111011'
```
25. Which register is allowed to be used as a pointer for register indirect addressing mode when accessing data stored in program ROM? Give the name and show how it is loaded.
26. Explain the role of the TABLAT register.
27. What is the size of the TBLPTR register? How much ROM space does it cover?
28. Give the SFR registers related to the TBLPTR.
29. Write a program to read the following message from ROM and place it in data RAM starting at 50:
```
             ORG 0x600
MYDATA       DB     "1-800-999-9999",0
```
30. Write a program to find y where $y = x^2 + 2x + 5$, and x is between 0 and 9.
31. Write a program to find y where $y = 20x + 5$, and x is between 0 and 9.
32. Write a program to read the following message from ROM and place it in data RAM starting at 40:
```
             ORG 0x700
MYDATA       DB     "The earth is but one country",0
```
33. True or false. The table read instruction works for all PIC18 family members.
34. True or false. The table write instruction works for PIC18 family members

with flash ROM.

35. Assume that the lower four bits of PORTB are connected to four switches. Write a program to send the following ASCII characters to a PORTC, based on the status of the switches:

0000	'0'
0001	'1'
0010	'2'
0011	'3'
0100	'4'
0101	'5'
0110	'6'
0111	'7'
1000	'8'
1001	'9'
1010	'A'
1011	'B'
1100	'C'
1101	'D'
1110	'E'
1111	'F'

SECTION 6.4: BIT-ADDRESSABILITY OF DATA RAM

36. Write a program to generate a square wave with 75% duty cycle on bit RB5.
37. Write a program to generate a square wave with 80% duty cycle on bit RC7.
38. Write a program to monitor RB4. When it goes HIGH, the program will generate a sound (square wave of 50% duty cycle) on pin RB7.
39. Write a program to monitor RC1. When it goes LOW, the program will send the value 55H to RD0.
40. What register does the carry flag belong to?
41. What bit address is assigned to the Z flag?
42. Which of the following instructions are valid? If valid, indicate which bit is altered.

    (a) BSF PORTB,1 (b) BSF PORTC.3   (c) BCF WREG,1
    (d) BCF 0x30,1 (e) BCF PORTD,0   (f) BST STATUS,C
    (g) CLRF WREG,3 (h) CLRF FSR0

43. "BTG PORTB, 0" is a(n) _____ (valid, invalid) instruction.
44. Which of the I/O ports of PORTB, PORTC, and PORTD are bit-addressable?
45. Which of the registers of the PIC18 are bit-addressable?
46. Give an instruction to clear the carry flag.
47. Show how would you check whether the C flag is HIGH.
48. Show how would you check whether the Z flag is HIGH.
49. Give the bit locations in the status register assigned to the flag bits C, Z, DC, and OV.
50. True or false. The bit addresses 0–7 are assigned to each RAM location of 000–FFFH.

51. True or false. SFR registers are not bit-addressable.
52. Write instructions to save the C flag bit in bit 4 of location 10.
53. Write instructions to save the DC flag bit in bit 2 of location 16H.
54. Write instructions to save the Z flag bit in bit 7 of location 12H.
55. Write instructions to see whether the D0 and D1 bits of register WREG are HIGH. If so, divide register WREG by 4.
56. Write a program to see whether the D7 bit of register WREG is HIGH. If so, send a message to the LCD stating that WREG has a negative number.
57. Write a program to set HIGH all the bits of RAM location 20H using the following methods:
    (a) byte addresses (b) bit addresses
58. Write a program to see whether the WREG register is divisible by 8.
59. Write a program to find the number of zeros in file register location 05.

SECTION 6.5: BANK SWITCHING IN THE PIC18

60. What addressing mode is used to access the SFRs?
61. What addressing mode is used to access the last 128 bytes of RAM in the PIC18?
62. Give the address range of the lower and the upper 128 bytes of the access bank.
63. In the PIC18, the SFRs use the same addresses across all members and they are from _____ to _____.
64. The PIC18 can have a maximum of _____ banks.
65. Explain the difference between these two instructions.
    (a) ADDWF MYREG, F, 1   (b) ADDWF MYREG, F, 0
66. Which addressing modes are used to access various banks?
67. Write a program to put 55H into RAM locations 1C0–1CFH.
68. Write a program to copy the contents of RAM locations 20–2FH to RAM locations 2D0–2DFH.
69. Explain the difference between these two instructions.
    (a) CLRF MYREG,F,1   (b) CLRF MYREG,F,0
70. Explain the difference between these two instructions.
    (a) SETF MYREG,W,1   (b) SETF MYREG,W,0
71. Explain the difference between these two instructions.
    (a) INCF MYREG,F,1   (b) INCF MYREG,F,0

SECTION 6.6: CHECKSUM AND ASCII SUBROUTINES

72. Find the checksum byte for the following ASCII message: "Hello"
73. True or false. If we add all bytes, including the checksum byte, and the result is 00H, there is no error in the data.
74. Write a program to (a) get the data "Hello, my fellow world citizens" from program ROM, (b) calculate the checksum byte, and (c) test the checksum byte for any data error.
75. To display data on LCD or PC monitors, it must be in _____ (BINARY, BCD, ASCII).

---

76. Assume that the lower 4 bits of PB are connected to four switches. Write a program to send the following ASCII characters to PD based on the status of the switches:

0000	'0'
0001	'1'
0010	'2'
0011	'3'
0100	'4'
0101	'5'
0110	'6'
0111	'7'
1000	'8'
1001	'9'

77. Write a program to convert a series of packed BCD numbers to ASCII. Assume that the packed BCD is located in ROM locations starting at 700H. Place the ASCII codes in RAM locations starting at 40H.

```
                 ORG  700H
MYDATA           DB   76H,87H,98H,43H
```

78. Write a program to convert a series of ASCII numbers to packed BCD. Assume that the ASCII data is located in ROM locations starting at 300H. Place the BCD data in RAM locations starting at 60H.

```
                 ORG  300H
MYDATA           DB   "87675649"
```

79. Write a program to get an 8-bit binary number from PD, convert it to ASCII, and save the result in RAM locations 40H, 41H, and 42H. What is the result if PD has 1000 1101 binary as input?

SECTION 6.7: MACROS AND MODULES

80. Give two advantages of macros.
81. Which uses more program ROM space: a macro or a module?
82. Give three reasons to write programs with modules.
83. If a label or parameter is not defined in a given module, it must be declared as

_____.

84. If a label or parameter is used by other modules, it must be declared as _____ in the present module.
85. Repeat Problem 79 using macros and modules.

ANSWERS TO REVIEW QUESTIONS

SECTION 6.1: IMMEDIATE AND DIRECT ADDRESSING MODES

1. No
2. MOVLW B'10000000'
3. PIC does not allow us to move a literal value directly to file register locations.
4. True
5. False

SECTION 6.2: REGISTER INDIRECT ADDRESSING MODE

1. Direct. The memory location address is 0x40.
2. The lower 8 bits of the 12-bit address of the data RAM file register. The address is 0FE9H.
3. The upper 4 bits of the 12-bit address of the data RAM file register. The address is 0FEAH.
4. 12-bit
5. FSR0, FSR1, and FSR2

SECTION 6.3: LOOK-UP TABLE AND TABLE PROCESSING

1. TBLPTR
2. TBLPTR
3. TABLAT
4. 21-bit, 2 MB
5. TBLPTR
6. In TBLRD*+ the element is read first, and then TBLPTR is incremented, while in TBLRD+*, TBLPTR is incremented first, and then the element is fetched.
7. False. Only with flash ROM.

SECTION 6.4: BIT-ADDRESSABILITY OF DATA RAM

1. True
2. True
3. True
4. a, b, c, and d
5. All of them
6. BTFSS 0x03,1
7. (a) It sets to HIGH bit 1 of RAM location 20H.
   (b) It clears bit 7 of RAM location 32H.
   (c) It sets to HIGH bit 2 of RAM location 12H.
   (d) It sets to HIGH bit 4 of PORTB.
   (e) It sets to HIGH bit 1 of the status register.
8. BCF STATUS,C

SECTION 6.5: BANK SWITCHING IN THE PIC18

1. True
2. True
3. True
4. F80–FFFH
5.
```
   MYREG EQU    0x2
   MOVLB 0x02           ;load 2 into BSR (use bank 2)
   MOVLW 0x99           ;WREG = 99h
   MOVWF MYREG,1
```

6.
```
MYREG EQU    0x08
MOVLB 0x4            ;load 4 into BSR (use bank 4)
MOVLW 0x55           ;WREG=55h
MOVWF MYREG,1
```

7. True
8. 8-bit, 4

SECTION 6.6: CHECKSUM AND ASCII SUBROUTINES

1. The 35H and 37H give 57H in BCD. The 39H and 34H give 94H.
2. The ASCII data is 32H, 30H, 30H, 35H, while 05 and 20H are for BCD.
3. No. To make it BCD, the radix must be set to hex.
4. 33H and 33H
5. 242 or 32, 34H and 32H
6. ROM
7. 88H + 99H + AAH + BBH + CCH + DDH = 42FH. Dropping the carries we have 2FH, and its 2's complement is D1H.
8. False

SECTION 6.7: MACROS AND MODULES

1. Macro programming can save the programmer time by allowing a set of frequently repeated instructions to be invoked within the program with a single line. This can also make the code easier to read.
2. The three parts of a macro are the MACRO directive, the body, and the ENDM directive.
3. The macro definition is the list of statements the macro will perform. It begins with the MACRO directive and ends with the ENDM directive. The macro is invoked whenever it is called from within an Assembly language program. The macro is expanded when the Assembly program replaces the line invoking the macro with the Assembly language code in the body of the macro.
4. False. A label that is to be local to a macro must be declared local with the LOCAL directive.
5. GLOBAL
6. EXTERN

CHAPTER 7

PIC PROGRAMMING IN C

OBJECTIVES

Upon completion of this chapter, you will be able to:

>>   Examine C data types for the PIC18
>>   Code C18 programs for time delay and I/O operations
>>   Code C18 programs for I/O bit manipulation
>>   Code C18 programs for logic and arithmetic operations
>>   Code C18 programs for ASCII and BCD data conversion
>>   Code C18 programs for binary (hex) to decimal conversion
>>   Code C18 programs for data serialization
>>   Understand C18 C compiler RAM and ROM allocation

Why program the PIC18 in C?

Compilers produce hex files that we download into the ROM of the micro-controller. The size of the hex file produced by the compiler is one of the main concerns of microcontroller programmers for two reasons:

1. Microcontrollers have limited on-chip ROM.
2. The code space for the PIC18 is limited to 2M.

How does the choice of programming language affect the compiled pro-gram size? While Assembly language produces a hex file that is much smaller than C, programming in Assembly language is often tedious and time consuming. On the other hand, C programming is less time consuming and much easier to write, but the hex file size produced is much larger than if we used Assembly language. The following are some of the major reasons for writing programs in C instead of Assembly:

1. It is easier and less time consuming to write in C than in Assembly.
2. C is easier to modify and update.
3. You can use code available in function libraries.
4. C code is portable to other microcontrollers with little or no modification.

Several third-party companies develop C compilers for the PIC microcon-troller. Our goal is not to recommend one over another, but to provide you with the fundamentals of C programming for the PIC18. You can use the compiler of your choice for the chapter examples and programs. For this book we have chosen Microchip's C18 C compiler to integrate with MPLAB IDE. Microchip has a stu-dent version of the C18 C compiler available for download from their web site. See http://www.MicroDigitalEd.com for tutorials on the C18 C compiler and MPLAB simulator.

C programming for the PIC18 is the main topic of this chapter. In Section 7.1, we discuss data types, and time delays. I/O programming is shown in Section 7.2. The logic operations AND, OR, XOR, inverter, and shift are discussed in Section 7.3. Section 7.4 describes ASCII and BCD conversions and checksums. In Section 7.5, data serialization for the PIC18 is shown. In Section 7.6, we show how the C18 C compiler uses program ROM for data storage. We will examine C18 data RAM allocation in Section 7.7.

SECTION 7.1: DATA TYPES AND TIME DELAYS IN C

In this section we first discuss C data types for the PIC18 and then provide code for time delay functions.

C data types for the PIC18

One of the goals of C18 programmers is to create smaller hex files, so it is worthwhile to re-examine C data types for C18. In other words, a good under-standing of C data types for the C18 can help programmers to create smaller hex files. In this section we focus on the specific C data types that are most useful and widely used for the PIC18 microcontroller. Table 7-1 shows data types and sizes.

Table 7-1: Some Data Types Widely Used by C18

Data Type	Size in Bits	Data Range/Usage
unsigned char	8-bit	0 to 255
char	8-bit	−128 to +127
unsigned int	16-bit	0 to 65,535
int	16-bit	−32,768 to +32,767
unsigned short	16-bit	0 to 65,535
short	16-bit	−32,768 to +32,767
unsigned short long	24-bit	0 to 16,777,215
short long	24-bit	−8,388,608 to +8,388,607
unsigned long	32-bit	0 to 4,294,967,295
long	32-bit	−2,147,483,648 to +2,147,483,648

Unsigned char

Because the PIC18 is an 8-bit microcontroller, the character data type is the most natural choice for many applications. The unsigned char is an 8-bit data type that takes a value in the range of 0–255 (00–FFH). It is one of the most widely used data types for the PIC18. In many situations, such as setting a counter value, where there is no need for signed data, we should use the unsigned char instead of the signed char. Remember that C compilers use the signed char as the default unless we put the keyword *unsigned* in front of the char (see Example 7-1). We can also use the unsigned char data type for a string of ASCII characters, including extended ASCII characters. Example 7-2 shows a string of ASCII characters. See Example 7-3 for toggling ports.

In declaring variables, we must pay careful attention to the size of the data and try to use unsigned char instead of int if possible. Because the PIC18 microcontroller has a limited number of registers and data RAM locations, using int in place of char can lead to a larger-size hex file. Such misuse of data types in compilers such as Microsoft Visual C++ for x86 IBM PCs is not a significant issue.

Example 7-1

Write a C18 program to send values 00–FF to Port B.
Solution:
```
#include <P18F458.h>          //for TRISB and PORTB declarations
void main(void)
  {
    unsigned char z;
    TRISB = 0;                //make Port B an output
    for(z=0;z<=255;z++)
      PORTB = z;
    while(1);                 //NEEDED IF RUNNING IN HARDWARE
  }
```
Run the above program on your simulator to see how Port B displays values 00–FFH in binary. Notice that "while(1)" is needed if this program is running in hardware.

Example 7-2

Write a C18 program to send hex values for ASCII characters of 0, 1, 2, 3, 4, 5, A, B, C, and D to Port B.

Solution:
```
#include <P18F458.h>
void main(void)
  {
    unsigned char mynum[]= "012345ABCD";//data is stored in RAM
    unsigned char z;
    TRISB = 0;              //make Port B an output
    for(z=0;z<10;z++)
      PORTB = mynum[z];
    while(1);               //stay here forever
  }
```
Run the above program on your simulator to see how Port B displays values 30H, 31H, 32H, 33H, 34H, 35H, 41H, 42H, 43H, and 44H (the hex values for ASCII 0, 1, 2, etc.). Notice that the last statement "while(1)" is needed only if we run the program in hardware. This is like "GOTO $" or "BRA $" in Assembly language.

Example 7-3

Write a C18 program to toggle all the bits of Port B continuously.

Solution:

```
// Toggle PB forever
#include <P18F458.h>
void main(void)
  {
    TRISB = 0;           //make Port B an output
    for(;;)              //repeat forever
      {
        PORTB = 0x55;   //0x indicates the data is in hex (binary)
        PORTB = 0xAA;
      }
  }
```

Run the above program on your simulator to see how Port B toggles continuously.

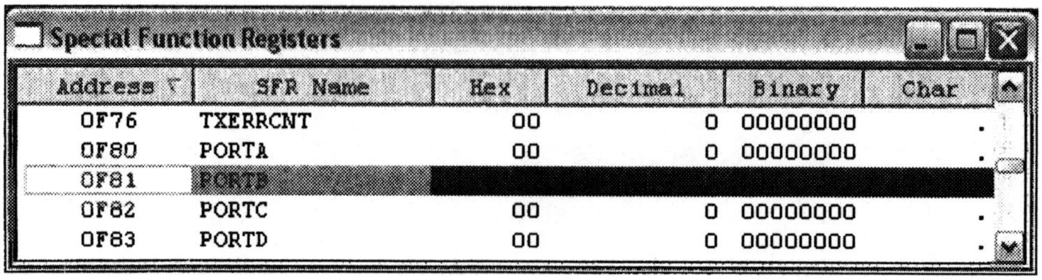

Special Function Registers						
Address	SFR Name	Hex	Decimal	Binary	Char	
0F76	TXERRCNT	00	0	00000000	.	
0F80	PORTA	00	0	00000000	.	
0F81	PORTB					
0F82	PORTC	00	0	00000000	.	
0F83	PORTD	00	0	00000000	.	

Figure 7-1. Examining the Special Function Registers Using MPLAB

Signed char

The signed char is an 8-bit data type that uses the most significant bit (D7 of D7–D0) to represent the – or + value. As a result, we have only 7 bits for the magnitude of the signed number, giving us values from –128 to +127. In situations where + and – are needed to represent a given quantity such as temperature, the use of the signed char data type is necessary.

Again, notice that if we do not use the keyword *unsigned,* the default is the signed value. For that reason we should stick with the unsigned char unless the data needs to be represented as signed numbers.

Example 7-4

Write a C18 program to send values of –4 to +4 to Port B.

Solution:
```
//sign numbers
#include <P18F458.h>
void main(void)
  {
    char mynum[]= {+1,-1,+2,-2,+3,-3,+4,-4};
    unsigned char z;
    TRISB = 0;                    //make Port B an output
    for(z=0;z<8;z++)
      PORTB = mynum[z];
    while(1);                     //stay here forever
  }
```

Run the above program on your simulator to see how PORTB displays values of 1, FFH, 2, FEH, 3, FDH, 4, and FCH (the hex values for +1, –1, +2, –2, etc.). See Chapter 5 for discussion of signed numbers.

Unsigned int

The unsigned int is a 16-bit data type that takes a value in the range of 0 to 65,535 (0000–FFFFH). In the PIC18, unsigned int is used to define 16-bit variables such as memory addresses. It is also used to set counter values of more than 256. Because the PIC18 is an 8-bit microcontroller and the int data type takes two bytes of RAM, we must not use the int data type unless we have to. Because registers and memory accesses are in 8-bit chunks, the misuse of int variables will result in a larger hex file. Such misuse is not a problem in PCs with 512 megabytes of memory, the 32-bit Pentium's registers and memory accesses, and a bus speed of 133 MHz. For PIC18 programming, however, do not use signed int in places where unsigned char will do the job. Of course, the compiler will not generate an error for this misuse, but the overhead in hex file size will be noticeable. Also, in situations where there is no need for signed data (such as setting counter values), we should use unsigned int instead of signed int. This gives a much wider range for data declaration. Again, remember that the C compiler uses signed int as the default unless we specify the keyword *unsigned.*

Signed int

Signed int is a 16-bit data type that uses the most significant bit (D15 of D15–D0) to represent the – or + value. As a result, we have only 15 bits for the magnitude of the number, or values from –32,768 to +32,767.

Other data types

The unsigned int is limited to values 0–65,535 (0000–FFFFH). The C18 C compiler supports both short long and long data types, if we want values greater than 16-bit. See Table 7-1. The short long value is 24 bits wide, while the long value is 32 bits wide.

Example 7-5

Write a C18 program to toggle all bits of Port B 50,000 times.

Solution:
```
#include <P18F458.h>
void main(void)
  {
     unsigned int z;
     TRISB = 0;                //make Port B an output
     for(z=0;z<=50000;z++)
        {
          PORTB = 0x55;
          PORTB = 0xAA;
        }
     while(1);                 //stay here forever
  }
```

Run the above program on your simulator to see how Port B toggles continuously. Notice that the maximum value for unsigned int is 65,535.

Example 7-6

Write a C18 program to toggle all bits of Port B 100,000 times.

Solution:

```
//toggle PB 100,00 times
#include <P18F458.h>
void main(void)
  {
     unsigned short long z;
     unsigned int x;
     TRISB = 0;                //make Port B an output
     for(z=0;z<=100000;z++)
        {
          PORTB = 0x55;
          PORTB = 0xAA;
        }
     while(1);                 //stay here forever
  }
```

Time delay

There are two ways to create a time delay in C18:

1. Using a simple for loop
2. Using the PIC18 timers

In either case, when we write a time delay we must use the oscilloscope to measure the duration of our time delay. Next, we use the for loop to create time delays. The use of the PIC18 timer to create time delays is postponed until Chapter 9.

In creating a time delay using a for loop, we must be mindful of two factors that can affect the accuracy of the delay:

1. The crystal frequency connected to the OSC1–OSC2 input pins is the most important factor in the time delay calculation. The duration of the clock period for the instruction cycle is a function of this crystal frequency.
2. The second factor that affects the time delay is the compiler used to compile the C program. When we program in Assembly language, we can control the exact instructions and their sequences used in the delay subroutine. In the case of C programs, it is the C compiler that converts the C statements and functions to Assembly language instructions. As a result, different compilers produce different code. In other words, if we compile a given C program with different compilers, each compiler produces different-size hex code.

For the above reasons, when we write time delays for C, we must use the oscilloscope to measure the exact duration. Look at Examples 7-7 and 7-8.

Example 7-7

Write a C18 program to toggle all the bits of Port B ports continuously with a 250 ms delay. Assume that the system is PIC18F458 with XTAL = 10 MHz.

Solution:

```
#include <P18F458.h>
void MSDelay(unsigned int);
void main(void)
  {
     TRISB = 0;                //make Port B an output
     while(1)                  //repeat forever
       {
           PORTB = 0x55;
           MSDelay(250);
           PORTB = 0xAA;
           MSDelay(250);
       }
  }

void MSDelay(unsigned int itime)
  {
     unsigned int i; unsigned char j;
     for(i=0;i<itime;i++)
       for(j=0;j<165;j++);
  }
```

Example 7-8

Write a C18 program to toggle all the bits of Port C and Port D continuously with a 250 ms delay.

Solution:

```c
//this program is tested for the PIC18F458 with XTAL = 10 MHz
#include <P18F458.h>
void MSDelay(unsigned int);
void main(void)
  {
    TRISC = 0;
    TRISD = 0;                  //make Ports C and D output
    while(1)                    //another way to do it forever
      {
        PORTC = 0x55;
        PORTD = 0x55;
        MSDelay(250);
        PORTC = 0xAA;
        PORTD = 0xAA;
        MSDelay(250);
      }
  }
void MSDelay(unsigned int itime)
  {
    unsigned int i; unsigned char j;
    for(i=0;i<itime;i++)
      for(j=0;j<165;j++);
  }
```

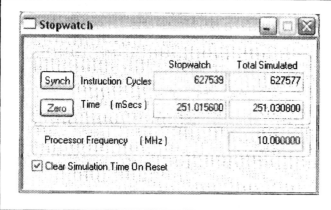

MPLAB's simulator has a stopwatch function that allows us to view the time delay before we program the microcontroller.

Figure 7-2. Time Delay Measurement for Example 7-8 Using MPLAB

Review Questions

1. Give the magnitude of the unsigned char and signed char data types.
2. Give the magnitude of the unsigned int and signed int data types.
3. If we are declaring a variable for a person's age, we should use the ___ data type.
4. True or false. Using a for loop to create a time delay is not recommended if you want your code be portable to other PIC18-based systems.
5. Give two factors that can affect the delay size.

SECTION 7.2: I/O PROGRAMMING IN C

In this section we look at C programming of the I/O ports for the PIC18. We look at both byte and bit I/O programming.

Byte size I/O

As we stated in Chapter 4, ports PORTA–PORTD are byte accessible. We use the PORTA–PORTD labels as defined in the C18 header file. See Example 7-9. Examine the next few examples to get a better understanding of how ports are accessed in C18.

Example 7-9

LEDs are connected to bits in Port B and Port C. Write a C18 program that shows the count from 0 to FFH (0000 0000 to 1111 1111 in binary) on the LEDs.

Solution:

```
#include <P18F458.h>
#define LED PORTC              //notice how we can define Port C
void main(void)
   {
    TRISB = 0;                 //make Port B an output
    TRISC = 0;                 //make Port C an output
    PORTB = 00;                //clear Port B
    LED = 0;                   //clear Port C
    for(;;)                    //repeat forever
       {
        PORTB++;               //increment Port B
        LED++;                 //increment Port C
       }
   }
```

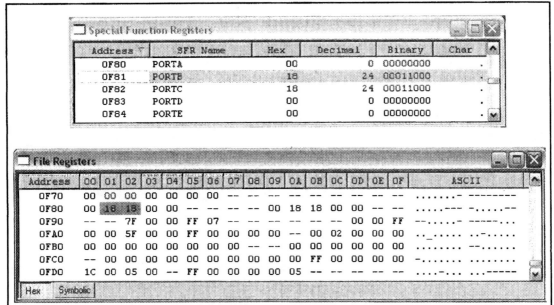

Figure 7-3. Example 7-9 Results After 24 Iterations

Example 7-10

Write a C18 program to get a byte of data from Port B, wait 1/2 second, and then send it to Port C.

Solution:

```
#include <P18F458.h>
void MSDelay(unsigned int);
void main(void)
  {
    unsigned char mybyte;
    TRISB = 0xFF;              //Port B as input
    TRISC = 0;                 //Port C as output
    while(1)
      {
        mybyte = PORTB;        //get a byte from Port B
        MSDelay(500);
        PORTC = mybyte;        //send it to Port C
      }
  }

void MSDelay(unsigned int itime)
  {
    unsigned int i;
    unsigned char j;
    for(i=0;i<itime;i++)
      for(j=0;j<165;j++);
  }
```

Example 7-11

Write a C18 program to get a byte of data from Port C. If it is less than 100, send it to Port B; otherwise, send it to Port D.

Solution:

```
#include <P18F458.h>
void main(void)
  {
    unsigned char mybyte;
    TRISC = 0xFF;              //make Port C an input
    TRISB = 0;
    TRISD = 0;                 //both Port B and D as output
    while(1)
      {
        mybyte = PORTC;        //get a byte from PORTC
        if(mybyte < 100)
          PORTB = mybyte;      //send it to PORTB if less than 100
        else
          PORTD = mybyte;      //send it to PORTD if more than 100
      }
  }
```

Bit-addressable I/O programming

The I/O ports of PIC18 are bit-addressable. We can access a single bit without disturbing the rest of the port. We use `PORTxbits.Rxy` to access a single bit of Portx, where x is the port A, B, C, or D, and y is the bit (0–7) of that port. For example, PORTBbits.RB7 indicates PORTB.7. We access the TRISx registers in the same way where TRISBbits.TRISB7 indicates the D7 of the TRISB. Table 7-2 shows the single-bit addresses of a PIC18. Study the next few examples to become familiar with the syntax.

Table 7-2: Single-Bit Addresses of PIC18F458/4580 Ports

PORTA	PORTB	PORTC	PORTD	PORTE	Port's Bit
RA0	RB0	RC0	RD0	RE0	D0
RA1	RB1	RC1	RD1	RE1	D1
RA2	RB2	RC2	RD2	RE2	D2
RA3	RB3	RC3	RD3		D3
RA4	RB4	RC4	RD4		D4
RA5	RB5	RC5	RD5		D5
	RB6	RC6	RD6		D6
	RB7	RC7	RD7		D7

PORT bits structure

Figure 7-4 shows the structure for the Port B bits as given by the C18 C compiler. You can find the structure of the ports in the microcontroller header file.

```
extern volatile near unsigned char        PORTB;
extern volatile near union {
  struct {
    unsigned RB0:1;
    unsigned RB1:1;
    unsigned RB2:1;
    unsigned RB3:1;
    unsigned RB4:1;
    unsigned RB5:1;
    unsigned RB6:1;
    unsigned RB7:1;
  };
  struct {
    unsigned INT0:1;
    unsigned INT1:1;
    unsigned CANTX:1;
    unsigned CANRX:1;
    unsigned :1;
    unsigned PGM:1;
    unsigned PGC:1;
    unsigned PGD:1;
  };
} PORTBbits;
```

Figure 7-4. Port B Bit Structure

Example 7-12

Write a C18 program to toggle only bit RB4 continuously without disturbing the rest of the bits of Port B.

Solution:

```c
#include <P18F458.h>
#define mybit PORTBbits.RB4          //declare single bit
void main(void)
   {
    TRISBbits.TRISB4=0;              //make RB4 an output
    while(1)
      {
        mybit = 1;                   //turn on RB4
        mybit = 0;                   //turn off RB4
      }
   }
```

Example 7-13

Write a C18 program to monitor bit PC5. If it is HIGH, send 55H to Port B; otherwise, send AAH to Port D.

Solution:

```c
#include <P18F458.h>
#define mybit PORTCbits.RC5          //notice single-bit declaration
void main(void)
   {
    TRISCbits.TRISC5 = 1;            //RC5 as input
    TRISD = 0;                       //Ports C and D output
    while(1)
      {
        if(mybit == 1)
           PORTD = 0x55;
        else
           PORTD = 0xAA;
      }
   }
```

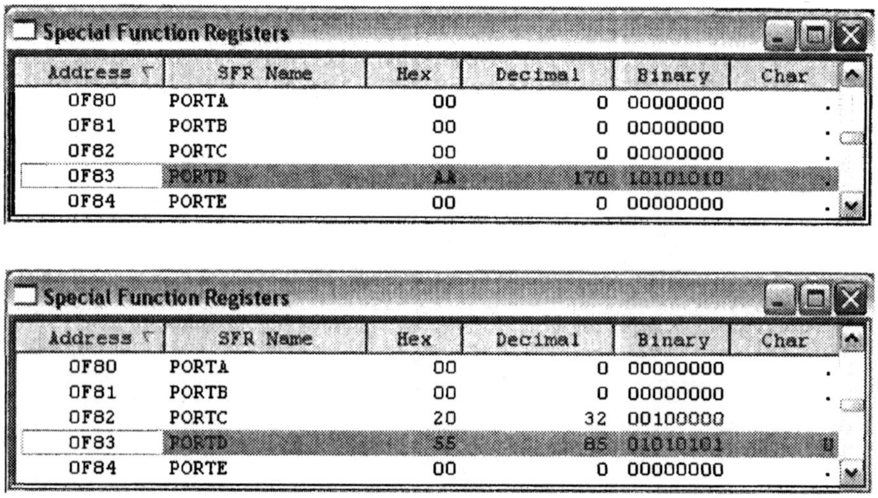

Special Function Registers

Address	SFR Name	Hex	Decimal	Binary	Char
0F80	PORTA	00	0	00000000	.
0F81	PORTB	00	0	00000000	.
0F82	PORTC	00	0	00000000	.
0F83	PORTD	AA	170	10101010	
0F84	PORTE	00	0	00000000	.

Special Function Registers

Address	SFR Name	Hex	Decimal	Binary	Char
0F80	PORTA	00	0	00000000	.
0F81	PORTB	00	0	00000000	.
0F82	PORTC	20	32	00100000	
0F83	PORTD	55	85	01010101	U
0F84	PORTE	00	0	00000000	.

Figure 7-5. Example 7-13 Results on MPLAB Simulator

Example 7-14

A door sensor is connected to the RB1 pin, and a buzzer is connected to RC7. Write a C18 program to monitor the door sensor, and when it opens, sound the buzzer. You can sound the buzzer by sending a square wave of a few hundred Hz frequency to it.

Solution:

```c
#include <P18F458.h>
void MSDelay(unsigned int);
#define Dsensor PORTBbits.RB1
#define buzzer PORTCbits.RC7
void main(void)
  {
    TRISBbits.TRISB1 = 1;          //PORTB.1 as an input
    TRISCbits.TRISC7 = 0;          //make PORTC.7 an output

    while(Dsensor == 1)
      {
        buzzer = 0;
        MSDelay(200);
        buzzer = 1;
        MSDelay(200);
      }
    while(1);                      //stay here forever
  }

void MSDelay(unsigned int itime)
  {
    unsigned int i;
    unsigned char j;
    for(i=0;i<itime;i++)
      for(j=0;j<165;j++);
  }
```

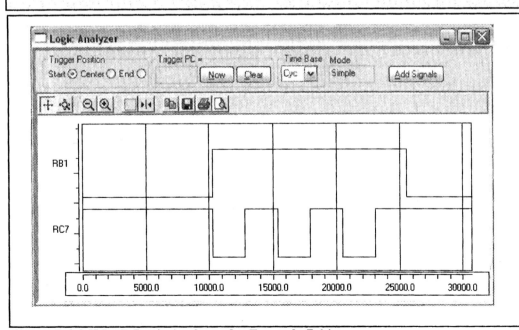

Figure 7-6. MPLAB Logic Analyzer for Example 7-14

Example 7-15

The data pins of an LCD are connected to Port B. The information is latched into the LCD whenever its Enable pin goes from HIGH to LOW. Write a C18 program to send "The Earth is but One Country" to this LCD.

Solution:
```
#include <P18F458.h>
#define LCDData PORTB                  //LCDData declaration
#define En PORTCbits.RC2               //the Enable pin
void main(void)
  {
    unsigned char message[] = "The Earth is but One Country";
    unsigned char z;
    TRISB = 0;                         //Port B as output
    TRISCbits.TRISC2 = 0;              //PortC.2 as output
    for(z=0;z<28;z++)                  //send all the 28 characters
      {
       LCDData = message[z];
       En=1;                           //a HIGH-
       En=0;                      //-to-LOW pulse to latch the LCD data
      }
    while(1);                          //stay here forever
  }
```

Run the above program on your simulator to see how PORTB displays each character of the message. Meanwhile, monitor bit PC.2 after each character is issued.

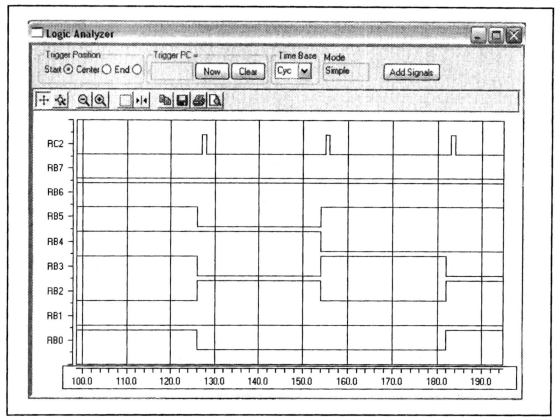

Figure 7-7. MPLAB Logic Analyzer Results for Example 7-15

Example 7-16

Write a C18 program to toggle all the bits of Port B, Port C, and continuously with a 250 ms delay.

Solution:

```c
#include <P18F458.h>
void MSDelay(unsigned int);
void main(void)
  {
    TRISB = 0;
    TRISC = 0;
    TRISD = 0;
    while(1)                      //do it forever
      {
        PORTB = 0x55;
        PORTC = 0x55;
        PORTD = 0x55;
        MSDelay(250);             //250 ms delay
        PORTB = 0xAA;
        PORTC = 0xAA;
        PORTD = 0xAA;
        MSDelay(250);
      }
  }

void MSDelay(unsigned int itime)
  {
    unsigned int i;
    unsigned char j;
    for(i=0;i<itime;i++)
      for(j=0;j<165;j++);
  }
```

Example 7-17

Write a C18 program to turn bit 5 of Port B on and off 50,000 times.

Solution:

```c
#include <P18F458.h>
#define MYBIT PORTBbits.RB5
void main(void)
  {
    unsigned int z;
    TRISBbits.TRISB5 = 0;           //make PORTB.5 an output
    for(z=0;z<50000;z++)
      {
        MYBIT = 1;
        MYBIT = 0;
      }
    while(1);                       //stay here forever
  }
```

Example 7-18

Write a C18 progra... the status of bit RB0, and send it to RC7 continuously.
Solution:

```c
#include <P18...h>
#define inbit  PORTBbits.RB0
#define outbit  PORTCbits.RC7
void main(...
{
    TRIS...s.TRISB0 = 1;        //make RB0 an input
    TRIS...its.TRISC7 = 0;      //make RC7 an output
    TRI...(1)
    whi...
    {
        outbit = inbit;        //get a bit from RB0
                               //and send it to RC7
    }
}
```

```
1:                #include <P18F458.h>
2:                #define inbit   PORTBbits.RB0
3:                #define outbit  PORTCbits.RC7
4:                void main(void)
5:                    {
6:                    TRISBbits.TRISB0 = 1;           //make RB0 an input
0000E2  8093      BSF 0xf93, 0, ACCESS
7:                    TRISCbits.TRISC7 = 0;           //make RC7 an output
0000E4  9E94      BCF 0xf94, 0x7, ACCESS
8:                    while(1)
0000F2  D7F9      BRA 0xe6
9:                        {
10:                       outbit = inbit;            //get bit from RB0
0000E6  5081      MOVF 0xf81, W, ACCESS
0000E8  0B01      ANDLW 0x1
0000EA  E002      BZ 0xf0
0000EC  8E82      BSF 0xf82, 0x7, ACCESS
0000EE  D001      BRA 0xf2
0000F0  9E82      BCF 0xf82, 0x7, ACCESS
11:                                                  //and send it to RC7
12:                        }
13:                    }
0000F4  0012   RETURN 0
```

Figure 7-8. Disassembly of Example 7-18

Review Questions

1. The address of PORTB is _____.
2. Write a short program that toggles all bits of PORTC.
3. Write a short program that toggles only bit 0 of PORTB.
4. True or false. All bits of PORTB are bit addressable.
5. True or false. All bits of TRISB are bit addressable.

SECTION 7.3: LOGIC OPERATIONS IN C

One of the most important and powerful features of the C language is its ability to perform bit manipulation. Because many books on C do not cover this important topic, it is appropriate to discuss it in this section. This section describes the action of bit-wise logic operators and provides some examples of how they are used.

Table 7-3: Bit-wise Logic Operators for C

		AND	OR	EX-OR	Inverter
A	B	A&B	A|B	A^B	Y=~B
0	0	0	0	0	1
0	1	0	1	1	0
1	0	0	1	1	
1	1	1	1	0	

Bit-wise operators in C

While every C programmer is familiar with the logical operators AND (&&), OR (||), and NOT (!), many C programmers are less familiar with the bit-wise operators AND (&), OR (|), EX-OR (^), inverter (~), shift right (>>), and shift left (<<). These bit-wise operators are widely used in software engineering for embedded systems and control; consequently, their understanding and mastery are critical in microprocessor-based system design and interfacing. See Table 7-3.

The following shows some examples using the C bit-wise operators:

1. 0x35 & 0x0F = 0x05   /* ANDing */
2. 0x04 | 0x68 = 0x6C   /* ORing: */
3. 0x54 ^ 0x78 = 0x2C   /* XORing */
4. ~0x55 = 0xAA   /* Inverting 55H */

Bit-wise shift operation in C

There are two bit-wise shift operators in C: (1) shift right (>>), and (2) shift left (<<).
Their format in C is as follows:
data >> number of bits to be shifted right
data << number of bits to be shifted left

The following shows some examples of shift operators in C:
1. 0x9A >> 3 = 0x13   /* shifting right 3 times */
2. 0x77 >> 4 = 0x07   /* shifting right 4 times */
3. 0x6 << 4 = 0x60   /* shifting left 4 times */

Study Examples 7-19 through 7-22. These show how the bit-wise operators are used in C.

Example 7-19

Run the following program on your simulator and examine the results.

Solution:

```c
#include <P18F458.h>
void main (void)
    {
       TRISB = 0;                  //make Ports B, C,
       TRISC = 0;                  //and D output ports
       TRISD = 0;
       PORTB = 0x35 & 0x0F;        //ANDing
       PORTC = 0x04 | 0x68;        //ORing
       PORTD = 0x54 ^ 0x78;        //XORing
       PORTB = ~0x55;              //inverting
       PORTC = 0x9A >> 3;          //shifting right 3 times
       PORTD = 0x77 >> 4;          //shifting right 4 times
       PORTB = 0x6 << 4;           //shifting left 4 times
       while(1);                   //stay here forever
    }
```

Example 7-20

Write a C18 program to toggle all the bits of Port B and Port C continuously with a 250 ms delay. Use the inverting operator.

Solution:

```c
#include <P18F458.h>
void MSDelay(unsigned int);
void main(void)
    {
       TRISB = 0;
       TRISC = 0;                  //make Ports B and C output
       PORTB = 0x55;
       PORTC = 0xAA;
       while(1)
         {
           PORTB = ~PORTB;
           PORTC = ~PORTC;
           MSDelay(250);
         }
    }

void MSDelay(unsigned int itime)
    {
       unsigned int i;
       unsigned char j;
       for(i=0;i<itime;i++)
         for(j=0;j<165;j++);
    }
```

Example 7-21

Rewrite the C18 program to toggle all the bits of Port B, Port C, and Port D continuously with a 250 ms delay. Use the EX-OR operator.

Solution:

```c
#include <P18F458.h>
void MSDelay(unsigned int);
void main(void)
  {
    TRISB = 0;
    TRISC = 0;
    TRISD = 0;                  //make Ports B,C, and D output
    PORTB=0x55;
    PORTC=0x55;
    PORTD=0x55;
    while(1)
      {
        PORTB=PORTB^0xFF;
        PORTC=PORTC^0xFF;
        PORTD=PORTD^0xFF;
        MSDelay(250);
      }
  }

void MSDelay(unsigned int itime)
  {
    unsigned int i;
    unsigned char j;
    for(i=0;i<itime;i++)
      for(j=0;j<165;j++);
  }
```

Example 7-22

Rewrite the C18 program to get bit RB0 and send it to RC7 after inverting it.

Solution:

```c
#include <P18F4550.h>
#define inbit PORTBbits.RB0
#define outbit PORTCbits.RC7
void main(void)
  {
    TRISBbits.TRISB0 = 1;           //make PORTB.0 an input
    TRISCbits.TRISC7 = 0;           //make PORTC.7 an output
    while(1)
      {
        outbit = ~inbit;            //get a bit from RB0
      }
  }
```

Example 7-23

Write a PIC C18 program to read the RB0 and RB1 bits and issue an ASCII character to PD according to the following table:

RB1	RB0	
0	0	send '0' to PORTD (notice ASCII '0' is 0x30)
0	1	send '1' to PORTD
1	0	send '2' to PORTD
1	1	send '3' to PORTD

Solution:

```
#include <P18F458.h>
void main(void)
  {
    unsigned char z;
    TRISB = 0xFF;                    //make Port B an input
    TRISD = 0;                       //make Port D an output
    while(1)                         //repeat forever
      {
        z = PORTB;                   //read PORTB
        z = z & 0x3;                 //mask the unused bits
        switch(z)                    //make decision
          {
            case(0):
              {
                PORTD = '0';         //issue ASCII 0
                break;
              }
            case(1):
              {
                PORTD = '1';         //issue ASCII 1
                break;
              }
            case(2):
              {
                PORTD = '2';         //issue ASCII 2
                break;
              }
            case(3):
              {
                PORTD = '3';         //issue ASCII 3
                break;
              }
          }
      }
  }
```

	Special Function Registers					
Address ▽	SFR Name	Hex	Decimal	Binary	Char	
0F80	PORTA	00	0	00000000	.	
0F81	PORTB	02	2	00000010	.	
0F82	PORTC	00	0	00000000	.	
0F83	PORTD	32	50	00110010	2	
0F84	PORTE	00	0	00000000	.	

Review Questions

1. Find the content of PORTB after the following C code in each case:
    (a) `PORTB=0x37&0xCA;`
    (b) `PORTB=0x37|0xCA;`
    (c) `PORTB=0x37^0xCA;`
2. To mask certain bits we must AND them with _____.
3. To set high certain bits we must OR them with _____.
4. EX-ORing a value with itself results in _____.
5. Find the contents of PORTC after execution of the following code:
    ```
    PORTC = 0;
    PORTC = PORTC | 0x99;
    PORTC = ~PORTC;
    ```

SECTION 7.4: DATA CONVERSION PROGRAMS IN C

Recall that BCD numbers were discussed in Chapters 5 and 6. As stated there, many newer microcontrollers have a real-time clock (RTC) where the time and date are kept even when the power is off. Very often the RTC provides the time and date in packed BCD. To display them, however, it must convert them to ASCII. In this section we show the application of logic and rotate instructions in the conversion of BCD and ASCII.

ASCII numbers

On ASCII keyboards, when the key "0" is activated, "011 0000" (30H) is provided to the computer. Similarly, 31H (011 0001) is provided for the key "1", and so on, as shown in Table 7-4.

Table 7-4: ASCII Code for Digits 0–9

Key	ASCII (hex)	Binary	BCD (unpacked)
0	30	011 0000	0000 0000
1	31	011 0001	0000 0001
2	32	011 0010	0000 0010
3	33	011 0011	0000 0011
4	34	011 0100	0000 0100
5	35	011 0101	0000 0101
6	36	011 0110	0000 0110
7	37	011 0111	0000 0111
8	38	011 1000	0000 1000
9	39	011 1001	0000 1001

Packed BCD to ASCII conversion

The RTC provides the time of day (hour, minute, second) and the date (year, month, day) continuously, regardless of whether the power is on or off. This data is provided in packed BCD, however. To convert packed BCD to ASCII, you must first convert it to unpacked BCD. Then the unpacked BCD is tagged with 011 0000 (30H). The following demonstrates converting from packed BCD to ASCII. See also Example 7-24.

```
Packed BCD      Unpacked BCD          ASCII
0x29             0x02, 0x09           0x32, 0x39
00101001         00000010,00001001    00110010,00111001
```

ASCII to packed BCD conversion

To convert ASCII to packed BCD, you first convert it to unpacked BCD (to get rid of the 3), and then combine to make packed BCD. For example, 4 and 7 on the keyboard give 34H and 37H, respectively. The goal is to produce 47H or "0100 0111", which is packed BCD.

```
Key    ASCII      Unpacked BCD      Packed BCD
4       34         00000100
7       37         00000111         01000111 or 47H
```

After this conversion, the packed BCD numbers are processed and the result will be in packed BCD format. Chapter 16 discusses the RTC chip and uses the BCD and ASCII conversion programs shown in Examples 7-24 and 7-25.

Example 7-24

Write a C18 program to convert packed BCD 0x29 to ASCII and display the bytes on PORTB and PORTC.

Solution:

```c
#include <P18F458.h>
void main(void)
   {
     unsigned char x, y, z;
     unsigned char mybyte = 0x29;
     TRISB = 0;
     TRISC = 0;                    //make Ports B and C output
     x = mybyte & 0x0F;            //mask upper 4 bits
     PORTB = x | 0x30;             //make it ASCII
     y = mybyte & 0xF0;            //mask lower 4 bits
     y = y >> 4;                   //shift it to lower 4 bits
     PORTC = y | 0x30;             //make it ASCII
   }
```

Address	SFR Name	Hex	Decimal	Binary	Char
0F80	PORTA	00	0	00000000	.
0F81	PORTB	39	57	00111001	9
0F82	PORTC	32	50	00110010	2
0F83	PORTD	00	0	00000000	.
0F84	PORTE	00	0	00000000	.

Figure 7-9. Result of Example 7-24 on MPLAB Simulator.

Example 7-25

Write a C18 program to convert ASCII digits of '4' and '7' to packed BCD and display it on PORTB.

Solution:

```
#include <P18F458.h>
void main(void)
   {
    unsigned char bcdbyte;
    unsigned char w = '4';
    unsigned char z = '7';
    TRISB = 0;              //make Port B an output
    w = w & 0x0F;           //mask 3
    w = w << 4;             //shift left to make upper BCD digit
    z = z & 0x0F;           //mask 3
    bcdbyte = w | z;        //combine to make packed BCD
    PORTB = bcdbyte;
   }
```

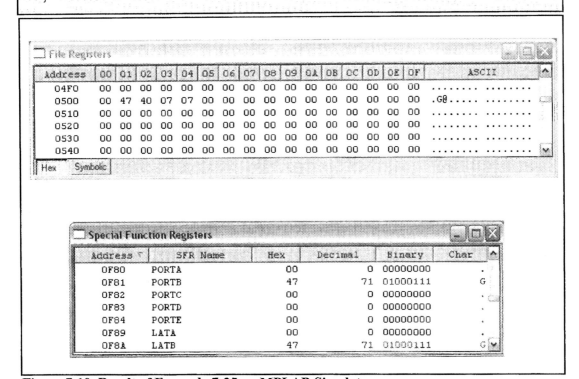

Figure 7-10. Result of Example 7-25 on MPLAB Simulator

Checksum byte in ROM

To ensure the integrity of ROM contents, every system must perform the checksum calculation. The checksum will detect any corruption of the contents of ROM. One of the causes of ROM corruption is current surge, either when the system is turned on or during operation. To ensure data integrity in ROM, the checksum process uses what is called a *checksum byte*. The checksum byte is an extra byte that is tagged to the end of a series of bytes of data. To calculate the checksum byte of a series of bytes of data, the following steps can be taken:

1. Add the bytes together and drop the carries.
2. Take the 2's complement of the total sum. This is the checksum byte, which becomes the last byte of the series.

To perform the checksum operation, add all the bytes, including the checksum byte. The result must be zero. If it is not zero, one or more bytes of data have been changed (corrupted). To clarify these important concepts, see Examples 7-26 through 7-28.

Example 7-26

Assume that we have 4 bytes of hexadecimal data: 25H, 62H, 3FH, and 52H.
(a) Find the checksum byte, (b) perform the checksum operation to ensure data integrity, and (c) if the second byte, 62H, has been changed to 22H, show how checksum detects the error.

Solution:

(a)   Find the checksum byte.
```
        25H
    +   62H
    +   3FH
    +   52H
       118H  (dropping carry of 1 and taking 2's complement, we get E8H)
```

(b)   Perform the checksum operation to ensure data integrity.
```
        25H
    +   62H
    +   3FH
    +   52H
    +   E8H
       200H  (dropping the carries we get 00, which means data is not corrupted)
```

(c)   If the second byte, 62H, has been changed to 22H, show how checksum detects the error.
```
        25H
    +   22H
    +   3FH
    +   52H
    +   E8H
       1C0H  (dropping the carry, we get C0H, which means data is corrupted)
```

Example 7-27

Write a C18 program to calculate the checksum byte for the data given in Example 7-26.

Solution:

```
#include <P18F458.h>
void main(void)
   {
   unsigned char mydata[] = {0x25,0x62,0x3F,0x52};
   unsigned char sum = 0;
   unsigned char x;
   unsigned char chksumbyte;
   TRISB = 0;
   TRISC = 0;                     //make Ports B and C output
   for(x=0;x<4;x++)
      {
      PORTC = mydata[x];          //issue each byte to PORTC
      sum = sum + mydata[x];      //add them together
      PORTB = sum;                //issue the sum to PORTB
      }
   chksumbyte = ~sum + 1;         //make 2's complement (invert +1)
   PORTB = chksumbyte;            //show the checksum byte
   }
```

Single-step the above program on the MPLAB simulator and examine the contents of PORTB and PORTC. Notice that each byte is put on PORTC as they are added.

Example 7-28

Write a C18 program to perform step (b) of Example 7-26. If the data is good, send ASCII character 'G' to PORTD. Otherwise, send 'B' to PORTD.

Solution:

```
#include <P18F458.h>
void main(void)
   {
   unsigned char mydata[] = {0x25,0x62,0x3F,0x52,0xE8};
   unsigned char chksum = 0;
   unsigned char x;
   TRISD = 0;                     //make Port D an output
   for(x=0;x<5;x++)
      chksum = chksum + mydata[x];   //add them together
   if(chksum == 0)
      PORTD = 'G';
   else
      PORTD = 'B';
   }
```

Change one or two values in the mydata array and simulate the program to see the results.

CHAPTER 7: PIC PROGRAMMING IN C

Binary (hex) to decimal and ASCII conversion in C18

The printf function is part of the standard I/O library in C and can do many things including converting data from binary (hex) to decimal, or vice versa. But printf takes a lot of memory space and increases your hex file substantially. For this reason, in systems based on the PIC18 microcontroller, it is better to write our own conversion function instead of using printf.

One of the most widely used conversions is binary to decimal conversion. In devices such as ADCs (Analog-to-Digital Converters), the data is provided to the microcontroller in binary. In some RTCs, the time and dates are also provided in binary. In order to display binary data, we need to convert it to decimal and then to ASCII. Because the hexadecimal format is a convenient way of representing binary data, we refer to the binary data as hex. The binary data 00–FFH converted to decimal will give us 000 to 255. One way to do that is to divide it by 10 and keep the remainder, as was shown in Chapters 5 and 6. For example, 11111101 or FDH is 253 in decimal. The following is one version of an algorithm for conversion of hex (binary) to decimal:

Hex	Quotient	Remainder
FD/0A	19	3 (low digit) LSD
19/0A	2	5 (middle digit)
		2 (high digit) (MSD)

Example 7-29 shows the C program for that algorithm.

Example 7-29

Write a C18 program to convert 11111101 (FD hex) to decimal and display the digits on PORTB, PORTC, and PORTD.

Solution:

```
#include <P18F458.h>
void main(void)
  {
    unsigned char x, binbyte, d1, d2, d3;
    TRISB = 0;
    TRISC = 0;
    TRISD = 0;              //Ports B, C, and D output
    binbyte = 0xFD;         //binary (hex) byte
    x = binbyte / 10;       //divide by 10
    d1 = binbyte % 10;      //find remainder (LSD)
    d2 = x % 10;            //middle digit
    d3 = x / 10;            //most-significant digit (MSD)
    PORTB = d1;
    PORTC = d2;
    PORTD = d3;
  }
```

Review Questions

1. For the following decimal numbers, give the packed BCD and unpacked BCD representations:
   (a) 15   (b) 99
2. Show the binary and hex formats for "76" and its packed BCD version.
3. 67H in BCD when converted to ASCII is ____H and ____H.
4. Does the following convert unpacked BCD to ASCII?
   ```
   mydata=0x09+0x30;
   ```
5. Why is the use of packed BCD preferable to ASCII?
6. Which takes more memory space: packed BCD or ASCII?
7. In Question 6, which is more universal?
8. Find the checksum byte for the following values; 22H, 76H, 5FH, 8CH, 99H.
9. To test data integrity, we add the bytes together, including the checksum byte. The result must be equal to ____ if the data is not corrupted.
10. An ADC provides an output of 0010 0110. How do we display that on the screen?

SECTION 7.5: DATA SERIALIZATION IN C

Serializing data is a way of sending a byte of data one bit at a time through a single pin of a microcontroller. There are two ways to transfer a byte of data serially:

1. Using the serial port. In using the serial port, the programmer has very limited control over the sequence of data transfer. The details of serial port data transfer are discussed in Chapter 10.
2. The second method of serializing data is to transfer data one bit a time and control the sequence of data and spaces between them. In many new generations of devices such as LCD, ADC, and EEPROM, the serial versions are becoming popular because they take less space on a printed circuit board. Although we can use standards such as I²C and CAN, not all devices support such standards. For this reason we need to be familiar with data serialization using the C language.

Examine the next four examples to see how data serialization is done in C.

Example 7-30

Write a C18 program to send out the value 44H serially one bit at a time via RC0. The LSB should go out first.

Solution:

```
//Serializing data via RC0 (SHIFTING RIGHT)
#include <P18F458.h>
#define PC0 PORTCbits.RC0
void main(void)
  {
    unsigned char conbyte = 0x44;
    unsigned char regALSB;
    unsigned char x;
    regALSB = conbyte;
    TRISCbits.TRISC0 = 0;           //make RC0 an output
    for(x=0;x<8;x++)
      {
        PC0 = regALSB & 0x01;
        regALSB = regALSB >> 1;
      }
  }
```

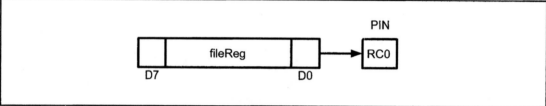

Figure 7-11. Shifting Bits Out (LSB Going First)

Example 7-31

Write a C18 program to send out the value 44H serially one bit at a time via RC0. The MSB should go out first.

Solution:

```
//Serializing data via RC0 (SHIFTING LEFT)
#include <P18F458.h>
#define PC0 PORTCbits.RC0
void main(void)
  {
    unsigned char conbyte = 0x88;
    unsigned char regAMSB;
    unsigned char x;
    regAMSB = conbyte;
    TRISCbits.TRISC0 = 0;           //make RC0 an output
    for(x=0;x<8;x++)
      {
        PC0 = (regAMSB >> 7) & 0x01;
        regAMSB = regAMSB << 1;
      }
  }
```

Example 7-32

Write a C18 program to bring in a byte of data serially one bit at a time via the RB0 pin. Place the byte on Port D. The LSB should come in first.

Solution:
```c
//Bringing in data via RB0 (SHIFTING RIGHT)
#include <P18F458.h>
#define PB0 PORTBbits.RB0
void main(void)
   {
      unsigned char x;
      unsigned char REGA=0;
      TRISBbits.TRISB0 = 1;      //RB0 as input
      TRISD = 0;                 //Port D as output
      for(x=0;x<8;x++)
        {
          REGA = REGA >> 1;
          REGA |= (PB0 & 0x01) << 7;
        }
      PORTD = REGA;
   }
```

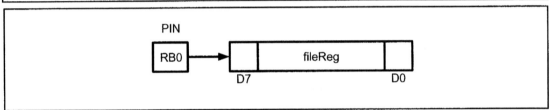

PIN

RB0 → fileReg

D7   D0

Figure 7-12. Shifting Bits in (Bring in LSB First)

Example 7-33

Write a C18 program to bring in a byte of data serially one bit at a time via the RB0 pin. The MSB should come in first.

Solution:

```c
//Bringing in data via RB0 (SHIFTING LEFT)
#include <P18F458.h>
#define PB0 PORTBbits.RB0
void main(void)
   {
      unsigned char x;
      unsigned char REGA=0;
      TRISBbits.TRISB0 = 1;      //RB0 as input
      TRISD = 0;                 //Port D as output
      for(x=0;x<8;x++)
        {
          REGA = REGA << 1;
          REGA |= PB0 & 0x01;
        }
      PORTD = REGA;
   }
```

SECTION 7.6: PROGRAM ROM ALLOCATION IN C18

Using program (code) space for predefined fixed data is a widely used option in the PIC18, as we saw in Chapter 6. In that chapter we saw how to use Assembly language instructions to access the data stored in the program code space. In this chapter, we explore the same concept using the C18 C compiler. We will also examine the far and near storage qualifier for ROM.

RAM data space vs. code data space

In the PIC18 we have two spaces in which to store data. They are as follows:

1. The 4096 bytes of data RAM space with address range 000–FFFH. As we have seen in previous chapters, many PIC18 chips have much less than 4096 bytes for the file register data RAM. We also have seen how we can read (from) or write (into) this RAM space directly or indirectly.

2. The 2M of code (program) space with addresses of 000000–1FFFFFH. This 2M bytes of on-chip ROM space is used for storing programs (opcodes) and therefore is directly under control of the program counter (PC). As we have seen in the previous chapters, many PIC18 chips have much less than

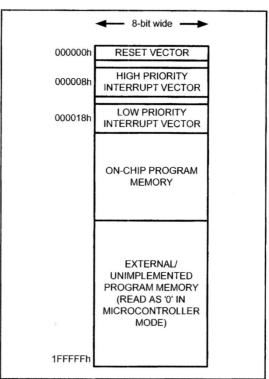

Figure 7-13. PIC18 Program ROM Space

2M of on-chip program ROM. We have also seen how to access the program ROM for the purpose of data storage using the TBLRD instruction (see Chapter 6). There is one problem with using this program code space for storage of fixed data: The more code space we use for data, the less is left for our program code. For example, if we have a PIC18 chip such as the PIC18F252 with only 4K of on-chip ROM, and we use 1K to store a look-up table, only 3K is left for the program. For some applications, this can be a problem. For this reason Microchip has added EEPROM memory to the PIC18 to be used for data storage. The EEPROM option of PIC18 is discussed in Chapter 14. Next, we will examine how the C18 compiler uses on-chip ROM space, and discuss how it places data into program ROM.

Table 7-5: Program ROM Size for Some PIC18F Family Members

	On-Chip Code ROM (Bytes)	Code Address Range (Hex)
PIC18F2220	4K	00000–00FFF
PIC18F2410	16K	00000–03FFF
PIC18F458/4580	32K	00000–07FFF
PIC18F6680	64K	00000–0FFFF
PIC18F8722	128K	00000–1FFFF

Allocating program space to data

In all our C18 examples so far, byte-size variables were stored in the data RAM. As we saw in Chapter 6, it is common practice to use the on-chip program ROM for the purpose of storing fixed data such as strings. This is specially useful since we have limited amount of file register data RAM. To make the C18 compiler use the program (code) ROM space for the fixed data, we use the keyword *rom* as shown in the following lines of C code:

```
rom char mynum[] = "Hello";        //use code space for data
rom char weekdays = 7, month= 12;  //use code space for data
```

The following code shows how to use program space for data in C18:

```
//Program 7-1
#include <P18F458.h>
rom const char mynum[]= "0123456789"; //uses program
                          //ROM space for fixed (constant) data
void main(void)
   {
     unsigned char z;
     TRISB = 0;           //make Port B an output
     for(z=0;z<10;z++)
       PORTB=mynum[z];
   }
```

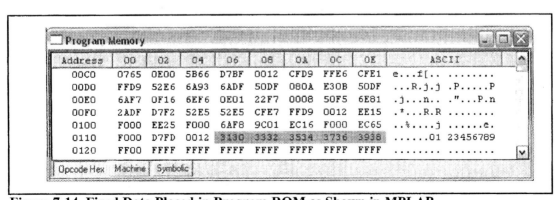

Figure 7-14. Fixed Data Placed in Program ROM as Shown in MPLAB

NEAR and FAR for code

As we have discussed earlier, the PIC18 microcontroller has a maximum of 2M of on-chip program ROM space. Not every family member, however, comes with that much on-chip program ROM. Some PIC18 chips come with as little as 4K and some come with 128K of program ROM. To make a more efficient use of the code space, the C18 compiler allows the use of near and far storage qualifiers to indicate in what region the data and code should be placed. The near qualifier is used to indicate that a program memory data variable is located in the first 64K of the program ROM. In order to indicate that a data variable in program ROM can be found anywhere in the 2M ROM space, we must use the *far* qualifier. See Table 7-6. Also see Program 7-2A. Note that the far storage qualifier is the default for the C18 if we do not specify it in our program.

Table 7-6: NEAR and FAR Usage for ROM

Storage qualifier	ROM
near	In program space of 0000–FFFFH (64 kB)
far	In program space of 000000–1FFFFFH (2 MB)

```
//Program 7-2A
#include <P18F458.h>
    near rom const char mydata[] = "HELLO"; //program ROM data
void main(void)
  {
    unsigned char z;
    TRISB = 0;                              //make Port B an output
    for(z=0;z<5;z++)
      PORTB = mydata[z];
  }
```

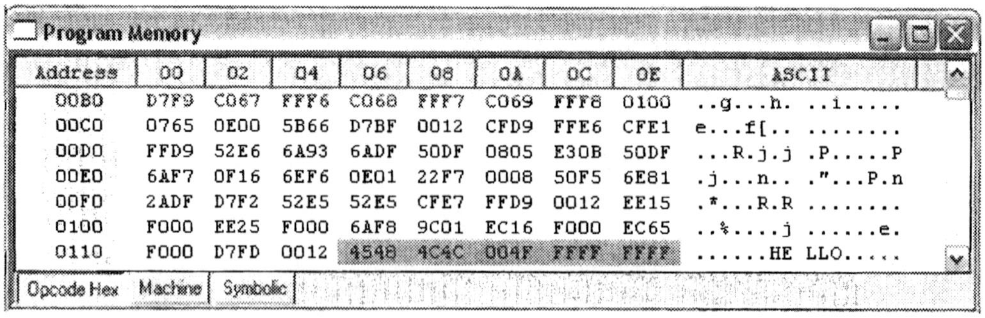

Address	00	02	04	06	08	0A	0C	0E	ASCII
00B0	D7F9	C067	FFF6	C068	FFF7	C069	FFF8	0100	..g...h. ..i.....
00C0	0765	0E00	5B66	D7BF	0012	CFD9	FFE6	CFE1	e...f[.. ....
00D0	FFD9	52E6	6A93	6ADF	50DF	0805	E30B	50DF	...R.j.j .P.....P
00E0	6AF7	0F16	6EF6	0E01	22F7	0008	50F5	6E81	.j...n.. ."...P.n
00F0	2ADF	D7F2	52E5	52E5	CFE7	FFD9	0012	EE15	.*....R.R ....
0100	F000	EE25	F000	6AF8	9C01	EC16	F000	EC65	..%....j ......e.
0110	F000	D7FD	0012	4548	4C4C	004F	FFFF	FFFF	......HE LLO.....

Opcode Hex Machine Symbolic

Notice 4 digits for address (0000–FFFF)

Figure 7-15. Using Near Storage Qualifier as Shown by MPLAB

In Program 7-2A, if we change the near to far and compile for the PIC18F8722 chip (which has 128K of program ROM), we have the following:

```
//Program 7-2B
#include <P18F8722.h>
   far rom const char mydata[]= "HELLO"; //program ROM data
void main(void)
  {
    unsigned char z;
    TRISB = 0;                             //make Port B an output
    for(z=0;z<5;z++)
      PORTB = mydata[z];
  }
```

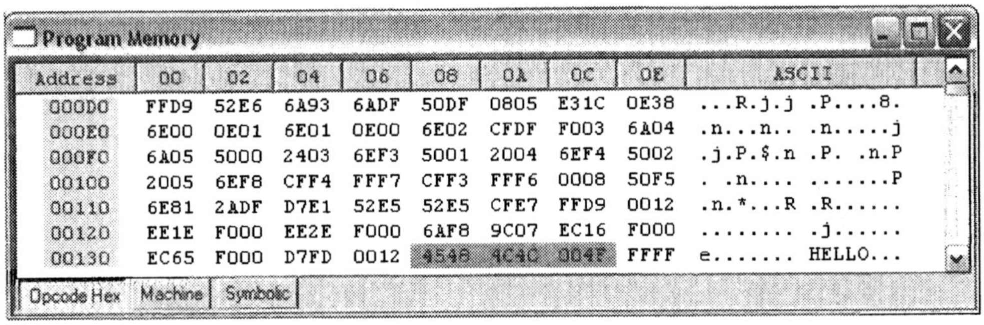

Program Memory
Address 00 02 04 06 08 0A 0C 0E ASCII
000D0 FFD9 52E6 6A93 6ADF 50DF 0805 E31C 0E38 ...R.j.j .P....8.
000E0 6E00 0E01 6E01 0E00 6E02 CFDF F003 6A04 .n...n.. .n.....j
000F0 6A05 5000 2403 6EF3 5001 2004 6EF4 5002 .j.P.$.n .P. .n.P
00100 2005 6EF8 CFF4 FFF7 CFF3 FFF6 0008 50F5 . .n.... ....P
00110 6E81 2ADF D7E1 52E5 52E5 CFE7 FFD9 0012 .n.*...R .R......
00120 EE1E F000 EE2E F000 6AF8 9C07 EC16 F000 .... .j......
00130 EC65 F000 D7FD 0012 4548 4C4C 004F FFFF e....... HELLO...
Opcode Hex
Machine
Symbolic

Notice 5 digits for hex address (00000–FFFFF)

Figure 7-16. Far Storage Qualifier as Shown in MPLAB

Pragma and allocating a fixed address to data and code

As we saw in Chapter 6, the MPLAB assembler allows us to place data or code at a specific ROM address using the ORG directive. To do the same thing in the C18 C compiler we use the #pragma section directive, where "section" is a portion of an application (code or data) that can be assigned an specific memory address location. In the case of the on-chip ROM program memory, we have two options: (1) code, and (2) romdata. The #pragma directive is used for the program because it contains executable instructions, while the #pragma romdata directive is used for fixed data such as strings and look-up tables. Next we explore the use of #pragma to allocate ROM addresses for the code and data.

Putting code in a specific ROM address

To place the code (containing executable instructions) at a specific address location of the program ROM, we use the #pragma code directive. Examine the Program 7-3 to see how the C code for the MSDelay function is placed at the ROM address of 0x300.

```
//Program 7-3
#include <P18F458.h>
#pragma code main = 0x50        //place the main at ROM addr 0x50
void MSDelay(unsigned int);
void main(void)
   {
     unsigned char mydata[] = "HELLO";
     unsigned char z;
     TRISB = 0;                 //make Port B an output
     for(z=0;z<5;z++)
       {
         PORTB = mydata[z];
         MSDelay(250);
       }
   }

#pragma code MSDelay = 0x300  //place delay at ROM addr 0x300
void MSDelay(unsigned int itime)
   {
     unsigned int i;
     unsigned char j;
     for(i=0;i<itime;i++)
        for(j=0;j<165;j++);
   }
```

Running the above programs on the MPLAB simulator and examining the program code space, we see that the main and MSDelay functions are located at the ROM addresses 0x50 and 0x300, respectively.

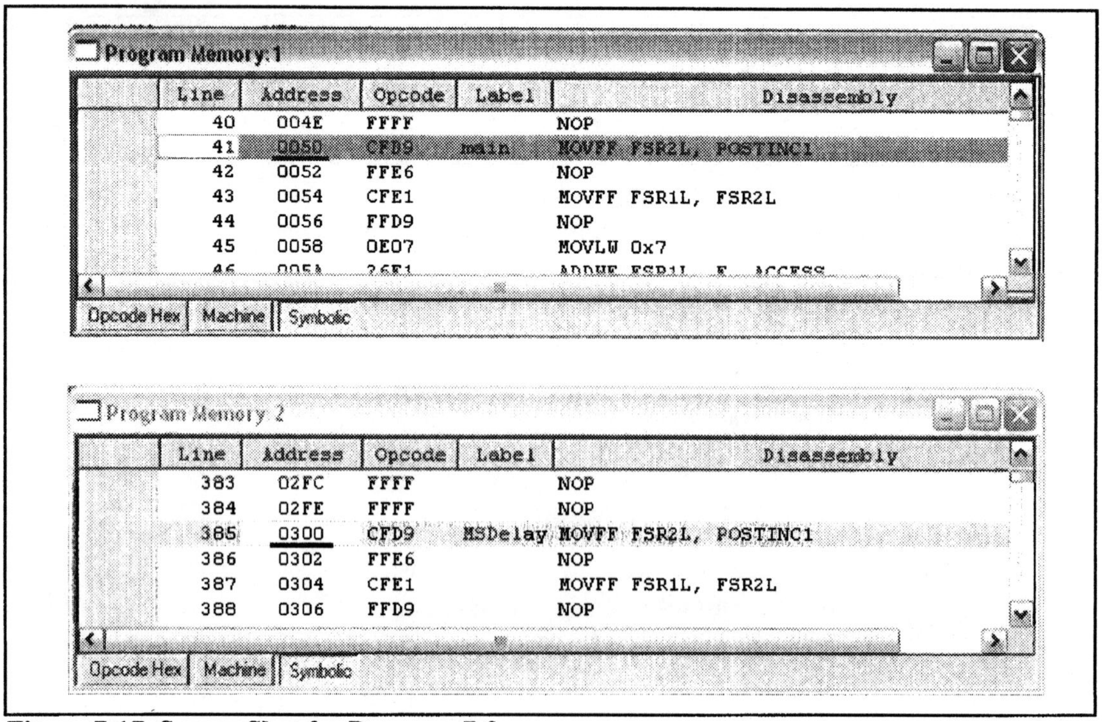

Figure 7-17. Screen Shot for Program 7-3

Putting data in a specific ROM address

To place the data (containing variables and constants, strings, and look-up tables) at a specific address of program ROM, we use the #pragma romdata directive. Examine the Program 7-4 to see how the C code assigns the program ROM address of 0x200 to the string "Hello".

```
//Program 7-4
#include <P18F458.h>
#pragma romdata mydata = 0x200   //place mydata at ROM addr 0x200
    near rom const char mydata[]= "HELLO";    //ROM data
void main(void)
  {
    unsigned char z;
    TRISB = 0;                     //make Port B an output
    for(z=0;z<5;z++)
      PORTB = mydata[z];
  }
```

Run the above program on the MPLAB simulator. Examine the program code space to see the string "HELLO" located at the ROM address starting at 0x50.

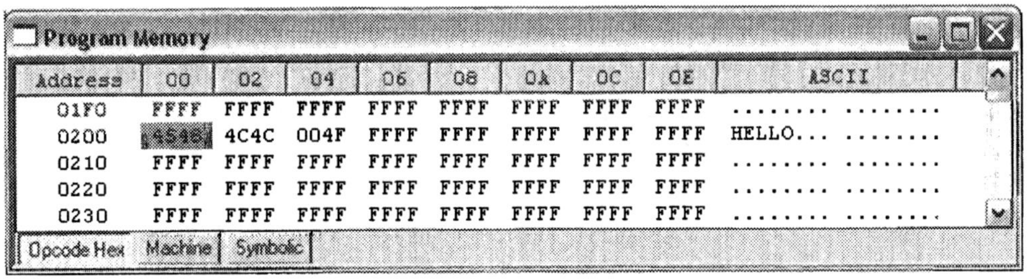

Figure 7-18. Screen Shot for Program 7-4

Review Questions

1. The PIC18 family has a maximum of _____ of program ROM space.
2. The PIC18F8722 has _____ K of program ROM
3. True or false. The program (code) ROM space can be used for data storage, but the data space cannot be used for code.
4. True or false. Using the program ROM space for data means the data is fixed and static.
5. If we have a message string with a size of over 1000 bytes, then we use _____ (program ROM, data RAM) to store it.

SECTION 7.7: DATA RAM ALLOCATION IN C18

In this chapter, we explore the data RAM file register usage and its allocation by the C18 C compiler. We will also explore the near and far data storage qualifiers for data RAM. In addition, we will examine how to place a given data or
stack at a fixed address using the C18 compiler.

RAM data space usage by the C18 C compiler

As we have seen in Chapters 2 through 6, PIC18 family members can have
a maximum of 4K of data RAM, but not all the members come with 4K of RAM.
As we have seen before, the data RAM can vary from 256 bytes to 4096 depending on the chip. That means all members of the PIC18 family come with at least
one bank of RAM, which is called the access bank. In Assembly language programming, the 128 bytes of the data RAM are used for the SFRs and the remaining RAM is used for the scratch pad. The C18 compiler does the same thing by
leaving the SFR region undisturbed and allocating the rest of the RAM to the stack
and the variables declared by the C program. See Program 7-5.

```
//Program 7-5
#include <P18F458.h>
void main(void)
  {
     unsigned char x=5,y=9;    //uses data RAM to store data
     unsigned char z;
     TRISB = 0;                //make Port B an output
     z = x + y;
     PORTB = z;
  }
```

Running the above program on the MPLAB simulator, we can examine the
data RAM space to locate *x*, *y*, and *z* as shown in Figure 7-19.

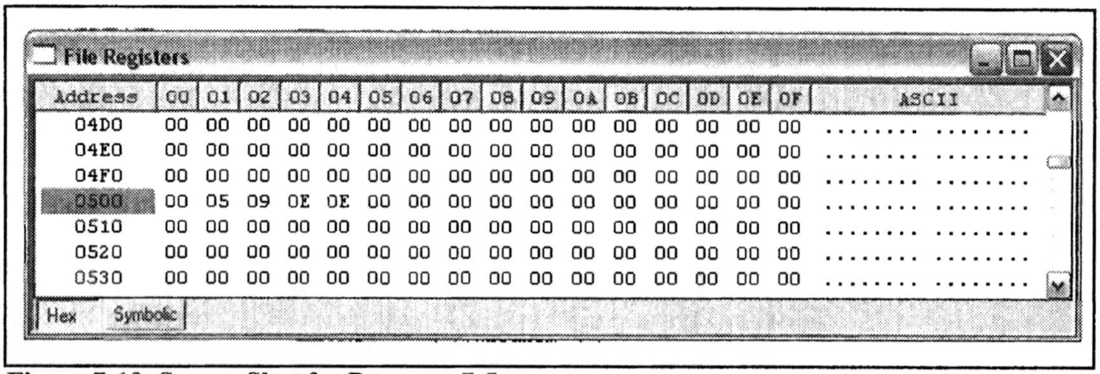

Figure 7-19. Screen Shot for Program 7-5

An array needs contiguous RAM locations for the array elements, which means the size of the array is limited to the size of data RAM in a given PIC18 chip. See Program 7-6 below.

```
//Program 7-6
#include <P18F458.h>
void main(void)
    {
    unsigned char mynum[]= "0123456789"; //uses RAM space
                                         //to store data
    unsigned char z;
    TRISB = 0;                           //make Port B an output
    for(z=0;z<10;z++)
      PORTB = mynum[z];
    }
```

Running the above program on the MPLAB simulator and examining the data RAM space, we can locate values 30H, 31H, 32H, ..., 41H, 42H, 43H, 44H, and so on (the hex values for ASCII '0', '1', '2', etc., as shown below).

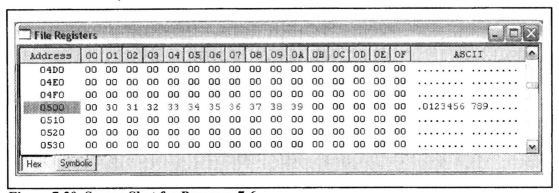

Figure 7-20. Screen Shot for Program 7-6

In the case of arrays with a large number of elements, such as 100, examine the code below:

```
//Program 7-7
#include <P18F458.h>
void main(void)
    {
    unsigned char mydata[100];    //100-byte space in RAM
    unsigned char x, z = 0xFF;
    TRISB = 0;                     //make Port B an output
    for(x=0;x<100;x++)
      {
        mydata[x] = z;             //save it in RAM
        PORTB = z;                 //give a copy to PORTB too
        z--;                       //count down
      }
    }
```

Running the above program on the MPLAB simulator, we can locate values FFH, FEH, FDH, and so on in the data RAM file register, as shown in Figure 7-21.

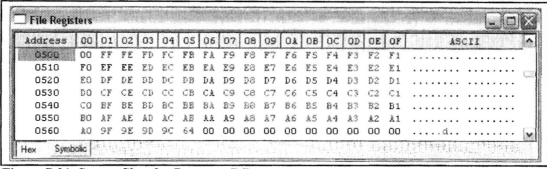

Figure 7-21. Screen Shot for Program 7-7

Change the size of the array and the targeted PIC18 chip (e.g., PIC18F252) and monitor the RAM space allocation by the C compiler.

NEAR and FAR for data

The C18 compiler has two storage qualifiers for data RAM allocation called near and far. The far and near qualifiers are used to indicate which sections of data RAM are to be used for the storage of declared variables. The keyword *near* will limit the C18 C compiler usage of RAM to the access bank for the data declaration, while the keyword *far* will put the entire data RAM at the disposal of C compilers. See Table 7-7. This means that programs written for the PIC18 chips with a limited data RAM cannot have too many arrays with a large number of elements. Using C18, compile and simulate Programs 7-8a and 7-8b to see the impact of *near* and *far* in RAM allocation. See Table 7-7.

Table 7-7: NEAR and FAR Usage for Data RAM

Storage qualifier	RAM
near	In access bank
far	Anywhere in data RAM file register (default)

```
//Program 7-8a
#include <P18F458.h>
near unsigned char mydata[100];        //100-byte space in RAM
void main(void)
  {
    unsigned char x, z = 0;
    TRISB = 0;                         //make Port B an output
    for(x=0;x<100;x++)
      {
        z--;                           //count down
        mydata[x] = z;                 //save it in RAM
        PORTB = z;                     //give a copy to PORTB too
      }
  }
```

```
//Program 7-8b
#include <P18F458.h>
void main(void)
  {
      far unsigned char mydata[100];    //100-byte space in RAM
      unsigned char x, z = 0;
      TRISB = 0;                        //make Port B an output
      for(x=0;x<100;x++)
        {
          z--;                          //count down
          mydata[x] = z;                //save it in RAM
          PORTB = z;                    //give a copy to PORTB too
        }
  }
```

Putting data in a specific RAM address

As we saw in Chapter 6, the MPLAB assembler allows us to place data at a specific RAM address using a combination of MOVLW and MOVWF instructions. To place data at a specific data RAM address in the C18 C compiler, we use the #pragma directive. In the last section we examined the use of the #pragma directive to set the ROM memory address. The #pragma directive can also be used to set the data RAM address. There are two options for #pragma when it is used for the data RAM: (1) idata, and (2) udata, where idata stands for initialized data and udata stands for uninitialized data. The idata (initialized data) and udata (uninitialized data) options are widely used by the C18 to assign an explicit address in RAM data. For example, the following code uses idata to place the data string "HELLO" at RAM address starting at 0x150:

```
//Program 7-9 (using idata)
#include <P18F458.h>
#pragma idata mydata = 0x150
unsigned char mydata[]= "HELLO";          //RAM data
void main(void)
  {
      unsigned char z;
      TRISB = 0;                            //make Port B an output
      for(z=0;z<5;z++)
        PORTB = mydata[z];
  }
```

We can verify the above concept by simulating the program on the MPLAB and examining the RAM at address 0x150.

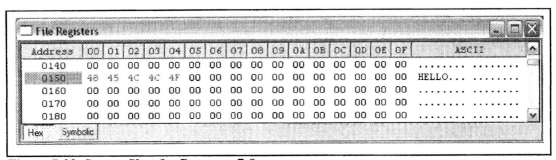

Figure 7-22. Screen Shot for Program 7-9

The following is a repeat of an earlier program and shows how to assign a fixed address of 0x200 using udata.

```
//Program 7-10 (using udata)
#include <P18F458.h>
#pragma udata mycount = 0x200          //assign RAM address 0x200
                                       //(bank 2)
far unsigned char mycount[100];        //100-byte space in RAM
void main(void)
  {
     unsigned char x,z=0;
     TRISB = 0;                        //make Port B an output
       for(x=0;x<100;x++)
         {
            z--;                       //count down
            mycount[x]=z;              //save it in RAM
            PORTB=z;                   //give a copy to PORTB too
         }
  }
```

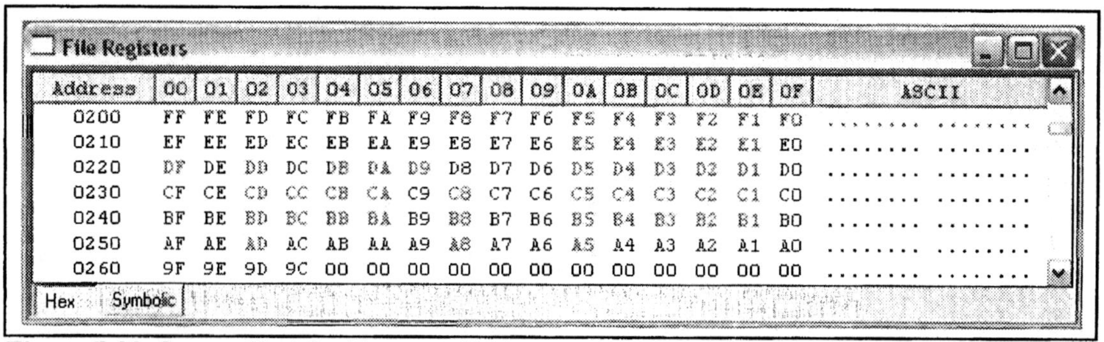

Address	00	01	02	03	04	05	06	07	08	09	0A	0B	0C	0D	0E	0F	ASCII
0200	FF	FE	FD	FC	FB	FA	F9	F8	F7	F6	F5	F4	F3	F2	F1	F0	....
0210	EF	EE	ED	EC	EB	EA	E9	E8	E7	E6	E5	E4	E3	E2	E1	E0	....
0220	DF	DE	DD	DC	DB	DA	D9	D8	D7	D6	D5	D4	D3	D2	D1	D0	....
0230	CF	CE	CD	CC	CB	CA	C9	C8	C7	C6	C5	C4	C3	C2	C1	C0	....
0240	BF	BE	BD	BC	BB	BA	B9	B8	B7	B6	B5	B4	B3	B2	B1	B0	....
0250	AF	AE	AD	AC	AB	AA	A9	A8	A7	A6	A5	A4	A3	A2	A1	A0	....
0260	9F	9E	9D	9C	00	00	00	00	00	00	00	00	00	00	00	00	....

Figure 7-23. Screen Shot for Program 7-10

The following program shows how to assign a fixed address using both udata and idata.

```
//Program 7-11 (assigning udata and idata to a fixed address)
#pragma idata x = 0x100   //assign fixed RAM address 0x100 to var x
unsigned char x=5;        //both data are initialized data
#pragma idata y = 0x101   //assign fixed RAM address 0x101 to var y
unsigned char y=9;        //both data are initialized data
#pragma udata z = 0x102   //assign fixed RAM address 0x102 to var z
unsigned char z;          //it is uninitialized data
#include <P18F458.h>
void main(void)
  {
     TRISB = 0;           //make Port B an output
     z = x + y;
     PORTB = z;
  }
```

Although assigning fixed addresses to a string of data in the data RAM can be justified, this practice is not recommended for individual variables, because it is the job of the compiler to assign addresses dynamically.

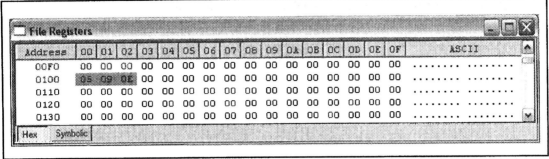

Figure 7-24. Screen Shot for Program 7-11

Overlay storage class

In an attempt to use the data space of the PIC18 more efficiently, the C18 compiler introduces the overlay storage class. The overlay conserves memory by allowing two variables to share the same physical address as long as they are not active at the same time. Compare the following two functions.

```
unsigned char proga(void)
  {
    overlay unsigned char x = 0;
    x = x + 1;
    return x;
  }
```

and

```
unsigned char progb(void)
  {
    overlay unsigned char y = 0;
    y = y + 2;
    return y;
  }
```

Because the *x* and *y* variables are not active at the same time, the C18 C compiler uses the same physical address location in the file register for both of them. If we remove the keyword *overlay* in the above programs, the C18 will assign two different locations to the *x* and *y* variables. The C18 will also use two different physical locations for the variables when the variables are dependent on each other and are both active. Look at the following cases.

```
unsigned char progc(void)
  {
    overlay unsigned char x = 0;
    x = progd()
    return x;
  }
```

and

```
unsigned char progd(void)
  {
    overlay unsigned char y = 0;
    y = y + 2;
    return y;
  }
```

In the previous programs, the C18 compiler assigned a separate RAM location to x and y, even though we used the keyword *overlay*. Because progc calls function progd, the variables are dependent on each other and they are both active at the same item. Note that the C18 C compiler supports all the ANSI C standard storage classes such as auto, extern, static, and so on. The overlay is a new storage class and applies to local variables.

To gain a better understanding of this concept, examine Example 7-34. It shows three different versions of a program that sends the string "HELLO" to Port B. Simulate each program with the C18 compiler and compare data storage methods. Also, compile each program and compare the hex file size to see the impact of the data storage method on the hex file size.

Example 7-34

Compare and contrast the following programs and discuss the advantages and disadvantages of each:

(a)
```c
#include <P18F458.h>
void main(void)
    {
        TRISB = 0;                  //make Port B an output
        PORTB = 'H';
        PORTB = 'E';
        PORTB = 'L';
        PORTB = 'L';
        PORTB = 'O';
    }
```

(b)
```c
#include <P18F458.h>
void main(void)
    {
        unsigned char mydata[] = "HELLO";
        unsigned char z;
        TRISB = 0;                  //make Port B an output
        for(z=0;z<5;z++)
            PORTB = mydata[z];
    }
```

(c)
```c
#include <P18F458.h>
void main(void)
    {
        rom unsigned char mydata[] = "HELLO";    //notice keyword rom
        unsigned char z;
        TRISB = 0;                  //make Port B an output
        for(z=0;z<5;z++)
            PORTB = mydata[z];
    }
```

Solution:

All the programs send out "HELLO" to PORTB one character at a time. They do the same thing in different ways. The first way is short and simple, but the individual characters are embedded into the program. If we change the characters, the whole program changes. This method also mixes the code and data together. The second one uses the RAM data space to store array elements; therefore, the size of the array is limited to file register size. The third one uses a separate area of the program code space for data. This allows the size of the array to be as big as you want provided that you have enough on-chip program ROM. The more program code space you use for data, however, the less space is left for your program code. Both the (b) and (c) programs are easily upgradable if we want to change the string itself or make it longer. That is not the case for program (a).

CHAPTER 7: PIC PROGRAMMING IN C

Review Questions

1. The PIC18 has a maximum of _____ of data RAM.
2. The PIC18F8722 has _____ of data RAM space.
3. True or false. The data space can be used for code.
4. Which space would you use to declare the following values for C18?
   (a) the number of days in a week
   (b) the number of months in a year
   (c) a counter for a delay
5. True or false. The near storage qualifier is used to place the variables in access RAM.

See the following web site for PIC18 C compilers:

http://www.microchip.com

The following web site has a tutorial for MPLAB and C18:

http://www.MicroDigitalEd.com

Running any of the C18 programs on the PIC18F hardware, the following points must be noted:

1. **Disable the WatchDog Timer in the configuration bits.**

2. **Place "while(1);" at the end of the program to prevent the program from executing again. This plays the role of "HERE BRA HERE" in Assembly language.**

SUMMARY

This chapter dealt with C18 programming, specifically I/O programming and time delays in C. We also showed the logic operators AND, OR, XOR, and complement. In addition, some applications for these operators were discussed. This chapter described BCD and ASCII formats and conversions in C. We also compared and contrasted the use of code space and RAM data space in C. The widely used technique of data serialization was also discussed.

PROBLEMS

SECTION 7.1: DATA TYPES AND TIME DELAYS IN C

1. Indicate what data type you would use for the following variables:
   (a) the temperature
   (b) the number of days in a week
   (c) the number of days in a year
   (d) the number of months in a year
   (e) the counter to keep the number of people getting on a bus
   (f) the counter to keep the number of people going to a class
   (g) an address of 64K RAM space
   (h) the age of a person
   (i) a string for a message to welcome people to a building
2. Give the hex value that is sent to the port for each of the following C statements:
   (a) PORTB=14;   (b) PORTB=0x18;   (c) PORTB='A';
   (d) PORTB=7;   (e) PORTB=32;   (f) PORTB=0x45;
   (g) PORTB=255;   (h) PORTB=0x0F;
3. Give two factors that can affect time delay code size in the PIC18 microcontroller.
4. Of the two factors in Problem 3, which can be set by the system designer?
5. Can the programmer set the number of clock cycles used to execute an instruction? Explain your answer.
6. Explain why various C compilers produce different hex file sizes.

SECTION 7.2: I/O PROGRAMMING IN C

7. What is the difference between the PORTBbits.RB4 and TRISBbits.TRISB4?
8. Write a C18 program to toggle all bits of PORTB every 200 ms.
9. Write a C18 program to toggle bits RB1 and RB7 every 200 ms.
10. Write a time delay function for 100 ms.
11. Write a C18 program to toggle only bit RB0 every 200 ms.
12. Write a C18 program to count up PORTB from 0–99 continuously.

SECTION 7.3: LOGIC OPERATIONS IN C

13. Indicate the data on the ports for each of the following:
    Note: The operations are independent of each other.
    (a) PORTB=0xF0&0x45;   (b) PORTB=0xF0&0x56;
    (c) PORTB=0xF0^0x76;   (d) PORTC=0xF0&0x90;
    (e) PORTC=0xF0^0x90;   (f) PORTC=0xF0|0x90;
    (g) PORTC=0xF0&0xFF;   (h) PORTC=0xF0|0x99;
    (i) PORTC=0xF0^0xEE;   (j) PORTC=0xF0^0xAA;
14. Find the contents of the port after each of the following operations:
    (a) PORTB=0x65&0x76;   (b) PORTB=0x70|0x6B;
    (c) PORTC=0x95^0xAA;   (d) PORTC=0x5D&0x78;
    (e) PORTC=0xC5|0x12;   (f) PORTD=0x6A^0x6E;
    (g) PORTB=0x37|0x26;

CHAPTER 7: PIC PROGRAMMING IN C   **295**

15. Find the port value after each of the following is executed:
    (a) PORTB=0x65>>2;   (b) PORTC=0x39<<2;
    (c) PORTB=0xD4>>3;   (d) PORTB=0xA7<<2;
16. Show the C code to swap 0x95 to make it 0x59.
17. Write a C program that finds the number of zeros in an 8-bit data item.
18. A stepper motor uses the following sequence of binary numbers to move the motor. How would you generate them in C18?
       1100,0110,0011,1001

SECTION 7.4: DATA CONVERSION PROGRAMS IN C

19. Write a program to convert the following series of packed BCD numbers to ASCII. Assume that the packed BCD is located in data RAM.
       76H,87H,98H,43H
20. Write a program to convert the following series of ASCII numbers to packed BCD. Assume that the ASCII data is located in data RAM.
       "8767"
21. Write a program to get an 8-bit binary number from PORTB, convert it to ASCII, and save the result if the input is packed BCD of 00–0x99. Assume that PORTB has 1000 1001 binary as input.

SECTION 7.6: PROGRAM ROM ALLOCATION IN C18

22. Indicate what type of memory (data RAM or code ROM space) you would use for the following variables:
    (a) the temperature
    (b) the number of days in a week
    (c) the number of days in a year
    (d) the number of months in a year
23. True or false. When using program ROM for data, the total size of the array should not exceed 256.
24. Why do we use the ROM code space for video game characters and shapes?
25. What is the advantage of using program ROM space for data?
26. What is the drawback of using program ROM space for data?
27. Write a C18 program to send your first and last names to PORTC. Use the program ROM space for the data.
28. What is the difference between far and near storage?
29. What is the difference between #pragma code and #pragma romdata?
30. In Problem 27, show how to place the last name at ROM address 0x200 and the first name at address 0x220.
31. Indicate the size of the program ROM space for each of the following chips:
    (a) PIC18F452/4520   (b) PIC18F458/4580 (c) PIC18F8722
32. In Problem 31, discuss what impact the ROM space has on your decision on the amount of the memory to be allocated to data.

SECTION 7.7: DATA RAM ALLOCATION IN C

33. Indicate what type of memory (data RAM, or code ROM space) you would use for the following variables:
    (a) the counter to keep the number of people getting on a bus
    (b) the counter to keep the number of people going to a class
    (c) an address of 64K RAM space
    (d) the age of a person
    (e) a string for a message to welcome people to building
34. Indicate the size of the data RAM space for each of the following chips:
    (a) PIC18F452/4520   (b) PIC18F458/4580 (c) PIC18F8722
35. Discuss why the total size of an array should be limited to 256 bytes if possible.
36. Why will we not use the data RAM space for video game characters and shapes?
37. What is the drawback of using RAM data space for fixed data?
38. What is the advantage of using data RAM space for variables?
39. What is the difference between #pragma udata and #pragma idata?
40. In Problem 27, show how to place the names at RAM address 0x300.
41. Explain when we use overlay for variables.
42. True or false. Overlay is used for variables that are not active at the same time.

ANSWERS TO REVIEW QUESTIONS

SECTION 7.1: DATA TYPES AND TIME DELAY IN C

1. 0 to 255 for unsigned char and –128 to +127 for signed char
2. 0 to 65,535 for unsigned int and –32,768 to +32,767 for signed int
3. Unsigned char
4. True
5. (a) Crystal frequency of PIC18 system
    (b) PIC18 machine cycle timing
    (c) compiler used for C

SECTION 7.2: I/O PROGRAMMING IN C

1. F81H
2.  ```
    void main()
       {
       TRISC = 0;
       PORTC = 0x55;
       PORTC = 0xAA;
       }
    ```
3.  ```
    #define PB0bit PORTBbits.RB0
    void main()
       {
       TRISBbits.TRISB0 = 0;
       PB0bit = 0;
       PB0bit = 1;
       }
    ```
4. True
5. True

SECTION 7.3: LOGIC OPERATIONS IN C

1. (a) 02
    (b) FFH
    (c) FDH
2. Zeros
3. One
4. All zeros
5. 66H

SECTION 7.4: DATA CONVERSION PROGRAMS IN C

1. (a) 15H = 0001 0101 packed BCD, 0000 0001,0000 0101 unpacked BCD
    (b) 99H = 1001 1001 packed BCD, 0000 1001,0000 1001 unpacked BCD
2. 3736H = 00110111 00110110B
    and in packed BCD we have 76H = 0111 0110B
3. 36, 37
4. Yes, because mydata = 0x39.
5. Space savings
6. ASCII   7. BCD
8. E4H   9. 0
10. First, convert from binary to decimal, then convert to ASCII, and then send results to the
    screen and we will see 038.

SECTION 7.6: PROGRAM ROM ALLOCATION IN C18

1. 2M
2. 128
3. True
4. True
5. Program ROM

SECTION 7.7: DATA RAM ALLOCATION IN C18

1. 4K
2. 4096 bytes
3. False
4. (a) ROM space, (b) ROM space, (c) RAM space
5. True

CHAPTER 8

PIC18F HARDWARE CONNECTION AND ROM LOADERS

OBJECTIVES

Upon completion of this chapter, you will be able to:

>> Explain the function of the reset pin of the PIC18F microcontroller
>> Show the hardware connection of the PIC18F chip
>> Show the use of a crystal oscillator for a clock source
>> Explain how to design a PIC18F-based system
>> Explain the role of brown-out reset voltage (BOR) in system reset
>> Explain the role of the CONFIG registers in PIC18-based systems
>> Show the design of the PIC Trainer
>> Code a test program in Assembly and C for testing the PIC18
>> Show how to download programs into the PIC18F system using Microchip PICkit 2
>> Explain the Intel hex file characteristics for 32-bit and 16-bit addresses

This chapter describes the process of physically connecting and testing PIC18F-based systems. In the first section we describe the functions of PIC18F458 pins. The configuration registers of the PIC18 and how they are set are explored in Section 8.2. In Section 8.3 we explain the characteristics of Intel hex files that are produced by MPLAB. In Section 8.4 we discuss the various methods of loading a program into the microcontroller. It also shows the hardware connection for a PIC18 Trainer using the PIC18F452/458 (PIC18F4520/4580) chips.

SECTION 8.1: PIC18F458/452 PIN CONNECTION

The PIC18F458 family members come in different packages, such as DIP (dual in-line package), QFP (quad flat package), and LLC (leadless chip carrier). They all have many pins that are dedicated to various functions such as I/O, ADC, timer, and interrupts. Note that Microchip provides an 18-pin version of the PIC18 family with a reduced number of I/O ports for less demanding applications. Because the vast majority of developers use the 40-pin chip, however, we will concentrate on that. Figure 8-1 shows the pins for the PIC18F458.

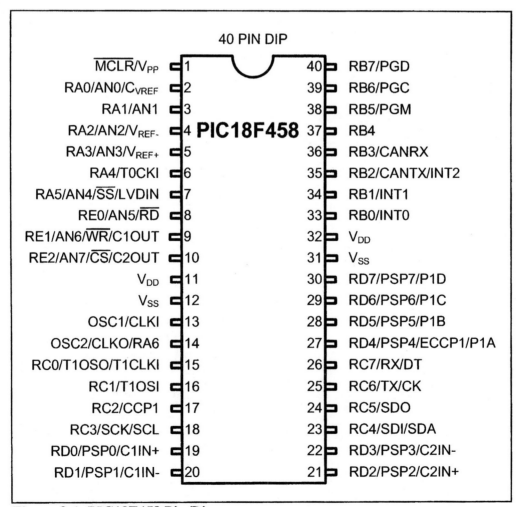

Figure 8-1. PIC18F458 Pin Diagram

Examining Figure 8-1, note that of the 40 pins, a total of 33 are set aside for the five ports A, B, C, D, and E, with their alternate functions. The rest of the pins are designated as Vdd, GND (Vss), OSC1, OSC2, and MCLR (master clear reset). Next, we describe the function of each pin.

Vdd (Vcc)

Two pins are used to provide supply voltage to the chip. The typical voltage source is +5V. Some PIC18F family members have lower voltage for Vdd pins in order to reduce the noise and power dissipation of the PIC system. We can choose other options for the Vdd voltage level by setting the bits in the configuration register. The configuration register for Vdd is discussed in the next section.

Vss (GND)

Two pins are also used for ground. In chips with 40 pins and more, it is common to have multiple pins for VCC and GND. This will help reduce the noise (ground bounce) in high-frequency systems, as discussed in Appendix C.

OSC1 and OSC2

The PIC18F has many options for the clock source. Most often a quartz crystal oscillator is connected to input pins OSC1 and OSC2. The quartz crystal oscillator connected to the OSC1 and OSC2 pins also needs two capacitors. One side of each capacitor is connected to the ground as shown in Figure 8-3. Note that PIC18F microcontrollers can have speeds of 0 Hz to 40 MHz.

We can choose options for the clock frequency by setting bits in the configuration register. The config register for the oscillator is discussed in the next section.

MCLR

Pin 1 (in the PIC18F458 40-pin DIP) is the MCLR (master clear reset) pin. It is an input and is active-LOW (normally HIGH). When a LOW pulse is applied to this pin, the microcontroller will reset and terminate all activities. This is often referred to as a *power-on reset (POR)*.

Program counter value upon reset

Activating a MCLR reset will cause all values in the registers to be lost. Table 8-1 provides a partial list of PIC18F registers and their values after power-on reset. From Table 8-1 we note that the value of the PC (program counter) is 0 upon reset, forcing the CPU to fetch the first opcode from ROM memory location 00000. This means that we must place the first byte of opcode in ROM location 0 because that is where the CPU expects to find the first instruction.

Table 8-1: RESET Values for Some PIC18 Registers

Register	Reset Value (hex)
PC	000000
WREG	00
SP	00
TRISA–TRISE	FF

Figures 8-2a and 8-2b show two ways of connecting the MCLR pin to the power-on reset circuitry. Figure 8-2b uses a momentary switch for reset circuitry. The most difficult time for any system is during the power-up. The CPU needs both a stable clock source and a stable voltage level to function properly. The PIC18 chips come with some features that help the reset process. We can choose these features by setting the bits in the configuration register. The configuration register for the reset pin is discussed in the next section. There are other sources of reset in the PIC18 family, and they are discussed in future chapters.

The pins discussed so far must be connected no matter which family member is used. They are the minimum pin connections that every PIC18 must have. See Figure 8-3.

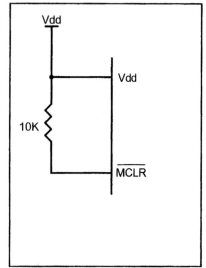

**Figure 8-2a. PIC18F458
Power-On Reset Circuit**

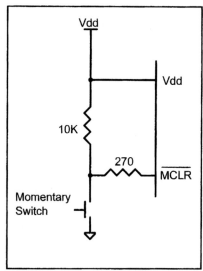

**Figure 8-2b. PIC18F458
Power-On Reset with a
Momentary Switch**

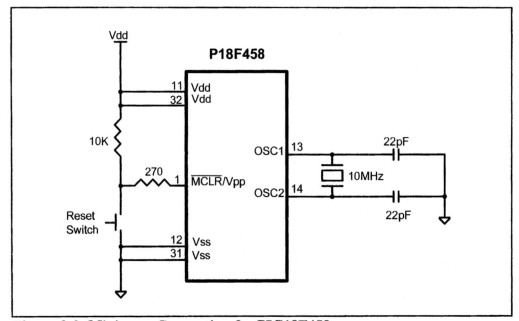

Figure 8-3. Minimum Connection for PIC18F458

The number of I/O ports varies among the PIC18 family members, as we saw in Chapter 4. The following is another look at them for the PIC18F458.

Ports A, B, C, D, and E

As shown in Figure 8-1 (and discussed in Chapter 4), the ports PORTA, PORTB, PORTC, PORTD, and PORTE use a total of 33 pins. All the ports upon RESET are configured as input, because TRISA–TRISE have the value FFH on them. Tables 8-2 through 8-5 provide summaries of features of ports PORTA–PORTE and their alternative functions. We will study the alternative functions of these pins in future chapters, as we discuss the PIC18 features.

Table 8-2: PORTA/PORTE Alternate Functions

Bit	Function
RA0	AN0/CVREF
RA1	AN1
RA2	AN2/VREF–
RA3	AN3/VREF+
RA4	T0CKI
RA5	AN4/SS/LVDIN
RA6	OSC2/CLKO
RE0	AN5/RD
RE1	AN6/WR/C1OU
RE2	AN7/CS/C2OUT

Table 8-3: PORTB Alternate Functions

Bit	Function
RB0	INT0
RB1	INT1
RB2	INT2/CANTX
RB3	CANRX
RB4	
RB5	PGM
RB6	PGC
RB7	PGD

Table 8-4: PORTC Alternate Functions

Bit	Function
RC0	T1OSO/T1CKI
RC1	T1OSI
RC2	CCP1
RC3	SCK/SCL
RC4	SDI/SDA
RC5	SDO
RC6	TX/CK
RC7	RX/DT

Table 8-5: PORTD Alternate Functions

Bit	Function
RD0	PSP0/C1IN+
RD1	PSP1/C1IN–
RD2	PSP2/C2IN+
RD3	PSP3/C2IN–
RD4	PSP4/ECCP1/P1A
RD5	PSP5/P1B
RD6	PSP6/P1C
RD7	PSP7/P1D

Review Questions

1. Which pin is used to reset the PIC18F458 chip?
2. Upon power-up, the program counter (PC) has a value of ____.
3. Upon power-up, the PIC18F458 fetches the first opcode from ROM address location _____.
4. MCLR is an active-_____ (LOW, HIGH) pin.
5. How many Vdd and Gnd pins are in the PIC18F458 chip?

SECTION 8.2: PIC18 CONFIGURATION REGISTERS

There are some features of the PIC18 that we can choose by programming the bits of the configuration registers. These features will reduce system cost by eliminating any need for external components. The configuration registers are located at the address starting at 300000H, as shown in Figure 8-4. Notice that the address 300000H is outside the range 000000–1FFFFFH, the address space belonging to program ROM. We write 8-bit values into the configuration register one byte at a time using the CONFIG directive in the source code, as we will see soon. In other words, we provide the values and register name in our application program and let the ROM programmer load them into the config register along with the application program itself. The configuration registers can be accessed from the user program using table reads and writes. In this section we examine some of the basic configuration registers such as reset, clock source, and Vdd voltage. MicroChip website provides the complete list of configuration registers for the PIC microcontrollers. For the configuration registers of a given member of the PIC18 family, see the "Configuration Register Settings Addendum" document on the Microchip web site. Table 8-6 gives a short description of the configuration registers. It must be noted that if a configuration register is incorrectly programmed, it can cause the system to fail. An example of this is changing the clock type connected to the microcontroller.

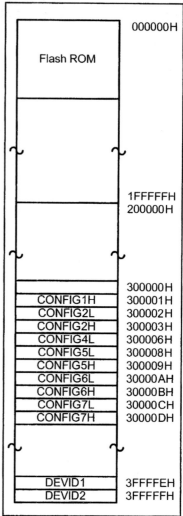

Figure 8-4. CONFIG Register Memory Map

Table 8-6: PIC18F458 Configuration Registers

Address (Hex)	Name	General Description
300001	CONFIG1H	Oscillator selection
300002	CONFIG2L	Brown out
300003	CONFIG2H	Watchdog enable
300006	CONFIG4L	Background debugger and ISCP
300008	CONFIG5L	Code protection
300009	CONFIG5H	EEPROM and boot block protection
30000A	CONFIG6L	Write protection
30000B	CONFIG6H	Write protection
30000C	CONFIG7L	Read protection
30000D	CONFIG7H	Boot block read protection
3FFFFE	DEVID1	Device ID and revision
3FFFFF	DEVID2	Device ID

CONFIG1H register and oscillator clock source

The CONFIG1H register is located at address 0x300001 and is set aside for the clock oscillator, as shown in Figure 8-5. The following is a description of the options for the CONFIG1H register.

U-0	U-0	R/P-1	U-0	U-0	R/P-1	R/P-1	R/P-1
—	—	OSCSEN	—	—	FOSC2	FOSC1	FOSC0
bit 7							bit 0

bit 7-6   **Unimplemented**: Read as '0'

bit 5   **OSCSEN**: Oscillator System Clock Switch Enable bit

　　　　1 = Oscillator system clock switch option is disabled (main oscillator is source)
　　　　0 = Oscillator system clock switch option is enabled (oscillator switching is enabled)

bit 4-3   **Unimplemented**: Read as '0'

bit 2-0   **FOSC2:FOSC0:** Oscillator Selection bits

　　　　111 = RC oscillator w/OSC2 configured as RA6
　　　　110 = HS oscillator with PLL enabled/clock frequency = (4 x Fosc)
　　　　101 = EC oscillator w/OSC2 configured as RA6
　　　　100 = EC oscillator w/OSC2 configured as divide-by-4 clock output
　　　　011 = RC oscillator
　　　　010 = HS oscillator
　　　　001 = XT oscillator
　　　　000 = LP oscillator

Legend:
R = Readable bit   P = Programmable bit   U = Unimplemented bit, read as '0'
-n = Value when device is unprogrammed   u = Unchanged from programmed state

Figure 8-5. CONFIG1H Register for Frequency Selection

FOSC2–FOSC0

The three bits of FOSC2, FOSC1, and FOSC0 are used to select the clock frequency to the CPU. The default choice is RC (111), which uses the on-chip oscillator with the help of an externally connected resistor and capacitor. In this option, all we have to do is to connect the OSC1 pin to the RC circuit. The values of R and C determine the clock speed. Providing clock to the CPU in this manner leaves the OSC2 (bit 6 of PORTA) available to be used as an I/O pin. We can use option 101 (EC: external clock) and provide an external clock source to the pin OSC1 and let A6 be used as an I/O pin. We can do the same thing with option 100 while OSC2 provides us with an Osc/4 frequency. This Osc/4 clock can be used to synchronize all the system activities with the CPU. The most widely used option is to connect the OSC1 and OSC2 pins to a crystal (or ceramic) oscillator, as shown in Figure 8-6. There are four choices for the crystal oscillator option. They are PPLHS, HS, XT, and LP. The main difference among them is the frequency range as shown in Table 8-7. The LP (low power) option uses the lowest power while the highest power consumption belongs to the PPLHS (phase lock loop high speed) option. Notice that the higher the frequency, the more power is dissipated by the CPU, as discussed in Appendix C. We use the HS (high speed) option for many of the circuits discussed in this textbook. If we connect pins OSC1–OSC2 to a 10 MHz crystal oscillator and choose the PPLHS option, then the CPU works on 40 MHz because the PPLHS uses phase lock loop to quadruple the clock source provided to the CPU. The PLLHS also has the highest power dissipation. Notice that the RC option (111) is the cheapest while the LP option (000) has the lowest power dissipation.

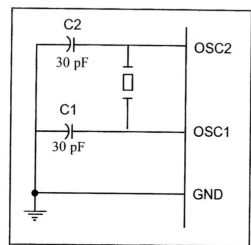

Figure 8-6a. OSC1–OSC2 Connection to Crystal Oscillator

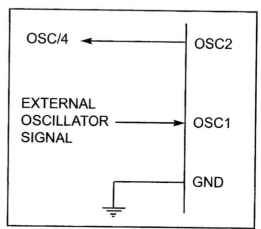

Figure 8-6b. OSC Connection to an External Clock Source

OSCSEN

The OSCSEN bit (D5) of CONFIG1H allows the CPU to switch to an internal clock source, which has a fixed frequency of 32 kHz. Switching the clock source from the external oscillator connected to the OSC1 and OSC2 pins to an internal 32 kHz clock source will reduce power dissipation to an absolute minimum in many systems running on *battery* power. Using this option, along with the LP option for the crystal frequency, can reduce CPU power consumption to the nanowatt range. Notice that this low-frequency 32 kHz clock source is in addition to the external clock source connected to the OSC1 and OSC2 pins. This secondary clock source of 32 kHz is independent of the OSC1–OSC2 clock source and will continue to provide the clock to the CPU in the event that the crystal frequency goes bad. In this textbook we disable this secondary clock source and use the OSC1–OSC2 oscillator as the main clock source.

Table 8-7: PIC18F458 Oscillator Frequency Choices and Capacitor Range

Osc choice	Crystal Freq	C1 range	C2 range
LP	32 kHz	33 pF	33 pF
LP	200 kHz	15 pF	15 pF
XT	200 kHz	47–65 pF	47–65 pF
XT	1 MHz	15 pF	15 pF
XT	4 MHz	15 pF	15 pF
HS	4 MHz	15 pF	15 pF
HS	8 MHz	15–33 pF	15–33 pF
HS	20 MHz	15–33 pF	15–33 pF
HS	25 MHz	15–33 pF	15–33 pF

OSCillator frequency and instruction clock cycle

We examined the instruction cycle time in Chapters 2 through 4 and showed how to create time delay subroutines. In PIC18 microcontrollers, the instruction cycle time is based on 1/4 of the clock source provided to the OSC pins. This is examined once more in Example 8-1.

Example 8-1

Find the instruction cycle time for the PIC18F458 chip with the following crystal oscillator connected to the OSC1 and OSC2 pins. The option for CONFIG1H is shown for each crystal speed.
(a) 4 MHz, XT   (b) 10 MHz, HS   (c) 20 MHz, HS

Solution:

All the options in CONFIG1H use 1/4 of the clock source for the instruction cycle time.

(a) 4 MHz/4 = 1 MHz and instruction cycle time is 1/1 MHz = 1 μs
(b) 10 MHz/4 = 2.5 MHz and instruction cycle time is 1/2.5 MHz = 0.4 μs = 400 ns
(c) 20 MHz/4 = 5 MHz and instruction cycle time is 1/5 MHz = 0.2 μs = 200 ns

If we use 10 MHz crystal oscillator speed and choose the HSPLL option (instead of HS), then the CPU has 40 MHz for the clock source. This means that the instruction cycle time is 1/10 MHz = 0.1 μs = 100 ns because 40 MHz/4 = 10 MHz.

Table 8-8: CONFIG1H Options Using CONFIG Directive in MPLAB

Oscillator Selection

OSC = LP	LP	Low Power
OSC = XT	XT	Crystal
OSC = HS	HS	High Speed
OSC = RC	RC	Resistor/Capacitor
OSC = EC	EC, OSC2 as Clock Out	External Clock
OSC = ECIO	EC, OSC2 as RA6	External Clock
OSC = HSPLL	HS-PLL Enabled	High Speed Phase Lock Loop
OSC = RCIO	EC, OSC2 as RA6	External Clock

Oscillator Switch Enable

OSCS = ON	Enabled
OSCS = OFF	Disabled

CONFIG directive

Table 8-8 shows the syntax choices for the CONFIG1H byte supported by the MPLAB. Whenever we load an application into the PIC18 program ROM, we need to load the CONFIG bytes into the configuration registers as well. This is done by using the CONFIG directive in the source program. In the source code, we use the CONFIG directive to set the CONFIG1H values according to Table 8-8 as shown below:

```
CONFIG OSC = HS              ;high-speed oscillator
CONFIG OSCS = OFF            ;disable Osc switch
```

Or, we can combine them into a single statement, as follows:

```
CONFIG OSC = HS, OSCS = OFF  ;oscillator, no Osc switch
```

CONFIG2L register and reset voltage

CONFIG2L is located at address 0x300002 and is set aside for the purpose of providing stable voltage and clock frequency during reset. See Figure 8-7. The most difficult time for a system is during power-up. The CPU needs both a stable clock source and a stable voltage level to function properly. Two internal timers help us achieve that: they are called the power-up timer (PWRT) and the oscillator start-up timer (OST). These two internal timers help to reduce the delay associated with the frequency and voltage sources during the power-up process. PWRT

U-0	U-0	U-0	U-0	R/P-1	R/P-1	R/P-1	R/P-1
—	—	—	—	BORV1	BORV0	BOREN	PWRTEN
bit 7							bit 0

bit 7-4   **Unimplemented**: Read as '0'

bit 3-2   **BORV1:BORV0:** Brown-out Reset voltage bits

        11 = VBOR set to 2.0V

        10 = VBOR set to 2.7V

        01 = VBOR set to 4.2V

        00 = VBOR set to 4.5V

bit 1   **BOREN:** Brown-out Reset Enable bit

        1 = Brown-out Reset enabled

        0 = Brown-out Reset disabled

bit 0   **PWRTEN:** Power-up Timer Enable bit

        1 = PWRT disabled

        0 = PWRT enabled

Legend:

R = Readable bit   P = Programmable bit   U = Unimplemented bit, read as '0'

-n = Value when device is unprogrammed   u = Unchanged from programmed state

Figure 8-7. CONFIG2L Configuration Register for Reset Voltage

provides a fixed delay during power-up, which keeps the CPU in the reset state until the power supply stabilizes. The OST timer does the same thing for the crystal oscillator. These two on-chip timers eliminate the need for external circuitry for voltage and frequency stabilization during the power-up. CONFIG2L allows us to set the voltage and frequency to keep the CPU in the reset state until both the clock and power supply are stable. Next, we discuss options for the bits for this important configuration register.

Table 8-9: CONFIG2L Selection for PIC18F458

Brown-out Voltage	
BORV = 45	4.5 V
BORV = 42	4.2 V
BORV = 27	2.7 V
BORV = 20	2.0 V

Power-up Timer	
PWRT = ON	Enabled
PWRT = OFF	Disabled

Brown-out Reset	
BOR = ON	Enabled
BOR = OFF	Disabled

BORV1:BORV0

Occasionally, the power source provided to the V_{CC} (V_{dd}) pin fluctuates, causing the CPU to malfunction. The PIC18 family has a provision for this which is called brown-out reset voltage. The brown-out reset voltage (BORV) bits in CONFIG2L allow us to set the minimum voltage for V_{dd}. If it falls below that, the CPU will go into the reset state and stop all activities. This is needed because the voltage connected to the V_{dd} (V_{CC}) pins can be set according to the oscillator frequency connected to the OSC1 and OSC2 pins. At the high frequency of 40 MHz with V_{dd} = 5 V, we set BORV to 4.5 V. That means that if V_{dd} falls below the BORV of 4.5 V, the CPU will go into the reset state and stop execution of programs without losing any data in registers. For a low-power system with a frequency of 2 MHz and below, we can connect the V_{dd} to 2 V and set BORV to 1.8 V. In such a situation if V_{dd} falls below 1.8 V, the CPU will go into the reset state, and when V_{dd} rises above that level it will come out of reset and continue the program execution. Therefore, the BORV1:BORV0 bits of CONFIG2L will be set according to the V_{dd} voltage supplied to the V_{dd} pins and the oscillator frequency connected to the OSC1 and OSC2 pins. In this book, we set BORV = 4.5 V because V_{dd} = 5 V and the crystal oscillator is 10 MHz.

BOREN

This will enable the option BORV1:BORV0 discussed above.

PWRTEN

This bit will enable the power-up timer (PWRT). The PWRT provides a fixed delay during power-up, which keeps the CPU in the reset state until the power supply is stabilized.

Table 8-9 provides the syntax options for CONFIG2L as supported by the

MPLAB. We use the CONFIG directive to set the values according to Table 8-9 as shown below:

```
CONFIG BORV=45          ;for Vdd = 5 V, OSC = 10 MHz
CONFIG PWRT = ON        ;use power-timer
CONFIG BOR=ON           ;enable BORV option
```

Or, we can combine them into a single statement as follows:

```
CONFIG BORV = 45, PWRT = ON,  BOR=ON
```

CONFIG2H register and watchdog timer

CONFIG2H is located at address 0x300003 and is set aside for the watchdog timer. In recent years, microcontrollers have come with a piece of hardware called a *watchdog timer*. We can use the watchdog timer to force the microcontroller into the known state of reset when the system is hung up or out of control due to execution of an incorrect sequence of codes. There are many uses for watchdog timers in embedded systems. One application is to use the watchdog timer to prevent a system from going into an infinite loop due to a software bug. Another application of the watchdog timer can be to catch events that cause the system to

U-0	U-0	U-0	U-0	R/P-1	R/P-1	R/P-1	R/P-1
—	—	—	—	WDTPS2	WDTPS1	WDTPS0	WDTEN

bit 7   bit 0

bit 7-4   **Unimplemented**: Read as '0'

bit 3-2   **WDTPS2:WDTPS0**: Watchdog Timer Postscale Select bits

              111 = 1:128

              110 = 1:64

              101 = 1:32

              100 = 1:16

              011 = 1:8

              010 = 1:4

              001 = 1:2

              000 = 1:1

              Note: The Watchdog Timer postscale select bits configuration used in the PIC 18FXXX devices has changed from the configuration used in the PIC 18CXXX devices.

bit 0   **WDTEN**: Watchdog Timer Enable bit

              1 = WDT enabled

              0 = WDT disabled (control is placed on the SWDTEN bit)

Legend:

R = Readable bit   P = Programmable bit   U = Unimplemented bit, read as '0'

-n = Value when device is unprogrammed   u = Unchanged from programmed state

Figure 8-8. CONFIG2H Configuration Register for Watchdog Timer

hang. These problems can happen due to corruption of the program ROM caused by a power surge, an electrically noisy environment, or inadvertent changes to the program counter. In such situations, the watchdog timer will force the system into a known state of reset, from which the system can recover. In some applications, the system can be put to sleep if there is no activity, thereby saving battery power. In such applications, one can use the watchdog timer to monitor the keyboard and, when there is activity on the keyboard, to awaken the system to process the information. Figure 8-8 shows the CONFIG2H register.

WDTEN

This bit will enable the watchdog timer.

WDTPS2:WDTPS0

The watchdog timer prescalar bits allow programming the WDT for up to 2 minutes. Appendix A discusses SLEEP instruction with the WDT.

For the applications in this textbook, we turn off the watchdog timer. We can turn off the watchdog timer in the MPLAB or use the CONFIG directive in the source code to set CONFIG2H values according to Table 8-10, as shown below:

Table 8-10: CONFIG2H Selection for PIC18F458

Watchdog Timer	
WDT = ON	Enabled
WDT = OFF	Disabled

```
CONFIG WDT = OFF
```

CONFIG4L register and background debugger

CONFIG4L is located at address 0x300006 and is set aside for the purpose of enabling the background debugger, among other things. See Figure 8-9. Table 8-11 shows the selection options for CONFIG4L. The following are its options.

DEBUG

If we connect the PIC18 system to an in-circuit debugger, then we lose the RB6 and RB7 pins of PORTB. By disabling the background debugger option in the CONFIG4L byte, we can use the RB6 and RB7 pins for general purpose I/O.

STVREN

The D0 bit is used for stack overflow. As we discussed in Chapter 3, the PIC18 has only 31 locations for the stack. By enabling the D0 bit we will cause the system go into a reset state if stack overflows (or underflows).

LVP

The D2 bit is set aside for the low-voltage in-circuit serial programming (ICSP)

Table 8-11: CONFIG4L Selection for PIC18F458

Background Debugger Enable	
DEBUG = ON	Enabled
DEBUG = OFF	Disabled

Low-Voltage ICSP	
LVP = ON	Enabled
LVP = OFF	Disabled

Stack Overflow Reset	
STVR = ON	Enabled
STVR = OFF	Disabled

via pin RB5. We can disable it and use the RB5 pin as an I/O.

Table 8-11 shows the CONFIG4L byte selection syntax used by the MPLAB assembler. For the applications in this textbook, we turn off all the options of debugger, LVP, and stack overflow using the CONFIG directive as follows:

```
CONFIG DEBUG = OFF, LVP = OFF, STVR = OFF
```

The first four CONFIG registers are the minimum number of registers that we need for any PIC18F452 or 458-based system. The rest of the CONFIG registers are dedicated to program and data protection. See the Microchip website.

```
  R/P-1      U-0      U-0      U-0      U-0     R/P-1     U-0     R/P-1
| DEBUG |  —   |   —   |   —   |   —   |  LVP  |   —   | STVREN |
   bit 7                                                          bit 0
```

bit 7   **DEBUG**: Background Debugger Enable bit

            1 = Background Debugger disabled. RB6 and RB7 configured as general purpose I/O
            0 = Background Debugger enabled. RB6 and RB7 are dedicated to In-Circuit Debug.

bit 6-3   **Unimplemented**: Read as '0'

bit 2   **LVP**: Low-Voltage ICSP Enable bit

            1 = Low-Voltage ICSP enabled
            0 = Low-Voltage ICSP disabled

bit 1   **Unimplemented**: Read as '0'

bit 0   **STVREN**: Stack Full/Underflow Reset Enable bit

            1 = Stack Full/Underflow will cause Reset
            0 = Stack Full/Underflow will not cause Reset

Legend:

R = Readable bit   C = Clearable bit   U = Unimplemented bit, read as '0'

-n = Value when device is unprogrammed   u = Unchanged from programmed state

Figure 8-9. CONFIG4L Configuration Register for Background Debugger

---

CONFIGURATION SETTINGS TO BE NOTED

1. Note that each member of the PIC18 family has its own values for the configuration registers. They are provided in a document called "PIC18 CONFIGURATION SETTINGS ADDENDUM" and can be found at the following web site:

   http://ww1.microchip.com/downloads/en/DeviceDoc/MPLAB_C18_Config_Settings_51537b.pdf

2. Microchip Corp. recommends using the CONFIG directive instead of __CONFIG for the PIC18 family. Although the __CONFIG directive (notice, __CONFIG has two undersigns) works with the PIC18, it is not recommended. According to Microchip, we should not use both of them in the same program.

The LIST directive

The LIST directive is another component used in the source code for a program that we intend to burn into the PIC ROM. The LIST directive informs the MPLAB assembler about some of the options, such as the Intel hex file format, the radix for data format, the printout of the source code, and so on. Table 8-12 provides some of the major options of the LIST directive used in this textbook.

For an example of how to use the LIST directive, look at the following setting.

```
LIST  P=18F458, F=INTHX32, MM=OFF, R=HEX, ST=OFF X=OFF
```

It must be noted that some of the options in Table 8-12 can be set by the MPLAB assembler itself. To ensure that they are set when we share source files, however, we use the LIST directive to set them.

Table 8-12: Some LIST Directive Options

B = nnn	Set tab space. (Default is 8.)
C = nnn	Set column width for the print-out. (Default is 132.)
F = format	Set the hex file output. The choices are INHX32, INHX8M, or INHX8S. Default is INHX8M. (See next section on this.)
MM = {ON/OFF}	Print memory map in list file. (Default is on.)
N = nnn	Set lines per page in the print-out. (Default is 60.)
P = type	Set microcontroller type. (Example: P = PIC18F458.)
R = radix	Set radix, the data format used throughout the source code. The options are hex, dec, and oct. (Default is hex.)
ST = {ON/OFF}	Print symbol table in list file. (Default is on.)
X = {ON/OFF}	Turn macro expansion on or off. (Default is on.)

Putting it all together

All the programs we showed in the first seven chapters were intended to be simulated. In order to create a ready-to-burn program, however, we must provide all the configuration register bytes, and set the desired options of the LIST directive in the source code before assembling and linking the program. By doing so, the hex output file provided by the MPLAB assembler can be burned into the program ROM of the PIC18 chip using a ROM burner. We can also send this hex file to anyone, knowing that it will work because all the configuration registers are already set. We can use the following skeleton source code for the programs that we intend to burn into ROM.

```
;skeleton of a PIC18 Assembly language program
        LIST  P=PIC18F458, F=INHX32, MM=OFF,  N=0, ST=OFF, R=HEX
        #include P18F458.INC
        CONFIG OSC=HS, OSCS=OFF          ;high-speed XTAL as clk src
        CONFIG WDT=OFF                   ;disable watchdog timer

        ;Brown-out Reset Volt = 4.5 V and Power-up Timer is on
        CONFIG BORV=45, PWRT=ON, BOR=ON

        ;no Background debug, no Reset if stack overflows
        ;and pin PB5 = I/O
        CONFIG DEBUG=OFF, LVP=OFF, STVR=OFF
        .......
        ORG 0
        .......
        .......
        .......
        END
```

As an example, examine the following program. It will toggle all the bits of PORTB with some delay in between the "on" and "off" states

```
;Test Program 8-1: Toggling PORTB for the PIC18F458 and
;XTAL = 10 MHz
        LIST P=PIC18F458, F=INHX32, N=0, ST=OFF, R=HEX
        #include P18F458.INC
        CONFIG OSC = HS, OSCS = OFF
        CONFIG WDT = OFF
        CONFIG BORV = 45, PWRT = ON, BOR = ON
        CONFIG DEBUG = OFF, LVP = OFF, STVR = OFF
R1      EQU    0x07
R2      EQU    0x08
R3      EQU    0x09


        ORG 0
        CLRF   TRISB        ;make Port B an output port
        MOVLW 0x55          ;WREG = 55h
        MOVWF PORTB         ;put 55h on port B pins
L3      COMF   PORTB,F      ;toggle bits of Port B
        CALL   QDELAY       ;quarter of a second delay
        BRA    L3           ;continue

;-----------1/4 SECOND DELAY
QDELAY
        MOVLW D'2'
        MOVWF R1
D1      MOVLW D'250'
        MOVWF R2
D2      MOVLW D'250'
        MOVWF R3
D3      NOP
        NOP
        DECF   R3, F
        BNZ    D3
        DECF   R2, F
        BNZ    D2
        DECF   R1, F
        BNZ    D1
        RETURN
        END
```

Setting the CONFIG registers in the MPLAB C18 C compiler

In Chapter 7 we covered C programming of the PIC18F using the C18 C compiler. Those programs were intended to be simulated. To create a burnable C program, we must ensure that the configuration registers are set. One way to do that is to use #pragma. We can use the following skeleton for C18 code for the programs that we intend to burn into ROM.

```
;skeleton of a PIC18 C18 C language program
    #pragma config OSC = HS, OSCS = OFF
    #pragma config BORV = 45, PWRT = ON, BOR = ON
    #pragma config WDT = OFF
    #pragma config DEBUG = OFF, LVP = OFF, STVR = OFF

    void main (void)
    {
    ......
    ......
    ......
    }
```

As an example, examine the following C18 program. It will toggle all the bits of PORTB with some delay between the "on" and "off" states.

```
;Test Program 8-2: Toggling PORTB for the PIC18F458 and
;XTAL = 10 MHz
    #pragma config OSC = HS, OSCS = OFF
    #pragma config PWRT = OFF, BOR = ON, BORV = 45
    #pragma config WDT = OFF
    #pragma config DEBUG = OFF, LVP = OFF , STVR = OFF
    #include <P18F458.h>

    void msdelay(unsigned int ms);
    void main(void)
      {
        TRISB = 0;                      //make Port B an output
        while(1)
          {
          PORTB = 0x55;
          msdelay(500);
          PORTB = 0xAA;
          msdelay(500);
          }
      }

    //this delay is for a 10 MHz clock
    void msdelay(unsigned int ms)
      {
      unsigned int x;
      unsigned char z;
      for(x=0;x<ms;x++)
        for(z=0;z<165;z++);
      }
```

Review Questions

1. A given PIC18F458-based system has a crystal frequency of 16 MHz with HS selected for the CONFIG1H. What is the instruction cycle time for the CPU?
2. Which address is used for the CONFIG1H register?
3. True or false. Upon power-up, both voltage and frequency are stable instantly.
4. The LP option for the OSC1–OSC2 frequency works for the frequency range of _____ to _____ kHz.
5. Which configuration register is used to disable the watchdog timer? What is its address?
6. True or false. Upon power-up, the power-up timer keeps the CPU in the reset state until the voltage source is stable.
7. True or false. The configuration registers are located within the ROM program address space.
8. True or false. The brown-out reset voltage (BORV) can be set at a lower voltage for a system with low crystal frequencies.
9. True or false. The higher the clock frequency for the system, the lower the power dissipation.
10. If we have the statement BORV = 42 in a given source code, what is the lowest V_{dd} voltage level at which the CPU goes into the reset state?

SECTION 8.3: EXPLAINING THE INTEL HEX FILE FOR PIC18

Intel hex file is a widely used file format designed to standardize the loading (transferring) of executable machine code into a ROM chip. Therefore, the loaders that come with every ROM burner (programmer) support the Intel hex file format. In many Windows-based assemblers such as MPLAB, the Intel hex file is produced according to the settings you set. In the PIC MPLAB environment, the object file is fed into the linker program to produce the Intel hex file. The hex file is used by the loader of an EPROM programmer such as the PICkit 2 programmer to transfer (load) the file into the ROM chip. The MPLAB assembler can produce three types of Intel hex files. They are (a) INHX8M, (b) INHX32, and (c) INHX8S. See Table 8-13. In this section we will explain each one with some examples.

Table 8-13: Intel Hex File Formats Produced by MPLAB (See http://www.microchip.com)

Format Name	Format Type	File Extension	Max. ROM Address
Intel Hex format	INHX8M	.hex	16-bit address
Intel Hex 32 format	INHX32	.hex	32-bit address
Intel Split Hex	INHX8S	.hxl and .hxh	16-bit address for each

Analyzing the Intel hex (INHX8M) file

We choose the Intel hex type of INHX8M, INHX32, or INHX8S by using the LIST directive or setting the options in the MPLAB assembler itself. If we do not choose one, the MPLAB assembler selects INHX32 by default. Next, we will analyze the hex file belonging to the list file for INHX8M. The INHX8M file is produced by the MPLAB assembler by choosing the INHX8M option in the LIST directive (or setting the MPLAB assembler). The file has the .hex extension. The INHX8M is used for PIC chips with program ROM space of up to 64K in size. To get Intel hex files for chips with more than 64K of program ROM space, we must use the INHX32 option. Figure 8-10 shows the Intel hex file of INHX8M for the test program whose list file was given earlier. Notice that we have chosen INHX8M in the LIST directive. Since the ROM burner (loader) uses the hex file to download the opcode into ROM, the hex file must provide the following: (1) the number of bytes of information to be loaded, (2) the information itself, and (3) the starting address where the information must be placed. Each line of the hex file consists of six parts as follows:

:BBAAAATTHHHHH.......HHHHCC

The following describes each part:
1. ":" Each line starts with a colon.
2. BB, the count byte. This tells the loader how many bytes are in the line. BB can range from 00 to 16 (10 in hex).
3. AAAA is for the address. This is a 16-bit address for the INHX8M. The loader places the first byte of data into this memory address.
4. TT is for type. This field is either 00 or 01. If it is 00, it means that there are more lines to come after this line. If it is 01, it means that this is the last line and the loading should stop after this line.
5. HH......H is the real information (data or code). There is a maximum of 16 bytes in this part. The loader places this information into successive memory locations of ROM. Because the PIC18 chips have 16-bit-wide program ROM space, the information in this field is presented as low byte followed by the high byte.

```
:10000000936A550E816E811E07EC00F0FCD7020E3C
:10001000076EFA0E086EFA0E096E0000000009065F
:0C002000FCE10806F8E10706F4E112001C
:0300010022020ECA
:010006008079
:060008000FC00FE00F40E5
:00000001FF


Separating the fields, we get the following:

:BB AAAA TT HHHHHHHHHHHHHHHHHHHHHHHHHHHHHHHH     CC
:10 0000 00 936A550E816E811E07EC00F0FCD7020E     3C
:10 0010 00 076EFA0E086EFA0E096E000000000906     5F
:0C 0020 00 FCE10806F8E10706F4E11200             1C

:03 0001 00 22020E                               CA
:01 0006 00 80                                   79
:06 0008 00 0FC00FE00F40                          E5
:00 0000 01                                       FF
```

Figure 8-10. Intel Hex File Test Program with the INHX8M Option

6. CC is a single byte. This last byte is the checksum byte of everything in that line. The checksum byte is used for error checking. Checksum bytes are discussed in detail in Chapters 6 and 7. Notice that the checksum byte at the end of each line represents the checksum byte for everything in that line and not just for the data portion.

Now, compare the data portion of the Intel hex file in Figure 8-10 with the information under the OBJ field of the .lst file in Figure 8-11. Notice that they are

```
LOC     OBJ           LINE

                      00003 LIST  P=PIC18F458,F=INHX8M,N=0,ST=OFF,R=HEX
                      00004 #include P18F458.INC

22 02 0E 80 0F        00005        CONFIG   OSC=HS, OSCS=OFF
C0 0F E0 0F 40
                      00006        CONFIG   BORV=45,PWRT=ON, BOR=ON
                      00007        CONFIG   WDT=OFF
                      00008        CONFIG   DEBUG=OFF, LVP=OFF, STVR=OFF
                      00009

00000007              00010 R1 EQU 0x07
00000008              00011 R2 EQU 0x08
00000009              00012 R3 EQU 0x09

000000               00014        ORG 0
000000 6A93          00015        CLRF   TRISB
000002 0E55          00016        MOVLW 0x55
000004 6E81          00017        MOVWF PORTB
000006 1E81          00018 L3     COMF   PORTB,F
000008 EC07 F000     00019        CALL   QDELAY
00000C D7FC          00020        BRA    L3

                     00023 ;-----------1/4 SECOND DELAY
00000E               00024 QDELAY
00000E 0E02          00025        MOVLW D'2'
000010 6E07          00026        MOVWF R1
000012 0EFA          00027 D1     MOVLW D'250'
000014 6E08          00028        MOVWF R2
000016 0EFA          00029 D2     MOVLW D'250'
000018 6E09          00030        MOVWF R3
00001A 0000          00031 D3     NOP
00001C 0000          00032        NOP
00001E 0609          00033        DECF   R3, F
000020 E1FC          00034        BNZ    D3
000022 0608          00035        DECF   R2, F
000024 E1F8          00036        BNZ    D2
000026 0607          00037        DECF   R1, F
000028 E1F4          00038        BNZ    D1
00002A 0012          00039        RETURN
                     00040        END
```

Figure 8-11. List File for Test Program with the INHX8M Option
(Comments and other lines are deleted for space and simplicity.)

identical, as they should be. The extra information is added by the Intel hex file format. You can run the C language version of the test program and verify its operation. Your C compiler will provide you both the .lst file and Intel hex file if you want to explore the Intel hex file concept.

Examine the next three examples to gain insight into the Intel hex file.

Example 8-2

From Figure 8-10, analyze the six parts of line 3.

Solution:

After the colon (:), we have 0C, which means that 12 bytes of data are in this line. 0020H is the address at which the data starts. Next, 00 means that this is not the last line of the record. Then the data, which is 12 bytes, is as follows: `FCE10806F8E10706F4E11200`. Finally, the last byte, 1C, is the checksum byte.

Example 8-3

Compare the data portion of the Intel hex file of Figure 8-10 with the opcodes in the list file of the test program given in Figure 8-11. Do they match?

Solution:

In the first line of Figure 8-10, the data portion starts with 936AH, where the low byte is followed by the high byte. That means it is 6A93, the opcode for the instruction "`CLRF TRISB`", as shown in the list file of Figure 8-11. The last byte of the data in line 3 is 1200, which is the opcode for the "`RETURN`" instruction in the list file.

Example 8-4

Verify the checksum byte for line 3 of Figure 8-10. Verify also that the information is not corrupted.

Solution:

0C + 20 + FC + E1 + 08 + 06 + F8 + E1 + 07 + 06 + F4 + E1 + 12 + 00 = 5E4 in hex. Dropping the carries (5) gives E4H, and its 2's complement is 1CH, which is the last byte of line 4.

If we add all the information in line 4, including the checksum byte, and drop the carries we should get 0C + 20 + FC + E1 + 08 + 06 + F8 + E1 + 07 + 06 + F4 + E1 + 12 + 00 = 600H.

Analyzing the Intel hex file of INHX32

For PIC chips with program ROM space of more than 64K, we must choose the INHX32 option. Figure 8-13 shows the Intel hex file for the test program (Figure 8-14) assembled with the INHX32 option instead of INHX8M. Notice that INHX8M is used for chips with a ROM size of 64K or less, while ROM chips with more than 64K use INHX32. The 32-bit space in the INHX32 is for the address. That means that the INHX32 can be used for ROM chips with address space of 1 byte to 4 gigabytes, because $2^{32} =$ 4 gigabytes, as shown in Figure 8-12.

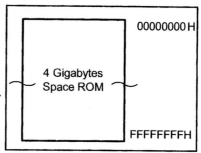

Figure 8-12. ROM Space for Chips with 32-bit Addressing

Notice that the MPLAB produces a file with the .hex extension for the INHX32, just like the INHX8M. The INHX32 is very similar to the INHX8M, except that the TT field has extra options to accommodate the 32-bit address of the ROM chip. As with the INHX8M, each line of the hex file consists of six parts.

:BBAAAATTHHHHH.......HHHHCC

The following describes each part.

1. ":" Each line starts with a colon.
2. BB, the count byte. This tells the loader how many bytes are in the line. BB can range from 00 to 16 (10 in hex).
3. AAAA is a A15–A0 address, or all zeros, depending on the TT status.
4. TT is for the record type. This field has four possibilities as follows:
   00 = Data record is in field HHHH
   01 = End of file record. It means this is the last line and the loading should stop after this line.
   02 = Segment address record.
   04 = Linear address record in the HHHH field (A31–A16 portion of A31–A0 is given in the HHHH field). Notice that if TT = 04, then the HHHH has the upper addresses of A31–A16 and AAAA = 0000.
5. HH......H is the real information (data, address, or code). There is a maximum of 16 bytes in this part. The loader places this information into successive memory locations of ROM. Because the PIC18 chips have 16-bit ROM, the information in this field is presented as low byte followed by the high byte.
6. CC is a single byte. This last byte is the checksum byte of everything in that line. The checksum byte is used for error checking. Checksum bytes were discussed in detail in Chapters 6 and 7. Notice that the checksum byte at the end of each line represents the checksum byte for everything in that line and not just for the data portion.

```
:020000040000FA
:0E000000936A550E816E811E78EC94F0FCD749
:020000040001F9
:1028F000020E076EFA0E086EFA0E096E0000000056
:0E2900000906FCE10806F8E10706F4E1120002
:020000040030CA
:0600010022020E830180C3
:06000800FFC0FFE0FF4015
:00000001FF


Separating the fields we get the following:

:BB  AAAA  TT  HHHHHHHHHHHHHHHHHHHHHHHHHHHHHHHH      CC
:02  0000  04  0000                                  FA
:0E  0000  00  936A550E816E811E78EC94F0FCD7          49
:02  0000  04  0001                                  F9
:10  28F0  00  020E076EFA0E086EFA0E096E00000000      56
:0E  2900  00  0906FCE10806F8E10706F4E11200          02
:02  0000  04  0030                                  CA
:06  0001  00  22020E830180                          C3
:06  0008  00  FFC0FFE0FF40                           15
:00  0000  01                                        FF
```

Figure 8-13. Intel Hex (INHX32) File Test Program as Provided by the Assembler

---

Example 8-5

From Figure 8-13, analyze the six parts of (a) line 3, and (b) line 4.

Solution:

(a) In line 3, after the colon (:), we have 02 which means that 2 bytes of data are in this line. The AAAA = 0000 and TT = 04 mean that the upper 16 bits of the address are provided by the HHHH field. That is exactly what we see where the 16-bit address of 000128F0H is given in the HHHH field, which is 0001.

(b) In line 4, after the colon (:), we have 10H (which is 16 in decimal) as the number of bytes in this line. The AAAA = 28F0 is the lower 16-bit address where information will be burned. Next, 00 means that this is not the last line of the record. Then the data, which is 16 bytes, is as follows: 020E076EFA0E086EFA0E096E00000000. Finally, the last byte, 56, is the checksum byte.

```
LOC       OBJ           LINE
                        00003 LIST P=PIC18F8720,F=INHX32,N=0,ST=OFF,R=HEX
                              ;INTX32 for > 64KB
                        00004   #include P18F8720.INC
                        00001        LIST
                01306      LIST
22 02 0E 83             00005        CONFIG OSC=HS, OSCS=OFF
01 80 FF C0 FF E0 FF 40
                        00006        CONFIG BORV=45,PWRT=ON, BOR=ON
                        00007        CONFIG WDT=OFF
                        00008        CONFIG DEBUG=OFF, LVP=OFF, STVR=OFF

  00000007              00010 R1 equ 0x07
  00000008              00011 R2 equ 0x08
  00000009              00012 R3 equ 0x09

000000                  00014        ORG  0
000000 6A93             00015        CLRF   TRISB
000002 0E55             00016        MOVLW  0x55      ;WREG = 55h
000004 6E81             00017        MOVWF  PORTB
000006 1E81             00018 L3     COMF   PORTB,F
000008 EC78 F094        00019        CALL   QDELAY
00000C D7FC             00020        BRA    L3

                        00023 ;-----------1/4 SECOND DELAY
0128F0                  00024        ORG 128F0H
0128F0                  00025 QDELAY
0128F0 0E02             00026        MOVLW D'2'
0128F2 6E07             00027        MOVWF R1
0128F4 0EFA             00028 D1     MOVLW D'250'
0128F6 6E08             00029        MOVWF R2
0128F8 0EFA             00030 D2     MOVLW D'250'
0128FA 6E09             00031        MOVWF R3
0128FC 0000             00032 D3     NOP
0128FE 0000             00033        NOP
012900 0609             00034        DECF   R3, F
012902 E1FC.            00035        BNZ    D3
012904 0608             00036        DECF   R2, F
012906 E1F8             00037        BNZ    D2
012908 0607             00038        DECF   R1, F
01290A E1F4             00039        BNZ    D1
01290C 0012             00040        RETURN
                        00041        END
```

Figure 8-14. List File for Test Program with INHX32 Option (Notice ORG address for QDELAY. Some of the comments and lines are deleted for clarity.)

Intel hex split file of INHX8S

The INHX8S option is called the Intel split hex format. When we choose the INHX8S option in the LIST directive, we get two files: .hxl for the low byte and and .hxh for the high byte. Because the PIC18 ROM is 16 bits wide, we have even addresses for the low byte and odd addresses for the high byte, as shown in Figure 8-15. The MPLAB assembler gives us this option because in many PIC18 chips with external memory we need to split the memory into odd and even banks to create a 16-bit wide ROM space. Note that ROM chips have pins D0–D7; there-fore, their organization is Nkx8 (e.g., 64kx8). That means we must burn the hex file into the even-bank ROM and the .hxh file into the ROM with odd addresses. Note that the .hxl and .hxh formats are the same as the INHX8M with the address space limited to 64K for each bank. That means that with split ROM, we can have a maximum of 128K of ROM with 64K for each bank.

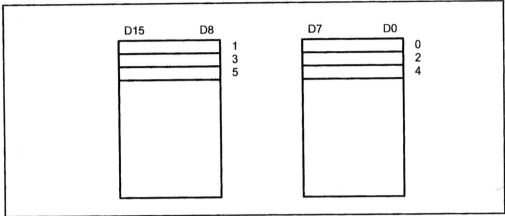

Figure 8-15. The Odd and Even Banks for External Memory of PIC18

Review Questions

1. True or false. The Intel hex file does not use the checksum byte method to ensure data integrity.
2. The first byte of a line in the Intel hex file represents ____.
3. The last byte of a line in the Intel hex file represents ____.
4. In the TT field of an Intel hex file, we have 00. What does it indicate?
5. Find the checksum byte for the following values: 22H, 76H, 5FH, 8CH, 99H.
6. In Question 5, add all the values and the checksum byte. What do you get?
7. True or false. In the TT field of the INHX32 file, we have 04, indicating the record is in the upper 16 bits of the 32-bit address.

SECTION 8.4: PIC18 TRAINER DESIGN AND LOADING

In this section, we discuss the connection for a simple PIC18-based train-er. We also show various ways of loading a hex file into the PIC microcontroller. Microchip has skillfully designed their microcontrollers for maximum flexibility of loading programs. The three primary ways to load a program are:

1. A device burner loads the program into the microcontroller separate from the system. This is useful on a manufacturing floor where a gang programmer is used to program many chips at one time. Most mainstream device burners sup-port the PIC families: Advin and EETools are two of the more popular compa-nies. Microchip supplies programmers for all their products; the PICkit 2 and PICSTART PLUS are just two examples. See Microchip's website for a com-plete list. You can also build your own device programmer. Doing this will

reduce the cost of purchasing a commercial programmer, which is usually expensive. Building a programmer from scratch is beyond the scope of this text; check the Internet for sites devoted to this.

The device programming method is straightforward: The chip is programmed before it is inserted into the circuit. Or, the chip can be removed and reprogrammed if it is in a socket. A ZIF (zero insertion force) socket is even quicker and less damaging than a standard socket. When removing and reinserting, we must observe ESD (electrostatic discharge) procedures. Although PIC devices are rugged, there is always a risk when handling them. Using this method allows all of the device's resources to be utilized in the design. No pins are shared, nor are internal resources of the chip used as in the other two methods. This allows the embedded designer to use the minimum board space for the design.

2. An in-circuit serial programmer (ICSP) allows the developer to program and debug their microcontroller while it is in the system. This is done by two wires with a system setup to accept this configuration. The Microchip ICD 2 is a wonderful device for debugging programs. This method also allows the manufacturer to install the devices unprogrammed on the board. Before shipping to customers, the microcontroller can be programmed with the most recent file.

In-circuit serial programming is excellent for designs that change or require periodic updating. The ICSP uses two pins, RB7 and RB6. These pins can be used as I/O after the device is programmed. The designer must make sure that these pins do not conflict with the programmer. MCLR also needs a 10 kΩ pull-up resistor for the ICSP. The ICD 2 also needs V_{dd} and Gnd. The designer must bring the pins to a header on the board so that the programmer can connect to it. Figure 8-16 shows the pin connections. The designer must weigh the pros and cons of these methods.

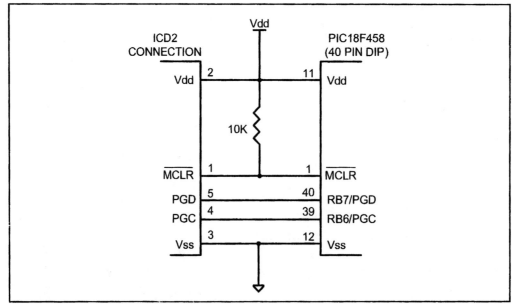

Figure 8-16. ICSP Connections

3. A boot loader is a piece of code burned into the microcontroller's program ROM. Its purpose is to communicate with the user's board to load the program. A boot loader can be written to communicate via a serial port, CAN port, USB port, or even a network connection. A boot loader can also be designed to debug a system, similar to the ICD. This method of programming is excellent for the developer who does not always have a device programmer or an ICD available. Microchip has several application notes on writing boot loaders on their website. The main drawback of the boot loader is that it does require a communication port and program code space on the microcontroller. Also, the boot loader has to be programmed into the device before it can be used, usually by one of the two previous ways.

The boot loader method is ideal for the developer who needs to quickly program and test code. This method also allows the update of devices in the field without the need of ICD tools. All one needs is a computer with a port that is compatible with the board. (The serial port is one of the most commonly used and discussed, but a CAN or USB boot loader can also be written.) This method also consumes the largest amount of resources. Code space must be reserved and protected, and external devices are needed to connect and communicate with the PC. Developing projects using this method really helps programmers test their code. For mature designs that do not change, the other two methods are better suited.

Next, we discuss the issues related to the ROM loader for PIC18F-based systems using chips such as the PIC18F458/4580 and PIC18F452/4520. We will also provide guidelines for design of a simple PIC18 Trainer. If you decide to wire-wrap one of these, make sure that you read Appendix B on wire wrapping.

PIC18F452/458-based Trainer

In systems based on a PIC18-type microcontroller, you need a ROM burner to burn your program into the microcontroller. For the PIC18F, the ROM burner can erase the Flash ROM in addition to burning a program into it. In the case of the PIC18C, you also need an EPROM erasure tool because it uses UV-EPROM. Before burning the PIC18C, you need to erase its contents, which takes approximately 20 minutes for UV-EPROM. For the PIC18F, this is not required because it has Flash ROM.

PIC18 Flash ROM size

While all PIC18 chips share the same features, they come with different amounts of on-chip ROM. Table 8-14 shows the on-chip ROM size for various

Table 8-14: PIC18 On-chip ROM Size and Address Space

	On-chip Code ROM (Bytes)	Code Address Range (Hex)
PIC18F2220	4K	00000–00FFF
PIC18F2410	16K	00000–03FFF
PIC18F458/4580	32K	00000–07FFF
PIC18F6680	64K	00000–0FFFF
PIC18F8722	128K	00000–1FFFF

PIC18 chips. Refer to the web site http://www.microchip.com for further information. Notice that while the PIC18F2220 comes with 4K of on-chip ROM, and the PIC18F2410 comes with 16K, the PIC18F458 has 32K of on-chip ROM. Also notice that the PIC18F458 is a substitute for the PIC18F452 with extra functions such as controlled area network (CAN).

Example 8-6

Find the ROM memory address of each of the following PIC chips:
(a) PIC18F2220 with 4 KB
(b) PIC18F2410 with 16 KB
(c) PIC18F458/4580 with 32 KB

Solution:

(a) With 4K of on-chip ROM memory space, we have 4096 bytes (4 × 1024 = 4096). This maps to address locations of 0000 to 0FFFH. Notice that 0 is always the first location.
(b) With 16K of on-chip ROM memory space, we have 16,384 bytes (16 × 1024 = 16,384), which gives 0000–3FFFH.
(c) With 32K we have 32,768 bytes (32 × 1024 = 32,768). Converting 32,768 to hex, we get 8000H; therefore, the memory space is 0000 to 7FFFH.

PIC18 Trainer connection

We selected the PIC18F458 for a PIC18-based Trainer because it allows you to easily wirewrap an inexpensive but powerful trainer to be used at work and home. Figure 8-17 shows the connection for the PIC18F-based system to be used with the PICkit 2 programmer.

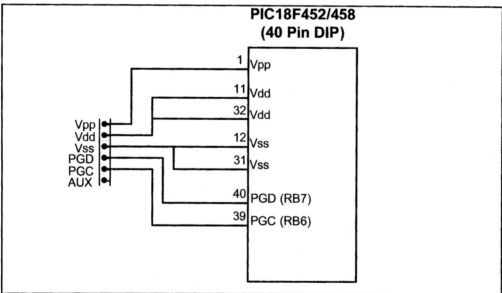

Figure 8-17. PIC18F Connection to PICkit 2 with 6-Pin Header
Note: This connection using the PICkit 2 header applies to all families of PIC microcontrollers. The only differences are the pin number and designation.

326

The PICkit 2 is an inexpensive programmer available from the Microchip website. The www.MicroDigitalEd.com web site shows the schematic for PIC18-based Trainer connection.

Downloading to the PIC18 Trainer

After we build our PIC18-based system, we can download the program into the Trainer using the PICkit 2's programmer utility. See Figure 8-18. Microchip is continuously updating MPLAB IDE to support PICkit 2 for programming of all PIC microcontrollers.

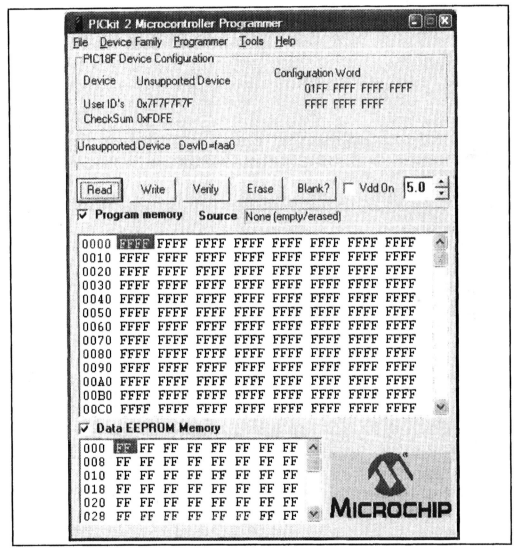

Figure 8-18. PICkit 2 Programmer Utility

Test program for the PIC18 in Assembly and C

To test your PIC18 hardware connection, we can run a simple test in which all the bits of PORTB toggle continuously with some delay between the "on" and "off" states. See Programs 8-3 and 8-3C. Notice in these programs that the time delay is based on a 10 MHz crystal. In developing your program, you can use the program shells provided in Figures 8-19 and 8-20.

Trainer Test Program in Assembly

```
;Program 8-3
        LIST P=PIC18F458, F=INHX32, N=0, ST=OFF, R=HEX
        #include P18F458.INC
        CONFIG OSC = HS, OSCS = OFF
        CONFIG WDT = OFF
        CONFIG BORV = 45, PWRT = ON, BOR = ON
        CONFIG DEBUG = OFF, LVP = OFF, STVR = OFF


R1      EQU     0x07
R2      EQU     0x08
R3      EQU     0x09


        ORG     0000H           ;note starting address
        CLRF    TRISB           ;make Port B an output port
        MOVLW   0x55            ;WREG = 55h
        MOVWF   PORTB           ;put 55h on port B pins
L3      COMF    PORTB,F         ;toggle bits of Port B
        CALL    QDELAY          ;quarter of a second delay
        BRA     L3              ;continue


;-----------1/4 SECOND DELAY
QDELAY
        MOVLW   D'2'
        MOVWF   R1
D1      MOVLW   D'250'
        MOVWF   R2
D2      MOVLW   D'250'
        MOVWF   R3
D3      NOP
        NOP
        DECF    R3, F
        BNZ     D3
        DECF    R2, F
        BNZ     D2
        DECF    R1, F
        BNZ     D1
        RETURN
        END
```

```
#include P18F458.INC
    CONFIG OSC = HS, OSCS = OFF
    CONFIG WDT = OFF
    CONFIG BORV = 45, PWRT = ON, BOR = ON
    CONFIG DEBUG = OFF, LVP = OFF, STVR = OFF


    ORG    0000H            ;start of user code space
                            ;begin user code




                            ;end of user code
    END
```

Figure 8-19: Shell of a Simple Assembly Language Program for MPLAB

Notice that the LIST directive is not used because it is one of the default settings in the MPLAB IDE.

Trainer Test Program in C

```
;Test Program 8-3C: Toggling PORTB for the PIC18F458/4580
;(452/4520) with XTAL = 10 MHz

      #pragma config OSC = HS, OSCS = OFF
      #pragma config PWRT = OFF, BOR = ON, BORV = 45
      #pragma config WDT = OFF, LVP = OFF
      #pragma config DEBUG = OFF, STVR = OFF

      #include <P18F458.h>

      void msdelay(unsigned int ms);
      void main(void)
        {
          TRISB = 0;                    //make Port B an output
          while(1)
            {
            PORTB = 0x55;
            msdelay(500);
            PORTB = 0xAA;
            msdelay(500);
            }

        }

      //this is for a 10 MHz clock
      void msdelay(unsigned int ms)
        {
        unsigned int x;
        unsigned char z;
        for(x=0;x<ms;x++)
          for(z=0;z<165;z++);
        }
```

```
#pragma config OSC = HS, OSCS = OFF
#pragma config PWRT = OFF, BOR = ON, BORV = 45
#pragma config WDT = OFF, LVP = OFF
#pragma config DEBUG = OFF, STVR = OFF

#include <P18F458.h>
void main(void)
  {





  }
```

Figure 8-20: Shell of a Simple C Language Program for MPLAB

Some troubleshooting tips

Running the test program on your PIC18F458-based trainer (or PIC18F452 system) should toggle all the I/O bits with some delay. If your wire-wrapped system does not work, follow these steps to find the problem:

1. With the power off, check your connection for all pins, especially V_{dd} and GND.

2. Check MCLR (pin 1) using an oscilloscope. When the system is powered up, pin 1 is HIGH. Upon pressing the momentary switch, it goes LOW. Make sure the momentary switch is connected properly.

3. Observe the OSC1 pin on the oscilloscope while the power is on. You should see a crude sine wave. This indicates that the crystal oscillator is operating.

4. If all the above steps pass inspection, check the content of the on-chip ROM. It must be the same as the opcodes provided by the .lst file. Your assembler produces the .lst file, which lists the opcodes and operands on the left side of the assembly instructions. This must match exactly the contents of your on-chip ROM if the proper steps were taken in burning and loading the program into the on-chip ROM.

Review Questions

1. Which method(s) to program the PIC microcontroller is/are the best for manfacturing of large-scale boards?
2. Which method(s) allow(s) for debugging a system?
3. Which method(s) would allow a small company to develop a prototype and test an embedded system for a variety of customers?
4. True or false. The PIC18C has Flash program ROM.
5. Which pin is used for reset in the PIC18F458/4580?
6. What is the status of the reset pin when it is not activated?
7. What kind of ROM is used in the PIC18F458/4580 chip?
8. True or false. The PIC18 can download the file into its ROM only if it is in Intel hex file format.
9. Give two reasons that the PIC18F is preferable over PIC18C chips.

See the following website for the PIC18 Trainer:

http://www.MicroDigitalEd.com

SUMMARY

This chapter began by describing the function of each pin of the PIC18F458. The CONFIG registers of the PIC18F458 were discussed. These CONFIG registers are located at address 300001H and beyond, which is out of the program ROM address range. They are burned into the PIC chip along with the application. We use CONFIG registers to enable features such as low power frequency and watchdog timer. We also explained the Intel hex file formats INHX8M and INHX32. We examined how the INHX32 format uses the 32-bit address of ROM while INHX8M files are used for 16-bit addresses. Then the design of the PIC18-based trainer was shown.

PROBLEMS

SECTION 8.1: PIC18F458/452 PIN CONNECTION

1. The PIC18F458 DIP package is a(n) _____-pin package.
2. Which pins are assigned to V_{CC} and GND?
3. In the PIC18F458, how many pins are designated as I/O port pins?
4. The crystal oscillator is connected to pins _____ and _____ .
5. If PIC18F458 is rated as 40 MHz, what is the maximum frequency that can be connected to it?
6. Indicate the pin number assigned to MCLR in the DIP package.
7. MCLR stands for _____ .
8. The MCLR pin is normally _____ (LOW, HIGH) and needs a _____ (LOW, HIGH) signal to be activated.
9. What are the contents of the PC (program counter) upon reset of the PIC18F458?
10. What are the contents of the SP register upon reset of the PIC18F458?
11. What are the contents of the WREG register upon reset of the PIC18F458?
12. What are the contents of the TRIS registers upon reset of the PIC18F458?
13. In PIC18F458, how many pins are set aside for the V_{dd}?
14. In PIC18F458, how many pins are set aside for the V_{ss} (Gnd)?
15. Which of the OSC pins are shared with the PORTA bit?
16. OSC1 and OSC2 are _____ (input, output) pins.
17. MCLR is an _____ (input, output) pin.
18. How many pins are designated as PORTA and what are those in the DIP package?
19. How many pins are designated as PORTB and what are those in the DIP package?
20. How many pins are designated as PORTC and what are those in the DIP package?
21. How many pins are designated as PORTD and what are those in the DIP package?

22. Upon reset, all the bits of ports are configured as _____ (input, output).
23. In the PIC18F458, which port has only 3 pins?
24. Which I/O pin of the PIC18F458 does not have an alternate function and can be used solely for I/O?

SECTION 8.2: PIC18 CONFIGURATION REGISTERS

25. True or false. For the PIC18F in reset state, the CPU does not execute any code.
26. True or false. When the system is powered up, the power-up timer (PWRT) and oscillator start-up timer (OST) keep the PIC18 in the reset state until the voltage and frequency are stable.
27. True or false. The power-up timer (PWRT) and oscillator start-up timer (OST) are components that we must add to the PIC18 externally.
28. True or false. The watchdog timer is a component that we must add to the PIC18 externally.
29. True or false. If we do not provide CONFIG values in our source code, PIC18 uses the default values for them.
30. True or false. The CONFIG registers use the same address space as program ROM.
31. Give the ROM address locations for CONFIG1H, CONFIG2L, CONFIG2H, and CONFIG4L.
32. The CONFIG registers are _____ bits wide.
33. Which CONFIG register is used to set the clock frequency for the PIC18F458?
34. Which CONFIG register is used to set the brown-out reset voltage for the PIC18F458?
35. Which CONFIG register is used to disble the watchdog timer for the PIC18F458?
36. If the brown-out reset voltage is set to 4.2 V, what does it mean to the system?
37. Show the CONFIG directive for a PIC18F458 system with the following options:
    (a) OSC1–OSC2 is connected to 20 MHz and it is the only source of the clock for the system.
    (b) The brown-out voltage is set for 4.2 V and the power-up timer is enabled.
    (c) No watchdog timer
    (d) No overflow on stack, no background debugger, and no LVP
38. For CONFIG1H, which option for the OSC frequency provides the lowest power dissipation?
39. Which CONFIG register is used to set the clock source for the PIC18F458?
40. Find the instruction cycle for the following crystal frequencies connected to OSC1 and OSC2. Assume that the HS option is chosen for all of them.
    (a) 12 MHz   (b) 20 MHz   (c) 25 MHz   (d) 30 MHz

SECTION 8.3: EXPLAINING THE INTEL HEX FILE FOR PIC18

41. True or false. The INHX32 option can be set by MPLAB without using the LIST directive.

42. True or false. The INHX32 option can be used for ROM sizes of more than 64 kilobytes.
43. True or false. The INHX8M option can be used for ROM sizes of more than 64 kilobytes.
44. True or false. The INHX8M option can be used for ROM sizes of less than 64 kilobytes.
45. True or false. The INHX32 option can be used for ROM of any size.
46. Analyze the six parts of line 1 of Figure 8-10.
47. Verify the checksum byte for line 1 of Figure 8-10. Verify also that the information is not corrupted.
48. Verify the checksum byte for line 2 of Figure 8-13. Verify also that the information is not corrupted.
49. What is the difference between the INHX8M and INHX32 hex files?
50. Analyze the INHX32 Intel hex file in Figure 8-13.

SECTION 8.4: PIC18 TRAINER DESIGN AND LOADING OPTIONS

51. True or false. Using the PICkit2, we must remove the PIC18F chip from the system and place it into the programmer.
52. True or false. The PICkit2 can only work with Flash chips.
53. Which of the following choices is the cheapest?
    (a) MPLAB ICD2   (b) PICkit2
54. Write a program to get 8-bit data from PORTB and send it to ports PORTC and PORTD.
55. Write a program to get 8-bit data from PORTD and send it to ports PORTB and PORTC.
56. Which pins of PORTB are PGD (program data) and PGC (program clock)?
57. At what program memory location does the PIC18F458 wake up upon reset? What is the implication of that?
58. Write a program to toggle all the bits of PORTB continuously
    (a) using AAH and 55H (b) using the COMF instruction.
59. What is the address of the last location of program ROM for the PIC18F458?
60. What is the address of the last location of program ROM for the PIC18F8722?
61. What is the address of the last location of program ROM for the PIC18F452?

ANSWERS TO REVIEW QUESTIONS

SECTION 8.1: PIC18F458/452 PIN CONNECTION

1. 1
2. 000000
3. 000000
4. LOW
5. Two pins for V_{dd} and 2 pins for Gnd

SECTION 8.2: PIC18 CONFIGURATION REGISTERS

1. 16 MHz/4 = 4 MHz and 1/4 MHz = 250 ns
2. 300001 hex

3. False
4. 0, 200
5. CONFIG2H, 300003H
6. True
7. False
8. True
9. False
10. 4.2 V

SECTION 8.3: EXPLAINING THE INTEL HEX FILE FOR PIC18

1. False
2. The number of bytes of data in the line
3. The checksum byte of all the bytes in that line
4. 00 means this is not the last line and that more lines of data follow.
5. 22H + 76H + 5FH + 8CH + 99H = 21CH. Dropping the carries we have 1CH and its 2's complement, which is E4H.
6. 22H + 76H + 5FH + 8CH + 99H + E4H = 300H. Dropping the carries we have 00, which means that the data is not corrupted.
7. True

SECTION 8.4: PIC18 TRAINER DESIGN AND LOADING OPTIONS

1 Device burner
2. In-circuit serial debugger
3. ICSP
4. False
5. Pin 1
6. HIGH
7. Flash
8. True
9. It can be used with ICSP and has a faster development time.

CHAPTER 9

PIC18 TIMER PROGRAMMING IN ASSEMBLY AND C

OBJECTIVES

Upon completion of this chapter, you will be able to:

>>   List the timers of the PIC18 and their associated registers
>>   Describe the various modes of the PIC18 timers
>>   Program the PIC18 timers in Assembly and C to generate time delays
>>   Program the PIC18 counters in Assembly and C as event counters

The PIC18 has two to five timers depending on the family member. They are referred to as Timers 0, 1, 2, 3, and 4. They can be used either as timers to generate a time delay or as counters to count events happening outside the microcontroller. In Section 9.1 we see how Timers 0 and 1 are used to generate time delays. In Section 9.2 we show how they are used as event counters. In Section 9.3 we use C language to program the PIC18 timers. Timers 2 and 3 are discussed in Section 9.4.

SECTION 9.1: PROGRAMMING TIMERS 0 AND 1

Every timer needs a clock pulse to tick. The clock source can be internal or external. If we use the internal clock source, then 1/4th of the frequency of the crystal oscillator on the OSC1 and OSC2 pins (Fosc/4) is fed into the timer. Therefore, it is used for time delay generation and for that reason is called a timer. By choosing the external clock option, we feed pulses through one of the PIC18's pins: this is called a counter. In this section we discuss the PIC18 timer and in the next section we program the timer as a counter.

Basic registers of the timer

Many of the PIC18 timers are 16 bits wide. Because the PIC18 has an 8-bit architecture, each 16-bit timer is accessed as two separate registers of low byte (TMRxL) and high byte (TMRxH). Each timer also has the TCON (timer control) register for setting modes of operation. Next, we discuss each timer separately.

Timer0 registers and programming

Timer0 can be used as an 8-bit or a 16-bit timer. The 16-bit register of Timer0 is accessed as low byte and high byte, as shown in Figure 9-1. The low-byte register is called TMR0L (Timer0 low byte) and the high-byte register is referred to as TMR0H (Timer0 high byte). These registers can be accessed like any other special function registers. For example, the instruction "MOVWF TMR0L" moves the value in WREG into TMR0L, the low byte of Timer0. These registers can also be read like any other register. For example, "MOVFF TMR0L, PORTB" copies TMR0L (low byte of Timer0) to PORTB.

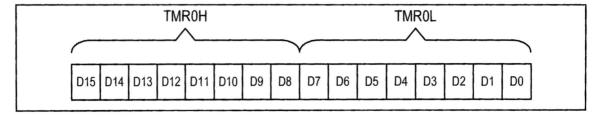

Figure 9-1. Timer 0 High and Low Registers

T0CON (Timer0 control) register

Each timer has a control register, called TCON, to set the various timer operation modes. T0CON is an 8-bit register used for control of Timer0. The bits for T0CON are shown in Figure 9-2.

TMR0ON	T08BIT	T0CS	T0SE	PSA	T0PS2	T0PS1	T0PS0

TMR0ON   D7   Timer0 ON and OFF control bit
                             1 = Enable (start) Timer0
                             0 = Stop Timer0

T08BIT   D6   Timer0 8-bit/16-bit selector bit
                             1 = Timer0 is configured as an 8-bit timer/counter.
                             0 = Timer0 is configured as a 16-bit timer/counter.

T0CS   D5   Timer0 clock source select bit
                             1 = External clock from RA4/T0CKI pin
                             0 = Internal clock (Fosc/4 from XTAL oscillator)

T0SE   D4   Timer0 source edge select bit
                             1 = Increment on H-to-L transition on T0CKI pin
                             0 = Increment on L-to-H transition on T0CKI pin

PSA   D3   Timer0 prescaler assignment bit
                             1 = Timer0 clock input bypasses prescaler.
                             0 = Timer0 clock input comes from prescaler output.

T0PS2:T0PS0 D2D1D0   Timer0 prescaler selector
               0 0 0 = 1:2   Prescale value (Fosc / 4 / 2)
               0 0 1 = 1:4   Prescale value (Fosc / 4 / 4)
               0 1 0 = 1:8   Prescale value (Fosc / 4 / 8)
               0 1 1 = 1:16   Prescale value (Fosc / 4 / 16)
               1 0 0 = 1:32   Prescale value (Fosc / 4 / 32)
               1 0 1 = 1:64   Prescale value (Fosc / 4 / 64)
               1 1 0 = 1:128 Prescale value (Fosc / 4 / 128)
               1 1 1 = 1:256 Prescale value (Fosc / 4 / 256)

Figure 9-2. T0CON (Timer0 Control) Register

T0CS (Timer0 clock source)

This bit in the T0CON register is used to decide whether the clock source is internal (Fosc/4) or external. If T0CS = 0, then the Fosc/4 is used as clock source. In this case, the timers are often used for time delay generation. See Example 9-1. If T0CS = 1, the clock source is external and comes from the RA4/T0CKI, which is pin 6 on the DIP package of PIC1818F4580/4520. When the clock source comes from an external source, the timer is used as an event counter. We will discuss that option in the next section. See Example 9-2.

Example 9-1

Find the value for T0CON if we want to program Timer0 in 16-bit mode, no prescaler. Use PIC18's Fosc/4 crystal oscillator for the clock source, increment on positive-edge.

Solution:

T0CON = 0000 1000   16-bit, Fosc/4 clock source, no prescaler, Timer0 off

---

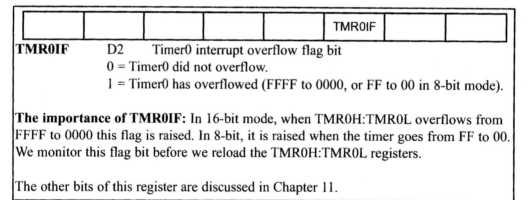
TMR0IF flag bit

Notice that the TMR0IF bit (Timer0 interrupt flag) is part of the INTCON (interrupt control) register. See Figure 9-3. The other options of the INTCON register are discussed in Chapter 11. As we will see, when the timer reaches its maximum value of FFFFH, it rolls over to 0000, and TMR0IF is set to 1 (see Figure 9-4). Chapter 11 shows how we can use TMR0IF to cause an interrupt. Next, we describe the 16-bit mode operation for Timer0.

					TMR0IF		

TMR0IF   D2   Timer0 interrupt overflow flag bit
                           0 = Timer0 did not overflow.
                           1 = Timer0 has overflowed (FFFF to 0000, or FF to 00 in 8-bit mode).

The importance of TMR0IF: In 16-bit mode, when TMR0H:TMR0L overflows from FFFF to 0000 this flag is raised. In 8-bit, it is raised when the timer goes from FF to 00. We monitor this flag bit before we reload the TMR0H:TMR0L registers.

The other bits of this register are discussed in Chapter 11.

Figure 9-3. INTCON (Interrupt Control Register) has the TMR0IF Flag

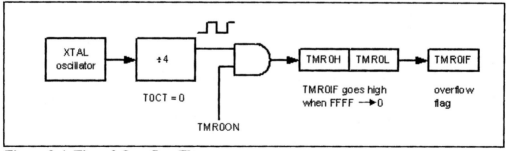

Figure 9-4. Timer0 Overflow Flag

16-bit timer programming

The following are the characteristics and operations of 16-bit mode:

1. It is a 16-bit timer; therefore, it allows values of 0000 to FFFFH to be loaded into the registers TMR0H and TMR0L.
2. After TMR0H and TMR0L are loaded with a 16-bit initial value, the timer must be started. This is done by "BSF T0CON, TMR0ON" for Timer0.
3. After the timer is started, it starts to count up. It counts up until it reaches its limit of FFFFH. When it rolls over from FFFFH to 0000, it sets HIGH a flag bit called TMR0IF (timer interrupt flag, which is part of the INTCON register). This timer flag can be monitored. When this timer flag is raised, one option would be to stop the timer.
4. After the timer reaches its limit and rolls over, in order to repeat the process, the registers TMR0H and TMR0L must be reloaded with the original value, and the TMR0IF flag must be reset to 0 for the next round.

Steps to program Timer0 in 16-bit mode

To generate a time delay using the Timer0 mode 16, the following steps are taken:

1. Load the value into the T0CON register indicating which mode (8-bit or 16-bit) is to be used and the selected prescaler option.
2. Load register TMR0H followed by register TMR0L with initial count values.
3. Start the timer with the instruction "BSF T0CON, TMR0ON".
4. Keep monitoring the timer flag (TMR0IF) to see if it is raised. Get out of the loop when TMR0IF becomes high.
5. Stop the timer with the instruction "BCF T0CON, TMR0ON".
6. Clear the TMR0IF flag for the next round.
7. Go back to Step 2 to load TMR0H and TMR0L again.

To clarify the above steps, see Example 9-3. To calculate the exact time delay and the square wave frequency generated on pin PB5, we need to know the XTAL frequency. See Example 9-4 and Example 9-5.

Notice in Figure 9-5 that we should load **TMR0H first**, and **then load TMR0L**, because the value for TMR0H is kept in a temporary register and written to TMR0H when TMR0L is loaded. This will prevent any error in counting if the TMR0ON flag is set HIGH.

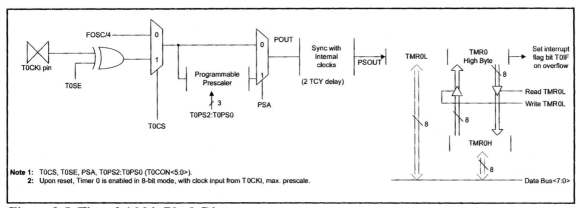

Figure 9-5. Timer0 16-bit Block Diagram

Example 9-3

In the following program, we are creating a square wave of 50% duty cycle (with equal portions high and low) on the PORTB.5 bit. Timer0 is used to generate the time delay. Analyze the program.

```
        BCF    TRISB,5            ;PB5 as an output
        MOVLW  0x08              ;Timer0,16-bit,int clk,no prescale
        MOVWF  T0CON             ;load T0CON reg.
HERE    MOVLW  0xFF              ;TMR0H = FFH, the high byte
        MOVWF  TMR0H             ;load Timer0 high byte
        MOVLW  0xF2              ;TMR0L = F2H, the low byte
        MOVWF  TMR0L             ;load Timer0 low byte
        BCF    INTCON, TMR0IF    ;clear timer interrupt flag bit
        BTG    PORTB,5           ;toggle PB5
        BSF    T0CON, TMR0ON     ;start Timer0
AGAIN   BTFSS  INTCON, TMR0IF    ;monitor Timer0 flag until
        BRA    AGAIN             ;it rolls over
        BCF    T0CON, TMR0ON     ;stop Timer0
        BRA    HERE              ;load TH, TL again
```

Solution:

In the above program notice the following steps:

1. T0CON is loaded.
2. FFF2H is loaded into TMR0H–TMR0L.
3. The Timer0 interrupt flag is cleared by the "BCF INTCON, TMR0IF" instruction.
4. PORTB.5 is toggled for the high and low portions of the pulse.
5. Timer0 is started by the "BSF T0CON, TMR0ON" instruction.
6. Timer0 counts up with the passing of each clock, which is provided by the crystal oscillator. As the timer counts up, it goes through the states of FFF3, FFF4, FFF5, FFF6, FFF7, FFF8, FFF9, FFFA, FFFB, and so on until it reaches FFFFH. One more clock rolls it to 0, raising the Timer0 flag (TMR0IF = 1). At that point, the "BTFSS INTCON, TMR0IF" instruction bypasses the "BRA AGAIN" instruction.
7. Timer0 is stopped by the instruction "BCF T0CON, TMR0ON", and the process is repeated.

Notice that to repeat the process, we must reload the TMR0L and TMR0H registers, and start the timer again.

FFF2   FFF3   FFF4   FFFF   0000

TMR0IF=0   TMR0IF=0   TMR0IF=0   TMR0IF=0   TMR0IF=1

Example 9-4

In Example 9-3, calculate the amount of time delay generated by the timer. Assume that XTAL = 10 MHz.

Solution:

The timer works with the Fosc/4 clock; therefore, we have 10 MHz / 4 = 2.5 MHz as the timer frequency. As a result, each clock has a period of T = 1 /2.5 MHz = 0.4 μs. In other words, Timer0 counts up each 0.4 μs resulting in delay = number of counts × 0.4 μs.

The number of counts for the rollover is FFFFH − FFF2H = 0DH (13 decimal). However, we add one to 13 because of the extra clock needed when it rolls over from FFFF to 0 and raises the TMR0IF flag. This gives 14 × 0.4 μs = 5.6 μs for half the pulse. For the entire period the time delay generated by the timer is T = 2 × 5.6 μs = 11.2 μs.

Example 9-5

Calculate the frequency of the square wave generated on pin PORTB.5.

Solution:

To get a more accurate timing, we need to add clock cycles due to the instructions in the loop.

```
                                              Cycles
            BCF    TRISB,5
            MOVLW  0x08
            MOVWF  T0CON
            BCF    INTCON, TMR0IF

HERE        MOVLW  0xFF                          1
            MOVWF  TMR0H                         1
            MOVLW  -D'48'                        1
            MOVWF  TMR0L                         1
            CALL   DELAY                         1
            BTG    PORTB,5                       1
            BRA    HERE                          1
;————————delay using Timer0
DELAY       BSF    T0CON, TMR0ON                 1
AGAIN       BTFSS  INTCON, TMR0IF                1
            BRA    AGAIN                         1
            BCF    T0CON, TMR0ON                 1
            BCF    INTCON, TMR0IF                1
            RETURN                               1
                                                ——
                                                13
```

T = 2 × (48 + 13) × 0.4 μs = 48.8 μs and F = 20.491 kHz.

We can develop a formula for delay calculations using 16-bit mode of the timer for a crystal frequency of XTAL = 10 MHz. This is given in Figure 9-6. The scientific calculator in the Accessory directory of Microsoft Windows can help you to find the TMR0H, TMR0L values. This calculator supports decimal, hex, and binary calculations. See Examples 9-6 and 9-7.

(a) in hex	(b) in decimal
(FFFF - YYXX + 1) × 0.4 μs where YYXX are the TMR0H, TMR0L initial values respectively. Notice that YYXX values are in hex.	Convert YYXX values of the TMR0H, TMR0L register to decimal to get a NNNNN decimal number, then (65536 - NNNNN) × 0.4 μs

Figure 9-6. Timer Delay Calculation for XTAL = 10 MHz with No Prescaler

---

Example 9-6

Find the delay generated by Timer0 in the following code, using both of the methods of Figure 9-6. Do not include the overhead due to instructions.

```
        BCF    TRISB,5          ;PB5 as an output
        MOVLW  0x80             ;Timer0,16-bit,int clk, no prescale
        MOVWF  T0CON
        BCF    INTCON,TMR0IF    ;clear Timer0 interrupt
HERE    MOVLW  0xB8             ;TMR0H = B8, the high byte
        MOVWF  TMR0H
        MOVLW  0x3E             ;TMR0L = 3E, the low byte
        MOVWF  TMR0L
        BSF    T0CON, TMR0ON    ;start Timer0
AGAIN   BTFSS  INTCON, TMR0IF   ;monitor Timer0 flag until
        BRA    AGAIN            ;it rolls over
        BCF    T0CON, TMR0ON    ;stop Timer0
        BCF    INTCON, TMR0IF   ;clear Timer0 interrupt
        BTG    PORTB,5          ;toggle PB5
        BRA    HERE             ;load TH, TL again
```

Solution:

(a) (FFFF – B83E + 1) = 47C2H = 18,370 in decimal and 18,370 × 0.4 μs = 7.348 ms.

(b) Because TMR0H : TMR0L = B83EH = 47166 (in decimal) we have 65,536 – 47,166 = 18,370. This means that the timer counts from B83EH to FFFFH. This plus rolling over to 0 goes through a total of 18,370 clock cycles, where each clock is 0.4 μs in duration. Therefore, we have 18,370 × 0.4 μs = 7.348 ms as the width of the pulse.

Example 9-7

Find the frequency of the square wave generated by the following program if XTAL = 10 MHz. In your calculation do not include the overhead due to instructions in the loop.

```
        BCF    TRISB,5          ;PB5 as an output
        MOVLW  0x08             ;Timer0,16-bit,int clk,no prescale
        MOVWF  T0CON            ;load T0CON reg.
HERE    MOVLW  0x76             ;TMR0H = 76H, the high byte
        MOVWF  TMR0H            ;load Timer0 high byte
        MOVLW  0x34             ;TMR0L = 34H, the low byte
        MOVWF  TMR0L            ;load Timer0 low byte
        BCF    INTCON,TMR0IF    ;clear timer interrupt flag bit
        CALL   DELAY
        BTG    PORTB,5          ;toggle PB5
        BRA    HERE             ;load TH, TL again
;----------delay using Timer0
DELAY   BSF    T0CON,TMR0ON     ;start Timer0
AGAIN   BTFSS  INTCON,TMR0IF    ;monitor Timer0 flag until
        BRA    AGAIN            ;it rolls over
        BCF    T0CON,TMR0ON     ;stop Timer0
        RETURN
```

Solution:

Because FFFFH – 7634H = 89CBH + 1 = 89CCH and 89CCH = 35,276 clock count, $35{,}276 \times 0.4\ \mu s = 14.11$ ms and frequency = $1/(14.11$ ms $\times 2) = 35.434$ Hz. In this calculation, the overhead due to all the instructions in the loop is not included.

Finding values to be loaded into the timer

Assuming that we know the amount of timer delay we need, the question is how to find the values needed for the TMR0H and TMR0L registers. To calculate the values to be loaded into the TMR0L and TMR0H registers, look at Examples 9-8 and 9-9, where we use a crystal frequency of 10 MHz for the PIC18 system.

Assuming XTAL = 10 MHz and no prescaler we can use the following steps for finding the TMR0H and TMR0L registers' values:

1. Divide the desired time delay by 0.4 μs.
2. Perform 65,536 – n, where n is the decimal value we got in Step 1.
3. Convert the result of Step 2 to hex, where $yyxx$ is the initial hex value to be loaded into the timer's registers.
4. Set TMR0L = xx and TMR0H = yy.

Example 9-8

Assuming that XTAL = 10 MHz, write a program to generate a square wave with a period of 10 ms on pin PORTB.3.

Solution:

For a square wave with T = 10 ms we must have a time delay of 5 ms. Because XTAL = 10 MHz, the counter counts up every 0.4 μs. This means that we need 5 ms / 0.4 μs = 12,500 clocks. 65,536 – 12,500 = 53,036 = CF2CH. Therefore, we have TMR0H = CF and TMR0L = 2C.

```
        BCF    TRISB,3            ;PB3 as an output
        MOVLW  0x08              ;Timer0,16-bit,int clk,no prescale
        MOVWF  T0CON             ;load T0CON reg
HERE    MOVLW  0xCF              ;TMR0H = CFH, the high byte
        MOVWF  TMR0H             ;load Timer0 high byte
        MOVLW  0x2C              ;TMR0L = 2CH, the low byte
        MOVWF  TMR0L             ;load Timer0 low byte
        BCF    INTCON,TMR0IF     ;clear timer interrupt flag bit
        CALL   DELAY
        BTG    PORTB,3           ;toggle PB3
        BRA    HERE              ;load TH, TL again
;------------delay using Timer0
DELAY   BSF    T0CON,TMR0ON      ;start Timer0
AGAIN   BTFSS  INTCON,TMR0IF     ;monitor Timer0 flag until
        BRA    AGAIN             ;it rolls over
        BCF    T0CON,TMR0ON      ;stop Timer0
        RETURN
```

Example 9-9

Assuming that XTAL = 10 MHz, modify the program in Example 9-8 to generate a square wave of 2 kHz frequency on pin PORTB.3.

Solution:

Look at the following steps.
(a) T = 1 / F = 1 / 2 kHz = 500 μs the period of the square wave.
(b) 1/2 of it for the high and low portions of the pulse is 250 μs.
(c) 250 μs / 0.4 μs = 625 and 65,536 – 625 = 64,911, which in hex is FD8FH.
(d) TMR0L = 8FH and TMR0H = FDH, all in hex.

Example 9-10

Modify TMR0L and TMR0H in Example 9-8 to get the largest time delay possible. Find the delay in ms. In your calculation, exclude the overhead due to the instructions in the loop.

Solution:

To get the largest delay we make TL and TH both 0. This will count up from 0000 to FFFFH and then roll over to zero.

```
        BCF    TRISB,3          ;PB3 as an output
        MOVLW  0x80             ;Timer0,16-bit,int clk,no prescale
        MOVWF  T0CON            ;load T0CON reg.
HERE    CLRF   TMR0H
        CLRF   TMR0L            ;TH = TL = 0
        BCF    INTCON,TMR0IF    ;clear timer interrupt flag bit
        CALL   DELAY
        BTG    PORTB,3          ;toggle PB3
        BRA    HERE             ;load TH, TL again
;------------delay using Timer0
DELAY   BSF    T0CON,TMR0ON     ;start Timer0
AGAIN   BTFSS  INTCON,TMR0IF    ;monitor Timer0 flag until
        BRA    AGAIN            ;it rolls over
        BCF    T0CON,TMR0ON     ;stop Timer0
        RETURN
```

Making TMR0H and TMR0L both zero means that the timer will count from 0000 to FFFFH, and then roll over to raise the TMR0IF flag. As a result, it goes through a total of 65,536 states. Therefore, we have delay = $(65{,}536 - 0) \times 0.4 \ \mu s = 26.214$ ms. That gives us the smallest frequency of $1 / (2 \times 26.214 \ ms) = 1 / (52.428 \ ms) = 19.073$ Hz.

Using the Windows calculator to find TH, TL

The scientific calculator in Microsoft Windows is a handy and easy-to-use tool to find the TMR0H, TMR0L values. Assume that we would like to find the TMR0H, TMR0L values for a time delay that uses 35,000 clocks of 0.4 μs. The following steps show the calculation:

1. Bring up the scientific calculator in MS Windows and select decimal.
2. Enter 35,000.
3. Select hex. This converts 35,000 to hex, which is 88B8H.
4. Select +/- to give –35,000 decimal (7748H).
5. The lowest two digits (48) of this hex value are for TMR0L and the next two (77) are for TMR0H. We ignore all the Fs on the left because our number is 16-bit data.

Prescaler and generating a large time delay

As we have seen in the examples so far, the size of the time delay depends on two factors, (a) the crystal frequency, and (b) the timer's 16-bit register. Both of these factors are beyond the control of the PIC18 programmer. We saw in Example 9-10 that the largest time delay is achieved by making both TMR0H and TMR0L zero. What if that is not enough? We can use the prescaler option in the T0CON register to increase the delay by reducing the period. The prescaler option of T0CON allows us to divide the instruction clock by a factor of 2 to 256 as was shown in Figure 9-2.

As we have seen so far, with no prescaler enabled, the crystal oscillator frequency is divided by 4 (Fosc/4) and then fed into Timer0. If we enable the prescaler bit in the T0CON register, however, then we can divide the instruction clock (Fosc/4) further before it is fed into Timer0. The lower 3 bits of the T0CON register give the options of the number we can divide by. As shown in Figure 9-2, this number can be 2, 4, 8, 16, 32, 64, and so on. Notice that the lowest number is 2 and the highest number is 256. Examine Examples 9-11 through 9-15 to see how the prescaler options are programmed.

Example 9-11

Find the value for T0CON if we want to program Timer0 in 16-bit mode with a prescaler of 64 and use internal clock (Fosc/4) for the clock source, positive-edge.
Solution:

From Figure 9-2 we have T0CON = 0000 0101; 16-bit mode, XTAL clock source, prescaler of 64.

Example 9-12

Find the timer's clock frequency and its period for various PIC18-based systems, with the following crystal frequencies. Assume that a prescaler of 1:64 is used.
(a) `10 MHz`   (b) `16 MHz`

Solution:

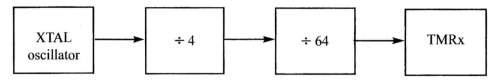

(a) $1/4 \times 10$ MHz = 2.5 MHz and $1/64 \times 2.5$ MHz = 39062.5 Hz due to 1:64 prescaler and T = 1/39062.5Hz = 25.6 μs

(b) $1/4 \times 16$ MHz = 4 MHz and $1/64 \times 4$ MHz = 62500 Hz due to prescaler and T = 1/62500 Hz = 16 μs

---

Example 9-13

Examine the following program and find the time delay in seconds. Exclude the overhead due to the instructions in the loop. Assume XTAL = 10 MHz.

```
        BCF    TRISB,2          ;PB2 as an output
        MOVLW  0x05             ;Timer0,16-bit,int clk,prescaler 64
        MOVWF  TOCON            ;load TOCON reg.
HERE    MOVLW  0x01             ;TMR0H = 01H, the high byte
        MOVWF  TMR0H            ;load Timer0 high byte
        MOVLW  0x08             ;TMR0L = 08H, the low byte
        MOVWF  TMR0L            ;load Timer0 low byte
        BCF    INTCON,TMR0IF    ;clear timer interrupt flag bit
        CALL   DELAY
        BTG    PORTB,2          ;toggle PB2
        BRA    HERE             ;load TH, TL again
;-------------delay using Timer0
DELAY   BSF    TOCON,TMR0ON     ;start Timer0
AGAIN   BTFSS  INTCON,TMR0IF    ;monitor Timer0 flag until
        BRA    AGAIN            ;it rolls over
        BCF    TOCON,TMR0ON     ;stop Timer0
        RETURN
```

Solution:

TMR0H:TMR0L = 0108H = 264 in decimal and 65,536 − 264 = 65,272. Now 65,272 × 64 × 0.4 μs = 1.671 seconds, or from Example 9-12, we have 65,272 × 25.6 μs = 1.671 seconds.

Example 9-14

Assume XTAL = 10 MHz. (a) Find the clock period fed into Timer0 if a prescaler option of 256 is chosen. (b) Show what is the largest time delay we can get using this prescaler option.

Solution:

(a) 1/4 × 10 MHz = 2.5 MHz and 1/256 × 2.5 MHz = 9765.625 Hz due to 1:256 prescaler and T = 1/9765.625 Hz = 1024 μs = 1.024 ms

(b) To get the largest delay, we make TMR0L and TMR0H both 0. Making TMR0H and TMR0L both zero means that the timer will count from 0000 to FFFFH, and then roll over to raise the TMR0IF flag. As a result, it goes through a total of 65,536 states. Therefore, we have delay = (65,536 − 0) × 1024 μs = 67,108,864 μs = 67.108864 seconds.

Example 9-15

Assuming XTAL = 10 MHz, write a program to generate a square wave of 50 Hz frequency on pin PORTB.7. Use Timer0, 16-bit mode, with prescaler = 128.

Solution:

Look at the following steps:
(a) T = 1 / 50 Hz = 20 ms, the period of the square wave.
(b) 1/2 of it for the high and low portions of the pulse = 10 ms
(c) 10 ms / 0.4 μs / 128 = 195 and 65,536 − 195 = 65,341 in decimal, and in hex it is FF3DH.
(d) TL = 3D and TH = FF (hex)

```
        BCF    TRISB,7             ;PB7 as an output
        MOVLW  0x06                ;Timer0,16-bit,int clk,128 prescale
        MOVWF  T0CON               ;load T0CON reg.
HERE    MOVLW  0xFF                ;TMR0H = FF, the high byte
        MOVWF  TMR0H               ;load Timer0 high byte
        MOVLW  0x3D                ;TMR0L = 3DH, the low byte
        MOVWF  TMR0L               ;load Timer0 low byte
        BCF    INTCON, TMR0IF      ;clear timer interrupt flag bit
        BTG    PORTB,7             ;toggle PB7
        BSF    T0CON, TMR0ON       ;start Timer0
AGAIN   BTFSS  INTCON, TMR0IF      ;monitor Timer0 flag until
        BRA    AGAIN               ;it rolls over
        BCF    T0CON, TMR0ON       ;stop Timer0
        BRA    HERE                ;load TH, TL again
```

8-bit mode programming of Timer0

Timer0 can also be used in 8-bit mode. The 8-bit mode allows only values of 00 to FFH to be loaded into the timer's register TMRL0. After the timer is started, it starts to count up by incrementing the TMR0L register. It counts up until it reaches its limit of FFH. When it rolls over from FFH to 00, it sets HIGH the TMR0IF. See Figure 9-7.

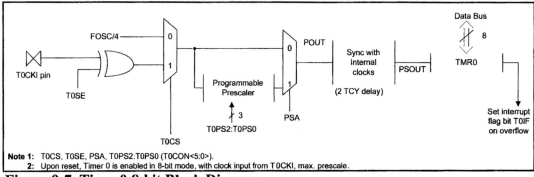

Note 1: T0CS, T0SE, PSA, T0PS2:T0PS0 (T0CON<5:0>).
    2: Upon reset, Timer 0 is enabled in 8-bit mode, with clock input from T0CKI, max. prescale.

Figure 9-7. Timer0 8-bit Block Diagram

Steps to program 8-bit mode of Timer0

To generate a time delay using Timer0 in 8-bit mode, take the following steps:

1. Load the T0CON value register indicating 8-bit mode is selected.
2. Load the TMR0L registers with the initial count value.
3. Start the timer.
4. Keep monitoring the timer flag (TMR0IF) to see if it is raised. Get out of the loop when TMR0IF becomes HIGH.
5. Stop the timer with the instruction "BCF T0CON, TMR0ON".
6. Clear the TMR0IF flag for the next round.
7. Go back to Step 2 to load TMR0L again.

Notice that when we choose the 8-bit option, only the TMR0L register is used and the TMR0H has a zero value during the count up. To clarify the above steps, see Examples 9-16 and 9-17.

Example 9-16

Assuming that XTAL = 10 MHz, find (a) the frequency of the square wave generated on pin PORTB.0 in the following program, and (b) the smallest frequency achievable in this program, and the TH value to do that.

```
        BCF    TRISB,0          ;PB0 as an output
        MOVLW  0x48             ;Timer0,8-bit,int clk,no prescaler
        MOVWF  T0CON            ;load T0CON reg.
        BCF    INTCON,TMR0IF    ;clear timer interrupt flag bit
HERE    MOVLW  0x5              ;TMR0L = 5, the low byte
        MOVWF  TMR0L            ;load Timer0 byte
        CALL   DELAY
        BTG    PORTB,0          ;toggle PB0
        BRA    HERE             ;load TL again
;-----------delay using Timer0
DELAY   BSF    T0CON,TMR0ON     ;start Timer0
AGAIN   BTFSS  INTCON,TMR0IF    ;monitor Timer0 flag until
        BRA    AGAIN            ;it rolls over
        BCF    T0CON,TMR0ON     ;stop Timer0
        BCF    INTCON,TMR0IF    ;clear Timer0 interrupt flag bit
        RETURN
```

Solution:

(a) Now (256 – 05) = 251 × 0.4 µs = 100.4 µs is the high portion of the pulse. Because it is a 50% duty cycle square wave, the period T is twice that; as a result T = 2 × 100.4 µs = 200.8 µs, and the frequency = 4.98 kHz.

(b) To get the smallest frequency, we need the largest T, and that is achieved when TMR0H = 00. In that case, we have T = 2 × 256 × 0.4 µs = 204.8 µs and the frequency = 1 / 204.8 µs = 4,882.8 Hz.

Example 9-17

Assume XTAL = 10 MHz. (a) Find the clock period fed into Timer0 if the prescaler option of 256 is chosen. (b) Show what is the largest time delay we can get using this prescaler option.

Solution:

(a) $1/4 \times 10$ MHz = 2.5 MHz and $1/256 \times 2.5$ MHz = 9765.625 Hz due to 1:256 prescaler and T = 1/9765.625 Hz = 1024 μs

(b) To get the largest delay, we make TMR0L = 0. Making TMR0L zero means that the timer will count from 00 to FFH, and then roll over to raise the TMR0IFF flag. As a result, it goes through a total of 256 states. Therefore, we have delay = (256 – 0) × 1024 μs = 262,144 μs = 0.262144 second.

Assemblers and negative values

Because the timer is in 8-bit mode, we can let the assembler calculate the value for TMR0H. For example, in "MOVLW, -D'100'", the assembler will calculate the –100 = 9C and make WREG = 9C in hex. This makes our job easier. See Examples 9-18 and 9-19.

Example 9-18

Assuming that we are programming the timers for 8-bit mode, find the value (in hex) loaded into TMR0L for each of the following cases.

(a)   MOVLW -D'200'   (b)   MOVLW -D'60'   (c)   MOVLW -D'12'
        MOVWF TMR0L   MOVWF TMR0L   MOVWF TMR0L

Solution:

You can use the Windows scientific calculator to verify the results provided by the assembler. In the Windows calculator, select decimal and enter 200. Then select hex, then +/– to get the negative value. The following is what we get.

Decimal	2's complement (TMR0L value)
–200	38H
–60	C4H
–12	F4H

Example 9-19

Find (a) the frequency of the square wave generated in the following code, and (b) the duty cycle of this wave. Assume XTAL = 10 MHz

```
        BCF    TRISB,3            ;PB3 as an output
        BCF    INTCON,TMR0IF      ;clear timer interrupt flag bit
        MOVLW  0x48               ;Timer0,8-bit,int clk,no prescaler
        MOVWF  T0CON              ;load T0CON reg.
HERE    MOVLW  -D'150'            ;loading negative value
        MOVWF  TMR0L              ;load Timer0 byte
        BSF    PORTB,3            ;PB3 = 1
        CALL   DELAY
        MOVWF  TMR0L              ;reload Timer0 byte
        CALL   DELAY
        BCF    PORTB,3            ;PB3 = 0
        MOVWF  TMR0L              ;reload Timer0 byte
        CALL   DELAY
        BRA    HERE               ;load TH, TL again
;------------delay using Timer0
DELAY   BSF    T0CON,TMR0ON       ;start Timer0
AGAIN   BTFSS  INTCON,TMR0IF      ;monitor Timer0 flag until
        BRA    AGAIN              ;it rolls over
        BCF    T0CON,TMR0ON       ;stop Timer0
        BCF    INTCON,TMR0IF      ;clear timer interrupt flag bit
        RETURN
```

Solution:

For the TMR0L value in 8-bit mode, the conversion is done by the assembler as long as we enter a negative number. This also makes the calculation easy. Because we are using 150 clocks, we have time for the DELAY subroutine = 150×0.4 μs = 60 μs. The high portion of the pulse is twice the size of the low portion (66% duty cycle). Therefore, we have: T = high portion + low portion = 2×60 μs + 60 μs = 180 μs and frequency = 5.555555 kHz.

Another version of this program could be as follows:

```
        BCF    TRISB,3            ;PB3 as an output
        BCF    INTCON,TMR0IF      ;clear timer interrupt flag bit
        MOVLW  0x48               ;Timer0,8-bit,int clk,no prescaler
        MOVWF  T0CON              ;load T0CON reg.
HERE    BSF    PORTB,3            ;PB3 = 1
        CALL   DELAY
        CALL   DELAY
        BCF    PORTB,3            ;PB3 = 0
        CALL   DELAY
        BRA    HERE               ;load TH, TL again
;------------delay using Timer0
DELAY   MOVLW  -D'150'            ;loading negative value
        MOVWF  TMR0L              ;load Timer0 byte
        BSF    T0CON,TMR0ON       ;start Timer0
AGAIN   BTFSS  INTCON,TMR0IF      ;monitor Timer0 flag until
        BRA    AGAIN              ;it rolls over
        BCF    T0CON,TMR0ON       ;stop Timer0
        BCF    INTCON,TMR0IF      ;clear timer interrupt flag bit
        RETURN
```

Timer1 programming

Timer1 is a 16-bit timer, and its 16-bit register is split into two bytes, referred to as TMR1L (Timer1 low byte) and TMR1H (Timer1 high byte). See Figure 9-8. Timer1 can be programmed in 16-bit mode only and unlike Timer0, it does not support 8-bit mode. Timer1 also has the T1CON (Timer 1 control) register in addition to the TMR1IF (Timer1 interrupt flag). The TMR1IF flag bit goes HIGH when TMR1H:TMR1L overflows from FFFF to 0000. Timer1 also has the prescaler option, but it only supports factors of 1:1, 1:2, 1:4, and 1:8. See Figure 9-9 for the Timer1 block diagram and Figure 9-10 for T1CON register options. The PIR1 register contains the TMR1IF flags. See Figure 9-11.

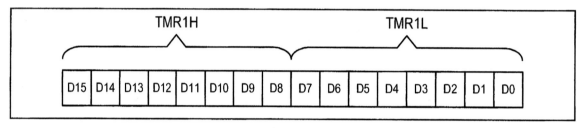

Figure 9-8. Timer1 High and Low Registers

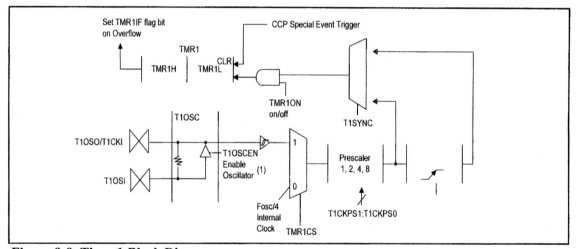

Figure 9-9. Timer1 Block Diagram

Examples 9-20 and 9-21 show how to program Timer1. Notice that in many of the time delay calculations, we have ignored the clocks caused by the overhead instructions in the loop. To get a more accurate time delay, and hence frequency, you need to include them. If you use a digital scope and you don't get exactly the same frequency as we have calculated, it is because of the overhead associated with those instructions.

In this section, we used the PIC18 timer for time delay generation. However, a more powerful and creative way to use these timers is as event counters. We discuss this use of the counter next.

RD16	---	T1CKPS1	T1CKPS0	T1OSCEN	T1SYNC	TMR1CS	TMR1ON

RD16   D7   16-bit read/write enable bit
                                  1 = Timer1 16-bit is accessible in one 16-bit operation.
                                  0 = Timer1 16-bit is accessible in two 8-bit operations.

                    D6   Not used

T1CKPS2:T1CKPS0   D5 D4 Timer1 prescaler selector
                                0 0 = 1:1   Prescale value
                                0 1 = 1:2   Prescale value
                                1 0 = 1:4   Prescale value
                                1 1 = 1:8   Prescale value

T1OSCEN   D3   Timer1 oscillator enable bit
                                    1 = Timer1 oscillator is enabled.
                                    0 = Timer1 oscillator is shutoff.

T1SYNC   D2   Timer1 synchronization (used only when TMR1CS = 1 for
                                    counter mode to synchronize external clock input)
                                    If TMR1CS = 0 this bit is not used.

TMR1CS   D1   Timer1 clock source select bit
                                    1 = External clock from pin RC0/T1CKI
                                    0 = Internal clock (Fosc/4 from XTAL)

TMR1ON   D0   Timer1 ON and OFF control bit
                                    1 = Enable (start) Timer1
                                    0 = Stop Timer1

Figure 9-10. T1CON (Timer 1 Control) Register

							TMR1IF

TMR1IF   D1   Timer1 Interrupt overflow flag bit
                            0 = Timer1 did not overflow.
                            1 = Timer1 has overflowed (FFFF to 0000).

The importance of TMR1IF: When TMR1H:TMR1L overflows from FFFF to 0000, this flag is raised. We monitor this flag bit before we reload the TMR1H:TMR1L registers.

The other bits of this register are discussed in Chapter 11.

Figure 9-11. PIR1 (Interrupt Control Register 1) Contains the TMR1IF Flag

Example 9-20

Find the frequency of the square wave generated by the following program if XTAL = 10 MHz. In your calculation do not include the overhead due to instructions in the loop.

```
        BCF     TRISB,5         ;make PB5 an output
        MOVLW   0x0             ;Timer1,16-bit,int clk,no prescale
        MOVWF   T1CON           ;load T0CON reg
HERE    MOVLW   0x76            ;TMR1H = 76H, the high byte
        MOVWF   TMR1H           ;load Timer1 high byte
        MOVLW   0x34            ;TMR1L = 34H, the low byte
        MOVWF   TMR1L           ;load Timer1 low byte
        BCF     PIR1,TMR1IF     ;clear timer interrupt flag bit
        CALL    DELAY
        BTG     PORTB,RB5       ;toggle PB5
        BRA     HERE            ;load TH, TL again
;———————delay using Timer1
DELAY   BSF     T1CON,TMR1ON    ;start Timer1
AGAIN   BTFSS   PIR1,TMR1IF     ;monitor Timer1 flag until
        BRA     AGAIN           ;it rolls over
        BCF     PIR1,TMR1ON     ;stop Timer1
        RETURN
```

Solution:

Because FFFFH – 7634H = 89CBH + 1 = 89CCH and 89CCH = 35276 clock count 35276 × 0.4 µs = 14.11 ms and frequency = 1 / (14.11 ms × 2) = 35.434 Hz. In this calculation, the overhead due to all the instructions in the loop is not included. Calculation is the same as Example 9-7.

Example 9-21

Assuming XTAL = 10 MHz, write a program to generate a square wave of 50 Hz frequency on pin PORTB.5. Use Timer1 in 16-bit mode with the maximum prescaler allowed.

Solution:

Because FFFFH – F3CBH = C34H + 1 = C35H and C35H = 3125 clock count 3125 × 8 × 0.4 µs = 10 ms and frequency = 1 / (2 × 10 ms) = 50 Hz. In this calculation, the overhead due to all the instructions in the loop is not included.

```
        BCF     TRISB,5         ;make PB5 an output
        MOVLW   0x30            ;Timer1,16-bit,int clk,prescale 1:8
        MOVWF   T1CON           ;load T1CON reg
HERE    MOVLW   0xF3            ;TMR1H = F3H, the high byte
        MOVWF   TMR1H           ;load Timer1 high byte
        MOVLW   0xCB            ;TMR1L = CBH, the low byte
        MOVWF   TMR1L           ;load Timer1 low byte
        BCF     PIR1,TMR1IF     ;clear timer interrupt flag bit
        CALL    DELAY
        BTG     PORTB,RB5       ;toggle PB5
        BRA     HERE            ;load TH, TL again
;———————delay using Timer1
DELAY   BSF     T1CON,TMR1ON    ;start Timer1
AGAIN   BTFSS   PIR1,TMR1IF     ;monitor Timer1 flag until
        BRA     AGAIN           ;it rolls over
        BCF     PIR1,TMR1ON     ;stop Timer1
        RETURN
```

Review Questions

1. How many timers do we have in the PIC18F458/4580?
2. True or false. Timer0 can be used only as a 16-bit timer.
3. True or false. Timer1 can be used only as a 16-bit timer.
4. True or false. The T0CON register is a bit-addressable register.
5. Indicate the selection made in the instruction "MOV T0CON, 0x08".
6. In 16-bit mode, the counter rolls over when the counter goes from _____ to _____.
7. In 8-bit mode, the counter rolls over when the counter goes from _____ to _____.
8. In the instruction "MOVLW -D'200'", find the hex value for WREG.
9. To get a 2-ms delay, what numbers should be loaded into TMR0H and TMR0L using 16-bit mode? Assume that XTAL = 10 MHz.
10. To get a 100-µs delay, what number should be loaded into the TMR0L register using 8-bit mode? Assume that XTAL = 10 MHz.

SECTION 9.2: COUNTER PROGRAMMING

In the last section, we used the timers of the PIC18 to generate time delays. These timers can also be used as counters to count events happening outside the PIC18. The use of the timer as an event counter is covered in this section. When the timer is used as a timer, the PIC18's crystal is used as the source of the frequency. When it is used as a counter, however, it is a pulse outside the PIC18 that increments the TH, TL registers. In counter mode, notice that registers such as T0CON, TMR0H, and TMR0L are the same as for the timer discussed in the last section; they even have the same names.

T0CS bit in T0CON register

Recall from the last section that the T0CS bit (Timer0 clock source) in the T0CON register decides the source of the clock for the timer. If T0CS = 0, the timer gets pulses from the crystal oscillator connected to the OSC1 and OSC2 pins (Fosc/4). In contrast, when T0CS = 1, the timer is used as a counter and gets its pulses from outside the PIC18. Therefore, when T0CS = 1, the counter counts up as pulses are fed from pin RA4 (PORTA.4). The pin is called T0CKI (Timer0 clock input). Notice that the pin belongs to Port A. In the case of Timer0, when T0CS = 1, pin RA4 (PORTA.4) provides the clock pulse and the counter counts up for each clock pulse coming from that pin. Similarly, for Timer1, when TMR1CS = 1, each clock pulse coming in from pin RC0 (PORTC.0) makes the counter count up. See Example 9-22.

In Example 9-23, we are using Timer1 as an event counter that counts up as clock pulses are fed into pin 3.5. These clock pulses could represent the number of people passing through an entrance, or the number of wheel rotations, or any other event that can be converted to pulses.

In Example 9-23, the TL data was displayed in binary. In Example 9-24, the TL registers are converted to ASCII to be displayed on an LCD.

As another example of the application of the timer with C/T = 1, we can feed an external square wave of 60 Hz frequency into the timer. The program will generate the second, the minute, and the hour out of this input frequency and display the result on an LCD. This will be a nice digital clock, but not a very accurate one.

Example 9-22

Find the value for T0CON if we want to program Timer0 as an 8-bit mode counter, no prescaler. Use an external clock for the clock source and increment on the positive edge.

Solution:

T0CON   = 0110 1000   8-bit, external clock source, no prescaler.

Example 9-23

Assuming that clock pulses are fed into pin T0CKI, write a program for counter 0 in 8-bit mode to count the pulses and display the state of the TMR0L count on PORTB.

Solution:
```
        BSF   TRISA,RA4         ;PORTA.4 as an input for clock
        CLRF  TRISB             ;PORTB as an output
        MOVLW 0x68              ;Timer0, 8-bit,ext clk,no prescale
        MOVWF T0CON             ;load T0CON reg
HERE    MOVLW 0x0               ;TMR0L = 0
        MOVWF TMR0L             ;load Timer0
        BCF   INTCON,TMR0IF     ;clear timer interrupt flag bit
        BSF   T0CON,TMR0ON      ;start Timer0
AGAIN   MOVFF TMR0L,PORTB       ;display the count on PORTB
        BTFSS INTCON,TMR0IF     ;monitor Timer0 flag until
        BRA   AGAIN             ;it rolls over
        BCF   T0CON,TMR0ON      ;stop Timer0
        GOTO  HERE
```

PORTB is connected to 8 LEDs and input T0CKI to pulse.

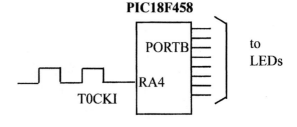

Using external crystal for Timer1 clock

Timer1 has two options when it comes to using the external clock source. It uses either the clock fed into the T1CKI pin or the clock from a crystal connected to the T1OSI and T1OSO pins, as shown in Figure 9-9. Generally, a 32-kHz crystal is connected to the T1OSI and T1OSO pins and is used for saving power during SLEEP mode because the SLEEP instruction does not disable Timer1. Notice that this 32-kHz crystal connected to the T1OSCI and T1OSO pins is in addition to the main crystal connected to the OSC1 and OSC2 pins. The PIC18 uses the main crystal to execute CPU instruction clock cycles among other things, and when the CPU goes into SLEEP mode, the main crystal is shut down to save power. The alternate 32-kHz crystal connected to pins T1OSO and T1OSI provides clock to Timer1 during SLEEP mode, while the main crystal is shut down. This allows the use of the timer to implement an on-chip RTC (real-time clock). Chapter 16 shows how to connect an external RTC to the PIC18. Notice that in order to use the alternate external clock source for Timer1, we must choose the external clock source option of TMR1CS = 1, in addition to enabling the T1OSCEN bit (T1OSCEN = 1) in the T1CON register, as shown in Figure 9-10. Study Examples 9-23 through 9-27 to see how timers are used as counters.

Before we finish this section, we need to state an important point. You might think monitoring the TMR0IF and TMR1IF flags is a waste of the microcontroller's time. You are right. There is a solution to this: the use of interrupts. Using interrupts enables us to do other things with the microcontroller. When a timer Interrupt flag such as TMR0IF is raised it will inform us. This important and powerful feature of the PIC18 is discussed in Chapter 11.

Example 9-24

Assume that a 1-Hz frequency pulse is connected to input for Timer0 (pin T0CKI). Write a program to display counter 0 on PORTB, PORTC, and PORTD in decimal. Set the initial value of TMR0L to –60.

Solution:

To display the TMR0L count on an LCD, we must convert 8-bit binary data to ASCII. See Chapter 5 for data conversion.

```
NUME        EQU    0x00        ;RAM loc for NUME
QU          EQU    0x20        ;RAM loc for quotient
RMND_L      EQU    0x30        ;the least significant digit loc
RMND_M      EQU    0x31        ;the middle significant digit loc
RMND_H      EQU    0x32        ;the most significant digit loc
MYDEN       EQU    D'10'       ;value for divide by 10
       BSF    TRISA,RA4        ;RA4 as an input
       MOVLW 0x68              ;Timer0,8-bit, ext clk,no prescale
       MOVWF T0CON             ;load T0CON reg
HERE   MOVLW 0x0               ;TMR0L = 0
       MOVWF TMR0L             ;load Timer0
       BCF    INTCON,TMR0IF    ;clear timer interrupt flag bit
       BSF    T0CON,TMR0ON     ;start Timer0
AGAIN  MOVF  TMR0L,W           ;save the count in WREG
       CALL  BIN_ASC_CON
       BTFSS INTCON,TMR0IF     ;monitor Timer0 flag until
       BRA    AGAIN            ;it rolls over
       BCF    T0CON,TMR0ON     ;stop Timer0
       GOTO  HERE


;converting 8-bit binary to decimal
BIN_DEC_CON
       MOVFF PORTB,WREG
       MOVWF NUME              ;load numerator
       MOVLW MYDEN             ;WREG = 10, the denominator
       CLRF  QU                ;clear quotient
D_1    INCF  QU                ;inc quotient for every subtract
       SUBWF NUME              ;subtract WREG from NUME value
       BC    D_1               ;if positive go back
       ADDWF NUME              ;once too many, first digit
       DECF  QU                ;once too many for quotient
       MOVFF NUME,RMND_L       ;save the first digit
       MOVFF QU,NUME           ;repeat the process one more time
       CLRF  QU                ;clear QU
D_2    INCF  QU
       SUBWF NUME              ;subtract WREG from NUME value
       BC    D_2
       ADDWF NUME              ;once too many
       DECF  QU
       MOVFF NUME,RMND_M       ;2nd digit
       MOVFF QU,RMND_H         ;3rd digit
       RETURN
```

PIC18

PB

PC

PD

RA4

1 Hz clock   T0CKI

In order to display the data on LCD, the decimal number must be converted to ASCII. See Chapter 6.

Example 9-25

Assume that a 16-Hz frequency pulse is connected to input for Timer0 (pin T0CKI). Write a program to display the counter values of TMR0H and TMR0L on ports B and D. Set the initial values to 0. Use Timer0, 16-bit mode, and positive-edge clock. Show the program for (a) no prescaler, (b) prescaler of 1:16.

Solution:
(a)
```
        BSF    TRISA,RA4        ;RA4 as an input
        CLRF   TRISB            ;PORTB as an output
        CLRF   TRISD            ;PORTD as an output
        MOVLW  0x28             ;Timer 0,16-bit,ext clk,no prescale
        MOVWF  T0CON            ;load T0CON reg
HERE    MOVLW  0x0              ;TMR0H = 0
        MOVWF  TMR0H            ;load Timer0 high byte
        MOVLW  0x0              ;TMR0L = 0
        MOVWF  TMR0L            ;load Timer0 low byte
        BCF    INTCON,TMR0IF    ;clear timer interrupt flag bit
        BSF    T0CON,TMR0ON     ;start Timer0
AGAIN   MOVFF  TMR0H,PORTD      ;display high byte count
        MOVFF  TMR0L,PORTB      ;display low byte count
        BTFSS  INTCON,TMR0IF    ;monitor Timer0 flag until
        BRA    AGAIN            ;it rolls over
        BCF    T0CON,TMR0ON     ;stop Timer0
        GOTO   HERE
```
(b)
```
        BSF    TRISA,RA4        ;RA4 as an input
        CLRF   TRISB            ;PORTB as an output
        CLRF   TRISD            ;PORTD as an output
        MOVLW  0x23             ;T0,16-bit,ext clk,prescale of 1:16
        MOVWF  T0CON            ;load T0CON reg
HERE    MOVLW  0x0              ;TMR0H = 0
        MOVWF  TMR0H            ;load Timer0 High byte
        MOVLW  0x0              ;TMR0L = 0
        MOVWF  TMR0L            ;load Timer0 low byte
        BCF    INTCON,TMR0IF    ;clear timer interrupt flag bit
        BSF    T0CON,TMR0ON     ;start Timer0
AGAIN   MOVFF  TMR0H,PORTD      ;display high byte count
        MOVFF  TMR0L,PORTB      ;display low byte count
        BTFSS  INTCON,TMR0IF    ;monitor Timer0 flag until
        BRA    AGAIN            ;it rolls over
        BCF    T0CON,TMR0ON     ;stop Timer0
        GOTO   HERE
```

Example 9-26

Assuming that clock pulses are fed into pin T0CKI and a buzzer is connected to pin PORTB.1, write a program for counter 0 in 8-bit mode to sound the buzzer every 100 pulses.

Solution:

To sound the buzzer every 100 pulses, we set the initial counter value to –100 (9C in hex), then the counter counts up until it reaches FF. Upon overflow, we can count the buzzer by toggling the PORTB.1 pin.

```
        BCF    TRISB,1          ;RB1 as an output
        BSF    TRISA,4          ;RA4 as an input for clock-in
        MOVLW 0x68             ;Timer0,8-bit,ext clk,no prescale
        MOVWF T0CON            ;load T0CON reg
        MOVLW -D'100'          ;TMR0L = 0
        MOVWF TMR0L            ;load Timer0
        BCF    INTCON,TMR0IF    ;clear timer interrupt flag bit
        BSF    T0CON,TMR0ON     ;start Timer0
AGAIN BTFSS INTCON,TMR0IF       ;monitor Timer0 flag until
        BRA    AGAIN            ;it rolls over
        BCF    T0CON,TMR0ON     ;stop Timer0
OVER  BTG    PORTB,1           ;sound the buzzer
        CALL  DELAY            ;quarter second delay
        GOTO  OVER             ;forever
```

Bit 1 of PORTB is connected to a buzzer and input T0CKI to a pulse.

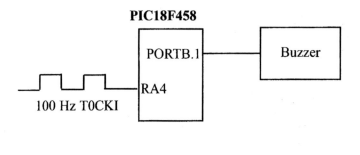

PIC18F458

100 Hz T0CKI

PORTB.1 — Buzzer

RA4

Example 9-27

Assume that a 1-Hz frequency pulse is connected to input for Timer1 (pin PORTC.0). Write a program to display the counter values of TMR1H and TMR1L on ports B and D. Set the initial values to 0. Use Timer1, 16-bit mode, no prescaler, and positive-edge clock.

Solution:

```
         BSF   TRISC,RC0      ;PC0 as an input
         CLRF  TRISB          ;PORTB as an output
         CLRF  TRISD          ;PORTD as an output
         MOVLW 0x02           ;Timer1,16-bit,ext clk,no prescale
         MOVWF T1CON          ;load T0CON reg
HERE     MOVLW 0x0            ;TMR1H = 0, the low byte
         MOVWF TMR1H          ;load Timer1 high byte
         MOVLW 0x0            ;TMR1L = 0, the low byte
         MOVWF TMR1L          ;load Timer1 low byte
         BCF   PIR1,TMR1IF    ;clear timer interrupt flag bit
         BSF   T1CON,TMR1ON   ;start Timer1
AGAIN    MOVFF TMR1H,PORTD    ;display high byte count
         MOVFF TMR1L,PORTB    ;display low byte count
         BTFSS PIR1,TMR1IF    ;monitor Timer1 flag until
         BRA   AGAIN          ;it rolls over
         BCF   PIR1,TMR1ON    ;stop Timer1
         GOTO  HERE
```

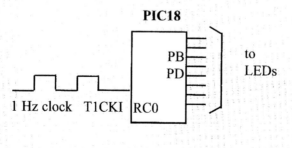

Review Questions

1. What provides the clock pulses to PIC18 timers if T0CS = 0?
2. What provides the clock pulses to PIC18 timers if T0CS = 1?
3. Does the discussion in Section 9.1 apply to timers if T0CS = 1?
4. To allow RC0 to be used as an input for the Timer1 clock, what must be done, and why?
5. Do we have a choice of counting up on the positive or negative edge of the clock?

SECTION 9.3: PROGRAMMING TIMERS 0 AND 1 IN C

In Chapter 7 we showed some examples of C programming for the PIC18. In this section we show C programming for the PIC18 timers. As we saw in the examples in Chapter 7, the general-purpose registers of the PIC18 are under the control of the C compiler and are not accessed directly by C statements. All of the SFRs, however, are accessible directly using C statements. As an example of accessing the SFRs directly, we saw how to access ports PORTB–PORTD in Chapter 7. Next, we discuss how to access the PIC18 timers directly using the C18 C compiler.

Accessing timer registers in C

In C18 we can access timer registers such as TMR0H, TMR0L, and T0CON directly using the PIC18Fxxx.h header file. This is shown in Example 9-28. Example 9-28 also shows how to access the TMR0ON and TMR0IF flag bits. Notice that all the SFR registers are bit-accessible.

Calculating delay length using timers

As we saw in the last two sections, the delay length depends on two factors: (a) the crystal frequency, (b) the prescaler factor. The third factor in the delay size is the C compiler because various C compilers generate different hex code sizes. Study Examples 9-28 through 9-33 and verify them using an oscilloscope.

Example 9-28

Write a C18 program to toggle all the bits of PORTB continuously with some delay. Use Timer0, 16-bit mode, and no prescaler options to generate the delay.

Solution:

```c
#include <p18f4580.h>
void T0Delay(void);
void main(void)
  {
  TRISB=0;                       //PORTB output port
  while(1)                       //repeat forever
    {
      PORTB=0x55;                //toggle all bits of Port B
      T0Delay();                 //delay size unknown
      PORTB=0xAA;                //toggle all bits of Port B
      T0Delay();
    }
  }

void T0Delay()
  {
    T0CON=0x08;                  //Timer0, 16-bit mode, no prescaler
    TMR0H=0x35;                  //load TH0
    TMR0L=0x00;                  //load TL0
    T0CONbits.TMR0ON=1;              //turn on T0
    while(INTCONbits.TMR0IF==0);     //wait for TF0 to roll over
    T0CONbits.TMR0ON=0;              //turn off T0
    INTCONbits.TMR0IF=0;         //clear TF0
  }
```

Example 9-29

Write a C18 program to toggle only the PORTB.4 bit continuously every 50 ms. Use Timer0, 16-bit mode, the 1:4 prescaler to create the delay. Assume XTAL = 10 MHz.

Solution:

```c
#include <p18f4580.h>
void T0Delay(void);
#define mybit PORTBbits.RB4
void main(void)
  {
    TRISBbits.TRISB4=0;
    while(1)
      {
        mybit^=1;                       //toggle PORTB.4
        T0Delay();                      //Timer0, mode 1 (16-bit)
      }
  }

void T0Delay()
  {
    T0CON=0x01;                 //Timer0, 16-bit mode, 1:4 prescaler
    TMR0H=0x85;                         //load TH0
    TMR0L=0xEE;                         //load TL0
    T0CONbits.TMR0ON=1;                 //turn on Timer0
    while(INTCONbits.TMR0IF==0);        //wait for TF0 to roll over
    T0CONbits.TMR0ON=0;                 //turn off Timer0
    INTCONbits.TMR0IF=0;                //clear TF0
  }
```

FFFFh–85EEH = 7A11H = 31249 + 1 = 31250

Timer delay = 31250 × 4 × 0.4 μs = 50 ms

Example 9-30

Write a C18 program to generate a frequency of 2 Hz only on pin PORTB.5. Use Timer0, 8-bit mode to create the delay.

Solution:

```c
#include <p18f4580.h>
void T0M8Delay(void);
#define mybit PORTBbits.RB5
void main(void)
    {
        unsigned char x,y;
        TRISBbits.TRISB5 = 0;
        while(1)
            {
                mybit^=1;                   //toggle PortB.5
                for(x=0;x<250;x++)          //due to for loop overhead
                    for(y=0;y<35;y++)       //we put 35 and not 39
                        T0M8Delay();
            }
    }
void T0M8Delay()
    {
        TOCON=0x45;                 //Timer0, 16-bit mode, prescaler 1:64
        TMR0L=-1;                           //load TL0
        T0CONbits.TMR0ON=1;                 //turn on T0
        while(INTCONbits.TMR0IF==0); //wait for TF0 to roll over
        T0CONbits.TMR0ON=0;                 //turn off T0
        INTCONbits.TMR0IF=0;                //clear TF0
    }
```

$256 - 255 = 1$

$1 \times 64 \times 0.4 \ \mu s = 25.6 \ \mu s$

$25.6 \ \mu s \times 250 \times \mathbf{39} = 0.2496$ by calculation.

$F = 1 / (2 \times 0.2496 \ s) = 1 / 0.4992 \ s = 2 \ Hz$

The scope output, however, does not give us this result. This is due to overhead of the for loop in C. To correct this problem, we put 35 instead of 39.

Example 9-31

Write a C18 program to generate a frequency of 250 Hz on all bits of PORTC. Use Timer0, 16-bit mode, and no prescaler to create the frequency. Assume XTAL = 10 MHz.

Solution:

```c
#include <p18f4580.h>
void T0Delay(void);
void main(void)
  {
    unsigned char x;
    TRISC=0;                 //PORTC output port
    PORTC=0x55;
    while(1)
      {
        PORTC=~PORTC;        //toggle all bits of Port C
        for(x=0;x<20;x++)
          T0Delay();
      }
  }

void T0Delay()
  {
    T0CON=0x0;         //Timer 0, 16-bit mode, no prescaler
    TMR0H=0xFF;        //load TH0
    TMR0L=0x06;        //load TL0
    T0CONbits.TMR0ON=1;            //turn on T0
    while(INTCONbits.TMR0IF==0); //wait for TF0 to roll over
    T0CONbits.TMR0ON=0;            //turn off T0
    INTCONbits.TMR0IF=0;           //clear TF0
  }
```

FF06H = 65286 in decimal

65536 − 65286 = 250

$250 \times 0.4\ \mu s = 0.1$ ms and 20×0.1 ms = 2 ms

$T = 1 / (2 \times 2\ ms) = 1 / 4\ ms = 250$ Hz

Another way is:

$T = 1 / 250$ Hz = 0.004 second and one half is 0.002 second

0.002 second $/ 0.4\ \mu s = 5000$

$5000 / 20 = 250$ because the for loop is set to 20.

Example 9-32

A switch is connected to pin PORTB.7. Write a C18 program to monitor SW and create
the following frequencies on pin PORTB.0:

SW = 0:   500 Hz

SW = 1:   750 Hz

Use Timer0 with prescaler for both of them.

Solution:

```c
#include <p18f4580.h>
#define mybit PORTBbits.RB0
#define SW PORTBbits.RB7
void T0PSDelay(unsigned char);
void main(void)
    {
    TRISBbits.TRISB7=1;//make PB.7 an input
     TRISBbits.TRISB0=0;//make PB.0 an output
    SW=1;
    while(1)
       {
         mybit^=1;        //toggle PB.0
         if(SW==0)        //check switch
            T0PSDelay(0);
         else
            T0PSDelay(1);
       }
    }
void T0PSDelay(unsigned char c)
    {
    T0CON=0x05;          //Timer 0, 16-bit mode, prescaler 1:64
    if(c==0)
       {
       TMR0H=0xFF;             //load TH0
        TMR0L=0xD9;            //load TL0
       }
    else
       {
       TMR0H=0xFF;             //load TH0
       TMR0L=0xE6;             //load TL0
       }
    T0CONbits.TMR0ON=1;                 //turn on T0
    while(INTCONbits.TMR0IF==0);   //wait for TF0 to roll over
    T0CONbits.TMR0ON=0;                 //turn off T0
    INTCONbits.TMR0IF=0;                //clear TF0
    }
```

FFD9H = 65497

65536–65497 = 39

$39 \times 64 \times 0.4 \ \mu s = 998 \ \mu s$

$1 / (998 \ \mu s \times 2) = 501$ Hz

FFE6H = 65510

65536–65510 = 26

$26 \times 64 \times 0.4 \ \mu s = 666 \ \mu s$

$1 / (666 \ \mu s \times 2) = 751$ Hz

Use the scope and modify TH:TL to get an exact frequency.

Example 9-33

Write a C18 program to create a frequency of 2500 Hz on pin PORTB.1. Use Timer1 to create the delay.

Solution:

```c
#include <p18f4580.h>
void T1Delay(void);
#define mybit PORTBbits.RB1

void main(void)
  {
     TRISBbits.TRISB1 = 0;
   while(1)
      {
        mybit^=1;              //toggle PB.1
        T1Delay();
      }
  }

void T1Delay()
  {
    T1CON=0x0;          //Timer1, 16-bit mode, no prescaler
    TMR1H=0xFE;         //load TH1
    TMR1L=0x0C;         //load TL1
    T1CONbits.TMR1ON=1;             //turn on T1
    while(PIR1bits.TMR1IF==0);   //wait for TF1 to roll over
    T1CONbits.TMR1ON=0;             //turn off T1
    PIR1bits.TMR1IF=0;              //clear TF1
  }
```

$1 / 2500 \text{ Hz} = 400 \text{ μs}$

$400 \text{ μs} / 2 = 200 \text{ μs}$

$200 \text{ μs} / 0.4 \text{ μs} = 500$

$65536 - 500 = 65036 = \text{FE0CH}$

C programming of Timers 0 and 1 as counters

In Section 9.2 we showed how to use Timers 0 and 1 as event counters. Timers can be used as counters if we provide pulses from outside the chip instead of using the frequency of the crystal oscillator as the clock source. By feeding pulses to the T0CKI (RA4) and T1CKI (RC0) pins, we turn Timer0 and Timer1 into Counter 0 and Counter 1, respectively. Study Examples 9-34 through 9-37 to see how Timers 0 and 1 are programmed as counters using C language.

Example 9-34

Assume that a 1-Hz external clock is being fed into pin T0CKI (RA4). Write a C18 program for Counter0 in 8-bit mode to count up and display the state of the TMR0L count on PORTB. Start the count at 0H.

Solution:

```c
#include <p18f4580.h>

void main(void)
  {
    TRISAbits.TRISA4=1;//make RA4/T0CKI an input
     TRISB=0;
    T0CON=0x68;            //Counter 0, 8-bit mode, no prescaler
    TMR0L=0;                          //set count to 0

    while(1)                          //repeat forever
      {
        do
         {
           T0CONbits.TMR0ON=1;        //turn on T0
            PORTB=TMR0L;              //place value on pins
         }
         while(INTCONbits.TMR0IF==0); //wait for TF0 to roll over
         T0CONbits.TMR0ON=0;          //turn off T0
         INTCONbits.TMR0IF=0;         //clear TF0
      }
  }
```

PORTB is connected to 8 LEDs.
T0CKI (RA4) is connected to a
1-Hz external clock.

PIC18

1 Hz   T0CKI

Example 9-35

Assume that a 1-Hz external clock is being fed into pin T0CKI (RA4). Write a C program for Counter 0 in mode 1 (16-bit) to count the pulses and display the TMR0H and TMR0L registers on PORTD and PORTB, respectively.

Solution:

```c
#include <p18f4580.h>

void main(void)
  {
    TRISAbits.TRISA4=1;               //make RA4 an input for T0CKI
    TRISB=0;                          //PORTB output port
    TRISD=0;                          //PORTD output port
    T0CON=0x25;         //Timer0, 16-bit mode, prescaler 1:64
    TMR0H=0;                          //set count to 0
    TMR0L=0;                          //set count to 0

    while(1)                          //repeat forever
      {
        do
          {
            T0CONbits.TMR0ON=1;   //turn on T0
            PORTB=TMR0L;          //
            PORTD=TMR0H;          //place value on pins
          }
        while(INTCONbits.TMR0IF==0);//wait for rollover

        T0CONbits.TMR0ON=0;           //turn off T0
        INTCONbits.TMR0IF=0;          //clear TF0
      }
  }
```

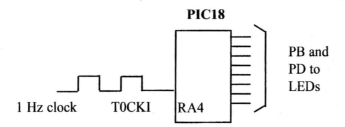

PIC18

1 Hz clock   T0CKI   RA4

PB and
PD to
LEDs

Example 9-36

Assume that a 64-Hz external clock is being fed into pin T0CKI (RA4). Write a C program for Counter 0 in 8-bit mode to display the count in ASCII. The 8-bit binary count must be converted to ASCII. Display the ASCII digits (in binary) on PORTB, PORTC, and PORTD, where PORTB has the least significant digit. Set the initial value of TMR0L to 0.

Solution:

To display the TMR0L count, we must convert 8-bit binary data to ASCII. See Chapter 7 for data conversion. The ASCII values will be shown in binary. For example, '9' will show as 00111001 on the ports.

```c
#include <p18f4580.h>
void BinToASCII(unsigned char);
void main()
  {
    unsigned char value;
    TRISAbits.TRISA4=1;            //make RA4 an input
     TRISB=0;                      //make PORTB an output
     TRISC=0;                      //make PORTC an output
     TRISD=0;                      //make PORTD an output
     TMR0L=0;
     T0CON=0x65;         //Counter 0, 8-bit mode, prescaler 1:64

    while(1)
      {
        do
          {
            T0CONbits.TMR0ON=1;   //turn on T0
            value=TMR0L;
            BinToASCII(value);
          }
    while(INTCONbits.TMR0IF==0); //wait for TF0 to roll over
    T0CONbits.TMR0ON=0;             //turn off T0
    INTCONbits.TMR0IF=0;            //clear TF0
      }
  }

void BinToASCII(unsigned char value)         //see Chapter 7
  {
    unsigned char x,d1,d2,d3;
    x=value/10;
    d1=value%10;
    d2=x%10;
    d3=x/10;
    PORTB=0x30 | d1;
    PORTC=0x30 | d2;
    PORTD=0x30 | d3;
  }
```

CHAPTER 9: PIC18 TIMER PROGRAMMING IN ASSEMBLY AND C

Example 9-37

Assume that a 60-Hz external clock is being fed into pin T0CKI (RA4). Write a C program for Counter 0 in 8-bit mode to display the seconds and minutes on PORTB and PORTD, respectively.

Solution:

```c
#include <p18f4580.h>

void ToTime(unsigned char);
void main()
  {
    unsigned char sec;
     TRISB=TRISD=0;                //PORTB,D outputs
    T0CON=0x68;                    //Timer 0, no prescaler
     TMR0L=-60;                    //sec = 60 pulses
    while(1)
      {
        do
         {
            T0CONbits.TMR0ON=1;    //turn on T0
            sec=TMR0L;
            ToTime(sec);
         }
    while(INTCONbits.TMR0IF==0); //wait for TF0 to roll over
    T0CONbits.TMR0ON=0;            //turn off T0
    INTCONbits.TMR0IF=0;           //clear TF0
      }
  }

void ToTime(unsigned char value)
  {
    unsigned char sec, min;
    min = value / 60;
    sec = value % 60;
    PORTB = sec;
    PORTD = min;
  }
```

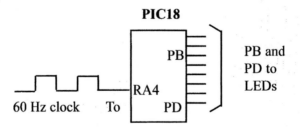

By using 60 Hz, we can generate seconds, minutes, and hours.

SECTION 9.4: PROGRAMMING TIMERS 2 AND 3

In this section we examine Timers 2 and 3 of the PIC18 family and show how to program them in both Assembly and C.

Timer2 programming

Timer2 is an 8-bit timer. The 8-bit register of Timer2 is called TMR2. Timer2 also has an 8-bit register called the period register (PR2). We can set the PR2 register to a fixed value and Timer2 will increment from 00 until it matches the value in PR2. At that point, the equal signal will raise the TMR2IF flag and reset TMR2 to 00. The clock source for Timer2 is Fosc/4 with the options of both prescaler and postscaler, as shown in Figure 9-12. Notice from Figure 9-12 that there is no external clock source for Timer2. In other words, it cannot be used as a counter. Examine the next few examples to learn the programming syntax for Timer2. See Figures 9-12 and 9-13.

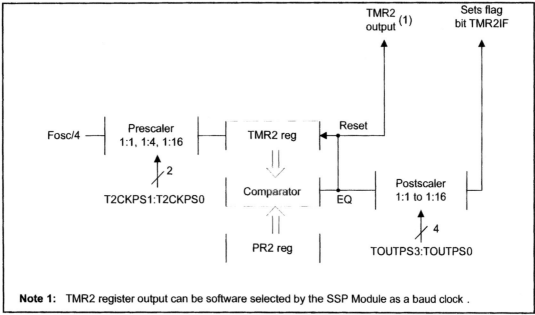

Note 1: TMR2 register output can be software selected by the SSP Module as a baud clock .

Figure 9-12 Timer2 Block Diagram

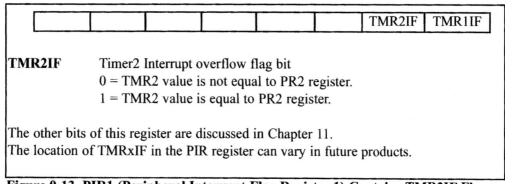

Figure 9-13. PIR1 (Peripheral Interrupt Flag Register 1) Contains TMR2IF Flag

TOUTPS3	TOUTPS2	TOUTPS1	TOUTPS0	TMR2ON	T2CKPS1	T2CKPS0

D7   Not used

TOUTPS3:TOUTPS0 D6–D3 Timer2 Output Postcale Select bits
00 0 0 = 1:1   Postscale value
00 0 1 = 1:2   Postscale value
00 1 0 = 1:3   Postscale value
00 1 1 = 1:4   Postscale value

11 1 0 = 1:15   Postscale value
11 1 1 = 1:16   Postscale value

TMR2ON   D2   Timer2 ON and OFF Control bit
1 = Enable (Start) Timer2
0 = Stop Timer2
T2CKPS1:T2CKPS0 D1–D0 Timer2 Clock Prescale Select bits
0 0 = Prescale is 1
0 1 = Prescale is 4
1 x = Prescale is 16

Figure 9-14. T2CON (Timer2 Control) Register

Example 9-38

Assuming that XTAL = 10 MHz, write a program to turn on pin PORTB4 when TMR2 reaches value 100 (decimal).

Solution:

Because XTAL = 10 MHz, TMR2 counts up every 0.4 μs. Therefore, when you have TMR2H = PR2 = 100, PORTB4 will be turned on.

```
        BCF    TRISB,4            ;make PORTB4 an output
        BCF    PORTB,4            ;turn off PORTB4
        MOVLW  0x0               ;Timer2, no prescale or postscale
        MOVWF  T2CON             ;load T2CON reg
        MOVLW  0x0               ;TMR2 = 0
        MOVWF  TMR2              ;load Timer2
        MOVLW  D'100'            ;PR2 = 100, the period register
        MOVWF  PR2               ;load PR2
        BCF    PIR1,TMR2IF       ;clear timer interrupt flag
        BSF    T2CON,TMR2ON      ;start Timer2
AGAIN   BTFSS  PIR1,TMR2IF       ;monitor Timer2 flag
        BRA    AGAIN             ;
        BSF    PORTB,4           ;turn on PORTB4
        BCF    T2CON,TMR2ON      ;stop Timer2
HERE    BRA    HERE
```

Example 9-39

Using the prescaler and postscaler, find the longest time delay that we can create using Timer2. Assume that XTAL = 10 MHz.

Solution:

We can create the longest time delay by making PR2 = 255. When TMR2 reaches value 255 (decimal), it toggles a pin.

```
        BCF    TRISB,4           ;make PORTB4 an output
        BCF    PORTB,4           ;turn off PORTB4
        MOVLW B'01111011'        ;Timer2, prescale = 16,postscale = 16
        MOVWF T2CON              ;load T2CON reg
        MOVLW 0x0                ;TMR2 = 0
        MOVWF TMR2               ;load Timer2
HERE    MOVLW D'255'             ;PR2 = 255, the period register
        MOVWF PR2                ;load PR2
        BCF    PIR1,TMR2IF       ;clear timer interrupt flag bit
        BSF    T2CON,TMR2ON      ;start Timer2
AGAIN BTFSS PIR1,TMR2IF          ;monitor Timer2 flag
        BRA    AGAIN             ;
        BTG    PORTB,4           ;turn on PORTB4
        BCF    T2CON,TMR2ON      ;stop Timer2
        BRA    HERE
```

Because XTAL = 10 MHz, TMR2 counts up every 0.4 μs. Therefore, when you have TMR2H = PR2 = 255, RB4 will be turned on and off every 52 ms because 255×0.4 μs $\times 16 \times 16 = 26.112$ ms.

Example 9-40

Assuming that XTAL = 10 MHz, write a C18 program to turn on pin PORTB4 when TMR2 reaches value 100. This is a repeat of Example 9-13 in C18.

Solution:

```c
#include <p18f4580.h>
#define mybit PORTBbits.RB4
void main(void)
  {
      TRISBbits.TRISB4=0;    //PORTB4 as output
      T2CON=0x0;             //Timer2, no prescaler/postscaler
      TMR2=0x00;             //TMR2 = 0
      mybit=0;               //PB.4 = 0
      PR2=100;               //load period register 2
      T2CONbits.TMR2ON=1;    //turn on T0
      while(PIR1bits.TMR2IF==0); //wait for TMR2IF to be raised
      mybit=1;               //PB.4 = 0
      T2CONbits.TMR2ON=0;    //turn off T2
      PIR1bits.TMR2IF=0;     //clear TF0
      while(1);              //stay here
  }
```

Example 9-41

Using the prescaler and postscaler, find the longest time delay that we can create using Timer2. Assume that XTAL = 10 MHz. This is a repeat of Example 9-39 in C18.

Solution:

```
#include <p18f4580.h>
#define mybit PORTBbits.RB4
void main(void)
   {
     TRISBbits.TRISB4=0;
     T2CON=0x7B;              //Timer2, prescaler = 16,postscaler = 16
     TMR2=0x00;                       //TMR2 = 0
       while(1)
        {
       PR2=255;                   //load period register 2
       T2CONbits.TMR2ON=1;        //turn on T2
       while(PIR1bits.TMR2IF==0); //wait for TMR2IF to be raised
       mybit=~mybit;              //toggle PORTB4
       T2CONbits.TMR2ON=0;        //turn off T2
       PIR1bits.TMR2IF=0;         //clear TF0
        }
   }
```

Because XTAL = 10 MHz, TMR2 counts up every 0.4 μs. Therefore, when TMR2H = PR2 = 255, PORTB4 will be turned on and off every 52 ms because 255 × 0.4 μs × 16 × 16 = 26.112 ms.

Timer3 Programming

Timer3 is a 16-bit timer that can be used as a timer or counter. Its 16-bit register is split into two bytes referred to as TMR3L and TMR3H. (See Figure 9-17.) Timer3 can be programmed in 16-bit mode only and does not support 8-bit mode. We select various options of Timer3 using the T3CON (Timer3 Control) register, as shown in Figure 9-15. Timer3 has the prescaler options of 1:1, 1:2, 1:4, and 1:8, as shown in Figure 9-15. Figure 9-15 also shows the bits related to the CCP (compare/capture pulse-width-modulation) feature of the PIC18. CCP is a widely used feature of the PIC18, and we will discuss it in Chapters 15 and 17. In Chapter 15, we will see how to use the CCP feature along with the interrupt to measure the pulse width. Pulse width modulation (PWM) is an important concept used in DC motor control. We will examine it in detail in Chapter 17. Because Timer3 is a 16-bit timer, the TMR3IF flag bit goes HIGH when TMR3H:TMR3L overflows from FFFF to 0000. The TMR3IF (Timer3 Interrupt flag) is part of the PIR2 register, as shown in Figure 9-16.

Examine the next few examples to learn the programming syntax for Timer3.

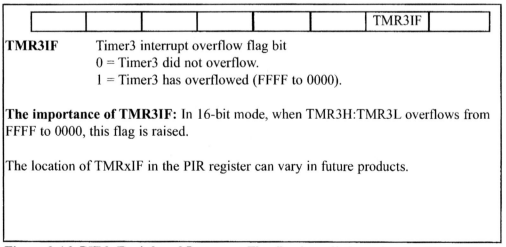

RD16	T3CCP2	T3CKPS1	T3CKPS0	T3CCP1	T3SYNC	TMR3CS	TMR3ON

RD16 D7 16-bit read/write enable bit
 1 = Timer3 16-bit is accessible in one 16-bit operation.
 0 = Timer3 16-bit is accessible in two 8-bit operations.

T3CCP2:T3CCP1 D6 D3 Timer3 and Timer1 to CPPx Enable bits
 0 0 = Timer1 is the clock source for compare/capture of the CCP module.
 0 1 = Timer3 is the clock source for compare/capture of the CCP2.
   Timer1 is the clock source for compare/capture of the CCP1.
 1 x = Timer3 is the clock source for compare/capture of the CCP module.

T3CKPS1:T3CKPS0 D5 D4 Timer3 Input Clock Prescaler Selector
 0 0 = 1:1 Prescale value
 0 1 = 1:2 Prescale value
 1 0 = 1:4 Prescale value
 1 1 = 1:8 Prescale value

T1SYNC D2 Timer3 external clock input synchronization control bit
   Used only when TMR3CS = 1 and clock comes from an
   external source. If TMR3CS = 0, this bit is not used.
   1 = Do not synchronize external clock input
   0 = Synchronize external clock input

TMR3CS D1 Timer3 clock source select bit
   1 = External clock from pin T1OSI or T1CKI
   0 = Internal clock (Fosc/4)

TMR3ON D0 Timer3 On and Off control bit
   1 = Enable (start) Timer1
   0 = Stop Timer1

Figure 9-15. T3CON (Timer3 Control) Register

						TMR3IF	

TMR3IF Timer3 interrupt overflow flag bit
   0 = Timer3 did not overflow.
   1 = Timer3 has overflowed (FFFF to 0000).

The importance of TMR3IF: In 16-bit mode, when TMR3H:TMR3L overflows from
FFFF to 0000, this flag is raised.

The location of TMRxIF in the PIR register can vary in future products.

**Figure 9-16. PIR2 (Peripheral Interrupt Flag Register 2) Contains the TMR3IF
Flag**

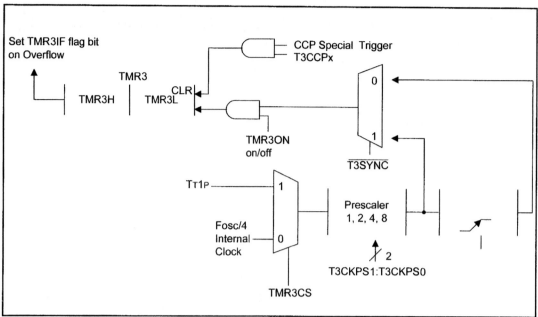

Figure 9-17. Timer3 Block Diagram

Example 9-42

Find the frequency of the square wave generated by the following program if XTAL = 10 MHz. In your calculation do not include overhead due to instructions in the loop.

```
        BCF    TRISB,5        ;PB5 as an output
        MOVLW 0x0             ;Timer3,16-bit,int clk,no prescale
        MOVWF T3CON           ;load T3CON reg
HERE    MOVLW 0x76            ;TMR3H = 76H, the high byte
        MOVWF TMR3H           ;load Timer3 high byte
        MOVLW 0x34            ;TMR3L = 34H, the low byte
        MOVWF TMR3L           ;load Timer3 low byte
        BCF    PIR2,TMR3IF    ;clear timer interrupt flag bit
        CALL   DELAY
        BTG    PORTB,RB5      ;toggle PB5
        BRA    HERE           ;load TH, TL again
;————————delay using Timer3
DELAY BSF    T3CON,TMR3ON     ;start Timer3
AGAIN BTFSS PIR2,TMR3IF       ;monitor Timer3 flag until
        BRA    AGAIN          ;it rolls over
        BCF    T3CON,TMR3ON    ;stop Timer3
        RETURN
```

Solution:

Because FFFFH − 7634H = 89CBH + 1 = 89CCH and 89CCH = 35276 clock count, $35276 \times 0.4 \ \mu s = 14.11$ ms and frequency = $1 / (2 \times 14.11$ ms$) = 1 / 28.22$ ms = 35.434 Hz. In this calculation, the overhead due to all the instructions in the loop is not included. Notice that the calculation is the same as in Example 9-20.

Example 9-43

Assume XTAL = 10 MHz, write a program to generate a square wave of 50 Hz frequency on pin PORTB5. Use Timer3, 16-bit mode, with the maximum prescaler allowed.

Solution:

Because FFFFH − 9E58H = 61A7H + 1 = 61A8H and 61A8H = 25,000 clock count, 25000×0.4 µs = 1 ms and frequency = 1/2 (1 ms) = 50 Hz. In this calculation, the overhead due to all the instructions in the loop is not included.

```
        BCF    TRISB,5          ;PB5 as an output
        MOVLW 0x0               ;Timer3,16-bit,int clk,no prescale
        MOVWF T3CON             ;load T3CON reg.
HERE    MOVLW 0x9E              ;TMR3H = 9EH, the high byte
        MOVWF TMR3H             ;load Timer3 high byte
        MOVLW 0x58              ;TMR3L = 58H, the low byte
        MOVWF TMR3L             ;load Timer3 low byte
        BCF    PIR2,TMR3IF      ;clear Timer3 interrupt flag bit
        CALL   DELAY
        BTG    PORTB,RB5        ;toggle PB5
        BRA    HERE             ;
;------------delay using Timer3
DELAY BSF    T3CON,TMR3ON       ;start Timer3
AGAIN BTFSS PIR2,TMR3IF         ;monitor timer flag until
        BRA    AGAIN            ;it rolls over
        BCF    T3CON,TMR3ON     ;stop timer
        RETURN
```

T = 1 / 50 Hz = 2 ms

$1 / 2 \times 2$ ms = 1 ms   for high and low portions

1 ms / 0.4 µs = 25000   number of clock counts

65536 − 25000 = 40536 = 9E58H

Example 9-44

Assume that a 1-Hz frequency pulse is connected to the input for Timer3 (pin RC0).
Write a program to display the counter values TMR3H and TMR3L on ports B and D.
Set the initial values of to 0. Use no prescaler.

Solution:

```
        BSF   TRISC,0          ;PORTC.0 as input T1CLKI
        CLRF  TRISB            ;PORTB as an output
        CLRF  TRISD            ;PORTD as an output
        MOVLW 0x02             ;Timer3,16-bit,ext clk,no prescale
        MOVWF T3CON            ;load T0CON reg
HERE    MOVLW 0x0              ;TMR3H = 0, the low byte
        MOVWF TMR3H            ;load Timer3 high byte
        MOVLW 0x0              ;TMR1L = 0, the low byte
        MOVWF TMR3L            ;load Timer3 low byte
        BCF   PIR2,TMR3IF      ;clear timer interrupt flag bit
        BSF   T3CON,TMR3ON     ;start timer
AGAIN   MOVFF TMR3L,PORTB      ;display low byte count
        MOVFF TMR3H,PORTD      ;display high byte count
        BTFSS PIR2,TMR3IF      ;monitor Timer3 flag until
        BRA   AGAIN            ;it rolls over
        BCF   T3CON,TMR3ON     ;stop Timer3
        GOTO  HERE
```

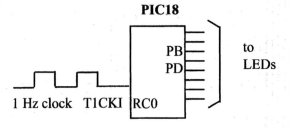

Example 9-45

Write a C18 program to create a frequency of 2500 Hz on pin PORTB.1. Use Timer3 to create the delay.

Solution:

```c
#include <p18f4580.h>
void T3Delay(void);
#define mybit PORTBbits.RB1

void main(void)
  {
     TRISBbits.TRISB1=0;          //PB1 as an output
     T3CON=0x00;                  //Timer3, 16-bit mode, no prescaler
     while(1)
       {
          mybit=~mybit;           //toggle PB.1
          T3Delay();
       }
  }

void T3Delay()
  {
     TMR3H=0xFE;                  //load TH3
     TMR3L=0x0C;                  //load TL3
     T3CONbits.TMR3ON=1;          //turn on T3
     while(PIR2bits.TMR3IF==0);   //wait for TF3 to roll over
     T3CONbits.TMR3ON=0;          //turn off T3
     PIR2bits.TMR3IF=0;           //clear TF3
  }
```

1 / 2500 Hz = 400 µs

400 µs / 2 = 200 µs

200 µs / 0.4 µs = 500

65536 − 500 = 65036 = FE0CH

Example 9-46

Assume that a 1-Hz external clock is being fed into pin T3 (RC0). Write a C18 program for Timer3 to be used as a counter. It should count the pulses and display the TMR3H and TMR3L registers on PORTD and PORTB, respectively.

Solution:

```c
#include <p18f4580.h>

void main(void)
  {
    TRISCbits.TRISC0=1;         //make RC0 an input for T1CKI
    TRISB = 0;                  //make PORTB an output
    TRISD = 0;                  //make PORTD an output
    T3CON=0x02;      //Timer1, 16-bit mode, no prescaler
    TMR3H=0;                    //set count to 0
    TMR3L=0;                    //set count to 0
    while(1)                    //repeat forever
      {
        do
          {
            T3CONbits.TMR3ON=1;  //turn on T3
            PORTB=TMR3L;         //place value on pins
            PORTD=TMR3H;         //
          }
      while(PIR2bits.TMR3IF==0); //wait for TF3 to roll over
      T3CONbits.TMR3ON=0;        //turn off T3
      PIR2bits.TMR3IF=0;         //clear TF3
          }
  }
```

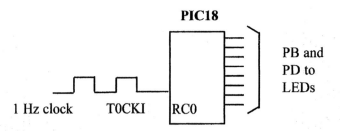

Review Questions

1. What provides the clock pulses to Timer2?
2. Indicate the selection made if T2CON = 0x00.
3. True or false. Timer2 cannot be used for a counter.
4. In Timer3, the counter rolls over when it goes from ____ to ____.
5. If we set PR2 = 200, state when TMR2IF is raised.
6. TMR2IF and TMR3IF are part of registers ____.

SUMMARY

The PIC18 can have up to four or more timers/counters, depending on the family member. When used as timers, they can generate time delays. When used as counters, they can serve as event counters. This chapter showed how to program the timers/counters for various modes.

Generally, the timers are accessed as two 8-bit registers, TMRLx and TMRHx. They can be used as a 16-bit timer, or as an 8-bit timer.

Each timer has its own TCON (Timer Control) register, allowing us to choose various operational modes. Among the modes are the prescaler and timer/counter options. When the timer is used as a timer, the PIC18's crystal is used as the source of the frequency (Fosc/4); however, when it is used as a counter, it is a pulse outside of the PIC18 that increments the TMRxH, TMRxL registers.

PROBLEMS

SECTION 9.1: PROGRAMMING TIMERS 0 AND 1

1. How many timers are in the PIC18F458?
2. Timer0 of the PIC18 is ____-bit, accessed as _____ and _____.
3. Timer1 of the PIC18 is ____-bit, accessed as _____ and _____.
4. Timer0 supports the highest prescaler value of _____.
5. Timer1 supports the highest prescaler value of _____.
6. The T0CON register is a(n) ___-bit register.
7. What is the job of the T0CON register?
8. True or false. T1CON is a bit-addressable register.
9. Find the T1CON value for 16-bit mode, no prescaler, timer oscillator off, with the clock coming from the PIC18's crystal.
10. Find the frequency and period used by the timer if the crystal attached to the PIC18 has the following values:
    (a) XTAL = 10 MHz   (b) XTAL = 20 MHz
    (c) XTAL = 24 MHz   (d) XTAL = 30 MHz
11. Indicate which register holds the TMRxIF (Timer Interrupt Flag) bit for each of the following timers:
    (a) Timer0   (b) Timer1

12. Indicate the rollover value (in hex and decimal) of the timer for each of the following modes:
    (a) 16-bit   (b) 8-bit
13. Indicate when the TMR0IF flag is raised for each of the following modes:
    (a) 16-bit   (b) 8-bit
14. True or false. Both Timer0 and Timer1 have their own timer interrupt flags.
15. True or false. Both Timer0 and Timer1 have their own timer start flags.
16. Assume that XTAL = 10 MHz. Find the TMR0H,TMR0L value needed to generate a time delay of 2 ms. Use 16-bit mode, no prescaler mode.
17. Assume that XTAL = 10 MHz. Find the TMR0H,TMR0L value needed to generate a time delay of 5 ms. Use 16-bit mode, and the largest prescaler possible.
18. Assume that XTAL = 10 MHz. Find the TMR1H,TMR1L value needed to generate a time delay of 2.5 ms. Use the largest prescaler possible.
19. Assume that XTAL = 10 MHz. Find the TMR1H,TMR1L value needed to generate a time delay of 0.2 ms. Use 16-bit mode, no prescaler mode.
20. Assume that XTAL = 20 MHz. Find the TMR1H,TMR1L value needed to generate a time delay of 2 ms. Use 16-bit mode, and the largest prescaler possible.
21. Assuming that XTAL = 10 MHz, and we are generating a square wave on pin RB7, find the lowest square wave frequency that we can generate using Timer0 in 16-bit mode.
22. Assuming that XTAL = 10 MHz, and we are generating a square wave on pin RB2, find the highest square wave frequency that we can generate using Timer0 in 16-bit mode.
23. Repeat Problems 21 and 22 for 8-bit mode.
24. In 8-bit mode, assuming that TMR0L = F1H, indicate which states Timer0 goes through until TMR0IF is raised. How many states is that?
25. Program Timer0 to generate a square wave of 1 kHz. Assume that XTAL = 10 MHz.
26. Program Timer1 to generate a square wave of 3 kHz. Assume that XTAL = 10 MHz. Use the largest prescaler possible.
27. State the differences between Timer0 and Timer1.
28. Find the value (in hex) loaded into WREG in each of the following:

(a)	MOVLW -D'12'	(b)	MOVLW -D'22'
(c)	MOVLW -D'34'	(d)	MOVLW -D'92'
(e)	MOVLW -D'120'	(f)	MOVLW -D'104'

SECTION 9.2: COUNTER PROGRAMMING

29. To use a timer as an event counter we must set the _____ bit in the T0CON register to _____ (LOW, HIGH).
30. Can we use both Timer0 and Timer1 as event counters?
31. For Counter 0, which pin is used for the input clock?
32. For Counter 1, which pin is used for the input clock?
33. Program Timer1 to be an event counter. Use 16-bit mode, and display the binary count on PORTB and PORTD continuously. Set the initial count to 20,000.
34. Program Timer0 to be an event counter. Use 8-bit mode and display the binary count on PORTB continuously. Set the initial count to 20.

35. The T1CON register is a(n) _____-bit register.
36. True or false. The T1CON register is not a bit-addressable register.

SECTION 9.3: PROGRAMMING TIMERS 0 AND 1 IN C

37. Program Timer0 in C to generate a square wave of 1 kHz. Assume that XTAL = 10 MHz.
38. Program Timer1 in C to generate a square wave of 1 kHz. Assume that XTAL = 10 MHz.
39. Program Timer0 in C to generate a square wave of 3 kHz. Assume that XTAL = 10 MHz.
40. Program Timer1 in C to generate a square wave of 3 kHz. Assume that XTAL = 10 MHz.
41. Program Timer1 in C to be an event counter. Use 16-bit mode and display the binary count on PORTB and PORTD continuously. Set the initial count to 20,000.
42. Program Timer0 in C to be an event counter. Use 8-bit mode and display the binary count on PORTD continuously. Set the initial count to 20.

SECTION 9.4: PROGRAMMING TIMERS 2 AND 3

43. Indicate the highest size of the prescaler supported for each of the following timers:
    (a) Timer2   (b) Timer3
44. Indicate the rollover value (in hex and decimal) of Timer3.
45. Indicate when the timer flag is raised for each of the following:
    (a) Timer2   (b) Timer3
46. True or false. The PR2 register of Timer2 is an 8-bit register.
47. True or false. Both Timer2 and Timer3 are 16-bit timers.
48. Assume that XTAL = 10 MHz. Find the TMR3H,TMR3L value needed to generate a time delay of 2 ms. Use no prescaler.
49. Assume that XTAL = 10 MHz. Find the TMR3H,TMR3L value needed to generate a time delay of 5 ms. Use the highest prescaler allowed.
50. Program Timer3 to be an event counter. Use 16-bit mode and display the binary count on PORTB and PORTD continuously. Set the initial count to 20,000.
51. Program Timer2 in Assembly to toggle pin RB3 when it counts up from 0 to 200. Assume that XTAL = 10 MHz.
52. Program Timer3 in C to generate a square wave of 3 kHz. Assume that XTAL = 10 MHz.
53. Program Timer2 in C to toggle pin RB3 when it counts up from 0 to 200. Assume that XTAL = 10 MHz.
54. Program Timer3 in C to generate a square wave of 1 kHz. Assume that XTAL = 10 MHz.

ANSWERS TO REVIEW QUESTIONS

SECTION 9.1: PROGRAMMING TIMERS 0 AND 1

1. 4
2. False
3. True
4. True
5. 0000 1000 indicates 16-bit mode, no prescaler, and using XTAL for frequency.
6. FFFFH, 0000
7. FFH, 00
8. −200 is 38H; therefore, WREG = 38H
9. 2 ms / 0.4 us = 5000, 65536 − 5000 = 60536 = EC78H, TMR0H = ECH and TMR0L = 78H.
10. 100 us / 0.4 us = 250, 256 − 250 = 06; therefore TMR0L = 06H.

SECTION 9.2: COUNTER PROGRAMMING

1. The crystal attached to the PIC18
2. The clock source for the timer comes from pin RA4 (PORTA4) .
3. Yes
4. We must configure the pin as input to allow the clocks to come in from an external source.
5. Yes

SECTION 9.4: PROGRAMMING TIMERS 2 AND 3

1. The crystal attached to the PIC18 (Fosc/4)
2. Prescale of 1, postscale of 1, and stop Timer2
3. True
4. FFFFH, 0
5. TMR2 counts up until it matches PR2. At that time TMR2IF is raised.
6. PIR1 and PIR2 respectively.

CHAPTER 10

PIC18 SERIAL PORT PROGRAMMING IN ASSEMBLY AND C

OBJECTIVES

Upon completion of this chapter, you will be able to:

>>   Contrast and compare serial versus parallel communication
>>   List the advantages of serial communication over parallel
>>   Explain serial communication protocol
>>   Contrast synchronous versus asynchronous communication
>>   Contrast half- versus full-duplex transmission
>>   Explain the process of data framing
>>   Describe data transfer rate and bps rate
>>   Define the RS232 standard
>>   Explain the use of the MAX232 and MAX233 chips
>>   Interface the PIC18 with an RS232 connector
>>   Discuss the baud rate of the PIC18
>>   Describe serial communication features of the PIC18
>>   Describe the main registers used by serial communication of the PIC18
>>   Program the PIC18 serial port in Assembly and C

Computers transfer data in two ways: parallel and serial. In parallel data transfers, often eight or more lines (wire conductors) are used to transfer data to a device that is only a few feet away. Devices that use parallel transfers include printers and hard disks; each uses cables with many wire strips. Although a lot of data can be transferred in a short amount of time by using many wires in parallel, the distance cannot be great. To transfer to a device located many meters away, the serial method is used. In serial communication, the data is sent one bit at a time, in contrast to parallel communication, in which the data is sent a byte or more at a time. Serial communication of the PIC18 is the topic of this chapter. The PIC18 has serial communication capability built into it, thereby making possible fast data transfer using only a few wires.

In this chapter we first discuss the basics of serial communication. In Section 10.2, PIC18 interfacing to RS232 connectors via MAX232 line drivers is discussed. Serial port programming of the PIC18 is discussed in Section 10.3. Section 10.4 covers PIC18 C programming for the serial port using the C18 compiler.

SECTION 10.1: BASICS OF SERIAL COMMUNICATION

When a microprocessor communicates with the outside world, it provides the data in byte-sized chunks. In some cases, such as printers, the information is simply grabbed from the 8-bit data bus and presented to the 8-bit data bus of the printer. This can work only if the cable is not too long, because long cables diminish and even distort signals. Furthermore, an 8-bit data path is expensive. For these reasons, serial communication is used for transferring data between two systems located at distances of hundreds of feet to millions of miles apart. Figure 10-1 diagrams serial versus parallel data transfers.

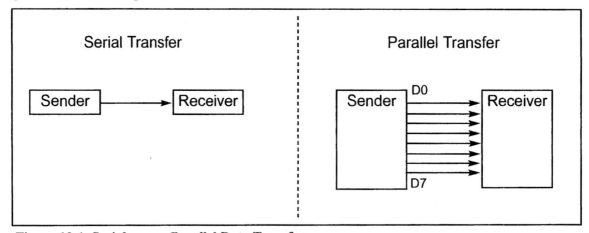

Figure 10-1. Serial versus Parallel Data Transfer

The fact that in a single data line is used in serial communication instead of the 8-bit data line of parallel communication makes serial transfer not only much cheaper but also enables two computers located in two different cities to communicate over the telephone.

For serial data communication to work, the byte of data must be converted to serial bits using a parallel-in-serial-out shift register; then it can be transmitted over a single data line. This also means that at the receiving end there must be a serial-in-parallel-out shift register to receive the serial data and pack them into a byte. Of course, if data is to be transferred on the telephone line, it must be converted from 0s and 1s to audio tones, which are sinusoidal signals. This conversion is performed by a peripheral device called a *modem*, which stands for "modulator/demodulator."

When the distance is short, the digital signal can be transferred as it is on a simple wire and requires no modulation. This is how IBM PC keyboards transfer data to the motherboard. For long-distance data transfers using communication lines such as a telephone, however, serial data communication requires a modem to *modulate* (convert from 0s and 1s to audio tones) and *demodulate* (convert from audio tones to 0s and 1s).

Serial data communication uses two methods, asynchronous and synchronous. The *synchronous* method transfers a block of data (characters) at a time whereas the *asynchronous* method ransfers a single byte at a time. It is possible to write software to use either of these methods, but the programs can be tedious and long. For this reason, special IC chips are made by many manufacturers for serial data communications. These chips are commonly referred to as UART (universal asynchronous receiver-transmitter) and USART (universal synchronous-asynchronous receiver-transmitter). The PIC18 chip has a built-in USART, which is discussed in detail in Section 10.3.

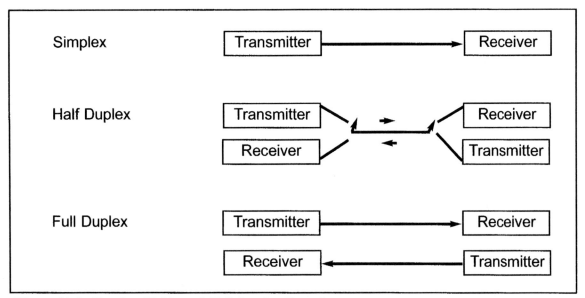

Figure 10-2. Simplex, Half-, and Full-Duplex Transfers

Half- and full-duplex transmission

In data transmission, if the data can be both transmitted and received, it is a *duplex* transmission. This is in contrast to *simplex* transmissions such as with printers, in which the computer only sends data. Duplex transmissions can be half or full duplex, depending on whether or not the data transfer can be simultaneous. If data is transmitted one way at a time, it is referred to as *half duplex*. If the data can go both ways at the same time, it is *full duplex*. Of course, full duplex requires two wire conductors for the data lines (in addition to the signal ground), one for transmission and one for reception, in order to transfer and receive data simultaneously. See Figure 10-2.

Asynchronous serial communication and data framing

The data coming in at the receiving end of the data line in a serial data transfer is all 0s and 1s; it is difficult to make sense of the data unless the sender and receiver agree on a set of rules, a *protocol*, on how the data is packed, how many bits constitute a character, and when the data begins and ends.

Start and stop bits

Asynchronous serial data communication is widely used for character-oriented transmissions, while block-oriented data transfers use the synchronous method. In the asynchronous method, each character is placed between start and stop bits. This is called *framing*. In data framing for asynchronous communications, the data, such as ASCII characters, are packed between a start bit and a stop bit. The start bit is always one bit but the stop bit can be one or two bits. The start bit is always a 0 (low) and the stop bit(s) is 1 (high). For example, look at Figure 10-3 in which the ASCII character "A" (8-bit binary 0100 0001) is framed between the start bit and a single stop bit. Notice that the LSB is sent out first.

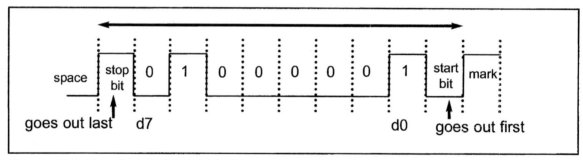

Figure 10-3. Framing ASCII 'A' (41H)

Notice in Figure 10-3 that when there is no transfer, the signal is 1 (high), which is referred to as *mark*. The 0 (low) is referred to as *space*. Notice that the transmission begins with a start bit followed by D0, the LSB, then the rest of the bits until the MSB (D7), and finally, the one stop bit indicating the end of the character "A".

In asynchronous serial communications, peripheral chips and modems can be programmed for data that is 7 or 8 bits wide. This is in addition to the number of stop bits, 1 or 2. While in older systems ASCII characters were 7-bit, in recent years, 8-bit data has become common due to the extended ASCII characters. In some older systems, due to the slowness of the receiving mechanical device, two stop bits were used to give the device sufficient time to organize itself before transmission of the next byte. In modern PCs, however, the use of one stop bit is standard. Assuming that we are transferring a text file of ASCII characters using 1 stop bit, we have a total of 10 bits for each character: 8 bits for the ASCII code, and 1 bit each for the start and stop bits. Therefore, each 8-bit character has an extra 2 bits, which gives 25% overhead.

In some systems, the parity bit of the character byte is included in the data frame in order to maintain data integrity. This means that for each character (7- or 8-bit, depending on the system) we have a single parity bit in addition to start and stop bits. The parity bit is odd or even. In the case of an odd parity bit the number of 1s in the data bits, including the parity bit, is odd. Similarly, in an even parity bit system the total number of bits, including the parity bit, is even. For example, the ASCII character "A", binary 0100 0001, has 0 for the even parity bit. UART chips allow programming of the parity bit for odd-, even-, and no-parity options.

Data transfer rate

The rate of data transfer in serial data communication is stated in *bps* (bits per second). Another widely used terminology for bps is *baud rate*. However, the baud and bps rates are not necessarily equal. This is because baud rate is the modem terminology and is defined as the number of signal changes per second. In modems, sometimes a single change of signal transfers several bits of data. As far as the conductor wire is concerned, the baud rate and bps are the same, and for this reason in this book we use the terms bps and baud interchangeably.

The data transfer rate of a given computer system depends on communication ports incorporated into that system. For example, the early IBM PC/XT could transfer data at the rate of 100 to 9600 bps. In recent years, however, Pentium-based PCs transfer data at rates as high as 56K. Note that in asynchronous serial data communication, the baud rate is generally limited to 100,000 bps.

RS232 standards

To allow compatibility among data communication equipment made by various manufacturers, an interfacing standard called RS232 was set by the Electronics Industries Association (EIA) in 1960. In 1963 it was modified and called RS232A. RS232B and RS232C were issued in 1965 and 1969, respectively. In this book we refer to it simply as RS232. Today, RS232 is the most widely used serial I/O interfacing standard. This standard is used in PCs and numerous types of equipment. Because the standard was set long before the advent of the TTL logic family, however, its input and output voltage levels are not TTL compatible. In RS232, a 1 is represented by −3 to −25 V, while a 0 bit is +3 to +25 V, making −3 to +3 undefined. For this reason, to connect any RS232 to a microcontroller system we must use voltage converters such as MAX232 to convert the TTL logic levels to the RS232 voltage level, and vice versa. MAX232 IC chips are commonly referred to as line drivers. RS232 connection to MAX232 is discussed in Section 10.2.

RS232 pins

Table 10-1 provides the pins and their labels for the RS232 cable, commonly referred to as the DB-25 connector. In labeling, DB-25P refers to the plug connector (male) and DB-25S is for the socket connector (female). See Figure 10-4.

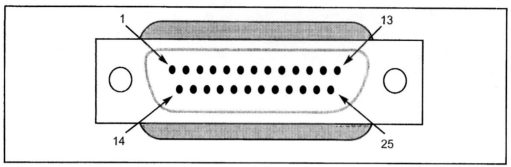

Figure 10-4. RS232 Connector DB-25

Because not all the pins are used in PC cables, IBM introduced the DB-9 version of the serial I/O standard, which uses only 9 pins, as shown in Table 10-2. The DB-9 pins are shown in Figure 10-5.

Data communication classification

Current terminology classifies data communication equipment as DTE (data terminal equipment) or DCE (data communication equipment). DTE refers to terminals and computers that send and receive data, while DCE refers to communication equipment, such as modems, that are responsible for transferring the data. Notice that all the RS232 pin function definitions of Tables 10-1 and 10-2 are from the DTE point of view.

The simplest connection between a PC and a microcontroller requires a minimum of three pins, TX, RX, and ground, as shown in Figure 10-6. Notice in that figure that the RX and TX pins are interchanged.

Examining RS232 handshaking signals

To ensure fast and reliable data transmission between two devices, the data transfer must be coordinated. Just as in the case of the printer, because the receiving device may have no room for the data in serial data communication, there must be a way to inform the sender to stop sending data. Many of the pins of the RS-232 connector are used for handshaking signals. Their description is provided below only as a reference, and they can be bypassed because they are not supported by the PIC18 UART chip.

Table 10-1: RS232 Pins (DB-25)

Pin	Description
1	Protective ground
2	Transmitted data (TxD)
3	Received data (RxD)
4	Request to send ($\overline{\text{RTS}}$)
5	Clear to send ($\overline{\text{CTS}}$)
6	Data set ready ($\overline{\text{DSR}}$)
7	Signal ground (GND)
8	Data carrier detect ($\overline{\text{DCD}}$)
9/10	Reserved for data testing
11	Unassigned
12	Secondary data carrier detect
13	Secondary clear to send
14	Secondary transmitted data
15	Transmit signal element timing
16	Secondary received data
17	Receive signal element timing
18	Unassigned
19	Secondary request to send
20	Data terminal ready ($\overline{\text{DTR}}$)
21	Signal quality detector
22	Ring indicator
23	Data signal rate select
24	Transmit signal element timing
25	Unassigned

1. DTR (data terminal ready). When the terminal (or a PC COM port) is turned on, after going through a self-test, it sends out signal DTR to indicate that it is ready for communication. If there is something wrong with the COM port, this signal will not be activated. This is an active-LOW signal and can be used to inform the modem that the computer is alive and kicking. This is an output pin from DTE (PC COM port) and an input to the modem.

2. DSR (data set ready). When the DCE (modem) is turned on and has gone through the self-test, it asserts DSR to indicate that it is ready to communicate. Thus, it is an output from the modem (DCE) and an input to the PC (DTE). This is an active-LOW signal. If for any reason the modem cannot make a connection to the telephone, this signal remains inactive, indicating to the PC (or terminal) that it cannot accept or send data.

3. RTS (request to send). When the DTE device (such as a PC) has a byte to transmit, it asserts RTS to signal the modem that it has a byte of data to transmit. RTS is an active-LOW output from the DTE and an input to the modem.

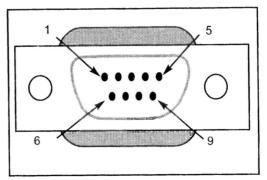

Figure 10-5. DB-9 9-Pin Connector

Table 10-2: IBM PC DB-9 Signals

Pin	Description
1	Data carrier detect ($\overline{\text{DCD}}$)
2	Received data (RxD)
3	Transmitted data (TxD)
4	Data terminal ready (DTR)
5	Signal ground (GND)
6	Data set ready ($\overline{\text{DSR}}$)
7	Request to send ($\overline{\text{RTS}}$)
8	Clear to send ($\overline{\text{CTS}}$)
9	Ring indicator (RI)

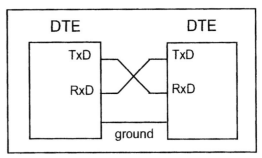

Figure 10-6. Null Modem Connection

4. CTS (clear to send). In response to RTS, when the modem has room to store the data it is to receive, it sends out signal CTS to the DTE (PC) to indicate that it can receive the data now. This input signal to the DTE is used by the DTE to start transmission.

5. DCD (data carrier detect). The modem asserts signal DCD to inform the DTE (PC) that a valid carrier has been detected and that contact between it and the other modem is established. Therefore, DCD is an output from the modem and an input to the PC (DTE).

6. RI (ring indicator). An output from the modem (DCE) and an input to a PC (DTE) indicates that the telephone is ringing. RI goes on and off in synchronization with the ringing sound. Of the six handshake signals, this is the least often used because modems take care of answering the phone. If in a given system the PC is in charge of answering the phone, however, this signal can be used.

From the above description, PC and modem communication can be summarized as follows: While signals DTR and DSR are used by the PC and modem, respectively, to indicate that they are alive and well, it is RTS and CTS that actually control the flow of data. When the PC wants to send data it asserts RTS, and in response, the modem, if it is ready (has room) to accept the data, sends back CTS. If, for lack of room, the modem does not activate CTS, the PC will deassert DTR and try again. RTS and CTS are also referred to as hardware control flow signals.

This concludes the description of the most important pins of the RS232 handshake signals plus TX, RX, and ground. Ground is also referred to as SG (signal ground).

IBM PC/compatible COM ports

IBM PC/compatible computers based on x86 (8086, 286, 386, 486, and all Pentiums) microprocessors used to have two COM ports. Both COM ports were RS232-type connectors. Many PCs used one each of the DB-25 and DB-9 RS232 connectors. The COM ports were designated as COM 1 and COM 2. In recent years, one of these has been replaced with the USB port, and COM 1 is the only serial port available, if any. We can connect the PIC18 serial port to the COM 1 port of a PC for serial communication experiments. In the absence of a COM port, we can use a COM-to-USB converter module.

With this background in serial communication, we are ready to look at the PIC18. In the next section we discuss the physical connection of the PIC18 and RS232 connector, and in Section 10.3 we see how to program the PIC18 serial communication port.

Review Questions

1. The transfer of data using parallel lines is _____ (faster, slower) but _____ (more expensive, less expensive).
2. True or false. Sending data to a printer is duplex.
3. True or false. In full duplex we must have two data lines, one for transfer and one for receive.
4. The start and stop bits are used in the _____ (synchronous, asynchronous) method.
5. Assuming that we are transmitting the ASCII letter "E" (0100 0101 in binary) with no parity bit and one stop bit, show the sequence of bits transferred serially.
6. In Question 5, find the overhead due to framing.
7. Calculate the time it takes to transfer 10,000 characters as in Question 5 if we use 9600 bps. What percentage of time is wasted due to overhead?
8. True or false. RS232 is not TTL compatible.
9. What voltage levels are used for binary 0 in RS232?
10. True or false. The PIC18 has a built-in UART.
11. On the back of x86 PCs, we normally have ____ COM port connectors.
12. The PC COM ports are designated by DOS and Windows as _____ and _____.

SECTION 10.2: PIC18 CONNECTION TO RS232

In this section, the details of the physical connections of the PIC18 to RS232 connectors are given. As stated in Section 10.1, the RS232 standard is not TTL compatible; therefore, a line driver such as the MAX232 chip is required to convert RS232 voltage levels to TTL levels, and vice versa. The interfacing of PIC18 with RS232 connectors via the MAX232 chip is the main topic of this section.

RX and TX pins in the PIC18

The PIC18 has two pins that are used specifically for transferring and receiving data serially. These two pins are called TX and RX and are part of the PORTC group (RC6 and RC7) of the 40-pin package. Pin 25 of the PIC18 (RC7) is assigned to TX and pin 26 (RC6) is designated as RX. These pins are TTL compatible; therefore, they require a line driver to make them RS232 compatible. One such line driver is the MAX232 chip. This is discussed next.

MAX232

Because the RS232 is not compatible with today's microprocessors and microcontrollers, we need a line driver (voltage converter) to convert the RS232's signals to TTL voltage levels that will be acceptable to the PIC18's TX and RX pins. One example of such a converter is MAX232 from Maxim Corp. (www.maxim-ic.com). The MAX232 converts from RS232 voltage levels to TTL voltage levels, and vice versa. One advantage of the MAX232 chip is that it uses a +5 V power source, which is the same as the source voltage for the PIC18. In other words, with a single +5 V power supply we can power both the PIC18 and MAX232, with no need for the dual power supplies that are common in many older systems.

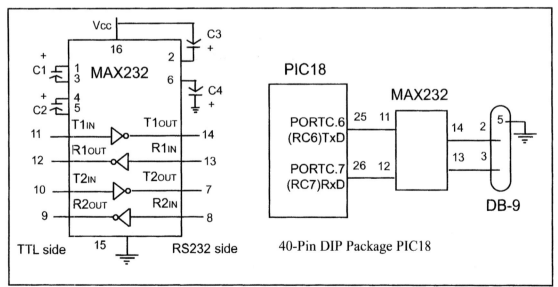

Figure 10-7. (a) Inside MAX232 and (b) its Connection to the PIC18 (Null Modem)

The MAX232 has two sets of line drivers for transferring and receiving data, as shown in Figure 10-7. The line drivers used for TX are called T1 and T2, while the line drivers for RX are designated as R1 and R2. In many applications only one of each is used. For example, T1 and R1 are used together for TX and RX of the PIC18, and the second set is left unused. Notice in MAX232 that the T1 line driver has a designation of T1in and T1out on pin numbers 11 and 14, respectively. The T1in pin is the TTL side and is connected to TX of the microcontroller, while T1out is the RS232 side that is connected to the RX pin of the RS232 DB connector. The R1 line driver has a designation of R1in and R1out on pin numbers 13 and 12, respectively. The R1in (pin 13) is the RS232 side that is connected to the TX pin of the RS232 DB connector, and R1out (pin 12) is the TTL side that is connected to the RX pin of the microcontroller. See Figure 10-7. Notice the null modem connection where RX for one is TX for the other.

MAX232 requires four capacitors ranging from 1 to 22 μF. The most widely used value for these capacitors is 22 μF.

MAX233

To save board space, some designers use the MAX233 chip from Maxim. The MAX233 performs the same job as the MAX232 but eliminates the need for capacitors. However, the MAX233 chip is much more expensive than the MAX232. Notice that MAX233 and MAX232 are not pin compatible. You cannot take a MAX232 out of a board and replace it with a MAX233. See Figure 10-8 for MAX233 with no capacitor used.

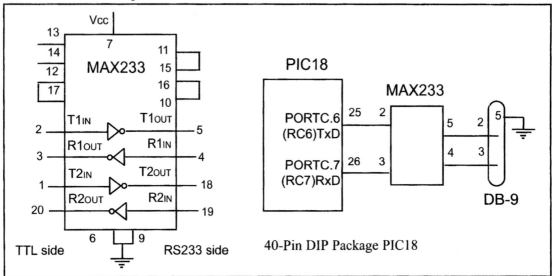

Figure 10-8. (a) Inside MAX233 and (b) Its Connection to the PIC18 (Null Modem)

Review Questions

1. True or false. The PC COM port connector is the RS232 type.
2. Which pins of the PIC18 are set aside for serial communication, and what are their functions?
3. What are line drivers such as MAX 232 used for?
4. MAX232 can support _____ lines for TX and _____ lines for RX.
5. What is the advantage of the MAX233 over the MAX232 chip?

SECTION 10.3: PIC18 SERIAL PORT PROGRAMMING IN ASSEMBLY

In this section we discuss the serial communication registers of the PIC18 and show how to program them to transfer and receive data using asynchronous mode. The USART (universal synchronous asynchronous receiver) in the PIC18 has both the synchronous and asynchronous features. The synchronous mode can be used to transfer data between the PIC and external peripherals such as ADC and EEPROMs. The asynchronous mode is the one we will use to connect the PIC18-based system to the IBM PC serial port for the purpose of full-duplex serial data transfer. In this section we examine the asynchronous mode only. In the PIC microcontroller six major registers are associated with the UART that we deal with in this chapter. They are (a) SPBGR (serial port baud rate generator), (b) TXREG (Transfer register), (c) RCREG (Receive register), (d) TXSTA (transmit status and control register), (e) RCSTA (receive status and control register), and (f) PIR1 (peripheral interrupt request register1). We examine each of them and show how they are used in full-duplex serial data communication.

SPBRG register and baud rate in the PIC18

Because IBM PC/compatible computers are so widely used to communicate with PIC18-based systems, we will emphasize serial communications of the PIC18 with the COM port of the PC. Some of the baud rates supported by PC HyperTerminal are listed in Table 10-3. You can examine these baud rates by going to the Microsoft Windows HyperTerminal program and clicking on the Communication Settings option. The PIC18 transfers and receives data serially at many different baud rates. The baud rate in the PIC18 is programmable. This is done with the help of the 8-bit register called SPBRG. For a given crystal frequency, the value loaded into the SPBRG decides the baud rate. The relation between the value loaded into SPBRG and the Fosc (frequency of oscillator connected to the OSC1 and OSC2 pins) is dictated by the following formula:

Table 10-3: Some PC Baud Rates in HyperTerminal

HyperTerminal
1,200
2,400
4,800
9,600
19,200
38,400
57,600
115,200

$$\text{Desired Baud Rate} = \text{Fosc}/(64X + 64) = \text{Fosc}/64(X + 1)$$

where X is the value we load into the SPBGR register. Assuming that Fosc = 10 MHz, we have the following:

$$\text{Desired Baud Rate} = \text{Fosc}/64(X + 1) = 10 \text{ MHz}/64(X + 1) = 6250 \text{ Hz}/(X + 1)$$

To get the X value for different baud rates we can solve the equation as follows:

$$X = (156250/\text{Desired Baud Rate}) - 1$$

Table 10-4 shows the X values for the different baud rates if Fosc = 10 MHz. Another way to understand the SPBRG values in Table 10-4 is to look at them from the perspective of the instruction cycle time. As we discussed in previous chapters, the PIC18 divides the crystal frequency (Fosc) by 4 to get the instruction cycle time frequency. In the case of XTAL = 10 MHz, the instruction cycle frequency is 2.5 MHz. The PIC18's UART circuitry divides the instruction cycle frequency by 16 once more before it is used by an internal timer to set the baud rate. Therefore, 2.5 MHz divided by 16 gives 156,250 Hz. This is the number we use to find the SPBRG value shown in Table 10-4. Example 10-1 shows how to verify the data in Table 10-4. Table 10-5 shows the SPBRG values with the crystal frequency of 4 MHz (Fosc = 4 MHz).

Example 10-1

With Fosc = 10 MHz, find the BGRP value needed to have the following baud rates:
(a) 9600　　　(b) 4800　　　(c) 2400　　　(d) 1200

Solution:

Because Fosc = 10 MHz, we have 10 MHz/4 = 2.5 MHz for the instruction cycle frequency. This is divided by 16 once more before it is used by UART. Therefore, we have 2.5 MHz/16= 156250 Hz and X = (156250 Hz/Desired Baud Rate) – 1:

(a) (156250/ 9600) – 1 = 16.27 – 1 = 15.27 = 15 = F (hex) is loaded into SPBRG
(b) (156250/ 4800) – 1 = 32.55 – 1 = 31.55 = 32 = 20 (hex) is loaded into SPBRG
(c) (156250/ 2400) – 1 = 65.1 – 1 = 64.1 = 64 = 40 (hex) is loaded into SPBRG
(d) (156250/ 1200) – 1 = 130.2 – 1 = 129.2 = 129 = 81 (hex) is loaded into SPBRG

Notice that dividing the instruction cycle frequency by 16 is the setting upon Reset. We can get a higher baud rate with the same crystal by changing this default setting. This is done by making bit BRGH = 1 in the TXSTA register. This is explained at the end of this section.

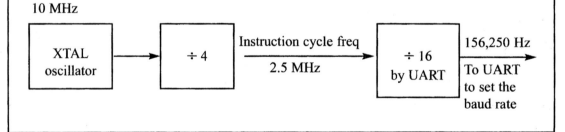

Table 10-4: SPBRG Values for Various Baud Rates (Fosc = 10 MHz, BRGH = 0)

Baud Rate	SPBRG (Decimal Value)	SPBRG (Hex Value)
38400	3	3
19200	7	7
9600	15	F
4800	32	20
2400	64	40
1200	129	81

Note: For Fosc = 10 MHz we have SPBRG = (156,250/BaudRate) – 1

Table 10-5: SPBRG Values for Various Baud Rates (Fosc = 4 MHz, BRGH = 0)

Baud Rate	SPBRG (Decimal Value)	SPBRG (Hex Value)
19200	2	2
9600	5	5
4800	12	0C
2400	25	19
1200	51	33

Note: For Fosc = 4 MHz we have 4 MHz/4 = 1 MHz for instruction cycle freq. The frequency used by the UART is 1 MHz/16 = 62,500 Hz. That means SPBRG = (62500/Baud Rate) – 1

TXREG register

TXREG is another 8-bit register used for serial communication in the PIC18. For a byte of data to be transferred via the TX pin, it must be placed in the TXREG register. TXREG is a special function register (SFR) and can be accessed like any other register in the PIC18. Look at the following examples of how this register is accessed:

```
MOVLW  0x41    ;WREG=41H, ASCII for letter 'A'
MOVWF  TXREG   ;copy WREG into TXREG

MOVFF  PORTB,TXREG ;copy PORTB contents into TXREG
```

The moment a byte is written into TXREG, it is fetched into a register called TSR (transmit shift register). The TSR frames the 8-bit data with the start and stop bits and the 10-bit data is transferred serially via the TX pin. Notice that while TXREG is accessible by the programmer, TSR is not accessible and is strictly for internal use.

RCREG register

Similarly, when the bits are received serially via the RX pin, the PIC18 deframes them by eliminating the stop and start bits, making a byte out of the data received, and then placing it in the RCREG register. The following code will dump the received byte into PORTB:

```
MOVFF  RCREG,PORTB ;copy RXREG to PORTB
```

TXSTA (transmit status and control register)

The TXSTA register is an 8-bit register used to select the synchronous/asynchronous modes and data framing size, among other things. Figure 10-9 describes various bits of the TXSTA register. In this textbook we use the asynchronous mode with a data size of 8 bits. The BRGH bit is used to select a higher speed for transmission. The default is lower baud rate transmission. We will examine the higher transmission rate at the end of this chapter. Notice that D6 of the TXSTA register determines the framing of data by specifying the number of bits per character. We use an 8-bit data size. There are some applications for the 9-bit in which the ninth bit can be used as an address.

CSRC	TX9	TXEN	SYNC	0	BRGH	TRMT	TX9D

CSRC D7   Clock Source Select (not used in asynchronous mode, therefore D7 = 0.)

TX9   D6   9-bit Transmit Enable

           1 = Select 9-bit transmission

           0 = Select 8-bit transmission (We use this option, therefore D6 = 0.)

TXEN D5   Transmit Enable

           1 = Transmit Enabled

           0 = Transmit Disabled

           We turn "on" and "off" this bit in order to start or stop data transfer.

SYNC D4   USART mode Select (We use asynchronous mode, therefore D4 = 0.)

           1 = Synchronous

           0 = Asynchronous

0   D3

BRGH D2   High Baud Rate Select

           0 = Low Speed (Default)

           1 = High Speed

           We can double the baud rate with the same Fosc. See the end of this section for further discussion on this bit.

TRMT D1   Transmit Shift Register (TSR) Status

           1 = TSR empty

           0 = TSR full

The importance of the TSR register. To transfer a byte of data serially, we write it into TXREG. The TSR (transmit shift register) is an internal register whose job is to get the data from the TXREG, frame it with the start and stop bits, and send it out one bit at a time via the TX pin. When the last bit, which is the stop bit, is transmitted, the TRMT flag is raised to indicate that it is empty and ready for the next byte. When TSR fetches the data from TXREG, it clears the TRMT flag to indicate it is full. Notice that TSR is a parallel-in-serial-out shift register and is not accessible to the programmer. We can only write to TXREG. Whenever the TSR is empty, it gets its data from TXREG and clears the TXREG register immediately, so it does not send out the same data twice.

TXD9 D0   9th bit of Transmit Data (Because we use the 8-bit option, we make

           D0 = 0)

           Can be used as an address/data or a parity bit in some applications

Figure 10-9. TXSTA: Transmit Status and Control Register

RCSTA (receive status and control register)

The RCSTA register is an 8-bit register used to enable the serial port to receive data, among other things. Figure 10-10 describes various bits of the RCSTA register. In this section we use the 8-bit data frame.

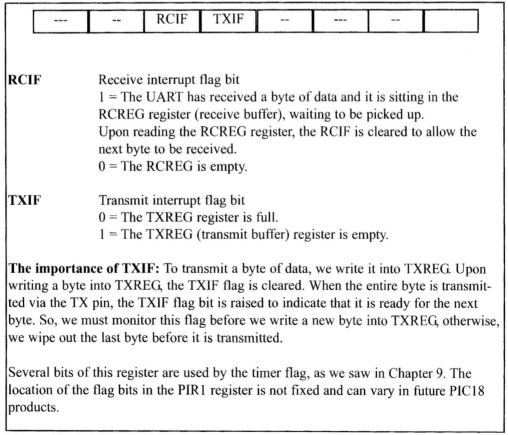

SPEN	RX9	SREN	CREN	ADDE	FERR	OERR	RX9D

SPEN D7   Serial port enable bit
                   1 = Serial port enabled, which makes TX and RX pins as serial port pins
                   0 = Serial port disabled
RX9   D6   9-bit Receive enable bit
                   1 = Select 9-bit reception
                   0 = Select 8-bit reception (We use this option; therefore, D6 = 0.)
SREN D5   Single receive enable bit (not used in asynchronous mode D5 = 0)
CREN D4   Continuous receive enable bit
                   1 = Enable continuous Receive (in asynchronous mode)
                   0 = Disable continuous Receive (in asynchronous mode)
ADDEN D3   Address delete enable bit (Because used with the 9-bit data frame D3 = 0)
FERR D2   Framing error bit
                   1 = Framing error
                   0 = No Framing error
OERR D1   Overrun error bit
                   1 = Overrun error
                   0 = No overrun error
TXD9 D0   9th bit of Receive data (Because we use the 8-bit option, we make D0 = 0)
                   Can be used as an address/data or a parity bit in some applications.

Figure 10-10. RCSTA: Receive Status and Control Register

---	--	RCIF	TXIF	--	---	--	

RCIF   Receive interrupt flag bit
                   1 = The UART has received a byte of data and it is sitting in the
                   RCREG register (receive buffer), waiting to be picked up.
                   Upon reading the RCREG register, the RCIF is cleared to allow the
                   next byte to be received.
                   0 = The RCREG is empty.

TXIF   Transmit interrupt flag bit
                   0 = The TXREG register is full.
                   1 = The TXREG (transmit buffer) register is empty.

The importance of TXIF: To transmit a byte of data, we write it into TXREG. Upon writing a byte into TXREG, the TXIF flag is cleared. When the entire byte is transmitted via the TX pin, the TXIF flag bit is raised to indicate that it is ready for the next byte. So, we must monitor this flag before we write a new byte into TXREG, otherwise, we wipe out the last byte before it is transmitted.

Several bits of this register are used by the timer flag, as we saw in Chapter 9. The location of the flag bits in the PIR1 register is not fixed and can vary in future PIC18 products.

Figure 10-11. PIR1 (Peripheral Interrupt Register 1)

PIR1 (peripheral interrupt request register 1)

In Chapter 9, we saw how some of the bits of PIR1 are used by the timers. Two of the PIR1 register bits are used by the UART. They are TXIF (transmit interrupt flag) and RCIF (receive interrupt flag). See Figure 10-11. We monitor (poll) the TXIF flag bit to make sure that all the bits of the last byte are transmitted before we write another byte into the TXREG. By the same logic, we monitor the RCIF flag to see if a byte of data has come in yet. In Chapter 11 we will see how these flags are used with interrupts instead of polling. Next we will examine how TXIF flags are used in serial data transfer.

Programming the PIC18 to transfer data serially

In programming the PIC18 to transfer character bytes serially, the following steps must be taken:

1. The TXSTA register is loaded with the value 20H, indicating asynchronous mode with 8-bit data frame, low baud rate, and transmit enabled.
2. Make TX pin of PORTC (RC6) an output for data to come out of the PIC.
3. The SPBRG is loaded with one of the values in Table 10-4 (or Table 10-5 if Fosc = 4 MHz) to set the baud rate for serial data transfer.
4. SPEN bit in the RCSTA register is set HIGH to enable the serial port of the PIC18.
5. The character byte to be transmitted serially is written into the TXREG register.
6. Monitor the TXIF bit of the PIR1 register to make sure UART is ready for next byte.
7. To transfer the next character, go to Step 5.

Example 10-2 shows the program to transfer data serially at 9600 baud. Example 10-3 shows how to transfer "YES" continuously.

Example 10-2

Write a program for the PIC18 to transfer the letter 'G' serially at 9600 baud, continuously. Assume XTAL = 10 MHz.

Solution:

```
        MOVLW B'00100000'  ;enable transmit and choose low baud rate
        MOVWF TXSTA        ;write to reg
        MOVLW D'15'        ;9600 bps (Fosc / (64 * Speed) - 1)
        MOVWF SPBRG        ;write to reg
        BCF TRISC, TX      ;make TX pin of PORTC an output pin
        BSF RCSTA, SPEN    ;enable the entire serial port of PIC18
OVER    MOVLW A'G'         ;ASCII letter 'G' to be transferred
S1      BTFSS PIR1, TXIF   ;wait until the last bit is gone
        BRA S1             ;stay in loop
        MOVWF TXREG        ;load the value to be transferred
        BRA OVER           ;keep sending letter 'G'
```

Example 10-3

Write a program to transmit the message "YES" serially at 9600 baud, 8-bit data, and 1 stop bit. Do this forever.

Solution:

```
        MOVLW B'00100000'        ;enable transmit and choose low baud
        MOVWF TXSTA              ;write to reg
        MOVLW D'15'              ;9600 bps (Fosc / (64 * Speed) - 1)
        MOVWF SPBRG              ;write to reg
        BCF TRISC, TX            ;make TX pin of PORTC an output pin
        BSF RCSTA, SPEN          ;enable the serial port
OVER    MOVLW A'Y'               ;ASCII letter 'Y' to be transferred
        CALL  TRANS
        MOVLW A'E'               ;ASCII letter 'E' to be transferred
        CALL  TRANS
        MOVLW A'S'               ;ASCII letter 'S' to be transferred
        CALL  TRANS
        MOVLW 0x0                ;NULL to purge the buffer
        CALL TRANS
        BRA   OVER               ;keep doing it
TRANS ;----serial data transfer subroutine
S1      BTFSS PIR1, TXIF         ;wait until the last bit is gone
        BRA S1                   ;stay in loop
        MOVWF TXREG              ;load the value to be transmitted
        RETURN                   ;return to caller
```

Importance of the TXIF flag

To understand the importance of the role of TXIF, look at the following sequence of steps that the PIC18 goes through in transmitting a character via TX:

1. The byte character to be transmitted is written into the TXREG register.
2. The TXIF flag is set to 1 internally to indicate that TXREG has a byte and will not accept another byte until this one is transmitted.
3. The TSR (Transmit Shift Register) reads the byte from TXREG and begins to transfer the byte starting with the start bit.
4. The TXIF is cleared to indicate that the last byte is being transmitted and TXREG is ready to accept another byte.
5. The 8-bit character is transferred one bit at a time.
6. By monitoring the TXIF flag, we make sure that we are not overloading the TXREG register. If we write another byte into the TXREG register before the TSR has fetched the last one, the old byte could be lost before it is transmitted.

From the above discussion we conclude that by checking the TXIF flag bit, we know whether or not the PIC18 is ready to transfer another byte. The TXIF flag bit can be checked by the instruction "BTFSS PIR1, TXIF" or we can use an interrupt, as we will see in Chapter 11. In Chapter 11 we will show how to use interrupts to transfer data serially, and avoid tying down the microcontroller with instructions such as "BTFSS PIR1, TXIF".

Programming the PIC18 to receive data serially

In programming the PIC18 to receive character bytes serially, the following steps must be taken:

1. The RCSTA register is loaded with the value 90H, to enable the continuous receive in addition to the 8-bit data size option.
2. The TXSTA register is loaded with the value 00H to choose the low baud rate option.
3. SPBRG is loaded with one of the values in Table 10-4 to set the baud rate (assuming XTAL = 10 MHz).
4. Make the RX pin of PORTC (RC7) an input for data to come into the PIC18.
5. The RCIF flag bit of the PIR1 register is monitored for a HIGH to see if an entire character has been received yet.
6. When RCIF is raised, the RCREG register has the byte. Its contents are moved into a safe place.
7. To receive the next character, go to Step 5.

Example 10-4 shows the coding of the above steps.

Example 10-4

Program the PIC18 to receive bytes of data serially and put them on PORTB. Set the baud rate at 9600, 8-bit data, and 1 stop bit.

Solution:

```
        MOVLW  B'10010000'          ;enable receive and serial port itself
        MOVWF  RCSTA                ;write to reg
        MOVLW  D'15'                ;9600 bps (Fosc / (64 * Speed) - 1)
        MOVWF  SPBRG                ;write to reg
        BSF    TRISC, RX            ;make RX pin of PORTC an input pin
        CLRF   TRISB                ;make port B an output port
;get a byte from serial port and place it on PORTB
R1      BTFSS PIR1, RCIF            ;check for ready
        BRA    R1                   ;stay in loop
        MOVFF  RCREG,  PORTB        ;save value into PORTB
        BRA    R1                   ;keep doing that
```

Importance of the RCIF flag bit

In receiving bits via its RX pin, the PIC18 goes through the following steps:

1. It receives the start bit indicating that the next bit is the first bit of the character byte it is about to receive.
2. The 8-bit character is received one bit at time. When the last bit is received, a byte is formed and placed in RCREG.
3. The stop bit is received. It is during receiving the stop bit that the PIC18 makes RCIF = 1, indicating that an entire character byte has been received and must

be picked up before it gets overwritten by another incoming character.

4. By checking the RCIF flag bit when it is raised, we know that a character has been received and is sitting in the RCREG register. We copy the RCREG contents to a safe place in some other register or memory before it is lost.

5. After the RCREG contents are read (copied) into a safe place, the RCIF flag bit is forced to 0 by the UART itself. This allows the next received character byte to be placed in RCREG, and also prevents the same byte from being picked up multiple times.

From the above discussion we conclude that by checking the RCIFI flag bit we know whether or not the PIC18 has received a character byte. If we fail to copy RCREG into a safe place, we risk the loss of the received byte. More importantly, note that the RCIF flag bit is raised by the PIC18, and it is also cleared by the CPU when the data in the RCREG is picked up. Note also that if we copy RCREG into a safe place before the RCIF flag bit is raised, we risk copying garbage. The RCIF flag bit can be checked by the instruction "BTFSS PIR1, RCIF" or by using an interrupt, as we will see in Chapter 11.

Quadrupling the baud rate in the PIC18

There are two ways to increase the baud rate of data transfer in the PIC18:

1. Use a higher-frequency crystal.
2. Change a bit in the TXSTA register, as shown below.

Option 1 is not feasible in many situations because the system crystal is fixed. Therefore, we will explore option 2. There is a software way to quadruple the baud rate of the PIC18 while the crystal frequency stays the same. This is done with the BRGH bit of the TXSTA register. When the PIC18 is powered up, D2 (BRGH bit) of the TXSTA register is zero. We can set it to high by software and thereby quadruple the baud rate.

To see how the baud rate is quadrupled with this method, we show the role of the BRGH bit (D2 bit of the TXSTA register), which can be 0 or 1. We discuss each case.

Baud rates for BRGH = 0

When BRGH = 0, the PIC18 divides Fosc/4 (crystal frequency) by 16 once more and uses that frequency for UART to set the baud rate. In the case of XTAL = 10 MHz we have:

```
Instruction cycle freq. = 10 MHz / 4 = 2.5 kHz
and
2.5 MHz / 16 = 156,250 Hz because BRGH = 0
```

This is the frequency used by UART to set the baud rate. This has been the basis of all the examples so far because it is the default when the PIC18 is powered up. The baud rate for BRGH = 0 was listed in Table 10-4 and Table 10-5.

Baud rates for BRGH = 1

With the fixed crystal frequency, we can quadruple the baud rate by making BRGH = 1. When the BRGH bit (D2 of the TXSTA register) is set to 1, Fosc/4 of XTAL is divided by 4 (instead of 16) once more, and that is the frequency used by UART to set the baud rate. In the case of XTAL = 10 MHz, we have:

```
Instruction cycle freq. = 10 MHz / 4 = 2.5 MHz
and
2.5 MHz / 4 = 625000 Hz because BHRG = 1
```

This is the frequency used by UART to set the baud rate if BHRH = 1.

Table 10-8 shows that the values loaded into SPBREG are the same for both cases; however, the baud rates are quadrupled when BRGH = 1. Look at Examples 10-5 through 10-7 to clarify the data given in Tables 10-6 and 10-7.

Table 10-6: SPBRG Values for Various Baud Rates (Fosc = 10MHz and BRGH = 1)

Baud Rate	SPBRG (Decimal Value)	SPBRG (Hex Value)
57600	10	0A
38400	15	0F
19200	32	20
9600	64	40
4800	129	81

Note: For Fosc = 10 MHz we have SPBRG = (625000/Baud Rate) – 1

Example 10-5

Find the SPBRG value (in both decimal and hex) to set the baud rate to each of the following:
(a) 9600 if BRGH = 1   (b) 4800 if BRGH = 1
Assume that XTAL = 10 MHz.

Solution:
With XTAL = 10 MHz, Fosc/4 = 2.5 MHz. Because BRGH = 1, we have UART frequency = 2.5 MHz/4 = 625,000 Hz.
(a) (625,500 / 9600) – 1 = 64; therefore, SPBRG = 64 or SPBRG = 40H (in hex).
(b) (625,500 / 4800) – 1 = 129; therefore, SPBRG = 129 or SPBRG = 81H (in hex).

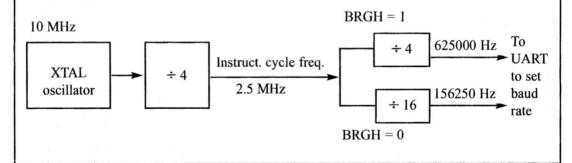

Table 10-7: SPBRG Values for Various Baud Rates (XTAL = 10 MHz)

	BRGH = 0	BRGH = 1
Baud Rate	**SPBRG (Decimal)**	**SPBRG (Decimal)**
57600	2	10
38400	3	15
19200	7	32
9,600	15	64
4,800	32	129
SPBRG = (156250/Baud rate) − 1		SPBRG = (625000/Baud rate) − 1

Table 10-8: SPBRG Values vs. Baud Rates for BRGH = 0 and BRGH = 1 (XTAL = 10 MHz)

	BRGH = 0	BRGH = 1
SPBRG (Decimal)	**Baud Rate**	**Baud Rate**
15	9600	38400
32	4800	19200
64	2400	9600

Table 10-9: SPBRG Values for Various Baud Rates (XTAL = 4 MHz)

	BRGH = 0	BRGH = 1
Baud Rate	**SPBRG (Decimal)**	**SPBRG (Decimal)**
19200	3	12
9,600	6	25
4,800	12	51
2,400	25	103
SPBRG = (62500/Baud rate) − 1		SPBRG = (250000/Baud rate) − 1

Example 10-6

Write a program for the PIC18 to transfer the letter 'G' serially at 57600 baud, continuously. Assume XTAL = 10 MHz. Use the BRGH = 1 mode

Solution:

```
        MOVLW B'00100100'   ;enable transmit and choose high baud rate
        MOVWF TXSTA         ;write to reg
        MOVLW D'10'         ;57600 bps (Fosc / (16 * Speed) - 1)
        MOVWF SPBRG         ;write to reg
        BCF TRISC, TX       ;make TX pin of PORTC an output pin
        BSF RCSTA, SPEN     ;enable the entire Serial port of PIC18
OVER    MOVLW A'G'          ;ASCII letter 'G' to be transferred
S1      BTFSS PIR1, TXIF    ;wait until the last bit is gone
        BRA S1              ;stay in loop
        MOVWF TXREG         ;load the value to be transferred
        BRA OVER            ;keep sending letter 'G'
```

Baud rate error calculation

In calculating the baud rate we have used the integer number for the SPBRG register values because PIC microcontrollers can only use integer values. By dropping the decimal portion of the calculated values we run the risk of introducing error into the baud rate. There are several ways to calculate this error. One way would be to use the following formula.

Error = (Calculated value for the SPBRG – Integer part)/Integer part

For example, with the XTAL = 10 MHz and BRGH = 0 we have the following for the 9600 baud rate:

SPBRG value = (156250/9600) – 1 = 16.27 – 1 = 15.27 = 15
and the error is
$$(15.27 - 15)/16 = 1.7\%$$
Another way to calculate the the error rate is as follows:

Error = (calculated baud rate – desired baud rate) / desired baud rate

---

Example 10-7

Assuming XTAL = 10 MHz, calculate the baud rate error for the following:
(a) 2400   (b) 1200   (c) 19200   (d) 57600
Use the BRGH = 0 mode.

Solution:

(a) SPBRG Value = (156250/2400) – 1 = 65.1 – 1 = 64.1 = 64

   Error = (64.1 – 64)/ 65 = 0.15%

(b) SPBRG Value (156250/1200) – 1 = 130.2 – 1 = 129.2 = 129

   Error = (129.2 – 129)/130 = 0.15%

(c) SPBRG Value (156250/19200) – 1 = 8.138 – 1 = 7.138 = 7

   Error = (7.138 – 7)/8 = 1.7%

(d) SPBRG Value (156250/57600) – 1 = 2.71 – 1 = 1.71 = 1

   Error = (1.71 – 1)/2 = 35%

Such an error rate is too high. Let's round up the number to see what happens.
   Error = (3 – 2.7)/3 = 10% This means we use SPBRG = 2 instead of SPBRG = 1.

---

where the desired baud rate is calculated using $X = ((\text{Fosc/Desired Baud rate})64) - 1$ and then the integer X (value loaded into SPBRG reg) is used for the calculated baud rate as follows:

calculated baud rate $= \text{Fosc}/(64(X + 1))$ (for BRGH = 0)

For XTAL = 10 MHz and 9600 baud rate, we got X = 15. Therefore, we get the calculated baud rate of $10 \text{ MHz}/(64(15 + 1)) = 9765$. Now the error is calculated as follows:

Error = (9765 − 9600)/9600 = 1.7%

which is the same as what we got earlier using the other method. Tables 10-10 and 10-11 provide the error rates for SPBRG values of 10 MHz and 4 MHz crystal frequencies, respectively. Compare Examples 10-7 and 10-8 to see how to calculate the error rates two different ways.

Example 10-8

Assuming XTAL = 10 MHz, calculate the baud rate error for the following:
(a) 2400   (b) 1200
Assume BRGH = 0
Solution:

(a) SPBRG Value = (156250/2400) − 1 = 65.1 − 1 = 64.1 = 64
and calculated baud rate is 156250/(64 + 1) = 2403
       Error = (2403 − 2400)/2400 = 0.12%

(b) SPBRG Value (156250/1200) − 1 = 130.2 − 1 = 129.2 = 129
where the calculated baud rate is 156250/(129 + 1) = 1202
       Error = (1202 − 1200)/1200 = 0.16%

Table 10-10: SPBRG Values for Various Baud Rates (XTAL = 10 MHz)

	BRGH = 0		BRGH = 1	
Baud Rate	SPBRG	Error	SPBRG	Error
38400	3	1.5%	15	1.7%
19200	7	1.7%	32	1.3%
9,600	15	1.7%	64	0.15%
4,800	32	1.3%	129	0.15%

SPBRG = (156250/Baud rate) − 1   SPBRG = (625000/Baud rate) − 1

Table 10-11: SPBRG Values for Various Baud Rates (XTAL = 4 MHz)

	BRGH = 0		BRGH = 1	
Baud Rate	SPBRG	Error	SPBRG	Error
19200	2	8.3%	12	0.15%
9,600	6	8%	25	0.15%
4,800	12	0.15%	51	0.15%
2,400	25	0.16%	103	0.16%

SPBRG = (62500/Baud rate) − 1   SPBRG = (250000/Baud rate) − 1

Examine the next few examples to master the topic of PIC18 serial port programming.

Example 10-9

Assume a switch is connected to pin RD7. Write a program to monitor its status and send two messages to the serial port continuously as follows:
SW = 0 send "NO"
SW = 1 send "YES"
Assume XTAL = 10 MHz, and set the baud rate to 9,600.

Solution:

```
        BSF TRISD,7         ;PORTD.7 as in input for SW
        MOVLW 0x20          ;enable transmit and choose low baud rate
        MOVWF TXSTA         ;write to reg
        MOVLW D'15'         ;9600 bps (Fosc / (64 * Speed) - 1)
        MOVWF SPBRG         ;write to reg
        BCF   TRISC, TX     ;make TX pin of PORTC an output pin
        BSF   RCSTA, SPEN   ;enable the entire serial port of PIC18
OVER    BTFSS PORTD,7
        BRA   NEXT
        MOVLW high(MESS1)   ;if SW = 0 display "NO"
        MOVWF TBLPTRH
        MOVLW low(MESS1)
        MOVWF TBLPTRL
FN      TBLRD*+             ;read the character
        MOVF  TABLAT,W
        BZ    EXIT          ;check for end of line
        CALL  SENDCOM       ;send character to serial port
        BRA   FN            ;repeat
NEXT    MOVLW high(MESS2)   ;if SW = 1 display "YES"
        MOVWF TBLPTRH
        MOVLW low(MESS2)
        MOVWF TBLPTRL
LN      TBLRD*+             ;read the character
        MOVF  TABLAT,W      ;Z = 1 if NULL
        BZ    EXIT          ;check for end of line
        CALL  SENDCOM       ;send character to serial port
        BRA   LN            ;repeat
EXIT    MOVLW 0x20          ;send space
        CALL  SENDCOM
        GOTO  OVER
;-------------
SENDCOM
S1      BTFSS PIR1, TXIF    ;wait until the last bit is gone
        BRA   S1            ;stay in loop
        MOVWF TXREG         ;load the value to be transferred
        RETURN              ;return to caller
;-----------------
MESS1 DB    "NO",0
MESS2 DB    "YES",0
```

Example 10-10

Write a program to send the message "The Earth is but One Country" to the serial port continuously. Assume a SW is connected to pin RB0. Monitor its status and set the baud rate as follows:
SW = 0 9600 baud rate
SW = 1 38400 baud rate
Assume XTAL = 10 MHz.

Solution:

As shown in Table 10-8, we can quadruple the baud rate by changing the BRGH bit of the TXSTA register.

```
        BSF   TRISB,0      ;PORTB.0 as in input for SW
        BCF   TRISC, TX    ;make TX pin of PORTC an output pin
        BSF   RCSTA, SPEN  ;enable the entire serial port of PIC18
        MOVLW 0x20         ;transmit at low baud rate
        MOVWF TXSTA        ;write to reg
        MOVLW D'15'        ;9600 bps (Fosc / (64 * Speed) - 1)
        MOVWF SPBRG        ;write to reg
OVER    BTFSC PORTB,0      ;test bit PORTB.0 and skip if LOW
        BSF   TXSTA,BRGH   ;transmit at high rate  by making BRGH = 1
        MOVLW upper(MESSAGE)
        MOVWF TBLPTRU
        MOVLW high(MESSAGE)
        MOVWF TBLPTRH
        MOVLW low(MESSAGE)
        MOVWF TBLPTRL
NEXT    TBLRD*+            ;read the character
        MOVF  TABLAT,W     ;place it in WREG
        BZ    OVER         ;if end of line, start over
        CALL  SENDCOM      ;send char to serial port
        BRA   NEXT         ;repeat for the next character
;-------------
SENDCOM
S1      BTFSS PIR1, TXIF   ;wait until the last bit is gone
        BRA   S1           ;stay in loop
        MOVWF TXREG        ;load the value to be transmitted
        RETURN             ;return to caller
;--------------------
MESSAGE DB  "The Earth is but One Country",0
```

Transmit and receive

Assume that the PIC18 serial port is connected to the COM port of the IBM PC, and we are using the HyperTerminal program on the PC to send and receive data serially. The ports PORTB and PORTD of the PIC18 are connected to LEDs and switches, respectively. Program 10-1 shows a PIC18 program with the following parts: (a) sends the message "YES" once to the PC screen, (b) gets data on switches and transmits it via the serial port to the PC's screen, and (c) receives any key press sent by HyperTerminal and puts it on LEDs. The program performs part (a) once, but parts (b) and (c) continuously. It uses the 9600 baud rate for XTAL = 10 MHz.

```
;Program 10-1 Transmit and Receive
      ORG 0
   ;initialize the serial ports for both transmit and receive
      MOVLW B'00100100'        ;enable transmit, choose high baud
      MOVWF TXSTA              ;write to reg
      MOVLW B'10010000'        ;enable receive, serial port itself
      MOVWF RCSTA              ;write to reg
      MOVLW D'15'              ;9600 bps (Fosc / (64 * Speed) - 1)
      MOVWF SPBRG              ;write to reg
      BSF   RCSTA, SPEN        ;enable the serial port itself
      BCF   TRISC, TX          ;make TX pin of PORTC an output
      BSF   TRISC, RX          ;make RX pin of PORTC an input
      CLRF  TRISB              ;make port B an output port
      SETF  TRISD              ;make port D an input port
   ;send the message "YES"
      MOVLW 'Y'                ;ASCII letter 'Y' to be transferred
      CALL  TRANS
      MOVLW 'E'                ;ASCII letter 'E' to be transferred
      CALL  TRANS
      MOVLW 'S'                ;ASCII letter 'S' to be transferred
      CALL  TRANS
   ;get a byte from switches and transmit data to PC screen
OVER  MOVF PORTD,W             ;get a byte from SW of PORTD
      CALL  TRANS              ;transmit it via serial port
   ;as keys are pressed on PC receive data and put it on LEDs
      CALL  RECV               ;receive the byte from serial port
      MOVWF PORTB              ;display it on LEDS of PORTB
      BRA   OVER               ;keep doing it
;--serial transfer (WREG needs the byte to be transmitted)
TRANS
S1    BTFSS PIR1, TXIF         ;wait until the last bit is gone
      BRA   S1                 ;
      MOVWF TXREG              ;load the value to be transferred
      RETURN                   ;return to caller
;----serial data receive subroutine (WREG = received byte)
```

```
RECV  BTFSS PIR1, RCIF        ;check for ready
      BRA   RECV              ;stay in loop
      MOVF RCREG,W            ;save value in WREG
      RETURN
```

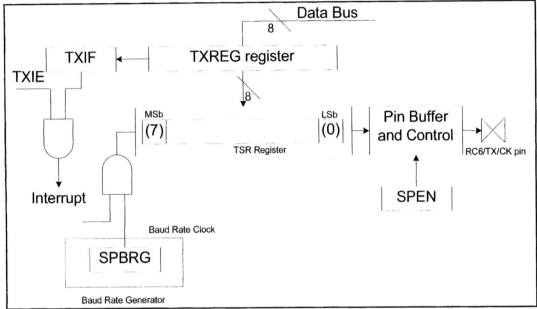

Figure 10-12. Simplified USART Transmit Block Diagram

Interrupt-based data transfer

By now you might have noticed that it is a waste of the microcontroller's time to poll the TXIF and RXIF flags. In order to avoid wasting the microcontroller's time we use interrupts instead of polling. In Chapter 11, we will show how to use interrupts to program the PIC18's serial communication port.

Review Questions

1. Which register of the PIC18 is used to set the baud rate?
2. If XTAL = 10 MHz, what frequency is used by the UART to set the baud rate (assuming default mode)?
3. Which bit of the TXSTA register is used to set the low or high baud rate?
4. With XTAL = 10 MHz, what value should be loaded into SPBRG to have a 9600 baud rate? Give the answer in both decimal and hex.
5. To transmit a byte of data serially, it must be placed in register _____.
6. TXSTA stands for _____ and it is a(n) ____-bit register.
7. Which register is used to set the data frame size?
8. True or false. TXSTA is a bit-addressable register.
9. When is TXIF raised? When is it cleared?
10. Which register has the BRGH bit, and what is its status when the PIC18 is powered up?

SECTION 10.4: PIC18 SERIAL PORT PROGRAMMING IN C

This section shows C programming of the serial ports for the PIC18 chip.

Transmitting and receiving data in PIC18 C

As we saw in Chapter 7, all the special function registers (SFRs) of the PIC18 are accessible directly in C18 compilers by using the appropriate header file. Examples 10-11 through 10-15 show how to program the serial port in PIC18 C. Connect your PIC18 Trainer to the PC's COM port and use HyperTerminal to test the operation of these examples. Notice that Examples 10-11 through 10-15 are C versions of the Assembly programs in the last section.

Example 10-11

Write a C program for the PIC18 to transfer the letter 'G' serially at 9600 baud, continuously. Use 8-bit data and 1 stop bit. Assume XTAL = 10 MHz.

Solution:

```
#include <P18F4580.h>
void main(void)
  {
    TXSTA=0x20;               //choose low baud rate,8-bit
    SPBRG=15;                 //9600 baud rate/ XTAL = 10 MHz
    TXSTAbits.TXEN=1;
    RCSTAbits.SPEN=1;

    while(1)
      {
        TXREG='G';            //place value in buffer
        while(PIR1bits.TXIF==0);    //wait until all gone
      }
  }
```

Review Questions

1. True or false. All the SFR registers of PIC18 are accessible in the C18 C compiler.
2. True or false. C18 compilers support the bit-addressable registers of the PIC18.
3. True or false. The TXIF flag is cleared the moment we write a character to the TXREG register.
4. Which register is used to set the baud rate?
5. To which register does the BRGH bit belong, and what is its role?

414

Example 10-12

Write a PIC18 C program to transfer the message "YES" serially at 9600 baud, 8-bit data, and 1 stop bit. Do this continuously.

Solution:
```c
#include <P18F458.h>
void SerTx(unsigned char);
void main(void)
  {
    TXSTA=0x20;              //choose low baud rate,8-bit
    SPBRG=15;                //9600 baud rate/ XTAL = 10 MHz
    TXSTAbits.TXEN=1;
    RCSTAbits.SPEN=1;
    while(1)
      {
        SerTx('Y');
        SerTx('E');
        SerTx('S');
      }
  }
void SerTx(unsigned char c)
  {
    while(PIR1bits.TXIF==0);   //wait until transmitted
    TXREG=c;                   //place character in buffer
  }
```

Example 10-13

Program the PIC18 in C to receive bytes of data serially and put them on PORTB. Set the baud rate at 9600, 8-bit data, and 1 stop bit.

Solution:
```c
#include <P18F458.h>
void main (void)
  {
    TRISB = 0;                //PORTB an output
    RCSTA=0x90;               //enable serial port and receiver
    SPBRG=15;                 //9600 baud rate/ XTAL = 10 MHz
    while(1)                  //repeat forever
      {
        while(PIR1bits.RCIF==0);  //wait to receive
        PORTB=RCREG;              //save value
      }
  }
```

Example 10-14

Write an C18 program to send two different strings to the serial port. Assuming that SW is connected to pin PORTB.5, monitor its status and make a decision as follows:
SW = 0: send your first name
SW = 1: send your last name
Assume XTAL = 10 MHz, baud rate of 9600, and 8-bit data.

Solution:

```
#include <P18F458.h>
#define MYSW PORTBbits.RB5          //INPUT SWITCH
void main(void)
  {
    unsigned char z;
    unsigned char fname[]="ALI";
    unsigned char lname[]="SMITH";
    TRISBbits.TRISB5 = 1;  //an input
    TXSTA=0x20;                 //choose low baud rate, 8-bit
    SPBRG=15;                   //9600 baud rate/ XTAL = 10 MHz
    TXSTAbits.TXEN=1;
    RCSTAbits.SPEN=1;
    if(MYSW==0)                          //check switch
      {
        for(z=0;z<3;z++)                 //write name
          {
            while(PIR1bits.TXIF==0); //wait for transmit
            TXREG=fname[z];              //place char in buffer
          }
      }
    else
      {
        for(z=0;z<5;z++)                 //write name
          {
            while(PIR1bits.TXIF==0); //wait for transmit
            TXREG=lname[z];              //place value in buffer
          }
      }
    while(1);
  }
```

Example 10-15

Write a PIC18 C program to send the two messages "Normal Speed" and "High Speed" to the serial port. Assuming that SW is connected to pin PORTB.0, monitor its status and set the baud rate as follows:

SW = 0 9600 baud rate

SW = 1 38400 baud rate

Assume that XTAL = 10 MHz for both cases.

Solution:

```c
#include <P18F458.h>
#define MYSW PORTBbits.RB5          //INPUT SWITCH
void main(void)
  {
    unsigned char z;
    unsigned char Mess1[]="Normal Speed";
    unsigned char Mess2[]="High Speed";
    TRISBbits.TRISB5 = 1;  //an input
    TXSTA=0x20;                     //choose low baud rate, 8-bit
    SPBRG=15;                       //9600 baud rate/ XTAL = 10 MHz
    TXSTAbits.TXEN=1;
    RCSTAbits.SPEN=1;
    if(MYSW==0)
      {
        for(z=0;z<12;z++)
          {
            while(PIR1bits.TXIF==0); //wait for transmit
            TXREG=Mess1[z];          //place value in buffer
          }
      }
    else
      {
        TXSTA=TXSTA|0x4;                //for high speed
        for(z=0;z<10;z++)
          {
            while(PIR1bits.TXIF==0); //wait for transmit
            TXREG=Mess2[z];             //place value in buffer
          }
      }
    while(1);
  }
```

SUMMARY

This chapter began with an introduction to the fundamentals of serial communication. Serial communication, in which data is sent one bit a time, is used in situations where data is sent over significant distances because in parallel communication, where data is sent a byte or more a time, great distances can cause distortion of the data. Serial communication has the additional advantage of allowing transmission over phone lines. Serial communication uses two methods: synchronous and asynchronous. In synchronous communication, data is sent in blocks of bytes; in asynchronous, data is sent one byte at a time. Data communication can be simplex (can send but cannot receive), half duplex (can send and receive, but not at the same time), or full duplex (can send and receive at the same time). RS232 is a standard for serial communication connectors.

The PIC18's UART was discussed. We showed how to interface the PIC18 with an RS232 connector and change the baud rate of the PIC18. In addition, we described the serial communication features of the PIC18, and programmed the PIC18 for serial data communication. We also showed how to program the serial port of the PIC18 chip in Assembly and C.

PROBLEMS

SECTION 10.1: BASICS OF SERIAL COMMUNICATION

1. Which is more expensive, parallel or serial data transfer?
2. True or false. 0- and 5-V digital pulses can be transferred on the telephone without being converted (modulated).
3. Show the framing of the letter ASCII 'Z' (0101 1010), no parity, 1 stop bit.
4. If there is no data transfer and the line is high, it is called _____ (mark, space).
5. True or false. The stop bit can be 1, 2, or none at all.
6. Calculate the overhead percentage if the data size is 7, 1 stop bit, no parity.
7. True or false. RS232 voltage specification is TTL compatible.
8. What is the function of the MAX 232 chip?
9. True or false. DB-25 and DB-9 are pin compatible for the first 9 pins.
10. How many pins of the RS232 are used by the IBM serial cable, and why?
11. True or false. The longer the cable, the higher the data transfer baud rate.
12. State the absolute minimum number of signals needed to transfer data between two PCs connected serially. What are those?
13. If two PCs are connected through the RS232 without the modem, both are configured as a _____ (DTE, DCE) -to- _____ (DTE, DCE) connection.
14. State the nine most important signals of the RS232.
15. Calculate the total number of bits transferred if 200 pages of ASCII data are sent using asynchronous serial data transfer. Assume a data size of 8 bits, 1 stop bit, and no parity. Assume each page has 80x25 of text characters.
16. In Problem 15, how long will the data transfer take if the baud rate is 9,600?

17. The MAX232 DIP package has _____ pins.
18. For the MAX232, indicate the V_{CC} and GND pins.
19. The MAX233 DIP package has _____ pins.
20. For the MAX233, indicate the V_{CC} and GND pins.
21. Is the MAX232 pin compatible with the MAX233?
22. State the advantages and disadvantages of the MAX232 and MAX233.
23. MAX232/233 has _____ line driver(s) for the RX wire.
24. MAX232/233 has _____ line driver(s) for the TX wire.
25. Show the connection of pins TX and RX of the PIC18 to a DB-9 RS232 connector via the second set of line drivers of MAX232.
26. Show the connection of the TX and RX pins of the PIC18 to a DB-9 RS232 connector via the second set of line drivers of MAX233.
27. What is the advantage of the MAX233 over the MAX232 chip?
28. Which pins of the PIC18 are set aside for serial communication, and what are their functions?

SECTION 10.3: PIC18 SERIAL PORT PROGRAMMING IN ASSEMBLY

29. Which of the following baud rates are supported by the HyperTerminal program in PC?
    (a) 4,800   (b) 3,600   (c) 9,600
    (d) 1,800   (e) 1,200   (f) 19,200
30. Which timer of the PIC18 is used for baud rate programming?
31. Which bit of the TXSTA is used for baud rate speed?
32. What is the role of the TXREG register in serial data transfer?
33. TXREG is a(n) _____-bit register.
34. What is the role of the TXSTA register in serial data transfer?
35. TXSTA is a(n) _____-bit register.
36. For XTAL = 10 MHz, find the SPBRG value (in both decimal and hex) for each of the following baud rates.
    (a) 9,600   (b) 4,800   (c) 1,200
37. What is the baud rate if we use SPBRG = 15 to program the baud rate? Assume XTAL = 10 MHz.
38. Write a PIC18 program to transfer serially the letter 'Z' continuously at 1,200 baud rate. Assume XTAL = 10 MHz.
39. Write a PIC18 program to transfer serially the message "The earth is but one country and mankind its citizens" continuously at 57,600 baud rate. Assume XTAL = 10 MHz.
40. When is the TXIF flag bit raised or cleared?
41. When is the RCIF flag bit raised or cleared?
42. To which register do RCIF and TXIF belong? Is that register bit-addressable?
43. What is the role of the SPEN bit in the RCSTA register?
44. In a given situation we cannot accept reception of any serial data. How do you block such a reception with a single instruction?

45. To which register does the BRGH bit belong? State its role in rate of data transfer.
46. Is the BRGH bit HIGH or LOW when the PIC18 is powered up?
47. Find the SPBRG for the following baud rates if XTAL = 16 MHz and BRGH = 0.
    (a) 9600   (b) 19200
    (c) 38400   (d) 57600
48. Find the SPBRG for the following baud rates if XTAL = 16 MHz and BRGH = 1.
    (a) 9600   (b) 19200
    (c) 38400   (d) 57600
49. Find the SPBRG for the following baud rates if XTAL = 20 MHz and BRGH = 0.
    (a) 9600   (b) 19200
    (c) 38400   (d) 57600
50. Find the SPBRG for the following baud rates if XTAL = 20 MHz and BRGH = 1.
    (a) 9600   (b) 19200
    (c) 38400   (d) 57600
51. Find the baud rate error for Problem 47.
52. Find the baud rate error for Problem 48.

SECTION 10.4: PIC18 SERIAL PORT PROGRAMMING IN C

53. Write an PIC18 C program to transfer serially the letter 'Z' continuously at 1,200 baud rate.
54. Write an PIC18 C program to transfer serially the message "The earth is but one country and mankind its citizens" continuously at 57,600 baud rate.

ANSWERS TO REVIEW QUESTIONS

SECTION 10.1: BASICS OF SERIAL COMMUNICATION

1. Faster, more expensive
2. False; it is simplex.
3. True
4. Asynchronous
5. With 0100 0101 binary the bits are transmitted in the sequence:
   (a) 0 (start bit) (b) 1 (c) 0 (d) 1 (e) 0 (f) 0 (g) 0 (h) 1 (i) 0 (j) 1 (stop bit)
6. 2 bits (one for the start bit and one for the stop bit). Therefore, for each 8-bit character, a total of 10 bits is transferred.
7. $10000 \times 10 = 100000$ total bits transmitted. $100000 / 9600 = 10.4$ seconds; $2 / 10 = 20\%$.
8. True
9. +3 to +25 V
10. True
11. One
12. COM 1, COM 2

SECTION 10.2: PIC18 CONNECTION TO RS232

1. True
2. Pins RC6 and RC7. Pin RC6 is for TX and pin RC7 for RX.
3. They are used for converting from RS232 voltage levels to TTL voltage levels and vice versa.
4. Two, two
5. It does not need the four capacitors that MAX232 must have.

SECTION 10.3: PIC18 SERIAL PORT PROGRAMMING IN ASSEMBLY

1. SPBRG
2. 156,250 Hz
3. BRGH
4. 15 in decimal (or F in hex) because 156,250/9600 – 1 = 15
5. TXREG
6. Transmit Status and Control Register, 8
7. TXSTA
8. True
9. It is raised during transfer of the stop bit. It is cleared when we write a byte to TXREG to be transmitted.
10. TXSTA; it is low upon power-on reset.

SECTION 10.4: PIC18 SERIAL PORT PROGRAMMING IN C

1. True
2. True
3. True
4. SPBRG
5. TXSTA. It allows us to quadruple the baud rate with the same crystal frequency.

CHAPTER 11

INTERRUPT PROGRAMMING IN ASSEMBLY AND C

OBJECTIVES

Upon completion of this chapter, you will be able to:

>>   Contrast and compare interrupts versus polling
>>   Explain the purpose of the ISR (interrupt service routine)
>>   List all the major interrupts of the PIC18
>>   Explain the purpose of the interrupt vector table
>>   Enable or disable PIC18 interrupts
>>   Program the PIC18 timers using interrupts
>>   Describe the external hardware interrupts of the PIC18
>>   Program the PIC18 for interrupt-based serial communication
>>   Define the interrupt priority of the PIC18
>>   Program PIC interrupts in C

In this chapter we explore the concept of the interrupt and interrupt programming. In Section 11.1, the basics of PIC18 interrupts are discussed. In Section 11.2, interrupts belonging to timers are discussed. External hardware interrupts are discussed in Section 11.3, while the interrupt related to serial communication is presented in Section 11.4. In Section 11.5, we cover the interrupt associated with PORTB. In Section 11.6, we cover interrupt priority. Throughout this chapter, we provide examples in both Assembly and C.

SECTION 11.1: PIC18 INTERRUPTS

In this section, first we examine the difference between polling and interrupts and then describe the various interrupts of the PIC18.

Interrupts vs. polling

A single microcontroller can serve several devices. There are two methods by which devices receive service from the microcontroller: interrupts or polling. In the *interrupt* method, whenever any device needs the microcontroller's service, the device notifies it by sending an interrupt signal. Upon receiving an interrupt signal, the microcontroller stops whatever it is doing and serves the device. The program associated with the interrupt is called the *interrupt service routine* (ISR) or *interrupt handler*. In *polling*, the microcontroller continuously monitors the status of a given device; when the status condition is met, it performs the service. After that, it moves on to monitor the next device until each one is serviced. Although polling can monitor the status of several devices and serve each of them as certain conditions are met, it is not an efficient use of the microcontroller. The advantage of interrupts is that the microcontroller can serve many devices (not all at the same time, of course); each device can get the attention of the microcontroller based on the priority assigned to it. The polling method cannot assign priority because it checks all devices in a round-robin fashion. More importantly, in the interrupt method the microcontroller can also ignore (mask) a device request for service. This also is not possible with the polling method. The most important reason that the interrupt method is preferable is that the polling method wastes much of the microcontroller's time by polling devices that do not need service. So interrupts are used to avoid tying down the microcontroller. For example, in discussing timers in Chapter 9 we used the bit test instruction "BTFSS TMR0IF" and waited until the timer rolled over, and while we were waiting we could not do anything else. That is a waste of microcontroller time that could have been used to perform some useful tasks. In the case of the timer, if we use the interrupt method, the microcontroller can go about doing other tasks, and when the TMR0IF flag is raised, the timer will interrupt the microcontroller in whatever it is doing.

Interrupt service routine

For every interrupt, there must be an interrupt service routine (ISR), or interrupt handler. When an interrupt is invoked, the microcontroller runs the interrupt service routine. Generally, in most microprocessors, for every interrupt there is a fixed location in memory that holds the address of its ISR. The group of memory locations set aside to hold the addresses of ISRs is called the *interrupt vector table.* In the case of the PIC18, there are only two locations for the interrupt vector table, locations 0008 and 0018, as shown in Table 11-1. We will discuss the difference between these two in Section 11.6 when we cover interrupt priority.

Table 11-1: Interrupt Vector Table for the PIC18

Interrupt	ROM Location (Hex)
Power-on Reset	0000
High Priority Interrupt	0008 (Default upon power-on reset)
Low Priority Interrupt	0018 (See Section 11.6)

Steps in executing an interrupt

Upon activation of an interrupt, the microcontroller goes through the following steps:

1. It finishes the instruction it is executing and saves the address of the next instruction (program counter) on the stack.
2. It jumps to a fixed location in memory called the interrupt vector table. The interrupt vector table directs the microcontroller to the address of the interrupt service routine (ISR).
3. The microcontroller gets the address of the ISR from the interrupt vector table and jumps to it. It starts to execute the interrupt service subroutine until it reaches the last instruction of the subroutine, which is RETFIE (return from interrupt exit).
4. Upon executing the RETFIE instruction, the microcontroller returns to the place where it was interrupted. First, it gets the program counter (PC) address from the stack by popping the top bytes of the stack into the PC. Then it starts to execute from that address.

Notice from Step 4 the critical role of the stack. For this reason, we must be careful in manipulating the stack contents in the ISR. Specifically, in the ISR, just as in any CALL subroutine, the number of pushes and pops must be equal.

Sources of interrupts in the PIC18

There are many sources of interrupts in the PIC18, depending on which peripheral is incorporated into the chip. The following are some of the most widely used sources of interrupts in the PIC18:
1. There is an interrupt set aside for each of the timers, Timers 0, 1, 2, and so on. See Section 11.2.
2. Three interrupts are set aside for external hardware interrupts. Pins RB0 (PORTB.0), RB1 (PORTB.1), and RB2 (PORTB.2) are for the external hardware interrupts INT0, INT1, and INT2, respectively. See Section 11.3.
3. Serial communication's USART has two interrupts, one for receive and another for transmit. See Section 11.4.
4. The PORTB-Change interrupt. See Section 11.5.
5. The ADC (analog-to-digital converter). See Chapter 13.
6. The CCP (compare capture pulse-width-modulation). See Chapters 15 and 17.

The PIC18 has many more interrupts than the above list shows. We will cover them throughout the book as we study the peripherals of the PIC18. Notice in Table 11-1 that a limited number of bytes is set aside for high-priority interrupts. For example, a total of 8 bytes, from location 0008 to 000017H, are set aside for high-priority interrupts. Normally, the service routine for an interrupt is too long to fit in the memory space allocated. For that reason, a GOTO instruction is placed in the vector table to point to the address of the ISR. The rest of the bytes allocated to the interrupt are unused. In upcoming sections of this chapter, we will see many examples of interrupt programming that clarify these concepts.

From Table 11-1, also notice that only 8 bytes of ROM space are assigned to the reset pin. They are ROM address locations 0–7. For this reason, in our program we put the GOTO as the first instruction and redirect the processor away from the interrupt vector table, as shown in Figure 11-1. In the next section we will see how this works in the context of some examples.

```
            ORG   0     ;wake-up ROM reset location
            GOTO MAIN  ;bypass interrupt vector table

;---- the wake-up program
            ORG   100H
MAIN:       .... ;enable interrupt flags
            ....
            END
```

Figure 11-1. Redirecting the PIC18 from the Interrupt Vector Table at Power-Up

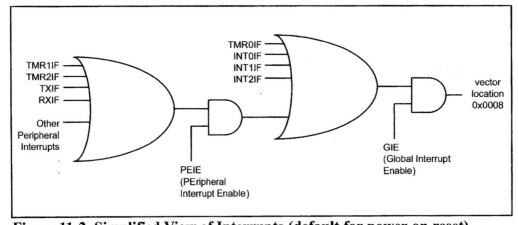

Figure 11-2. Simplified View of Interrupts (default for power-on reset)

Enabling and disabling an interrupt

Upon reset, all interrupts are disabled (masked), meaning that none will be responded to by the microcontroller if they are activated. The interrupts must be enabled (unmasked) by software in order for the microcontroller to respond to them. The D7 bit of the INTCON (Interrupt Control) register is responsible for enabling and disabling the interrupts globally. Figure 11-3 shows the INTCON register. The GIE bit makes the job of disabling all the interrupts easy. With a single instruction (BCF INTCON,GIE), we can make GIE = 0 during the operation of a critical task. See Figure 11-2.

Steps in enabling an interrupt

To enable any one of the interrupts, we take the following steps:

1. Bit D7 (GIE) of the INTCON register must be set to HIGH to allow the interrupts to happen. This is done with the "BSF INTCON, GIE" instruction.
2. If GIE = 1, each interrupt is enabled by setting to HIGH the interrupt enable (IE) flag bit for that interrupt. Because there are a large number of interrupts in the PIC18, we have many registers holding the interrupt enable bit. Figure 11-2 shows that the INTCON has interrupt enable bits for Timer0 (TMR0IE) and external interrupt 0 (INT0IE). As we study each of peripherals throughout the book we will examine the registers holding the interrupt enable bits. It must be noted that if GIE = 0, no interrupt will be responded to, even if the corresponding interrupt enable bit is high. To understand this important point look at Example 11-1.
3. As shown in Figures 11-2 and 11-3, for some of the peripheral interrupts such as TMR1IF, TMR2IF, and TXIF, we have to enable the PEIE flag in addition to the GIE bit.

D7   D0

| GIE | | TMR0IE | INT0IE | | | | |

GIE (Global Interrupt Enable)
GIE = 0   Disables all interrupts. If GIE = 0, no interrupt is acknowledged, even if they are enabled individually.
If GIE = 1, interrupts are allowed to happen. Each interrupt source is enabled by setting the corresponding interrupt enable bit.

TMR0IE   Timer0 interrupt enable
                    = 0 Disables Timer0 overflow interrupt
                    = 1 Enables Timer0 overflow interrupt
INT0IE   Enables or disables external interrupt 0
                    = 0 Disables external interrupt 0
                    = 1 Enables external interrupt 0
These bits, along with the GIE, must be set high for an interrupt to be responded to. Upon activation of the interrupt, the GIE bit is cleared by the PIC18 itself to make sure another interrupt cannot interrupt the microcontroller while it is servicing the current one. At the end of the ISR, the RETFIE instruction will make GIE = 1 to allow another interrupt to come in.
PEIE (PEripheral Interrupt Enable)
For many of the peripherals, such as Timers 1, 2, .. and the serial port, we must enable this bit in addition to the GIE bit. (See Figure 11-2.)

Figure 11-3. INTCON (Interrupt Control) Register

Example 11-1

Show the instructions to (a) enable (unmask) the Timer0 interrupt and external hardware interrupt 0 (INT0), and (b) disable (mask) the Timer0 interrupt, then (c) show how to disable (mask) all the interrupts with a single instruction.

Solution:

(a)

```
BSF  INTCON,TMR0IE   ;enable(unmask) Timer0 interrupt
BSF  INTCON,INT0IE   ;enable external interrupt 1(INT0)
BSF  INTCON,GIE ;allow interrupts to come in
```

We can perform the above actions with the following two instructions:

```
MOVLW B'10110000'    ;GIE = 1, TMR0IF = 1,INTIF0 = 1
MOVWF INTCON         ;load the INTCON reg
```

(b)

```
BCF  INTCON,TMR0IE ;mask (disable) Timer0 interrupt
```

(c)

```
BCF  INTCON,GIE      ;mask all interrupts globally
```

Review Questions

1. Of the interrupt and polling methods, which one avoids tying down the micro-controller?
2. Give the name of the interrupts in the INTCON register.
3. Upon power-on reset of the PIC18, what memory area is assigned to the interrupt vector table? Can the programmer change the memory space assigned to the table?
4. What is the content of D7 (GIE) of the INTCON register upon reset, and what does it mean?
5. Show the instruction needed to enable the TMR0 interrupt.
6. What address in the interrupt vector table is assigned to high-priority and low-priority interrupts?

SECTION 11.2: PROGRAMMING TIMER INTERRUPTS

In Chapter 9 we discussed how to use Timers 0, 1, 2, and 3 with the polling method. In this section we use interrupts to program the PIC18 timers. Please review Chapter 9 before you study this section.

Rollover timer flag and interrupt

In Chapter 9 we stated that the timer flag is raised when the timer rolls over. In that chapter, we also showed how to monitor the timer flag with the instruction "BTFSS TMR0IF". In polling TMR0IF, we have to wait until TMR0IF is raised. The problem with this method is that the microcontroller is tied down waiting for TMR0IF to be raised, and cannot do anything else. Using interrupts avoids tying down the controller. If the timer interrupt in the interrupt register is enabled, TMR0IF is raised whenever the timer rolls over and the microcontroller jumps to the interrupt vector table to service the ISR. In this way, the microcontroller can do other things until it is notified that the timer has rolled over. To use an interrupt in place of polling, first we must enable the interrupt because all the interrupts are masked upon power-on reset. The TMRxIE bit enables the interrupt for a given timer. TMRxIE bits are held by various registers as shown in Table 11-2. In the case of Timer0, the INTCON register (Figure 11-4) contains the TMR0IE bit, and PIE1 (peripheral interrupt enable) holds the TMR1IE bit. See Figure 11-5 and Program 11-1.

Table 11-2: Timer Interrupt Flag Bits and Associated Registers

Interrupt	Flag Bit	Register		Enable Bit	Register
Timer0	TMR0IF	INTCON		TMR0IE	INTCON
Timer1	TMR1IF	PIR1		TMR1IE	PIE1
Timer2	TMR2IF	PIR1		TMR2IE	PIE1
Timer3	TMR3IF	PIR3		TMR3IE	PIE2

		TMR0IE		TMR0IF		

Figure 11-4. INTCON Register with Timer0 Interrupt Enable and Interrupt Flag

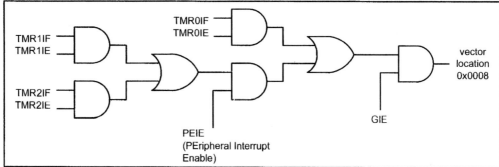

Figure 11-5. The Role of Timer Interrupt Enable Flag (TMRxIE)
Note: The TMRxIP (timer interrupt priority) flag is not shown. TMRxIP is used to force the interrupt to land at vector location 0x0018. See Section 11.6.

Notice the following points about Program 11-1:

1. We must avoid using the memory space allocated to the interrupt vector table. Therefore, we place all the initialization codes in memory starting at an address such as 100H. The GOTO instruction is the first instruction that the PIC18 executes when it is awakened at address 00000 upon power-on reset (POR). The GOTO instruction at address 00000 redirects the controller away from the interrupt vector table.
2. In the MAIN program, we enable (unmask) the Timer0 interrupt with instruction "BSF INTCON, TMR0IE" followed by the instruction "BSF INT-CON, GIE" to enable all interrupts globally.
3. In the MAIN program, we initialize the Timer0 register and then enter an infinite loop to keep the CPU busy. This could be a real-world application being executed by the CPU. In this case, the loop gets data from PORTC and sends it to PORTD. While the PORTC data is brought in and issued to PORTD continuously, the TMR0IF flag is raised as soon as Timer0 rolls over, and the microcontroller gets out of the loop and goes to 00008H to execute the ISR associated with Timer0. At this point, the PIC18 clears the GIE bit (D7 of INTCON) to indicate that it is currently serving an interrupt and cannot be interrupted again; in other words, no interrupt inside the interrupt. In Section 11.6, we show how to allow an interrupt inside an interrupt.
4. The ISR for Timer0 is located starting at memory location 00200H because it is too large to fit into address space 08–17H, the address allocated to high-priority interrupts.
5. In the ISR for Timer0, notice that the "BCF INTCON, TMR0IF" instruction is needed before the RETFIE instruction. This will ensure that a single interrupt is serviced once and is not recognized as multiple interrupts.
6. RETFIE must be the last instruction of the ISR. Upon execution of the RET-FIE instruction, the PIC18 automatically enables the GIE (D7 of the INTCON register) to indicate that it can accept new interrupts.

Program 11-1: For this program, we assume that PORTC is connected to 8 switches and PORTD to 8 LEDs. This program uses Timer0 to generate a square wave on pin PORTB.5, while at the same time data is being transferred from PORTC to PORTD.

```
;Program 11-1
        ORG   0000H
        GOTO MAIN         ;bypass interrupt vector table
    ;--on default all interrupts land at address 00008
        ORG   0008H       ;interrupt vector table
        BTFSS INTCON,TMR0IF   ;Timer0 interrupt?
        RETFIE                ;No. Then return to main
        GOTO  T0_ISR          ;Yes. Then go Timer0 ISR
    ;--main program for initialization and keeping CPU busy
        ORG   00100H      ;after vector table space
    MAIN BCF   TRISB,5     ;PB5 as an output
        CLRF TRISD        ;make PORTD output
        SETF TRISC        ;make PORTC input
```

```
        MOVLW  0x08         ;Timer0,16-bit,
                            ;no prescale,internal clk
        MOVWF  T0CON        ;load T0CON reg
        MOVLW  0xFF         ;TMR0H = FFH, the high byte
        MOVWF  TMR0H        ;load Timer0 high byte
        MOVLW  0xF2         ;TMR0L = F2H, the low byte
        MOVWF  TMR0L        ;load Timer0 low byte
        BCF INTCON,TMR0IF;clear timer interrupt flag bit
        BSF T0CON,TMR0ON        ;start Timer0
        BSF INTCON,TMR0IE      ;enable Timer 0 interrupt
        BSF INTCON,GIE   ;enable interrupts globally
;--keeping CPU busy waiting for interrupt
OVER  MOVFF PORTC,PORTD  ;send data from PORTC to PORTD
        BRA OVER                ;stay in this loop forever
;-------------------------ISR for Timer 0
T0_ISR
        ORG 200H
        MOVLW  0xFF         ;TMR0H = FFH, the high byte
        MOVWF  TMR0H        ;load Timer0 high byte
        MOVLW  0xF2         ;TMR0L = F2H, the low byte
        MOVWF  TMR0L        ;load Timer0 low byte
        BTG   PORTB,5       ;toggle RB5
        BCF INTCON,TMR0IF   ;clear timer interrupt flag bit
EXIT  RETFIE ;return from interrupt (See Example 11-2)
        END
```

Example 11-2

What is the difference between the RETURN and RETFIE instructions? Explain why we cannot use RETURN instead of RETFIE as the last instruction of an ISR.

Solution:

Both perform the same actions of popping off the top bytes of the stack into the program counter, and making the PIC18 return to where it left off. However, RETFIE also performs the additional task of clearing the GIE flag, indicating that the servicing of the interrupt is over and the PIC18 now can accept a new interrupt. If you use RETURN instead of RETFIE as the last instruction of the interrupt service routine, you simply block any new interrupt after the first interrupt, because the GIE would indicate that the interrupt is still being serviced.

In Program 11-1, the Timer0 ISR (interrupt service routine) was too long to be placed in memory locations allocated to the high interrupt (addresses of 0008–00017H). There was enough space, however, to test to see which interrupt was the cause of landing at the 0008 address. Very often, we go from address 0008 to another address with a larger space to check the source of the interrupt, given the fact that the PIC18 has so many interrupts and we have limited space at address 0008. See Program 11-2.

Program 11-2 uses Timer0 and Timer1 interrupts to generate square waves on pins RB1 and RB7 respectively, while data is being transferred from PORTC to PORTD.

```
;Program 11-2
          ORG    0000H
          GOTO   MAIN          ;bypass interrupt vector table
;--on default all interrupts land at address 00008
          ORG    0008H         ;interrupt vector table
          GOTO CHK_INT         ;go to an address with more space
;--check to see the source of interrupt
          ORG    0040H         ;we got here from 0008
CHK_INT
          BTFSC INTCON,TMR0IF   ;Is it Timer0 interrupt?
          BRA   T0_ISR         ;Yes. Then branch to T0_ISR
          BTFSC PIR1,TMR1IF     ;Is it Timer1 interrupt?
          BRA   T1_ISR         ;Yes. Then branch to T1_ISR
          RETFIE               ;No. Then return to main
;--main program for initialization and keeping CPU busy
          ORG    0100H  ;somewhere after vector table space
MAIN  BCF   TRISB,1    ;PB1 as an output
      BCF   TRISB,7    ;PB7 as an output
      CLRF TRISD       ;make PORTD output
      SETF TRISC       ;make PORTC input
      MOVLW 0x08       ;Timer0,16-bit,
                       ;no prescale,internal clk
      MOVWF T0CON      ;load T0CON reg
      MOVLW 0xFF       ;TMR0H = FFH, the high byte
      MOVWF TMR0H      ;load Timer0 high byte
      MOVLW 0xF2       ;TMR0L = F2H, the low byte
      MOVWF TMR0L      ;load Timer0 low byte
      BCF INTCON,TMR0IF;clear Timer0 interrupt flag bit
      MOVLW 0x0        ;Timer1,16-bit,
                       ;no prescale,internal clk
      MOVWF T1CON      ;load T1CON reg
      MOVLW 0xFF       ;TMR1H = FFH, the high byte
      MOVWF TMR1H      ;load Timer0 high byte
      MOVLW 0xF2       ;TMR1L = F2H, the low byte
      MOVWF TMR1L      ;load Timer1 low byte
      BCF PIR1,TMR1IF  ;clear Timer1 interrupt flag bit
      BSF INTCON,TMR0IE    ;enable Timer0 interrupt
      BSF PIE1,TMR1IE      ;enable Timer1 interrupt
      BSF INTCON,PEIE  ;enable peripheral interrupts
      BSF INTCON,GIE       ;enable interrupts globally
      BSF T0CON,TMR0ON     ;start Timer0
      BSF T1CON,TMR1ON     ;start Timer1
;--keeping CPU busy waiting for interrupt
OVER  MOVFF PORTC,PORTD ;send data from PORTC to PORTD
      BRA OVER               ;stay in this loop forever
;------------------------ISR for Timer 0
T0_ISR
      ORG 200H
```

```
        MOVLW  0xFF        ;TMR0H = FFH, the high byte
        MOVWF  TMR0H       ;load Timer0 high byte
        MOVLW 0xF2         ;TMR0L = F2H, the low byte
        MOVWF  TMR0L       ;load Timer0 low byte
        BTG    PORTB,1     ;toggle PB1
        BCF INTCON,TMR0IF  ;clear timer interrupt flag bit
        GOTO CHK_INT
;--------------------------ISR for Timer1
T1_ISR
        ORG 300H
        MOVLW  0xFF        ;TMR1H = FFH, the high byte
        MOVWF  TMR1H       ;load Timer0 high byte
        MOVLW 0xF2         ;TMR1L = F2H, the low byte
        MOVWF  TMR1L       ;load Timer1 low byte
        BTG PORTB,7
        BCF PIR1,TMR1IF    ;clear Timer1 interrupt flag bit
        GOTO CHK_INT
        END
```

Notice that the addresses 0040H, 0100H, 00200H, and 0300H that we used in Program 11-2 are all arbitrary and can be changed to any addresses we want. The only addresses that we have no choice on are the power-on reset location of 0000 and high-priority address of 0008 because they were fixed at the time of the PIC18 design.

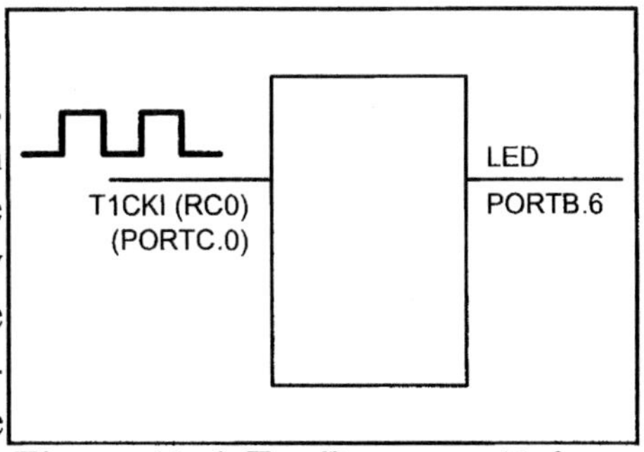

Figure 11-6. For Program 11-3

Program 11-3 has two interrupts: (1) PORTD counts up everytime Timer0 over-flows. It uses 16-bit mode of Timer0 with the largest prescale possible; (2) 1-Hz pulse is fed into Timer1 where Timer1 is used as counter and counts up. Whenever the count reaches 200, it will toggle the pin PORTB.6.

```
;Program 11-3
        ORG    0000H
        GOTO   MAIN          ;bypass interrupt vector table
;---on default all interrupts land at address 00008
        ORG    0008H
        GOTO CHK_INT
;----------------find the interrupt source
        ORG    0040H
CHK_INT
        BTFSC INTCON,TMR0IF   ;Is it Timer0 interrupt?
        BRA    T0_ISR         ;Yes. Then branch to T0_ISR
        BTFSC PIR1,TMR1IF     ;Is it Timer1 interrupt?
        BRA    T1_ISR         ;Yes. Then branch to T1_ISR
        RETFIE                ;No. Then return to main
;---the main program for initialization
        ORG    00100H         ;after vector table space
MAIN BSF   TRISC,T13CKI       ;PORTC.0 as an input
        CLRF TRISD            ;make PORTD output
        BCF TRISB,6           ;make RB6 output
```

```
        MOVLW  0x08         ;16-bit, prescale = 256,
                            ;internal clk
        MOVWF  T0CON        ;load T0CON reg
        MOVLW  0x00         ;TMR0H = 00H, the high byte
        MOVWF  TMR0H        ;load Timer0 high byte
        MOVLW 0x00          ;TMR0L = 0, the low byte
        MOVWF TMR0L         ;load Timer0 low byte
        BCF INTCON,TMR0IF   ;clear timer interrupt flag bit
        MOVLW 0x6           ;Timer1, no prescale,
                            ;ext. clock
        MOVWF  T1CON        ;load T1CON reg
        MOVLW  D'255'       ;TMR1H = 255
        MOVWF  TMR1H        ;load Timer1 high byte
        MOVLW  -D'200'      ;TMR1L = 0
        MOVWF  TMR1L        ;load Timer1 low byte
        BCF PIR1,TMR1IF     ;clear timer interrupt flag bit
        BSF T0CON,TMR0ON        ;start Timer0
        BSF T1CON,TMR1ON        ;start Timer1
        BSF INTCON,TMR0IE       ;enable Timer0 interrupt
        BSF PIE1,TMR1IE         ;enable Timer1 interrupt
        BSF INTCON,PEIE    ;enable peripheral interrupts
        BSF INTCON,GIE     ;enable interrupts globally
OVER    BRA OVER           ;stay in this loop forever
;-------------------------ISR for Timer0
T0_ISR
        ORG 200H
        INCF PORTD         ;increment PORTD
        MOVLW 0x00         ;TMR0L = 0, the low byte
        MOVWF TMR0L        ;load Timer0 low byte
        MOVLW  0x00        ;TMR0H = 00, the high byte
        MOVWF  TMR0H       ;load Timer0 high byte
        BCF INTCON,TMR0IF  ;clear timer interrupt flag bit
        GOTO CHK_INT
;-------------------------ISR for Timer2
T1_ISR
        ORG 300H
        BTG PORTB,6        ;toggle PORTC.6
        MOVLW D'255'       ;TMR1H = 255
        MOVWF TMR1H        ;load Timer1 high byte
        MOVLW -D'200'      ;TMR1L = 0
        MOVWF TMR1L        ;load Timer1 low byte
        BCF PIR1,TMR1IF    ;clear Timer1 interrupt flag bit
        GOTO CHK_INT
        END
```

**Notice in Programs 11-2 and 11-3 that we use the "GOTO CHK_INT" instruction
instead of RETFIE as the last instruction of the ISR. This is because we are checking for activation of multiple interrupts.**

PIC18 interrupt programming in C using C18 compiler

In Chapter 7, we discussed how the C18 compiler uses "#pragma code" to place code at a specific ROM address. Because the C18 does not place an ISR at the interrupt vector table automatically, we must use Assembly language instruction GOTO at the interrupt vector to transfer control to the ISR. This is done as follows:

```
#pragma code high_vector =0x0008  // High-priority interrupt location
void My_HiVect_Int (void)
{
_asm
GOTO my_isr
_endasm
}
#pragma code                        // End of code
```

Now we redirect it from address location 00008 to another program to find the source of the interrupt and finally to the ISR. This is done with the help of the keyword **interrupt** as follows:

```
#pragma interrupt my_isr    //interrupt is reserved keyword
void my_isr (void)          //used for high-priority interrupt
{

//C18 places RETFIE here automatically due to
//interrupt keyword
}
```

Note that "pragma", "code", and "interrupts" are reserved keywords while the choice of all other labels is up to us. Examine Programs 11-2C and 11-3C. They are the C versions of Programs 11-2 and 11-3.

Program 11-2C uses Timer0 and Timer1 interrupts to generate square waves on pins RB1 and RB7, respectively, while data is being transferred from PORTC to PORTD. This is a C version of Program 11-2.

```
//Program 11-2C (C version of Program 11-2)
#include <p18F458.h>
#define myPB1bit PORTBbits.RB1
#define myPB7bit PORTBbits.RB7

void T0_ISR(void);
void T1_ISR(void);

#pragma interrupt chk_isr  //used for high-priority
                           //interrupt only
void chk_isr (void)
{
```

```
        if (INTCONbits.TMR0IF==1)    //Timer0 causes interrupt?
            T0_ISR();                //Yes. Execute Timer0 ISR
        if(PIR1bits.TMR1IF==1)       //Or was it Timer1?
            T1_ISR();                // Yes. Execute Timer1 ISR
}

#pragma code My_HiPrio_Int=0x08//high-priority interrupt
void My_HiPrio_Int (void)
{
   _asm
     GOTO chk_isr
   _endasm
}
#pragma code
void main(void)
   {
     TRISBbits.TRISB1=0;      //RB1 = OUTPUT
     TRISBbits.TRISB7=0;      //RB7 = OUTPUT
     TRISC = 255;            //PORTC = INPUT
     TRISD = 0;            //PORTD = OUTPUT
     T0CON=0x0;            //Timer 0, 16-bit mode, no prescaler
     TMR0H=0x35;           //load TH0
     TMR0L=0x00;           //load TL0
     T1CON=0x88;           //Timer 1, 16-bit mode, no prescaler
     TMR1H=0x35;           //load TH1
     TMR1L=0x00;           //load TL1
     INTCONbits.TMR0IF=0;    //clear TF0
     PIR1bits.TMR1IF=0;      //clear TF1
     INTCONbits.TMR0IE=1;    //enable Timer0 interrupt
     INTCONbits.TMR0IE=1;    //enable Timer1 interrupt
     T0CONbits.TMR0ON=1;     //turn on Timer0
     T1CONbits.TMR1ON=1;     //turn on Timer1
     INTCONbits.PEIE=1;//enable all peripheral interrupts
     INTCONbits.GIE=1; //enable all interrupts globally
   while(1)      //keep looping until interrupt comes
     {
       PORTD=PORTC;    //send data from PORTC to PORTD
     }
   }

void T0_ISR(void)
   {
     myPB1bit=~myPB1bit;     //toggle PORTB.1
     TMR0H=0x35;           //load TH0
     TMR0L=0x00;           //load TL0
     INTCONbits.TMR0IF=0;    //clear TF0
   }

void T1_ISR(void)
   {
     myPB7bit=~myPB7bit;     //toggle PORTB.7
```

```
    TMR1H=0x35;              //load TH0
    TMR1L=0x00;              //load TL0
    PIR1bits.TMR1IF=0;       //clear TF1
}
```

Program 11-3C shows the C version of Program 11-3.

Program 11-3C has two interrupts: (1) PORTC counts up every time Timer0 over-flows. It uses the 16-bit mode of Timer0 with the largest prescale possible; (2) a 1-Hz pulse is fed into Timer1 where Timer1 is used as a counter and counts up. Whenever the count reaches 200, it will toggle pin RB6.

```
//Program 11-3C
#include <p18F458.h>
#define myPB6bit PORTBbits.RB6

void chk_isr(void);
void T0_ISR(void);
void T1_ISR(void);
#pragma interrupt chk_isr //for high-priority interrupt only
void chk_isr (void)
{
if (INTCONbits.TMR0IF==1)   //Timer0 causes interrupt?
T0_ISR( );                  //Yes. Execute Timer0 program
    if(PIR1bits.TMR1IF==1)//Or was it Timer2?
    T1_ISR();               //Yes. Execute Timer2 program
    }

#pragma code My_HiPrio_Int=0x0008 //high-priority interrupt
void My_HiPrio_Int (void)
{
_asm
GOTO chk_isr
_endasm
}
#pragma code

void main(void)
  {
    TRISBbits.TRISB6=0;     //RB6 = OUTPUT
    TRISCbits.TRISC0=1;     //PORTC0 = INPUT
    TRISD=0;
    T0CON=0x08;             //Timer0, 16-bit mode,
                            //no prescaler
    TMR0H=0;                //load Timer0 high byte
    TMR0L=0;                //load Timer0 low byte
    T1CON=0x06;             //Timer 2, no prescaler
    TMR1H=255;              //load Timer1 high byte
    TMR1L=-200;             //load Timer1 low byte
    INTCONbits.TMR0IF=0;    //clear TF0
    PIR1bits.TMR1IF=0;      //clear TF1
```

```
        INTCONbits.TMR0IE=1;   //enable Timer0 interrupt
        PIE1bits.TMR1IE=1;     //enable Timer1 interrupt
        T0CONbits.TMR0ON=1;    //turn on Timer0
        T1CONbits.TMR1ON=1;    //turn on Timer1
        INTCONbits.PEIE=1;//enable all peripheral interrupts
        INTCONbits.GIE=1;      //enable all interrupts globally
        while(1);  //keep looping until interrupt comes
    }

void T0_ISR(void)
    {
        PORTD++;                   //count up PORTD
        TMR0H=0;                   //load Timer0 high byte
        TMR0L=0;                   //load Timer0 low byte
        INTCONbits.TMR0IF=0;       //clear TF0
    }

void T1_ISR(void)
    {
        myPB6bit=~myPB6bit;        //toggle PB.6
        TMR1H=255;                 //load Timer1 high byte
        TMR1L=-200;                //load Timer1 low byte
        PIR1bits.TMR2IF=0;         //clear TF1
    }
```

Review Questions

1. True or false. A unique address in the interrupt vector table is assigned to each of Timer0–Timer3.
2. Upon power-on reset, what address in the interrupt vector table is assigned to the high-priority interrupt?
3. Which register does TMR1IE belong to? Show how it is enabled.
4. Assume that Timer1 is programmed in 8-bit mode, TMR1L = F5H, and the TMR1IF bit is enabled. Explain how the interrupt for the timer works.
5. True or false. The last two instructions of the ISR for Timer0 are:
   ```
   BCF   INTCON,TMR0IF
   RETFIE
   ```

SECTION 11.3: PROGRAMMING EXTERNAL HARDWARE INTERRUPTS

The PIC18 has three external hardware interrupts. Pins RB0 (PORTB.0), RB1 (PORTB.1), and RB2 (PORTB.2), designated as INT0, INT1, and INT2 respectively, are used as external hardware interrupts. Upon activation of these pins, the PIC18 gets interrupted in whatever it is doing and jumps to the vector table to perform the interrupt service routine. In this section we study these three external hardware interrupts of the PIC18 with some examples in both Assembly and C.

External interrupts INT0, INT1, and INT2

There are three external hardware interrupts in the PIC18: INT0, INT1, and INT2. They are located on pins RB0, RB1, and RB2, respectively. See Figures 11-7 and 11-8. On default, all three hardware interrupts are directed to vector table location 0008H, unless we specify otherwise. They must be enabled before they can take effect. This is done using the INTxIE bit. The registers associated with INTxIE bits are shown in Table 11-3. For example, the instruction "BSF INTCON, INT0IE" enables INT0. The INT0 is a *positive-edge-triggered interrupt*, which means, when a low-to-high signal is applied to pin RB0 (PORTB.0), the INT0IF is raised, causing the controller to be interrupted. The raising of INT0IF forces the PIC18 to jump to location 0008H in the vector table to service the ISR. In Table 11-3, notice the INTxIF bits and the registers they belong to. Upon power-on reset, the PIC18 makes INT0, INT1, and INT2 rising (positive) edge-triggered interrupts. To make them falling (negative) edge-triggered interrupts, we must program the INTEDGx bits, as we will see shortly.

Examine Program 11-4 and its C version, Program 11-4C, to gain insight into external hardware interrupts.

Table 11-3: Hardware Interrupt Flag Bits and Associated Registers

Interrupt (Pin)	Flag bit	Register	Enable bit	Register
INT0 (RB0)	INT0IF	INTCON	INT0IE	INTCON
INT1 (RB1)	INT1IF	INTCON3	INT1IE	INTCON3
INT2 (RB2)	INT2IF	INTCON3	INT2IE	INTCON3

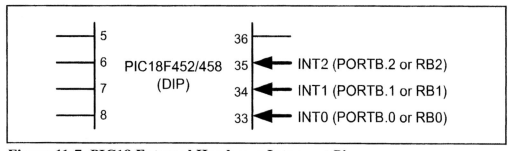

Figure 11-7. PIC18 External Hardware Interrupt Pins

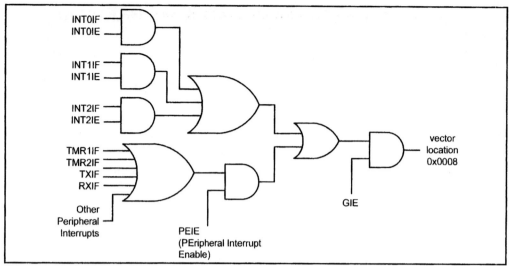

Figure 11-8. INT0–INT2 Hardware Interrupts

Program 11-4 connects a switch to INT0 and an LED to pin RB7. In this program, every time INT0 is activated, it toggles the LED, while at the same time data is being transferred from PORTC to PORTD.

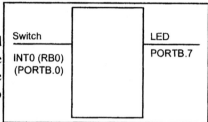

Figure 11-9. For Program 11-4

```
;Program 11-4
      ORG    0000H
      GOTO   MAIN              ;bypass interrupt vector table
;--on default all interrupts go to to address 00008
      ORG    0008H             ;interrupt vector table
      BTFSS INTCON,INT0IF      ;Did we get here due to INT0?
      RETFIE                   ;No. Then return to main
      GOTO  INT0_ISR           ;Yes. Then go INT0 ISR
;;--the main program for initialization
      ORG    00100H
MAIN  BCF    TRISB,7           ;PB7 as an output
      BSF TRISB,INT0           ;make INT0 an input pin
      CLRF TRISD               ;make PORTD output
      SETF TRISC               ;make PORTC input
      BSF INTCON,INT0IE        ;enable INT0 interrupt
      BSF INTCON,GIE           ;enable interrupts globally
OVER  MOVFF PORTC,PORTD        ;send data from PORTC to PORTD
      BRA OVER                 ;stay in this loop forever
;-------------------------ISR for INT0
INT0_ISR
      ORG 200H
      BTG PORTB,7              ;toggle PB7
      BCF INTCON,INT0IF        ;clear INT0 interrupt flag bit
      RETFIE                   ;return from ISR
      END
```

Look at Program 11-4. When a rising edge of the signal is applied to pin INT0, the LED will toggle. In this example, to toggle the LED again, the INT0 pulse must be brought back LOW and then forced HIGH to create a rising edge to activate the interrupt.

```c
//Program 11-4C (This is the C version of Program 11-4)
    #include <p18F4580.h>
    #define mybit PORTBbits.RB7
    void chk_isr(void);
    void INT0_ISR(void);
    #pragma interrupt chk_isr//used for high-priority int
    void chk_isr (void)
    {
    if (INTCONbits.INT0IF==1)  //INT0 caused interrupt?
    INT0_ISR( );                //Yes. Execute INT0 program
    }
    #pragma code My_HiPrio_Int=0x08 //high-priority
                                //interrupt location
    void My_HiPrio_Int (void)
    {
    _asm
    GOTO chk_isr
    _endasm
    }
    #pragma code
    void main(void)
      {
        TRISBbits.TRISB7=0;   //RB7 = OUTPUT
        TRISBbits.TRISB0=1;   //INT0 = INPUT
        TRISC = 0xFF;         //PORTC = INPUT
        TRISD = 0;            //PORTD = OUTPUT
        INTCONbits.INT0IF=0;  //clear TF1
        INTCONbits.INT0IE=1;  //enable Timer0 interrupt
        INTCONbits.GIE=1;     //enable all interrupts
      while(1)      //keep looping until interrupt comes
        {
        PORTD=PORTC;
        }
      }
    void INT0_ISR(void)
      {
        mybit=~mybit;
        INTCONbits.INT0IF=0;            //clear INT0 flag
      }
```

Negative edge-triggered interrupts

Upon power-on reset, the PIC18 makes INT0, INT1, and INT2 positive (rising) edge-triggered interrupts. To make any of them a negative (falling) edge-triggered interrupt, we must program the corresponding bit called INTEDGx, where x can be 0, 1, or 2. The INTCON2 register holds, among other bits, the INTEDG0, INTEDG1, and INTEDG flag bits as shown in Figure 11-10. INTEDG0, INTEDG1, and INTEDG2 are bits D4, D5, and D6 of the INTCON2 register, respectively, as shown in Figure 11-10. The status of these bits determines the negative or positive edge-triggered mode of the hardware interrupts. Upon reset, INTEDGx bits are all 1s, meaning that the external hardware interrupts are positive edge-triggered. By making the INTEDG0 bit LOW, the external hardware interrupts of INT0 become *negative edge-triggered interrupts*. For example, the instruction "BSF INTCON2, INTEDG1" makes INTEDG1 a negative edge-triggered interrupt, in which, when a high-to-low signal is applied to pin RB1 (PORTB.1), the controller will be interrupted and forced to jump to location 0008H in the vector table to service the ISR (assuming that the GIE and INT0IE bits are enabled). This is shown in Program 11-5. Its C version is shown in Program 11-5C.

	INTEDG0	INTEDG1	INTEDG2				

INTEDGx   External Hardware Interrupt Edge trigger bit
0 = Interrupt on negative (falling) edge
1 = Interrupt on positive (rising) edge (Default for power-on reset)

Figure 11-10. INTCON2 Register INTEDG Allows Positive or Negative Edge Trigger

In Program 11-5 we assume that pin RB1 (INT1) is connected to a pulse generator and the pin RB7 is connected to an LED. The program will toggle the LED on the falling edge of the pulse. In other words, the LED is turned on and off at the same rate as the pulses are applied to the INT1 pin.

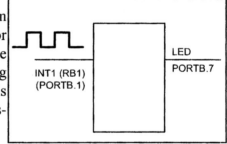

Figure 11-11. For Program 11-5

```
;Program 11-5
          ORG   0000H
          GOTO  MAIN          ;bypass interrupt vector table
      ;--on default all interrupts go to to address 00008
          ORG   0008H         ;interrupt vector table
          BTFSS INTCON3,INT1IF ;Did we get here due to
                              ;INT1 interrupt?
          RETFIE              ;No. Then return to main
          GOTO  INT1_ISR      ;Yes. Then go INT1 ISR
      ;--the main program for initialization
          ORG   00100H
     MAIN BCF   TRISB,7       ;PB7 as an output
          BSF   TRISB,INT1 ;make INT1 an input pin
```

442

```
                 BSF   INTCON3,INT1IE    ;enable INT1 interrupt
                 BCF INTCON2,INTEDG1      ;make it negative
                                          ;edge-triggered
                 BSF INTCON,GIE           ;enable interrupts globally
          OVER BRA OVER                   ;stay in this loop forever
          ;------------------------ISR for INT1
          INT1_ISR
                 ORG 200H
                 BTG PORTB,7              ;toggle on RB7
                 BCF INTCON3,INT1IF ;clear INT1 interrupt flag bit
                 RETFIE
                 END


//Program 11-5C (This is the C version of Program 11-5)
     #include <p18F4580.h>
     #define mybit PORTBbits.RB7
     void chk_isr(void);
     void INT1_ISR(void);
     #pragma code My_HiPrio_Int =0x0008    //high-priority
interrupt location
     void My_HiPrio_Int (void)
     {
     _asm
     GOTO chk_isr
     _endasm
     }
     #pragma code
     #pragma interrupt chk_isr  //used for high-priority
                                //interrupt only
     void chk_isr (void)
     {
     if (INTCON3bits.INT1IF==1) //INT1 causes interrupt?
     INT1_ISR( );    //Yes. Execute INT1 program
         }
     void main(void)
       {
          TRISBbits.TRISB7=0;   //RB7 = OUTPUT
          TRISBbits.TRISB1=1;   //INT1 = INPUT
          INTCON3bits.INT1IF=0; //clear INT1
          INTCON3bits.INT1IE=1; //enable INT1 interrupt
          INTCON2bits.INTEDG1=0;//make it negative edge
          INTCONbits.GIE=1;     //enable all interrupts
          while(1);  //keep looping until interrupt comes
       }

     void INT1_ISR(void)
       {
          mybit=~mybit;
          INTCON3bits.INT1IF=0;         //clear INT1 flag
       }
```

Sampling the edge-triggered interrupt

Before ending this section, we need to answer the question of how often the edge-triggered interrupt is sampled. In edge-triggered interrupts, the external source must be held HIGH for at least two instruction cycles, and then held LOW for at least two instruction cycles to ensure that the transition is seen by the microcontroller.

The rising edge (or the falling edge) is latched by the PIC18 and is held by the INTxIF bits. The INT0IF, INT1IF, and INTIF2 bits hold the latched rising (or falling, depending on the INTEDGx bit) edge of pins RB0–RB2. The INT0IF–INT2IF bits function as interrupt-in-service flags. When an interrupt-in-service flag is raised, it indicates to the external world that the interrupt is being serviced and no new interrupt on this INT*n* pin will be responded to until this service is finished. This is just like the busy signal you get when calling a telephone number that is in use. Regarding the INT0IF–INT2IF one more point must be emphasized. The point is that before the ISRs are finished (that is, before execution of instruction RETFIE), these bits (INT0IF–INT2IF) must be cleared, indicating that the interrupt is finished and the PIC18 is ready to respond to another interrupt on that pin. For another interrupt to be recognized, the pin must go back to a logic LOW state and be brought back HIGH to be considered a positive edge-triggered interrupt.

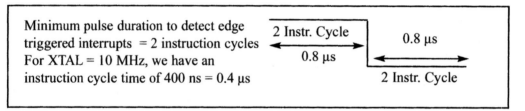

```
Minimum pulse duration to detect edge
triggered interrupts  = 2 instruction cycles       2 Instr. Cycle        0.8 µs
For XTAL = 10 MHz, we have an                    ←——————→
instruction cycle time of 400 ns = 0.4 µs            0.8 µs          ←——————→
                                                                     2 Instr. Cycle
```

Review Questions

1. True or false. Upon reset, all external hardware interrupts INT0–INT2 go to the interrupt vector table address of 0008.
2. For PIC18F458, what pins are assigned to INT0–INT2?
3. Show how to enable the INT1.
4. Assume that the INT0IE bit for the external hardware interrupt INT0 is enabled. Explain how this interrupt works when it is activated.
5. True or false. Upon reset, the external hardware interrupt is negative edge-triggered.
6. How do we make sure that a single interrupt is not recognized as multiple interrupts?
7. True or false. The last two instructions of the ISR for INT0 are:

```
        BCF    INTCON2,INT0IF
        RETFIE
```

SECTION 11.4: PROGRAMMING THE SERIAL COMMUNI-CATION INTERRUPTS

In Chapter 10 we studied the serial communication of the PIC18. All examples in that chapter used the polling method. In this section we explore interrupt-based serial communication, which allows the PIC18 to do many things, in addition to sending and receiving data from the serial communication port.

RCIF and TXIF flags and interrupts

As you may recall from Chapter 10, TXIF (transfer interrupt) is raised when the last bit of the framed data, the stop bit, is transferred, indicating that the TXREG register is ready to transfer the next byte. RCIF (received interrupt) is raised when the entire frame of data, including the stop bit, is received. In other words, when the RCREG register has a byte, RCIF is raised to indicate that the received byte needs to be picked up before it is lost (overrun) by new incoming serial data. As far as serial communication is concerned, all the above concepts apply equally when using either polling or an interrupt. The only difference is in how the serial communication needs are served. In the polling method, we wait for the flag (TXIF or RCIF) to be raised; while we wait we cannot do anything else. In the interrupt method, we are notified when the PIC18 has received a byte, or is ready to send the next byte; we can do other things while the serial communication needs are served.

In the PIC18 two interrupts are set aside for serial communication. One interrupt is used for send and the other for receive. If the corresponding interrupt bit of TXIE or RCIE is enabled, when TXIF or RCIF is raised the PIC18 gets interrupted and jumps to memory address location 0008H to execute the ISR.

Table 11-4: Serial Port Interrupt Flag Bits and their Associated Registers

Interrupt	Flag bit	Register	Enable bit	Register
TXIF (Transmit)	TXIF	PIR1	TXIE	PIE1
RCIF (Receive)	RCIF	PIR1	RCIE	PIE1

		RCIE	TXIE				

Figure 11-12. PIE1 Register Bits Holding TXIE and RCIE

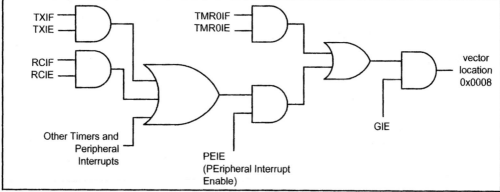

Figure 11-13: Serial Interrupt Enable Flags

Use of serial COM in the PIC18

In the vast majority of applications, the serial interrupt is used mainly for receiving data and is seldom used for sending data serially. This is like receiving a telephone call, where we need a ring to be notified of an incoming call. If we need to make a phone call there are other ways to remind ourselves and so no need for ringing. In receiving the phone call, however, we must respond immediately no matter what we are doing or we will miss the call. Similarly, we use the serial interrupt to receive incoming data so that it is not lost. Look at Program 11-6. Notice that the last instruction of the ISR is RETFIE and there is no clearing of the TXIF flag, since it is done by writing a byte to TXREG.

In Figure 11-13, notice the role of PEIE (PEripheral Interrupt Enable) in allowing serial communication interrupts and other interrupts to come in. This is in addition to the GIE bit discussed in Section 11-1.

For Program 11-6 we assume an 8-bit switch is connected to PORTD. In this program, the PIC18 reads data from PORTD and writes it to TXREG continuously to be transmitted serially. We assume that XTAL = 10 MHz. The baud rate is set at 9600.

```
;Program 11-6
        ORG    0000H
        GOTO  MAIN          ;bypass interrupt vector table
    ;--on default all interrupts go to to address 00008
        ORG    0008H         ;interrupt vector table
        BTFSC PIR1,TXIF  ;Is interrupt due to transmit?
        BRA TX_ISR         ;Yes. Then go to ISR
        RETFIE              ;No. Then return
        ORG 0040H
    TX_ISR                  ;service routine for TXIF
        MOVWFF PORTD,TXREG;load new value, clear TXIF
        RETFIE              ;then return to main
    ;--the main program for initialization
        ORG    00100H
    MAIN  SETF TRISD       ;make PORTD input
        MOVLW 0x20          ;enable transmit and choose
                            ;low baud
        MOVWF TXSTA         ;write to reg
        MOVLW D'15'         ;9600 bps
                            ;(Fosc / (64 * Speed) - 1)
        MOVWF SPBRG         ;write to reg
        BCF TRISC, TX       ;make TX pin of PORTC an
                            ;output pin
        BSF RCSTA, SPEN    ;enable the serial port
        BSF PIE1,TXIE      ;enable TX interrupt
        BSF INTCON,PEIE    ;enable peripheral interrupts
        BSF INTCON,GIE     ;enable interrupts globally
    OVER  BRA OVER         ;stay in this loop forever
        END
```

Program 11-7 is a modification of Program 11-6 with receive interrupt. In this program, the PIC18 gets data from PORTD and sends it to TXREG continuously while incoming data from the serial port is sent to PORTB. We assume that XTAL = 10 MHz and the baud rate = 9600. This program can be verified by connecting your PICTrainer to the serial port of the x86 IBM PC and using HyperTerminal to send and receive data between the PIC Trainer and the IBM PC.

```
;Program 11-7
      ORG   0000H
      GOTO  MAIN              ;bypass interrupt vector table
;--on default all interrupts go to to address 00008
      ORG   0008H             ;interrupt vector table
HI_ISR BTFSC PIR1,TXIF        ;is it TX interrupt?
      BRA   TX_ISR            ;Yes. Then branch to TX_ISR
      BTFSC PIR1,RCIF         ;Is it RC interrupt?
      BRA   RC_ISR            ;Yes. Then branch to RC_ISR
      RETFIE                  ;No. Then return to main
TX_ISR MOVFF PORTD,TXREG      ;loading TXREG clears TXIF
      GOTO  HI_ISR
RC_ISR
      MOVFF RCREG,PORTB       ;copy received data to PORTB
      GOTO  HI_ISR
;--the main program for initialization
      ORG   00100H
MAIN  CLRF TRISB              ;PORTB as an output
      SETF TRISD              ;make PORTD input
      MOVLW 0x20              ;enable transmit and choose low baud
      MOVWF TXSTA             ;write to reg
      MOVLW D'15'             ;9600 bps (Fosc / (64 * Speed) - 1)
      MOVWF SPBRG             ;write to reg
      BCF TRISC,TX            ;make TX pin of PORTC an output pin
      BSF TRISC,RX            ;make RCV pin of PORTC an input pin
      MOVLW 0x90              ;enable receive and serial port
      MOVWF RCSTA             ;write to reg
      BSF PIE1,TXIE           ;enable TX interrupt
      BSF PIE1,RCIE           ;enable receive interrupt
      BSF INTCON,PEIE         ;enable peripheral interrupts
      BSF INTCON,GIE          ;enable interrupts globally
OVER  BRA OVER                ;stay in this loop forever
      END
```

```c
//Program 11-7C (This is the C version of Program 11-7)
#include <p18F458.h>
void chk_isr(void);
void TX_ISR(void);
void RC_ISR(void);
#pragma code My_HiPrio_Int=0x08 //high-priority interrupt
void My_HiPrio_Int (void)
{
    _asm
        GOTO chk_isr
    _endasm
```

```
}
#pragma code
#pragma interrupt chk_isr//used for high-priority interrupt
void chk_isr (void)
{
if (PIR1bits.TXIF==1)          //Transmit caused interrupt?
  TX_ISR( );                   //Yes. Execute Transmit program
if (PIR1bits.RCIF==1)          //Receive caused interrupt?
  RC_ISR( );                   //Yes. Execute Receive program
}
void main(void)
  {
      TRISD = 0xFF;            //PORTD = INPUT
      TRISB = 0;              //PORTB = OUTPUT
      TRISCbits.TRISC6=0;    //TX pin = OUTPUT
      TRISCbits.TRISC7=1;    //RCV pin = INPUT
      TXSTA=0x20;            //choose low baud rate, 8-bit
      SPBRG=15;             //9600 baud rate/ XTAL = 10 MHz
      RCSTAbits.CREN=1;
      RCSTAbits.SPEN=1;
      TXSTAbits.TXEN=1;
      PIE1bits.RCIE=1;        //enable RCV interrupt
      PIE1bits.TXIE=1;        //enable TX interrupt
      INTCONbits.PEIE=1;      //enable peripheral interrupts
      INTCONbits.GIE=1;       //enable all interrupts globally
      while(1);              //keep looping until interrupt comes
  }
void TX_ISR(void)
  {
      TXREG=PORTD;
  }
void RC_ISR(void)
  {
      PORTB=RCREG;
  }
}
```

Review Questions

1. True or false. All interrupts, including the TXIF and RXIF, are directed to a single location in the interrupt vector table.
2. What address in the interrupt vector table is assigned to the serial interrupt?
3. Which register do the TXIF and RXIF flags belong to? Show how they are enabled.
4. Assume that the RCIF bit is enabled. Explain how this interrupt gets activated and its actions upon activation.
5. True or false. Upon reset, the serial interrupts are active and ready to go.
6. True or false. The last two instructions of the ISR for the receive interrupt are:
        BCF   RIR1,RCIF
        RETFIE
7. Answer Question 6 for the transmit interrupt.

SECTION 11.5: PORTB-CHANGE INTERRUPT

The four pins of the PORTB (RB4–RB7) can cause an interrupt when any changes are detected on any one of them. They are referred to as "PORTB-Change interrupt" to distinguish them from the INT0–INT2 interrupts, which are also located on PORTB (RB0–RB2). See Figure 11-15. The PORTB-Change interrupt has a single interrupt flag called RBIF and is located in the INTCON register. This is shown in Figure 11-14. In Figure 11-14, also notice the RBIE bit for enabling the PORTB-Change interrupt. In Section 11.3 we discussed the external hardware interrupts of INT0, INT1, and INT2. Notice the following differences between the PORTB-Change interrupt and INT0–INT2 interrupts:

(a) Each of the INT0–INT2 interrupts has its own pin and is independent of the others. These interrupts use pins PORTB.0 (RB0), PORTB.1 (RB1), and PORTB.2 (RB2), respectively. The PORTB-change interrupt uses all four of the PORTB pins RB4–PB7 and is considered to be a single interrupt even though it can use up to four pins.

(b) While each of the INT0–INT2 interrupts has its own flag, and is independent of the others, there is only a single flag for the PORTB-Change interrupt.

(c) While each of the INT0–INT2 interrupts can be programmed to trigger on the negative or positive edge, the PORTB-Change interrupt causes an interrupt if any of its pins changes status from HIGH to LOW, or LOW to HIGH. See Figure 11-16.

PORTB-Change is widely used in keypad interfacing as we will see in Chapter 12. Another way to use the PORTB-Change interrupt is shown in Program 11-8. In that program, we assume a door sensor is connected to pin RB4 and upon opening or closing the door, the buzzer will sound. See Figure 11-17.

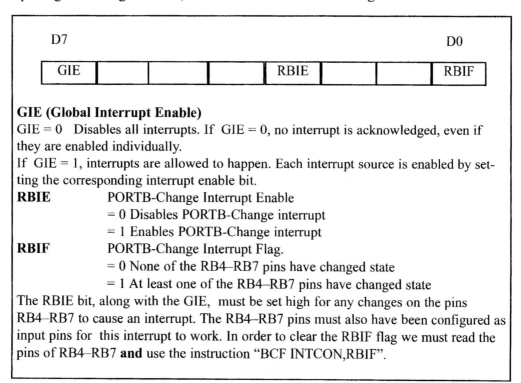

Figure 11-14. INTCON (Interrupt Control) Register

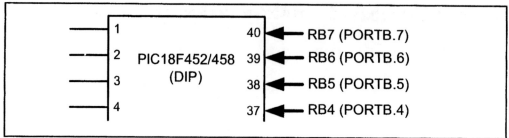

Figure 11-15. PORTB-Change Interrupt Pins

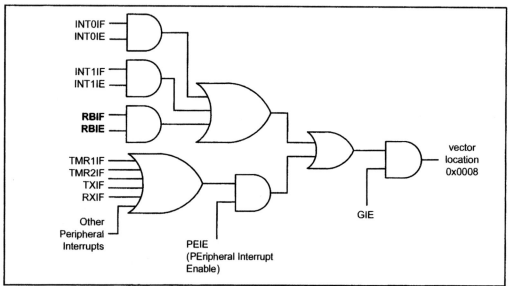

Figure 11-16. PORTB-Change Interrupt (RBIF)

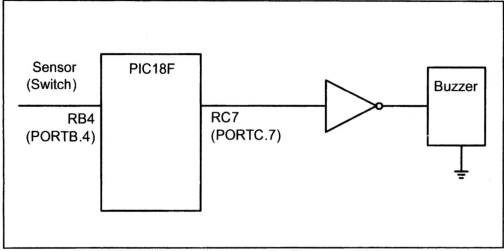

Figure 11-17. PORTB-Change Interrupt for Program 11-8

For Program 11-8 we have connected a door sensor to pin RB4 and a buzzer to pin RC7. In this program, every time the door is opened, it sounds the buzzer by sending it a square wave frequency.

```
;Program 11-8
MYREG EQU   0x20        ;set aside a couple of registers
DELRG EQU 0x80          ;for buzzer time delay
      ORG   0000H
      GOTO  MAIN        ;bypass interrupt vector table
;--on default all interrupts go to to address 00008
      ORG   0008H       ;interrupt vector table
      BTFSS INTCON,RBIF ;Did we get here due to RBIF?
      RETFIE            ;No. Then return to main
      GOTO  PB_ISR      ;Yes. Then go ISR
;--the main program for initialization
      ORG   00100H
MAIN  BCF   TRISC,7     ;PORTC.7 as an output for buzzer
      BSF   TRISB,4     ;PORTB.4 as an input for interrupt
      BSF   INTCON,RBIE ;enable PORTB-Change interrupt
      BSF   INTCON,GIE  ;enable interrupts globally
OVER  BRA OVER          ;stay in this loop forever
;------------------------ISR for PORTB-Change INT
PB_ISR
      ORG 200H
      MOVF  PORTB,W     ;we must read PORTB
      MOVLW D'250'      ;for delay
      MOVWF MYREG
BUZZ  BTG PORTC,7       ;toggle PC7 for the buzzer
      MOVLW D'255'      ;for delay
      MOVWF DELRG
DELAY DECF DELRG,F
      BNZ DELAY         ;keep sounding the buzzer
      DECF MYREG,F
      BNZ BUZZ
      BCF INTCON,RBIF   ;and clear RBIF interrupt flag bit
      RETFIE
      END
```

Notice for the PORTB-Change interrupt, there is no need to enable the PEIE; however, we still need to enable the GIE bit.

It must be noted again that, while the INT0–INT2 interrupts each have their own interrupt flags, there is only a single interrupt flag (RBIF) for all the four pins of the RB4–RB7. Examine Program 11-9. For this program, we assume that each of the pins RB4 and RB5 is connected to an external switch. Upon activation of the SW, an LED reflects the status. See Figure 11-18.

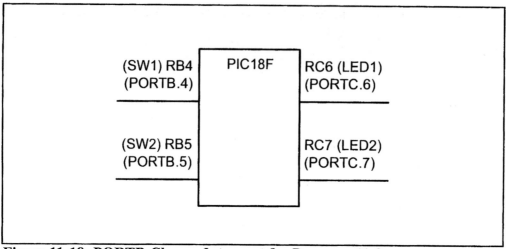

Figure 11-18: PORTB-Change Interrupt for Program 11-9

For Program 11-9 we have connected SW1 and SW2 to pins RB4 and RB5 respectively. In this program, the activation of SW1 and SW2 will result in changing the state of LED1 and LED2 respectively.

```
;Program 11-9
        ORG   0000H
        GOTO  MAIN  ;bypass interrupt vector table
;--on default all interrupts go to to address 00008
        ORG   0008H      ;interrupt vector table
        BTFSS INTCON,RBIF;Did we get here due to RBIF?
        RETFIE           ;No. Then return to main
        GOTO  PB_ISR     ;Yes. Then go ISR
;--the main program for initialization
        ORG   0100H
MAIN  BCF TRISC,4      ;PC4 as an output
      BCF TRISC,5      ;PC5 as an output
      BSF TRISB,4 ;PB4 as an input for the interrupt
      BSF TRISB,5 ;PB5 as an input for the interrupt
      BSF INTCON,RBIE ;enable PORTB interrupt
      BSF INTCON,GIE  ;enable interrupts globally
OVER  BRA OVER         ;stay in this loop forever
;------------------------ISR for PORTB_Change
PB_ISR
      ORG 200H
      MOVFF PORTB,W ;get the status of switches
      ANDLW 0x30 ;mask unneeded bits
      MOVFF W,PORTC ;update LEDs
      BCF INTCON,RBIF ;clear RBIF interrupt flag bit
      RETFIE
      END
```

```
//Program 11-9C (This is the C version of Program 11-9)
#include <p18F458.h>
#define LED1 PORTCbits.RC4
#define LED2 PORTCbits.RC5

void chk_isr(void);
void RBINT_ISR(void);

#pragma code My_HiPrio_Int =0x0008 //high-priority int
void My_HiPrio_Int (void)
{
    _asm
        GOTO chk_isr
    _endasm
}
#pragma code
#pragma interrupt chk_isr  //used for high-priority int
void chk_isr (void)
{
    if (INTCONbits.RBIF==1)    //RBIF caused interrupt?
    RBINT_ISR( );              //Yes. Execute ISR program
}
void main(void)
  {
    TRISCbits.TRISC4=0;    //RC4 = OUTPUT
    TRISCbits.TRISC5=0;    //RC5 = OUTPUT
    TRISBbits.TRISB4 = 1; //RB4 = INPUT for interrupt
    TRISBbits.TRISB5 = 1; //RB5 = INPUT for interrupt
    INTCONbits.RBIF=0;     //clear RBIF
    INTCONbits.RBIE=1;     //enable RB interrupt
    INTCONbits.GIE=1;      //enable all interrupts globally
    while(1);  //keep looping until interrupt comes
  }
void RBINT_ISR(void)
  {
    LED1=PORTBbits.RB4;
    LED2=PORTBbits.RB5;
    INTCONbits.RBIF=0;     //clear RBIF flag
  }
```

Review Questions

1. True or false. There is a single interrupt for each of the PORTB pins.
2. What address in the interrupt vector table is assigned to the PORTB-Change interrupt?
3. Which register do the RBIF and RBIE flags belong to? Show how RBIE is enabled.
4. Give the last two instructions of the ISR for the PORTB-Change interrupt.
5. True or false. Upon reset, the RBIF interrupt is active and ready to go.

SECTION 11.6: INTERRUPT PRIORITY IN THE PIC18

The next topic that we must deal with is what happens if two interrupts are activated at the same time? Which of these two interrupts is responded to first? Interrupt priority is the main topic of discussion in this section.

Setting interrupt priority

In the PIC18 microcontroller, there are only two levels of interrupt priority: (a) low level, and (b) high level. While address 0008 is assigned to high-priority interrupts, the low-priority interrupts are directed to address 00018 in the interrupt vector table. See Table 11-5. Upon power-on reset, all interrupts are automatically designated as high priority and will go to address 00008H. This is done to make the PIC18 compatible with the earlier generation of PIC microcontrollers such as PIC16xxx. We can make the PIC18 a two-level priority system by way of programming the IPEN (interrupt priority enable) bit in the RCON register. Figure 11-19 shows the IPEN bit of the RCON register. Upon power-on reset, the IPEN bit contains 0, making the PIC18 a single priority level chip, just like the PIC16xxx. To make the PIC18 a two-level priority system, we must first set the IPEN bit to HIGH. It is only after making IPEN = 1 that we can assign a low priority to any of the interrupts by programming the bits called IP (interrupt priority). Figure 11-20 shows IPR1 (interrupt priority register) with the IP bits for TXIP, RCIP, TMR1IP, and TMR2IP. If IPEN = 1, then the IP bit will take effect and will assign a given interrupt a low priority. As a result of assigning a low priority to a given interrupt, it will land at the address 0018 instead of 0008 in the interrupt vector table. The IP (interrupt priority) bit along with the IF (interrupt flag) and IE (interrupt enable) bits will complete all the flags needed to program the interrupts for the PIC18. Table 11-6 shows the three flags and the registers they belong to for some of the interrupts used in this chapter. In Table 11-6, notice the absence of the INT0 priority flag. The INT0 has only one priority and that is high priority. That means all the PIC18 interrupts can be assigned a low or high priority level, except the external hardware interrupt of INT0. Study Figures 11-22 through 11-25 very carefully. When examining these figures, the following point must be noted. By making IPEN = 1, we enable the interrupt priority feature. Now we must also enable two bits to enable the interrupts: (a) We must set GIEH = 1. The GIEH bit is part of the INTCON register (Figure 11-21) and is the same as GIE, which we have used in previous sections. In this regard there is no difference between the priority and no-priority systems. (b) The second bit we must set high is GIEL (part of INTCON). Making GIEL = 1 will enable all the interrupts whose IP = 0. That means all the interrupts that have been given the low priority will be forced to vector location 00018H.

Table 11-5: Interrupt Vector Table for the PIC18

Interrupt	ROM Location (Hex)
Power-on-Reset	0000
High-priority Interrupt	0008 (Default upon power-on reset)
Low-priority Interrupt	0018 (Selected with IP bit)

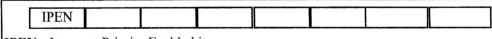

IPEN							

IPEN   Interrupt Priority Enable bit
   0 = All the interrupts are directed to the vector location 0008 (default).
   1 = Interrupts can be assigned a low or high priority.

The importance of IPEN: Upon power-on reset, all the interrupts of PIC18 are directed to location 0008, making it a single-priority system, just like PIC16xxx. To prioritize the PIC18 interrupts into low- and high-level priorities, we must make IPEN = 1. When IPEN = 1, we can assign either a low or a high priority to any of the interrupts by manipulating the corresponding bit in the IPR (interrupt priority register) for that interrupt. When interrupt priority is enabled (IPEN = 1), we must set both the GIEH and GIEL bits to high in order to enable the interrupts globally. Notice in Figure 11-21 that GIE is the same as GIEH.

Figure 11-19. RCON Register. IPEN Allows Prioritizing the Interrupt into 2 Levels

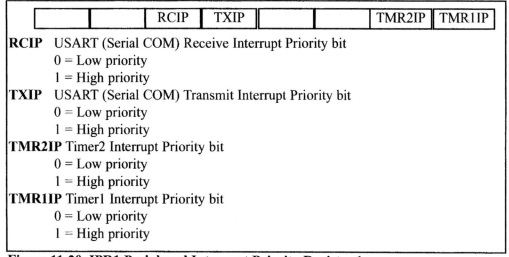

		RCIP	TXIP			TMR2IP	TMR1IP

RCIP   USART (Serial COM) Receive Interrupt Priority bit
   0 = Low priority
   1 = High priority
TXIP   USART (Serial COM) Transmit Interrupt Priority bit
   0 = Low priority
   1 = High priority
TMR2IP Timer2 Interrupt Priority bit
   0 = Low priority
   1 = High priority
TMR1IP Timer1 Interrupt Priority bit
   0 = Low priority
   1 = High priority

Figure 11-20. IPR1 Peripheral Interrupt Priority Register 1

Table 11-6: Interrupt Flag Bits for PIC18 Timers

Interrupt	Flag bit (Register)	Enable bit (Register)	Priority (Register)
Timer0	TMR0IF (INTCON)	TMR0IE (INTCON)	TMR0IP (INTCON2)
Timer1	TMR1IF (PIR1)	TMR1IE (PIE1)	TMR1IP (IPR1)
Timer2	TMR2IF (PIR1)	TMR2IE (PIE1)	TMR2IP (IPR1)
Timer3	TMR3IF (PIR3)	TMR3IE (PIE2)	TMR3IP (IPR2)
INT1	INT1IF (PIR1)	INT1IE (PIE1)	INT1IP (INTCON3)
INT2	INT2IF (PIR1)	INT2IE (PIE1)	INT2IP (INTCON)
TXIF	TXIF (PIR1)	TXIE (PIE1)	TXIP (IPR1)
RCIF	RCIF (PIR1)	RCIE (PIE1)	RCIP (IPR1)
RB INT	RBIF (INTCON)	RBIE (INTCON)	RBIP (INTCON2)

Note: INT0 has only the high-level priority.

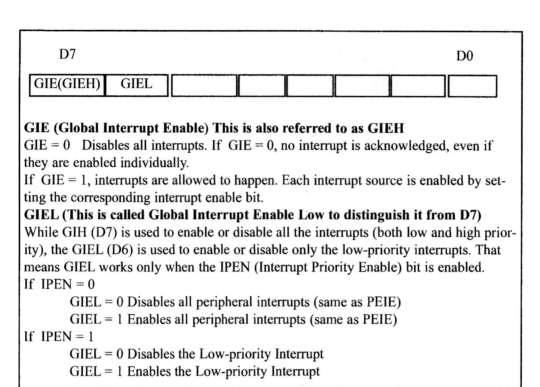

GIE (Global Interrupt Enable) This is also referred to as GIEH

GIE = 0   Disables all interrupts. If GIE = 0, no interrupt is acknowledged, even if they are enabled individually.

If GIE = 1, interrupts are allowed to happen. Each interrupt source is enabled by setting the corresponding interrupt enable bit.

GIEL (This is called Global Interrupt Enable Low to distinguish it from D7)

While GIH (D7) is used to enable or disable all the interrupts (both low and high priority), the GIEL (D6) is used to enable or disable only the low-priority interrupts. That means GIEL works only when the IPEN (Interrupt Priority Enable) bit is enabled.

If IPEN = 0

   GIEL = 0 Disables all peripheral interrupts (same as PEIE)

   GIEL = 1 Enables all peripheral interrupts (same as PEIE)

If IPEN = 1

   GIEL = 0 Disables the Low-priority Interrupt

   GIEL = 1 Enables the Low-priority Interrupt

Figure 11-21. INTCON (Interrupt Control) Register

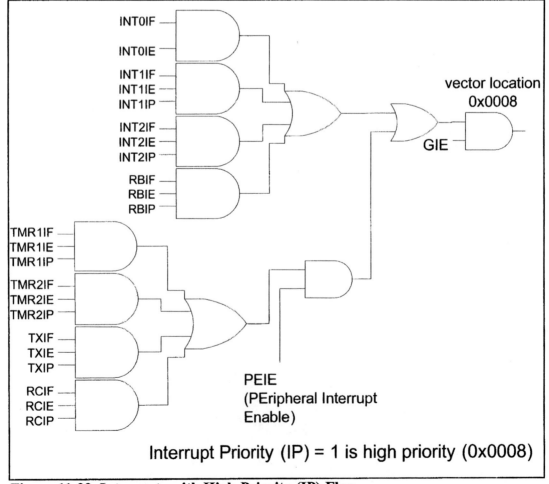

Figure 11-22. Interrupts with High-Priority (IP) Flag

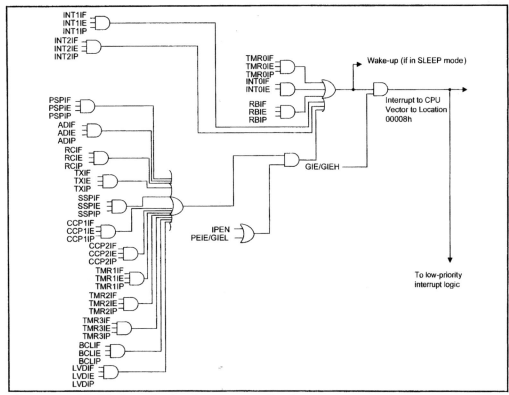

Figure 11-23. High-Priority Interrupts (Redrawn from PIC18 Manual)

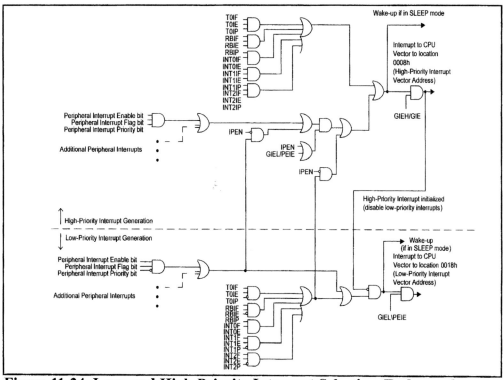

Figure 11-24. Low- and High-Priority Interrupt Selection (Redrawn from PIC18 Manual)

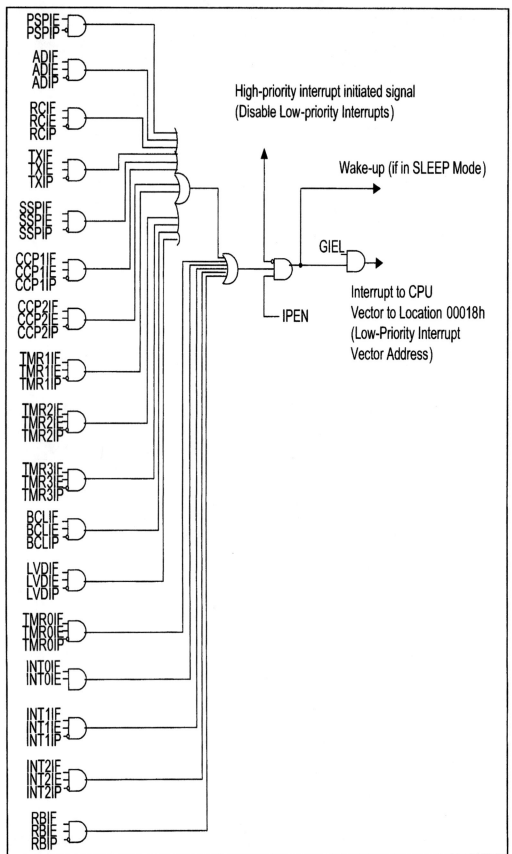

Figure 11-25. Low-Priority Interrupt Selection with IP Flag (Redrawn from PIC18 Manual)

Program 11-10 uses Timer0 and Timer1 interrupts to generate square waves on pins RB1 and RB7 respectively, while data is being transferred from PORTC to PORTD. This is a repeat of Program 11-2, except Timer1 has been assigned to low priority.

```
;Program 11-10
     ORG    0000H
     GOTO MAIN          ;bypass interrupt vector table
;--high-priority interrupts go to address 00008
     ORG    0008H       ;high-priority interrupt vector table
     BTFSC INTCON,TMR0IF  ;Is it Timer0 interrupt?
     BRA T0_ISR           ;Yes. Then branch to T0_ISR
     RETFIE 0x01          ;No. Then fast return to main
;--low-priority interrupts go to address 00018
     ORG    0018H       ;low-priority interrupt vector table
     BTFSC PIR1,TMR1IF    ;Is it Timer1 interrupt?
     BRA T1_ISR           ;Yes. Then branch to T1_ISR
     RETFIE               ;No. Then return to main

;-main program for initialization and keeping CPU busy
     ORG    0100H ;somewhere after vector table space
MAIN BCF    TRISB,1   ;PB1 as an output
     BCF    TRISB,7   ;PB7 as an output
     CLRF TRISD       ;make PORTD output
     SETF TRISC       ;make PORTC input
     MOVLW 0x08       ;Timer0, 16-bit, no prescale,
                      ;internal clk
     MOVWF T0CON      ;load T0CON reg
     MOVLW  0xFF      ;TMR0H = FFH, the high byte
     MOVWF  TMR0H     ;load Timer0 high byte
     MOVLW 0x00       ;TMR0L = 00H, the low byte
     MOVWF TMR0L      ;load Timer0 low byte
     BCF INTCON,TMR0IF ;clear Timer0 interrupt flag bit
     BSF INTCON,TMR0IE ;enable Timer0 interrupt
     MOVLW 0x0        ;Timer1, 16-bit, no prescale,
                      ;internal clk
     MOVWF T1CON      ;load T1CON reg
     MOVLW  0xFF      ;TMR1H = FFH, the high byte
     MOVWF  TMR1H     ;load Timer0 high byte
     MOVLW 0x00       ;TMR1L = 00H, the low byte
     MOVWF TMR1L      ;load Timer1 low byte
     BCF PIR1,TMR1IF  ;clear Timer1 interrupt flag bit
     BSF PIE1,TMR1IE  ;enable Timer1 interrupt
     BCF IPR1,TMR1IP  ;make Timer1 low-priority interrupt
     BSF    RCON,IPEN ;enable priority levels
     BSF INTCON,GIEL
     BSF INTCON,GIEH ;enable interrupts globally
     BSF T1CON,TMR1ON ;start Timer1
     BSF T0CON,TMR0ON ;start Timer0
;--keeping CPU busy waiting for interrupt
OVER MOVFF PORTC,PORTD ;send data from PORTC to PORTD
     BRA OVER          ;stay in this loop forever
```

```
;-------------------------ISR for Timer0
T0_ISR
      ORG 200H
      MOVLW  0xFF       ;TMR0H = FFH, the high byte
      MOVWF  TMR0H      ;load Timer0 high byte
      MOVLW 0x00        ;TMR0L = 00H, the low byte
      MOVWF TMR0L       ;load Timer0 low byte
      BTG   PORTB,1     ;toggle RB1
      BCF INTCON,TMR0IF ;clear timer interrupt flag bit
      RETFIE 0x01
;------------------------ISR for Timer1
T1_ISR
      ORG 300H
      MOVLW  0xFF       ;TMR1H = FFH, the high byte
      MOVWF  TMR1H      ;load Timer0 high byte
      MOVLW 0x00        ;TMR1L = 00H, the low byte
      MOVWF TMR1L       ;load Timer1 low byte
      BTG PORTB,7
      BCF PIR1,TMR1IF   ;clear Timer1 interrupt flag bit
      RETFIE
      END
```

Program 11-11 has four interrupts. It uses Timer0 and Timer1 interrupts to generate square waves on pins RC0 and RC1 respectively. It also uses the transmit and receive interrupts to send and receive data serially. Data from PORTD is transmitted and the received byte is placed on PORTB. Timer0 and receive ISRs have the high priority while Timer1 and send ISRs are assigned a low priority level.

```
;Program 11-11
      ORG   0000H
      GOTO  MAIN        ;bypass interrupt vector table
;--high-priority interrupts go to address 00008
      ORG   0008H       ;need to redirect because not
                        ;enough space
      GOTO CHK_HI_PRIO
;--no need to redirect because we have plenty of space
      ORG   00018       ;low-priority interrupt vector table
      BTFSC PIR1,TMR1IF     ;is it Timer1 interrupt?
      BRA   T1_ISR      ;Yes. Then branch to T1_ISR
      BTFSC PIR1,TXIF   ;Did we get here due to TxD?
      BRA   TX_ISR      ;Yes. Then branch to TX_ISR
      RETFIE            ;No. Then return to main
;-----------------
CHK_HI_PRIO       ORG   0x50
      BTFSC INTCON,TMR0IF    ;Is it Timer0 interrupt?
      BRA   T0_ISR      ;Yes. Then branch to T0_ISR
      BTFSC PIR1,RCIF   ;Did we get here due to RCV int?
      BRA RC_ISR        ;Yes. Then branch to RC_ISR
      RETFIE 0x01       ;No. Then return to main
```

```
;—main program for initialization and keeping CPU busy
      ORG   0100H         ;somewhere after vector table space
MAIN  BCF   TRISC,RC0
      BCF   TRISC,RC1
      CLRF  TRISB         ;make PORTB output
      SETF  TRISD         ;make PORTD input
      MOVLW 0x08          ;Timer0, 16-bit, no prescale,
                          ;internal clk
      MOVWF T0CON         ;load T0CON reg
      MOVLW 0xFF          ;TMR0H = FFH, the high byte
      MOVWF TMR0H         ;load Timer0 high byte
      MOVLW 0x00          ;TMR0L = 00H, the low byte
      MOVWF TMR0L         ;load Timer0 low byte
      BCF   INTCON,TMR0IF ;clear Timer0 interrupt flag bit
      MOVLW 0x0           ;Timer1, 16-bit, no prescale,
                          ;internal clk
      MOVWF T1CON         ;load T1CON reg
      MOVLW 0xFF          ;TMR1H = FFH, the high byte
      MOVWF TMR1H         ;load Timer0 high byte
      MOVLW 0x00          ;TMR1L = 00H, the low byte
      MOVWF TMR1L         ;load Timer1 low byte
      BCF   PIR1,TMR1IF   ;clear Timer1 interrupt flag bit
      MOVLW 0x20          ;enable transmit and choose low baud
      MOVWF TXSTA         ;write to reg
      MOVLW D'15'         ;9600 bps (Fosc / (64 * Speed) - 1)
      MOVWF SPBRG         ;write to reg
      MOVLW 0x90          ;enable receive and serial port
      MOVWF RCSTA         ;write to reg
      BCF   TRISC, TX     ;make TX pin of PORTC an output pin
      BSF   TRISC, RX     ;make RCV pin of PORTC an input pin
      BSF   RCON,IPEN
      BSF   PIE1,RCIE
      BSF   PIE1,TXIE     ;enable TX interrupt
      BSF   INTCON,TMR0IE    ;enable Timer0 interrupt
      BSF   PIE1,TMR1IE   ;enable Timer1 interrupt
      BCF   IPR1,TMR1IP   ;make Timer1 a low-priority interrupt
      BCF   IPR1,TXIP     ;make Transmit a low-priority interrupt
      BSF   T0CON,TMR0ON     ;start Timer0
      BSF   T1CON,TMR1ON     ;start Timer1
      BSF   INTCON,GIEL      ;enable low-priority interrupts
      BSF   INTCON,GIEH      ;enable high-priority interrupts
;--keeping CPU busy waiting for interrupt
OVER  BRA   OVER          ;stay in this loop forever

;------------------------ISR for Timer0
T0_ISR      ORG 200H
      MOVLW 0xFF          ;TMR0H = FFH, the high byte
      MOVWF TMR0H         ;load Timer0 high byte
      MOVLW 0x00          ;TMR0L = 00H, the low byte
      MOVWF TMR0L         ;load Timer0 low byte
      BTG   PORTC,0       ;toggle RB1
```

```
        BCF INTCON,TMR0IF ;clear timer interrupt flag bit
        RETFIE 0x01
;--------------------------ISR for Timer1
T1_ISR       ORG 300H
        MOVLW  0xFF       ;TMR1H = FFH, the high byte
        MOVWF  TMR1H      ;load Timer0 high byte
        MOVLW 0x00        ;TMR1L = 00H, the low byte
        MOVWF TMR1L       ;load Timer1 low byte
        BTG PORTC,1
        BCF PIR1,TMR1IF ;clear Timer1 interrupt flag bit
        RETFIE
;----------------Transmit ISR
TX_ISR
        BCF PIR1,TXIF ;clear TX interrupt flag bit
        MOVFF PORTD,TXREG
        RETFIE
;------------
RC_ISR
        MOVFF RCREG,PORTB  ;copy received data to PORTD
        BCF   PIR1, RCIF      ;clear RCIF
        RETFIE 1
        END
```

Example 11-3

For Program 11-11 (or Program 11-11C), discuss what happens: (a) if interrupt RCIF is activated when the PIC18 is serving the Timer1 interrupt, (b) Timer 1 is activated when Timer0 is being served, (c) RCIF and TMR0IF and TMR1IF are activated at the same time.

Solution:

In Program 11-11, notice that the Receive (RCIF) and Timer0 (TMR0IF) interrupts are assigned to high priority while the Transmit (TXIF) and Timer1 (TMR1IF) interrupts have low priority. As a result we have the following:

(a) if the RCIF is activated during the execution of the Timer1 ISR, the Receive interrupt comes in and its ISR is executed first because it has a higher priority. After it is finished, the PIC18 goes back and finishes the Timer1 ISR.

(b) if the TMR1IF is activated during the execution of the Timer0 ISR, it is ignored because it has lower priority. After the Timer0 ISR is finished, the PIC18 will execute the Timer1 ISR.

(c) If all three, RCIF, TMR0IF, and TMR1IF, are activated at the same time, the Received ISR and and Timer0 ISR are taken care of first because they are assigned to high priority. Between the Receive and Timer0 interrupts, the Timer0 ISR is served first due to the programming sequence we have set in the interrupt vector table for the high-priority interrupts. That means if these three interrupts are activated at the same time, they are executed in the following sequence: Timer0 ISR, Receive ISR, and Timer1 ISR.

Low-priority interrupt programming in C

As we saw in the last four sections, the C18 compiler uses the reserved keyword **interrupt** to designate an interrupt as high priority. To assign low priority level to a given interrupt, it uses the keyword **interruptlow**. See Table 11-7. This is shown in Program 11-11C, which is a repeat of Program 11-11 in C.

Table 11-7: Interrupt Vector Table for the PIC18 using C18 Syntax

Interrupt	ROM Location	C18 keyword
High-priority Interrupt	0x0008 (Default)	**interrupt**
Low-priority Interrupt	0x0018 (Selected with IP bit)	**interruptlow**

Program 11-11C has four interrupts. It uses Timer0 and Timer 1 interrupts to generate square waves on pins RC1 and RC7, respectively. It also uses the transmit and receive interrupts to send and receive data serially. Data from PORTB is transmitted and the received byte is placed on PORTD. Timer0 and receive ISRs have the high priority while Timer1 and transmit ISRs are assigned low priority level.

```
//Program 11-11C (This is the C version of Program 11-11)
#include <p18F458.h>
#define myPC0bit PORTCbits.RC0
#define myPC1bit PORTCbits.RC1

void chk_isr(void);
void chk_low_isr(void);
void T0_ISR(void);
void T1_ISR(void);
void TX_ISR(void);
void RC_ISR(void);

#pragma code My_HiPrio_Int =0x0008 //high-priority int
void My_HiPrio_Int (void)
{
_asm
GOTO chk_isr
_endasm
}

#pragma code My_Lo_Prio_Int =0x00018 //low-priority int
void My_Lo_Prio_Int (void)
{
    _asm
        GOTO chk_low_isr
    _endasm
}

#pragma interruptlow chk_low_isr //used for low-priority
void chk_low_isr (void)
{
```

```c
        if(PIR1bits.TMR1IF==1)//Timer1 causes interrupt?
          T1_ISR();              //Yes. Execute Timer1 ISR
        if (PIR1bits.TXIF==1) //Transmit causes interrupt?
          TX_ISR( );             //Yes. Execute Transmit ISR
}

#pragma interrupt chk_isr//used for high-priority interrupt
void chk_isr (void)
{
        if (PIR1bits.TMR1IF==1)//Timer0 causes interrupt?
            T0_ISR( );          //Yes. Execute Timer0 ISR
        if (PIR1bits.RCIF==1) //Receiver causes interrupt?
            RC_ISR( );          //Yes. Execute Receiver ISR
}

void main(void)
   {
        TRISCbits.TRISC0=0;    //RC0 = OUTPUT
        TRISCbits.TRISC1=0;    //RC1 = OUTPUT
        TRISD = 255;           //PORTD = INPUT
        TRISB = 0;             //PORTB = OUTPUT
        T0CON=0x08;            //Timer0, 16-bit mode,
                               //no prescaler
        TMR0H=0xFF;            //load TH0
        TMR0L=0x00;            //load TL0
        INTCONbits.TMR0IF=0;   //clear TF1
        T1CON=0x0; //Timer 1, 16-bit mode, no prescaler
        TMR1H=0xFF;            //load TH1
        TMR1L=0x00;            //load TL1
        PIR1bits.TMR1IF=0;     //clear TF1
        TXSTA=0x20;            //choose low baud rate,8-bit
        SPBRG=15;              //9600 baud rate/ XTAL = 10 MHz
        RCSTAbits.CREN=1;
        RCSTAbits.SPEN=1;
        TRISCbits.TRISC6=0;    //TX pin = OUTPUT
        TRISCbits.TRISC7=1;    //RCV pin = INPUT
        RCONbits.IPEN=1;
        PIE1bits.RCIE=1;
        PIE1bits.TXIE=1;       //enable TX interrupt
        INTCONbits.TMR0IE=1;
        PIE1bits.TMR1IE=1;
        IPR1bits.TMR1IP=0;     //make Timer1 a low-priority
        IPR1bits.TXIP=0;       //make TX a low-priority
        T0CONbits.TMR0ON=1;    //turn on T0
        T1CONbits.TMR1ON=1;    //turn on T1
        INTCONbits.GIEL=1;     //enable low-priority interrupts
        INTCONbits.GIEH=1;//enable high-priority interrupts
        while(1);              //keep looping until interrupt comes
   }
```

```
//-------------ISR for Timer0
void T0_ISR(void)
   {
       TMR0H=0xFF;              //load TH0
       TMR0L=0x00;              //load TL0
       myPC0bit=~myPC0bit;      //toggle RB1
       INTCONbits.TMR0IF=0;     //clear TF0
   }
//-------------ISR for Timer1
void T1_ISR(void)
   {
       TMR1H=0xFF;              //load TH0
       TMR1L=0x00;              //load TL0
       PIR1bits.TMR1IF=0;       //clear TF1
       myPC1bit=~myPC1bit;      //toggle RB1
   }
//-------------ISR for Transmit
void TX_ISR(void)
   {
       TXREG=PORTD;             //clear Tx Interrupt flag
   }
//-------------ISR for Receive
void RC_ISR(void)
   {
       PORTB=RCREG;
       RCSTAbits.CREN=0;//clear CREN to clear any error
       RCSTAbits.CREN=1;//set CREN for continuous reception
   }
```

Interrupt inside an interrupt

What happens if the PIC18 is executing an ISR belonging to an interrupt and another interrupt is activated? In such cases, a high-priority interrupt can interrupt a low-priority interrupt. This is an interrupt inside an interrupt. In the PIC18 a low-priority interrupt can be interrupted by a higher-priority interrupt, but not by another low-priority interrupt. Although all the interrupts are latched and kept internally, no low-priority interrupt can get the immediate attention of the CPU until the PIC18 has finished servicing all the high-priority interrupts. The GIE (which is also called GIEH) and GIEL bits play an important role in the process of the interrupt inside the interrupt. Regarding the interrupt inside an interrupt concept, the following points must be emphasized:

1. When a high-priority interrupt is vectored into address 0008H, the GIE bit is disabled (GIEH = 0), thereby blocking another interrupt (low or high) from coming in. The RETFIE instruction at the end of the ISR will enable the GIE (GIE = 1) automatically, which allows interrupts to come in again. If we want to allow another high-priority interrupt to come in during the execution of the current ISR, then we must make GIE = 1 at the beginning of the current ISR.
2. When a low-priority interrupt is vectored into address 0018H, the GIEL bit is disabled (GIEL = 0), thereby blocking another low-priority interrupt from

coming in. The RETFIE instruction at the end of the ISR will enable the GIEL (GIEL = 1) automatically, which allows low-priority interrupts to come in again. Notice that the low-priority interrupt cannot block a high-priority interrupt from coming in during the execution of the current low-priority ISR because GIEH is still set to one (GIEH = 1).

3. When two or more interrupts have the same priority level. In this case, they are serviced according to the sequence by which the program checks them in the interrupt vector table. We saw many examples of that in this chapter. See Example 11-3.

Fast context saving in task switching

In many applications, such as multitasking real-time operating systems (RTOS), the CPU brings in one task (job or process) at a time and executes it before it moves to the next one. In executing each task, which is often organized as the interrupt service routine, access to all the resources of the CPU is critical in performing the task in a timely manner. In early CPUs, the limited number of registers forced programmers to save the entire contents of the CPU on the stack before execution of the new task. This saving of the CPU contents before switching to a new task is called *context saving* (or *context switching*). The use of the stack as a place to save the CPU's contents is tedious, time consuming, and slow. For this reason some CPUs such as x86 microprocessors have instructions such as PUSHA (Push All) and POPA (Pop All), which will push and pop all the main registers onto the stack with a single instruction. Because the PIC18 has numerous general purpose registers, there is no need for using the stack to save the CPU's general purpose registers. However, each task generally needs the key registers of WREG, BSR, and STATUS. For that reason the PIC18 automatically saves these three registers internally in shadow registers when a high-priority interrupt is activated. This way, the three key registers of the main task are saved internally. To restore the original contents of these three key registers, one must use instruction "RETFIE 0x01" instead of "RETFIE" at the end of the high-priority ISR. The "RETFIE 0x01" is called fast context saving in PIC18 literature. Regarding fast context saving in the PIC18, two important points must be noted:

1. It is not available for the low-priority interrupts, and works only for high-priority interrupts. That means that when a low-priority interrupt is activated, there is no fast context saving and we must save these three registers at the beginning of the low-priority ISR, if they are being used by the low-priority ISR.

2. The shadow registers keeping these three key registers have a depth of one, meaning that there is only one of them. For that reason, the fast context saving works only when a high-priority ISR is activated during the main subroutine. If two or three high-priority interrupts are activated at the same time, only the first ISR can use the fast context saving because the depth of shadow registers is only one. In that case, the second and third ISRs must save these key registers at the beginning of the body of their ISRs. This should not be difficult because we know the sequence by which the ISRs are executed, as we saw in many examples in this chapter.

Interrupt latency

The time from the moment an interrupt is activated to the moment the CPU starts to execute the code at the vector address of 0008H (or 0x0018H) is called the *interrupt latency*. This latency can be anywhere from 2 to 4 instruction cycle times depending on whether the source of the interrupt is an internal (e.g., timers) or external hardware (e.g., hardware INTx and PORTB-Change) interrupt. The duration of an interrupt latency can also be affected by the type of the instruction in which the CPU was executing when the interrupt comes in. It takes slightly longer in cases where the instruction being executed lasts for two instruction cycles (e.g., MOVFF reg,reg) compared to the instructions that last for only one instruction cycle time (e.g., ADDWL). See PIC18 for the timing data sheet.

Triggering the interrupt by software

Sometimes when we need to test an ISR by way of simulation. This can be done with simple instructions to set the interrupts HIGH and thereby cause the PIC18 to jump to the interrupt vector table. For example, if the TMR1IE bit for Timer 1 is set, an instruction such as "BSF INTCON, TMR1IF" will interrupt the PIC18 in whatever it is doing and force it to jump to the interrupt vector table. In other words, we do not need to wait for Timer 1 to roll over to have an interrupt. We can cause an interrupt with an instruction that raises the interrupt flag.

Review Questions

1. True or false. Upon reset, all interrupts have the same priority.
2. Which bit of what register is used to enable the interrupt priority option in the PIC18? Is it a bit-addressable register?
3. Which register has the TXIP bit? Show how to assign it low priority.
4. Assume that INT0 and INT1 have the same low priority. Explain what happens if both INT0 and INT1 are activated at the same time. Also assume that INT0 is checked first in the program for the interrupt vector table.
5. Explain what happens if a higher-priority interrupt is activated while the PIC18 is serving a lower-priority interrupt (i.e., executing a lower-priority ISR).

SUMMARY

An interrupt is an external or internal event that interrupts the microcontroller to inform it that a device needs its service. Every interrupt has a program associated with it called the ISR, or interrupt service routine. The PIC18 has many sources of interrupts, depending on the family members. Some of the most widely used interrupts are for the timers, external hardware interrupts, and serial communication. When an interrupt is activated, the IF (Interrupt flag) bit is raised.

The PIC18 can be programmed to enable (unmask) or disable (mask) an interrupt, which is done with the help of the GIE (global interrupt enable) and IE (interrupt enable) bits. The PIC18 has two levels of priority, low and high. Upon power-on reset, all the interrupts are designated as high priority and are directed to address 0008 in the interrupt vector table. This default setting can be altered with the help of the IP (interrupt priority) bits. By programming the IP bit, we can make an interrupt a low priority and force it to land at address 0x00018 in the interrupt vector table. This chapter also showed how to program PIC18 interrupts in both Assembly and C languages.

PROBLEMS

SECTION 11.1: PIC18 INTERRUPTS

1. Which technique, interrupt or polling, avoids tying down the microcontroller?
2. List some of the interrupt sources in the PIC18.
3. In the PIC18 what memory area is assigned to the interrupt vector table?
4. True or false. The PIC18 programmer cannot change the memory address location assigned to the interrupt vector table.
5. What memory address in the interrupt vector table is assigned to low-priority interrupts?
6. What memory address in the interrupt vector table is assigned to high-priority interrupts?
7. Do we have a memory address in the interrupt vector table assigned to the Timer0 interrupt?
8. Do we have a memory address in the interrupt vector table assigned to the INT1 interrupt?
9. To which register does the GIE bit belong?
10. Why do we put a GOTO instruction at address 0?
11. What is the state of the GIE bit upon power-on reset, and what does it mean?
12. Show the instruction to enable the INT0 interrupt.
13. Show the instruction to enable the Timer0 interrupt.
14. The TMR0IE bit belongs to register_____.
15. How many bytes of address space in the interrupt vector table are assigned to the high-priority interrupt?
16. How many bytes of address space in the interrupt vector table are assigned to the low-priority interrupt?
17. To put the entire interrupt service routine in the interrupt vector table for high priority, it must be no more than _____ bytes in size.
18. True or false. The INTCON register is not a bit-addressable register.
19. With a single instruction, show how to disable all the interrupts.
20. With a single instruction, show how to disable the INT0 interrupt.
21. True or false. Upon reset, all interrupts are enabled by the PIC18.
22. In the PIC18, how many bytes of ROM space are assigned to the reset?

SECTION 11.2: PROGRAMMING TIMER INTERRUPTS

23. True or false. For each of Timer 0 and Timer1, there is a unique address in the interrupt vector table.
24. What address in the interrupt vector table is assigned to Timer1?
25. Show how to enable the Timer0 interrupt.
26. Which bit of INTCON belongs to the Timer0 interrupt? Show how it is enabled.

27. Assume that Timer0 is programmed in 8-bit mode, TMR0H = F0H, and the TMR0IE bit is enabled. Explain how the interrupt for the timer works.
28. True or false. The last two instructions of the ISR for Timer1 are:
```
        BCF   PIR1,TMR1IF
        RETFIE
```
29. Assume that Timer1 is programmed for 16-bit mode, TMR1H = FFH, TMR1L = F8H, and the TMR1IE bit is enabled. Explain how the interrupt is activated.
30. If Timer 1 is programmed for interrupts in 8-bit mode, explain when the interrupt is activated.
31. Write a program using the Timer0 interrupt to create a square wave of 1 Hz on pin RB7 while data from PORTC is being sent to PORTD. Assume XTAL = 10 MHz.
32. Write a program using the Timer1 interrupt to create a square wave of 3 kHz on pin RB7 while data from PORTC is being sent to PORTD. Assume XTAL = 10 MHz.

SECTION 11.3: PROGRAMMING EXTERNAL HARDWARE INTERRUPTS

33. True or false. An address location is assigned to each of the external hardware interrupts INT0, INT1, and INT2.
34. What address in the interrupt vector table is assigned to INT0, INT1 and INT2? How about the pin numbers on PORTB?
35. To which register does the INT0IE bit belong? Show how it is enabled.
36. To which register does the INT1IE bit belong? Show how it is enabled.
37. Show how to enable all three external hardware interrupts.
38. Assume that the INT0IE bit for external hardware interrupt INT0 is enabled and is negative edge-triggered. Explain how this interrupt works when it is activated.
39. True or false. Upon reset, all the external hardware interrupts are negative edge-triggered.
40. In Question 38, how do we make sure that a single interrupt is not recognized as multiple interrupts?
41. The INT0IF bit belongs to the _____ register.
42. The INT1IF bit belongs to the _____ register.
43. True or false. The last two instructions of the ISR for INT1 are:
```
        BCF   INTCON3,INT1IF
        RETFIE
```
44. Explain the role of INT0IF and INT0IE in the execution of external interrupt 0.
45. Explain the role of INT1IF and INT1IE in the execution of external interrupt 1.
46. Assume that the INT1IE bit for external hardware interrupt INT1 is enabled and is positive edge-triggered. Explain how this interrupt works when it is activated. How can we make sure that a single interrupt is not interpreted as multiple interrupts?
47. True or false. INT0–INT2 are part of the PEIE group.
48. True or false. Upon power-on reset, all of INT0–INT2 are positive edge-triggered.
49. Explain the difference between positive and negative edge-triggered interrupts.

50. How do we make the hardware interrupt negative edge-triggered?
51. True or false. INT0–INT2 must be configured as an input pin for a hardware interrupt to come in.
52. Which register holds the INTEDGx bits?

SECTION 11.4: PROGRAMMING THE SERIAL COMMUNICATION INTERRUPTS
and
SECTION 11.5: PORTB-CHANGE INTERRUPT

53. True or false. Two separate interrupts are assigned to each of the interrupts, TXIF and RCIF.
54. Upon power-on reset, what address in the interrupt vector table is assigned to the serial interrupt? How many bytes are assigned to it?
55. To which register does the TXIF belong? Show how it is enabled.
56. Assume that the TXIE bit for the serial interrupt is enabled. Explain how this interrupt gets activated and also explain its working upon activation.
57. True or false. Upon reset, serial interrupts are blocked.
58. True or false. The last two instructions of the ISR for the transmit interrupt are:

```
        BCF     PIR1,TXIF
        RETFIE
```

59. State how the RCIF is cleared.
60. Assuming that the TXIE bit is set when TXIF is raised, what happens subsequently?
61. Assuming that the RCIE bit is set when RCIF is raised, what happens subsequently?
62. Write a program using interrupts to get data serially and send it to PORTD while at the same time any changes on PORTB.4 will cause the LED connected to PORTC.7 to toggle.
63. Provide the following information for the PORTB-Change interrupt.
    (a) the flag associated with the PORTB-Change interrupt
    (b) the register to which these flag belong
    (c) the difference between the PORTB-Change and INT0–INT2 interrupts
    (d) the pins that are part of the PORTB-Change interrupt

SECTION 11.6: INTERRUPT PRIORITY IN THE PIC18

64. True or false. Upon reset, all interrupts have high priority.
65. What register enables the interrupt priority in the PIC18 ? Explain its role.
66. Which register has the INT0IP bit? Show how to assign it low priority.
67. Which register has TMR1IP bit? Show how to assign it low priority.
68. Which register has the INT1IP bit? Show how to assign it low priority.
69. Assume that INT1IP and INT2IP are both 0s. Explain what happens if both INT1IF and INT2IF are activated at the same time.
70. Assume that TMR0IP and TMR1IP are both 0s. Explain what happens if both TMR0IF and TMR1IF are activated at the same time.
71. If both TMR0IP and TMR1IP are set to HIGH, what happens if both are activated at the same time?

72. If both INT1IP and INT2IP in the IP are set to HIGH, what happens if both are activated at the same time?

73. Explain what happens if a low-priority interrupt is activated while the PIC18 is serving a high-priority interrupt.

74. Explain what happens if a high-priority interrupt is activated while the PIC18 is serving a low-priority interrupt.

75. Explain the role of the GIEH bit in masking and unmasking the interrupts.

76. True or false. In PIC18, an interrupt inside an interrupt is not allowed.

77. Explain the role of the GIEL bit in masking and unmasking interrupts.

78. Explain the role of RETFIE in enabling the GIEL bit.

79. Explain the difference between the "RETFIE" and "RETFIE 1" instructions.

80. Explain the concept of fast context saving in PIC.

ANSWERS TO REVIEW QUESTIONS

SECTION 11.1: PIC18 INTERRUPTS

1. Interrupts
2. INT0 and TMR0
3. Address locations 0x0008 to 0x00017. No. It is set when the processor is designed.
4. GIE = 0 means that all interrupts are masked, and as a result no interrupts will be responded to by the PIC18.
5. Assuming GIE = 1, we need "BSF INTCON,TMR0IE".
6. 0008 for the high-priority interrupts and 0x0018 for the low-priority interrupts.

SECTION 11.2: PROGRAMMING TIMER INTERRUPTS

1. False. There is a single address for all the timers, Timer0, Timer1, and so on.
2. 0008H
3. PIE1 and "BSF PIE1,TMR1IE" will enable the Timer1 interrupt.
4. After Timer1 is started, the timer will count up from F5H to FFH on its own while the PIC18 is executing other tasks. Upon rolling over from FFH to 00, the TMR1IF flag is raised, which will interrupt the PIC18 in whatever it is doing and forces it to jump to memory location 0008 to execute the ISR belonging to this interrupt.
5. True

SECTION 11.3: PROGRAMMING EXTERNAL HARDWARE INTERRUPTS

1. True
2. Bits RB0 (PORTB.0), RB1 (PORTB.1), and RB2 (PORTB.2)
3. BSF INTCON3,INT1IE
4. Upon application of a low-to-high pulse to pin RB0, the PIC18 is interrupted in whatever it is doing and jumps to ROM location 0008H to execute the ISR.
5. False
6. When the CPU jumps to ROM location 0008 to execute the ISR, the GIE becomes 0, effectively blocking another interrupt from the same source. The last two instructions of the ISR are "BCF INCON, INTOIF" followed by "RETFIE". While the first instruction will clear the previous request for interrupt, the second one will make GIE = 1, allowing a new interrupt to come in from the same source. That can happen only if a new low-to-high pulse is applied to the pin.
7. True

SECTION 11.4: PROGRAMMING THE SERIAL COMMUNICATION INTERRUPTS

1. True. There is only one interrupt for all all interrupts including the transfer and receive.
2. 0x0008 for high-priority interrupts and 0x0018 for low-priority interrupts.
3. "BSF PIE1,TXIE" will enable the send interrupt and "BSF PIE1,RCIE" will enable the receive interrupt.
4. The RCIF (received interrupt flag) is raised when the entire frame of data, including the stop bit, is received. As a result the received byte is delivered to the RCREG register and the PIC18 jumps to memory location 0008H to execute the ISR belonging to this interrupt. In the serial COM interrupt service routine, we must save the RCREG content before it is lost by the incoming data.
5. False
6. True
7. BCF RIR1,TXIF
   RETFIE

SECTION 11.5: PORTB-CHANGE INTERRUPT

1. False
2. All interrupts, including the PORTB-Change interrupt, go to location 0008 on default.
3. INTCON, and we enable it with the instruction "BSF INTCON, RBIF"
4. BCF INTCON,RBIF
   RETFIE
5. False

SECTION 11.6: INTERRUPT PRIORITY IN THE PIC18

1. True
2. IPEN bit of the RCON register. Yes, it is bit-addressable.
3. IPR1 and the instruction "BCF IPR1,TXIP" will do it.
4. If both are activated at the same time, INT0 is serviced first because it is checked first. After INT0 is serviced, INT1 is serviced.
5. We have an interrupt inside an interrupt, meaning that the lower-priority interrupt is put on hold and the higher one is serviced. After servicing this higher-priority interrupt, the PIC18 resumes servicing the lower-priority ISR.

CHAPTER 12

LCD AND KEYBOARD INTERFACING

OBJECTIVES

Upon completion of this chapter, you will be able to:

>> Describe the functions of the pins of a typical LCD
>> List instruction command codes for programming an LCD
>> Interface an LCD to the PIC18
>> Program an LCD in Assembly and C
>> Explain the basic operation of a keyboard
>> Describe the key press and detection mechanisms
>> Interface a 4 × 4 keypad to the PIC18 using C and Assembly

This chapter explores some real-world applications of the PIC18. We explain how to interface the PIC18 to devices such as an LCD and a keyboard. In Section 12.1, we show LCD interfacing with the PIC18. In Section 12.2, keyboard interfacing with the PIC18 is shown. We use C and Assembly for both sections.

SECTION 12.1: LCD INTERFACING

This section describes the operation modes of LCDs, then describes how to program and interface an LCD to a PIC18 using Assembly and C.

LCD operation

In recent years the LCD has been finding widespread use replacing LEDs (seven-segment LEDs or other multisegment LEDs). This is due to the following reasons:

1. The declining prices of LCDs.
2. The ability to display numbers, characters, and graphics. This is in contrast to LEDs, which are limited to numbers and a few characters.
3. Incorporation of a refreshing controller into the LCD, thereby relieving the CPU of the task of refreshing the LCD. In contrast, the LED must be refreshed by the CPU (or in some other way) to keep displaying the data.
4. Ease of programming for characters and graphics.

LCD pin descriptions

The LCD discussed in this section has 14 pins. The function of each pin is given in Table 12-1. Figure 12-1 shows the pin positions for various LCDs.

V_{CC}, V_{SS}, and V_{EE}

While V_{CC} and V_{SS} provide +5 V and ground, respectively, V_{EE} is used for controlling LCD contrast.

RS, register select

There are two very important registers inside the LCD. The RS pin is used for their selection as follows. If RS = 0, the instruction command code register is selected, allowing the user to send a command such as clear display, cursor at home, and so on. If RS = 1 the data register is selected, allowing the user to send data to be displayed on the LCD.

R/W, read/write

R/W input allows the user to write information to the LCD or read information from it. R/W = 1 when reading; R/W = 0 when writing.

Table 12-1: Pin Descriptions for LCD

Pin	Symbol	I/O	Description
1	V_{SS}	--	Ground
2	V_{CC}	--	+5 V power supply
3	V_{EE}	--	Power supply to control contrast
4	RS	I	RS = 0 to select command register, RS = 1 to select data register
5	R/W	I	R/W = 0 for write, R/W = 1 for read
6	E	I/O	Enable
7	DB0	I/O	The 8-bit data bus
8	DB1	I/O	The 8-bit data bus
9	DB2	I/O	The 8-bit data bus
10	DB3	I/O	The 8-bit data bus
11	DB4	I/O	The 8-bit data bus
12	DB5	I/O	The 8-bit data bus
13	DB6	I/O	The 8-bit data bus
14	DB7	I/O	The 8-bit data bus

E, enable

The enable pin is used by the LCD to latch information presented to its data pins. When data is supplied to data pins, a high-to-low pulse must be applied to the En pin in order for the LCD to latch in the data present at the data pins. This pulse must be a minimum of 450 ns wide. In this book we call this delay the SDELAY (short delay) to distinguish it from other delays.

D0–D7

The 8-bit data pins, D0–D7, are used to send information to the LCD or read the contents of the LCD's internal registers.

To display letters and numbers, we send ASCII codes for the letters A–Z, a–z, and numbers 0–9 to these pins while making RS = 1.

There are also instruction command codes that can be sent to the LCD to clear the display or force the cursor to the home position or blink the cursor. Table 12-2 lists the instruction command codes. To send any of the commands listed in Table 12-2 to the LCD, make pin RS = 0. For data, make RS = 1. Then send a high-to-low pulse to the E pin to enable the internal latch of the LCD. There are two ways to send characters (command/data) to the LCD: (1) use a delay before sending the next one, (2) use the busy flag to see if the LCD is ready for the next one.

Table 12-2: LCD Command Codes

Code (Hex)	Command to LCD Instruction Register
1	Clear display screen
2	Return home
4	Decrement cursor (shift cursor to left)
6	Increment cursor (shift cursor to right)
5	Shift display right
7	Shift display left
8	Display off, cursor off
A	Display off, cursor on
C	Display on, cursor off
E	Display on, cursor blinking
F	Display on, cursor blinking
10	Shift cursor position to left
14	Shift cursor position to right
18	Shift the entire display to the left
1C	Shift the entire display to the right
80	Force cursor to beginning of 1st line
C0	Force cursor to beginning of 2nd line
38	2 lines and 5x7 matrix

Note: This table is extracted from Table 12-4.

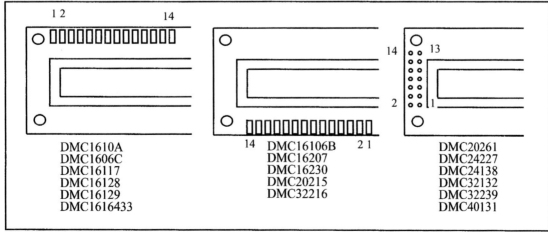

DMC1610A
DMC1606C
DMC16117
DMC16128
DMC16129
DMC1616433

DMC16106B
DMC16207
DMC16230
DMC20215
DMC32216

DMC20261
DMC24227
DMC24138
DMC32132
DMC32239
DMC40131

Figure 12-1. Pin Positions for Various LCDs from Optrex

Sending commands and data to LCDs with a time delay

Program 12-1 shows how to send characters (command/data) to the LCD without checking the busy flag. Notice that we need to wait 5–10 ms (DELAY) between issuing each character to the LCD. We call this delay simply DELAY. In programming an LCD, we also need a long delay for the power-up process. We call it LDELAY (long delay). SDELAY (short delay) is used to make the En signal wide enough for the LCD's enable input. See Chapter 3 for delays.

Figure 12-2 shows the LCD connections to the microcontroller.

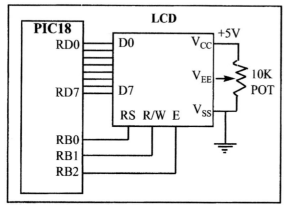

Figure 12-2. LCD Connections

```
;Program 12-1: Using delay before sending data/command
LCD_DATA   EQU  PORTD         ;LCD data pins RD0-RD7
LCD_CTRL   EQU  PORTB         ;LCD control pins
RS         EQU  RB0           ;RS pin of LCD
RW         EQU  RB1           ;R/W pin of LCD
EN         EQU  RB2           ;E pin of LCD
           CLRF   TRISD       ;PORTD = Output
           CLRF   TRISB       ;PORTB = Output
           BCF    LCD_CTRL,EN ;enable idle low
           CALL   LDELAY      ;wait for initialization
           MOVLW  0x38        ;init. LCD 2 lines, 5x7 matrix
           CALL   COMNWRT     ;call command subroutine
           CALL   LDELAY      ;initialization hold
           MOVLW  0x0E        ;display on, cursor on
           CALL   COMNWRT     ;call command subroutine
           CALL   DELAY       ;give LCD some time
           MOVLW  0x01        ;clear LCD
           CALL   COMNWRT     ;call command subroutine
           CALL   DELAY       ;give LCD some time
           MOVLW  0x06        ;shift cursor right
           CALL   COMNWRT     ;call command subroutine
           CALL   DELAY       ;give LCD some time
           MOVLW  0x84        ;cursor at line 1, pos. 4
           CALL   COMNWRT     ;call command subroutine
           CALL   DELAY       ;give LCD some time
           MOVLW  A'N'        ;display letter 'N'
           CALL   DATAWRT     ;call display subroutine
           CALL   DELAY       ;give LCD some time
           MOVLW  A'O'        ;display letter 'O'
```

```
               CALL    DATAWRT       ;call display subroutine
AGAIN          BTG     LCD_CTRL,0
               BRA     AGAIN         ;stay here
COMNWRT                              ;send command to LCD
               MOVWF   LCD_DATA      ;copy WREG to LCD DATA pin
               BCF     LCD_CTRL,RS   ;RS = 0 for command
               BCF     LCD_CTRL,RW   ;R/W = 0 for write
               BSF     LCD_CTRL,EN   ;E = 1 for high pulse
               CALL    SDELAY        ;make a wide En pulse
               BCF     LCD_CTRL,EN   ;E = 0 for H-to-L pulse
               RETURN
DATAWRT                              ;write data to LCD
               MOVWF   LCD_DATA      ;copy WREG to LCD DATA pin
               BSF     LCD_CTRL,RS   ;RS = 1 for data
               BCF     LCD_CTRL,RW   ;R/W = 0 for write
               BSF     LCD_CTRL,EN   ;E = 1 for high pulse
               CALL    SDELAY        ;make a wide En pulse
               BCF     LCD_CTRL,EN   ;E = 0 for H-to-L pulse
               RETURN
;look in previous chapters for delay routines
               END
```

Sending command or data to the LCD using busy flag

We use RS = 0 to read the busy flag bit to see if the LCD is ready to receive information. The busy flag is D7, and can be read when R/W = 1 and RS = 0, as follows: if R/W = 1, RS = 0. When D7 = 1 (busy flag = 1), the LCD is busy taking care of internal operations and will not accept any new information. When D7 = 0, the LCD is ready to receive new information.

This is shown in Program 12-2.

```
;Program 12-2: Check busy flag before sending
;data or command to LCD (See Fig. 12-2)
LCD_DATA   EQU PORTD             ;LCD data pins RD0-RD7
LCD_CTRL   EQU PORTB             ;LCD control pins
RS         EQU RB0               ;RS pin of LCD
RW         EQU RB1               ;R/W pin of LCD
EN         EQU RB2               ;E pin of LCD
           CLRF    TRISD         ;PORTD = Output
           CLRF    TRISB         ;PORTB = Output
           BCF     LCD_CTRL,EN   ;enable idle low
           CALL    LDELAY        ;long delay (250 ms) for power-up
           MOVLW   0x38          ;init. LCD 2 lines, 5x7 char
           CALL    COMMAND       ;issue command
           CALL    LDELAY        ;initialization hold
           MOVLW   0x0E          ;LCD on, cursor on
           CALL    COMMAND       ;issue command
           CALL    READY         ;Is LCD ready?
           MOVLW   0x01          ;clear LCD command
           CALL    COMMAND       ;issue command
           CALL    READY         ;Is LCD ready?
           MOVLW   0x06          ;shift cursor right
```

```
          CALL      COMMAND        ;issue command
          CALL      READY          ;Is LCD ready?
          MOVLW     0x86           ;cursor: line 1, pos. 6
          CALL      COMMAND        ;command subroutine
          CALL      READY          ;Is LCD ready?
          MOVLW     A'N'           ;display letter 'N'
          CALL      DATA_DISPLAY
          CALL      READY          ;Is LCD ready?
          MOVLW     A'O'           ;display letter 'O'
          CALL      DATA_DISPLAY
HERE      BRA       HERE           ;STAY HERE
;-------------------------------------------------
COMMAND   MOVWF     LCD_DATA       ;issue command code
          BCF       LCD_CTRL,RS    ;RS = 0 for command
          BCF       LCD_CTRL,RW    ;R/W = 0 for write
          BSF       LCD_CTRL,EN    ;E = 1 for high pulse
          CALL      SDELAY         ;make a wide En pulse
          BCF       LCD_CTRL,EN    ;E = 0 for H-to-L pulse
          RETURN
;-------------------------------------------------
DATA_DISPLAY MOVWF  LCD_DATA       ;copy WREG to LCD DATA pin
          BSF       LCD_CTRL,RS    ;RS = 1 for data
          BCF       LCD_CTRL,RW    ;R/W = 0 for write
          BSF       LCD_CTRL,EN    ;E = 1 for high pulse
          CALL      SDELAY         ;make a wide En pulse
          BCF       LCD_CTRL,EN    ;E = 0 for H-to-L pulse
          RETURN
;-------------------------------------------------
READY     SETF      TRISD          ;make PORTD input port for LCD data
          BCF       LCD_CTRL,RS    ;RS = 0 access command reg
          BSF       LCD_CTRL,RW    ;R/W = 1 read command reg
;read command reg and check busy flag
BACK      BSF       LCD_CTRL,EN    ;E = 0 for L-to-H pulse
          CALL      SDELAY         ;make a wide En pulse
          BCF       LCD_CTRL,EN    ;E = 1 L-to-H pulse
          BTFSC     LCD_DATA,7     ;stay until busy flag = 0
          BRA       BACK
          CLRF      TRISD   ;make PORTD output port for LCD data
          RETURN
;look in previous chapters for delay routines
          END
```

Notice in Program 12-2 that the busy flag is D7 of the command register. To read the command register, we make R/W = 1 and RS = 0, and a L-to-H pulse for the E pin will provide us the command register. After reading the command register, if bit D7 (the busy flag) is HIGH, the LCD is busy and no information (command or data) should be issued to it. Only when D7 = 0 can we send data or commands to the LCD. Notice that no time delays are used in this method because we are checking the busy flag before issuing commands or data to the LCD. Contrast the read and write timing for the LCD in Figures 12-3 and 12-4. Note that the E line is negative edge-triggered for the write while it is positive edge-trig-gered for the read.

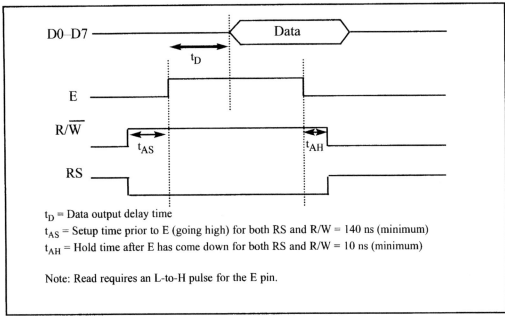

t_D = Data output delay time

t_{AS} = Setup time prior to E (going high) for both RS and R/W = 140 ns (minimum)

t_{AH} = Hold time after E has come down for both RS and R/W = 10 ns (minimum)

Note: Read requires an L-to-H pulse for the E pin.

Figure 12-3. LCD Timing for Read (L-to-H for E line)

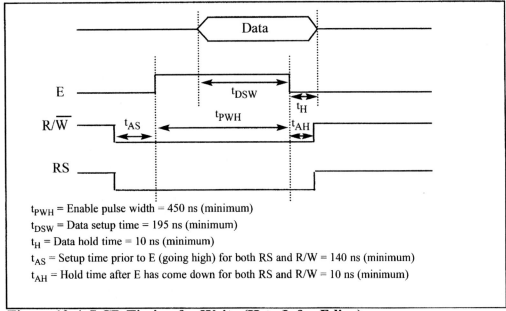

t_{PWH} = Enable pulse width = 450 ns (minimum)

t_{DSW} = Data setup time = 195 ns (minimum)

t_H = Data hold time = 10 ns (minimum)

t_{AS} = Setup time prior to E (going high) for both RS and R/W = 140 ns (minimum)

t_{AH} = Hold time after E has come down for both RS and R/W = 10 ns (minimum)

Figure 12-4. LCD Timing for Write (H-to-L for E line)

LCD data sheet

In the LCD, one can put data at any location. The following shows address locations and how they are accessed.

RS	R/W	DB7	DB6	DB5	DB4	DB3	DB2	DB1	DB0
0	0	1	A	A	A	A	A	A	A

where AAAAAAA = 0000000 to 0100111 for line 1 and AAAAAAA = 1000000 to 1100111 for line 2. See Table 12-3.

The upper address range can go as high as 0100111 for the 40-character-wide LCD, while for the 20-character-wide LCD it goes up to 010011 (19 decimal = 10011 binary). Notice that the upper-range 0100111 (binary) = 39 decimal, which corresponds to locations 0 to 39 for the LCDs of 40x2 size.

From the above discussion we can get the addresses of cursor positions for various sizes of LCDs. See Figure 12-5 for the cursor addresses for common types of LCDs. Note that all the addresses are in hex. Table 12-4 provides a detailed list of LCD commands and instructions. (Table 12-2 is extracted from this table.)

Table 12-3: LCD Addressing

	DB7	DB6	DB5	DB4	DB3	DB2	DB1	DB0
Line 1 (min)	1	0	0	0	0	0	0	0
Line 1 (max)	1	0	1	0	0	1	1	1
Line 2 (min)	1	1	0	0	0	0	0	0
Line 2 (max)	1	1	1	0	0	1	1	1

16 x 2 LCD	80	81	82	83	84	85	86 through 8F
	C0	C1	C2	C3	C4	C5	C6 through CF
20 x 1 LCD	80	81	82	83	through 93
20 x 2 LCD	80	81	82	83	through 93
	C0	C1	C2	C3	through D3
20 x 4 LCD	80	81	82	83	through 93
	C0	C1	C2	C3	through D3
	94	95	96	97	through A7
	D4	D5	D6	D7	through E7
40 x 2 LCD	80	81	82	83	through A7
	C0	C1	C2	C3	through E7

Note: All data is in hex.

Figure 12-5. Cursor Addresses for Some LCDs

Table 12-4: List of LCD Instructions

Instruction	RS	R/W	DB7	DB6	DB5	DB4	DB3	DB2	DB1	DB0	Description	Execution Time (Max)
Clear Display	0	0	0	0	0	0	0	0	0	1	Clears entire display and sets DD RAM address 0 in address counter	1.64 ms
Return Home	0	0	0	0	0	0	0	0	1	-	Sets DD RAM address 0 as address counter. Also returns display being shifted to original position. DD RAM contents remain unchanged.	1.64 ms
Entry Mode Set	0	0	0	0	0	0	0	1	1/D	S	Sets cursor move direction and specifies shift of display. These operations are performed during data write and read.	40 μs
Display On/ Off Control	0	0	0	0	0	0	1	D	C	B	Sets On/Off of entire display (D), cursor On/Off (C), and blink of cursor position character (B).	40 μs
Cursor or Display Shift	0	0	0	0	0	1	S/C	R/L	-	-	Moves cursor and shifts display without changing DD RAM contents.	40 μs
Function Set	0	0	0	0	1	DL	N	F	-	-	Sets interface data length (DL), number of display lines (L), and character font (F).	40 μs
Set CG RAM Address	0	0	0	1	AGC						Sets CG RAM address. CG RAM data is sent and received after this setting.	40 μs
Set DD RAM Address	0	0	1	ADD							Sets DD RAM address. DD RAM data is sent and received after this setting.	40 μs
Read Busy Flag & Address	0	1	BF	AC							Reads busy flag (BF) indicating internal operation is being performed and reads address counter contents.	40 μs
Write Data CG or DD RAM	1	0	Write Data								Writes data into DD or CG RAM.	40 μs
Read Data CG or DD RAM	1	1	Read Data								Reads data from DD or CG RAM.	40 μs

Notes:

1. Execution times are maximum times when fcp or fosc is 250 kHz.
2. Execution time changes when frequency changes.
   (e.g., when fcp or fosc is 270 kHz: 40 μs × 250 / 270 = 37 μs.)
3. Abbreviations:

DD RAM	Display data RAM
CG RAM	Character generator RAM
ACC	CG RAM address
ADD	DD RAM address, corresponds to cursor address
AC	Address counter used for both DD and CG RAM addresses.

1/D = 1	Increment	1/D = 0	Decrement
S = 1	Accompanies display shift		
S/C = 1	Display shift;	S/C = 0	Cursor move
R/L = 1	Shift to the right;	R/L = 0	Shift to the left
DL = 1	8 bits, DL = 0: 4 bits		
N = 1	1 line, N = 0: 1 line		
F = 1	5 × 10 dots, F = 0: 5 × 7 dots		
BF = 1	Internal operation;	BF = 0	Can accept instruction

Sending information to LCD using the TBLRD instruction

Program 12-3 shows how to use the TBLRD instruction to send data and commands to an LCD.

For a PIC18 C version of LCD programming see Program 12-1C and Program 12-2C.

```
;Program 12-3: Using TableRead
;PORTD = D0-D7, RB0 = RS, RB1 = R/W, RB2 = E pins
LCD_DATA    EQU PORTD          ;LCD data pins RD0-RD7
LCD_CTRL    EQU PORTB          ;LCD control pins
RS          EQU RB0            ;RS pin of LCD
RW          EQU RB1            ;R/W pin of LCD
EN          EQU RB2            ;E pin of LCD
            CLRF    TRISD      ;PORTD = Output
            CLRF    TRISB      ;PORTB = Output
            BCF     LCD_CTRL,EN ;enable idle low
            CALL    LDELAY     ;long delay (250 ms) for power-up
            MOVLW   upper(MYCOM)
            MOVWF   TBLPTRU
            MOVLW   high(MYCOM)
            MOVWF   TBLPTRH
            MOVLW   low(MYCOM)
            MOVWF   TBLPTRL
C1          TBLRD*+
            MOVF TABLAT,W       ;give it to WREG
            IORLW 0x0           ;Is it the end of command?
            BZ      SEND_DAT    ;if yes then go to display data
            CALL    COMNWRT     ;call command subroutine
            CALL    DELAY       ;give LCD some time
            BRA     C1
```

```
SEND_DAT    MOVLW   upper(MYDATA)
            MOVWF   TBLPTRU
            MOVLW   high(MYDATA)
            MOVWF   TBLPTRH
            MOVLW   low(MYDATA)
            MOVWF   TBLPTRL
DT1         TBLRD*+
            MOVF TABLAT,W          ;give it to WREG
            IORLW   0x0        ;Is it the end of data string?
            BZ      OVER           ;if yes then exit
            CALL    DATAWRT        ;call DATA subroutine
            CALL    DELAY          ;give LCD some time
            BRA     DT1
OVER        BRA     OVER           ;stay here
COMNWRT                            ;send command to LCD
            MOVWF   LCD_DATA       ;copy WREG to LCD DATA pin
            BCF     LCD_CTRL,RS    ;RS = 0 for command
            BCF     LCD_CTRL,RW    ;R/W = 0 for write
            BSF     LCD_CTRL,EN    ;E = 1 for high pulse
            CALL    SDELAY         ;make a wide En pulse
            BCF     LCD_CTRL,EN    ;E = 0 for H-to-L pulse
            RETURN
DATAWRT                            ;write data to LCD
            MOVWF   LCD_DATA       ;copy WREG to LCD DATA pin
            BSF     LCD_CTRL,RS    ;RS = 1 for data
            BCF     LCD_CTRL,RW    ;R/W = 0 for write
            BSF     LCD_CTRL,EN    ;E = 1 for high pulse
            CALL    SDELAY         ;make a wide En pulse
            BCF     LCD_CTRL,EN    ;E = 0 for H-to-L pulse
            RETURN
            ORG     500H
MYCOM       DB 0x38,0x0E,0x01,0x06,0x84,0;commands and null
MYDATA      DB "HELLO",0           ;data and null
;look in previous chapters for delay routines
            END
```

This C18 program sends letters 'M', 'D', and 'E' to the LCD using delays.

```c
//Program 12-1C: This is the C version of Program 12-1.
#include <P18F4580.h>
#define ldata PORTD            //PORTD = LCD data pins (Fig. 12-2)
#define rs PORTBbits.RB0       //rs = PORTB.0
#define rw PORTBbits.RB1       //rw = PORTB.1
#define en PORTBbits.RB2       //en = PORTB.2

void main()
  {
    TRISD = 0;                 //both ports B and D as output
    TRISB = 0;
    en = 0;                    //enable idle low
    MSDelay(250);
    lcdcmd(0x38);              //init. LCD 2 lines, 5x7 matrix
    MSDelay(250);
    lcdcmd(0x0E);              //display on, cursor on
```

```
        MSDelay(15);
        lcdcmd(0x01);              //clear LCD
        MSDelay(15);
        lcdcmd(0x06);              //shift cursor right
        MSDelay(15);
        lcdcmd(0x86);              //line 1, position 6
        MSDelay(15);
        lcddata('M');              //display letter 'M'
        MSDelay(15);
        lcddata('D');              //display letter 'D'
        MSDelay(15);
        lcddata('E');              //display letter 'E'
    }

void lcdcmd(unsigned char value)
    {
        ldata = value;             //put the value on the pins
        rs = 0;
        rw = 0;
        en = 1;                    //strobe the enable pin
        MSDelay(1);
        en = 0;
    }

void lcddata(unsigned char value)
    {
        ldata = value;             //put the value on the pins
        rs = 1;
        rw = 0;
        en = 1;                    //strobe the enable pin
        MSDelay(1);
        en = 0;
    }

void MSDelay(unsigned int itime)
    {
        unsigned int i, j;
        for(i=0;i<itime;i++)
          for(j=0;j<135;j++);
    }
```

The following is the C version of Program 12-2, using the busy flag method.

```
//Program 12-2C. C version of Program 12-2
#include <P18F458.h>
#define ldata PORTD    //PORTD = LCD data pins (Fig. 12-2)
#define rs PORTBbits.RB0   //rs = PORTB.0
#define rw PORTBbits.RB1   //rw = PORTB.1
#define en PORTBbits.RB2   //en = PORTB.2
#define busy PORTDbits.RD7 //busy = PORTD.7

void main()
    {
        TRISD = 0;                 //both ports B and D as output
```

```
        TRISB = 0;
        en = 0;                  //enable idle low
        MSDelay(250);            //long delay
        lcdcmd(0x38);
        MSDelay(250);            //long delay
        lcdcmd(0x0E);
        lcdready();              //check the LCD busy flag
        lcdcmd(0x01);
        lcdready();              //check the LCD busy flag
        lcdcmd(0x06);
        lcdready();              //check the LCD busy flag
        lcdcmd(0x86);            //line 1, position 6
        lcdready();              //check the LCD busy flag
        lcddata('M');
        lcdready();              //check the LCD busy flag
        lcddata('D');
        lcdready();              //check the LCD busy flag
        lcddata('E');
    }

void lcdcmd(unsigned char value)
    {
        ldata = value;          //put the value on the pins
        rs = 0;
        rw = 0;
        en = 1;                 //strobe the enable pin
        MSDelay(1);
        en = 0;
    }

void lcddata(unsigned char value)
    {
        ldata = value;          //put the value on the pins
        rs = 1;
        rw = 0;
        en = 1;                 //strobe the enable pin
        MSDelay(1);
        en = 0;
    }

void lcdready()
    {
        TRISD = 0xFF;           //make PORTD an input
        rs = 0;
        rw = 1;
        do                      //wait here for busy flag
          {
            en = 1;             //strobe the enable pin
            MSDelay(1);
            en = 0;
          }while(busy==1);
        TRISD = 0;
    }

void MSDelay(unsigned int itime)
    {
        unsigned int i, j;
        for(i=0;i<itime;i++)
          for(j=0;j<135;j++);
    }
```

```
//Program 12-3C: C version of Program 12-3 Displaying Data in ROM
#include <P18F458.h>
#define ldata PORTD        //PORTD = LCD data pins (Fig. 12-2)
#define rs PORTBbits.RB0             //rs = PORTB.0
#define rw PORTBbits.RB1             //rw = PORTB.1
#define en PORTBbits.RB2             //en = PORTB.2

#pragma romdata mycom = 0x300       //command at ROM addr 0x300
far rom const char mycom[] = {0x0E,0x01,0x06,0x84};

#pragma romdata mydata = 0x320      //data at ROM addr 0x320
far rom const char mydata[] = "HELLO";

void main()
  {
    unsigned char z=0;
    TRISD = 0;              //both ports B and D as output
    TRISB = 0;
    en = 0;                 //enable idle low
    MSDelay(250);
    lcdcmd(0x38);
    MSDelay(250);
    //send out the commands
    for(;z<4;z++)
      {
        lcdcmd(mycom[z]);
        MSDelay(15);
      }
    //send out the data
    for(z=0;z<5;z++)
      {
        lcddata(mydata[z]);
        MSDelay(15);
      }
    while(1); //infinite loop
  }

void lcdcmd(unsigned char value)
  {
    ldata = value;          //put the value on the pins
    rs = 0;
    rw = 0;
    en = 1;                 //strobe the enable pin
    MSDelay(1);
    en = 0;
  }

void lcddata(unsigned char value)
  {
    ldata = value;          //put the value on the pins
    rs = 1;
    rw = 0;
```

```
        en = 1;                    //strobe the enable pin
        MSDelay(1);
        en = 0;
    }
void MSDelay(unsigned int itime)
    {
        unsigned int i, j;
        for(i=0;i<itime;i++)
            for(j=0;j<135;j++);
    }
```

Review Questions

1. The RS pin is an _____ (input, output) pin for the LCD.
2. The E pin is an _____ (input, output) pin for the LCD.
3. The E pin requires an _____ (H-to-L, L-to-H) pulse to latch in information at the data pins of the LCD.
4. For the LCD to recognize information at the data pins as data, RS must be set to _____ (HIGH, LOW).
5. Give the command codes for line 1, first character, and line 2, first character.

SECTION 12.2: KEYBOARD INTERFACING

Keyboards and LCDs are the most widely used input/output devices and a basic understanding of them is essential. In this section, we first discuss keyboard fundamentals, along with key press detection and key identification mechanisms. Then we show how a keyboard is interfaced to a PIC18.

Interfacing the keyboard to the PIC18

At the lowest level, keyboards are organized in a matrix of rows and columns. The CPU accesses both rows and columns through ports; therefore, with two 8-bit ports, an 8×8 matrix of keys can be connected to a microprocessor. When a key is pressed, a row and a column make a contact; otherwise, there is no connection between rows and columns. In IBM PC keyboards, a single microcontroller takes care of hardware and software interfacing of the keyboard. In such systems, programs stored in the ROM of the microcontroller scan the keys continuously, identify which one has been activated, and present it to the motherboard. See Example 12-3. In programming for keypad interfacing we must have two processes: (a) key press detection, and (b) key identification. There are two ways by which the PIC18 can perform key press detection: (1) the interrupt method, and (2) the scanning method. In the PIC18, the PORTB-Change interrupt can be used to implement the interrupt-based key press detection. Next we explain the interrupt method.

Interrupt method of key press detection

Figure 12-6 shows a 4×4 matrix keypad connected to PORTB. The rows are connected to PORTB.Low (RB3–RB0) and the columns are connected to PORTB.High (RB7–RB4), which is the PORTB-Change interrupt. As we dis-

From Figure 12-6, identify the row and column of the pressed key for the following:
RB3– RB0 = 1110 for the row, RB7–RB4 = 1011 for the column
Solution:

From Figure 12-6, the row and column can be used to identify the key. The row belongs
to RB0 and the column belongs to RB6; therefore, key number 2 was pressed.

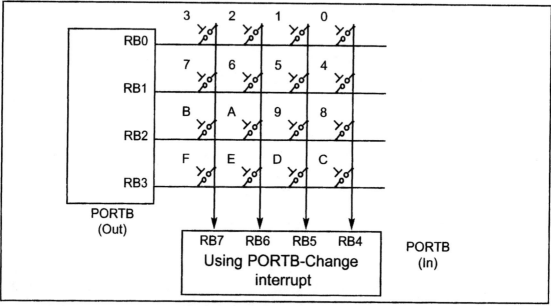

Figure 12-6. Matrix Keyboard Connection to Ports

cussed in Chapter 11, any changes on the RB7–RB4 pins will cause an interrupt
indicating a key press. Examine Program 12-4, which goes through the following
stages:

1. To make sure that the preceeding key has been released, 0s are output to all
   rows at once, and the columns are read and checked repeatedly until all the
   columns are HIGH. When all columns are found to be HIGH, the program
   waits for a short amount of time before it goes to the next stage of waiting for
   a key to be pressed.

2. To see if any key is pressed, the columns are connected to the PORTB-Change
   interrupt. Therefore, any key press will cause an interrupt and the microcontroller will execute the ISR. The ISR must do two things: (a) ensure that the
   first key press detection was not erroneous due to spike noise, and (b) wait 20
   ms to prevent the same key press from being interpreted as multiple key press-
   es. See Figure 12-8 for keyboard debounce.

3. To detect which row the key press belongs to, the microcontroller grounds one
   row at a time, reading the columns each time. If it finds that all columns are
   HIGH, this means that the key press cannot belong to that row; therefore, it
   grounds the next row and continues until it finds the row the key press belongs

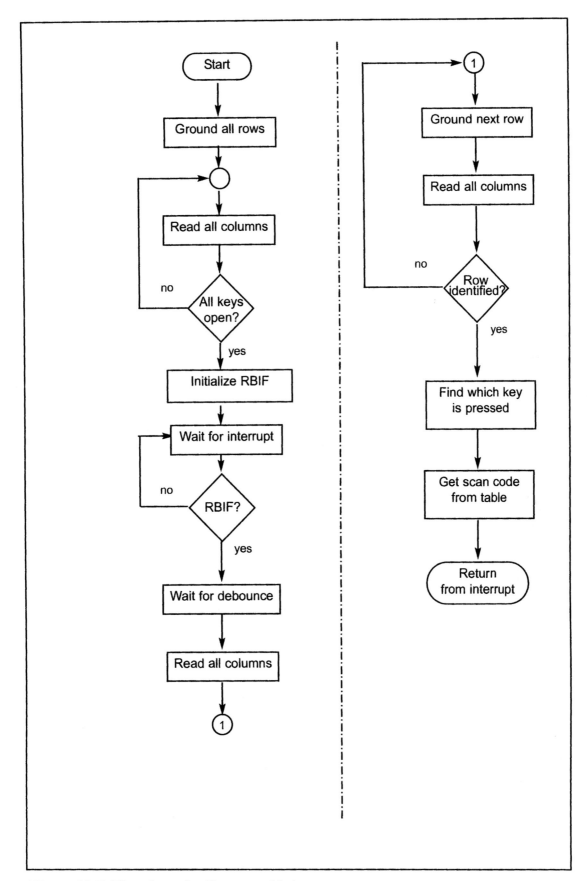

Figure 12-7. Flowchart for Program 12-4

to. Upon finding the row that the key press belongs to, it sets up the starting address for the look-up table holding the scan codes (or the ASCII value) for that row and goes to the next stage to identify the key.

4. To identify the key press, the microcontroller rotates the column bits, one bit at a time, into the carry flag and checks to see if it is LOW. Upon finding the zero, it pulls out the ASCII code for that key from the look-up table; otherwise, it increments the pointer to point to the next element of the look-up table. Figure 12-7 flowcharts this process.

The look-up table method shown in Program 12-4 can be modified to work with any matrix up to 8 × 4. Figure 12-7 provides the flowchart for Program 12-4 for scanning and identifying the pressed key.

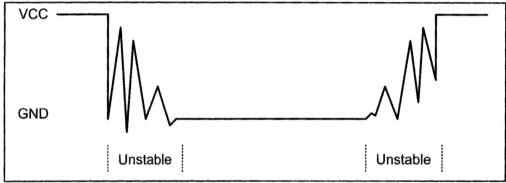

Figure 12-8. Keyboard Debounce

Examine Program 12-4. Notice in that program that the interrupt detects the key press. Then it is the job of the ISR to identify to which key the key press belongs (key identification). Program 12-4C is a C18 version of Program 12-4. In the Assembly version (12-4), the character is placed on PORTD, while in the C18 version (12-4C), it is sent to the serial port to be displayed on the monitor.

Program 12-4: This program waits for a key press on PORTB, then places the character on PORTD. We assume the following for this program:
RB3–RB0 connected to rows
RB7–RB4 connected to columns

```
D15mH       EQU    D'100'        ;15 ms delay high byte of value
D15mL       EQU    D'255'        ;low byte of value
COL         EQU    0x08          ;holds the column found
DR15mH      EQU    0x09          ;registers for 15 ms delay
DR15mL      EQU    0x0A          ;
;-----------------------------------------------
            ORG    0x000000
RESET_ISR   GOTO   MAIN          ;jump over interrupt table
            ORG    0x000008
HI_ISR      BTFSC  INTCON,RBIF   ;Was it a PORTB change?
            BRA    RBIF_ISR      ;yes then go to ISR
            RETFIE               ;else return
```

```
;-----------program for initialization
MAIN        CLRF    TRISD       ;make PORTD output port
            BCF     INTCON2,RBPU;enable PORTB pull-up resistors
            MOVLW   0xF0        ;make PORTB high input ports
            MOVWF   TRISB       ;make PORTB low output ports
            MOVWF   PORTB       ;ground all rows
KEYOPEN     CPFSEQ  PORTB       ;are all keys open
            GOTO    KEYOPEN     ;wait until keypad ready
            MOVLW   upper(KCODE0)
            MOVWF   TBLPTRU     ;load upper byte of TBLPTR
            MOVLW   high(KCODE0)
            MOVWF   TBLPTRH     ;load high byte of TBLPTR
            BSF     INTCON,RBIE ;enable PORTB change interrupt
            BSF     INTCON,GIE  ;enable all interrupts globally
LOOP        GOTO    LOOP        ;wait for key press
;-----------key identification ISR
RBIF_ISR    CALL    DELAY       ;wait for debounce
            MOVFF   PORTB,COL   ;get the column of key press
            MOVLW   0xFE
            MOVWF   PORTB       ;ground row 0
            CPFSEQ  PORTB       ;Did PORTB change?
            BRA     ROW0        ;yes then row 0
            MOVLW   0xFC
            MOVWF   PORTB       ;ground row 1
            CPFSEQ  PORTB       ;Did PORTB change?
            BRA     ROW1        ;yes then row 1
            MOVLW   0xFB
            MOVWF   PORTB       ;ground row 2
            CPFSEQ  PORTB       ;Did PORTB change?
            BRA     ROW2        ;yes then row 2
            MOVLW   0xF7
            MOVWF   PORTB       ;ground row 3
            CPFSEQ  PORTB       ;Did PORTB change?
            BRA     ROW3        ;yes then row 3
            GOTO    BAD_RBIF    ;no then key press too short
ROW0        MOVLW   low(KCODE0) ;set TBLPTR = start of row 0
            BRA     FIND        ;find the column
ROW1        MOVLW   low(KCODE1) ;set TBLPTR = start of row 1
            BRA     FIND        ;find the column
ROW2        MOVLW   low(KCODE2) ;set TBLPTR = start of row 2
            BRA     FIND        ;find the column
ROW3        MOVLW   low(KCODE3) ;set TBLPTR = start of row 3
FIND        MOVWF   TBLPTRL     ;load low byte of TBLPTR
            MOVLW   0xF0
            XORWF   COL         ;invert high nibble
            SWAPF   COL,F       ;bring to low nibble
AGAIN       RRCF    COL         ;rotate to find column
            BC      MATCH       ;column found, get the ASCII code
            INCF    TBLPTRL     ;else point to next col. address
```

```
                BRA    AGAIN          ;keep searching
MATCH           TBLRD*+               ;get ASCII code from table
                MOVFF TABLAT,PORTD;display pressed key on PORTD
WAIT            MOVLW 0xF0
                MOVWF PORTB           ;reset PORTB
                CPFSEQ PORTB          ;Did PORTB change?
                BRA    WAIT           ;yes then wait for key release
                BCF    INTCON,RBIF    ;clear PORTB, change flag
                RETFIE                ;return and wait for key press
BAD_RBIF        MOVLW 0x00            ;return null
                GOTO  WAIT            ;wait for key release
;------------delay
DELAY:          MOVLW D15mH           ;high byte of delay
                MOVWF DR15mH          ;store in register
D2:             MOVLW D15mL           ;low byte of delay
                MOVWF DR15mL          ;store in register
D1:             DECF  DR15mL,F        ;stay until DR15mL becomes 0
                BNZ    D1
                DECF  DR15mH,F        ;loop until all DR15m = 0x0000
                BNZ    D2
                RETURN
;------------key scancode look-up table
                ORG    300H
KCODE0:         DB    '0','1','2','3'    ;ROW 0
KCODE1:         DB    '4','5','6','7'    ;ROW 1
KCODE2:         DB    '8','9','A','B'    ;ROW 2
KCODE3:         DB    'C','D','E','F'    ;ROW 3
                END
```

Program 12-4C shows keypad programming in PIC18 C.

Program 12-4C: This C18 program reads the keypad and sends the result to the serial port. We assume the following for this program.
RB0–RB3 connected to rows
RB4–RB7 connected to columns
Serial port is set for 9600 baud (10 MHz XTAL), 8-bit mode, and 1 stop bit.

```c
#include <p18f458.h>
void SerTX(unsigned char x);
void RBIF_ISR(void);
void MSDelay(unsigned int millisecs);
unsigned char keypad[4][4] = {'0','1','2','3',
                              '4','5','6','7',
                              '8','9','A','B',
                              'C','D','E','F'};

#pragma code My_HiPrio_Int =0x0008 //high-priority interrupt
void My_HiPrio_Int (void)
{
  _asm
    GOTO chk_isr
  _endasm
```

```
}
#pragma code

#pragma interrupt chk_isr    //which ISR
void chk_isr (void)
{
  if (INTCONbits.RBIF==1)     //RBIF caused interrupt?
  RBIF_ISR( );                //yes go to RBIF_ISR
}
#pragma code

void main()
{
  TRISD=0;                    //make PORTD output port
  INTCON2bits.RBPU=0;         //enable PORTB pull-up resistors
  TRISB=0xF0;            //PORTB low as output and high as input
  PORTB=0xF0;                 //clear PORTB low
  while(PORTB!=0xF0);         //wait until key not pressed
  TXSTA=0x20;                 //choose low baud rate, 8-bit
  SPBRG=15;                   //9600 baud rate, XTAL = 10 MHz
  TXSTAbits.TXEN=1;           //enable transmit
  RCSTAbits.SPEN=1;           //enable serial port
  INTCONbits.RBIE=1;          //enable PORTB interrupt on change
  INTCONbits.GIE=1;           //enable interrupts globally
  while(1);                   //wait until key press
}
void RBIF_ISR(void)           //finds the key pressed
{
  unsigned char temp,COL=0,ROW=4;
  MSDelay(15);
  temp = PORTB;               //get column
  temp ^= 0xF0;               //invert high nibble
  if(!temp) return;           //if false alarm return
  while(temp<<=1) COL++;      //find the column
  PORTB = 0xFE;               //ground row 0
  if(PORTB != 0xFE)           //Did high nibble change?
    ROW = 0;                  //yes then row 0
  else {                      //try next row
    PORTB = 0xFD;             //ground row 1
    if(PORTB != 0xFD)         //Did high nibble change?
      ROW = 1;                //yes then row 1
    else {                    //try next row
      PORTB = 0xFB;           //ground row 2
      if(PORTB != 0xFB)       //Did high nibble change?
        ROW = 2;              //yes then row 2
      else {                  //try last row
        PORTB = 0xF7;         //ground row 3
          if(PORTB != 0xF7)   //Did high nibble change?
            ROW = 3;          //yes then row 3
      }
    }
  }
  if(ROW<4)                   //Did we find a valid row?
    SerTX(keypad[ROW][COL]);  //then send character
  while(PORTB!=0xF0) PORTB=0xF0;   //wait for release
  INTCONbits.RBIF=0;          //reset flag
}
void SerTX(unsigned char x)   //sends character
{
  while(PIR1bits.TXIF!=1);    //wait until ready
  TXREG=x;                    //send character out serial port
}
```

```
void MSDelay(unsigned int millisecs)
{
    unsigned int i, j;
    for(i=0;i<millisecs;i++)
        for(j=0;j<135;j++);
}
```

Scanning method for key press detection

Another method for key press detection is by scanning. In this method, to detect a pressed key, the microcontroller grounds all rows by providing 0 to the output latch, then it reads the columns. If the data read from the columns are equal to 1111, no key has been pressed and the process continues until a key press is detected. If one of the column bits has a zero, however, this means that a key press has occurred. After a key press is detected, the microcontroller will go through the process of identifying the key. Starting with the top row, the microcontroller grounds it by providing a LOW to the first row only; then it reads the columns. If the data read is all 1s, no key in that row is activated and the process is moved to the next row. It grounds the next row, reads the columns, and checks for any zero. This process continues until the row is identified. After identification of the row in which the key has been pressed, the next task is to find out which column the pressed key belongs to. This should be easy since the microcontroller knows at any time which row and column are being accessed. Figure 12-9 shows the flowchart for this method. The program implementation is left to the reader.

Some IC chips, such as National Semiconductor's MM74C923, incorporate keyboard scanning and decoding all in one chip. Such chips use combinations of counters and logic gates (no microcontroller) to implement the underlying concepts presented in this section.

Review Questions

1. True or false. To see if any key is pressed, all rows are grounded.
2. If RB7–RB4 = 0111 is the data read from the columns, which column does the pressed key belong to?
3. True or false. Key press detection and key identification require two different processes.
4. In Figure 12-6, if the rows are RB3–RB0 = 1110 and the columns are RB7–RB4 = 1110, which key is pressed?
5. True or false. To identify the pressed key, one row at a time is grounded.

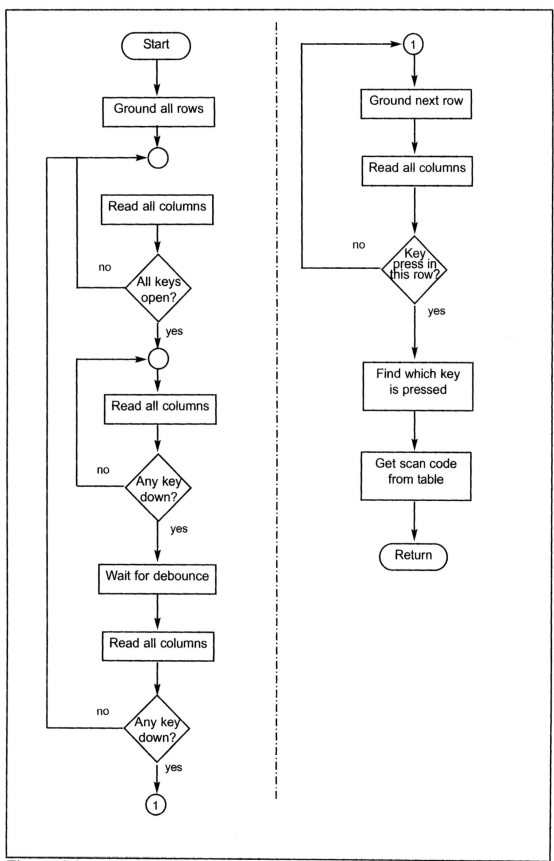

Figure 12-9. Flowchart of Scanning Method for Key Press Detection

SUMMARY

This chapter showed how to interface real-world devices such as LCDs and keypads to the PIC18. First, we described the operation modes of LCDs, then described how to program the LCD by sending data or commands to it via its interface to the PIC18.

Keyboards are one of the most widely used input devices for PIC18 projects. This chapter also described the operation of keyboards, including key press detection and key identification mechanisms. Then the PIC18 was shown interfacing with a keyboard. PIC18 programs were written to return the ASCII code for the pressed key.

PROBLEMS

SECTION 12.1: LCD INTERFACING

1. The LCD discussed in this section has _____ (4, 8) data pins.
2. Describe the function of pins E, R/W, and RS in the LCD.
3. What is the difference between the V_{CC} and V_{EE} pins on the LCD?
4. "Clear LCD" is a _____ (command code, data item) and its value is ___ hex.
5. What is the hex value of the command code for "display on, cursor on"?
6. Give the state of RS, E, and R/W when sending a command code to the LCD.
7. Give the state of RS, E, and R/W when sending data character 'Z' to the LCD.
8. Which of the following is needed on the E pin in order for a command code (or data) to be latched in by the LCD?
   (a) H-to-L pulse (b) L-to-H pulse
9. True or false. For the above to work, the value of the command code (data) must already be at the D0–D7 pins.
10. There are two methods of sending streams of characters to the LCD: (1) checking the busy flag, or (2) putting some time delay between sending each character without checking the busy flag. Explain the difference and the advantages and disadvantages of each method. Also explain how we monitor the busy flag.
11. For a 16×2 LCD, the location of the last character of line 1 is 8FH (its command code). Show how this value was calculated.
12. For a 16×2 LCD, the location of the first character of line 2 is C0H (its command code). Show how this value was calculated.
13. For a 20×2 LCD, the location of the last character of line 2 is 93H (its command code). Show how this value was calculated.
14. For a 20×2 LCD, the location of the third character of line 2 is C2H (its command code). Show how this value was calculated.
15. For a 40×2 LCD, the location of the last character of line 1 is A7H (its command code). Show how this value was calculated.
16. For a 40×2 LCD, the location of the last character of line 2 is E7H (its command code). Show how this value was calculated.

17. Show the value (in hex) for the command code for the 10th location, line 1 on a 20 × 2 LCD. Show calculations.

18. Show the value (in hex) for the command code for the 20th location, line 2 on a 40 × 2 LCD. Show calculations.

19. Rewrite the COMNWRT subroutine. Assume connections RC4 = RS, RC5 = R/W, RC6 = E.

20. Repeat Problem 19 for the data write subroutine. Send the string "Hello" to the LCD by checking the busy flag. Use the instruction TBLRD.

SECTION 12.2: KEYBOARD INTERFACING

21. In reading the columns of a keyboard matrix, if no key is pressed we should get all _____ (1s, 0s).

22. In Program 12-4, to detect the key press, which of the following is performed?
    (a) PORTB-Change interrupt   (b) grounding one row at time

23. In Figure 12-6, to identify the key pressed, which of the following is grounded?
    (a) all rows   (b) one row at time   (c) both (a) and (b)

24. For Figure 12-6, indicate the key press for RB7–RB4 = 0111, RB3–RB0 = 1110.

25. Indicate an advantage and a disadvantage of using an IC chip instead of a microcontroller for keyboard scanning and decoding.

26. What is the best compromise for the answer to Problem 25?

ANSWERS TO REVIEW QUESTIONS

SECTION 12.1: LCD INTERFACING

1.   Input
2.   Input
3.   H-to-L
4.   HIGH
5.   80H and C0H

SECTION 12.2: KEYBOARD INTERFACING

1.   True
2.   Column 3
3.   True
4.   0
5.   True

CHAPTER 13

ADC, DAC, AND SENSOR INTERFACING

OBJECTIVES

Upon completion of this chapter, you will be able to:

>> Discuss the ADC (analog-to-digital converter) section of the PIC18 chip
>> Interface temperature sensors to the PIC18
>> Explain the process of data acquisition using ADC
>> Describe factors to consider in selecting an ADC chip
>> Program the PIC18's ADC in C and Assembly
>> Describe the basic operation of a DAC (digital-to-analog converter) chip
>> Interface a DAC chip to the PIC18
>> Program a DAC chip to produce a sine wave on an oscilloscope
>> Program DAC chips in PIC18 C and Assembly
>> Explain the function of precision IC temperature sensors
>> Describe signal conditioning and its role in data acquisition

This chapter explores more real-world devices such as ADCs (analog-to-digital converters), DACs (digital-to-analog converters), and sensors. We will also explain how to interface the PIC18 to these devices. In Section 13.1, we describe analog-to-digital converter (ADC) chips. We will program the ADC portion of the PIC18 chip in Section 13.2. The characteristics of DAC chips are discussed in Section 13.3. In Section 13.4, we show the interfacing of sensors and discuss the issue of signal conditioning.

SECTION 13.1: ADC CHARACTERISTICS

This section will explore ADC programming in PIC18 chips. First, we describe some general aspects of the ADC itself, then show how to program the ADC portion of the PIC18 in both Assembly and C.

ADC devices

Analog-to-digital converters are among the most widely used devices for data acquisition. Digital computers use binary (discrete) values, but in the physical world everything is analog (continuous). Temperature, pressure (wind or liquid), humidity, and velocity are a few examples of physical quantities that we deal with every day. A physical quantity is converted to electrical (voltage, current) signals using a device called a *transducer*. Transducers are also referred to as *sensors*. Sensors for temperature, velocity, pressure, light, and many other natural quantities produce an output that is voltage (or current). Therefore, we need an analog-to-digital converter to translate the analog signals to digital numbers so that the microcontroller can read and process them. See Figures 13-1 and 13-2.

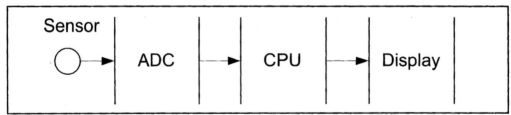

Figure 13-1. Microcontroller Connection to Sensor via ADC

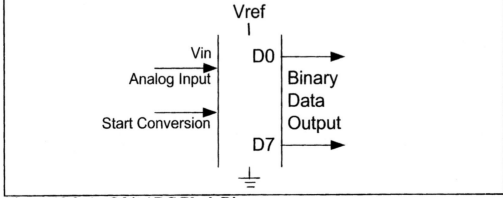

Figure 13-2. An 8-bit ADC Block Diagram

Table 13-1: Resolution versus Step Size for ADC (Vref = 5 V)

n-bit	Number of steps	Step size (mV)
8	256	5/256 = 19.53
10	1,024	5/1,024 = 4.88
12	4,096	5/4,096 = 1.2
16	65,536	5/65,536 = 0.076

Notes: V_{CC} = 5 V

Step size (resolution) is the smallest change that can be discerned by an ADC.

Some of the major characteristics of the ADC are as follows:

Resolution

ADC has *n*-bit resolution, where *n* can be 8, 10, 12, 16, or even 24 bits. The higher-resolution ADC provides a smaller step size, where *step size* is the smallest change that can be discerned by an ADC. Some widely used resolutions for ADCs are shown in Table 13-1. Although the resolution of an ADC chip is decided at the time of its design and cannot be changed, we can control the step size with the help of what is called Vref. This is discussed below.

Conversion time

In addition to resolution, conversion time is another major factor in judging an ADC. *Conversion time* is defined as the time it takes the ADC to convert the analog input to a digital (binary) number. The conversion time is dictated by the clock source connected to the ADC in addition to the method used for data conversion and technology used in the fabrication of the ADC chip such as MOS or TTL technology.

V_{ref}

V_{ref} is an input voltage used for the reference voltage. The voltage connected to this pin, along with the resolution of the ADC chip, dictate the step size. For an 8-bit ADC, the step size is $V_{ref}/256$ because it is an 8-bit ADC, and 2 to the power of 8 gives us 256 steps. See Table 13-1. For example, if the analog input range needs to be 0 to 4 volts, V_{ref} is connected to 4 volts. That gives 4 V/256 = 15.62 mV for the step size of an 8-bit ADC. In another case, if we need a step size of 10 mV for an 8-bit ADC, then V_{ref} = 2.56 V, because 2.56 V/256 = 10 mV. For

Table 13-2: V_{ref} Relation to V_{in} Range for an 8-bit ADC

V_{ref} (V)	V_{in} (V)	Step Size (mV)
5.00	0 to 5	5/256 = 19.53
4.0	0 to 4	4/256 = 15.62
3.0	0 to 3	3/256 = 11.71
2.56	0 to 2.56	2.56/256 = 10
2.0	0 to 2	2/256 = 7.81
1.28	0 to 1.28	1.28/256 = 5
1	0 to 1	1/256 = 3.90

Table 13-3: V_{ref} Relation to V_{in} Range for an 10-bit ADC

V_{ref} (V)	V_{in} (V)	Step Size (mV)
5.00	0 to 5	5/1,024 = 4.88
4.096	0 to 4.096	4.096/1,024 = 4
3.0	0 to 3	3/1,024 = 2.93
2.56	0 to 2.56	2.56/1,024 = 2.5
2.048	0 to 2.048	2.048/1,024 = 2
1.28	0 to 1.28	1/1,024 = 1.25
1.024	0 to 1.024	1.024/1,024 = 1

the 10-bit ADC, if the Vref = 5V, then the step size is 4.88 mV as shown in Table 13-1. Tables 13-2 and 13-3 show the relationship between the Vref and step size for the 8- and 10-bit ADCs, respectively. In some applications, we need the differential reference voltage where Vref = Vref (+) − Vref (−). Often the Vref (−) pin is connected to ground and the Vref (+) pin is used as the Vref.

Digital data output

In an 8-bit ADC we have an 8-bit digital data output of D0–D7 while in the 10-bit ADC the data output is D0–D9. To calculate the output voltage, we use the following formula:

$$D_{out} = \frac{V_{in}}{step\ size}$$

where Dout = digital data output (in decimal), Vin = analog input voltage, and step size (resolution) is the smallest change, which is Vref/256 for an 8-bit ADC. See Example 13-1. This data is brought out of the ADC chip either one bit at a time (serially), or in one chunk, using a parallel line of outputs. This is discussed next.

Example 13-1

For an 8-bit ADC, we have V_{ref} = 2.56 V. Calculate the D0–D7 output if the analog input is: (a) 1.7 V, and (b) 2.1 V.
Solution:

Because the step size is 2.56/256 = 10 mV, we have the following:
(a) D_{out} = 1.7 V/10 mV = 170 in decimal, which gives us 10101011 in binary for D7–D0.

(b) D_{out} = 2.1 V/10 mV = 210 in decimal, which gives us 11010010 in binary for D7–D0.

Parallel versus serial ADC

The ADC chips are either parallel or serial. In parallel ADC, we have 8 or more pins dedicated to bringing out the binary data, but in serial ADC we have only one pin for data out. That means that inside the serial ADC, there is a paral-

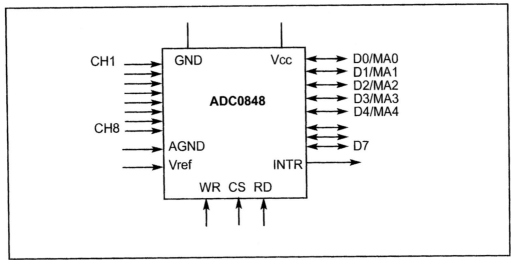

Figure 13-3. ADC0848 Parallel ADC Block Diagram

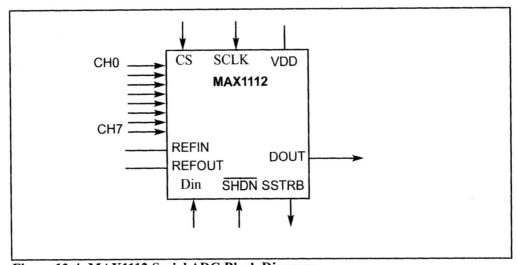

Figure 13-4. MAX1112 Serial ADC Block Diagram

lel-in-serial-out shift register responsible for sending out the binary data one bit at a time. The D0–D7 data pins of the 8-bit ADC provide an 8-bit parallel data path between the ADC chip and the CPU. In the case of the 16-bit parallel ADC chip, we need 16 pins for the data path. In order to save pins, many 12- and 16-bit ADCs use pins D0–D7 to send out the upper and lower bytes of the binary data. In recent years, for many applications where space is a critical issue, using such a large number of pins for data is not feasible. For this reason, serial devices such as the serial ADC are becoming widely used. While the serial ADCs use fewer pins and their smaller packages take much less space on the printed circuit board, more CPU time is needed to get the converted data from the ADC because the CPU must get data one bit at a time, instead of in one single read operation as with the parallel ADC. ADC848 is an example of a parallel ADC with 8 pins for the data output, while the MAX1112 is an example of a serial ADC with a single pin for Dout. Figures 13-3 and 13-4 show the block diagram for ADC848 and MAX1112, respectively.

Analog input channels

Many data acquisition applications need more than one ADC. For this reason, we see ADC chips with 2, 4, 8, or even 16 channels on a single chip. Multiplexing of analog inputs is widely used as shown in the ADC848 and MAX1112. In these chips, we have 8 channels of analog inputs, allowing us to monitor multiple quantities such as temperature, pressure, heat, and so on. PIC18 microcontroller chips come with 5 to 15 ADC channels, depending on the family member. The PIC18 ADC feature is discussed in the next section.

Start conversion and end-of-conversion signals

The fact that we have multiple analog input channels and a single digital output register makes it necessary for start conversion (SC) and end-of-conversion (EOC) signals. When SC is activated, the ADC starts converting the analog input value of Vin to an n-bit digital number. The amount of time it takes to convert varies depending on the conversion method as was explained earlier. When the data conversion is complete, the end-of-conversion signal notifies the CPU that the converted data is ready to be picked up.

From the discussion we conclude that the following steps must be followed for data conversion by an ADC chip:

1. Select a channel.
2. Activate the start conversion (SC) signal to start the conversion of analog input.
3. Keep monitoring the end-of-conversion (EOC) signal.
4. After the EOC has been activated, we read data out of the ADC chip.

Review Questions

1. Give two factors that affect the step size calculation.
2. The ADC0848 is a(n) _____-bit converter.
3. True or false. While the ADC0848 has 8 pins for D_{OUT}, the MAX1112 has only one D_{OUT} pin.
4. Indicate the number of analog input channels for each of the following ADC chips.
   (a) ADC0848   (b) MAX1112
5. Find the step size for an 8-bit ADC, if Vref = 1.28 V
6. For question 5, calculate the D0–D7 output if the analog input is: (a) 0.7 V, and (b) 1 V.

SECTION 13.2: ADC PROGRAMMING IN THE PIC18

Because the ADC is widely used in data acquisition, in recent years an increasing number of microcontrollers have an on-chip ADC peripheral, just like timers and USART. An on-chip ADC eliminates the need for an external ADC connection, which leaves more pins for other I/O activities. The vast majority of the PIC18 chips come with 8 channels of ADC, and some PIC18s have as many as 16 channels of ADCs. In this section we discuss the ADC feature of the PIC18452/458 and show how it is programmed in both Assembly and C.

PIC18F452/458 ADC features programming

The ADC peripheral of the PIC18 has the following characteristics:
(a) It is a 10-bit ADC.
(b) It can have 5 to 15 channels of analog input channels, depending on the family member. In PIC18452/458, pins RA0–RA7 of PORTA are used for the 8 analog channels. See Figures 13-5A and 13-5B.
(c) The converted output binary data is held by two special function registers called ADRESL (A/D Result Low) and ADRESH (A/D Result High).

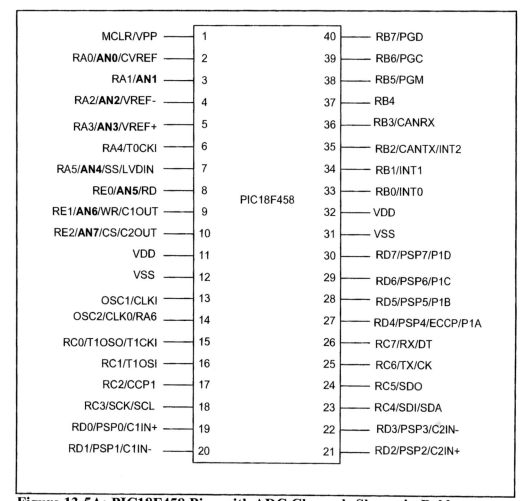

Figure 13-5A: PIC18F458 Pins with ADC Channels Shown in Bold

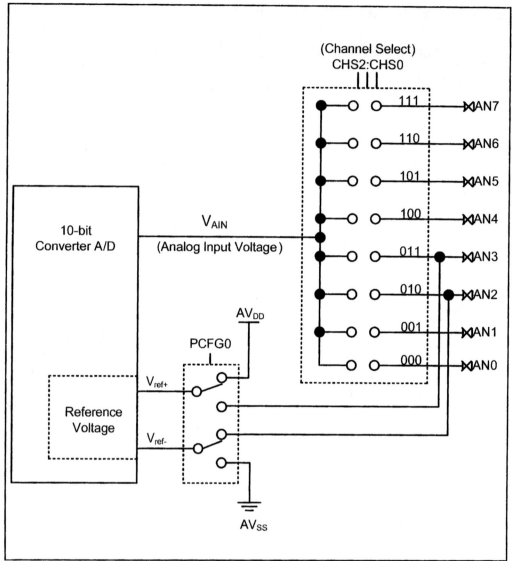

Figure 13-5B: PIC18 ADC Channel and Reference Selection

(d) Because the ADRESH:ADRESL registers give us 16 bits and the ADC data out is only 10 bits wide, 6 bits of the 16 are unused. We have the option of making either the upper 6 bits or the lower 6 bits unused.

(e) We have the option of using Vdd (Vcc), the voltage source of the PIC18 chip itself, as the Vref or connecting it to an external voltage source for the Vref.

(f) The conversion time is dictated by the Fosc of crystal frequency connected to the OSCs pins. While the Fosc for PIC18 can be as high as 40 MHz, the conversion time can not be shorter than 1.6 ms.

(g) It allows the implementation of the differential Vref voltage using the Vref(+) and Vref(−) pins, where Vref = Vref (+) − Vref (−).

Many of the above features can be programmed by way of ADCON0 (A/D control register 0) and ADCON1 (A/D control register 1), as we will see next.

ADCON0 register

The ADCON0 register is used to set the conversion time and select the analog input channel among other things. Figure 13-6 shows the ADCON0 register. In order to reduce the power consumption of the PIC18, the ADC feature is turned off when the microcontroller is powered up. We turn on the ADC with the ADON bit of the ADCON0 register, as shown in Figure 13-6. The other important bit is the GO/DONE bit. We use this bit to start conversion and monitor it to see if conversion has ended. Notice in ADCCON0 that not all family members have all the 8 analog input channels. The conversion time is set with the ADCS bits. While ADCS1 and ADCS0 are held by the ADCON0 register, ADCS2 is part of the ADCON1 register. This is discussed next.

ADCS1	ADCS0	CHS2	CHS1	CHS0	GO/DONE	--	ADON

ADCS2 (from ADCON1)	ADCS1	ADCS0	Conversion Clock Source
0	0	0	Fosc/2
0	0	1	Fosc/8
0	1	0	Fosc/32
0	1	1	Internal RC used for clock source
1	0	0	Fosc/4
1	0	1	Fosc/16
1	1	0	Fosc/64
1	1	1	Internal RC used for clock source

CHS2	CHS1	CHS0	CHANNEL SELECTION
0	0	0	CHAN0 (AN0)
0	0	1	CHAN1 (AN1)
0	1	0	CHAN2 (AN2)
0	1	1	CHAN3 (AN3)
1	0	0	CHAN4 (AN4)
1	0	1	CHAN5 (AN5) not implemented on 28-pin PIC18
1	1	0	CHAN6 (AN6) not implemented on 28-pin PIC18
1	1	1	CHAN7 (AN7) not implemented on 28-pin PIC18

GO/DONE A/D conversion status bit.
    1 = A/D conversion is in progress. This is used as start conversion, which means after the conversion is complete, it will go LOW to indicate the end-of-conversion.
    0 = A/D conversion is complete and digital data is available in registers ADRESH and ADRESL.

ADON A/D on bit
    0 = A/D part of the PIC18 is off and consumes no power. This is the default and we should leave it off for applications in which ADC is not used.
    1 = A/D feature is powered up.

Figure 13-6. ADCON0 (A/D Control Register 0)

ADCON1 register

Another major register of the PIC18's ADC feature is ADCON1. The ADCON1 register is used to select the Vref voltage among other things. It is shown in Figure 13-7. After the A/D conversion is complete, the result sits in registers ADRESL (A/D Result Low Byte) and ADRESH (A/D Result High Byte). The ADFM bit of the ADCON1 is used for making it right-justified or left-justified because we need only 10 bits of the 16. See Figure 13-8.

ADFM	ADCS2	--	--	PCFG3	PCFG2	PCFG1	PCFG0

ADFM A/D Result format select bit
> 1 = Right justified: The 10-bit result is in the ADRESL register and the lower 2 bits of ADRESH. That means the 6 most significant bits of the ADRESH register are all 0s.
> 0 = Left justified: The 10-bit result is in the ADRESL register and the upper 2 bits of ADRESL. That means the 6 least significant bits of the ADRESL register are all 0s.

ADCS2 A/D Clock Select bit 2. This bit along with the ADCS1 and ADCS0 bits of the ADCON0 register decide the conversion clock for the ADC. The default value for ADCS2 is 0, which means setting the ADCS0 and ADCS1 values of ADCON0 can give us clock conversion of Fosc/2, Fosc/8, and Fosc/32. See the ADCON0 register.

PCFGs: A/D Port Configuration Control bits:

PCFGs	AN7	AN6	AN5	AN4	AN3	AN2	AN1	AN0	Vref+	Vref-	C/R
0 0 0 0	A	A	A	A	A	A	A	A	Vdd	Vss	8/0
0 0 0 1	A	A	A	A	Vref+	A	A	A	AN3	Vss	7/1
0 0 1 0	D	D	D	A	A	A	A	A	Vdd	Vss	5/0
0 0 1 1	D	D	D	A	Vref+	A	A	A	AN3	Vss	4/1
0 1 0 0	D	D	D	D	A	D	A	A	Vdd	Vss	3/0
0 1 0 1	D	D	D	D	Vref+	D	A	A	AN3	Vss	2/1
0 1 1 x	D	D	D	D	D	D	D	D	-	-	0/0
1 0 0 0	A	A	A	A	Vref+	Vref–	A	A	AN3	AN2	6/2
1 0 0 1	D	D	A	A	A	A	A	A	Vdd	Vss	6/0
1 0 1 0	D	D	A	A	Vref+	A	A	A	AN3	Vss	5/1
1 0 1 1	D	D	A	A	Vref+	Vref–	A	A	AN3	AN2	4/2
1 1 0 0	D	D	D	A	Vref+	Vref–	A	A	AN3	AN2	3/2
1 1 0 1	D	D	D	D	Vref+	Vref–	A	A	AN3	AN2	2/2
1 1 1 0	D	D	D	D	D	D	D	A	Vdd	Vss	1/0
1 1 1 1	D	D	D	D	Vref+	Vref–	D	A	AN3	AN2	1/2

A = Analog input, D = Digital I/O
C/R = # of analog input channels / # of pins used for A/D voltage reference
The default is option 0000, which gives us 8 channels of analog input and uses the Vdd of PIC18 as Vref.

Figure 13-7. ADCON1 (A/D Control Register 1)

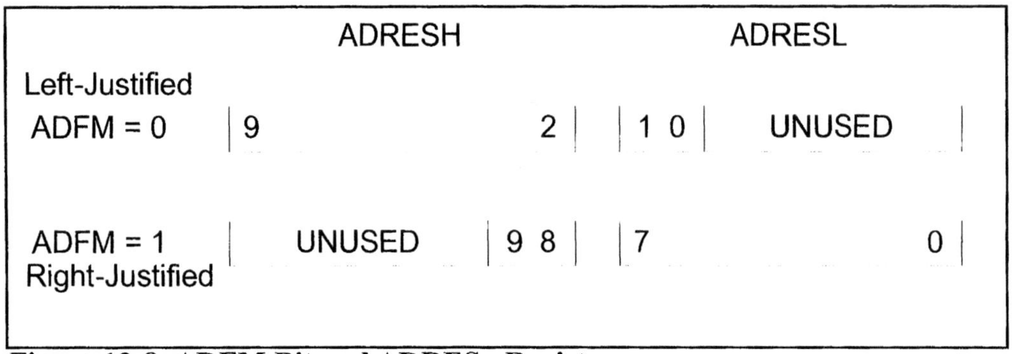

	ADRESH		ADRESL	
Left-Justified ADFM = 0	|9	2|	|1 0|	UNUSED |
ADFM = 1 Right-Justified	| UNUSED	|9 8|	|7	0|

Figure 13-8. ADFM Bit and ADRESx Registers

The port configuration for A/D channels is handled by the PCFG (A/D port configuration) bits. While in chips such as the PIC18452/458, we can have up to 8 channels of analog input, not all applications need that many ADC inputs. The PORTA pins of RA0–RA3 and RA5 and RE0–RE2 of PORTE are used for the analog input channels. With PCFG = 0110, we can use all the pins of the PORTA as the digital I/O. The default is PCFG = 0000, which allows us to use all 8 pins for analog inputs. In that case Vref = Vdd, the same voltage source used by the PIC18 chip itself. In many applications we need Vref other than Vdd. The AN3 pin can be used as an external source of voltage for Vref. For example, option PCFG = 0101 allows us to use two channels for analog inputs, AN3 = Vref, and the other 5 pins for the digital I/O. In this case the Vss (Gnd) pin of the PIC18 is used for the Vref (–). See Examples 13-2 and 13-3.

Example 13-2

For a PIC18-based system, we have V_{ref} = Vdd = 5 V. Find (a) the step size, and (b) the ADCON1 value if we need 3 channels. Assume that the ADRESH:ADRESL registers are right justified.

Solution:

(a) The step size is 5/1,024 = 4.8 mV.
(b) ADCON1 = 1x000100 because option 100 gives us 3 analog input channels. The x = ADCS2 is decided by the conversion speed.

Example 13-3

For a PIC18-based system, we have V_{ref} = 2.56 V. Find (a) the step size, and (b) the ADCON1 value if we need 3 channels. Assume that the ADRESH:ADRESL registers are right justified.

Solution:

(a) The step size is 2.56/1,024 = 2.5 mV.
(b) ADCON1 = 1x000011 because option 0011 gives us 3 analog input channels where x = ADCS2 is decided by the conversion speed.

Calculating A/D conversion time

By using the ADCS (A/D clock source) bits of both the ADCON0 and ADCON1 registers we can set the A/D conversion time. The conversion time is defined in terms of Tad, where Tad is the conversion time per bit. To calculate the Tad, we can select a conversion clock source of Fosc/2, Fosc/4, Fosc/8, Fosc/16, Fosc/32, or Fosc/64, where Fosc is the speed of the crystal frequency connected to the PIC18 chip. For the PIC18, the conversion time is 12 times the Tad. Notice that the Tad cannot be faster than 1.6 ms. Look at Examples 13-4 and 13-5 for clarification.

We can also use the the internal RC oscillator for the conversion clock source, instead of the Fosc of the external crystal oscillator. In that case the Tad is typically 4–6 μs and conversion time is 12×6 μs = 72 μs.

Another timing factor that we must pay attention to is the acquisition time (Tacq). After an A/D channel is selected, we must allow some time for the sample-and-hold capacitor (C hold) to charge fully to the input voltage level present at the channel. It is only after the elapsing of this acquisition time that the A/D conversion can be started. Although many factors (e.g., Vdd and temperature) affect the duration of Tacq, we can use a typical value of 15 μs. In some newer generations of the PIC18, we have the option of controlling the exact time of Tacq by programming the internal register ADCON2. In the PIC18F452/458, we have only the ADCON0 and ADCON1 registers. See Example 13-6.

Example 13-4

A PIC18 is connected to the 10 MHz crystal oscillator. Calculate the conversion time for all options of ADCS bits in both the ADCON0 and ADCON1 registers.

Solution:

The options for the conversion clock source for both ADCON0 and ADCON1 are as follows:
(a) For Fosc/2, we have 10 MHz / 2 = 5 MHz.
Tad = 1 / 5 MHz = 200 ns. Invalid because it is faster than 1.6 μs.
(b) For Fosc/4, we have 10 MHz / 4 = 2.5 MHz.
Tad = 1 / 2.5 MHz = 400 ns. Invalid because it is faster than 1.6 μs.
(c) For Fosc/8, we have 10 MHz / 8 = 1.25 MHz.
Tad = 1 / 2.5 MHz = 800 ns. Invalid because it is faster than 1.6 μs.
(d) For Fosc/16, we have 10 MHz / 16 = 625 kHz.
Tad =1 / 625 kHz = 1.6 μs. The conversion time = 12×1.6 μs = 19.2 μs
(e) For Fosc/32, we have 10 MHz / 32 = 312.5 kHz.
Tad = 1 / 312.5 kHz = 3.2 μs. The conversion time = 12×3.2 μs = 38.4 μs
(f) For Fosc/64, we have 10 MHz / 64 = 156.25 kHz.
Tad = 1 / 156.25 kHz = 6.4 μs. The conversion time = 12×6.4 μs = 76.8 μs
Notice that for the Fosc/4, Fosc/16, and Fosc/64 selections, we must use the ADSC2 bit in the ADCON1 register, in addition to the ADCS bits in the ADCON0 register.

Example 13-5

A PIC18 is connected to the 4 MHz crystal oscillator. Calculate the conversion time if we want to use only the ADCS bits of the ADCON0 register.

Solution:

The options for the conversion clock source available in the ADCON0 register are as follows:
(a) For Fosc/2, we have 4 MHz / 2 = 2 MHz.
Tad = 1 / 2 MHz = 400 ns. Invalid because it is faster than 1.6 μs.
(b) For Fosc/8, we have 4 MHz / 8 = 500 kHz.
Tad = 1 / 500 kHz = 2 μs. The conversion time = 12 × 2 μs = 24 μs
(c) For Fosc/32, we have 4 MHz / 32 = 125 kHz.
Tad = 1 / 125 kHz = 8 μs. The conversion time = 12 × 8 μs = 96 μs

Example 13-6

Find the values for the ADCON0 and ADCON1 registers for the following options: (a) channel AN0 as analog input, (b) Vref+ = Vdd, Vref– = Vss, (c) Fosc/64, (d) A/D result is right justified, and (e) A/D module is on.

Solution:

From Figure 13-6, we have ADCON0 = 10000x1. With x = 0 we have 10000001.
From Figure 13-7, we have ADCON1 = 11xx1110. With x = 0 we have 11001110.

Steps in programming the A/D converter using polling

To program the A/D converter of the PIC18, the following steps must be taken:

1. Turn on the ADC module of the PIC18 because it is disabled upon power-on reset to save power. We can use the "BSF ADCON0,ADON" instruction.
2. Make the pin for the selected ADC channel an input pin. We use "BSF TRISA,x." or "BSF TRISE,x" where x is the channel number.
3. Select voltage reference and A/C input channels. We use registers ADCON0 and ADCON1.
4. Select the conversion speed. We use registers ADCON0 and ADCON1.
5. Wait for the required acquisition time.
6. Activate the start conversion bit of GO/DONE.
7. Wait for the conversion to be completed by polling the end-of-conversion (GO/DONE) bit.
8. After the GO/DONE bit has gone LOW, read the ADRESL and ADRESH registers to get the digital data output.
9. Go back to step 5.

Programming PIC18F458 ADC in Assembly

The Assembly language Program 13-1 illustrates the steps for ADC conversion shown above. The C version of the program is shown in Program 13-1C.

```
Program 13-1: This program gets data from channel 0 (RA0)
of ADC and displays the result on PORTC and PORTD. This is
done every quarter of second.
;Program 13-1
      ORG    0000H
      CLRF   TRISC           ;make PORTC an output
      CLRF   TRISD           ;make PORTD an output
      BSF    TRISA,0 ;make RA0 an input for analog input
      MOVLW  0x81            ;Fosc/64, channel 0, A/D is on
      MOVWF  ADCON0
      MOVLW  0xCE ;right justified, Fosc/64, AN0 = analog
      MOVWF  ADCON1
OVER  CALL   DELAY ;wait for Tacq (sample and hold time)
      BSF    ADCON0,GO       ;start conversion
BACK  BTFSC  ADCON0,DONE     ;keep polling end-of-conversion
      BRA    BACK            ;wait for end of conversion
      MOVFF  ADRESL,PORTC    ;give the low byte to PORTC
      MOVFF  ADRESH,PORTD    ;give the high byte to PORTD
      CALL   QSEC_DELAY
      BRA    OVER            ;keep repeating it
      END
```

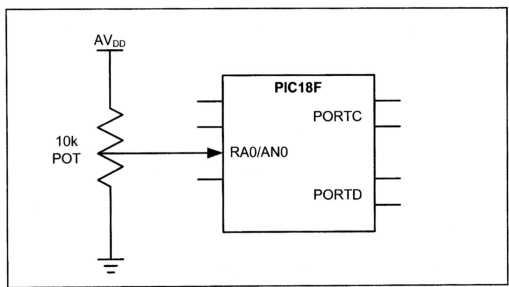

Figure 13-9. A/D Connection for Program 13-1

Programming PIC18F458 A/D in C

Program 13-1C is the C version of the ADC conversion for Program 13-1.

Program 13-1C: This program gets data from channel 0 (RA0) of ADC and displays the result on PORTC and PORTD. This is done every quarter of second. This is the C version of Program 13-1.
//Program 13-1C

```c
void main(void)
    {
        TRISC=0;                //make PORTC output port
        TRISD=0;                //make PORTD output port
        TRISAbits.TRISA0=0;     //RA0 = INPUT for analog input
        ADCON0 = 0x81;  //Fosc/64, channel 0, A/D is on
        ADCON1 = 0xCE;          //right justified, Fosc/64,
                                //AN0 = analog
        while(1)
        {
         DELAY(1); //give A/D channel time to sample
         ADCON0bits.GO = 1;     //start converting
         while(ADCON0bits.DONE == 1);
         PORTC=ADRESL;          //display low byte on PORTC
         PORTD=ADRESH;          //display high byte on PORTD
         DELAY(250);            //wait for one quarter of a
                                //second before trying again
        }
    }
```

Programming A/D converter using interrupts

In Chapter 11, we showed how to use interrupts instead of polling to avoid tying down the microcontroller. To program the A/D using the interrupt method, we need to set HIGH the ADIE (A/D interrupt enable) flag. If ADIE = 1, then upon the completion of the conversion, the ADIF (A/D interrupt flag) becomes HIGH, which will force the CPU to jump to read binary outputs. Table 13-4 shows to which registers these two flags belong.

Table 13-4: A/D Converter Interrupt Flag Bits and their Registers

Interrupt	Flag bit	Register	Enable bit	Register
ADIF (ADC)	ADIF	PIR1	ADIE	PIE1

Note: Upon power-on reset, the A/D is assigned to high-priority interrupt (vector address of 0008). We can use the ADIP bit of the IPR1 register to assign low priority to it, which will land at vector address 00018H. See Chapter 11.

```
;Program 13-2
      ORG   0000H
      GOTO MAIN          ;bypass interrupt vector table
;--on default all interrupts go to to address 00008
      ORG   0008H        ;interrupt vector table
      BTFSS PIR1,ADIF ;Did we get here due to A/D int?
      RETFIE             ;No. Then return to main
      GOTO  AD_ISR       ;Yes. Then go INT0 ISR
;--the main program for initialization
      ORG   00100H
MAIN CLRF   TRISC        ;make PORTC an output
      CLRF   TRISD       ;make PORTD an output
      BSF    TRISA,0 ;make RA0 an input pin for analog input
      MOVLW 0x81         ;Fosc/64, channel 0, A/D is on
      MOVWF ADCON0
      MOVLW 0xCE  ;right justified, Fosc/64, AN0 = analog
      MOVWF ADCON1
      BCF PIR1,ADIF      ;clear ADIF for the first round
      BSF PIE1,ADIE      ;enable A/D interrupt
      BSF INTCON,PEIE ;enable peripheral interrupts
      BSF INTCON,GIE  ;enable interrupts globally
OVER  CALL DELAY         ;wait for Tacq (sample and hold time)
      BSF   ADCON0,GO    ;start conversion
      BRA OVER           ;stay in this loop forever
;-----A/D Converter ISR
AD_ISR
      ORG 200H
      MOVFF ADRESL,PORTC    ;give the low byte to PORTC
      MOVFF ADRESH,PORTD    ;give the high byte to PORTD
      CALL QSEC_DELAY
      BCF PIR1,ADIF    ;clear ADIF interrupt flag bit
      RETFIE
      END


//Program 13-2C (This is the C version of Program 13-2)

#include <PIC18F458.h>

#pragma code My_HiPrio_Int=0x0008 //high-priority interrupt
void My_HiPrio_Int (void)
{
  chk_isr();
}
#pragma code              //end high-priority interrupt
#pragma interrupt chk_isr  //Which interrupt?
void chk_isr (void)
{
  if (PIR1bits.ADIF==1)   //A/D caused interrupt?
  AD_ISR( );              //Yes. Execute INT0 program
}
```

```
void main(void)
{
  TRISC=0;                       //make PORTC output port
  TRISD=0;                       //make PORTD output port
  TRISAbits.TRISA0=0;            //RA0 = INPUT for analog input
  ADCON0 = 0x81;                 //Fosc/64, channel 0, A/D is on
  ADCON1 = 0xCE; //right justified, Fosc/64, AN0 = analog
  PIR1bits.ADIF=0;               //clear A/D interrupt flag
  PIE1bits.ADIE=1;               //enable A/D interrupt
  INTCONbits.PEIE=1;             //enable peripheral interrupts
  INTCONbits.GIE=1;              //enable all interrupts globally
  while(1)               //keep looping until interrupt comes
  {
    DELAY(1);
    ADCON0bits.GO = 1;     //start converting
  }
}
//----------A/D ISR
void AD_ISR(void)
  {
    PORTC=ADRESL;              //display low byte on PORTC
    PORTD=ADRESH;              //display high byte on
    DELAY(250);                //wait for one quarter of a
                               //second before trying again
    PIR1bits.ADIF=0;           //clear A/D interrupt flag
  }
```

Review Questions

1. Give the main factor affecting the step size of A/D in PIC18.
2. The A/D of PIC18 is a(n) _____-bit converter.
3. True or false. The A/D of PIC18 has pins for D_{OUT}.
4. True or false. A/D in the PIC18 is an off-chip module.
5. Find the step size for an PIC18 ADC, if Vref = 1.024 V.
6. For problem 5, calculate the D0–D9 output if the analog input is: (a) 0.7 V, and (b) 1 V.
7. Indicate the number of available analog input channels for each of the following options in the ADCON0 register:
   (a) PCFG = 0100   (b) PCFG = 1001
8. True or false. The conversion time is equal to $12 \times$ Tad.
9. The minimum Tad allowed is _____ μs.
10. Which bit is used to poll for the end of conversion?

SECTION 13.3: DAC INTERFACING

This section will show how to interface a DAC (digital-to-analog converter) to the PIC18. Then we demonstrate how to generate a sine wave on the scope using the DAC.

Digital-to-analog converter (DAC)

The digital-to-analog converter (DAC) is a device widely used to convert digital pulses to analog signals. In this section we discuss the basics of interfacing a DAC to the PIC18.

Recall from your digital electronics course the two methods of creating a DAC: binary weighted and R/2R ladder. The vast majority of integrated circuit DACs, including the MC1408 (DAC0808) used in this section, use the R/2R method because it can achieve a much higher degree of precision. The first criterion for judging a DAC is its resolution, which is a function of the number of binary inputs. The common ones are 8, 10, and 12 bits. The number of data bit inputs decides the resolution of the DAC because the number of analog output levels is equal to 2^n, where n is the number of data bit inputs. Therefore, an 8-input DAC such as the DAC0808 provides 256 discrete voltage (or current) levels of output. See Figure 13-10. Similarly, the 12-bit DAC provides 4,096 discrete voltage levels. There are also 16-bit DACs, but they are more expensive.

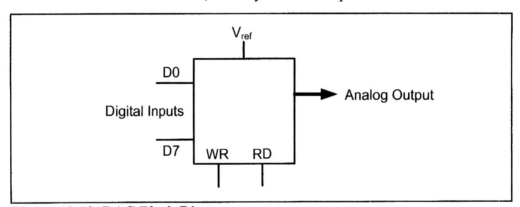

Figure 13-10. DAC Block Diagram

MC1408 DAC (or DAC0808)

In the MC1408 (DAC0808), the digital inputs are converted to current (I_{out}), and by connecting a resistor to the I_{out} pin, we convert the result to voltage. The total current provided by the I_{out} pin is a function of the binary numbers at the D0–D7 inputs of the DAC0808 and the reference current (I_{ref}), and is as follows:

$$I_{out} = I_{ref} \left(\frac{D7}{2} + \frac{D6}{4} + \frac{D5}{8} + \frac{D4}{16} + \frac{D3}{32} + \frac{D2}{64} + \frac{D1}{128} + \frac{D0}{256} \right)$$

where D0 is the LSB, D7 is the MSB for the inputs, and I_{ref} is the input current that must be applied to pin 14. The I_{ref} current is generally set to 2.0 mA. Figure 13-11 shows the generation of current reference (setting $I_{ref} = 2$ mA) by using the

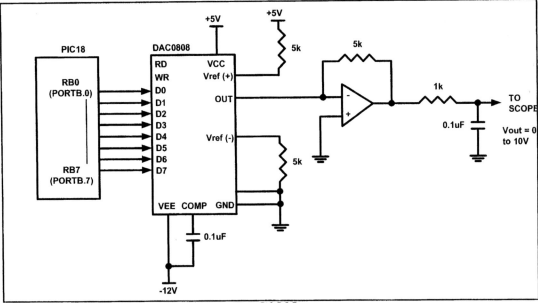

Figure 13-11. PIC18 Connection to DAC0808

standard 5 V power supply. Now assuming that I_{ref} = 2 mA, if all the inputs to the DAC are high, the maximum output current is 1.99 mA (verify this for yourself).

Converting I_{out} to voltage in DAC0808

Ideally we connect the output pin I_{out} to a resistor, convert this current to voltage, and monitor the output on the scope. In real life, however, this can cause inaccuracy because the input resistance of the load where it is connected will also affect the output voltage. For this reason, the I_{ref} current output is isolated by connecting it to an op-amp such as the 741 with R_f = 5 kOhms for the feedback resistor. Assuming that R = 5 kOhms, by changing the binary input, the output voltage changes as shown in Example 13-7.

Example 13-7
Assuming that R = 5 kOhms and I_{ref} = 2 mA, calculate V_{out} for the following binary inputs: (a) 10011001 binary (99H)   (b) 11001000 (C8H) **Solution:** (a) I_{out} = 2 mA (153/256) = 1.195 mA and V_{out} = 1.195 mA × 5K = 5.975 V (b) I_{out} = 2 mA (200/256) = 1.562 mA and V_{out} = 1.562 mA × 5K = 7.8125 V

Generating a sine wave

Example 13-8 shows how to generate a stair-step ramp. To generate a sine wave, we first need a table whose values represent the magnitude of the sine of angles between 0 and 360 degrees. The values for the sine function vary from –1.0 to +1.0 for 0- to 360-degree angles. Therefore, the table values are integer num-

Example 13-8

In order to generate a stair-step ramp, set up the circuit in Figure 13-11 and connect the output to an oscilloscope. Then write a program to send data to the DAC to generate a stair-step ramp.

Solution:

```
        CLRF  TRISB       ;PORTB as output
        CLRF  PORTB       ;clear PORTB
AGAIN:  INCF  PORTB,F     ;count from 0 to FFH send it to DAC
        RCALL DELAY       ;let DAC recover
        BRA   AGAIN
```

bers representing the voltage magnitude for the sine of theta. This method ensures that only integer numbers are output to the DAC by the PIC18 microcontroller. Table 13-5 shows the angles, the sine values, the voltage magnitudes, and the integer values representing the voltage magnitude for each angle (with 30-degree increments). To generate Table 13-5, we assumed a full-scale voltage of 10 V for DAC output (as designed in Figure 13-11). Full-scale output of the DAC is achieved when all the data inputs of the DAC are HIGH. Therefore, to achieve the full-scale 10 V output, we use the following equation.

$$V_{out} = 5\ V + (5 \times \sin\theta)$$

V_{out} of DAC for various angles is calculated and shown in Table 13-5. See Example 13-9 for verification of the calculations.

Table 13-5: Angle versus Voltage Magnitude for Sine Wave

Angle θ (degrees)	Sin θ	V_{out} (Voltage Magnitude) 5 V + (5 V × sin θ)	Values Sent to DAC (decimal) (Voltage Mag. × 25.6)
0	0	5	128
30	0.5	7.5	192
60	0.866	9.33	238
90	1.0	10	255
120	0.866	9.33	238
150	0.5	7.5	192
180	0	5	128
210	−0.5	2.5	64
240	−0.866	0.669	17
270	−1.0	0	0
300	−0.866	0.669	17
330	−0.5	2.5	64
360	0	5	128

Example 13-9

Verify the values given for the following angles: (a) 30° (b) 60°.

Solution:

(a) $V_{out} = 5\ V + (5\ V \times \sin \theta) = 5\ V + 5 \times \sin 30° = 5\ V + 5 \times 0.5 = 7.5\ V$
   DAC input value $= 7.5\ V \times 25.6 = 192$ (decimal)

(b) $V_{out} = 5\ V + (5\ V \times \sin \theta) = 5\ V + 5 \times \sin 60° = 5\ V + 5 \times 0.866 = 9.33\ V$
   DAC input value $= 9.33\ V \times 25.6 = 238$ (decimal)

To find the value sent to the DAC for various angles, we simply multiply the V_{out} voltage by 25.60 because there are 256 steps and full-scale V_{out} is 10 volts. Therefore, 256 steps / 10 V = 25.6 steps per volt. To further clarify this, look at the following code. This program sends the values to the DAC continuously (in an infinite loop) to produce a crude sine wave. See Figure 13-12.

```
;Program 13-3
OVER  MOVLW  upper(TABLE)
      MOVWF  TBLPTRU
      MOVLW  high(TABLE)
      MOVWF  TBLPTRH
      MOVLW  low(TABLE)
      MOVWF  TBLPTRL
      CLRF   TRISB
AGAIN TBLRD*
      MOVF   TABLAT,W
      XORLW  0x0
      BZ     OVER
      MOVWF  PORTB
      INCF   TBLPTRL,F
      BRA    AGAIN
      ORG    0x250
TABLE:     DB D'128',D'192',D'238',D'255',D'238',D'192'
      DB D'128',D'64',D'17',D'1',D'17',D'64',D'0'
      END

      ;to get a better looking sine wave, regenerate
      ;Table 13-5 for 2-degree angles
```

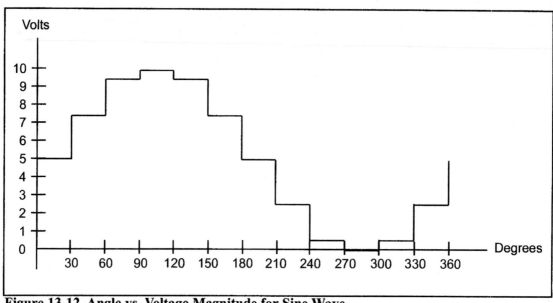

Figure 13-12. Angle vs. Voltage Magnitude for Sine Wave

Programming DAC in C

```c
//Program 13-3C. This is the C version of Program 13-3.
#include <p18F458.h>
rom const unsigned char WAVEVALUE[12] ={128,192,238,255,
                  238,192,128,64,
                  17,0,17,64};
void main()
  {
    unsigned char x;
    TRISB=0;
    while(1)
      {
      for(x=0;x<12;x++)
        PORTB = WAVEVALUE[x];
      }
  }
```

Review Questions

1. In a DAC, input is _____ (digital, analog) and output is _____ (digital, analog).
2. In an ADC, input is _____ (digital, analog) and output is _____ (digital, analog).
3. DAC0808 is a(n) ____-bit D-to-A converter.
4. (a) The output of DAC0808 is in _____ (current, voltage).
   (b) True or false. The output of DAC0808 is ideal to drive a motor.

SECTION 13.4: SENSOR INTERFACING AND SIGNAL CONDITIONING

This section will show how to interface sensors to the microcontroller. We examine some popular temperature sensors and then discuss the issue of signal conditioning. Although we concentrate on temperature sensors, the principles discussed in this section are the same for other types of sensors such as light and pressure sensors.

Temperature sensors

Transducers convert physical data such as temperature, light intensity, flow, and speed to electrical signals. Depending on the transducer, the output produced is in the form of voltage, current, resistance, or capacitance. For example, temperature is converted to electrical signals using a transducer called a *thermistor*. A thermistor responds to temperature change by changing resistance, but its response is not linear, as seen in Table 13-6.

Table 13-6: Thermistor Resistance vs. Temperature

Temperature (C)	Tf (K ohms)
0	29.490
25	10.000
50	3.893
75	1.700
100	0.817

From William Kleitz, *Digital Electronics*

The complexity associated with writing software for such nonlinear devices has led many manufacturers to market a linear temperature sensor. Simple and widely used linear temperature sensors include the LM34 and LM35 series from National Semiconductor Corp. They are discussed next.

LM34 and LM35 temperature sensors

The sensors of the LM34 series are precision integrated-circuit temperature sensors whose output voltage is linearly proportional to the Fahrenheit temperature. See Table 13-7. The LM34 requires no external calibration because it is internally calibrated. It outputs 10 mV for each degree of Fahrenheit temperature. Table 13-7 is a selection guide for the LM34.

The LM35 series sensors are precision integrated-circuit temperature sensors whose output voltage is linearly proportional to the Celsius (centigrade) tem-

Table 13-7: LM34 Temperature Sensor Series Selection Guide

Part Scale	Temperature Range	Accuracy	Output
LM34A	−50 F to +300 F	+2.0 F	10 mV/F
LM34	−50 F to +300 F	+3.0 F	10 mV/F
LM34CA	−40 F to +230 F	+2.0 F	10 mV/F
LM34C	−40 F to +230 F	+3.0 F	10 mV/F
LM34D	−32 F to +212 F	+4.0 F	10 mV/F

Note: Temperature range is in degrees Fahrenheit.

Table 13-8: LM35 Temperature Sensor Series Selection Guide

Part	Temperature Range	Accuracy	Output Scale
LM35A	−55 C to +150 C	+1.0 C	10 mV/C
LM35	−55 C to +150 C	+1.5 C	10 mV/C
LM35CA	−40 C to +110 C	+1.0 C	10 mV/C
LM35C	−40 C to +110 C	+1.5 C	10 mV/C
LM35D	0 C to +100 C	+2.0 C	10 mV/C

Note: Temperature range is in degrees Celsius.

perature. The LM35 requires no external calibration because it is internally calibrated. It outputs 10 mV for each degree of centigrade temperature. Table 13-8 is the selection guide for the LM35. (For further information see http://www.national.com.)

Signal conditioning and interfacing the LM35 to the PIC18

Signal conditioning is widely used in the world of data acquisition. The most common transducers produce an output in the form of voltage, current, charge, capacitance, and resistance. We need to convert these signals to voltage, however, in order to send input to an A-to-D converter. This conversion (modification) is commonly called *signal conditioning*. See Figure 13-13. Signal conditioning can be a current-to-voltage conversion or a signal amplification. For example, the thermistor changes resistance with temperature. The change of resistance must be translated into voltages in order to be of any use to an ADC. Look at the case of connecting an LM34 to an ADC of the PIC18F458. The A/D has 10-bit resolution with a maximum of 1,024 steps and the LM34 (or LM35) produces 10 mV for every degree of temperature change. Now, if we use the step size of 10 mV, the V_{out} will be 10,240 mV (10.24 V) for full-scale output. This is not acceptable even though the maximum temperature sensed by the LM34 is 300 degrees F, and the highest output for the A/D we will get is 3,000 mV (3.00 V). Now, if we use the step size of 2.5 mV, the V_{out} will be 1,024 × 2.5 mV = 2,560 mV (2.56 V) for full-scale output. That means we must set V_{ref} = 2.56 V. This makes the binary output number for the A/D 4 times the real temperature (10 mV/2.5 mV = 4). We can scale it by dividing it by 4 to get the real number for temperature. See Table 13-9.

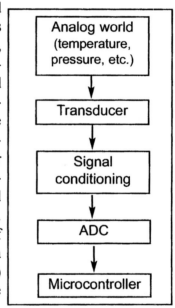

Figure 13-13. Getting Data From the Analog World

Figure 13-14 shows the connection of a temperature sensor to the PIC18F458. Notice that we use the LM336-2.5 zener diode to fix the voltage across the 10K pot at 2.5 volts. The use of the LM336-2.5 should overcome any fluctuations in the power supply.

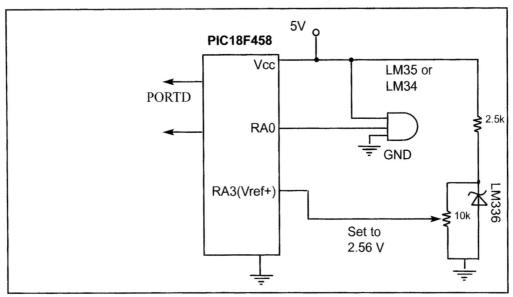

Figure 13-14. PIC18F458 Connection to Temperature Sensor

Table 13-9: Temperature vs. V_{out} for PIC18 with Vref = 2.56 V (SS = 2.5 mV)

Temp. (F)	V_{in} (mV)	#of steps	Binary V_{out} (b9–b0)	Temp. in Binary
0	0	0	00 00000000	00000000
1	10	4	00 00000100	00000100
2	20	8	00 00001000	00000010
3	30	12	00 00001100	00000011
10	100	20	00 00101000	00001010
20	200	80	00 01010000	00010100
30	300	120	00 01111000	00011110
40	400	160	00 10100000	00101000
50	500	200	00 11001000	00110010
60	600	240	00 11110000	00111100
70	700	300	01 00101100	01001011
80	800	320	01 01000000	01010000
90	900	360	01 01101000	01011010
100	1000	400	01 10010000	01100100

Example 13-10

In Table 13-9, verify the PIC output for a temperature of 70 degrees. Find values in the PIC18 A/D registers of ADRESL and ADRESH.

Solution:

The step size is 2.56/1,024 = 2.5 mV because Vref = 2.56 V.
For the 70 degrees temperature we have 700 mV output because the LM34 provides 10 mV output for every degree. Now, the number of steps are 700 mV/2.5 mV = 300 in decimal. Now 300 = 0101000000 in binary and the PIC18 A/D output registers have ADRESL = 0100000 and ADRESH = 00000001.

Reading and displaying temperature

Programs 13-4 and 13-4C show code for reading and displaying temperature in both Assembly and C respectively.

The programs correspond to Figure 13-14. Regarding these two programs, the following points must be noted:

(1) The LM34 (or LM35) is connected to channel 0 (RA0 pin).

(2) The channel AN3 (RA3 pin) is connected to the Vref of 2.56 V. That makes PCFG = 0010 for the ADCON1 register.

(3) The 10-bit output of the A/D is divided by 4 to get the real temperature.

The algorithm is as follows: (a) Shift right the ADRESL 2 bits, (b) rotate the ADRESH 2 bits, and (c) OR the ADRESH with ADRESL together to get the 8-bit output for temperature.

```
;Program 13-4
;this program reads the sensor and displays it on PORTD
L_Byte      SET 0x20    ;set a location 0x20 for L_Byte
H_Byte      SET 0x21    ;set a location 0x21 for H_Byte
BIN_TEMP    SET 0x22    ;set a location 0x22 for BIN_TEMP
      CLRF  TRISD         ;make PORTD an output
      BSF   TRISA,0 ;make RA0 an input pin for analog volt
      BSF   TRISA,3 ;make RA3 an input pin for Vref volt
      MOVLW 0x81          ;Fosc/64, channel 0, A/D is on
      MOVWF ADCON0
      MOVLW 0xC5          ;right justified, Fosc/64,
      MOVWF ADCON1        ;AN0 = analog, AN3 = Vref+
OVER  CALL  DELAY       ;wait for Tacq (sample and hold time)
      BSF   ADCON0,GO     ;start conversion
BACK  BTFSC ADCON0,DONE;keep polling end-of-conversion(EOC)
      BRA   BACK          ;wait for end-of-conversion
      MOVFF ADRESL,L_Byte ;save the low byte
      MOVFF ADRESH,H_Byte ;save the high byte
      CALL  ALGO_10_to_8  ;make it an 8-bit value
      MOVFF BIN_TEMP,PORTD ;display the temp on PORTD
      BRA   OVER          ;keep repeating it
;----------------
ALGO_10_to_8
      RRNCF L_Byte,F        ;rotate right twice
      RRNCF L_Byte,W
      ANDLW 0x3F            ;mask the upper 2 bits
      MOVWF L_Byte
      RRNCF H_Byte,F    ;rotate right through carry twice
      RRNCF H_Byte,W
      ANDLW 0xC0           ;mask the lower 6 bits
      IORWF L_Byte,W       ;combine low and high
      MOVWF BIN_TEMP
      RETURN
;-----------

//Program 13-4C
void main(void)
{
   unsigned char L_Byte,H_Byte,Bin_Temp;
   TRISD=0;               //make PORTD output port
   TRISAbits.TRISA0=1; //RA0 = INPUT for analog input
```

```
TRISAbits.TRISA2=1; //RA2 = INPUT for vref input
ADCON0 = 0x81;       //Fosc/64, channel 0, A/D is on
ADCON1 = 0xC5;       //right justified, Fosc/64,
                     //AN0 = analog, AN3 = Vref+
while(1)
{
  MSDelay(1);        //give A/D channel time to sample
  ADCON0bits.GO = 1;     //start converting
  while(ADCON0bits.DONE == 1);      //wait for EOC
  L_Byte=ADRESL;     //save the low byte
  H_Byte=ADRESH;     //save the high byte
  L_Byte>>=2;        //shift right
  L_Byte&=0x3F;      //mask the upper 2 bits
  H_Byte<<=6;        //shift left 6 times
  H_Byte&=0xC0;      //mask the lower 6 bits
  Bin_Temp= L_Byte|H_Byte;
  PORTD=Bin_Temp;
}
}
```

Review Questions

1. True or false. The transducer must be connected to signal conditioning circuitry before it is sent to the ADC.
2. The LM35 provides _____ mV for each degree of _____ (Fahrenheit, Celsius) temperature.
3. The LM34 provides _____ mV for each degree of _____ (Fahrenheit, Celsius) temperature.
4. Why do we set the V_{ref} of the PIC to 2.56 V if the analog input is connected to the LM35?
5. In Question 4, what is the temperature if the ADC output is 0011 1001?

SUMMARY

This chapter showed how to interface real-world devices such as DAC chips, ADC chips, and sensors to the PIC. First, we discussed both parallel and serial ADC chips, then described how the ADC module inside the PIC18 works and explained how to program it in both Assembly and C. Next we explored the DAC chip, and showed how to interface it to the PIC. In the last section we studied sensors. We also discussed the relation between the analog world and a digital device, and described signal conditioning, an essential feature of data acquisition systems.

PROBLEMS

SECTION 13.1: ADC CHARACTERISTICS

1. True or false. Sensor output is in analog.
2. True or false. A 10-bit ADC has 10-bit digital output.
3. True or false. ADC0848 is an 8-bit ADC.

4. True or false. MAX1112 is a 10-bit ADC.
5. True or false. An ADC with 8 channels of analog input must have 8 pins, one for each analog input.
6. True or false. For a serial ADC, it takes a longer time to get the converted digital data out of the chip.
7. True or false. ADC0848 has 4 channels of analog input.
8. True or false. MAX1112 has 8 channels of analog input.
9. True or false. ADC0848 is a serial ADC.
10. True or false. MAX1112 is a parallel ADC.
11. Which of the following ADC sizes provides the best resolution?
    (a) 8-bit   (b) 10-bit   (c) 12-bit   (d) 16-bit   (e) They are all the same.
12. In Question 11, which provides the smallest step size?
13. Calculate the step size for the following ADCs, if V_{ref} is 5 V:
    (a) 8-bit   (b) 10-bit   (c) 12-bit   (d) 16-bit
14. With V_{ref} = 1.28 V, find the V_{in} for the following outputs:
    (a) D7–D0 = 11111111   (b) D7–D0 = 10011001   (c) D7–D0 = 1101100
15. In the ADC0848, what should be the Vref value if we want a step size of 5 mV?
16. With Vref+ = 2.56 V and Vref– = Gnd, find the V_{in} for the following outputs:
    (a) D7–D0 = 11111111   (b) D7–D0 = 10011001   (c) D7–D0 = 01101100

SECTION 13.2: ADC PROGRAMMING IN THE PIC18

17. True or false. The PIC18F452/458 has an on-chip A/D converter.
18. True or false. A/D of the PIC18 is an 8-bit ADC.
19. True or false. PIC18F452/458 has 8 channels of analog input.
20. True or false. The unused analog pins of the PIC18F452/458 can be used for I/O pins.
21. True or false. The A/D conversion speed in the PIC18F452/458 depends on the crystal frequency.
22. True or false. Upon power-on reset, the A/D module of the PIC18F452/458 is turned on and ready to go.
23. True or false. The A/D module of the PIC18F452/458 has an external pin for the start-conversion signal.
24. True or false. The A/D module of the PIC18F452/458 can convert only one channel at a time.
25. True or false. The A/D module of the PIC18F452/458 can have multiple external Vref+ at any given time.
26. True or false. The A/D module of the PIC18F452/458 can use the Vdd for Vref+.
27. In the A/D of PIC18 what happens to the converted analog data? How do we know that the ADC is ready to provide us the data?
28. In the A/D of PIC18 what happens to the old data if we start conversion again before we pick up the last data?
29. Assume Vref– = Gnd. For the A/D of PIC18, find the step size for each of the following V_{ref+}:
    (a) V_{ref} = 1.024 V   (b) V_{ref} = 2.048 V   (c) V_{ref} = 2.56 V

30. In the PIC18, what should be the Vref value if we want a step size of 2 mV?
31. In the PIC18, what should be the Vref value if we want a step size of 3 mV?
32. With a step size of 1 mV, what is the analog input voltage if all outputs are 1?
33. With V_{ref} = 1.024 V, find the V_{in} for the following outputs:

    (a) D9–D0 = 0011111111 (b) D9–D0 = 0010011000 (c) D9–D0 = 0011010000
34. In the A/D of PIC18, what should be the Vref value if we want a step size of 4 mV?
35. With Vref+ = 2.56 V and Vref– = Gnd, find the V_{in} for the following outputs:.

    (a) D9–D0 = 1111111111 (b) D9–D0 = 1000000001 (c) D9–D0 = 1100110000
36. Find the conversion time for the following cases if XTAL = 8 MHz:

    (a) Fosc/2   (b) Fosc/4   (c) Fosc/8   (d) Fosc/16   (e) Fosc/32
37. Find the conversion time for the following cases if XTAL = 12 MHz:

    (a) Fosc/8   (b) Fosc/16   (c) Fosc/32   (d) Fosc/64
38. How do we start conversion in the PIC18?
39. How do we recognize the end of conversion in the PIC18?
40. The PIC18F452/458 can have a minimum of _____ channels of analog input.
41. In the PIC18F452/458, what ports are used for the analog channels?
42. Which register of the PIC18 is used to designate the number of A/D channels?
43. Which register of the PIC18 is used to select the A/D's conversion speed?
44. Which register of the PIC18 is used to select the analog channel to be converted?
45. Find the value for the ADCON0 register if we want Fosc/8, channel 0, and ADON on.
46. Find the value for the ADCON1 register if we want Fosc/64, 3 channels of analog input, and right-justified output.
47. Find the value for the ADCON0 register if we want Fosc/2, channel 2, and ADON off.
48. Find the value for the ADCON1 register if we want Fosc/32, 2 channels of analog input with external source for Vref+, and left-justified output.
49. Give the name of the interrupt flags for the A/D of the PIC18F452/458. State to which register they belong.
50. Upon power-on reset, the A/D of the PIC18F452/458 is given (low, high) priority.

SECTION 13.3: DAC INTERFACING

51. True or false. DAC0808 is the same as DAC1408.
52. Find the number of discrete voltages provided by the *n*-bit DAC for the following:

    (a) *n* = 8 (b) *n* = 10 (c) *n* = 12
53. For DAC1408, if I_{ref} = 2 mA, show how to get the I_{out} of 1.99 when all inputs are HIGH.
54. Find the I_{out} for the following inputs. Assume I_{ref} = 2 mA for DAC0808.

    (a) 10011001   (b) 11001100   (c) 11101110
    (d) 00100010   (e) 00001001   (f) 10001000

---

55. To get a smaller step, we need a DAC with _____ (more, fewer) digital inputs.
56. To get full-scale output, what should be the inputs for DAC?

SECTION 13.4: SENSOR INTERFACING AND SIGNAL CONDITIONING

57. What does it mean when a given sensor is said to have a linear output?
58. The LM34 sensor produces _____ mV for each degree of temperature.
59. What is signal conditioning?
60. What is the purpose of the LM336 Zener diode around the pot setting the V_{ref} in Figure 13-14?

ANSWERS TO REVIEW QUESTIONS

SECTION 13.1: ADC CHARACTERISTICS

1. Number of steps and Vref voltage
2. 8
3. True
4. (a) 8   (b) 8
5. 1.28 V/256 = 5 mV
6. (a) 0.7 V/ 5 mV= 140 in decimal and D7–D0 = 10001100 in binary.
    (a) 1 V/ 5 mV= 200 in decimal and D7–D0 = 11001000 in binary.

SECTION 13.2: ADC PROGRAMMING IN THE PIC18

1. Vref
2. 10
3. False
4. False
5. 1 mV
6. (a) 700 mV (1010111100), (b) 1000 mV (1111101000)
7. (a) 2 channels (b) 6 channels
8. True
9. 1.6
10. DONE bit of the ADCON0 register

SECTION 13.3: DAC INTERFACING

1. Digital, analog
2. Analog, digital
3. 8
4. (a) current   (b) true

SECTION 13.3: SENSOR INTERFACING AND SIGNAL CONDITIONING

1. True
2. 10, Celsius
3. 10, Fahrenheit
4. Using the 8-bit part of the 10-bit ADC, it gives us 256 steps, and 2.56 V/256 = 10 mV. The LM35 produces 10 mV for each degree of temperature, which matches the ADC's step size.
5. 00111001 = 57, which indicates it is 57 degrees.

CHAPTER 14

USING FLASH AND EEPROM MEMORIES FOR DATA STORAGE

OBJECTIVES

Upon completion of this chapter, you will be able to:

>> Contrast and compare various types of semiconductor memories
   in terms of their capacity, organization, and access time
>> Describe the relationship between the number of memory locations
   on a chip, the number of data pins, and the chip capacity
>> Define Flash ROM memory and describe its use in PIC18-based systems
>> Contrast and compare PROM, EPROM, UV-EPROM, EEPROM,
   Flash memory EPROM, and mask ROM memories
>> Code PIC18 Assembly and C programs for writing data into PIC18
   Flash memory space
>> Code PIC18 Assembly and C programs for erasing the Flash memory
   in PIC18
>> Explain how to write data to EEPROM memory of the PIC18
>> Explain how to read data from EEPROM memory of the PIC18

In this chapter we discuss how to access data stored in both Flash and EEPROM memories of the PIC18F. In Section 14.1 we study semiconductor memory concepts with emphasis on different types of ROM. In Section 14.2, the writing of data into PIC18F Flash memory is discussed. The accessing of EEPROM in the PIC18 is explored in Section 14.3.

SECTION 14.1: SEMICONDUCTOR MEMORY

In this section we discuss various types of semiconductor memories and their characteristics such as capacity, organization, and access time. In the design of all microprocessor-based systems, semiconductor memories are used as primary storage for code and data. Semiconductor memories are connected directly to the CPU and are the memory that the CPU first asks for information (code and data). For this reason, semiconductor memories are sometimes referred to as *primary memory*. The most widely used semiconductor memories are ROM and RAM. Before we discuss different types of RAM and ROM, we discuss some important terminology common to all semiconductor memories, such as capacity, organization, and speed.

Memory capacity

The number of bits that a semiconductor memory chip can store is called chip *capacity*. It can be in units of Kbits (kilobits), Mbits (megabits), and so on. This must be distinguished from the storage capacity of computer systems. While the memory capacity of a memory IC chip is always given in bits, the storage capacity of a computer system is given in bytes. For example, an article in a technical journal may state that the 128M chip has become popular. In that case, it is understood, although it is not mentioned, that 128M means 128 megabits because the article is referring to an IC memory chip. However, if an advertisement states that a computer comes with 128M memory, it is understood that 128M means 128 megabytes because it is referring to a computer system.

Memory organization

Memory chips are organized into a number of locations within the IC. Each location can hold 1 bit, 4 bits, 8 bits, or even 16 bits, depending on how it is designed internally. The number of bits that each location within the memory chip can hold is always equal to the number of data pins on the chip. How many locations exist inside a memory chip? That depends on the number of address pins. The number of locations within a memory IC always equals 2 to the power of the number of address pins. Therefore, the total number of bits that a memory chip can store is equal to the number of locations times the number of data bits per location. To summarize:

1. A memory chip contains 2^x locations, where x is the number of address pins.
2. Each location contains y bits, where y is the number of data pins on the chip.
3. The entire chip will contain $2^x \times y$ bits, where x is the number of address pins and y is the number of data pins on the chip.

Speed

Table 14-1: Powers of 2

x	2^x
10	1K
11	2K
12	4K
13	8K
14	16K
15	32K
16	64K
17	128K
18	256K
19	512K
20	1M
21	2M
22	4M
23	8M
24	16M
25	32M
26	64M
27	128M

One of the most important characteristics of a memory chip is the speed at which its data can be accessed. To access the data, the address is presented to the address pins, the READ pin is activated, and, after a certain amount of time has elapsed, the data shows up at the data pins. The shorter this elapsed time, the better, and consequently, the more expensive the memory chip. The speed of the memory chip is commonly referred to as its *access time*. The access time of memory chips varies from a few nanoseconds to hundreds of nanoseconds, depending on the IC technology used in the design and fabrication process.

The three important memory characteristics of capacity, organization, and access time will be explored extensively in this chapter. Table 14-1 serves as a reference for the calculation of memory characteristics. Examples 14-1 and 14-2 demonstrate these concepts.

Example 14-1

A given memory chip has 12 address pins and 4 data pins. Find:
(a) the organization, and (b) the capacity.

Solution:

(a) This memory chip has 4,096 locations (2^{12} = 4,096), and each location can hold 4 bits of data. This gives an organization of 4,096 × 4, often represented as 4Kx4.
(b) The capacity is equal to 16K bits because there is a total of 4K locations and each location can hold 4 bits of data.

Example 14-2

A 512K memory chip has 8 pins for data. Find:
(a) the organization, and (b) the number of address pins for this memory chip.

Solution:

(a) A memory chip with 8 data pins means that each location within the chip can hold 8 bits of data. To find the number of locations within this memory chip, divide the capacity by the number of data pins. 512K/8 = 64K; therefore, the organization for this memory chip is 64Kx8.
(b) The chip has 16 address lines because 2^{16} = 64K.

ROM (read-only memory)

ROM is a type of memory that does not lose its contents when the power is turned off. For this reason, ROM is also called *nonvolatile* memory. There are different types of read-only memory, such as PROM, EPROM, EEPROM, flash EPROM, and mask ROM. Each is explained below.

PROM (programmable ROM) and OTP

PROM refers to the kind of ROM that the user can burn information into. In other words, PROM is a user-programmable memory. For every bit of the PROM, there exists a fuse. PROM is programmed by blowing the fuses. If the information burned into PROM is wrong, that PROM must be discarded because its internal fuses are permanently blown. For this reason, PROM is also referred to as OTP (one-time programmable). Programming ROM, also called *burning* ROM, requires special equipment called a ROM burner or ROM programmer.

EPROM (erasable programmable ROM) and UV-EPROM

EPROM was invented to allow changes in the contents of PROM after it is burned. In EPROM, one can program the memory chip and erase it thousands of times. This is especially necessary during development of the prototype of a microprocessor-based project. A widely used EPROM is called UV-EPROM, where UV stands for ultraviolet. The only problem with UV-EPROM is that erasing its contents can take up to 20 minutes. All UV-EPROM chips have a window through which the programmer can shine ultraviolet (UV) radiation to erase its contents. For this reason, EPROM is also referred to as UV-erasable EPROM or simply UV-EPROM. Figure 14-1 shows the pins for a UV-EPROM chip.

To program a UV-EPROM chip, the following steps must be taken:

1. Its contents must be erased. To erase a chip, remove it from its socket on the system board and place it in EPROM erasure equipment to expose it to UV radiation for 15–20 minutes.
2. Program the chip. To program a UV-EPROM chip, place it in the ROM burner (programmer). To burn code or data into EPROM, the ROM burner uses 12.5 volts or higher, depending on the EPROM type. This voltage is referred to as V_{PP} in the UV-EPROM data sheet.
3. Place the chip back into its socket on the system board.

As can be seen from the above steps, not only is there an EPROM programmer (burner), but there is also separate EPROM erasure equipment. The main problem, and indeed the major disadvantage, of UV-EPROM is that it cannot be erased and programmed while it is in the system board. To find a solution to this problem, EEPROM was invented.

Notice the patterns of the IC numbers in Table 14-2. For example, part number 27128-25 refers to UV-EPROM that has a capacity of 128K bits and an access time of 250 nanoseconds. The capacity of the memory chip is indicated in the part number and the access time is given with a zero dropped. In part numbers, C refers to CMOS technology. Notice that 27XX refers to UV-EPROM chips.

Table 14-2: Some UV-EPROM Chips

Part #	Capacity	Org.	Access	Pins	V_{PP}
2716	16K	2Kx8	450 ns	24	25 V
2732	32K	4Kx8	450 ns	24	25 V
2732A-20	32K	4Kx8	200 ns	24	21 V
27C32-1	32K	4Kx8	450 ns	24	12.5 V CMOS
2764-20	64K	8Kx8	200 ns	28	21 V
2764A-20	64K	8Kx8	200 ns	28	12.5 V
27C64-12	64K	8Kx8	120 ns	28	12.5 V CMOS
27128-25	128K	16Kx8	250 ns	28	21 V
27C128-12	128K	16Kx8	120 ns	28	12.5 V CMOS
27256-25	256K	32Kx8	250 ns	28	12.5 V
27C256-15	256K	32Kx8	150 ns	28	12.5 V CMOS
27512-25	512K	64Kx8	250 ns	28	12.5 V
27C512-15	512K	64Kx8	150 ns	28	12.5 V CMOS
27C010-15	1,024K	128Kx8	150 ns	32	12.5 V CMOS
27C020-15	2,048K	256Kx8	150 ns	32	12.5 V CMOS
27C040-15	4,096K	512Kx8	150 ns	32	12.5 V CMOS

Example 14-3

For ROM chip 27128, find the number of data and address pins.

Solution:

The 27128 has a capacity of 128K bits. It has 16Kx8 organization (all ROMs have 8 data pins), which indicates that there are 8 pins for data and 14 pins for address ($2^{14} = 16K$).

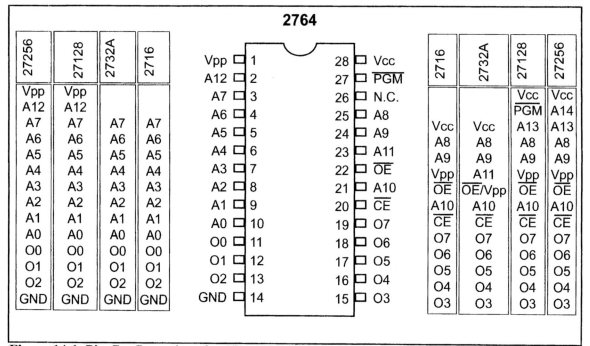

Figure 14-1. Pin Configurations for 27xx ROM Family

EEPROM (electrically erasable programmable ROM)

EEPROM has several advantages over EPROM, such as the fact that its method of erasure is electrical and therefore instant, as opposed to the 20-minute erasure time required for UV-EPROM. In addition, in EEPROM one can select the byte to be erased, in contrast to UV-EPROM, in which the entire contents of ROM are erased. The main advantage of EEPROM is that one can program and erase its contents while it is still in the system board. It does not require physical removal of the memory chip from its socket. In other words, unlike UV-EPROM, EEP-ROM does not require an external erasure and programming device. To utilize EEPROM fully, the designer must incorporate the circuitry to program the EEP-ROM into the system board. In general, the cost per bit for EEPROM is much higher than for UV-EPROM. In Section 14.3 we show how to access the PIC18 on-chip EEPROM.

Table 14-3: Some EEPROM and Flash Chips

EEPROMs

Part No.	Capacity	Org.	Speed	Pins	V_{PP}
2816A-25	16K	2Kx8	250 ns	24	5 V
2864A	64K	8Kx8	250 ns	28	5 V
28C64A-25	64K	8Kx8	250 ns	28	5 V CMOS
28C256-15	256K	32Kx8	150 ns	28	5 V
28C256-25	256K	32Kx8	250 ns	28	5 V CMOS

Flash

Part No.	Capacity	Org.	Speed	Pins	V_{PP}
28F256-20	256K	32Kx8	200 ns	32	12 V CMOS
28F010-15	1,024K	128Kx8	150 ns	32	12 V CMOS
28F020-15	2,048K	256Kx8	150 ns	32	12 V CMOS

Flash memory EPROM

Since the early 1990s, flash EPROM has become a popular user-programmable memory chip, and for good reasons. First, the erasure of the entire contents takes less than a second, or one might say in a flash, hence its name, flash memory. In addition, the erasure method is electrical, and for this reason it is sometimes referred to as flash EEPROM. To avoid confusion, it is commonly called flash memory. The major difference between EEPROM and flash memory is that when flash memory's contents are erased (or written to), the entire device is erased, in contrast to EEPROM, where one can erase a desired section or byte. In recent decades, Flash memory contents are divided into blocks and the erasure (or write) is done block by block. Unlike EEPROM, Flash memory has no byte erasure (or write) option. Because Flash memory can be programmed while it is in its socket on the system board, it has replaced the UV-EPROM for the storage of BIOS ROM of the PC. Nowadays, Flash memory is widely used for mass storage devices such as PDAs, cell phones, USB memory sticks, and MP3 players. Some computer scientists believe that Flash memory will replace the hard disk as a mass storage medium. This would increase the performance of the computer tremendously, because Flash memory is semiconductor memory with access time in the range of 100 ns compared with disk access time in the range of tens of milliseconds. For this to happen, flash memory's program/erase cycles must become infinite, just like hard disks. Program/erase cycle refers to the number of times that a chip can be erased and programmed (written to) before it becomes unreliable. At this time, the program/erase cycle is in the 100,000s for Flash and EEPROM, in the 1,000s for UV-EPROM, and infinite for RAM and hard disks.

Mask ROM

Mask ROM refers to a kind of ROM in which the contents are programmed by the IC manufacturer. In other words, it is not a user-programmable ROM. The term *mask* is used in IC fabrication. Because the burning process is costly, mask ROM is used when the needed volume is high (hundreds of thousands) and it is absolutely certain that the contents will not change. It is common practice to use UV-EPROM or flash for the development phase of a project; and only after the code/data have been finalized is the mask version of the product ordered. The main advantage of mask ROM is its cost, because it is significantly cheaper than other kinds of ROM; but if an error is found in the data/code, the entire batch must be thrown away. Many manufacturers of 8051 microcontrollers support the mask ROM version of the 8051. Note that all ROM memories have 8 bits for data pins; therefore, the organization is x8.

RAM (random access memory)

RAM memory is called *volatile* memory because cutting off the power to the IC results in the loss of data. Sometimes RAM is also referred to as RAWM (read and write memory), in contrast to ROM, which cannot be written to. There are three types of RAM: static RAM (SRAM), NV-RAM (nonvolatile RAM), and dynamic RAM (DRAM). Each is explained separately.

SRAM (static RAM)

Storage cells in static RAM memory are made of flip-flops and therefore do not require refreshing in order to keep their data. This is in contrast to DRAM, discussed below. The problem with the use of flip-flops for storage cells is that each cell requires at least 6 transistors to build, and the cell holds only 1 bit of data. In recent years, the cells have been made of 4 transistors, which still is too many. The use of 4-transistor cells plus the use of CMOS technology has given birth to a high-capacity SRAM, but its capacity is far below that of DRAM. Table 14-4 shows some examples of SRAM. Figure 14-2 shows the pin

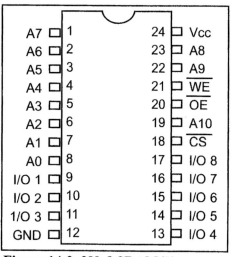

Figure 14-2. 2Kx8 SRAM Pins

---

diagram for an SRAM chip. In Figure 14-2, notice that WE is write enable, and OE is output enable, for read and write signals, respectively.

Table 14-4: Some SRAM and NV-RAM Chips

SRAM

Part No.	Capacity	Org.	Speed	Pins	V_{PP}
6116P-1	16K	2Kx8	100 ns	24	CMOS
6116P-2	16K	2Kx8	120 ns	24	CMOS
6116P-3	16K	2Kx8	150 ns	24	CMOS
6116LP-1	16K	2Kx8	100 ns	24	Low-power CMOS
6116LP-2	16K	2Kx8	120 ns	24	Low-power CMOS
6116LP-3	16K	2Kx8	150 ns	24	Low-power CMOS
6264P-10	64K	8Kx8	100 ns	28	CMOS
6264LP-70	64K	8Kx8	70 ns	28	Low-power CMOS
6264LP-12	64K	8Kx8	120 ns	28	Low-power CMOS
62256LP-10	256K	32Kx8	100 ns	28	Low-power CMOS
62256LP-12	256K	32Kx8	120 ns	28	Low-power CMOS

NV-RAM from Dallas Semiconductor

Part No.	Capacity	Org.	Speed	Pins	V_{PP}
DS1220Y-150	16K	2Kx8	150 ns	24	
DS1225AB-150	64K	8Kx8	150 ns	28	
DS1230Y-85	256K	32Kx8	85 ns	28	

NV-RAM (nonvolatile RAM)

Whereas SRAM is volatile, there is a new type of nonvolatile RAM called NV-RAM. Like other RAMs, it allows the CPU to read and write to it, but when the power is turned off the contents are not lost. NV-RAM combines the best of RAM and ROM: the read and write ability of RAM, plus the nonvolatility of ROM. To retain its contents, every NV-RAM chip internally is made of the following components:

1. It uses extremely power-efficient (very low power consumption) SRAM cells built out of CMOS.
2. It uses an internal lithium battery as a backup energy source.
3. It uses an intelligent control circuitry. The main job of this control circuitry is to monitor the V_{CC} pin constantly to detect loss of the external power supply. If the power to the V_{CC} pin falls below out-of-tolerance conditions, the control circuitry switches automatically to its internal power source, the lithium battery. The internal lithium power source is used to retain the NV-RAM contents only when the external power source is off.

It must be emphasized that all three of the components above are incorporated into a single IC chip, and for this reason nonvolatile RAM is a very expensive type of RAM as far as cost per bit is concerned. Offsetting the cost, however, is the fact that it can retain its contents up to ten years after the power has been turned off and allows one to read and write in exactly the same way as SRAM. See Table 14-4 for NV-RAM parts made by Dallas Semiconductor.

DRAM (dynamic RAM)

Since the early days of the computer, the need for huge, inexpensive read/write memory has been a major preoccupation of computer designers. In 1970, Intel Corporation introduced the first dynamic RAM (random access memory). Its density (capacity) was 1,024 bits and it used a capacitor to store each bit. Using a capacitor to store data cuts down the number of transistors needed to build the cell; however, the capacitor requires constant refreshing due to leakage. This is in contrast to SRAM (static RAM), whose individual cells are made of flip-flops. Because each bit in SRAM uses a single flip-flop, and each flip-flop requires 6 transistors, SRAM has much larger memory cells and consequently lower density. The use of capacitors as storage cells in DRAM results in much smaller net memory cell size.

The advantages and disadvantages of DRAM memory can be summarized as follows. The major advantages are high density (capacity), cheaper cost per bit, and lower power consumption per bit. The disadvantage is that it must be refreshed periodically because the capacitor cell loses its charge; furthermore, while DRAM is being refreshed, the data cannot be accessed. This is in contrast to SRAM's flip-flops, which retain data as long as the power is on, do not need to be refreshed, and whose contents can be accessed at any time. Since 1970, the capacity of DRAM has exploded. After the 1K-bit (1,024-bit) chip came the 4K-bit in 1973, and then the 16K chip in 1976. The 1980s saw the introduction of 64K, 256K, and finally 1M and 4M memory chips. The 1990s saw 16M, 64M, 256M, and the beginning of 1G-bit DRAM chips. In the 2000s, 2G-bit chips are standard, and as the fabrication process gets smaller, larger memory chips will be rolling off the manufacturing line. Keep in mind that when talking about IC memory chips, the capacity is always assumed to be in bits. Therefore, a 1M chip means a 1-megabit chip and a 256K chip means a 256-kilobit memory chip. When talking about the memory of a computer system, however, it is always assumed to be in bytes.

Packaging issue in DRAM

In DRAM there is a problem in packing a large number of cells into a single chip with the normal number of pins assigned to addresses. For example, a 64K chip (64Kx1) must have 16 address lines and 1 data line, requiring 16 pins to send in the address if the conventional method is used. This is in addition to V_{CC} power, ground, and read/write control pins. Using the conventional method of data access, the large number of pins defeats the purpose of high density and small packaging, so dearly cherished by IC designers. Therefore, to reduce the number of pins needed for addresses, multiplexing/demultiplexing is used. The method used is to split the address in half and send in each half of the address through the same pins, thereby requiring fewer address pins. See Table 14-5. Internally, the DRAM structure is divided into a square of rows and columns. The first half of the address is called the row and the second half is called the column. For example, in the case of DRAM of 64Kx1 organization, the first half of the address is sent in through the 8 pins A0–A7, and by activating RAS (row address strobe), the internal latches inside DRAM grab the first half of the address. After that, the second half of the address is sent in through the same pins, and by activating CAS (column address strobe), the internal latches inside DRAM latch the second half of the address. This results in using 8 pins for addresses plus RAS and CAS, for a total of 10 pins, instead of the 16 pins that would be

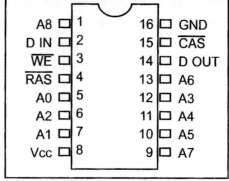

Figure 14-3. 256Kx1 DRAM

required without multiplexing. To access a bit of data from DRAM, both row and column addresses must be provided. For this concept to work, there must be a 2-by-1 multiplexer outside the DRAM circuitry and a demultiplexer inside every DRAM chip. Due to the complexities associated with DRAM interfacing (RAS, CAS, the need for multiplexer and refreshing circuitry), there are DRAM controllers designed to make DRAM interfacing much easier. However, many small microcontroller-based projects that do not require much RAM (usually less than 64K bytes) use SRAM of types EEPROM and NV-RAM, instead of DRAM.

Table 14-5: Some Widely Used DRAMs

Part No.	Speed	Capacity	Org.	Pins
4164-15	150 ns	64K	64Kx1	16
41464-8	80 ns	256K	64Kx4	18
41256-15	150 ns	256K	256Kx1	16
41256-6	60 ns	256K	256Kx1	16
414256-10	100 ns	1M	256Kx1	20
511000P-8	80 ns	1M	1Mx1	18
514100-7	70 ns	4M	4Mx1	20

DRAM organization

In the discussion of ROM, we noted that all of them have 8 pins for data. This is not the case for DRAM memory chips, which can have x1, x4, x8, or x16 organizations. See Example 14-4.

Example 14-4

Discuss the number of pins set aside for addresses in each of the following memory chips: (a) 16Kx4 DRAM   (b) 16Kx4 SRAM

Solution:

Because $2^{14} = 16K$:
(a) For DRAM we have 7 pins (A0–A6) for the address pins and 2 pins for RAS and CAS.
(b) For SRAM we have 14 pins for address and no pins for RAS and CAS because they are associated only with DRAM. In both cases we have 4 pins for the data bus.

In memory chips, the data pins are also called I/O. In some DRAMs there are separate D_{in} and D_{out} pins. Figure 14-3 shows a 256Kx1 DRAM chip with pins A0–A8 for address, RAS and CAS, WE (write enable), and data in and data out, as well as power and ground.

Review Questions

1. The speed of semiconductor memory is in the range of
   (a) microseconds (b) milliseconds
   (c) nanoseconds   (d) picoseconds
2. Find the organization and chip capacity for each ROM with the indicated number of address and data pins:
   (a) 14 address, 8 data   (b) 16 address, 8 data (c) 12 address, 8 data
3. Find the organization and chip capacity for each RAM with the indicated number of address and data pins:
   (a) 11 address, 1 data SRAM   (b) 13 address, 4 data SRAM
   (c) 17 address, 8 data SRAM   (d) 8 address, 4 data DRAM
   (e) 9 address, 1 data DRAM   (f) 9 address, 4 data DRAM
4. Find the capacity and number of pins set aside for address and data for memory chips with the following organizations:
   (a) 16Kx4 SRAM   (b) 32Kx8 EPROM   (c) 1Mx1 DRAM
   (d) 256Kx4 SRAM   (e) 64Kx8 EEPROM   (f) 1Mx4 DRAM
5. Which of the following is (are) volatile memory?
   (a) EEPROM   (b) SRAM   (c) DRAM   (d) NV-RAM

SECTION 14.2: ERASING AND WRITING TO FLASH IN THE PIC18F

The PIC18F comes with three types of memory (a) SRAM, (b) Flash, and (c) EEPROM. The SRAM is for general purpose usage including function registers, as we have seen throughout the book. The EEPROM is used for storing data only. While the Flash memory is used primarily to store program (code), we can also use it for storing fixed data such as look-up tables as we have seen throughout the book. In Chapter 6 we discussed how to use the TBLRD instruction to read the fixed data stored in program Flash. In this section, we discuss how to write to Flash memory. In the next section, we discuss how to access the EEPROM memory in the PIC18.

There are two ways to store (write) information (code or data) to the Flash memory or erase its content: (a) using an external Flash programmer (burner) such as PICSTART, and (b) using instructions such as TBLWR. In this section, we show how to use the TBLWR (table write) instruction to write to Flash memory. We will also show how to erase the contents of Flash memory. Due to similarities between the TBLRD and TBLWR instructions, it is very helpful to understand the material in Section 6.3 of Chapter 6, where we showed how to use the TBLRD instruction to read data stored in the Flash ROM.

Using TBLWR to write data to Flash

There are some major similarities between the TBLRD and TBLWR instructions. From Chapter 6, recall that in using the TBLRD instruction, we use the TBLPTR register as pointer to the data in Flash and the TABLAT register as a temporary place to store the data fetched from Flash. In the same way, the TBLWR instruction writes data held in the TABLAT register to the Flash ROM location whose address is pointed to by the TBLPTR register. In terms of autoincrement /autodecrement, the TBLRD and TBLWR instructions are exactly the same. See Table 14-6.

Table 14-6: PIC18 Table Write Instructions

Instruction	Function	Description
TBLWT*	Table Write	After write, TBLPTR stays the same
TBLWT*+	Table Write with post-inc	Write and increment TBLPTR
TBLWT*-	Table Write with post-dec	Write and decrement TBLPTR
TBLWT+*	Table Write with pre-inc	Increment TBLPTR and then write

There is a major difference between the TBLRD and TBLWR instructions. While the TBLRD reads individual bytes from Flash, the TBLWR writes a block of 8 bytes to Flash. The TBLRD instruction reads one byte at a time from the Flash into the TABLAT registers, which means we must save the contents of the TAB-LAT before the next read wipes it out. The TBLWR instruction uses what is called short write and long write to write to Flash. In the short write, we use the TBLWR instruction to write a block of 8 bytes of data into 8 TABLAT registers one byte at a time. After the short write is done, we use the long write to actually store (write, or one might say burn) the entire block of 8 bytes into the Flash. The long write is done with the help of a register called EECON1, shown in Figure 14-4. Notice that the EECON1 register is used for both Flash and EEPROM memory, as we will see in the next section. Also contrast the difference between the Flash and EEPROM memory. In Flash memory, the write or erase process is done on a block of data, while in EEPROM we can write or erase one byte at a time, which means it is byte-accessible memory. Readings for both Flash and EEPROM memories are in byte sizes. The block size for Flash memory varies among the Flash memories depending on their size and intended application. The block size for write/erase in the PIC184580 is 8 bytes, while in other Flash memories it is 64 or 256 bytes. The breaking of the PIC18F Flash into blocks of 8 bytes means the memory addresses must be on the 8-byte boundaries. This means that the lower three bits of the address A21–A0 of the Flash ROM location must be all zeros. See Figure 14-5.

EEPGD	CFGS	--	FREE	WRERR	WERN	WR#	RD#

EEPGD Flash Program or Data EEPROM Memory select bit
    1 = Access Program Flash memory
    0 = Access Data EEPROM memory

CFGS Flash Program/Data EE or Configuration Select bit
    1 = Access Configuration Registers
    0 = Access Program Flash or Data EEPROM memory

FREE Flash Row Erase Enable bit
    1 = Erase the Program Flash memory row addresses by TBLPTR on the next
    WR command (this bit is cleared when the Erase operation is completed)
    0 = Perform write only

WERR Write Error Flag bit
    1 = A write operation is prematurely terminated
    0 = The write operation is completed

WREN Write Enable bit
    1 = Allows write cycle
    0 = Inhibits write to the EEPROM or Flash memory

WR# Write Control bit. This is an an active-LOW signal used for both Flash and
    EEPROM. We can only make it HIGH by software and the PIC will make it
    LOW automatically when the write cycle is completed
    1 = initiates the write cycle to Flash or EEPROM (also used for initiating
    Erase / write cycle).
    0 = Write cycle is completed

RD# Read Control bit. This is an an active-LOW signal used by EEPROM only. We
    can only make it HIGH by software and the PIC will make it LOW
    automatically when the read cycle is completed.
    1 = Initiates the read cycle to EEPROM
    0 = Does not initiate an EEPROM read

Figure 14-4. EECON1 (EEPROM Control Register, also used for Flash)

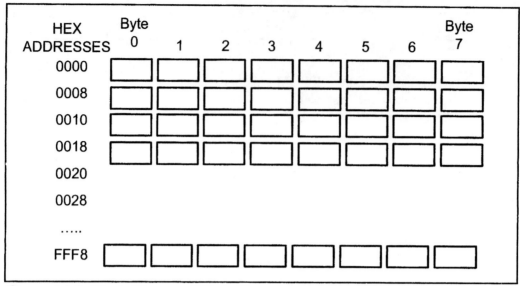

Figure 14-5. Flash Memory 8-Byte Boundaries

Although there are 8 TABLAT registers for the short write, they are not accessible individually. These 8 TABLAT registers are internal and are used solely for the purpose of the short write using the TBLWRT instruction. Compare Figures 14-6 and 14-7 to contrast reading and writing to the Flash memory.

Steps in writing to Flash memory

Assuming that an area of Flash memory is erased, we can use the following steps to write a block of 8 bytes of data to the Flash memory:

(1) Load the TBLPTR registers with the address of the first byte being written.

(2) Using the TBLWR instruction, write 8 bytes of data to the TABLAT registers one after another. This is the end of the short write.

(3) Set the EECON1 register for the write operation by setting (a) EEPGD = 1, (b) CFGS = 0, and (c) WREN = 1.

(4) Disable all interrupts globally with "BCF INTCON, GIE".

(5) Write 55H to the EECON2 dummy register. This is the start of a long write.

(6) Write AAH to the EECON2 dummy register.

(7) Set WR# to 1 with the instruction "BSF EECON1, WE". With WE = 1, the write cycle begins.

(8) It will take about 2 ms to finish writing the 8 bytes to Flash. During this write cycle, the CPU is stalled and will not allow fetching any opcode. Upon completion of the write cycle, the WE# bit will go back low automatically to indicate that the write cycle is finished. This step concludes the end of long-write cycle.

(9) Reenable the interrupts globally with "BSF INTCON, GIE".

Notice from step 4 that we must disable the interrupts to prevent any interruption of the write cycle (long write). If writing to Flash is interrupted by the reset pin (MCLR) or the WDT (watch dog timer), the WERR (write error) bit of the

EECON1 will go HIGH to indicate that. The good thing is the EEPGD bit of the EECON1 remains HIGH, allowing us to fix the error by rewriting the data to Flash. The EECON2 register does not exist physically and cannot be accessed. It is used exclusively for the purpose of writing/erasing the Flash/EEPROM memory. Program 14-1 shows how to write 8 bytes of data to Flash locations starting at address 400H. After writing the bytes, we read and display them on PORTB one byte at a time to verify the write operation.

The C language version of Program 14-1 is given at the end of this section.

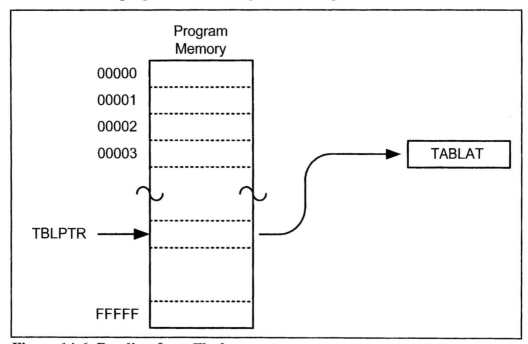

Figure 14-6. Reading from Flash

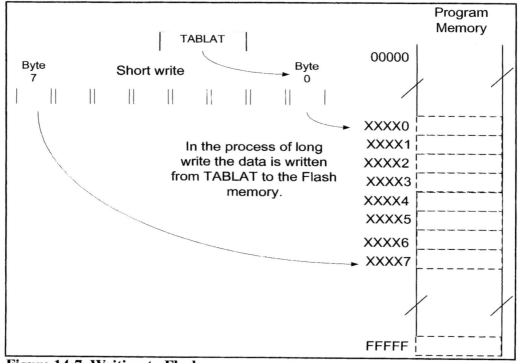

Figure 14-7. Writing to Flash

Program 14-1 (a) writes the message "GOOD BYE" to Flash memory starting at location 400H, and (b) reads the data from Flash and places it in PORTB one byte at a time.

```
;Program 14-1
COUNT       EQU    0x20
      MOVLW     0x00
      MOVWF     TBLPTRL     ;load the low byte of address
      MOVLW     0x04
      MOVWF     TBLPTRH;load the high byte of address
      ;start a short write
      MOVLW     A'G'        ;load the 'G' byte into WREG
      MOVWF     TABLAT      ;move it to TABLATch reg
      TBLWT*+   ;perform short write increment address
      MOVLW     A'O'        ;load the 'O' byte into WREG
      MOVWF     TABLAT      ;move it to TABLATch reg
      TBLWT*+   ;perform short write increment address
      MOVLW     A'O'        ;load the 'O' byte into WREG
      MOVWF     TABLAT      ;move it to TABLATch reg
      TBLWT*+               ;perform short write
      MOVLW     A'D'        ;load the 'D' byte into WREG
      MOVWF     TABLAT      ;move it to TABLATch reg
      TBLWT*+               ;perform short write
      MOVLW     A' '        ;load the space into WREG
      MOVWF     TABLAT      ;move it to TABLATch reg
      TBLWT*+               ;perform short write
      MOVLW     A'B'        ;load the 'B' byte into WREG
      MOVWF     TABLAT      ;move it to TABLATch reg
      TBLWT*+               ;perform short write
      MOVLW     A'Y'        ;load the 'Y' byte into WREG
      MOVWF     TABLAT      ;move it to TABLATch reg
      TBLWT*+               ;perform short write
      MOVLW     A'E'        ;load the 'E' byte into WREG
      MOVWF     TABLAT      ;move it to TABLATch reg
      TBLWT*+               ;perform short write
;start the long write cycle (write to Flash itself)
      MOVLW     0x00
      MOVWF     TBLPTRL     ;load the low byte of address
      MOVLW     0x04
      MOVWF     TBLPTRH;load the high byte of address
      BSF       EECON1,EEPGD    ;point to Flash memory
      BCF       EECON1,CFGS     ;access Flash program
      BSF       EECON1,WREN     ;enable write
      BCF       INTCON,GIE      ;disable all interrupts
      MOVLW     55H             ;wreg = 55h
      MOVWF     EECON2          ;write to dummy reg
      MOVLW     0AAH            ;wreg = aah
      MOVWF     EECON2          ;write to dummy reg
```

```
        BSF        EECON1,WR          ;now write it to Flash
        NOP                           ;wait
        BSF        INTCON,GIE         ;enable all interrupts
        BCF        EECON1,WREN        ;disable write to memory
;read them back one byte at a time and examine the
;bytes on PORTB
        MOVLW      0x00
        MOVWF      TBLPTRL;reload the low byte of address
        MOVLW      0x04
        MOVWF      TBLPTRH;reload the high byte of address
        CLRF       TRISB       ;PORTB an output port
        MOVLW      0x8         ;counter = 8
        MOVWF      COUNT
OVER    TBLRD*+    ;read the byte to TABLAT and increment
        MOVFF      TABLAT,PORTB    ;send it to PORTB
        CALL       DELAY       ;wait enough to see the byte
        DECF COUNT,F           ;decrement counter
        DECFSZ     COUNT,F
        BRA        OVER        ;continue for all the bytes
```

If the size of the block is not 8 bytes, then the rest of the unused block will be untouched. See Program 14-2.

Program 14-2 (a) writes the message "HELLO" to Flash memory starting at location 450H, and (b) reads the data from Flash and places it in PORTB one byte at a time.

```
;Program 14-2
COUNT       EQU  0x20
        MOVLW      0x50
        MOVWF      TBLPTRL     ;load the low byte of address
        MOVLW      0x04
        MOVWF      TBLPTRH;load the high byte of address
        ;start a short write
        MOVLW      A'H'        ;load the 'H' byte into WREG
        MOVWF      TABLAT      ;move it to TABLATch reg
        TBLWT*+    ;perform short write and increment
        MOVLW      A'E'        ;load the 'E' byte into WREG
        MOVWF      TABLAT      ;move it to TABLATch reg
        TBLWT*+    ;perform short write and increment
        MOVLW      A'L'        ;load the 'L' byte into WREG
        MOVWF      TABLAT      ;move it to TABLATch reg
        TBLWT*+    ;perform short write
        MOVLW      A'L'        ;load the 'L' byte into WREG
        MOVWF      TABLAT      ;move it to TABLATch reg
        TBLWT*+    ;perform short write
        MOVLW      A'O'        ;load the 'O' byte into WREG
        MOVWF      TABLAT      ;move it to TABLATch reg
```

```
        TBLWT*+     ;perform short write
;start the long write cycle (write to Flash itself)
        MOVLW       0x50
        MOVWF       TBLPTRL     ;load the low byte of address
        MOVLW       0x04
        MOVWF       TBLPTRH     ;load the high byte of address
        BSF         EECON1,EEPGD    ;point to Flash memory
        BCF         EECON1,CFGS     ;access Flash program
        BSF         EECON1,WREN     ;enable write
        BCF         INTCON,GIE      ;disable all interrupts
        MOVLW       55H             ;wreg = 55h
        MOVWF       EECON2          ;write to dummy reg
        MOVLW       0AAH            ;wreg = aah
        MOVWF       EECON2          ;write to dummy reg
        BSF         EECON1,WR       ;now write it to Flash
        NOP                         ;wait
        BSF         INTCON,GIE      ;enable all interrupts
        BCF         EECON1,WREN     ;disable write to memory
;read them back one byte at a time and examine the
;bytes on PORTB
        MOVLW       0x50
        MOVWF       TBLPTRL ;reload the low byte of address
        MOVLW       0x04
        MOVWF       TBLPTRH;reload the high byte of address
        CLRF        TRISB           ;PORTB an output port
        MOVLW       0x05            ;counter = 5
        MOVWF       COUNT
OVER    TBLRD*+                     ;read the byte and increment
        MOVFF       TABLAT,PORTB    ;send it to PORTB
        CALL        DELAY           ;wait enough to see the byte
        DECF COUNT,F                ;dec counter
        DECFSZ      COUNT,F
        BRA         OVER            ;continue for all the bytes
```

Program 14-3 (a) transfers a block of data from the code space of the PIC18 chip into RAM, (b) then writes the same data from RAM to Flash, and (c) reads the data from new Flash locations and sends it to the serial port of the PIC18 one byte at a time.

```
;Program 14-3
COUNT       EQU     0x0B
BUFRAM      EQU     0x20
        MOVLW D'8'              ;number of bytes to retrieve
        MOVWF COUNT
        MOVLW high (BUFRAM) ;point to buffer
        MOVWF FSR0H
```

```
        MOVLW low (BUFRAM)
        MOVWF FSR0L
        MOVLW upper (CODE_DATA)    ;load TBLPTR
        MOVWF TBLPTRU
        MOVLW high (CODE_DATA)
        MOVWF TBLPTRH
        MOVLW low (CODE_DATA)
        MOVWF TBLPTRL
    ;retrieve the data from program memory
READ_BLOCK
        TBLRD*+             ;read into TABLAT, and increment
        MOVF TABLAT, W              ;get data
        MOVWF POSTINC0             ;store data
        DECFSZ COUNT               ;done?
        BRA READ_BLOCK             ;repeat
        MOVLW upper (NEW_DATA)     ;load TBLPTR
        MOVWF TBLPTRU
        MOVLW high (NEW_DATA)
        MOVWF TBLPTRH
        MOVLW low (NEW_DATA)
        MOVWF TBLPTRL
        MOVLW high (BUFRAM)        ;point to buffer
        MOVWF FSR0H
        MOVLW low (BUFRAM)
        MOVWF FSR0L
        MOVLW 8                    ;number of bytes in RAM
        MOVWF COUNT
    ;move the data back to program memory
WRITE_BACK
        MOVF POSTINC0, W           ;get a byte from RAM
        MOVWF TABLAT   ;store the byte in table latch
        TBLWT*+                    ;perform a short write
        DECFSZ COUNT        ;loop until buffers are full
        BRA WRITE_BACK
        MOVLW upper (NEW_DATA)  ;load TBLPTR
        MOVWF TBLPTRU
        MOVLW high (NEW_DATA)
        MOVWF TBLPTRH
        MOVLW low (NEW_DATA)
        MOVWF TBLPTRL
        BSF EECON1, EEPGD ;point to Flash program memory
        BCF EECON1, CFGS    ;access Flash program memory
        BSF EECON1, WREN    ;enable write to memory
        BCF INTCON, GIE     ;disable interrupts
        MOVLW 55h           ;write 55h
        MOVWF EECON2
        MOVLW 0AAh          ;write 0AAh
```

```
        MOVWF  EECON2           ;start program (CPU stall)
        BSF EECON1, WR
        NOP
        BSF INTCON, GIE      ;re-enable interrupts
        BCF EECON1, WREN     ;disable write to memory
;read them back one byte at a time and send serially
        BSF   TRISD,7        ;PORTD.7 as in input
        MOVLW 0x20   ;enable transmit and low baud rate
        MOVWF      TXSTA      ;write to reg
        BCF        PIR1,TXIF
        MOVLW      D'15'      ;9600 bps (Fosc/(64*Speed)-1)
        MOVWF      SPBRG      ;write to reg
        BCF   TRISC, TX ;make TX pin of PORTC an output
        BSF   RCSTA, SPEN ;enable the entire serial port
        MOVLW 8                ;number of bytes in RAM
        MOVWF  COUNT
        MOVLW upper (NEW_DATA) ;load TBLPTR
        MOVWF TBLPTRU
        MOVLW high (NEW_DATA)
        MOVWF TBLPTRH
        MOVLW low (NEW_DATA)
        MOVWF TBLPTRL
        CLRF TRISB             ;PORTB an output port
        MOVLW      0x8         ;counter = 8
        MOVWF      COUNT
LN      TBLRD*+                ;read the character
        MOVF       TABLAT,W
R1      BTFSS PIR1, TXIF ;wait until the last bit is gone
        BRA  R1                ;stay in loop
        MOVWF      TXREG ;load the value to be transmitted
        DECFSZ COUNT           ;loop until buffers are full
        BRA  LN                ;repeat
```

Steps in erasing Flash memory

Although we can use external Flash programmers to erase the Flash memory contents, the PIC18 allows us to write a program to erase the Flash memory. The erasure process works on block-size, not byte-size data. The minimum block size for the erasure is 64 bytes. That means the lowest 6 bits of addresses are all zeros, making them 64-byte block boundaries. We can use the following steps to erase a single 64-byte block of Flash memory:

1. Load the TBLPTR registers with the address of the block being erased.
2. Set the EECON1 register for the erase operation by setting
   (a) EEPGD = 1, (b) CFGS = 0, (c) WREBN = 1, and (d) FREE = 1.
3. Disable all interrupts globally using "BCF INTCON, GIE".
4. Write 55H to the EECON2 dummy register.

5. Write AAH to the EECON2 dummy register.
6. Set WR# to 1 with the instruction "BSF EECON1,WE". With WE = 1, the erase cycle begins.
7. It will take about 2 ms to finish erasing the block of 64 bytes. During this erase cycle, the CPU is installed and will not allow fetching of any opcode. Upon completion of the erase cycle, the WE# bit will go back HIGH automatically to indicate the erase cycle is finished.
8. Reenable the interrupts globally using "BCF INTCON, GIE".

Program 14-4 shows how to erase the 64-byte block.

```
;Program 14-4: This program erases the Flash
;memory starting at location 0x500.
      ORG   0
      MOVLW     upper(MYDATA)
      MOVWF     TBLPTRH          ;load the upper address
      MOVLW     high(MYDATA)
      MOVWF     TBLPTRH   ;load the high byte of address
      MOVLW     low(MYDATA)
      MOVWF     TBLPTRL    ;load the low byte of address
      BSF       EECON1,EEPGD   ;point to Flash memory
      BCF       EECON1,CFGS    ;access Flash program
      BSF       EECON1,WREN    ;enable write
      BSF   EECON1, FREE    ;enable row erase operation
      BCF       INTCON,GIE     ;disable all interrupts
      MOVLW     55H            ;wreg = 55h
      MOVWF     EECON2         ;write to dummy reg
      MOVLW     0AAH           ;wreg = aah
      MOVWF     EECON2         ;write to dummy reg
      BSF       EECON1,WR      ;now write it to Flash
      NOP                      ;wait
      BSF       INTCON,GIE     ;enable all interrupts
      BCF       EECON1,WREN    ;disable write to memory
HERE BRA         HERE
      ORG 500H
MYDATA data  "ABCDEFGH"
      END
```

Examine Program 14-5. It combines erasing, writing, and reading of the Flash memory.

```
;Program 14-5: This program erases the message of
;"GOOD BYE" from Flash addresses 0x1200 and replaces
;it with "HELLO".
      MOVLW     upper(MYDATA)
      MOVWF     TBLPTRU    ;load the upper address
      MOVLW     high(MYDATA)
```

```
        MOVWF     TBLPTRH   ;load the high byte of address
        MOVLW     low(MYDATA)
        MOVWF     TBLPTRL    ;load the low byte of address
        BSF       EECON1,EEPGD   ;point to Flash memory
        BCF       EECON1,CFGS    ;access Flash program
        BSF       EECON1,WREN    ;enable write
        BSF   EECON1, FREE    ;enable row erase operation
        BCF       INTCON,GIE     ;disable all interrupts
        MOVLW     55H            ;wreg = 55h
        MOVWF     EECON2         ;write to dummy reg
        MOVLW     0AAH           ;wreg = aah
        MOVWF     EECON2         ;write to dummy reg
        BSF       EECON1,WR      ;now write it to Flash
        NOP                      ;wait
        BSF       INTCON,GIE     ;enable all interrupts
        BCF       EECON1,WREN    ;disable write to memory
        MOVLW     upper(MYDATA)
        MOVWF     TBLPTRU        ;load the upper address
        MOVLW     high(MYDATA)
        MOVWF     TBLPTRH   ;load the high byte of address
        MOVLW     low(MYDATA)
        MOVWF     TBLPTRL    ;load the low byte of address
;start a short write
        MOVLW     A'H'           ;load the byte into WREG
        MOVWF     TABLAT         ;move it to TABLATch reg
        TBLWT*+    ;perform short write and increment
        MOVLW     A'E'           ;load the byte into WREG
        MOVWF     TABLAT         ;move it to TABLATch reg
        TBLWT*+    ;perform short write and increment
        MOVLW     A'L'           ;load the byte into WREG
        MOVWF     TABLAT         ;move it to TABLATch reg
        TBLWT*+                  ;perform short write
        MOVLW     A'L'           ;load the byte into WREG
        MOVWF     TABLAT         ;move it to TABLATch reg
        TBLWT*+                  ;perform short write
        MOVLW     A'O'           ;load the byte into WREG
        MOVWF     TABLAT         ;move it to TABLATch reg
        TBLWT*+                  ;perform short write
;start the long write cycle (write to Flash itself)
        BSF       EECON1,EEPGD   ;point to Flash memory
        BCF       EECON1,CFGS    ;
        BSF       EECON1,WREN    ;enable write
        BCF       INTCON,GIE     ;disable all interrupts
        MOVLW     55H            ;wreg = 55h
        MOVWF     EECON2         ;write to dummy reg
        MOVLW     0AAH           ;wreg = aah
        MOVWF     EECON2         ;write to dummy reg
```

```
        BSF          EECON1,WR          ;now write it to Flash
        NOP                             ;wait
        BSF          INTCON,GIE         ;enable all interrupts
        BCF          EECON1,WREN        ;disable write to memory
;read them back one byte at a time and examine the
;bytes on PORTB
        MOVLW        upper(MYDATA)
        MOVWF        TBLPTRU            ;load the upper address
        MOVLW        high(MYDATA)
        MOVWF        TBLPTRH  ;load the high byte of address
        MOVLW        low(MYDATA)
        MOVWF        TBLPTRL    ;load the low byte of address
        CLRF TRISB                      ;PORTB an output port
        MOVLW        0x05               ;counter = 5
        MOVWF        COUNT
OVER TBLRD*+     ;read byte to TABLAT and point to next
        MOVFF        TABLAT,PORTB    ;send it to PORTB
        CALL         DELAY ;wait enough to see byte on PORTB
        DECF         COUNT,F            ;decrement counter
        BNZ          OVER         ;continue for all the bytes
        ORG 1200H
MYDATA data "GOOD BYE"
        END
```

Erasing and writing to Flash memory in C

Programs 14-6C through 14-8C are the C versions of earlier programs.

```
/*Program 14-6C: This C program (a) writes the mes-
sage "GOOD BYE" to Flash memory starting at location
400H, (b) reads the data from Flash and places it in
PORTB one byte at a time. */

#include <p18Cxxx.h>
void Delay(unsigned int itime);

void main()
{
    unsigned char x;
    //write to program memory
    TBLPTR = (short long)0x0400;  //load TBLPTR
    TABLAT='G';                   //load in TABLAT
    _asm TBLWTPOSTINC _endasm     //short write
    TABLAT='O';                   //load in TABLAT
    _asm TBLWTPOSTINC _endasm     //short write
    TABLAT='O';                   //load in TABLAT
    _asm TBLWTPOSTINC _endasm     //short write
```

```
        TABLAT='D';                    //load in TABLAT
        _asm TBLWTPOSTINC _endasm      //short write
        TABLAT=' ';                    //load in TABLAT
        _asm TBLWTPOSTINC _endasm      //short write
        TABLAT='B';                    //load in TABLAT
        _asm TBLWTPOSTINC _endasm      //short write
        TABLAT='Y';                    //load in TABLAT
        _asm TBLWTPOSTINC _endasm      //short write
        TABLAT='E';                    //load in TABLAT
        _asm TBLWTPOSTINC _endasm      //short write

        //long write
        TBLPTR = (short long)0x0400;  //reload TBLPTR
        EECON1bits.EEPGD=1;
        EECON1bits.CFGS=0;
        EECON1bits.WREN=1;
        INTCONbits.GIE=0;
        EECON2=0x55;
        EECON2=0xAA;
        EECON1bits.WR=1;
        _asm NOP _endasm
        INTCONbits.GIE=1;
        EECON1bits.WREN=0;

        //read from program memory send to PORTB
        TBLPTR = (short long)0x0400;  //reload TBLPTR
        for(x=0;x<8;x++){
            _asm TBLRDPOSTINC _endasm
            PORTB=TABLAT;
            Delay(250);
        }
}

//Program 14-7C: This C program erases the Flash
//memory starting at location 0x500.

#include <p18Cxxx.h>

#pragma romdata const_table = 0x500
const rom char my_const_array[10] = "GOOD BYE";
#pragma romdata
void main()
{

  //erase program memory
  TBLPTR = (short long)0x0500;      //load TBLPTR
  EECON1bits.EEPGD=1;
```

```
      EECON1bits.CFGS=0;
      EECON1bits.WREN=1;
      EECON1bits.FREE=1;
      INTCONbits.GIE=0;
      EECON2=0x55;
      EECON2=0xAA;
      EECON1bits.WR=1;
      _asm NOP _endasm
      INTCONbits.GIE=1;
      EECON1bits.WREN=0;
}


//Program 14-8C: This C program erases the message of
//"GOOD BYE" from Flash addresses 0x1200 and replaces
//it with "HELLO".

#include <p18Cxxx.h>

void Delay(unsigned int itime);

#pragma romdata const_table = 0x1200
const rom char my_const_array[10] = "GOOD BYE";
#pragma romdata
void main()
{
   unsigned char x;

   //erase program memory
   TBLPTR = (short long)0x1200;      //load TBLPTR
   EECON1bits.EEPGD=1;
   EECON1bits.CFGS=0;
   EECON1bits.WREN=1;
   EECON1bits.FREE=1;
   INTCONbits.GIE=0;
   EECON2=0x55;
   EECON2=0xAA;
   EECON1bits.WR=1;
   _asm NOP _endasm
   INTCONbits.GIE=1;
   EECON1bits.WREN=0;

   TBLPTR = (short long)0x1200;      //load TBLPTR
   TABLAT='H';                       //load in TABLAT
   _asm TBLWTPOSTINC _endasm         //short write
   TABLAT='E';                       //load in TABLAT
   _asm TBLWTPOSTINC _endasm         //short write
   TABLAT='L';                       //load in TABLAT
```

```
 _asm TBLWTPOSTINC _endasm        //short write
TABLAT='L';                       //load in TABLAT
 _asm TBLWTPOSTINC _endasm        //short write
TABLAT='O';                       //load in TABLAT
 _asm TBLWTPOSTINC _endasm        //short write

//long write
TBLPTR = (short long)0x1200;      //reload TBLPTR
EECON1bits.EEPGD=1;
EECON1bits.CFGS=0;
EECON1bits.WREN=1;
INTCONbits.GIE=0;
EECON2=0x55;
EECON2=0xAA;
EECON1bits.WR=1;
 _asm NOP _endasm
INTCONbits.GIE=1;
EECON1bits.WREN=0;

//read from program memory send to PORTB
TBLPTR = (short long)0x1200;      //reload TBLPTR
for(x=0;x<8;x++){
   _asm TBLRDPOSTINC _endasm
   PORTB=TABLAT;
   Delay(250);
}
}
```

Review Questions

1. True or false. The PIC18F Flash memory can be used for both program code and data.
2. True or false. The PIC18F SRAM memory can be used for both program code and data.
3. True or false. In the PIC18F, writing to Flash is not allowed.
4. True or false. Reading from Flash memory is in byte size, while writing to it is in block size.
5. True or false. During the long write, the CPU keeps fetching and executing the instructions.
6. What is the size of the block for writing to Flash memory in the PIC18F458?
7. What is the size of the block for erasing the Flash memory in the PIC18F458?

SECTION 14.3: READING AND WRITING TO DATA EEP-ROM IN THE PIC18

The vast majority of the members of the PIC18 family come with some EEPROM memory. The amount varies from 256 bytes to a few K depending on the family member. For example, the PIC18F4520 has 256 bytes of EEPROM while PIC184585 has only 1,024 bytes. Table 14-7 shows some of the family members and their EEPOM space. While the Flash memory in PIC18F can be used for storing both code and data, the EEPROM space is used exclusively for storing data. Of the three memory spaces that PIC18 has, the SRAM and EEPROM are used for data only while the Flash is used mainly for program and sometimes for fixed data storage. See Figure 14-8.

Table 14-7: EEPROM Size for Some PIC18 Chips

Part No.	On-chip Flash	On-chip RAM	On-chip EEPROM
PIC18F1220	4 KB	256 B	256 B
PIC18F1230	4 KB	256 B	128 B
PIC18F2410	16 KB	768 B	0 B
PIC18F4520	32 KB	1,536 B	256 B
PIC18F4580	32 KB	1,536 B	256 B
PIC18F4585	48 KB	3,328 B	1,024 B

Note: On-chip RAM does not include the SFR space.

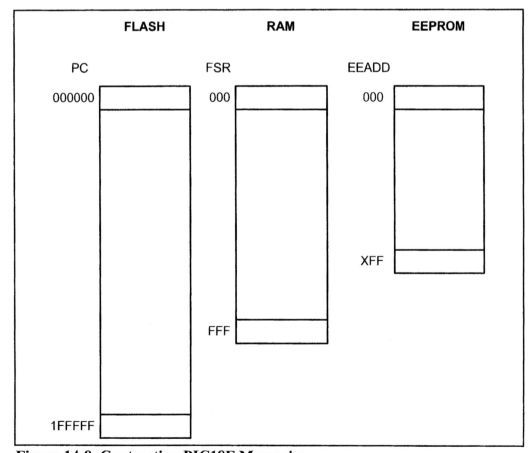

Figure 14-8. Contrasting PIC18F Memories

Writing data to EEPROM

There are four registers associated with the EEPROM. They are as follows:
(a) EEADR: An 8-bit register, used as pointer to EEPROM location.
(b) EEDATA: An 8-bit register, holds data to be written to EEPROM.
(c) EECON1: See Figure 14-4. Used by both EEPROM and Flash.
(d) EECON2: The dummy register. Used by both EEPROM and Flash.

Notice that the EEADR (EE address) register is only 8 bits wide in the PIC18F452/458. The 8-bit address gives us a total space of 256 bytes, which should cover the size of the EEPROM in these chips. In microcontroller chips such as PIC18F4585, which have 1,024 bytes of EEPROM, we have the low-byte and high-byte addresses for the EEADR and they are called EEADRL and EEADRH.

Steps in writing to EEPROM

To write a byte of data to a location in the EEPROM memory, we go through the following steps:

1. Load the EEADR registers with the address of the EEPROM location we want to write the data byte to.
2. Load the EEDATA registers with the data byte we want to write to EEP ROM.
3. Set the EECON1 register for the EEPROM write by making
   (a) EEPGD = 0, (b) CFGS = 0, and (c) WREN = 1.
4. Disable all interrupts globally using "BCF INTCON, GIE".
5. Write 55H to the EECON2 dummy register.
6. Write AAH to the EECON2 dummy register.
7. Set WR# to 1 with the instruction "BSF EECON1,WE".
   With WE = 1, the write cycle begins.
8. Upon completion of the write cycle, the WE# bit will be cleared automatically to indicate that the write cycle is finished.
9. Re-enable the interrupts globally using "BCF INTCON, GIE".
10. The WREN bit should be cleared to prevent an accidental write to the EEPROM by some runaway program.

In the above steps notice the last one. It is important to make WREN = 0, because the PIC18 will not do that automatically. The following program writes a single ASCII letter of 'H' to EEPROM address 10H.

```
MOVLW    0x10  ;starts at location 10H of EEPROM
MOVWF    EEADRD      ;load the EEPROM  address
MOVLW    A'H'           ;load the byte into WREG
MOVWF    EEDATA         ;move it to EEDATA reg
BCF      EECON1,EEPGD   ;point to EEPROM memory
BCF      EECON1,CFGS    ;
BSF      EECON1,WREN    ;enable write
BCF      INTCON,GIE     ;disable all interrupts
MOVLW    0x55           ;wreg = 55h
```

```
        MOVWF       EECON2          ;write to dummy reg
        MOVLW       0xAA            ;wreg = aah
        MOVWF       EECON2          ;write to dummy reg
        BSF         EECON1,WR       ;now write it to Flash
        BSF         INTCON,GIE      ;enable all interrupts
        BCF         EECON1,WREN     ;disable write to memory
```

Steps in reading from EEPROM

Reading a byte from the EEPROM memory is simple and straightforward as shown in the following steps:

1. Load the EEARD register with the address of the EEPROM location we want to read from.
2. Set the EECON1 register for the EEPROM read by making (a) EEPGD = 0, (b) CFGS = 0, and (c) RD = 1.
3. Within the next instruction cycle, the PIC18 will automatically fetch the data from the EEPROM location and place it in the EEDATA register. The only thing we have to do is to move data from the EEDTAT register to a safe place before we do another read. The following shows how to read a byte from EEPROM and place it in PORTB:

```
        MOVLW       0x10            ;read location 10H of EEPROM
        MOVWF       EEADR           ;load the EEPROM address
        BCF         EECON1,EEPGD    ;point to EEPROM memory
        BCF         EECON1,CFGS     ;
        BSF         EECON1,RD       ;enable read
        NOP   ;data is fetched from EEPROM to EEDATA reg
        MOVFF       EEDATA,PORTB    ;place the data in PORTB
```

Program 14-9 (a) writes the message "HELLO" to EEPROM memory starting at location 0, and (b) reads the data back from EEPROM and places it in PORTB one byte at a time.

```
;Program 14-9: Writing to EEPROM
        MOVLW 0x0       ;starts at location 0H of EEPROM
        MOVWF EEADR     ;load the EEPROM address
        MOVLW A'H'      ;load the byte into WREG
        MOVWF EEDATA    ;move it to EEDATA reg
        CALL EE_WRT
        INCF EEADR,F    ;point to next location
        MOVLW A'E'      ;load the byte into WREG
        MOVWF EEDATA    ;move it to EEDATA reg
        CALL EE_WRT
        INCF EEADR,F    ;point to next location
        MOVLW A'L'      ;load the byte into WREG
        MOVWF EEDATA    ;move it to EEDATA reg
        CALL EE_WRT
```

```
        INCF EEADR,F     ;point to next location
        MOVLW A'L'       ;load the byte into WREG
        MOVWF EEDATA     ;move it to EEDATA reg
        CALL EE_WRT
        INCF EEADR,F     ;point to next location
        MOVLW A'O'       ;load the byte into WREG
        MOVWF EEDATA     ;move it to EEDATA reg
        CALL EE_WRT
        INCF EEADR,F     ;point to next location
;read EEPROM one byte at a time and send it to
;PORTB
        MOVLW 0x0 ;starts at location 0H of EEPROM
        MOVWF EEADR           ;load the EEPROM address
        BCF   EECON1,EEPGD    ;point to EEPROM memory
        BCF   EECON1,CFGS     ;
        MOVLW 0x05            ;count = 5
        MOVWF COUNT
        CLRF TRISB            ;make PORTB output port
OVER BSF  EECON1,RD           ;enable read
        NOP
        MOVFF EEDATA,PORTB    ;read the data to PORTB
        CALL DELAY            ;wait
        INCF EEADR,F          ;point to next location
        DECF COUNT,F          ;decrement counter
        BNZ  OVER             ;keep repeating
HERE BRA  HERE
EE_WRT
        BCF   EECON1,EEPGD    ;point to EEPROM memory
        BCF   EECON1,CFGS     ;
        BSF   EECON1,WREN     ;enable write
        BCF   INTCON,GIE      ;disable all interrupts
        MOVLW 0x55            ;wreg = 55h
        MOVWF EECON2          ;write to dummy reg
        MOVLW 0xAA            ;wreg = aah
        MOVWF EECON2          ;write to dummy reg
        BSF   EECON1,WR       ;now write it to Flash
        BSF   INTCON,GIE      ;enable all interrupts
EE_WAIT    BTFSS PIR2,EEIF
        BRA   EE_WAIT
        BCF   PIR2,EEIF
        RETURN
```

Program 14-10 (a) moves a block of data from the code space of the PIC18 chip into EEPROM, and (b) then reads the same data from EEPROM and sends it to the serial port of the PIC18, one byte at a time.

```
#include p18f458.inc

COUNT      EQU   0x0B
BUFRAM     EQU   0x20

       MOVLW  D'8'              ;number of bytes to retrieve
       MOVWF  COUNT
       MOVLW  0H          ;starts at location 0H of EEPROM
       MOVWF  EEADR                ;load the EEPROM address
       MOVLW  upper (CODE_DATA)   ;load TBLPTR
       MOVWF  TBLPTRU
       MOVLW  high (CODE_DATA)
       MOVWF  TBLPTRH
       MOVLW  low (CODE_DATA)
       MOVWF  TBLPTRL
;retrieve the data from program memory
READ_BLOCK
       TBLRD*+            ;read into TABLAT, and increment
       MOVF TABLAT, W      ;get data
       MOVWF EEDATA        ;load data
       CALL EE_WRT         ;save data
       INCF EEADR,F        ;point to next location
       DECFSZ COUNT        ;done?
       BRA READ_BLOCK      ;repeat
;read them back one byte at a time and send to serial
;port
       BSF   TRISD,7       ;PORTD.7 as in input
       MOVLW 0x20 ;enable transmit and low baud rate
       MOVWF TXSTA         ;write to reg
       BCF   PIR1,TXIF
       MOVLW D'15'  ;9600 bps (Fosc / (64 * Speed) - 1)
       MOVWF SPBRG         ;write to reg
       BCF   TRISC, TX ;make TX pin of PORTC an output
       BSF   RCSTA, SPEN ;enable the entire serial port
       MOVLW 8            ;number of bytes in RAM
       MOVWF COUNT
       MOVLW     0x0  ;start at location 0H of EEPROM
       MOVWF     EEADR       ;load the EEPROM address
       CLRF TRISB         ;make PORTB an output port
       MOVLW     0x8      ;counter = 8
       MOVWF     COUNT
```

```
LN      CALL EE_RD              ;read the character
        CALL SENDCOM    ;send character to serial port
        INCF EEADR,F
        DECFSZ COUNT    ;loop until buffers are full
        BRA  LN                 ;repeat
HERE BRA HERE
;-------------
SENDCOM
S1   BTFSS PIR1, TXIF ;wait unil the last bit is gone
     BRA S1             ;stay in loop
     MOVWF TXREG        ;load the value to be transferred
     RETURN             ;return to caller
;-------------
EE_WRT
        BCF   EECON1,EEPGD      ;point to EEPROM memory
        BCF   EECON1,CFGS       ;
        BSF   EECON1,WREN       ;enable write
        BCF   INTCON,GIE        ;disable all interrupts
        MOVLW 0x55              ;wreg = 55h
        MOVWF EECON2            ;write to dummy reg
        MOVLW 0xAA              ;wreg = aah
        MOVWF EECON2            ;write to dummy reg
        BSF   EECON1,WR         ;now write it to Flash
        BSF   INTCON,GIE        ;enable all interrupts
EE_WAIT BTFSS PIR2,EEIF
        BRA   EE_WAIT
        BCF   PIR2,EEIF
        RETURN
;-------------
EE_RD
        BCF EECON1, EEPGD       ;point to DATA memory
        BCF EECON1, CFGS
        BSF EECON1, RD          ;EEPROM read
        MOVF EEDATA, W          ;W = EEDATA
        RETURN
;-------------
        ORG 0x0300
CODE_DATA
        DATA "MOVE ME"
        END
```

Accessing the EEPROM in C

Program 14-11C shows how to write and read the EEPROM memory in C language. This is the C version of an earlier program.

Program 14-11C (a) writes the message "YES" to EEPROM memory, and (b) then reads the same data from EEPROM and sends it to PORTB one byte at a time.

```c
//Program 14-11C
#include <p18F458.h>
void EE_WRT(void);
unsigned char EE_READ(void);
void Delay(unsigned int itime);
void main()
{
    unsigned char x;
    TRISB=0;            //make PORTB output
//write to EEPROM
    EEADR=0x0;          //EEPROM location
    EEDATA='Y';         //write this char to it
    EE_WRT();
    EEADR=0x1;
    EEDATA='E';
    EE_WRT();
    EEADR=0x2;
    EEDATA='S';
    EE_WRT();
    EECON1bits.WREN=0; //disable write
//read from EEPROM and place it on PORTB
    EECON1bits.RD=1;    //enable read
    EEADR =0x0;         //EEPROM location
    x=EE_READ();        //read data from EEPROM
    PORTB=x;            //place it on PORTB
    Delay(250);
    EEADR =0x1;         //EEPROM location
    x=EE_READ();
    PORTB=x;            //place it on PORTB
    Delay(250);
    EEADR =0x2;         //EEPROM location
    x=EE_READ();
    PORTB=x;            //place it on PORTB
    while(1);
}
void EE_WRT()
{
    EECON1bits.EEPGD=0;//point to EEPROM
    EECON1bits.CFGS=0;
    EECON1bits.WREN=1; //enable write
    INTCONbits.GIE=0;  //disable interrupts
    EECON2=0x55;
    EECON2=0xAA;
```

```
        EECON1bits.WR=1;
        INTCONbits.GIE=1;
        while(!PIR2bits.EEIF);
        PIR2bits.EEIF=0;
}

unsigned char EE_READ()
{
        EECON1bits.EEPGD=0;
        EECON1bits.CFGS=0;
        EECON1bits.RD=1;
        return(EEDATA);
}
```

Program 14-12C (a) transfers a block of data from Flash to RAM, (b) writes the block to EEPROM memory, and (b) then reads the same data from EEPROM and sends it to the serial port one byte at a time.

```
//Program 14-12C
#include <p18f458.h>
void EE_WRT(void);
unsigned char EE_READ(void);
void SerTx(unsigned char);

void main(){
    rom far char* RomPointer="MOVE ME";
    char RamString[7];
    unsigned char x,ch,k=sizeof(RomPointer);
    TXSTA=0x20;            //choose low baud rate,8-bit
    SPBRG=15;              //9600 baud rate, XTAL = 10 MHz
    TXSTAbits.TXEN=1;
    RCSTAbits.SPEN=1;

    //move the string to RAM
    for(x=0;x<7;x++){
        RamString[x]=RomPointer[x];
    }

    //move the string to EEPROM
    for(x=0;x<7;x++){
        EEADR=x;
        EEDATA=RamString[x];
        EE_WRT();
    }
    EECON1bits.WREN=0;//disable write

//read from EEPROM and send serially
    for(x=0;x<7;x++){
```

```
        EEADR=x;
        ch=EE_READ();
        SerTx(ch);
    }

   while(1);//infinite loop
}

void EE_WRT()
{
    EECON1bits.EEPGD=0;   //point to EEPROM
    EECON1bits.CFGS=0;
    EECON1bits.WREN=1;    //enable write
    INTCONbits.GIE=0;     //disable interrupts
    EECON2=0x55;
    EECON2=0xAA;
    EECON1bits.WR=1;
    INTCONbits.GIE=1;
    while(!PIR2bits.EEIF);
    PIR2bits.EEIF=0;
 }

unsigned char EE_READ()
{
    EECON1bits.EEPGD=0;   //point to EEPROM
    EECON1bits.CFGS=0;
    EECON1bits.RD=1;
    return(EEDATA);
}

void SerTx(unsigned char c)
{
    while(PIR1bits.TXIF==0);  //wait until transmitted
    TXREG=c;              //place character in buffer
}
```

Review Questions

1. True or false. The PIC18 EEPROM memory is used for both program code and data.
2. True or false. The PIC18F4580 has 1,024 bytes of EEPROM memory.
3. True or false. In the PIC18, EEPROM contents are lost when power is cut off to the chip.
4. True or false. In the PIC18, EEPROM memory is read and write memory.
5. True or false. Every PIC18F chip comes with 1 KB of EEPROM.
6. What is the advantage of the EEPROM over Flash?

SUMMARY

This chapter described memory interfacing with 8031/51-based systems. We began with an overview of semiconductor memories. Types of memories were compared in terms of their capacity, organization, and access time.

ROM (read-only memory) is nonvolatile memory typically used to store programs. The relative advantages of various types of ROM were described in this chapter, including PROM, EPROM, UV-EPROM, EEPROM, flash memory EPROM, and mask ROM.

RAM (random-access memory) is typically used to store data or programs. The relative advantages of its various types, including SRAM, NV-RAM, check-sum byte RAM, and DRAM, were discussed.

The Flash memory space of the PIC18 was discussed, and programs were written in both Assembly and C to access it. Finally, the EEPROM memory of the PIC18 chip was explored and we showed how to access it in both Assembly and C.

PROBLEMS

SECTION 14.1: SEMICONDUCTOR MEMORY

1. What is the difference in capacity between a 4M memory chip and 4M of computer memory?
2. True or false. The more address pins, the more memory locations are inside the chip. (Assume that the number of data pins is fixed.)
3. True or false. The more data pins, the more each location inside the chip will hold.
4. True or false. The more data pins, the higher the capacity of the memory chip.
5. True or false. The more data pins and address pins, the greater the capacity of the memory chip.
6. The speed of a memory chip is referred to as its _____.
7. True or false. The price of memory chips varies according to capacity and speed.
8. The main advantage of EEPROM over UV-EPROM is _____.
9. True or false. SRAM has a larger cell size than DRAM.
10. Which of the following, EPROM, DRAM, or SRAM, must be refreshed periodically?
11. Which memory is used for PC cache?
12. Which of the following, SRAM, UV-EPROM, NV-RAM, or DRAM, is volatile memory?
13. RAS and CAS are associated with which memory?
    (a) EPROM (b) SRAM (c) DRAM (d) all of the above
14. Which memory needs an external multiplexer?
    (a) EPROM (b) SRAM (c) DRAM (d) all of the above
15. Find the organization and capacity of memory chips with the following pins:
    (a) EEPROM A0–A14, D0–D7 (b) UV-EPROM A0–A12, D0–D7

(c) SRAM A0–A11, D0–D7   (d) SRAM A0–A12, D0–D7
(e) DRAM A0–A10, D0   (f) SRAM A0–A12, D0
(g) EEPROM A0–A11, D0–D7   (h) UV-EPROM A0–A10, D0–D7
(i) DRAM A0–A8, D0–D3   (j) DRAM A0– A7, D0–D7

16. Find the capacity, address, and data pins for the following memory organizations:
    (a) 16Kx8 ROM   (b) 32Kx8 ROM
    (c) 64Kx8 SRAM   (d) 256Kx8 EEPROM
    (e) 64Kx8 ROM   (f) 64Kx4 DRAM
    (g) 1Mx8 SRAM   (h) 4Mx4 DRAM
    (i) 64Kx8 NV-RAM

SECTION 14.2: ERASING AND WRITING TO FLASH IN THE PIC18F

17. True or false. The Flash memory in PIC18F is used primarily for the program code.
18. True or false. The Flash memory in PIC18F can be also used for storing fixed data.
19. True or false. The maximum memory space for program memory in PIC18F is 2M bytes.
20. True or false. Reading data from Flash memory can be done one byte at a time.
21. True or false. Writing data to Flash memory can be done one byte at a time.
22. True or false. Writing data to Flash memory must be done in blocks of 64 bytes.
23. True or false. Erasing of Flash memory can be done one byte at a time.
24. True or false. The use of the EECON2 register in writing/erasing of Flash memory must is optional.
25. What registers are used in reading the fixed data stored in Flash memory?
26. What registers are used in writing fixed data to Flash memory?
27. What registers are used in erasing the Flash memory?
28. What is the difference between the WREN and WR bits in the EECON1 register?
29. What registers are used by the TBLRD instruction?
30. What registers are used by the TBLWRT instruction?
31. Explain the difference between the short write and the long write in the PIC18.
32. During which write is the fetching of the opcode suspended by the CPU?
33. What is the size of the block of data for writing to Flash memory in PIC184580?
34. What is the size of the block of data for erasing the Flash memory in PIC184580?
35. Indicate all the addresses that have an 8-byte boundary:
    (a) 510H   (b) 512H   (c) 514H   (d) 516H   (e) 518H
    (f) 51AH   (g) 51CH   (h) 51EH
36. Indicate all the addresses that have a 64-byte boundary:
    (a) 500H   (b) 520H   (c) 540H   (d) 560H   (e) 580H
    (f) 5A0H   (g) 5C0H
37. Give the boundary addresses for the address range of 2000–2020H that can be

used in writing to Flash.

38. Give the boundary addresses for the address range of 2000–2100H that can be used in erasing of Flash memory.

39. Write a program to erase a section of Flash and then write the message "Hello World" to it.

40. For Problem 39, write a program to verify the write operation by reading it and sending it to the serial port one byte at a time

SECTION 14.3: READING AND WRITING TO DATA EEPROM IN THE PIC18F

41. True or false. The EEPROM memory in the PIC18F is used primarily for the program code.

42. True or false. The EEPROM memory in the PIC18F is used for data only.

43. True or false. Every PIC18F member has at least 256 bytes of EEPROM memory.

44. True or false. Reading data from EEPROM memory can be done one byte at a time.

45. True or false. Writing data to EEPROM memory can be done one byte at a time.

46. True or false. Writing data to EEPROM memory must be done in blocks of 64 bytes.

47. True or false. Erasing of data in EEPROM memory can be done one byte at a time.

48. True or false. The use of the EECON2 register in reading and writing of EEPROM memory is optional.

49. True or false. The EECON2 register is used by both the Flash and EEPROM memory write operation.

50. What registers are used in reading data from EEPROM memory?

51. What registers are used in writing data to EEPROM memory?

52. Give the major differences between Flash and EEPROM in the PIC18.

53. What is the size of the block of data for writing to EEPROM memory in the PIC18?

54. Which bits of the EECON1 are used by the read operation of the EEPROM?

55. Why do we disable the interrupts during the write cycle of Flash/EEPROM memory?

56. Why don't we disable the interrupts during the read cycle of Flash/EEPROM memory?

57. Write a program to write the message "Hello World" to EEPROM.

58. For Problem 57, write a program to verify the write operation by reading it and sending it to the serial port one byte at a time.

ANSWERS TO REVIEW QUESTIONS

SECTION 14.1: SEMICONDUCTOR MEMORY

1. c
2. (a) 16Kx8, 128K bits   (b) 64Kx8, 512K   (c) 4Kx8, 32K

3. (a) 2Kx1, 2K bits   (b) 8Kx4, 32K   (c) 128Kx8, 1M
   (d) 64Kx4, 256K   (e) 256Kx1, 256K   (f) 256Kx4, 1M
4. (a) 64K bits, 14 address, and 4 data   (b) 256K, 15 address, and 8 data
   (c) 1M, 10 address, and 1 data   (d) 1M, 18 address, and 4 data
   (e) 512K, 16 address, and 8 data   (f) 4M, 10 address, and 4 data
5. b, c

SECTION 14.2: ERASING AND WRITING TO FLASH IN THE PIC18F

1. True
2. False
3. False
4. True
5. False
6. 8 bytes
7. 64 bytes

SECTION 14.3: READING AND WRITING TO DATA EEPROM IN THE PIC18

1. False
2. False
3. False
4. True
5. False
6. In EEPROM we can write a single byte of data, while in Flash, we must write a block of data.

CHAPTER 15

CCP AND ECCP PROGRAMMING

This chapter discusses the capture/compare/pulse width modulation (CCP) features of the PIC18. In Section 15.1, we show the difference between standard and enhanced CCP modules. In Section 15.2, we describe the compare feature while Section 15.3 deals with the capture feature of the PIC18. The pulse width modulation (PWM) of the PIC18 is shown in Section 15.4. An overview of ECCP is given in Section 15.5. In all these sections we use both Assembly and C language programs to show these important features of the PIC18.

SECTION 15.1: STANDARD AND ENHANCED CCP MODULES

Depending on the family member, the PIC18 has anywhere between 0 and 5 CCP modules inside it. The multiple CCP modules are designated as CCP1, CCP2, CCP3, and so on (CCPx). In recent years, the PWM feature of the CCP has been enhanced greatly for better DC motor control, producng what is called enhanced CCP (ECCP). Therefore, a given family member can have two standard CCP modules and one or more ECCP modules, all on a single chip. See Table 15-1. The ECCP modules are discussed in Chapter 17.

CCP and timers

Table 15-1: PIC18 CCP and ECCP Modules

To program these CCP modules, we must understand PIC18 timers. Review timers in Chapter 9 before you embark on this chapter. Depending on the CCP feature used, the

Chip	# of CCP	# of ECCP
PIC18F2220	2	0
PIC18F4220	1	1
PIC18F452/4520	1	1
PIC18F458/4580	1	1
PIC18F65J10	2	3

timer usage is different. The allocation of the timers among the CCP features is shown in Table 15-2.

Table 15-2: PIC18 Usage of Timers

CCP mode	Timer
Capture	Timer1 or Timer3
Compare	Timer1 or Timer3
PWM	Timer2

Note: The T3CON register is used to choose the timer for the compare and capture modes.

The CCP registers

Each CCP module has three registers associated with it. They are as follows:

(a) CCPxCON is an 8-bit control register. We select one of the compare, capture, and PWM modes using this register. See Figure 15-1.

(b) and (c) CCPRxL and CCPxH form the low byte and the high byte of the 16-bit register. This 16-bit register can be used either as a 16-bit compare register, or a 16-bit capture register, or an 8-bit duty cycle register by the PWM, but not all at the same time. See Figures 15-1 and 5-2. The CCP1CON register selects the mode of operation.

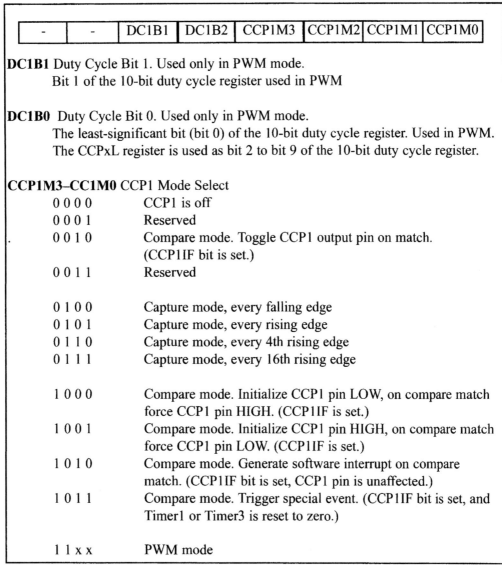

| - | - | DC1B1 | DC1B2 | CCP1M3 | CCP1M2 | CCP1M1 | CCP1M0 |

DC1B1 Duty Cycle Bit 1. Used only in PWM mode.
  Bit 1 of the 10-bit duty cycle register used in PWM

DC1B0 Duty Cycle Bit 0. Used only in PWM mode.
  The least-significant bit (bit 0) of the 10-bit duty cycle register. Used in PWM.
  The CCPxL register is used as bit 2 to bit 9 of the 10-bit duty cycle register.

CCP1M3–CC1M0 CCP1 Mode Select
  0 0 0 0   CCP1 is off
  0 0 0 1   Reserved
  0 0 1 0   Compare mode. Toggle CCP1 output pin on match.
             (CCP1IF bit is set.)
  0 0 1 1   Reserved

  0 1 0 0   Capture mode, every falling edge
  0 1 0 1   Capture mode, every rising edge
  0 1 1 0   Capture mode, every 4th rising edge
  0 1 1 1   Capture mode, every 16th rising edge

  1 0 0 0   Compare mode. Initialize CCP1 pin LOW, on compare match
             force CCP1 pin HIGH. (CCP1IF is set.)
  1 0 0 1   Compare mode. Initialize CCP1 pin HIGH, on compare match
             force CCP1 pin LOW. (CCP1IF is set.)
  1 0 1 0   Compare mode. Generate software interrupt on compare
             match. (CCP1IF bit is set, CCP1 pin is unaffected.)
  1 0 1 1   Compare mode. Trigger special event. (CCP1IF bit is set, and
             Timer1 or Timer3 is reset to zero.)

  1 1 x x   PWM mode

Figure 15-1. CCP1 Control Register. (This register selects one of the operation modes of Capture, Compare, or PWM.)

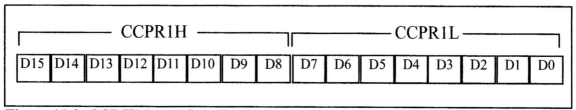

| CCPR1H | | | | | | | | CCPR1L | | | | | | | |
| D15 | D14 | D13 | D12 | D11 | D10 | D9 | D8 | D7 | D6 | D5 | D4 | D3 | D2 | D1 | D0 |

Figure 15-2. CCP High and Low Registers

CCP pins

Each CCP module has a single pin assigned to it. That means that a PIC18 family member with two standard CCP modules (e.g., PIC18F65J10) has two pins, one assigned to each of the CCPs. See Figure 15-3. In the case of the enhanced CCP (ECCP), although it has a single pin, we can program up to four pins to be used by the PWM feature of the ECCP, as we will see at the end of this chapter and in Chapter 17.

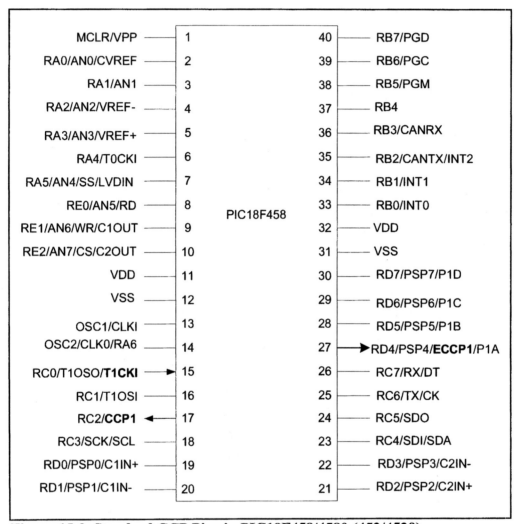

Figure 15-3. Standard CCP Pins in PIC18F458/4580 (452/4520)

Review Questions

1. True or false. The PIC18 chip can have multiple CCP modules inside a single chip.
2. True or false. The CCP1 register is a 16-bit register.
3. True or false. A single pin is associated with each of the standard CCP modules.
4. Give the pin number used for the standard CCP1 in the PIC18F452/458 (or PIC18F4520/4580) chip.

SECTION 15.2: COMPARE MODE PROGRAMMING

The Compare mode of the CCP module is selected using the select bits in the CCPxCON register. The Compare mode can cause an event outside the micro-controller. This event can be simply turning on a device connected to the CCP pin, or the start of an ADC conversion. This event is caused when the content of the Timer1 (or Timer3) register is equal to the 16-bit CCPR1H:CCPR1L register. To use the compare mode of the CCP, we must load both the 16-bit (CCP1H:CCP1L) and the Timer1 (or Timer3) register with some initial values. As Timer1 (or Timer3) counts up, its value is constantly compared with the CCPR1H:CCPR1L register and when a match occurs, the CCP1 pin can perform one of the following actions:

(a) Drive high the CCP1 pin
(b) Drive low the CCP1 pin
(c) Toggle the CCP1 pin
(d) Remain unaffected
(e) Trigger a special event with a hardware interrupt and clear the timer

We use the CCP1CON register to select one of the above actions. See Example 15-1. Note that upon match, the CCP1IF will also go HIGH. See Figure 15-5. Notice that for the above options of (a), (b), and (c) to work, the CCP pin must be configured as an output pin. From Figure 15-4 we use the T3CON register to select Timer1 or Timer3 for the Compare mode. In PIC18F452/458 (or their newer version, PIC18F4520/4580) chips with both CCP1 and ECCP1 modules on the chip, we can assign Timer1 to CCP1 and Timer3 to ECCP1, therefore making them work independently of each other. Also note that only option (e), the special event trigger, will cause Timer1 (or Timer3) to clear, while in other cases we must clear the timer.

Example 15-1

Using Figures 15-1 and 15-4, find the following:
(a) The CCP1CON register value for Compare mode if we want to toggle the CCP1 pin upon match
(b) The T3CON register value if we want to to use Timer3 for the Compare mode of CCP1 with no prescaler

Solution:

(a) From Figure 15-1 we have 00000010 (binary) or 0x20 for the CCP1CON register.

(b) From Figure 15-4 we have 01000010 (binary) or 0x42 for the T3CON register.

There are many applications for the compare feature. One application can be to count the number of people going through a door and closing the door when it reaches a certain number.

RD16	T3CCP2	T3CKPS1	T3CKPS0	T3CCP1	T3SYNC	TMR3CS	TMR3ON

RD16   D7   16-bit read/write enable bit
                         1 = Timer3 16-bit is accessible in one 16-bit operation.
                         0 = Timer3 16-bit is accessible in two 8-bit operations.

T3CCP2:T3CCP1   D6 D3 assigns Timer3 or Timer1 to CPP1 and CCP2 modules
                      CCP1   ECCP1 (or CCP2)

	CCP1	ECCP1 (or CCP2)	
0 0 =	Timer1	Timer1	(clock source for compare/capture)
0 1 =	Timer1	Timer3	(clock source for compare/capture)
1 x =	Timer3	Timer3	(clock source for compare/capture)

T3CKPS1:T3CKPS0 D5 D4 Timer3 Input Clock Prescaler Selector

0 0	= 1:1	Prescale value
0 1	= 1:2	Prescale value
1 0	= 1:4	Prescale value
1 1	= 1:8	Prescale value

T1SYNC   D2   Timer3 External Clock Input Synchronization Control bit
                         Used only when TMR3CS = 1 and clock comes from an
                         external source. If TMR3CS = 0, this bit is not used.
                         1 = Do not synchronize external clock input.
                         0 = Synchronize external clock input.

TMR3CS   D1   Timer 3 Clock Source Select bit
                         1 = External clock from pin RC0 (T1OSI or T1CKI)
                         0 = Internal clock (Fosc/4)

TMR3ON   D0   Timer3 ON and OFF Control bit
                         1 = Enable (start) Timer3
                         0 = Stop Timer3

Figure 15-4. T3CON (Timer 3 Control) Register

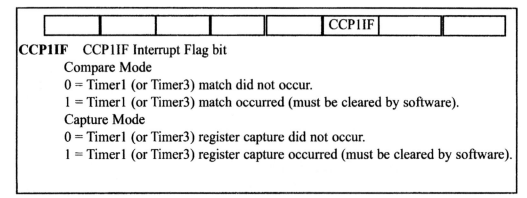

					CCP1IF		

CCP1IF   CCP1IF Interrupt Flag bit
        Compare Mode
        0 = Timer1 (or Timer3) match did not occur.
        1 = Timer1 (or Timer3) match occurred (must be cleared by software).
        Capture Mode
        0 = Timer1 (or Timer3) register capture did not occur.
        1 = Timer1 (or Timer3) register capture occurred (must be cleared by software).

Figure 15-5. PIR1 (Peripheral Interrupt flag register 1) Contains the CCP1IF Flag

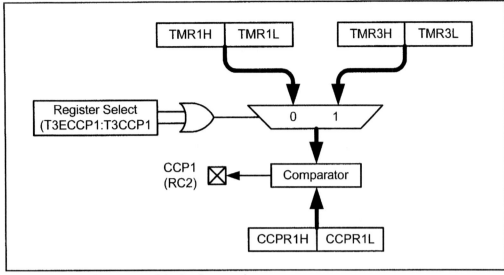

Figure 15-6. Compare Mode Operation

Steps for programming the Compare mode

The following steps are taken in programming the Compare mode:

1. Initialize the CCP1CON register for the compare option.
2. Initialize the T3CON register for Timer1 (or Timer3).
3. Initialize the CCPR1H:CCPR1L registers.
4. Make the CCP1 pin an output pin.
5. Initialize Timer1 (or Timer3) register values.
6. Start Timer1 (or Timer3).
7. Monitor the CCP1IF flag (or use an interrupt).

Program 15-1 shows an example of the Compare mode. It uses Timer3 as a counter and counts the number of pulses fed to Timer3. The pulses could be the number of people going into an elevator. When the count reaches 10, it toggles the LED connected to the CCP1 pin.

For Program 15-1 assume that a 1-Hz pulse is connected to the Timer3 pin and an LED is connected to the CCP1 pin. Timer3 is being used as a counter. Using the Compare mode, this Assembly language program will toggle the LED every 10 pulses.

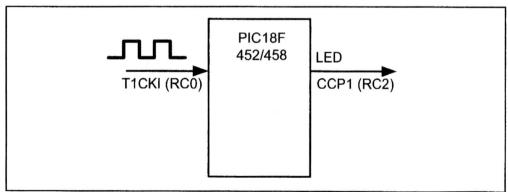

Figure 15-7. Drawing for Programs 15-1 and 15-1C

```
;Program 15-1
      MOVLW  0x02
      MOVWF  CCP1CON          ;Compare mode, toggle upon match
      MOVLW  0x42
      MOVWF  T3CON        ;Timer3 for Compare, 1:1 prescaler
      BCF    TRISC,CCP1    ;CCP1 pin as output
      BSF    TRISC,T1CKI   ;T3CLK pin as input pin
      MOVLW  D'10'
      MOVWF  CCPR1L        ;CCPR1L = 10
      MOVLW  0x0           ;CCPR1H = 0
      MOVWF  CCPR1H
OVER  CLRF   TMR3H         ;clear TMR3H
      CLRF   TMR3L         ;clear TMR3L
      BCF    PIR1,CCP1IF   ;clear CCP1IF
      BSF    T3CON,TMR3ON  ;start Timer3
B1    BTFSS  PIR1,CCP1IF
      BRA    B1
;---------CCP toggle CCP pin upon match
B2    BCF    T3CON,TMR3ON  ;stop Timer3
      GOTO   OVER          ;keep doing it


//Program 15-1C is a C version of Program 15-1
   CCP1CON=0x02;              //Compare mode, toggle upon match
   T3CON=0x42;         //Timer3 for Compare, 1:1 prescaler
   TRISCbits.TRISC2=0; //CCP1 pin an output
   TRISCbits.TRISC0=1;    //T3CLK pin an input
   CCPR1L=10;             //load CCPR1L
   CCPR1H=0;              //load CCPR1H
   while(1)
   {
     TMR3H=0;
     TMR3L=0;
     PIR1bits.CCP1IF=0;    //clear CCP1IF flag
     T3CONbits.TMR3ON=1;   //turn on Timer3
     while(PIR1bits.CCP1IF==0);//wait for CCP1IF
     //CCP toggles CCP pin upon match
     T3CONbits.TMR3ON=0;   //stop Timer3
   }
```

Examine Program 15-2: For this program we assume that the PIC18452/458 has Fosc = 10 MHz. It programs the CCP1 module in Compare mode to create a square wave with a period of 40 ms on the CCP1 pin continuously. The square wave has a 50% duty cycle, which means it is high 50% of each period. This is an example of how Timer1 is used in compare mode. See Figure 15-8. Because the timer uses the Fosc/4, we have 1/2.5 MHz = 0.4 μs for the clock. A 40 ms period gives us 20 ms for high and low portions of the square wave. Now 20 ms / 0.4 μs = 50,000 or C350 in hex. This is the value we load into CCPR1H:CCPR1L for the Compare mode.

RD16	—	T1CKPS1	T1CKPS0	T1OSCEN	T1SYNC	TMR1CS	TMR1ON

RD16 D7 16-bit read/write enable bit
1 = Timer1 16-bit is accessible in one 16-bit operation.
0 = Timer1 16-bit is accessible in two 8-bit operations.

D6 Not used

T1CKPS2:T1CKPS0 D5 D4 Timer1 prescaler selector
0 0 = 1:1   Prescale value
0 1 = 1:2   Prescale value
1 0 = 1:4   Prescale value
1 1 = 1:8   Prescale value

T1OSCEN D3 Timer1 oscillator enable bit
1 = Timer1 oscillator is enabled
0 = Timer1 oscillator is shut off

T1SYNC D2 Timer1 synchronization (used only when TMR1CS = 1 for counter mode to synchronize external clock input)
If TMR1CS = 0, this bit is not used.

TMR1CS D1 Timer1 clock source select bit
1 = External clock from pin RC0/T1CKI
0 = Internal clock (Fosc/4 from XTAL)

TMR1ON D0 Timer1 ON and OFF control bit
1 = Enable (start) Timer 1
0 = Stop Timer 1

Figure 15-8. T1CON (Timer 1 Control) Register

Program 15-2 creates a square wave with a 40 ms period and 50% duty cycle on CCP1 pin using the Compare mode.

```
;Program 15-2
     MOVLW 0x02
     MOVWF CCP1CON    ;Compare mode, toggle upon match
     MOVLW 0x0
     MOVWF T3CON      ;use Timer1 for Compare mode
     MOVLW 0x0
     MOVWF T1CON      ;Timer1, internal CLK, 1:1 prescale
     BCF   TRISC,CCP1    ;CCP1 pin as output
     MOVLW 0xC3
     MOVWF CCPR1H     ;CCPR1H = 0xC3
     MOVLW 0x50
```

```
        MOVWF  CCPR1L        ;CCPR1L = 0x50
OVER  CLRF   TMR1H         ;clear TMR1H
        CLRF   TMR1L         ;clear TMR1L
        BCF    PIR1,CCP1IF   ;clear CCP1IF
        BSF    T1CON,TMR1ON  ;start Timer1
B1      BTFSS  PIR1,CCP1IF
        BRA    B1
        ;CCP toggles CCP1 pin upon match
        BCF    T1CON,TMR1ON  ;stop Timer1
        GOTO   OVER          ;keep doing it
```

```c
//Program 15-2C is the C version of Program 15-2.
  CCP1CON=0x02;       //Compare mode, toggle upon match
  T3CON=0x0;          //Timer1 for Compare, 1:1 prescaler
  T1CON=0x0;          //Timer1 internal clk, 1:1 prescaler
  TRISCbits.TRISC2=0; //make CCP1 pin an output
  TRISCbits.TRISC0=1; //make T1CLK pin an input
  CCPR1H=0xC3;        //load CCPR1H
  CCPR1L=0x50;        //load CCPR1L
  while(1)
  {
    TMR1H=0;                        //clear Timer1
    TMR1L=0;
    PIR1bits.CCP1IF=0;              //clear CCP1IF flag
    T1CONbits.TMR1ON=1;             //turn on Timer1
    while(PIR1bits.CCP1IF==0);      //wait for CCP1IF
    //CCP toggles CCP1 pin upon match
    T1CONbits.TMR1ON=0;             //stop Timer1
  }
```

Review Questions

1. True or false. We can use any timers we want for the Compare mode.
2. True or false. There is a single pin associated with the Compare mode.
3. True or false. In Compare mode, the CCP pin must be configured as an input pin.
4. Which register is used to choose the timer for the Compare mode?

SECTION 15.3: CAPTURE MODE PROGRAMMING

We select Capture mode with the bit selection in the CCP1CON register. In Capture mode, an event at the CCP pin will cause the contents of the Timer1 (or Timer3) register to be loaded into the 16-bit CCPR1H:CCPR1L register. That means, for the Capture mode to work, the CCP pin must be configured as an input pin. The event that causes the contents of Timer1 (or Timer 3) to be captured into the CCPR1H:CCPR1L register can be a High-to-Low (falling-edge) pulse or Low-to-High (rising-edge) pulse. As far as the edge-triggering pulse is concerned, we have the following four options to choose from:

(a) every falling-edge pulse
(b) every rising-edge pulse
(c) every fourth rising-edge pulse
(d every 16th rising-edge pulse

One of the above options can be chosen by selection bits in the CCP1CON register. See Example 15-2. Notice that for any of the above options to work, the CCP pin must be configured as an input pin.

Example 15-2

Using Figures 15-1 and 15-4, find the following:
(a) The CCP1CON register value for Capture mode if we want to capture on the rising edge of every pulse
(b) The T3CON register value if we want to to use Timer3 for Capture mode of the CCP1 with no prescaler

Solution:

(a) From Figure 15-1, we have 00000101 (binary) or 0x05 for the CCP1CON register.
(b) From Figure 15-4, we have 01000010 (binary) or 0x42 for the T3CON register.

One application of Capture mode is measuring the frequency of an incoming pulse. See Program 15-3.

Steps for programming Capture mode

The following steps are used in programming Capture mode for measuring the period of a pulse:

1. Initialize the CCP1CON register for capture.
2. Make the CCP1 pin an input pin.
3. Initialize the T3CON register to select Timer1 or Timer3.
4. Read the Timer1 (or Timer3) register value on the first rising edge and save it.
5. Read the Timer1 (or Timer3) register value on the second rising edge and save it.
6. Subtract the value in step 4 from the value in step 5.

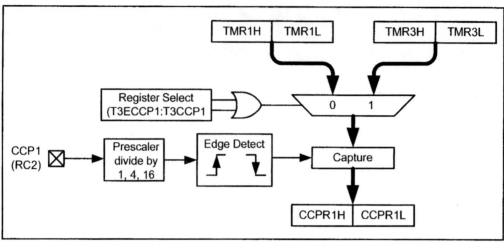

Figure 15-9. Capture Mode Operation

Measuring the period of a pulse

Program 15-3 shows an example of capture mode. See Figure 15-9. It measures the period of the pulse fed to the CCP pin. The measurement is in terms of the number of clock cycles, Tsclk $(1/(Fosc/4))$. See Figure 15-10.

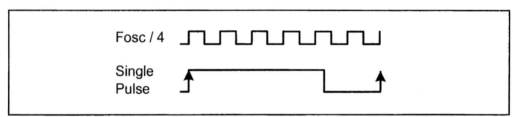

Figure 15-10. Measuring Pulse Period in Terms of Fosc/4 Clock Period

For Program 15-3 assume a pulse is being fed to the CCP1 pin. Using Capture mode, this Assembly language program measures the period of the pulse and puts the results on PORTB and PORTD. The measurement is in terms of the Fosc/4 clock period. See Figure 15-11.

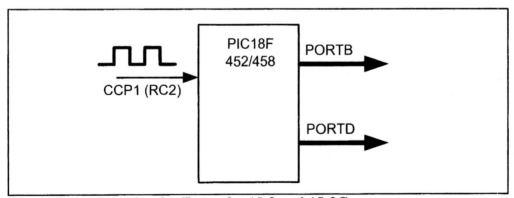

Figure 15-11. Drawing for Examples 15-3 and 15-3C.

```
;Program 15-3
        MOVLW   0x05
        MOVWF   CCP1CON         ;Capture mode rising edge
        MOVLW   0x0
        MOVWF   T3CON           ;Timer1 for Capture
        MOVLW   0x0
        MOVWF   T1CON       ;Timer1, internal CLK, 1:1 prescale
        CLRF    TRISB           ;make PORTB output port
        CLRF    TRISD           ;make PORTD output port
        BSF     TRISC,CCP1      ;make CCP1 pin an input
        MOVLW   0x0
        MOVWF   CCPR1H          ;CCPR1H = 0
        MOVWF   CCPR1L          ;CCPR1L = 0
OVER    CLRF    TMR1H           ;clear TMR1H
        CLRF    TMR1L           ;clear TMR1L
        BCF     PIR1,CCP1IF     ;clear CCP1IF
RE_1    BTFSS   PIR1,CCP1IF
        BRA     RE_1            ;stay here for 1st rising edge
        BSF     T1CON,TMR1ON    ;start Timer1
        BCF     PIR1,CCP1IF     ;clear CCP1IF for next
RE_2    BTFSS   PIR1,CCP1IF
        BRA     RE_2            ;stay here for 2nd rising edge
        BCF     T1CON,TMR1ON    ;stop Timer1
        MOVFF   TMR1L,PORTB     ;put low byte on PORTB
        MOVFF   TMR1H,PORTD     ;put high byte on PORTD
        GOTO    OVER            ;keep doing it


//Program 15-3C is the C version of Program 15-3.
    CCP1CON=0x05;       //Capture mode on every rising edge
    T3CON=0x0;                  //Timer1 for capture
    T1CON=0x0;          //Timer1 internal clk, 1:1 prescaler
    TRISB=0;                    //make PORTB output port
    TRISD=0;                    //make PORTD output port
    TRISCbits.TRISC2=1;     //make CCP1 pin an input
    CCPR1L=0;                   //CCPR1L = 0
    CCPR1H=0;                   //CCPR1H = 0
  while(1)
    {
    TMR1H=0;                    //clear Timer1
    TMR1L=0;
    PIR1bits.CCP1IF=0;      //clear CCP1IF flag
    while(PIR1bits.CCP1IF==0);  //wait for 1st rising edge
    T1CONbits.TMR1ON=1;     //start Timer1
    PIR1bits.CCP1IF=0;      //clear CCPIF for next edge
    while(PIR1bits.CCP1IF==0);  //wait for 2nd rising edge
    T1CONbits.TMR1ON=0;     //stop Timer1
    PORTB=CCPR1L;
    PORTD=CCPR1H;               //display the clock count
    }
```

One problem in measuring the period in the above program is the rate of error due to overhead associated with the program. One way to reduce the effect of the overhead is to use every fourth or every sixteenth rising edge.

Measuring pulse width

One of the most widely used applications of Capture mode is measuring the pulse width. A large number of devices measure things such as distance, temperature, and so on, in which the quantity is provided in terms of the pulse width instead of traditional voltage or current. In these devices, the output is provided in pulse-width-modulated (PWM) form. In a device with PWM output, the output has a fixed frequency and the variable duty cycle provides the quantity we are measuring. For example, the MAX6666/6667 temperature sensors from Maxim Corp. "convert the ambient temperature into a ratiometric PWM output with temperature information contained in the duty cycle of the output square wave." According to their data sheets the output is a square wave with a nominal frequency of 35 Hz at +25°C. The output format is decoded as follows:

$$\text{Temperature (°C)} = 235 - (400 \times t1) / t2 \qquad \text{(Equation 15-1)}$$

where t1 is fixed with a typical value of 10 ms and t2 is modulated by the temperature. In the above formula, $T = t1 + t2$ where T is the period of the pulse, t1 is the high portion of the pulse, and t2 is the low portion, as shown in Figure 15-12. With t1 = 10 and t2 = 20 ms, we get temperature = $235 - (400 \times 10 \text{ ms}) / 20 \text{ ms}$ = $235 - 200 = 35$°C. Program 15-4 shows how to measure the duty cycle using Capture mode.

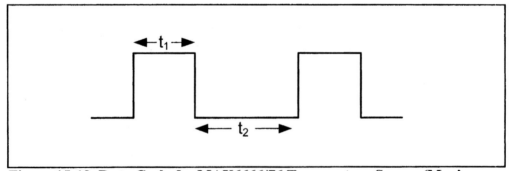

Figure 15-12. Duty Cycle for MAX6666/76 Temperature Sensor (Maxim Corp.)

```
;Program 15-4
FLAG EQU 0x10     ;flag register for steps in detection
DISP EQU 0x0      ;flag for capture complete
RF   EQU 0x1      ;flag for rising or falling edge
     ORG 0x0000
     GOTO MAIN
     ORG 0x0008
     BTFSC PIR1,CCP1IF    ;Is it CCP1?
     GOTO CCP_ISR         ;service CCP1
     RETFIE               ;else return
MAIN MOVLW 0x05
     MOVWF CCP1CON        ;Capture mode rising edge
     MOVLW 0x0
     MOVWF T3CON          ;Timer1 for Capture
     MOVLW 0x0
     MOVWF T1CON ;Timer1, internal CLK, 1:1 prescale
     CLRF  TRISB          ;make PORTB output port
     CLRF  TRISD          ;make PORTD output port
     BSF   TRISC,CCP1     ;make CCP1 pin an input
     CLRF  CCPR1H         ;CCPR1H = 0
     CLRF  CCPR1L         ;CCPR1L = 0
     BSF   PIE1,CCP1IE    ;enable CCP1 interrupt
     BSF   INTCON,PEIE    ;enable peripheral interrupt
     BSF   INTCON,GIE     ;enable all interrupts
OVER CLRF  TMR1H          ;clear TMR1H
     CLRF  TMR1L          ;clear TMR1L
WAIT BTFSS FLAG,DISP      ;Is capture complete?
     BRA WAIT             ;else wait
     BCF FLAG,DISP        ;clear flag for next capture
     MOVLW 0x03
     SUBWF TMR1L,F        ;subtract the overhead
     MOVFF TMR1L,PORTB    ;put low byte on PORTB
     MOVFF TMR1H,PORTD    ;put high byte on PORTD
     GOTO  OVER           ;keep doing it
CCP_ISR   BTFSS FLAG,RF   ;Is it rising edge?
     GOTO RISE_ISR        ;service rising edge
     GOTO FALL_ISR        ;else service falling edge
RISE_ISR  BSF  T1CON,TMR1ON    ;start Timer1
     BSF FLAG,RF          ;ready for falling edge
     BCF CCP1CON,0        ;detect falling edge
     BCF PIR1,CCP1IF ;clear interrupt
     RETFIE              ;return and wait for falling edge
FALL_ISR  BCF  T1CON,TMR1ON    ;stop Timer1
     BSF FLAG,DISP        ;capture complete
     BCF FLAG,RF          ;ready for rising edge
     BSF CCP1CON,0        ;detect rising edge
     BCF PIR1,CCP1IF ;clear interrupt
     RETFIE              ;return capture complete
     END
```

```
//Program 15-4C
#include "p18f458.h"

void CCP1_ISR(void);
void rising(void);
void falling(void);
unsigned char disp = 0;
unsigned char rf = 0;

#pragma interrupt chk_isr
void chk_isr (void)
{
  if (PIR1bits.CCP1IF==1)
    CCP1_ISR();
}

#pragma code My_HiPrio_Int=0x0008
void My_HiPrio_Int (void)
{
  _asm
  GOTO chk_isr
  _endasm
}

#pragma code
void main()
  {
    CCP1CON=0x05;           //Capture mode rising edge
    T3CON=0x0;              //Timer1 for Capture
    T1CON=0x0;          //Timer1, internal CLK, 1:1 prescale
    TRISB=0x0;              //make PORTB output port
    TRISD=0x0;              //make PORTD output port
    TRISCbits.TRISC2=1;     //make CCP1 pin an input
    CCPR1H=0x0;             //CCPR1H = 0
    CCPR1L=0x0;             //CCPR1L = 0
      PIE1bits.CCP1IE=1;   //enable CCP1 interrupt
      INTCONbits.PEIE=1;   //enable peripheral interrupt
      INTCONbits.GIE=1;    //enable all interrupts
      while(1)
      {
    TMR1H=0x0;              //clear TMR1H
        TMR1L=0x0;          //clear TMR1L
    while(disp==0);         //Is data ready to display?
        disp=0;             //clear the flag
        TMR1L-=15;          //subtract the overhead
        PORTB=TMR1L;        //put low byte on PORTB
        PORTD=TMR1H;        //put high byte on PORTD
      }
    }
```

```
void CCP1_ISR()
   {
      if(rf==0)  rising();
         else falling();
      }

void rising()
   {
      T1CONbits.TMR1ON=1;        //start Timer1
      rf=1;                      //ready for falling edge
      CCP1CONbits.CCP1M0=0;      //detect falling edge
      PIR1bits.CCP1IF=0;         //clear interrupt
   }

void falling()
   {
      T1CONbits.TMR1ON=0;        //stop Timer1
      disp=1;                    //capture complete
      rf=0;                      //ready for rising edge
      CCP1CONbits.CCP1M0=1;      //detect rising edge
      PIR1bits.CCP1IF=0;         //clear interrupt
      }
```

Notice that in the company's web site for data sheets the output for a given device is identified as analog (voltage or current) or PWM.

Review Questions

1. True or false. In Capture mode, the CCP pin must be configured as an input pin.
2. True or false. We can use only Timers 1 and 3 for Capture mode.
3. True or false. The timer's register values are transferred to CCPR1H:CCPR1L every time the CPU is reset.
4. True or false. After the timer's register values are transferred to CCPR1H:CCPR1L, the timer's registers are cleared.
5. Which register is used to choose the timer for Capture mode?

SECTION 15.4: PWM PROGRAMMING

Another feature of CCP is pulse width modulation (PWM). The PWM feature allows us to create pulses with variable widths. Although we can program timers to create PWM, the CCP module makes the programming of PWM much easier and less tedious. PWM is widely used in industrial controls such as DC motor controls, as we will see in Chapter 17. Indeed the PWM is so widely used that Microchip has enhanced the PWM capabilities of the newer generation of the PIC18 family members and has designated them as ECCP (enhanced CCP). We will study ECCP in the next section. The main difference between ECCP and standard CCP is the PWM capability. In creating pulses with variable widths for the PWM, two factors are important: The period of the pulse and its duty cycle. The duty cycle (DC) is the portion of the pulse that stays HIGH relative to the entire period. Very often the DC is stated in the form of percentages. For example, a pulse with a 4 ms period that stays HIGH for 1 ms has DC of 25% (1 ms / 4 ms = 25%), as shown in Figure 15-13.

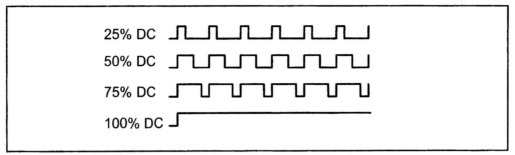

Figure 15-13. Period and Duty Cycle

The period of PWM

The CCP module uses Timer2 and its associated register, PR2, for the PWM time-base, which means that the frequency of the PWM is a fraction of the Fosc, the crystal frequency. It uses the PR2 register to set the PWM period as follows:

$$Tpwm = [(PR2) + 1] 4 \times N \times Tosc \qquad \text{(Equ. 15-2)}$$

where Tosc is the inverse of 1/Fosc, the crystal frequency, Tpwm is the desired PWM period, and N is the prescaler of 1, 4, or 16 set by the Timer2 control register (T2CON). Therefore, we can get the value for the PR2 register as follows:

$$PR2 = [Fosc / (Fpwm \times 4 \times N)] - 1 \qquad \text{(Equ. 15-3)}$$

From Equation 15-2, we can conclude that the maximum value for Tpwm can be achieved when N = 16 and PR2 = 255. Therefore, we have:

$$Tpwm = [(255) + 1] \times 4 \times 16 \times Tosc = 16,382 \ Tosc$$

which means that the minimum allowed Fpwm = Fosc/16,382.

Examine Examples 15-3 to 15-5 to see the calculation of the PWM period.

Example 15-3

Find the PR2 value and the prescaler needed to get the following PWM frequencies. Assume XTAL = 20 MHz.
(a) 1.22 kHz, (b) 4.88 kHz, (c) 78.125 kHz

Solution:

(a) PR2 value = [(20 MHz / (4 × 1.22 kHz)] − 1 = 4,097, which is larger than 255, the maximum value allowed for the PR2. Now choosing the prescaler of 16 we get
PR2 value = [(20 MHz / (4 × 1.22 kHz × 16)] − 1 = 255

(b) PR2 value = [(20 MHz / (4 × 4.88 kHz)] − 1 = 1,023, which is larger than 255, the maximum value allowed for the PR2. Now choosing the prescaler of 4 we get
PR2 value = [(20 MHz / (4 × 4.88 kHz × 4)] − 1 = 255

(c) PR2 value = [(20 MHz / (4 × 78.125 kHz)] − 1 = 63

Example 15-4

Find the PR2 value for the following PWM frequencies. Assume XTAL = 10 MHz and prescaler = 1.
(a) 10 kHz, (b) 25 kHz

Solution:

(b) PR2 value = [(10 MHz / (4 × 10 kHz × 1)] − 1 = 250 − 1 = 249
(c) PR2 value = [(10 MHz / 4 × 25 kHz × 1) − 1 = 100 − 1 = 99

Example 15-5

Find the minimum and maximum Fpwm frequency allowed for XTAL = 10 MHz. State the PR2 and prescaler values for the minimum and maximum Fpwm.

Solution:

We get the minimum Fpwm by making PR2 = 255 and prescaler = 16, which gives us 10 MHz / (4 × 16 × 256) = 610 Hz.

We get the maximum Fpwm by making PR2 = 1 and prescaler = 1, which gives us 10 MHz / (4 × 1 × 1) = 2.5 MHz.

The duty cycle of PWM

As stated earlier, the duty cycle of PWM is the portion of the pulse that stays HIGH relative to the entire period. To set the duty cycle, the CCP module uses the 10-bit register of DC1B9:DC1B0. The 10-bit register of DC1B9:DC1B0 is formed from 8 bits of CCPR1L and 2 bits from the CCP1CON register, where CCPR1L is the upper 8 bits and DC1B2:DC1B1 of the CCP1CON are the lower 2 bits of the 10-bit register. In reality, CCPRL1 is the main register for the duty cycle and the lower 2 bits of DC1B2:DC1B1 are for the decimal point portion of the duty cycle and are set as follows:

DC1B2	DC1B1	Decimal points
0	0	0
0	1	0.25
1	0	0.5
1	1	075

It must be noted that the value for the duty cycle register of the CCPR1L is always some percentage of the PR2 register. For example, if PR2 = 50, and we need a 20% duty cycle, then CCPRL1 = 10 because 20% × 50 = 10. In this case, DC1DB2:DC1B1 = 00. Now assume that we want a 25% duty cycle for the same PR2. Because 50 × 25% = 12.5, we make CCPRL1 = 12 and DC1B2:DC1B1 = 10 to take care of the 0.5 part. See Example 15-6 for further clarification.

Example 15-6

Find the values of registers PR2, CCP1RL, and DC1B2:DC1B1 for the following PWM frequencies if we want a 75% duty cycle. Assume XTAL = 10 MHz.
(a) 1 kHz b) 2.5 kHz

Solution:
(a)
Using the PR2 = Fosc / (4 × Fpwm × N) equation, we must set N = 16 for prescale. Therefore, we have
PR2 = [(10 MHz / (4 × 1 kHz × 16)] – 1 = 156 – 1 = 155 and because 155 × 75% = 116.25 we have CCPR1L = 116 and DC1B2:DC1B1 = 01 for the 0.25 portion.
(b)
Using the PR2 = Fosc / (4 × Fpwm × N) equation, we can set N = 4 for prescale. Therefore, we have
PR2 = [(10 MHz / (4 × 2.5 kHz × 4)] – 1 = 250 – 1 = 249 and because 249 × 75% = 186.75 we have CCPR1L = 186 and DC1B2:DC1B1 = 11 for the 0.75 portion.

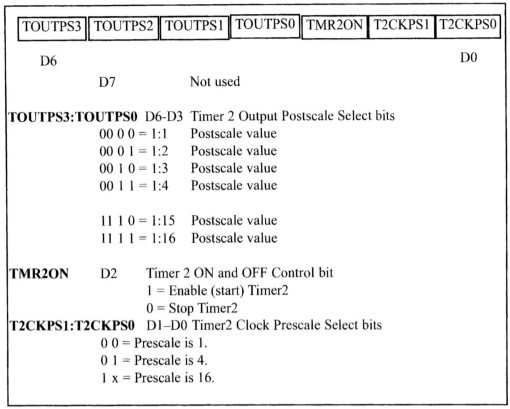

TOUTPS3	TOUTPS2	TOUTPS1	TOUTPS0	TMR2ON	T2CKPS1	T2CKPS0

   D6   D0

           D7   Not used

TOUTPS3:TOUTPS0 D6-D3 Timer 2 Output Postscale Select bits
            00 0 0 = 1:1   Postscale value
            00 0 1 = 1:2   Postscale value
            00 1 0 = 1:3   Postscale value
            00 1 1 = 1:4   Postscale value

            11 1 0 = 1:15 Postscale value
            11 1 1 = 1:16 Postscale value

TMR2ON   D2   Timer 2 ON and OFF Control bit
                      1 = Enable (start) Timer2
                      0 = Stop Timer2
T2CKPS1:T2CKPS0 D1–D0 Timer2 Clock Prescale Select bits
            0 0 = Prescale is 1.
            0 1 = Prescale is 4.
            1 x = Prescale is 16.

Figure 15-14. T2CON (Timer2 Control) Register

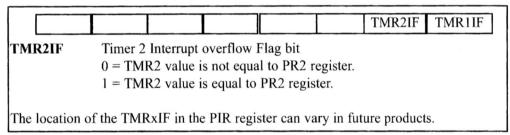

						TMR2IF	TMR1IF

TMR2IF   Timer 2 Interrupt overflow Flag bit
                    0 = TMR2 value is not equal to PR2 register.
                    1 = TMR2 value is equal to PR2 register.

The location of the TMRxIF in the PIR register can vary in future products.

Figure 15-15. PIR1 (Peripheral Interrupt Flag Register 1) Has the TMR2IF Flag

Steps in programming PWM

The following steps are taken to program the PWM feature of the CCP module:

1. Set the PWM period by writing to the PR2 register.
2. Set the PWM duty cycle by writing to CCPR1L for the higher 8 bits.
3. Set the CCP pin as an output.
4. Using the T2CON register, set the prescale value. See Figure 15-14.
5. Clear the TMR2 register.
6. Configure the CCP1CON register for PWM and set DC1B2:DC1B1 bits for the decimal portion of the duty cycle.
7. Start Timer2.

Examine Programs 15-5 and 15-5C to see how the PWM feature is programmed. These programs use the TMR2IF flag. See Figure 15-15.

Using data from Example 15-6, Program 15-5 will create a 2.5 kHz PWM frequency with a 75% duty cycle on the CCP1 pin.

```
;Program 15-5
      CLRF    CCP1CON            ;clear CCP1CON reg
      MOVLW   D'249'
      MOVWF   PR2
      MOVLW   D'186'             ;75% duty cycle
      MOVWF   CCPR1L
      BCF     TRISC,CCP1         ;make PWM pin an output
      MOVLW   0x01        ;Timer2, 4 prescale, no postscaler
      MOVWF   T2CON
      MOVLW   0x3C               ;PWM mode, 11 for DC1B1:B0
      MOVWF   CCP1CON
      CLRF    TMR2               ;clear Timer2
      BSF     T2CON,TMR2ON       ;turn on Timer2
AGAIN BCF     PIR1,TMR2IF        ;clear Timer2 flag
OVER  BTFSS   PIR1,TMR2IF        ;wait for end of period
      BRA     OVER
      GOTO    AGAIN              ;continue
```

```
//Program 15-5C is the C version of Program 15-5.
 CCP1CON=0;                 //clear CCP1CON reg
 PR2=249;
 CCPR1L=186;                //75% duty cycle
 TRISCbits.TRISC2=0;        //make PWM pin an output
 T2CON=0x01;           //Timer2, 4 prescale, no postscaler
 CCP1CON=0x3C;              //PWM mode, 11 for DC1B1:B0
 TMR2=0;                    //clear Timer2
 T2CONbits.TMR2ON=1;        //turn on Timer2
 while(1)
 {
    PIR1bits.TMR2IF=0;      //clear Timer2 flag
    while(PIR1bits.TMR2IF==0); //wait for end of period
 }
```

The role of CCPR1H in the process of creating the duty cycle must be noted. A copy of the duty cycle value in register CCPR1L is given to CCPR1H as soon as we start Timer2. Timer2 goes through the following stages in creating the PWM:

(a) The CCPR1L is loaded into CCPR1H and the CCP1 pin goes HIGH to start the beginning of the period.

(b) As TMR2 counts up, the TMR2 value is compared with both the CCPR1H and PR2 registers.

(c) When the TMR2 and CCPR1H (which is the same as CCPR1L) values are equal, the CCP pin is forced low. That ends the duty cycle portion of the period.

(d) The TMR2 keeps counting up until its value matches the PR2. At that point, the CCP pin goes high, indicating the end of one period and the beginning

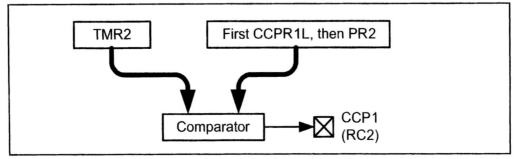

Figure 15-16. TMR2 and PR2 Role in Creating the Duty Cycle

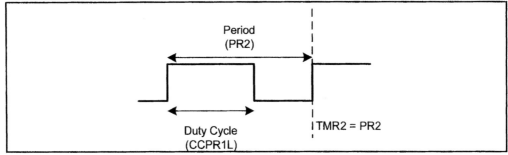

Figure 15-17. TMR2 Relation to CCPR1L and PR2 in PWM

of the next one. It also clears Timer2 for the next round. The CCPR1L is loaded into CCPR1H, and the process continues. See Figures 15-16 and 15-17.

Notice that because the CCPR1L is a fraction of PR2, Timer2 matches CCPR1L first before it matches PR2, unless we have a 100% duty cycle. In that case, Timer2 matches both CCPR1L and PR2 at same time because they have equal values for the 100% duty cycle.

Duty cycle and Fosc

The PIC18 datasheet gives the relation between the Fosc and duty cycle period as follows:

$$\text{Tdutycycle} = (\text{DC1B9:DC1B0 value}) \times \text{Tosc} \times N \qquad (\text{Equ. 15-4})$$

where Tosc = 1 / Fosc and N is the prescaler of 1, 4, or 16 set by the Timer2 control register. To get the value for the DC1B9:DC1B0 register, we can rearrange the above equation as follows:

$$\text{DC1B9:DC1B0} = [(\text{Fosc} / (\text{Fdutycycle} \times N)] \qquad (\text{Equ. 15-5})$$

To calculate the the maximum resolution (number of bits) that can be used for the PWM, the PIC manual gives the following equation:

$$\text{Maximum PWM Resolution (bits)} = \log(\text{Fosc} / \text{Fpwm}) / \log(2) \text{ bits.}$$

Notice that the maximum resolution is 10 bits.

Review Questions

1. True or false. Every standard CCP module has only one PWM pin.
2. How many standard CCP modules do we have in the PIC18F458/4580?
3. True or false. For CCP1, we must use PR2 to set the PWM period.
4. True or false. For CCP1, we must use CCPR1L to set the PWM duty cycle.
5. Which pin of the PIC18F458/4580 is used for PWM?
6. True or false. The duty cycle is always a fraction of the period, unless we want a 100% duty cycle.

SECTION 15.5: ECCP PROGRAMMING

A large number of the PIC18F family members come with ECCP (enhanced CCP) in addition to the standard CCP. While the standard CCP modules are called CCP1, CCP2, and so on, the ECCP modules are designated as ECCP1, ECCP2, and so on. Just like standard CCP, the ECCP has its own pins and registers. The PIC18F452/458 chip uses pin RD4 (PORTD.4) for the ECCP1 pin, while pin RC2 (PORTC.2) is used by the standard CCP1. See Figure 15-18. Figure 15-19 shows the ECCP1 control register.

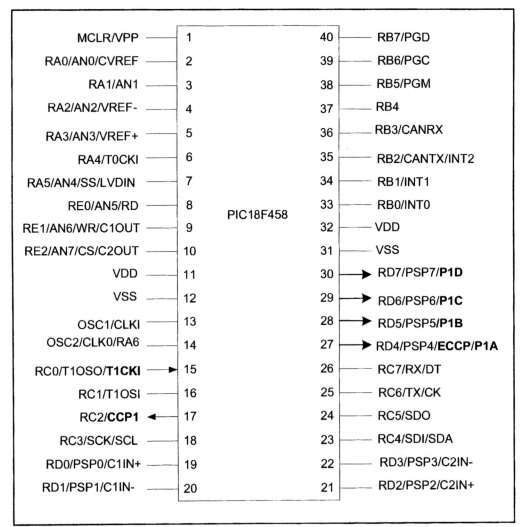

Figure 15-18. ECCP Pins for PWM in PIC18F458/4580 (452/4520)

EPWM1M1	EPWM1M0	EDC1B1	EDC1B0
D7			

ECCP1M3	ECCP1M2	ECCP1M1	ECCP1M0
			D0

EPWM1M1:EPWM1M0 PWM output pin configuration. It allows the use of a single pin for the capture/compare mode, or four pins for the PWM.

In compare/capture mode, only pin P1A (RD4) is used. In that case, there is no selection for these two bits.

In PWM mode, the options for these two bits are as follows:

00	P1A is used for a modulated output. P1B, P1C, and P1D are used as I/O.
01	Full-Bridge output forward. P1D modulated, P1A active. P1B and P1C inactive.
10	Half-Bridge output. P1A and P1D modulated with deadband control, P1C and P1D used as I/O.
01	Full-Bridge output reverse. P1B modulated, P1C active. P1A and P1D inactive.

EDC1B10:EDC1B1 PWM Duty Cycle least-significant bits. Used in PWM only.

The least-significant bits (Bit 1 and Bit 0) of the 10-bit duty cycle register are used in PWM. The ECCPR1L register is used as Bit 2 to Bit 9 of the 10-bit duty cycle register.

ECCP1M3–ECC1M0 ECCP1 Mode Select

0 0 0 0	ECCP1 is off
0 0 0 1	Reserved
0 0 1 0	Compare Mode. Toggle ECCP1 output pin on match. (ECCP1IF bit is set.)
0 0 1 1	Reserved
0 1 0 0	Capture Mode, every falling edge
0 1 0 1	Capture Mode, every rising edge
0 1 1 0	Capture Mode, every 4th rising edge
0 1 1 1	Capture Mode, every 16th rising edge
1 0 0 0	Compare Mode. Initialize ECCP1 pin LOW, on compare match, force CCP1 pin HIGH. (ECCP1IF is set.)
1 0 0 1	Compare Mode. Initialize CCP1 pin HIGH, on compare match, force CCP1 pin LOW. (ECCP1IF is set.)
1 0 1 0	Compare Mode. Generate software interrupt on compare match. (ECCP1IF bit is set, ECCP1 pin is unaffected.)
1 0 1 1	Compare Mode. Trigger special event (ECCP1IF bit is set, and Timer1 or Timer3 is reset to zero.)
1 1 0 0	PWM Mode; P1A, P1C active-HIGH; P1B and P1D active-HIGH
1 1 0 1	PWM Mode; P1A, P1C active-HIGH; P1B and P1D active-LOW
1 1 1 0	PWM Mode; P1A, P1C active-LOW; P1B and P1D active-HIGH
1 1 1 1	PWM Mode; P1A, P1C active-LOW; P1B and P1D active-LOW

Figure 15-19. ECCP1 Control Register. (This register selects one of the operation modes of Capture, Compare, or PWM of EECP1.)

The ECCP1 also has the registers of ECCPR1L, ECCPR1H, and ECCP-CON1. Register PIR2 has the ECCP1IF flag. See Figures 15-20 and 15-21. Just like the standard CCP, it uses Timer1, Timer2, and Timer3 to program the features of compare-capture and PWM. See Table 15-3.

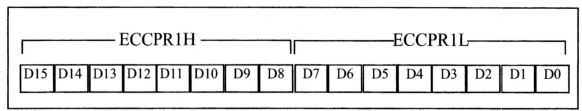

Figure 15-20. ECCP High and Low Registers

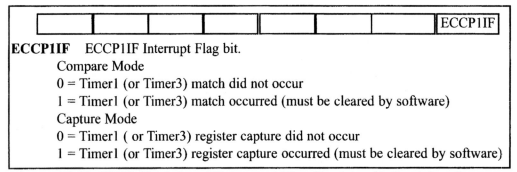

ECCP1IF ECCP1IF Interrupt Flag bit.
   Compare Mode
   0 = Timer1 (or Timer3) match did not occur
   1 = Timer1 (or Timer3) match occurred (must be cleared by software)
   Capture Mode
   0 = Timer1 (or Timer3) register capture did not occur
   1 = Timer1 (or Timer3) register capture occurred (must be cleared by software)

Figure 15-21. PIR2 (Peripheral Interrupt Flag Register 2) Contains the ECCP1IF Flag

Table 15-3: PIC18 Use of Timers for ECCP1

ECCP mode	Timer
Capture	Timer1 or Timer3
Compare	Timer1 or Timer3
PWM	Timer2

Steps for programming the Compare mode in ECCP

Programming the ECCP1 in compare mode is identical to the standard CCP, except we use the ECCP registers. The following steps are taken in programming the Compare mode for ECCP1:

1. Initialize the ECCP1CON register for the compare option.
2. Initialize the T3CON register for Timer1 (or Timer3).
3. Initialize the ECCPR1H:ECCPR1L registers.
4. Make the ECCP1 pin an output pin.
5. Initialize the Timer1 (or Timer3) register values.
6. Start Timer1 (or Timer3).
7. Monitor the ECCP1IF flag (or use an interrupt).

Program 15-6 shows an example of the Compare mode. It uses Timer3 as a counter and counts the number of pulses fed to Timer3. When the count reaches 20, it toggles the LED connected to the ECCP1 pin.

For Program 15-6 assume that a 1-Hz pulse is connected to the Timer3 pin and an LED is connected to the CCP1 pin. Timer3 is being used as a counter. Using the Compare mode, this Assembly language program will toggle the LED every 20 pulses.

```
;Program 15-6
        MOVLW   0x02
        MOVWF   ECCP1CON  ;Compare mode, toggle upon match
        MOVLW   0x42
        MOVWF   T3CON       ;Timer3 for Compare, 1:1 prescaler
        BCF     TRISD,ECCP1    ;ECCP pin as output
        BSF     TRISC,T1CKI    ;T3CLK pin as input pin
        MOVLW   D'20'
        MOVWF   ECCPR1L      ;ECCPR1L = 20
        MOVLW   0x0          ;ECCPR1H = 0
        MOVWF   ECCPR1H
OVER    CLRF    TMR3H        ;clear TMR3H
        CLRF    TMR3L        ;clear TMR3L
        BCF     PIR2,ECCP1IF ;clear ECCP1IF
        BSF     T3CON,TMR3ON ;start Timer3
B1      BTFSS   PIR2,ECCP1IF
        BRA     B1
;---------CCP toggle CCP pin upon match
B2      BCF     T3CON,TMR3ON ;stop Timer3
        GOTO    OVER         ;keep doing it

//Program 15-6C is the C version of Program 15-6.
    ECCP1CON=0x02;     //Compare mode, toggle upon match
    T3CON=0x42;        //Timer3 for Compare, 1:1 prescaler
    TRISDbits.TRISD4=0;     //make ECCP1 pin an output
    TRISCbits.TRISC0=1;     //make T3CLK pin an input
    ECCPR1L=20;        //load ECCPR1L
    ECCPR1H=0;         //load ECCPR1H
    while(1)
    {
      TMR3H=0;
      TMR3L=0;
      PIR2bits.ECCP1IF=0;               //clear ECCP1IF flag
      T3CONbits.TMR3ON=1;               //turn on Timer3
      while(PIR2bits.ECCP1IF==0);       //wait for CECP1IF
      //ECCP toggles ECCP pin upon match
      T3CONbits.TMR3ON=0;               //stop Timer3
    }
```

Steps for programming the Capture mode in ECCP

Programming the ECCP1 in capture mode is identical to the standard CCP, except that we use the ECCP registers. The following steps are taken in programming the Capture mode of ECCP1 for measuring the period of a pulse:

1. Initialize the ECCP1CON register for the Capture option.
2. Make the ECCP1 pin an input pin.
3. Initialize the T3CON register to select Timer1 or Timer3.
4. Read the Timer1 (or Timer3) register value on the first rising edge and save it.
5. Read the Timer1 (or Timer3) register value on the second rising edge and save it.
6. Subtract the value in step 4 from the value in step 3.

For Program 15-7 assume that a pulse is being fed to the ECCP1 pin. Using the Capture mode, this Assembly language program measures the period of the pulse and puts it on PORTB and PORTC. The measure is in terms of the Fosc/4 clock period.

```
;Program 15-7
        MOVLW   0x05
        MOVWF   ECCP1CON        ;Capture mode on rising edge
        MOVLW   0x0
        MOVWF   T3CON           ;Timer1 for capture
        MOVLW   0x0
        MOVWF   T1CON       ;Timer1, internal clk, 1:1 prescale
        CLRF    TRISB           ;make PORTB output port
        CLRF    TRISC           ;make PORTC output port
        BSF     TRISD,ECCP1     ;make ECCP1 pin an input
        MOVLW   0x0
        MOVWF   CCPR1H          ;ECCPR1H = 0
        MOVWF   CCPR1L          ;ECCPR1L = 0
OVER    CLRF    TMR1H           ;clear TMR1H
        CLRF    TMR1L           ;clear TMR1L
        BCF     PIR2,ECCP1IF    ;clear ECCP1IF
RE_1    BTFSS   PIR2,ECCP1IF
        BRA     RE_1            ;stay here for 1st rising edge
        BSF     T1CON,TMR1ON    ;start Timer1
        BCF     PIR2,ECCP1IF    ;clear ECCP1IF for next
RE_2    BTFSS   PIR2,ECCP1IF
        BRA     RE_2            ;stay here for 2nd rising edge
        BCF     T1CON,TMR1ON    ;stop Timer1
        MOVFF   TMR1L,PORTC     ;put low byte on PORTC
        MOVFF   TMR1H,PORTD     ;put high byte on PORTD
        GOTO    OVER            ;keep doing it

//Program 15-7C is the C version of Program 15-7.
        ECCP1CON=0x05;    //Capture mode on every rising edge
```

```
    T3CON=0x0;               //Timer1 for capture
    T1CON=0x0;           //Timer1, internal clk, 1:1 prescaler
    TRISC=0;                 //make PORTB output port
    TRISD=0;                 //make PORTD output port
    TRISDbits.TRISD4=1;      //make ECCP1 pin an input
    ECCPR1L=0;               //ECCPR1L = 0
    ECCPR1H=0;               //ECCPR1H = 0
  while(1)
    {
    TMR1H=0;                 //clear Timer1
    TMR1L=0;
    PIR2bits.ECCP1IF=0;      //clear ECCP1IF flag
    while(PIR2bits.ECCP1IF==0); //wait for 1st rising edge
    T1CONbits.TMR1ON=1;      //start Timer1
    PIR2bits.ECCP1IF=0;      //clear ECCPIF for next edge
    while(PIR2bits.ECCP1IF==0); //wait for 2nd rising edge
    T1CONbits.TMR1ON=0;      //stop Timer1
    PORTC=CCPR1L;
    PORTD=CCPR1H;            //display the clock count
    }
```

PWM features of ECCP

The main difference between the ECCP and standard CCP module is the PWM capability. The standard CCP allows only a single pin for PWM output. This is not enough for implementation of the H-Bridge used widely in DC motor control. As we will see in Chapter 17, we need four pins to drive the H-Bridge for DC motor control. The ECCP allows the use of four pins for the implementation of Full-Bridge or two pins for the Half-Bridge. The four pins used by the ECCP are shown in Table 15-4. In terms of the duty cycle calculation, ECCP1 is the same as CCP1. It uses the PR2 for the duty cycle.

Table 15-4: PIC18 Usage of Pins for ECCP1

ECCP mode	RD4	RD5	RD6	RD7
Compare/Capture	ECCP1	I/O	I/O	I/O
Dual Output PWM	P1A	P1B	I/O	I/O
Quad Output PWM	P1A	P1B	P1C	P1D

Note: I/O means they are used for input/output purpose or other functions associated with the pins.

Steps in programming PWM of ECCP

The following steps are taken to program the PWM feature of the ECCP module:

1. Set the PWM period by writing to the PR2 register.
2. Set the PWM duty cycle by writing to ECCPR1L for the higher 8 bits.

3. Set the ECCP pins as output.
4. Using the T2CON register, set the prescale value.
5. Clear the TMR2 register.
6. Configure the ECCP1CON register for PWM and set the EDC1B2:EDC1B1 bits for the decimal portion of the duty cycle.
7. Start Timer2.

Notice that in programming the compare/capture features, we can assign Timer1 to standard CCP1 and Timer3 to ECCP1 (or vice versa). For the PWM, however, there is only one register for setting the duty cycle. As a result, if we program the PWM feature for both CCP1 and ECCP1, then they will have the same period because there is only one PR2 to set the period. In Chapter 17 we will show how to use ECCP for DC motor control using all four pins in H-Bridge implementations.

Review Questions

1. True or false. Every ECCP module can use only one pin for PWM.
2. How many ECCP modules does the PIC18F458/4580 have?
3. True or false. For ECCP1, we must use PR2 to set the PWM period.
4. True or false. For ECCP1, we must use CCPR1L to set the PWM duty cycle.
5. Which pins of the PIC18F458/4580 are used for PWM?

SUMMARY

This chapter began by describing the CCP features of the PIC18 family. We discussed both the standard CCP and enhanced CCP (ECCP) modules and described each of the compare, capture, and PWM features. We showed how to use Timer1 or Timer3 as the time basis for the compare and capture modes. We also showed how PWM uses Timer2 to create the pulse width modulation.

PROBLEMS

SECTION 15.1: STANDARD AND ENHANCED CCP MODULES

1. True or false. Every member of the PIC18 family has an on-chip CCP module.
2. True or false. The PIC18F452/458 has only one standard CCP.
3. True or false. The PIC18F452/458 has only one ECCP module.
4. True or false. Each CCP module has a 16-bit register accessible as CCPRL and CCPRH.
5. True or false. Each CCP module has a single pin.
6. Give the number of standard and enhanced CCP (ECCP) modules in the PIC18F4520/4580.
7. Give the pin used for standard CCP in the 40-pin DIP package of the PIC18F458/4580.

8. True or false. We use register CCP1CON to choose the Compare mode.
9. True or false. We can use Timer0 and Timer2 for Compare mode.
10. True or false. To use Compare mode, we must make the CCP pin an output pin.
11. Which timers can be used for the Compare mode?
12. Assuming that we are using Timer1 for the Compare mode, indicate when the CCP pin is driven HIGH.
13. Which register holds the CCP flag bit?
14. Find the value for the CCP1CON register in compare mode if we want to drive HIGH the CCP pin upon match.
15. Find the value for the CCP1CON register in compare mode if we want to drive LOW the CCP pin upon match.
16. Find the value for the CCP1CON register in compare mode if we want to toggle the CCP pin upon match.
17. Rewrite Program 15-1 (or 15-1C) for Timer1.
18. Rewrite Program 15-1 (or 15-1C) for the count of 1000.
19. Rewrite Program 15-2 (or 15-2C) for Timer3.
20. Rewrite Program 15-2 (or 15-2C) to create a square wave with a frequency of 100 Hz.

SECTION 15.3: CAPTURE MODE PROGRAMMING

21. True or false. We use the CCP1CON register to choose the Capture mode.
22. True or false. We can use Timer0 and Timer2 for Capture mode.
23. True or false. To use Capture mode, we must make the CCP pin an output pin.
24. Which timers can be used for the capture mode?
25. Find the value for the CCP1CON register in capture mode if we want to capture on the falling edge.
26. Find the value for the CCP1CON register in capture mode if we want to capture every fourth rising edge.
27. Find the value for the T3CON register if we want to use Timer1 for capture mode.
28. Rewrite Program 15-3 (or 15-3C) for Timer3.

SECTION 15.4: PWM PROGRAMMING

29. True or false. We use the CCP1CON register to choose the PWM mode.
30. True or false. We can use Timer0 and Timer1 for the PWM mode.
31. True or false. To use PWM mode, we must make the CCP pin an output pin.
32. Which timer can be used for PWM mode for the standard CCP1?
33. Find the value for the CCP1CON register for PWM mode.
34. Of the CCPR1L and CCPR1H registers, which one is used to set the duty cycle?
35. Which register holds the DC1B2:DC1B1 bits?
36. What is the role of the DC1B2:DC1B1 bits in creating the duty cycle?
37. What is the value for the DC1B2:DC1B1 bits if we want 0.75 for the decimal

points part of the duty cycle?

38. In programming the PWM, the value loaded into the CCPRL1 is always a _____ (fraction, multiple) of the PR2 value.
39. Find the values of registers PR2, CCP1RL, and DCB1B2:DC1B1 bits for the PWM frequency of 2 kHz with 25% duty cycle. Assume XTAL = 10 MHz.
40. Find the values of registers PR2, CCP1RL, and DCB1B2:DC1B1 bits for the PWM frequency of 1.8 kHz with duty cycle of 25%. Assume XTAL = 10 MHz.
41. Find the values of registers PR2, CCP1RL, and DCB1B2:DC1B1 bits or the PWM frequency of 1.5 kHz with duty cycle of 25%. Assume XTAL = 10 MHz.
42. Find the values of registers PR2, CCP1RL, and DCB1B2:DC1B1 bits for the PWM frequency of 1.2 kHz with duty cycle of 25%. Assume XTAL = 10 MHz.

SECTION 15.5: ECCP PROGRAMMING

43. True or false. We use ECCP1CON to choose the PWM mode.
44. True or false. We can use Timer1 or Timer3 for the PWM mode in ECCP.
45. True or false. To use capture mode, we must make the ECCP pin an output pin.
46. Which timer can be used for the PWM mode for ECCP1?
47. Which register holds the ECCP1IF flag bit?
48. Find the value for the ECCP1CON register in compare mode if we want to drive HIGH the ECCP pin upon match.
49. In the PIC18F452/458, give the pin used for ECCP for compare/capture mode.
50. Which timers can be used for the compare mode in ECCP?
51. Which pins are used for PWM in ECCP1?
52. Find the value for the ECCP1CON register in compare mode if we want to drive HIGH the ECCP pin upon match.
53. Find the value for the ECCP1CON register in PWM mode if we want to have H-Bridge where P1A and P1C are active high and the rest are active low.
54. Of the ECCPR1L and ECCPR1H registers, which one is used to set the duty cycle?
55. Which register holds the EDC1B2:EDC1B1 bits?
56. What is role of the EDC1B2:EDC1B1 bits in creating duty cycle?
57. What is value for the EDC1B2:EDC1B1 bits if we want 0.5 for the decimal points part of the duty cycle?
58. In programming the PWM, the value loaded into ECCPRL1 is always a _____ (fraction, multiple) of the PR2 value.
59. Find the values of registers PR2, ECCP1RL, and EDCB1B2:EDC1B1 bits for the PWM frequency of 2 kHz with 25% duty cycle. Assume XTAL = 10 MHz.
60. Find the values of registers PR2, ECCP1RL, and EDCB1B2:DC1B1 bits for the PWM frequency of 1.8 kHz with duty cycle of 25%. Assume XTAL = 10 MHz.
61. Find the values of registers PR2, ECCP1RL, and EDCB1B2:DC1B1 bits for the PWM frequency of 1.5 kHz with duty cycle of 25%. Assume XTAL = 10 MHz.
62. Find the values of registers PR2, ECCP1RL, and EDCB1B2:DC1B1 bits for the PWM frequency of 1.2 kHz with duty cycle of 25%. Assume XTAL = 10 MHz.

ANSWERS TO REVIEW QUESTIONS

SECTION 15.1: STANDARD AND ENHANCED CCP MODULES

1. True
2. True
3. True
4. RC2 (PORTC.2)

SECTION 15.2: COMPARE MODE PROGRAMMING

1. False
2. True
3. True
4. T3CON

SECTION 15.3: CAPTURE MODE PROGRAMMING

1. True
2. True
3. False
4. False
5. T3CON

SECTION 15.4: PWM PROGRAMMING

1. True
2. One
3. True
4. True
5. RC2
6. True

SECTION 15.5: ECCP PROGRAMMING

1. False. Up to four pins.
2. One
3. True
4. False
5. RD4–RD7

CHAPTER 16

SPI PROTOCOL AND DS1306 RTC INTERFACING

OBJECTIVES

Upon completion of this chapter, you will be able to:

>> Understand the Serial Peripheral Interfacing (SPI) protocol
>> Explain how the SPI read and write operations work
>> Examine the SPI pins SDO, SDI, CE, and SCLK
>> Code programs in Assembly and C for SPI
>> Explain how the real-time clock (RTC) chip works
>> Explain the function of the DS1306 RTC pins
>> Explain the function of the DS1306 RTC registers
>> Understand the interfacing of the DS1306 RTC to the PIC18
>> Code programs to display time and date in Assembly and C
>> Explore and program the alarm and interrupt features of the RTC

This chapter discusses the SPI bus and shows the interfacing and programming of the DS1306 real-time clock (RTC), an SPI chip. In Section 16.1, we describe SPI bus connection and protocol. In Section 16.2, we describe the DS1306 RTC's pin functions and show its interfacing and programming with the PIC18. The C programming of DS1306 is shown in Section 16.3. The alarm feature of the DS1306 is discussed in Section 16.4.

SECTION 16.1: SPI BUS PROTOCOL

The SPI (serial peripheral interface) is a bus interface connection incorporated into many devices such as ADC, DAC, and EEPROM. In this section we examine the pins of the SPI bus and show how the read and write operations in the SPI work.

SPI bus

The SPI bus was originally started by Motorola Corp. (now Freescale), but in recent years has become a widely used standard adapted by many semiconductor chip companies. SPI devices use only 2 pins for data transfer, called SDI (Din) and SDO (Dout), instead of the 8 or more pins used in traditional buses. This reduction of data pins reduces the package size and power consumption drastically, making them ideal for many applications in which space is a major concern. The SPI bus has the SCLK (shift clock) pin to synchronize the data transfer between two chips. The last pin of the SPI bus is CE (chip enable), which is used to initiate and terminate the data transfer. These four pins, SDI, SDO, SCLK, and CE, make the SPI a 4-wire interface. See Figure 16-1. There is also a widely used standard called a 3-wire interface bus. In a 3-wire interface bus, we have SCLK and CE, and only a single pin for data transfer. The SPI 4-wire bus can become a 3-wire interface when the SDI and SDO data pins are tied together. However, there are some major differences between the SPI and 3-wire devices in the data transfer protocol. For that reason, a device must support the 3-wire protocol internally in order to be used as a 3-wire device. Many devices such as the DS1306 RTC (real-time clock) support both SPI and 3-wire protocols.

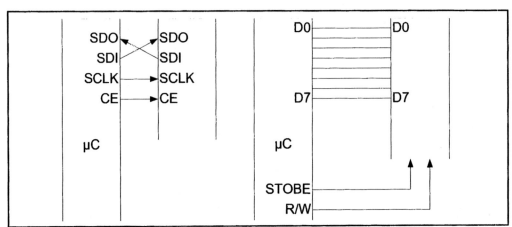

Figure 16-1. SPI Bus vs. Traditional Parallel Bus Connection to Microcontroller

SPI read and write protocol

In connecting a device with an SPI bus to a microcontroller, we use the microcontroller as the master while the SPI device acts as a slave. This means that the microcontroller generates the SCLK, which is fed to the SCLK pin of the SPI device. The SPI protocol uses SCLK to synchronize the transfer of information one bit at a time, where the most-significant bit (MSB) goes in first. During the transfer, the CE must stay HIGH. The information (address and data) is transferred between the microcontroller and the SPI device in groups of 8 bits, where the address byte is followed immediately by the data byte. To distinguish between the read and write, the D7 bit of the address byte is always 1 for write, while for the read, the D7 bit is LOW, as we will see next.

Steps for writing data to an SPI device

In accessing SPI devices, we have two modes of operation: single-byte and multibyte. We will explain each one separately.

Single-byte write

The following steps are used to send (write) data in single-byte mode for SPI devices, as shown in Figure 16-2:

1. Make CE = 1 to begin writing.
2. The 8-bit address is shifted in one bit at a time, with each edge of SCLK. Notice that A7 = 1 for the write operation, and the A7 bit goes in first.
3. After all 8 bits of the address are sent in, the SPI device expects to receive the data belonging to that address location immediately.
4. The 8-bit data is shifted in one bit at a time, with each edge of the SCLK.
5. Make CE = 0 to indicate the end of the write cycle.

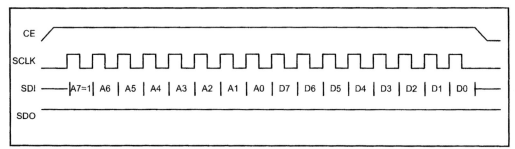

Figure 16-2. SPI Single-Byte Write Timing (Notice A7 = 1)

Multibyte burst write

Burst mode writing is an effective means of loading consecutive locations. In burst mode, we provide the address of the first location, followed by the data for that location. From then on, while CE = 1, consecutive bytes are written to consecutive memory locations. In this mode, the SPI device internally increments the address location as long as CE is HIGH. The following steps are used to send (write) multiple bytes of data in burst mode for SPI devices as shown in Figure 16-3:

1. Make CE = 1 to begin writing.
2. The 8-bit address of the first location is provided and shifted in one bit at a time, with each edge of SCLK. Notice that A7 = 1 for the write operation and the A7 bit goes in first.
3. The 8-bit data for the first location is provided and shifted in one bit at a time, with each edge of the SCLK. From then on, we simply provide consecutive bytes of data to be placed in consecutive memory locations. In the process, CE must stay high to indicate that this is a burst mode multibyte write operation.
4. Make CE = 0 to end writing.

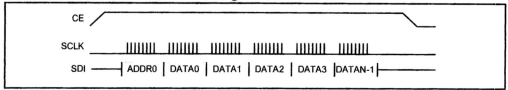

Figure 16-3. SPI Burst (MultiByte) Mode Writing

Steps for reading data from an SPI device

In reading SPI devices, we also have two modes of operation: single-byte and multibyte. We will explain each one separately.

Single-byte read

The following steps are used to get (read) data in single-byte mode from SPI devices as shown in Figure 16-4:

1. Make CE = 1 to begin writing.
2. The 8-bit address is shifted in one bit at a time, with each edge of SCLK. Notice that A7 = 0 for the read operation, and the A7 bit goes in first.
3. After all 8 bits of the address are sent in, the SPI device sends out data belonging to that location.
4. The 8-bit data is shifted out one bit at a time, with each edge of the SCLK.
5. Make CE = 0 to indicate the end of the read cycle.

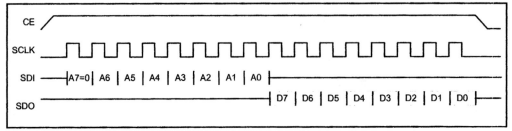

Figure 16-4. SPI Single-Byte Read Timing (Notice A7 = 0)

Multibyte burst read

Burst mode reading is an effective means of bringing out the contents of consecutive locations. In burst mode, we provide the address of the first location only. From then on, while CE = 1, consecutive bytes are brought out from consecutive memory locations. In this mode, the SPI device internally increments the address location as long as CE is HIGH. The following steps are used to get (read) multiple bytes of data in burst mode for SPI devices, as shown in Figure 16-5:

1. Make CE = 1 to begin reading.
2. The 8-bit address of the first location is provided and shifted in one bit at a time, with each edge of SCLK. Notice that A7 = 0 for the read operation, and the A7 bit goes in first.
3. The 8-bit data for the first location is shifted out one bit at a time, with each edge of the SCLK. From then on, we simply keep getting consecutive bytes of data belonging to consecutive memory locations. In the process, CE must stay HIGH to indicate that this is a burst mode multibyte read operation.
4. Make CE = 0 to end reading.

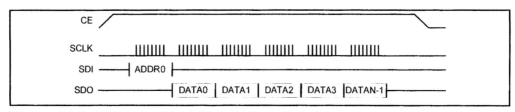

Figure 16-5. SPI Burst (MultiByte) Mode Reading

Review Questions

1. True or false. The SPI protocol writes and reads information in 8-bit chunks.
2. True or false. In SPI, the address is immediately followed by the data.
3. True or false. In an SPI write cycle, bit A7 of the address is LOW.
4. True or false. In an SPI write, the LSB goes in first.
5. State the difference between the single-byte and burst modes in terms of the CE signal.

SECTION 16.2: DS1306 RTC INTERFACING AND PROGRAMMING

The real-time clock (RTC) is a widely used device that provides accurate time and date information for many applications. Many systems such as the x86 IBM PC come with such a chip on the motherboard. The RTC chip in the IBM PC provides the time components of hour, minute, and second, in addition to the date/calendar components of year, month, and day. Many RTC chips use an internal battery, which keeps the time and date even when the power is off. Although some microcontrollers, such as the DS5000T, come with the RTC already embedded into the chip, we have to interface the vast majority of them to an external RTC chip. One of the most widely used RTC chips is the DS12887 from Dallas Semiconductor/Maxim Corp. This chip is found in the vast majority of x86 PCs. The original IBM PC/AT used the MC14618B RTC from Motorola. The DS12887 is the replacement for that chip. It uses an internal lithium battery to keep operating for over 10 years in the absence of external power. The DS12887 is a parallel RTC with 8 pins for the data bus. In this chapter, we interface and program the DS1306 RTC, which has an SPI bus. According to the DS1306 data sheet from Maxim, it keeps track of "seconds, minutes, hours, day of week, date, month, and year with leap-year compensation valid up to year 2099." The DS1306 RTC provides the above information in BCD format only. It supports both 12-hour and 24-hour clock modes, with AM and PM in the 12-hour mode. It does not support the Daylight Savings Time option. The DS1306 has a total of 128 bytes of nonvolatile RAM. It uses 28 bytes of RAM for clock/calendar and control registers, and the other 96 bytes of RAM are for general-purpose data storage. Next, we describe the pins of the DS1306. See Figure 16-6.

V_{CC2}
Pin 1 provides an external back-up supply voltage to the chip. This pin is connected to an external rechargeable power source. This option is called *trickle charge*. If this pin is not used, it must be grounded.

V_{bat}
Pin 2 can be connected to an external +3 V lithium battery, thereby providing the power source to the chip externally as back-up supply voltage. We must connect this pin to ground if it is not used.

V_{CC1}
Pin 16 is used as the primary external voltage supply to the chip. This primary external voltage source is generally set to +5 V. When V_{cc1} falls below the Vbat voltage level, the DS1306 switches to Vbat and the external lithium battery provides power to the RTC. According to the DS1306 data sheet "upon power-up, the device switches from Vbat to Vcc1 when Vcc1 is greater than Vbat+0.2 Volts." Because we can connect the standard 3 V lithium battery to the Vbat pin, the Vcc1 voltage level must remain above 3.2 V in order for the Vcc1 to remain as the primary voltage source to the chip. This nonvolatile capability of the RTC prevents

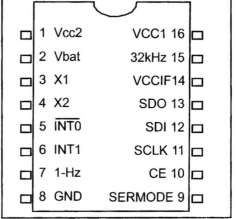

Figure 16-6. DS1306 RTC Chip (from Maxim/Dallas Semiconductor)

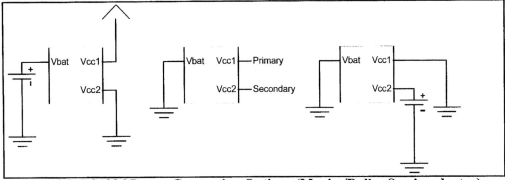

Figure 16-7. DS1306 Power Connection Options (Maxim/Dallas Semiconductor)

any loss of data. See Figure 16-7.

GND
Pin 8 is the ground.
SDI (Serial Din)
The SDI pin provides the path to bring data into the chip, one bit at a time.
SDO (Serial Dout)
The SDO pin provides the path to bring data out of the chip, one bit at a time.
32KHz
This is an output pin providing a 32.768 kHz frequency. This frequency is always present at the pin.
X1–X2
These are input pins that allow the DS1306 connection to an external crystal oscillator to provide the clock source to the chip. We must use the standard 32.768 kHz quartz crystal. The accuracy of the clock depends on the quality of this crystal oscillator. See Figure 16-8. Heat can cause a drift on the oscillator. To avoid this, we use the DS32KHZ chip, which automatically adjusts for temperature variations. Note that when using the DS32KHZ or similar clock generators, we only need to connect X1 because the X2 loopback is not required.
SCLK (serial clock)
An input pin is used for the serial clock to synchronize the data transfer between the DS1306 and the microcontroller.
1-Hz
An output pin provides a 1-Hz square wave frequency. The DS1306 creates the 1-Hz square wave automatically. To get this 1-Hz frequency to show up on the pin, however, we must enable the associated bit in the DS1306 control register.
CE
Chip enable is an input pin and an active-HIGH signal. During the read and write cycle time, CE must be high.
INT0#
Interrupt request is an output pin and an active-LOW signal. To use INT0, the interrupt-enable bit in the RTC control register must be set HIGH. The interrupt feature of the DS1306 is discussed in Section 16.4.
INT1
Interrupt request is an output pin and an active-HIGH signal. To use INT1, the interrupt-enable bit in the RTC control register must be set HIGH. The inter-

---

rupt feature of the DS1306 is discussed in Section 16.4.

SERMODE (serial mode selection)

Pin 9 is an input pin. If it is HIGH, then the SPI mode is selected. If it is connected to ground, the 3-wire mode is used. In our application, the SERMODE pin is connected to the V_{cc} pin because we program the 1306 chip using the SPI protocol.

V_{CCif}

Pin 14 is the interface logic power-supply input. This pin allows interfacing of the DS1306 with systems with 3 V logic in mixed supply systems. See the DS1306 data sheet if you are using a power source other than 5 V in your system.

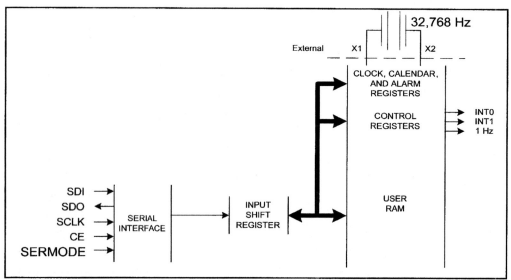

Figure 16-8. Simplified Block Diagram of DS1306 (Maxim/Dallas Semiconductor)

Importance of the WP bit in the Control register

As shown in Table 16-1, the Control register has an address of 8FH for write and 0FH for read. The most important bit in the Control register is the WP bit. The WP bit is undefined upon reset. In order to write to any of the registers of the DS1306, we must clear the WP bit first. See Figure 16-9. Upon powering up the DS1306, we have to clear the WP bit at least once. This means that after initializing the DS1306 we can write protect all the registers by making WP = 1.

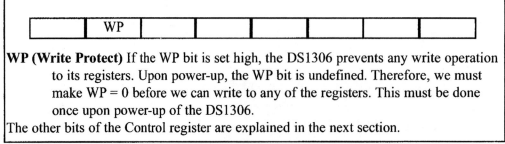

WP (Write Protect) If the WP bit is set high, the DS1306 prevents any write operation to its registers. Upon power-up, the WP bit is undefined. Therefore, we must make WP = 0 before we can write to any of the registers. This must be done once upon power-up of the DS1306.

The other bits of the Control register are explained in the next section.

Figure 16-9. WP Bit of DS1306 Control Register (write location address is 8FH)

Address map of the DS1306

The DS1306 has a total of 128 bytes of RAM space with addresses 00–7FH. The first fifteen locations, 00–0E, are set aside for RTC values of time, date, and alarm data. The next three bytes are used for the control and status registers. They are located at addresses 0F–11 in hex. The next 14 bytes from addresses 12H to 1FH are reserved and cannot be used. That leaves 96 bytes, from addresses 20H to 7FH, available for general purpose data storage. That means the entire 128 bytes of RAM are accessible directly for read or write except the addresses 12–1FH. Table 16-1 shows the address map of the DS1306. In this section we study the time and date. The alarm is examined in Section 16.4.

Table 16-1: Registers of the DS1306 (modified from datasheet)

HEX ADDRESS READ	WRITE	D7	D6	D5	D4	D3	D2	D1	D0	RANGE in HEX
0x00	0x80	0	10 SEC				SEC			00-59
0x01	0x81	0	10 MIN				MIN			00-59
0x02	0x82	0	24 / 12	20 HR P/A	10 HR		HOURS			00-23 01-12 P/A
0x03	0x83	0	0	0	0	0	DAY			01-07
0x04	0x84	0	0	10 DATE			DATE			01-31
0x05	0x85	0	0	10 MONTH			MONTH			01-12
0x06	0x86	0	10 YEAR				YEAR			00-99
0x07	0x87	M	10 SEC ALARM0				SEC ALARM0			00-59
0x08	0x88	M	10 MIN ALARM0				MIN ALARM0			00-59
0x09	0x89	M	24 / 12	20 HR P/A	10 HR		HOUR ALARM0			00-23 01-12 P/A
0x0A	0x08A	M	0	0	0	0	DAY ALARM0			01-07
0x0B	0x8B	M	10 SEC ALARM1				SEC ALARM1			00-59
0x0C	0x8C	M	10 MIN ALARM1				MIN ALARM1			00-59
0x0D	0x8D	M	24 / 12	20 HR P/A	10 HR		HOUR ALARM1			00-23 01-12 P/A
0x0E	0x8E	M	0	0	0	0	DAY ALARM1			01-07
0x0F	0x8F	CONTROL REGISTER								
0x10	0x90	STATUS REGISTER								
0x11	0x91	TRICKLE CHARGER REGISTER								
0x12-0x1F	0x92-0x9F	RESERVED								

Time and date address locations and modes

The byte addresses 0–6 are set aside for the time and date, as shown in Table 16-2. Table 16-2 is extracted from Table 16-1. It shows a summary of the address locations in read/write modes with data ranges for each location. The DS1306 provides data in BCD format only. Notice the data range for the hour mode. We can select 12-hour or 24-hour mode with bit 6 of hour location 02. When D6 = 1, the 12-hour mode is selected, and D6 = 0 provides us the 24-hour mode. In the 12-hour mode, we decide the AM and PM with the bit 5. If D5 = 0, the AM is selected and D5 = 1 is for the PM. Example 16-1 shows how to get the range of the data acceptable for the hour location.

Table 16-2: DS1306 Address Locations for Time and Date (extracted from Table 16-1)

Hex Address Location		Function	Data Range	Range in hex
Read	Write		BCD	
00	80	Seconds	00–59	00–59
01	81	Minutes	00–59	00–59
02	82	Hours, 12-Hour Mode	01–12	41–52 AM
		Hours, 12-Hour Mode	01–12	61–72 PM
		Hours, 24-Hour Mode	00–23	00–23
03	83	Day of the Week, Sun = 1	01–07	01–07
04	84	Day of the Month	01–31	01–31
05	85	Month	01–12	01–12
06	86	Year	00–99	00–99

Example 16-1

Using Table 16-1, verify the hour location values in Table 16-2.

Solution:

(a) For 24-hour mode, we have D6 = 0. Therefore, the range goes from 0000 0000 to 0010 0011, which is 00–23 in BCD.
(b) For 12-hour mode, we have D6 = 1 and D5 = 0 for AM. Therefore, the range goes from 0100 0001 to 0101 0010, which is 41–52 in BCD.
(c) For 12-hour mode, we have D6 = 1 and D5 = 1 for PMn. Therefore, the range goes from 0110 0001 to 0111 0010, which is 61–72 in BCD.

PIC18 interfacing to DS1306 using MSSP module

The DS1306 supports both SPI and 3-wire modes. In DS1306, we select the SPI mode by connecting the SERMODE pin to Vcc. If SERMODE = Gnd, then the 3-wire protocol is used. In this section, we use SPI mode only. The MSSP (Master Synchronous Serial Port) module inside the PIC18 supports SPI bus protocol. Three registers are associated with SPI of the MSSP module. They are SSP-BUF, SSPCON1, and SSPSTAT. To transfer a byte of data, we place it in SSPBUF. The SSPBUF register also holds the byte received via the SPI bus. Figures 16-10 and 16-11 show the other two major registers of the PIC18 for SPI interfacing. We use SSPCON1 to select the SPI mode operation of the PIC18. Notice that the SSPEN bit in the SSPCON1 register must be set to HIGH to allow the use of the PIC18 pins for SPI data bus protocol. We must also choose the SPI Master mode using the SSPM3:SSPM0 bits of SSPCON1. In our application, we will use Fosc/64 speed for best performance in data transfer between the PIC18 and the DS1306 RTC.

After the selection of SSPCON1, we must also select the proper bits for timing in the SSPSTAT register, as shown in Figure 16-11. In our application, we send data to an SPI device on the rising edge, and receive data from the SPI device in the middle of the SCLK clock pulse.

Because we are using the SPI feature of the PIC18 to communicate with our SPI device, we must use the designated pins for the SPI signals. They are RC2 (CE), RC3 (SCLK), RC4 (SDI), and RC5 (SDO), as shown in Figure 16-12.

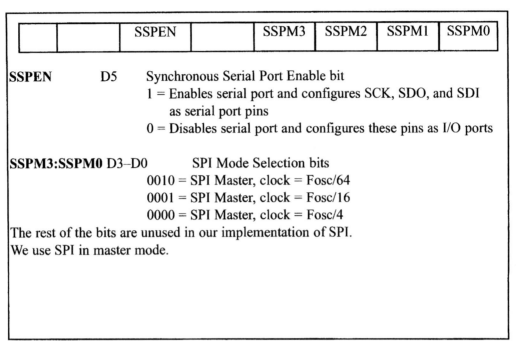

		SSPEN		SSPM3	SSPM2	SSPM1	SSPM0

SSPEN   D5   Synchronous Serial Port Enable bit
                        1 = Enables serial port and configures SCK, SDO, and SDI
                            as serial port pins
                        0 = Disables serial port and configures these pins as I/O ports

SSPM3:SSPM0 D3–D0   SPI Mode Selection bits
                        0010 = SPI Master, clock = Fosc/64
                        0001 = SPI Master, clock = Fosc/16
                        0000 = SPI Master, clock = Fosc/4
The rest of the bits are unused in our implementation of SPI.
We use SPI in master mode.

Figure 16-10. SSPCON1 - SSP Control Register 1
Note: Portion shown is used for SPI.

SMP	CKE						BF

SMP   D7   Sample bit
                      1 = Input data sampled at end of data output time
                      0 = Input data sampled at middle of data output time

CKE   D6   SPI Clock Edge Select bit
                      1 = Transmit occurs on transition from active to idle clock
                          state.
                      0 = Transmit occurs on transition from idle to active clock
                          state.

BF   D0   Buffer Full Status bit. Used for receive only.
                      1 = Receive complete, SSPBUF is full.
                      0 = Receive not complete, SSPBUF is empty.
The rest of the bits are used for I^2C module.

Figure 16-11. SSPSTAT - SSP Status Register
Note: Portion shown is used for SPI.

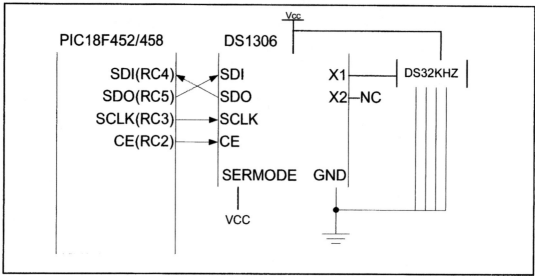

Figure 16-12. DS1306 Connection to PIC18

Note: For more accuracy, we use the DS32KHZ chip in place of a crystal.

Setting the time in Assembly

Program 16-1 initializes the clock at 16:58:55 using the 24-hour clock mode. It uses the single-byte operation for writing into the control register of the DS1306 and multibyte burst mode for writing seconds, minutes, and hours. Regarding the SPI subroutine in Program 16-1, we must note the following points:

1. In order for the PIC18 to transfer a byte of data using SPI protocol, it must be placed in SSPBUF.
2. After writing to SSPBUF, we must monitor the BF flag bit of the SSPSTAT register to ensure the entire byte has been transferred.
3. SSPBUF is also used as the destination for incoming data from an SPI device. This happens as data is being sent. The BF flag indicates that the entire byte has been received.

```
;Program 16-1: Setting the Time
      MOVLW 0x00
      MOVWF SSPSTAT ;read at middle, send on active edge
      MOVLW 0x22
      MOVWF SSPCON1 ;enable master SPI, Fosc / 64
      CLRF TRISC          ;make PORTC output
      BSF TRISC,SDI       ;except SDI
 ;--  send control byte to DS1306 in single-byte mode
      BSF   PORTC,RC2     ;make CE = 1 for single-byte
      CALL SDELAY
      MOVLW 0x8F                ;DS1306 control register address
      CALL SPI
      MOVLW 0x00                ;clear WP bit for write
```

```
        CALL SPI
        BCF PORTC,RC2
;make CE = 0 to end write (single-byte)
        CALL SDELAY
;-- send the data to DS1306 in burst mode
        BSF   PORTC,RC2   ;make CE = 1 (start multibyte write)
        MOVLW 0x80              ;seconds register address
        CALL SPI                ;send address
        MOVLW 0x55              ;55 seconds
        CALL SPI                ;send seconds
        MOVLW 0x58              ;58 minutes
        CALL SPI                ;send minutes
        MOVLW 0x16              ;24-hour clock at 16 hours
        CALL SPI                ;send hour
        BCF PORTC,RC2     ;make CE = 0 (end multibyte write)
;-- SPI write/read subroutine
SPI MOVWF SSPBUF              ;load SSPBUF for transfer
WAIT BTFSS SSPSTAT,BF        ;wait for all bits
        BRA WAIT
        MOVF SSPBUF,W            ;get the received byte
        RETURN                  ;return with byte in WREG
        END
```

Setting the date in Assembly

Program 16-2 shows how to set the date to October 19th, 2004.

```
;Program 16-2: Setting the Date
        MOVLW 0x00
        MOVWF SSPSTAT ;read at middle, send on active edge
        MOVLW 0x22
        MOVWF SSPCON1           ;master SPI enable, Fosc / 64
        CLRF TRISC             ;make PORTC output
        BSF TRISC,SDI          ;except SDI
        BSF   PORTC,RC2        ;enable the RTC
        MOVLW 0x8F             ;DS1306 control register address
        CALL SPI
        MOVLW 0x00             ;clear WP bit for write
        CALL SPI
        BCF PORTC,RC2          ;turn off RTC
;-- send the date to DS1306
        BSF   PORTC,RC2        ;enable the RTC
        MOVLW 0x84             ;date register address
        CALL SPI               ;send address
        MOVLW 0x19             ;19th of the month
        CALL SPI               ;send date
        MOVLW 0x10             ;October
        CALL SPI               ;send month
        MOVLW 0x04             ;2004
        CALL SPI               ;send year
        BCF PORTC,RC2          ;disable RTC
;-- SPI write/read subroutine
```

```
SPI MOVWF SSPBUF              ;load SSPBUF for transfer
WAIT BTFSS SSPSTAT,BF         ;wait for all bits
     BRA WAIT
     MOVF SSPBUF,W            ;get the received byte
     RETURN                  ;return with byte in WREG
     END
```

RTCs setting, reading, and displaying time and date

Program 16-3 is the complete Assembly code for setting, reading, and displaying the time and date. The times and dates are sent to the IBM PC screen via the serial port after they are converted from packed BCD to ASCII.

```
;Program 16-3
#include p18f458.inc
D1uL EQU D'2'                    ;1 microsecond delay byte
DR1uL EQU 0x0D              ;register for 1 microsecond delay
DAY  EQU 10H                     ;for day of the week
MON  EQU 11H                     ;fileReg starting with month
DAT  EQU 12H                     ;for day of the month
YR   EQU 13H                     ;for year
HR   EQU 14H                     ;for hour
MIN  EQU 15H                     ;for minutes
SEC  EQU 16H                     ;for seconds
CNT  EQU 20H                     ;for counter
TMP  EQU 21H                     ;for conversions
     MOVLW 0x00
     MOVWF SSPSTAT ;read at middle, send on active edge
     MOVLW 0x22
     MOVWF SSPCON1             ;master SPI enable, Fosc / 64
     CLRF TRISC                ;make PORTC output
     BSF TRISC,SDI             ;except SDI
     BSF TRISC,RX              ;and RX
;-- enable USART communication
     MOVLW B'00100000'         ;enable transmit and low baud
     MOVWF TXSTA               ;write to reg
     MOVLW D'15'            ;9600 bps (Fosc / (64 * Speed) - 1)
     MOVWF SPBRG               ;write to reg
     BCF TRISC, TX          ;make TX pin of PORTC an output pin
     BSF RCSTA, SPEN           ;enable the serial port
;-- start a new line for USART communications
     MOVLW 0x0A                ;form feed
     CALL TRANS
     MOVLW 0x0D                ;new line
     CALL TRANS
;-- send control byte to DS1306
     BSF  PORTC,RC2            ;enable the RTC
     CALL SDELAY
     MOVLW 0x8F                ;control register address
     CALL SPI
     MOVLW 0x00                ;clear WP bit for write
```

```
            CALL SPI
            BCF  PORTC,RC2        ;disable RTC
            CALL SDELAY
    ;-- send the time followed by date
            BSF  PORTC,RC2        ;enable the RTC
            MOVLW 0x80        ;seconds register address for write
            CALL SPI             ;send address
            MOVLW 0x55           ;55 seconds
            CALL SPI             ;send seconds
            MOVLW 0x58           ;58 minutes
            CALL SPI             ;send minutes
            MOVLW 0x16           ;24-hour clock at 16 hours
            CALL SPI             ;send hour
            MOVLW 0x3            ;Tuesday
            CALL SPI             ;send day of the week
            MOVLW 0x19           ;19th of the month
            CALL SPI             ;send day of the month
            MOVLW 0x10           ;October
            CALL SPI             ;send month
            MOVLW 0x04           ;2004
            CALL SPI             ;send year
            BCF  PORTC,RC2       ;disable RTC
            CALL SDELAY
    ;-- get the time and date from DS1306
    RDA     BSF  PORTC,RC2       ;enable the RTC
            CALL SDELAY
            MOVLW 0x00        ;seconds register address for read
            CALL SPI             ;send address to DS1306
            CALL SPI             ;start getting time/date
            MOVWF SEC            ;save the seconds
            CALL SPI             ;get the minutes
            MOVWF MIN            ;save the minutes
            CALL SPI             ;get the hour
            MOVWF HR             ;save the hour
            CALL SPI             ;get the day
            MOVWF DAY            ;save the day
            CALL SPI             ;get the date
            MOVWF DAT            ;save the date
            CALL SPI             ;get the month
            MOVWF MON            ;save the month
            CALL SPI             ;get the year
            MOVWF YR             ;save the year
            BCF  PORTC,RC2       ;disable RTC
    ;-- convert packed BCD to ASCII and display
            LFSR FSR0,0x11   ;address of fileReg for time/date
            MOVLW D'6'           ;6 bytes of data to display
            MOVWF CNT            ;set up the counter
    SND     MOVFF INDF0,TMP      ;get the data for high nibble
            MOVLW 0xF0           ;clear low nibble
            ANDWF TMP,F          ;keep in TMP register
            SWAPF TMP,F          ;switch high and low nibbles
```

```
        MOVLW 0x30          ;convert to ASCII
        IORWF TMP,W         ;put in WREG
        CALL TRANS          ;display the data
        MOVFF  POSTINC0,TMP ;get the data and point to next
        MOVLW 0x0F          ;clear high nibble
        ANDWF TMP,F         ;keep in TMP register
        MOVLW 0x30          ;convert to ASCII
        IORWF TMP,W         ;put in WREG
        CALL TRANS          ;display the data
        MOVLW ':'
        CALL TRANS
        DECFSZ CNT          ;Is it the last one?
        BRA SND             ;no
        MOVLW 0x0D          ;line feed
        CALL TRANS
        BRA RDA     ;keep reading time/date and display them
;-- SPI write/read subroutine
 SPI  MOVWF SSPBUF         ;load SSPBUF for transfer
 WAIT BTFSS SSPSTAT,BF     ;wait for all bits
      BRA WAIT
      MOVF SSPBUF,W        ;get the received byte
      RETURN               ;return with byte in WREG
;----serial data transfer subroutine
TRANS BTFSS PIR1, TXIF     ;wait until the last bit is gone
      BRA TRANS            ;stay in loop
      MOVWF TXREG    ;load the value to be transmitted
      RETURN               ;return to caller
;----short delay
SDELAY: MOVLW   D1uL       ;low byte of delay
        MOVWF DR1uL        ;store in register
DS1   DECF DR1uL,F         ;stay until DR1uL becomes 0
      BNZ   DS1
      RETURN
      END
```

Review Questions

1. True or false. All of the RAM contents of the DS1306 are nonvolatile.
2. How many bytes of RAM in the DS1306 are set aside for the clock and date?
3. How many bytes of RAM in the DS1306 are set aside for general-purpose applications?
4. True or false. The DS1306 has a single pin for Din.
5. Which pin of the DS1306 is used for Clock in SPI connection?
6. True or false. To use the DS1306 in SPI mode, we make SERMODE = GND.

SECTION 16.3: DS1306 RTC PROGRAMMING IN C

In this section, we program the DS1306 in PIC18 C language. Before you embark on this section, make sure you understand the basic concepts of the DS1306 chip covered in the first section.

Setting the time and date in C

Program 16-4C shows how to set the time and date for the DS1306 configuration in Figure 16-12.

```
//Program 16-4C : Setting time and date
#include <p18f458.h>
unsigned char SPI(unsigned char);
void SDELAY(int ms);
void main()
  {
    SSPSTAT = 0; //read at middle, send on active edge
    SSPCON1 = 0x22;         //master SPI enable, Fosc / 64
    TRISC = 0;              //make PORTC output
    TRISCbits.TRISC4 = 1; //except SDI
    TRISCbits.TRISC7 = 1; //and RX
    PORTCbits.RC2 = 1;     //enable the RTC
    SDELAY(1);
    SPI(0x8F);             //control register address
    SPI(0x00);             //clear WP bit for write
    PORTCbits.RC2 = 0;     //end of single-byte write
    SDELAY(1);
    PORTCbits.RC2 = 1;     //begin multibyte write
    SPI(0x80);             //seconds register address
    SPI(0x55);             //55 seconds
    SPI(0x58);             //58 minutes
    SPI(0x16);             //24-hour clock at 16 hours
    SPI(0x3);              //Tuesday
    SPI(0x19);             //19th of the month
    SPI(0x10);             //October
    SPI(0x04);             //2004
    PORTCbits.RC2 = 0;     //end multibyte write
    SDELAY(1);
    }
//-- SPI Write/Read subroutine
unsigned char SPI(unsigned char myByte)
  {
    SSPBUF = myByte;                //load SSPBUF for transfer
    while(!SSPSTATbits.BF);    //wait for all bits
    return SSPBUF;            //return with received byte
    }
```

Reading and displaying the time and date in C

Program 16-5C shows how to read the time, convert it to ASCII, and send it to the PC screen via the serial port.

```
//Program 16-5C : Reading and Displaying Time
#include <p18f458.h>
unsigned char SPI(unsigned char);
void TRANS(unsigned char);
void BCDtoASCIIandSEND(unsigned char);
void SDELAY(int ms);

void main()
  {
  unsigned char data[7];   //holds date and time
  unsigned char tmp;       //for BCD to ASCII conversion
  int i;
  SSPSTAT = 0;            //read at middle, send on active edge
  SSPCON1 = 0x22;          //master SPI enable, Fosc / 64
  TRISC   = 0;             //make PORTC output
  TRISCbits.TRISC4 = 1;    //except SDI
  TRISCbits.TRISC7 = 1;    //and RX
  TXSTA = 0x20;            //enable transmit and low baud
  SPBRG = 15;            //9600 bps (Fosc / (64 * Speed) - 1)
  RCSTAbits.SPEN = 1;      //enable the serial port
  TRANS(0x0A);             //form feed
  TRANS(0x0D);             //new line
//-- get the time and date from RTC and save them
  while(1)
     {
     PORTCbits.RC2 = 1;    //begin multibyte read
     SDELAY(1);
     SPI(0x00);            //seconds register address
     for(i=0;i<7;i++)
       {
       data[i] = SPI(0x00); //get time/date and save
       }
     PORTCbits.RC2 = 0;    //end of multibyte read
//-- convert time/date and display MM:DD:YY:HH:MM:SS
     BCDtoASCIIandSEND(data[5]); //the month
     BCDtoASCIIandSEND(data[4]); //the date
     BCDtoASCIIandSEND(data[6]); //the year
     BCDtoASCIIandSEND(data[2]); //the hour
     BCDtoASCIIandSEND(data[1]); //the minute
     BCDtoASCIIandSEND(data[0]); //the second
     TRANS(0x0D);             //new line
     }
  }
```

```
//-- SPI Write/Read
unsigned char SPI(unsigned char myByte)
  {
  SSPBUF = myByte;
  while(!SSPSTATbits.BF);
  return SSPBUF;
  }
void TRANS(unsigned char myChar) //serial data transfer
  {
  while(!PIR1bits.TXIF);
  TXREG = myChar;      //load the value to be transmitted
  }
void BCDtoASCIIandSEND(unsigned char myValue)
  {
  unsigned char tmp = myValue;
  tmp = tmp & 0xF0;         //mask lower nibble
  tmp = tmp >> 4;           //swap it
  tmp = tmp | 0x30;         //make it ASCII
  TRANS(tmp);               //display
  tmp = myValue;            //for other digit *
  tmp = tmp & 0x0F;         //mask upper nibble
  tmp = tmp | 0x30;         //make it ASCII
  TRANS(tmp);               //display
  TRANS(':');               //display separator
  }

void SDELAY(int ms)
  {
  unsigned int i, j;
  for(i=0;i<ms;i++)
    for(j=0;j<135;j++);
  }
```

Review Questions

1. True or false. All of the RAM contents of the DS1306 are volatile.
2. What locations of RAM in the DS1306 are set aside for the clock and date?
3. What locations of RAM in the DS1306 are set aside for general-purpose applications?
4. True or false. The DS1306 has a single pin for Dout.
5. True or false. CE is an output pin.
6. True or false. To use the DS1306 in SPI mode, we make SERMODE = VCC.

SECTION 16.4: ALARM AND INTERRUPT FEATURES OF THE DS1306

In this section, we program the alarm and interrupt features of the DS1306 chip using Assembly and C languages. These powerful features of the DS1306 can be very useful in many real-world applications. In the DS1306 there are two alarms, called Alarm0 and Alarm1, each with their own hardware interrupts. There is also a 1-Hz square wave output pin, which we discuss next. These features are accessed with the Control register shown in Figure 16-13.

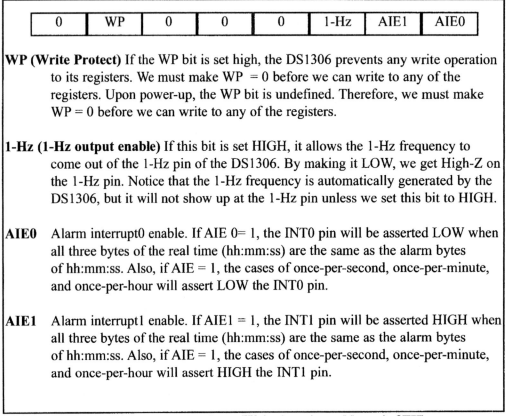

| 0 | WP | 0 | 0 | 0 | 1-Hz | AIE1 | AIE0 |

WP (Write Protect) If the WP bit is set high, the DS1306 prevents any write operation to its registers. We must make WP = 0 before we can write to any of the registers. Upon power-up, the WP bit is undefined. Therefore, we must make WP = 0 before we can write to any of the registers.

1-Hz (1-Hz output enable) If this bit is set HIGH, it allows the 1-Hz frequency to come out of the 1-Hz pin of the DS1306. By making it LOW, we get High-Z on the 1-Hz pin. Notice that the 1-Hz frequency is automatically generated by the DS1306, but it will not show up at the 1-Hz pin unless we set this bit to HIGH.

AIE0 Alarm interrupt0 enable. If AIE 0= 1, the INT0 pin will be asserted LOW when all three bytes of the real time (hh:mm:ss) are the same as the alarm bytes of hh:mm:ss. Also, if AIE = 1, the cases of once-per-second, once-per-minute, and once-per-hour will assert LOW the INT0 pin.

AIE1 Alarm interrupt1 enable. If AIE1 = 1, the INT1 pin will be asserted HIGH when all three bytes of the real time (hh:mm:ss) are the same as the alarm bytes of hh:mm:ss. Also, if AIE = 1, the cases of once-per-second, once-per-minute, and once-per-hour will assert HIGH the INT1 pin.

Figure 16-13. DS1306 Control Register (Write location address is 8FH)

Programming the 1-Hz feature

The 1-Hz pin of the DS1306 provides us a square wave output of 1-Hz frequency. Internally, the DS1306 generates the 1-Hz square wave automatically but it is blocked. We must enable the 1-Hz bit in the Control register to let it show up on the 1-Hz pin. This is shown below. Because we are writing to a single location, burst mode is not used.

```
      MOVLW 0x00
      MOVWF SSPSTAT    ;middle read, active edge send
      MOVLW 0x22
      MOVWF SSPCON1    ;master SPI enable, Fosc / 64
      CLRF  TRISC      ;make PORTC output
      BSF   TRISC,SDI  ;except SDI
;-- send control byte to enable write first (Figure 16-13)
      BSF   PORTC,RC2  ;enable the RTC
```

```
        CALL  SDELAY
        MOVLW  0x8F              ;Control register address
        CALL  SPI
        MOVLW  0x0               ;clear WP bit for write
        CALL  SPI
        BCF   PORTC,RC2          ;disable RTC
        CALL  SDELAY
;-- send control byte to enable 1 Hz signal after WP = 0
        BSF   PORTC,RC2          ;enable the RTC
        CALL  SDELAY
        MOVLW  0x8F              ;Control register address
        CALL  SPI
        MOVLW  0x04 ;enable 1 Hz signal in Control register
        CALL  SPI
        BCF   PORTC,RC2          ;disable RTC
        CALL  SDELAY
```

Alarm0, Alarm1, and interrupt

There are two time-of-day alarms in the DS1306 chip. They are referred to as Alarm0 and Alarm1. We can access Alarm0 by writing to its registers located at addresses 87H through 8AH, as shown in Table 16-3. Alarm1 is accessed by writing to its registers located at addresses 8BH through 8EH, as shown in Table 16-3. During each clock update, the RTC compares the clock registers and alarm registers. When the values stored in the timekeeping registers of 0, 1, and 2 match the values stored in the alarm registers, the corresponding alarm flag bit (IRQF0 or IRQF1) in the status register will go HIGH. See Figure 16-14. Because polling the IRQxF is too time-consuming, we can enable the AIEx bit in the Control register, and make it a hardware interrupt coming out of the INT0 and INT1 pins.

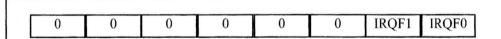

| 0 | 0 | 0 | 0 | 0 | 0 | IRQF1 | IRQF0 |

IRQF0 (Interrupt 0 Request Flag) The IRQF0 bit will go HIGH when all three bytes of the current real time (hh:mm:ss) are the same as the Alarm0 bytes of hh:mm:ss. Also, the cases of once-per-second, once-per-minute, and once-per-hour will assert HIGH the IRQ0 bit. We can use polling to see the status of IRQF0. However, in the Control register, if we make AIE0 = 1, IRQF0 will assert LOW the INT0 pin, making it a hardware interrupt. Any read or write of the Alarm0 registers will clear IRQF0.

IRQF1 (Interrupt 0 Request Flag) The IRQF1 bit will go HIGH when all three bytes of the current real time (hh:mm:ss) are the same as the Alarm1 bytes of hh:mm:ss. Also, the cases of once-per-second, once-per-minute, and once-per-hour will assert HIGH the IRQ1 bit. We can use polling to see the status of IRQF1. However, in the Control register, if we make AIE1 = 1, IRQF1 will assert HIGH the INT1 pin, making it a hardware interrupt. Any read or write of the Alarm1 registers will clear IRQF1.

Figure 16-14. Status Register (Read location address is 10H)

Table 16-3: DS1306 Address Locations for Time, Calendar, and Alarm

Hex Address Read	Write	Function	D7	Data Range BCD	Possible Hex Range
00H	80H	Seconds	0	00–59	00–59
01H	81H	Minute	0	00–59	00–59
02H	82H	Hours, 12-Hour Mode	0	01–12	41–52 AM
		Hours, 12-Hour Mode	0	01–12	61–72 PM
		Hours, 24-Hour Mode	0	00–23	00–23
03H	83H	Day of the Week, Sun = 1	0	01–07	01–07
04H	84H	Day of the Month	0	01–31	01–31
05H	85H	Month	0	01–12	01–12
06H	86H	Year	0	00–99	00–99
07H	87H	SEC Alarm0	0 or 1	00–59	00–59 or 89–A9
08H	88H	MIN Alarm0	0 or 1	00–59	00–59 or 89–A9
09H	89H	Hour Alarm0, 12-Hour	0 or 1	01–12	41–52 or C1–A2 AM
		Hour Alarm0, 12-Hour	0 or 1	01–12	61–72 or D1–F2 PM
		Hour Alarm0, 24-Hour	0 or 1	00–23	00–23 or 80–A3
0A	8A	Day Alarm0	0 or 1	1–7	01–07
0BH	8BH	SEC Alarm1	0 or 1	00–59	00–59 or 89–A9
0CH	8CH	MIN Alarm1	0 or 1	00–59	00–59 or 89–A9
0DH	8DH	Hour Alarm1, 12-Hour	0 or 1	01–12	41–52 or C1–A2 AM
		Hour Alarm1, 12-Hour	0 or 1	01–12	61–72 or D1–F2 PM
		Hour Alarm1, 24-Hour	0 or 1	00–23	00–23 or 80–A3
0EH	8EH	Day Alarm1	0 or 1	1–7	01–07
0FH	8FH	CONTROL REGISTER			
10H	90H	STATUS REGISTER			
11H	91H	TRICKLE REGISTER			
12–1FH	82–9FH	RESERVED			
20–7FH	A0–FFH	96-BYTE USER RAM			

Alarm and IRQ output pins

The alarm interrupts of INT0 and INT1 can be programmed to occur at rates of (a) once per week (b) once per day, (c) once per hour, (d) once per minute, and (f) once per second. Next, we look at each of these.

Once-per-day alarm

Table 16-3 shows the address locations belonging to the alarm seconds, alarm minutes, alarm hours, and alarm days. Notice the D7 bits of these locations. An alarm is generated every day when D7 of the day alarm location is set to HIGH. Therefore, to program the alarm for once-per-day, we must (a) write the desired time for the alarm into the hour, minute, and second of Alarm locations, and (b) set HIGH D7 of the alarm day. See Table 16-4. As the clock keeps the time, when all three bytes of hour, minute, and second for the real-time clock match the values in the alarm hour, minute, and second, the IRQxF flag bit in the Status register of the DS1306 will go high. We can poll the IRQxF bit in the Status register, which is a waste of microcontroller resources, or allow the hardware INTx pin to be activated upon matching the alarm time with the real time. It must be noted that in order to use the hardware INTx pin of the DS1306 for an alarm, the interrupt-enable bit for alarm in control register (AIEx) must be set HIGH. We will examine the process shortly.

Once-per-hour alarm

To program the alarm for once per hour, we must set HIGH D7 of both the day alarm and hour alarm registers. See Table 16-4.

Once-per-minute alarm

To program the alarm for once per minute, we must set HIGH D7 of all three, day alarm, hour alarm, and minute alarm locations. See Table 16-4.

Once-per-second alarm

To program the alarm for once per second, we must set HIGH D7 of all four locations of alarm day, alarm hour, alarm minute, and alarm second. See Table 16-4.

Once-per-week alarm

To program the alarm for once per week, we must clear D7 of all four locations of alarm day, alarm hour, alarm minute, and alarm second. See Table 16-4.

Table 16-4: DS1306 Time-of-day Alarm Mask Bits

Alarm Register Mask Bits (D7)

Seconds	Minutes	Hours	Days	Function
1	1	1	1	Alarm once per second
0	1	1	1	Alarm when seconds match (once-per-minute)
0	0	1	1	Alarm when minutes and seconds match (once-per-hour)
0	0	0	1	Alarm when hours, minutes, and seconds match (once-per-day)
0	0	0	0	Alarm when day, hours, minutes, and seconds match (once-per-week)

Example 16-2

Using Table 16-4, find the values we must place in the Alarm1 register if we want to have an alarm activated at 16:05:07, and from then on once-per-minute at 7 seconds past the minute.

Solution:

Because we use 24-hour clock, we have D6 = 0 for the HR register. Therefore, we have 1001 0110 for 16 in BCD. This means that we must put value 96H into register location 8D of the DS1306. Notice that D7 is 1, according to Table 16-4.

For the MIN register, we have 1000 0101 for 05 in BCD. This means that we must put value 85H into register location 8C of the DS1306. Notice that D7 is 1, according to Table 16-4.

For the SEC register we have 0000 0111 for 07 in BCD. This means that we must put value 07H into register location 8B of the DS1306. Notice that D7 is 0 according to Table 16-4.

For once-per-minute to work, we must make sure that D7 of Alarm1 day is also set to 1. See Table 16-4.

Using INT0 of DS1306 to activate the PIC18 interrupt

We can connect the INT0 bit of the DS1306 to the external interrupt pin of the PIC18 (INT0). See Figure 16-15. This allows us to perform a task once per day, once per minute, and so on. Example 16-2 shows the values needed for the Alarm0 registers. Program 16-6 uses the Alarm0 interrupt (INT0) to send the message "YES" to the serial port once per minute, at exactly 8 seconds past the minute.

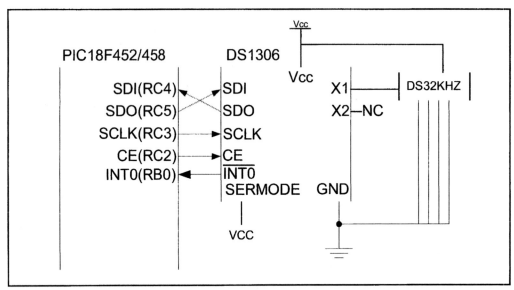

Figure 16-15. DS1306 Connection to PIC18 with Hardware INT0

```
;Program 16-6
D1uL EQU   D'2'             ;1 microsecond delay byte
DR1uL EQU  0x0D         ;register for 1 microsecond delay
     ORG   0x00
     BRA   MAIN            ;bypass INT vector table
     ORG   0x08
     BTFSC INTCON,INT0IF   ;Was it INT0?
     BRA   INT0_ISR        ;yes, go to INT0 ISR
     RETFIE
     ORG   0x28
;-- initialize SPI, INT0, and USART
MAIN CLRF  TRISC           ;make PORTC output
     BSF   TRISC,SDI       ;except SDI
     BSF   TRISC,RX        ;and RX
     BSF   TRISB,INT0      ;make RB0 input for interrupt
     MOVLW 0x00
     MOVWF SSPSTAT         ;middle read, active edge send
     MOVLW 0x22
     MOVWF SSPCON1         ;master SPI enable, Fosc / 64
     BCF   INTCON2,INTEDG0 ;make INT0 negative-edge
                           ;triggered
     BSF   INTCON,INT0IE   ;enable INT0
     MOVLW B'00100000';enable transmit and choose low baud
     MOVWF TXSTA           ;write to reg
     MOVLW D'15'      ;9600 bps (Fosc / (64 * Speed) - 1)
     MOVWF SPBRG          ;write to reg
```

```
        BCF    TRISC, TX   ;make TX pin of PORTC an output pin
        BSF    RCSTA, SPEN      ;enable the serial port
        BSF    INTCON,GIE      ;enable interrupts globally
;-- send control byte to enable write
        BSF    PORTC,RC2       ;enable the RTC
        CALL   SDELAY
        MOVLW  0x8F            ;control register
        CALL   SPI
        MOVLW  0x0             ;clear WP bit for write
        CALL   SPI
        BCF    PORTC,RC2       ;disable RTC
        CALL   SDELAY
;-- send the data
        BSF    PORTC,RC2  ;enable for multibyte write
        MOVLW  0x87            ;Alarm0 address
        CALL   SPI             ;send address
        MOVLW  0x08            ;alarm at 8 seconds
        CALL   SPI             ;send second
        MOVLW  0x80            ;once-per-minute
        CALL   SPI             ;send minute
        MOVLW  0x80            ;once-per-minute
        CALL   SPI             ;send hour
        MOVLW  0x80            ;once-per-minute
        CALL   SPI             ;send day
        BCF PORTC,RC2          ;end of multibyte write
        CALL   SDELAY
;-- send control byte to enable INT0
        BSF    PORTC,RC2       ;enable the RTC
        CALL   SDELAY
        MOVLW  0x8F            ;control register of DS1306
        CALL   SPI
        MOVLW  0x01            ;enable INT0 pin of DS1306
        CALL   SPI
        BCF PORTC,RC2          ;disable RTC
        CALL   SDELAY
LOOP BRA LOOP                  ;wait for interrupt
;-- service Alarm0
INT0_ISR
        BSF    PORTC,RC2       ;enable the RTC
        CALL   SDELAY
        MOVLW  0x8F            ;control register
        CALL   SPI
        MOVLW  0x04            ;1 Hz on, Alarm0 off
        CALL   SPI
        BCF PORTC,RC2          ;disable RTC
        CALL   SDELAY
;-- send Alarm0 seconds to reset alarm
        BSF    PORTC,RC2       ;enable the RTC
        CALL   SDELAY
        MOVLW  0x87            ;Alarm0 seconds register
        CALL   SPI
```

```asm
        MOVLW 0x08              ;at 8 seconds
        CALL SPI
        BCF PORTC,RC2           ;disable RTC
        CALL SDELAY
;-- begin displaying
        MOVLW upper(MESSAGE)
        MOVWF TBLPTRU
        MOVLW high(MESSAGE)
        MOVWF TBLPTRH
        MOVLW low(MESSAGE)
        MOVWF TBLPTRL
NEXT    TBLRD*+                 ;read the characters
        MOVF TABLAT,W           ;place it in WREG
        IORLW 0x0
        BZ    OVER              ;if end of line, start over
        CALL  TRANS             ;send char to serial port
        BRA   NEXT              ;repeat for the next character
;-- send control byte to enable INT0
OVER BSF   PORTC,RC2            ;enable the RTC
        CALL  SDELAY
        MOVLW 0x8F              ;control register
        CALL SPI
        MOVLW 0x01             ;1 Hz off, Alarm0 on
        CALL SPI
        BCF PORTC,RC2           ;disable RTC
        CALL  SDELAY
        BCF   INTCON,INT0IF
        RETFIE
;-- SPI subroutine
;-- serial data transfer subroutine
;-- delay for SPI communications
        RETURN;SEE PREVIOUS PROGRAMS FOR ABOVE SUBROUTINES
;--message to be displayed upon interrupt
MESSAGE:    DB    0x0A,0x0D,"Yes",0
        END
```

The following is the C version of the above program.

```c
//Program 16-6C
#include <p18f458.h>
//INSERT FUNCTION PROTOTYPES
#pragma interrupt chk_isr  //used for high priority inter-
rupt only
void chk_isr (void)
{
   if (INTCONbits.INT0IF==1)//INT0 caused interrupt?
   INT0_ISR( );               //Yes. Execute INT0 program
}
#pragma code My_HiPrio_Int=0x0008 //high-priority interrupt
void My_HiPrio_Int (void)
{
```

```
    _asm
    GOTO chk_isr
    _endasm
}
#pragma code
void main(void)
    {
//-- initialize SPI, INT0, and USART
    TRISC=0x90;         //make PORTC output, except SDI and RX
    TRISBbits.TRISB0=1;       //make RB0 input for interrupt
    SSPSTAT=0x0;              //middle read, active edge send
    SSPCON1=0x22;            //master SPI enable, Fosc / 64
    INTCON2bits.INTEDG0=0;   //make INT0 negative edge
                             //triggered
    INTCONbits.INT0IE=1;     //enable INT0
    TXSTA=0x20;           //enable transmit and choose low baud
    SPBRG=15;             //9600 bps (Fosc / (64 * Speed) - 1)
    RCSTAbits.SPEN=1;        //enable the serial port
    INTCONbits.GIE=1;        //enable interrupts globally
//-- send control byte to enable write
    PORTCbits.RC2=1;         //enable the RTC
    MSDelay(1);
    SPI(0x8F);               //control register address
    SPI(0x0);                //enable write
    PORTCbits.RC2=0;         //disable RTC
    MSDelay(1);
//-- send the data
    PORTCbits.RC2=1;         //enable the RTC
    MSDelay(1);
    SPI(0x87);               //Alarm0 address
    SPI(0x08);               //alarm at 8 seconds
    SPI(0x80);               //once-per-minute
    SPI(0x80);               //once-per-minute
    SPI(0x80);               //once-per-minute
    PORTCbits.RC2=0;         //disable RTC
    MSDelay(1);
//-- send control byte to enable INT0
    PORTCbits.RC2=1;         //enable the RTC
    MSDelay(1);
    SPI(0x8F);               //control register
    SPI(0x01);               //enable INT0
    PORTCbits.RC2=0;         //disable RTC
    MSDelay(1);
    while(1);                //wait for interrupt
    }
//-- service Alarm0
void INT0_ISR()
    {
    unsigned char mess[]={0x0D,0x0A,'Y','E','S',0};
    unsigned char i;
    PORTCbits.RC2=1;         //enable the RTC
```

```
     MSDelay(1);
     SPI(0x8F);              //control register
     SPI(0x04);              //1 Hz on, Alarm0 off
     PORTCbits.RC2=0;        //disable RTC
     MSDelay(1);
//-- send Alarm0 seconds to reset alarm
     PORTCbits.RC2=1;        //enable the RTC
     MSDelay(1);
     SPI(0x87);              //Alarm0 seconds register
     SPI(0x08);              //at 8 seconds
     PORTCbits.RC2=0;        //disable RTC
     MSDelay(1);
//-- begin sending the data
     for(i=0;mess[i]!=0;i++)
       TRANS(mess[i]);
//-- send control byte to enable INT0
     PORTCbits.RC2=1;        //enable the RTC
     MSDelay(1);
     SPI(0x8F);              //control register
     SPI(0x01);              //1 Hz offbits. Alarm0 on
     PORTCbits.RC2=0;        //turn off RTC
     INTCONbits.INT0IF=0;
     }
//--SEE PREVIOUS EXAMPLES FOR SUBROUTINES
```

In the last program, we send a message to the serial port to indicate that the alarm has occurred. We can use the 32 kHz output to sound an actual alarm. Because 32 kHz is too high a frequency for human ears, however, we can use multiple D flip flops to bring down the frequency. See Figure 16-16. The modification of Program 16-6 for Figure 16-16 is left to the reader.

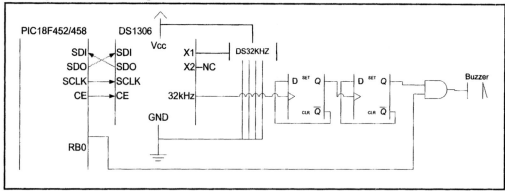

Figure 16-16. DS1306 Connection to PIC18 with Buzzer Control

Review Questions

1. Which bit of the Control register belongs to the 1-Hz pin?
2. True or false. The INT0 pin is an input for the DS1306.
3. True or false. The INT0 pin is active-LOW.
4. Which bit of the Control register belongs to the Alarm1 interrupt?
5. Give the address locations for Alarm1.

SUMMARY

This chapter began by describing the SPI bus connection and protocol. We also discussed the function of each pin of the DS1306 RTC chip. The DS1306 can be used to provide a real-time clock and dates for many applications. Various features of the RTC were explained, and numerous programming examples were given.

PROBLEMS

SECTION 16.1: SPI BUS PROTOCOL

1. True or false. The SPI bus needs an external clock.
2. True or false. The SPI CE is active-LOW.
3. True or false. The SPI bus has a single Din pin.
4. True or false. The SPI bus has multiple Dout pins.
5. True or false. When the SPI device is used as a slave, the SCLK is an input pin.
6. True or false. In SPI devices, data is transferred in 8-bit chunks.
7. True or false. In SPI devices, each bit of information (data, address) is transferred with a single clock pulse.
8. True or false. In SPI devices, the 8-bit data is followed by an 8-bit address.
9. In term of data pins, what is the difference between the SPI and 3-wire connections?
10. How does the SPI protocol distinguish between the read and write cycles?

SECTION 16.2: DS1306 RTC INTERFACING AND PROGRAMMING

11. The DS1306 DIP package is a(n) _____ -pin package.
12. Which pin is assigned as primary V_{cc}?
13. In the DS1306, how many pins are designated as address/data pins?
14. True or false. The DS1306 needs an external crystal oscillator.
15. True or false. The DS1306's crystal oscillator and heat affect the time-keeping accuracy.
16. In DS1306, what is the maximum year that it can provide?
17. Describe the functions of pins SDI, SDO, and SCLK.
18 CE is an _____ (input, output) pin.
19. The CE pin is normally _____ (LOW, HIGH) and needs a _____ (LOW, HIGH) signal to be activated.
20. Who keeps the contents of the DS1306 time and date registers if power to the primary V_{cc} pin is cut off?
21. Vbat pin stands for _____ and is an_____ (input, output) pin.
22. For the DS1306 chip, pin Vcc2 is connected to _____ (V_{cc}, GND).
23. SERMODE is an _____ (input, output) pin and it is connected to _____ for SPI mode.
24. Vcc1 is an _____ (input, output) pin and is connected to _____ voltage.
25. 1-Hz is an _____ (input, output).

---

26. INT0 is an _____ (input, output) pin.
27. 32KHz is an _____ (input, output) pin.
28. INT1 is an _____ (input, output) pin.
29. DS1306 has a total of _____ bytes of locations. Give the addresses for read and write operations.
30. What are the contents of the DS1306 time and date registers if power to the V_{cc} pin is lost?
31. What are the contents of the general-purpose RAM locations if power to the V_{cc1} is lost?
32. When does the DS1306 switch to a battery energy source?
33. What are the addresses assigned to the real-time clock (time) registers?
34. What are the addresses assigned to the calendar?
35. Which register is used to set the AM/PM mode? Give the bit location of that register.
36. Which register is used to set the 24-hour mode? Give the bit location of that register.
37. At what memory location does the DS1306 store the year 2007?
38. What is the address of the last location of RAM for the DS1306?
39. True or false. The DS1306 provides data in BCD format only.
40. Write a program to get the year data in BCD and send it to ports PORTB and PORTD.
41. Write a program to get the hour and minute data in BCD and send it to ports PORTB and PORTD.
42. Write a program to set the time to 9:15:05 PM.
43. Write a program to set the time to 22:47:19.
44. Write a program to set the date to May 14, 2009.
45. What are the roles of Vbat and Vcc2?

SECTION 16.3: DS1306 RTC PROGRAMMING IN C

46. Write a C program to display the time in AM/PM mode.
47. Write a C program to get the year data in BCD and send it to ports PORTB and PORTD.
48. Write a C program to get the hour and minute data and send it to ports PORTB and PORTD.
49. Write a C program to set the time to 9:15:05 PM.
50. Write a C program to set the time to 22:47:19.
51. Write a C program to set the date to May 14, 2009.
52. In Question 51, how does the RTC keep track of the century?

SECTION 16.4: ALARM AND INTERRUPT FEATURES OF THE DS1306

53. INT0 is an _____ (input, output) pin and active-_____ (LOW, HIGH).
54. 1-Hz is an _____ (input, output) pin.
55. Give the bit location of the Control register belonging to the alarm interrupt. Show how to enable it.

56. Give the bit location of the Control register belonging to the 1-Hz pin. Show how to enable it.
57. Give the bit location of the Status register belonging to the Alarm0 interrupt.
58. Give the bit location of the Status register belonging to the Alarm1 interrupt.
59. True or False. For the 32KHz output pin, the frequency is set and cannot be changed.
60. Give sources of interrupts that can activate the INT1 pin.
61. Why do we want to direct the AIE0 (Alarm0 flag) to an IRQ pin?
62. What is the difference between the IRQF0 and AIE0 bits?
63. What is the difference between the IRQF1 and AIE1 bits?
64. How do we allow the square wave to come out of the 1-Hz pin?
65. Which register is used to set the once-per-second Alarm1?
66. Explain how the IRQ1F pin is activated due to the once-per-minute alarm option.

ANSWERS TO REVIEW QUESTIONS

SECTION 16.1: SPI BUS PROTOCOL

1. True
2. True
3. False
4. False
5. In single-byte mode, after each byte, the CE pin must go LOW before the next cycle. In burst mode, the CE pin stays HIGH for the duration of the burst (multibyte) transfer.

SECTION 16.2: DS1306 RTC INTERFACING AND PROGRAMMING

1. True. Only if Vbat is connected to an external battery.
2. 7
3. 96
4. True
5. Pin 11 is SCLK.
6. False. SERMODE = Vcc

SECTION 16.3: DS1306 RTC PROGRAMMING IN C

1. True
2. 0–6
3. 20–7FH
4. True
5. False
6. False

SECTION 16.4: ALARM AND INTERRUPT FEATURES OF THE DS1306

1. Bit 2
2. False
3. True
4. Bit 1
5. Byte addresses of 0B–0E (in hex) for read and 8B–8E (in hex) for write

CHAPTER 17

MOTOR CONTROL: RELAY, PWM, DC, AND STEPPER MOTORS

OBJECTIVES

Upon completion of this chapter, you will be able to:

>> Describe the basic operation of a relay
>> Interface the PIC18 with a relay
>> Describe the basic operation of an optoisolator
>> Interface the PIC18 with an optoisolator
>> Describe the basic operation of a stepper motor
>> Interface the PIC18 with a stepper motor
>> Code PIC18 programs to control and operate a stepper motor
>> Define stepper motor operation in terms of step angle, steps per revolution, tooth pitch, rotation speed, and RPM
>> Describe the basic operation of a DC motor
>> Interface the PIC18 with a DC motor
>> Code PIC18 programs to control and operate a DC motor
>> Describe how PWM is used to control motor speed
>> Code CCP programs to control and operate a DC motor
>> Code ECCP programs to control and operate a DC motor

This chapter discusses motor control and shows PIC18 interfacing with relays, optoisolators, stepper motors, and DC motors. In Section 17.1, the basics of relays and optoisolators are described. Then we show their interfacing with the PIC18. In Section 17.2, stepper motor interfacing with the PIC18 is shown. The characteristics of DC motors are discussed in Section 17.3, along with their interfacing to the PIC18. We will also discuss the topic of PWM (pulse width modulation). In Section 17.4, the CCP feature of PIC18 is used to control DC motors, while the ECCP usage in motor control is shown in Section 17.5. We use both Assembly and C in our programming examples.

SECTION 17.1: RELAYS AND OPTOISOLATORS

This section begins with an overview of the basic operations of electromechanical relays, solid-state relays, reed switches, and optoisolators. Then we describe how to interface them to the PIC18. We use both Assembly and C language programs to demonstrate their control.

Electromechanical relays

A *relay* is an electrically controllable switch widely used in industrial controls, automobiles, and appliances. It allows the isolation of two separate sections of a system with two different voltage sources. For example, a +5 V system can be isolated from a 120 V system by placing a relay between them. One such relay is called an electromechanical (or electromagnetic) relay (EMR) as shown in Figure 17-1. The EMRs have three components: the coil, spring, and contacts. In Figure 17-1, a digital +5 V on the left side can control a 12 V motor on the right side without any physical contact between them. When current flows through the coil, a magnetic field is created around the coil (the coil is energized), which causes the armature to be attracted to the coil. The armature's contact acts like a switch and closes or opens the circuit. When the coil is not energized, a spring pulls the armature to its normal state of open or closed. In the block diagram for electomechanical relays (EMR) we do not show the spring, but it does exist internally. There are all types of relays for all kinds of applications. In choosing a relay the following characteristics need to be considered:

1. The contacts can be normally open (NO) or normally closed (NC). In the NC type, the contacts are closed when the coil is not energized. In the NO, the contacts are open when the coil is unenergized.
2. There can one or more contacts. For example, we can have SPST (single pole, single throw), SPDT (single pole, double throw), and DPDT (double pole, double throw) relays.
3. The voltage and current needed to energize the coil. The voltage can vary from a few volts to 50 volts, while the current can be from a few mA to 20 mA. The relay has a minimum voltage, below which the coil will not be energized. This minimum voltage is called the "pull-in" voltage. In the datasheet for relays we might not see current, but rather coil resistance. The V/R will give you the pull-in current. For example, if the coil voltage is 5 V, and the coil resistance is 500 ohms, we need a minimum of 10 mA (5 V/500 ohms = 10 mA) pull-in current.

4. The maximum DC/AC voltage and current that can be handled by the contacts. This is in the range of a few volts to hundreds of volts, while the current can be from a few amps to 40 A or more, depending on the relay. Notice the difference between this voltage/current specification and the voltage/current needed for energizing the coil. The fact that one can use such a small amount of voltage/current on one side to handle a large amount of voltage/current on the other side is what makes relays so widely used in industrial controls. Examine Table 17-1 for some relay characteristics.

Table 17-1: Selected DIP Relay Characteristics (www.Jameco.com)

Part No.	Contact Form	Coil Volts	Coil Ohms	Contact Volts-Current
106462CP	SPST-NO	5 VDC	500	100 VDC-0.5 A
138430CP	SPST-NO	5 VDC	500	100 VDC-0.5 A
106471CP	SPST-NO	12 VDC	1000	100 VDC-0.5 A
138448CP	SPST-NO	12 VDC	1000	100 VDC-0.5 A
129875CP	DPDT	5 VDC	62.5	30 VDC-1 A

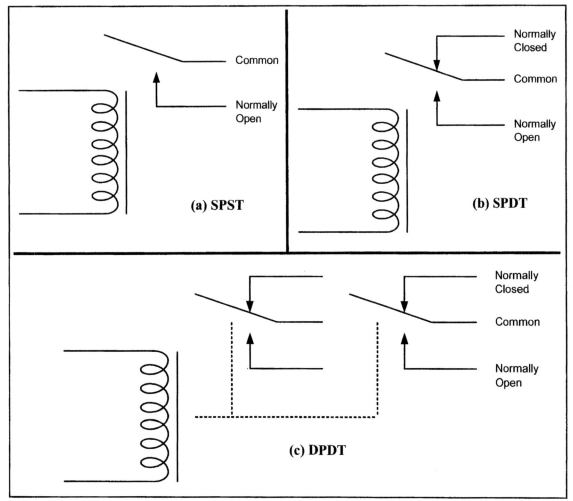

Figure 17-1. Relay Diagrams

Driving a relay

Digital systems and microcontroller pins lack sufficient current to drive the relay. While the relay's coil needs around 10 mA to be energized, the microcontroller's pin can provide a maximum of 1–2 mA current. For this reason, we place a driver, such as the ULN2803, or a power transistor between the microcontroller and the relay as shown in Figure 17-2.

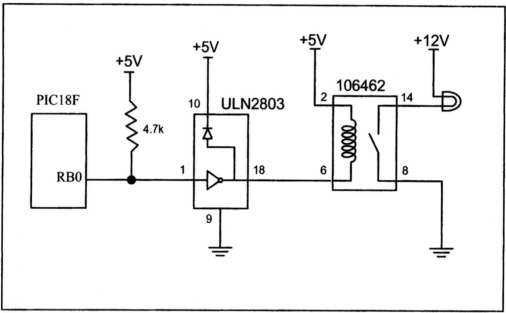

Figure 17-2. PIC18 Connection to Relay

Program 17-1 turns the lamp on and off shown in Figure 17-2 by energizing and de-energizing the relay every few ms.

```
;Program 17-1
R3    SET   0x20        ;set aside location 0x20 for R3
R4    SET   0x21        ;loc. 0x21 for R4
      ORG 0H
      BCF TRISB,0       ;PORTB.0 as output
OVER  BSF PORTB,0       ;turn on the lamp
      CALL DELAY
      BCF   PORTB,0     ;turn off the lamp
      CALL DELAY
      BRA OVER
DELAY MOVLW 0xFF
      MOVWF R4
D1    MOVLW 0xFF
      MOVWF R3
D2    NOP
      NOP
      DECF  R3,F
      BNZ   D2
      DECF  R4,F
      BNZ   D1
      RETURN
```

Solid-state relay

Another widely used relay is the solid-state relay. See Table 17-2. In this relay, there is no coil, spring, or mechanical contact switch. The entire relay is made out of semiconductor materials. Because no mechanical parts are involved in solid-state relays, their switching response time is much faster than that of electromechanical relays. Another advantage of the solid-state relay is its greater life expectancy. The life cycle for the electromechanical relay can vary from a few hundred thousand to a few million operations. Wear and tear on the contact points can cause the relay to malfunction after a while. Solid-state relays, however, have no such limitations. Extremely low input current and small packaging make solid-state relays ideal for microprocessor and logic control switching. They are widely used in controlling pumps, solenoids, alarms, and other power applications. Some solid-state relays have a phase control option, which is ideal for motor-speed control and light-dimming applications. Figure 17-3 shows control of a fan using a solid-state relay (SSR).

Table 17-2: Selected Solid-State Relay Characteristics (www.Jameco.com)

Part No.	Contact Style	Control Volts	Contact Volts	Contact Current
143058CP	SPST	4–32 VDC	240 VAC	3 A
139053CP	SPST	3–32 VDC	240 VAC	25 A
162341CP	SPST	3–32 VDC	240 VAC	10 A
172591CP	SPST	3–32 VDC	60 VDC	2 A
175222CP	SPST	3–32 VDC	60 VDC	4 A
176647CP	SPST	3–32 VDC	120 VDC	5 A

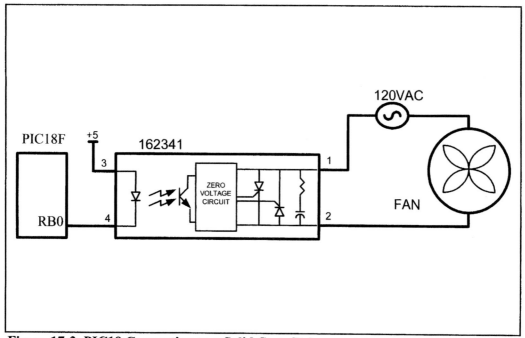

Figure 17-3. PIC18 Connection to a Solid-State Relay

Reed switch

Another popular switch is the reed switch. When the reed switch is placed in a magnetic field, the contact is closed. When the magnetic field is removed, the contact is forced open by its spring. See Figure 17-4. The reed switch is ideal for moist and marine environments where it can be submerged in fuel or water. Reed switches are also widely used in dirty and dusty atmospheres because they are tightly sealed.

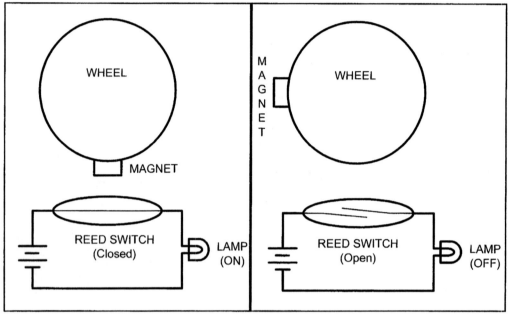

Figure 17-4. Reed Switch and Magnet Combination

Optoisolator

In some applications we use an optoisolator (also called optocoupler) to isolate two parts of a system. An example is driving a motor. Motors can produce what is called back EMF, a high-voltage spike produced by a sudden change of current as indicated in the V = Ldi/dt formula. In situations such as printed circuit board design, we can reduce the effect of this unwanted voltage spike (called ground bounce) by using decoupling capacitors (see Appendix C). In systems that have inductors (coil winding), such as motors, a decoupling capacitor or a diode will not do the job. In such cases we use optoisolators. An optoisolator has an LED (light-emitting diode) transmitter and a photosensor receiver, separated from each other by a gap. When current flows through the diode, it transmits a signal light across the gap and the receiver produces the same signal with the same phase but a different current and amplitude. See Figure 17-5. Optoisolators are also widely used in communication equipment such as modems. This device allows a computer to be connected to a telephone line without risk of damage from power surges. The gap between the transmitter and receiver of optoisolators prevents the electrical current surge from reaching the system.

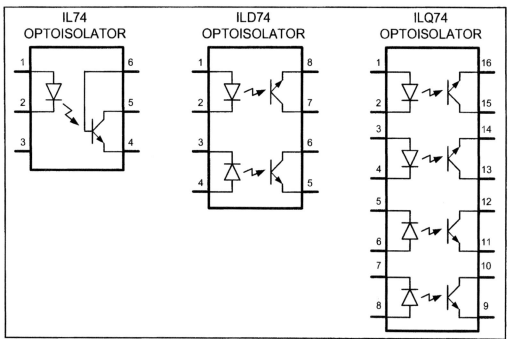

Figure 17-5. Optoisolator Package Examples

Interfacing an optoisolator

The optoisolator comes in a small IC package with four or more pins. There are also packages that contain more than one optoisolator. When placing an optoisolator between two circuits, we must use two separate voltage sources, one for each side, as shown in Figure 17-6. Unlike relays, no drivers need to be placed between the microcontroller/digital output and the optoisolators.

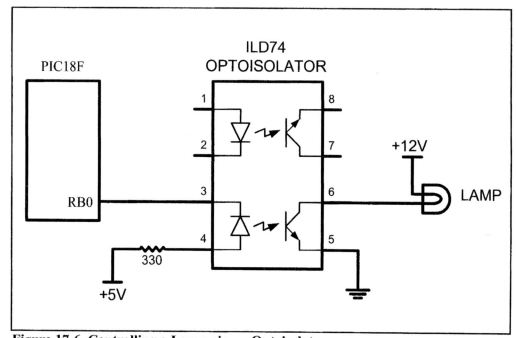

Figure 17-6. Controlling a Lamp via an Optoisolator

Review Questions

1. Give one application where would you use a relay.
2. Why do we place a driver between the microcontroller and the relay?
3. What is an NC relay?
4. Why are relays that use coils called electromechanical relays?
5. What is the advantage of a solid-state relay over EMR?
6. What is the advantage of an optoisolator over an EM relay?

SECTION 17.2: STEPPER MOTOR INTERFACING

This section begins with an overview of the basic operation of stepper motors. Then we describe how to interface a stepper motor to the PIC18. Finally, we use Assembly language programs to demonstrate control of the angle and direction of stepper motor rotation.

Stepper motors

A *stepper motor* is a widely used device that translates electrical pulses into mechanical movement. In applications such as disk drives, dot matrix printers, and robotics, the stepper motor is used for position control. Stepper motors commonly have a permanent magnet *rotor* (also called the *shaft*) surrounded by a *stator* (see Figure 17-7). There are also steppers called variable reluctance *stepper motors* that do not have a permanent magnet rotor. The most common stepper motors have four stator windings that are paired with a center-tapped common as shown in Figure 17-8. This type of stepper motor is commonly referred to as a *four-phase* or *unipolar* stepper motor. The center tap allows a change of current direction in

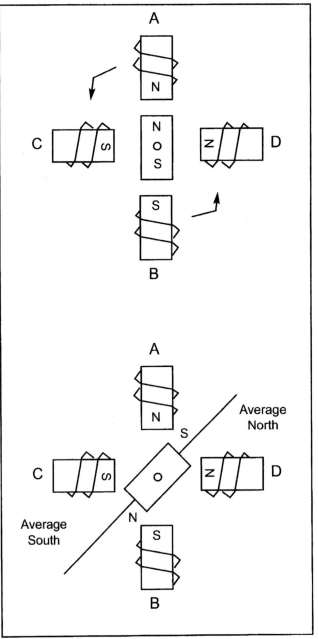

Figure 17-7. Rotor Alignment

each of two coils when a winding is grounded, thereby resulting in a polarity change of the stator. Notice that while a conventional motor shaft runs freely, the stepper motor shaft moves in a fixed repeatable increment, which allows one to move it to a precise position. This repeatable fixed movement is possible as a result of basic magnetic theory where poles of the same polarity repel and opposite poles attract. The direction of the rotation is dictated by the stator

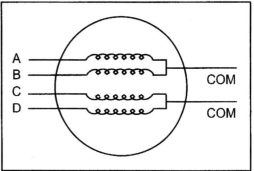

Figure 17-8. Stator Winding Configuration

poles. The stator poles are determined by the current sent through the wire coils. As the direction of the current is changed, the polarity is also changed causing the reverse motion of the rotor. The stepper motor discussed here has a total of six leads: four leads representing the four stator windings and two commons for the center-tapped leads. As the sequence of power is applied to each stator winding, the rotor will rotate. There are several widely used sequences, each of which has a different degree of precision. Table 17-3 shows a two-phase, four-step stepping sequence.

Note that although we can start with any of the sequences in Table 17-3, once we start we must continue in the proper order. For example, if we start with step 3 (0110), we must continue in the sequence of steps 4, 1, 2, etc.

Table 17-3: Normal Four-Step Sequence

Clockwise	Step #	Winding A	Winding B	Winding C	Winding D	Counter-clockwise
↓	1	1	0	0	1	↑
	2	1	1	0	0	
	3	0	1	1	0	
	4	0	0	1	1	

Step angle

How much movement is associated with a single step? This depends on the internal construction of the motor, in particular the number of teeth on the stator and the rotor. The *step angle* is the minimum degree of rotation associated with a single step. Various motors have different step angles. Table 17-4 shows some step angles for various motors. In Table 17-4, notice the term *steps per revolution*. This is the total number of steps needed to rotate one complete rotation or 360 degrees (e.g., 180 steps × 2 degrees = 360).

It must be noted that perhaps contrary to one's initial impression, a stepper motor does not need more terminal leads for the stator to achieve smaller steps. All the stepper motors

Table 17-4: Stepper Motor Step Angles

Step Angle	Steps per Revolution
0.72	500
1.8	200
2.0	180
2.5	144
5.0	72
7.5	48
15	24

discussed in this section have four leads for the stator winding and two COM wires for the center tap. Although some manufacturers set aside only one lead for the common signal instead of two, they always have four leads for the stators. See Example 17-1. Next we discuss some associated terminology in order to understand the stepper motor further.

Example 17-1

Describe the PIC18 connection to the stepper motor of Figure 17-9 and code a program to rotate it continuously.

Solution:

The following steps show the PIC18 connection to the stepper motor and its programming:

1. Use an ohmmeter to measure the resistance of the leads. This should identify which COM leads are connected to which winding leads.
2. The common wire(s) are connected to the positive side of the motor's power supply. In many motors, +5 V is sufficient.
3. The four leads of the stator winding are controlled by four bits of the PIC18 port (RB0–RB3). Because the PIC18 lacks sufficient current to drive the stepper motor windings, we must use a driver such as the ULN2003 to energize the stator. Instead of the ULN2003, we could have used transistors as drivers, as shown in Figure 17-11. However, notice that if transistors are used as drivers, we must also use diodes to take care of inductive current generated when the coil is turned off. One reason that using the ULN2003 is preferable to the use of transistors as drivers is that the ULN2003 has an internal diode to take care of back EMF.

```
MyReg       SET   0x30              ;loc 30H for MyReg
R2          SET   0x20              ;loc 20H for R2 Reg
            CLRF  TRISB             ;Port B as output
            MOVLW 0x66              ;load step sequence
            MOVWF MyReg
BACK        MOVFF MyReg,PORTB       ;issue sequence to motor
            RRNCF MyReg,F           ;rotate right clockwise
            CALL  DELAY             ;wait
            BRA   BACK              ;keep going
DELAY
            MOVLW 0xFF
            MOVWF R2
D1          NOP
            DECF  R2,F
            BNZ   D1
            RETURN
            END
```

Change the value of DELAY to set the speed of rotation.
We can use the single-bit instructions BSF and BCF instead of RRNCF to create the sequences.

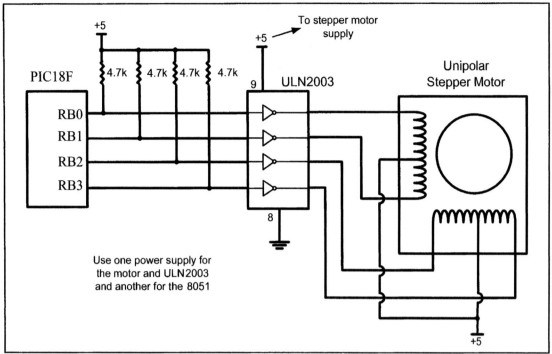

Figure 17-9. PIC18 Connection to Stepper Motor

Steps per second and rpm relation

The relation between rpm (revolutions per minute), steps per revolution, and steps per second is as follows.

$$Steps\ per\ second = \frac{rpm \times Steps\ per\ revolution}{60}$$

The 4-step sequence and number of teeth on rotor

The switching sequence shown earlier in Table 17-3 is called the 4-step switching sequence because after four steps the same two windings will be "ON". How much movement is associated with these four steps? After completing every four steps, the rotor moves only one tooth pitch. Therefore, in a stepper motor with 200 steps per revolution, the rotor has 50 teeth because $4 \times 50 = 200$ steps are needed to complete one revolution. This leads to the conclusion that the minimum step angle is always a function of the number of teeth on the rotor. In other words, the smaller the step angle, the more teeth the rotor passes. See Example 17-2.

Example 17-2

Give the number of times the four-step sequence in Table 17-3 must be applied to a stepper motor to make an 80-degree move if the motor has a 2-degree step angle.

Solution:

A motor with a 2-degree step angle has the following characteristics:

Step angle:	2 degrees	Steps per revolution:	180
Number of rotor teeth:	45	Movement per 4-step sequence:	8 degrees

To move the rotor 80 degrees, we need to send 10 consecutive 4-step sequences, because 10×4 steps $\times 2$ degrees $= 80$ degrees.

Looking at Example 17-2, one might wonder what happens if we want to move 45 degrees, because the steps are 2 degrees each. To allow for finer resolutions, all stepper motors allow what is called an *8-step* switching sequence. The 8-step sequence is also called *half-stepping,* because in the 8-step sequence each step is half of the normal step angle. For example, a motor with a 2-degree step angle can be used as a 1-degree step angle if the sequence of Table 17-5 is applied.

Table 17-5: Half-Step 8-Step Sequence

Clockwise	Step #	Winding A	Winding B	Winding C	Winding D	Counter-clockwise
	1	1	0	0	1	
	2	1	0	0	0	
	3	1	1	0	0	
	4	0	1	0	0	
	5	0	1	1	0	
	6	0	0	1	0	
	7	0	0	1	1	
	8	0	0	0	1	

Motor speed

The motor speed, measured in steps per second (steps/s), is a function of the switching rate. Notice in Example 17-1 that by changing the length of the time delay loop, we can achieve various rotation speeds.

Holding torque

The following is a definition of holding torque: "With the motor shaft at standstill or zero rpm condition, the amount of torque, from an external source, required to break away the shaft from its holding position. This is measured with rated voltage and current applied to the motor." The unit of torque is ounce-inch (or kg-cm).

Wave drive 4-step sequence

In addition to the 8-step and the 4-step sequences discussed earlier, there is another sequence called the wave drive 4-step sequence. It is shown in Table 17-6. Notice that the 8-step sequence of Table 17-5 is simply the combination of the wave drive 4-step and normal 4-step normal sequences shown in Tables 17-6 and 17-3, respectively. Experimenting with the wave drive 4-step sequence is left to the reader.

Table 17-6: Wave Drive 4-Step Sequence

Clockwise	Step #	Winding A	Winding B	Winding C	Winding D	Counter-clockwise
	1	1	0	0	0	
	2	0	1	0	0	
	3	0	0	1	0	
	4	0	0	0	1	

Table 17-7: Selected Stepper Motor Characteristics (www.Jameco.com)

Part No.	Step Angle	Drive System	Volts	Phase Resistance	Current
151861CP	7.5	unipolar	5 V	9 ohms	550 mA
171601CP	3.6	unipolar	7 V	20 ohms	350 mA
164056CP	7.5	bipolar	5 V	6 ohms	800 mA

Unipolar versus bipolar stepper motor interface

There are three common types of stepper motor interfacing: universal, unipolar, and bipolar. They can be identified by the number of connections to the motor. A universal stepper motor has eight, while the unipolar has six and the bipolar has four. The universal stepper motor can be configured for all three modes, while the unipolar can be either unipolar or bipolar. Obviously the bipolar cannot be configured for universal nor unipolar mode. Table 17-7 shows selected stepper motor characteristics. Figure 17-10 shows the basic internal connections of all three type of configurations.

Unipolar stepper motors can be controlled using the basic interfacing shown in Figure 17-11, whereas the bipolar stepper requires H-Bridge circuitry. Bipolar stepper motors require a higher operational current than the unipolar; the advantage of this is a higher holding torque.

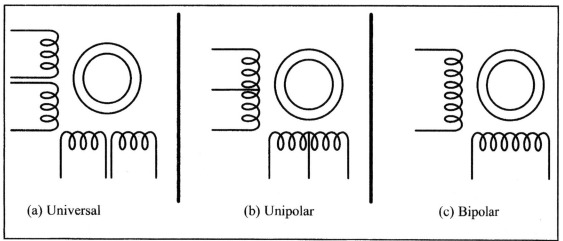

(a) Universal   (b) Unipolar   (c) Bipolar

Figure 17-10. Common Stepper Motor Types

Using transistors as drivers

Figure 17-11 shows an interface to a unipolar stepper motor using transistors. Diodes are used to reduce the back EMF spike created when the coils are energized and de-energized, similar to the electromechanical relays discussed earlier. TIP transistors can be used to supply higher current to the motor. Table 17-8 lists the common industrial Darlington transistors. These transistors can accommodate higher voltages and currents.

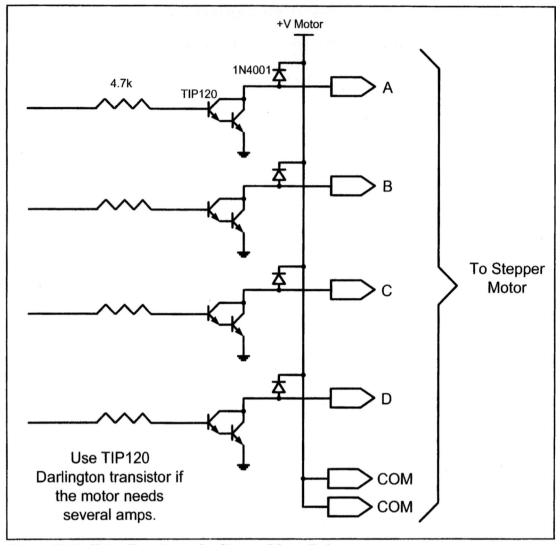

Figure 17-11. Using Transistors for Stepper Motor Driver

Table 17-8: Darlington Transistor Listing

NPN	PNP	Vceo (volts)	Ic (amps)	hfe (common)
TIP110	TIP115	60	2	1000
TIP111	TIP116	80	2	1000
TIP112	TIP117	100	2	1000
TIP120	TIP125	60	5	1000
TIP121	TIP126	80	5	1000
TIP122	TIP127	100	5	1000
TIP140	TIP145	60	10	1000
TIP141	TIP146	80	10	1000
TIP142	TIP147	100	10	1000

Controlling stepper motor via optoisolator

In the first section of this chapter we examined the optoisolator and its use. Optoisolators are widely used to isolate the stepper motor's EMF voltage and keep it from damaging the digital/microcontroller system. This is shown in Figure 17-12. See Examples 17-3 and 17-4.

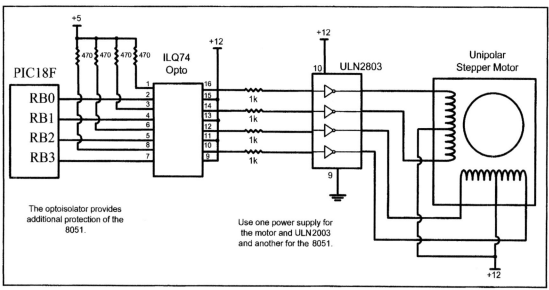

Figure 17-12. Controlling Stepper Motor via Optoisolator

Example 17-3

A switch is connected to pin RD7 (PORTD.7). Write a program to monitor the status of SW and perform the following:
(a) If SW = 0, the stepper motor moves clockwise.
(b) If SW = 1, the stepper motor moves counterclockwise.

Solution:

```
MyReg       SET   0x30              ;loc 30H for MyReg
            BSF   TRISD,RD7         ;RD7 as input pin
            CLRF  TRISB             ;Port B as output
            MOVLW 0x66              ;load step sequence
            MOVWF MyReg
BACK        BTFSS PORTD,RD7         ;check the SW
            BRA   OVER         ;It is high. Make it clockwise
            MOVFF MyReg,PORTB       ;issue sequence to motor
            RRNCF MyReg,F           ;rotate right clockwise
            CALL  DELAY             ;wait
            BRA   BACK              ;keep going
OVER        MOVFF MyReg,PORTB       ;issue sequence to motor
            RLNCF MyReg,F           ;rotate left clockwise
            CALL  DELAY             ;wait
            BRA   BACK              ;keep going
```

Stepper motor control with PIC18 C

The PIC18 C version of the stepper motor control is given below. In this program we could have used << (shift left) and >> (shift right) as was shown in Chapter 7.

```
#include <p18f458.h>
void main()
  {
    TRISB=0x0;           //PORTB as output
    while(1)
      {
         PORTB = 0x66;
         MSDelay(100);
         PORTB = 0xCC;
         MSDelay(100);
         PORTB = 0x99;
         MSDelay(100);
         PORTB = 0x33;
         MSDelay(100);
      }
  }
```

Example 17-4

A switch is connected to pin RD7. Write a C program to monitor the status of SW and perform the following:
(a) If SW = 0, the stepper motor moves clockwise.
(b) If SW = 1, the stepper motor moves counterclockwise.

Solution:

```
#include <p18f458.h>
#define SW PORTDbits.RD7
void MSDelay(int ms);
void main()
  {
    TRISD=0x80;          //RD7 as input pin
    TRISB=0x0;           //PORTB as output
    while(1)
      {
        if(SW == 0)
          {
           PORTB = 0x66;
           MSDelay(100);
           PORTB = 0xCC;
           MSDelay(100);
           PORTB = 0x99;
           MSDelay(100);
           PORTB = 0x33;
           MSDelay(100);
          }
        else
          {
           PORTB = 0x66;
           MSDelay(100);
           PORTB = 0x33;
```

Example 17-4 Cont.

```
        MSDelay(100);
        PORTB = 0x99;
        MSDelay(100);
        PORTB = 0xCC;
        MSDelay(100);
        }
    }
}

void MSDelay(unsigned int value)
    {
    unsigned int x, y;
    for(x=0;x<1275;x++)
        for(y=0;y<value;y++);
    }
```

Review Questions

1. Give the 4-step sequence of a stepper motor if we start with 0110.
2. A stepper motor with a step angle of 5 degrees has _____ steps per revolution.
3. Why do we put a driver between the microcontroller and the stepper motor?

SECTION 17.3: DC MOTOR INTERFACING AND PWM

This section begins with an overview of the basic operation of DC motors. Then we describe how to interface a DC motor to the PIC18. Finally, we use Assembly and C language programs to demonstrate the concept of pulse width modulation (PWM) and show how to control the speed and direction of a DC motor.

DC motors

A direct current (DC) motor is another widely used device that translates electrical pulses into mechanical movement. In the DC motor we have only + and − leads. Connecting them to a DC voltage source moves the motor in one direction. By reversing the polarity, the DC motor will move in the opposite direction. One can easily experiment with the DC motor. For example, small fans used in many motherboards to cool the CPU are run by DC motors. By connecting their leads to the + and − voltage source, the DC motor moves. While a stepper motor moves in steps of 1 to 15 degrees, the DC motor moves continuously. In a stepper motor, if we know the starting position we can easily count the number of steps the motor has moved and calculate the final position of the motor. This is not possible in a DC motor. The maximum speed of a DC motor is indicated in rpm and is given in the data sheet. The DC motor has two rpms: no-load and loaded. The manufacturer's data sheet gives the no-load rpm. The no-load rpm can be from a few thousand to tens of thousands. The rpm is reduced when moving a load and it decreases as the load is increased. For example, a drill turning a screw has a much lower rpm speed than when it is in the no-load situation. DC motors also have voltage and current ratings. The nominal voltage is the voltage for that motor under normal conditions, and can be from 1 to 150 V, depending on the motor. As we increase the voltage, the rpm goes up. The current rating is the current consumption when the nominal voltage is applied with no load, and can be from 25 mA to a few amps. As the load increases, the rpm is decreased, unless the current or voltage provided to the motor is increased, which in turn increases the torque. With a fixed voltage, as the load increases, the current (power) consumption of a DC motor is increased. If we overload the motor it will stall, and that can damage the motor due to the heat generated by high current consumption.

Unidirectional control

Figure 17-13 shows the DC motor rotation for clockwise (CW) and counterclockwise (CCW) rotations. See Table 17-9 for selected DC motors.

Table 17-9: Selected DC Motor Characteristics (www.Jameco.com)

Part No.	Nominal Volts	Volt Range	Current	RPM	Torque
154915CP	3 V	1.5–3 V	0.070 A	5,200	4.0 g-cm
154923CP	3 V	1.5–3 V	0.240 A	16,000	8.3 g-cm
177498CP	4.5 V	3–14 V	0.150 A	10,300	33.3 g-cm
181411CP	5 V	3–14 V	0.470 A	10,000	18.8 g-cm

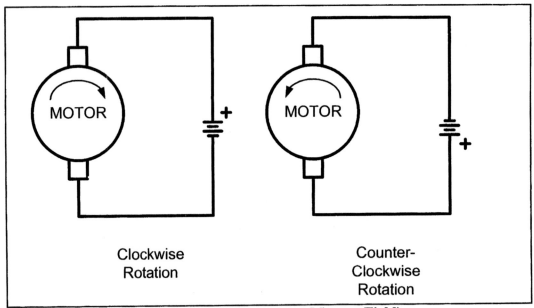

Figure 17-13. DC Motor Rotation (Permanent Magnet Field)

Bidirectional control

With the help of relays or some specially designed chips we can change the direction of the DC motor rotation. Figures 17-14 through 17-17 show the basic concepts of H-Bridge control of DC motors.

Figure 17-14 shows the connection of an H-Bridge using simple switches. All the switches are open, which does not allow the motor to turn.

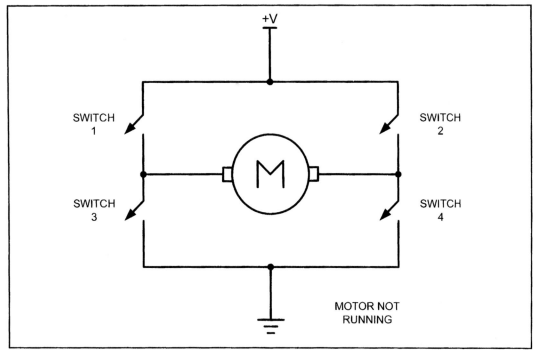

Figure 17-14. H-Bridge Motor Configuration

Figure 17-15 shows the switch configuration for turning the motor in one direction. When switches 1 and 4 are closed, current is allowed to pass through the motor.

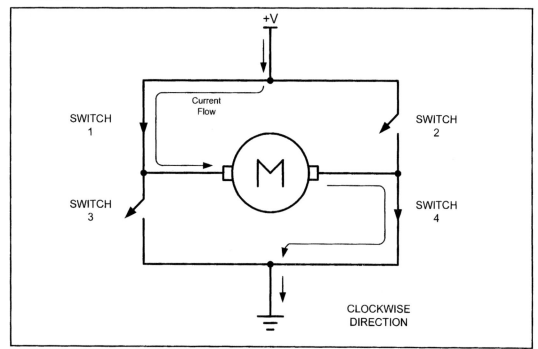

Figure 17-15. H-Bridge Motor Clockwise Configuration

Figure 17-16 shows the switch configuration for turning the motor in the opposite direction from the configuration of Figure 17-15. When switches 2 and 3 are closed, current is allowed to pass through the motor.

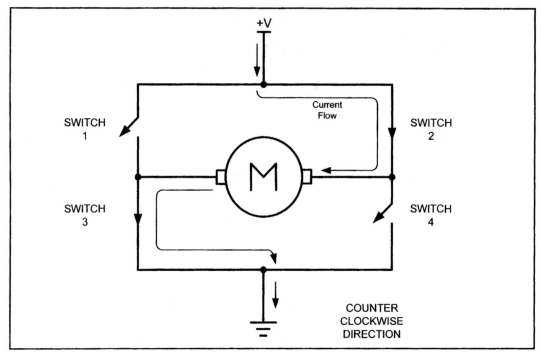

Figure 17-16. H-Bridge Motor Counterclockwise Configuration

Figure 17-17 shows an invalid configuration. Current flows directly to ground, creating a short circuit. The same effect occurs when switches 1 and 3 are closed or switches 2 and 4 are closed.

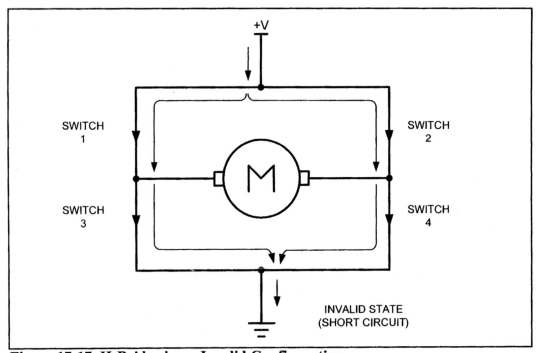

Figure 17-17. H-Bridge in an Invalid Configuration

Table 17-10 shows some of the logic configurations for the H-Bridge design.

H-Bridge control can be created using relays, transistors, or a single IC solution such as the L293. When using relays and transistors, you must ensure that invalid configurations do not occur.

Table 17-10: Some H-Bridge Logic Configurations for Figure 17-14

Motor Operation	SW1	SW2	SW3	SW4
Off	Open	Open	Open	Open
Clockwise	Closed	Open	Open	Closed
Counterclockwise	Open	Closed	Closed	Open
Invalid	Closed	Closed	Closed	Closed

Although we do not show the relay control of an H-Bridge, Example 17-5 shows a simple program to operate a basic H-Bridge.

Example 17-5

A switch is connected to pin RD7 (PORTD.7). Using a simulator, write a program to simulate the H-Bridge in Table 17-10. We must perform the following:
(a) If DIR = 0, the DC motor moves clockwise.
(b) If DIR = 1, the DC motor moves counterclockwise.

Solution:

```
          BCF   TRISB,0          ;PORTB.0 as output for switch 1
          BCF   TRISB,1          ;      .1   "         switch 2
          BCF   TRISB,2          ;      .2   "         switch 3
          BCF   TRISB,3          ;      .3   "         switch 4
          BSF   TRISD,7          ;make PORTD.7 an input DIR
MONITOR:
          BTFSS PORTD,7
          BRA   CLOCKWISE
          BSF   PORTB,0          ;switch 1
          BCF   PORTB,1          ;switch 2
          BCF   PORTB,2          ;switch 3
          BSF   PORTB,3          ;switch 4
          BRA   MONITOR
CLOCKWISE:
          BCF   PORTB,0          ;switch 1
          BSF   PORTB,1          ;switch 2
          BSF   PORTB,2          ;switch 3
          BCF   PORTB,3          ;switch 4
          BRA   MONITOR
```

View the results on your simulator. This example is for simulation only and should not be used on a connected system.

See http://www.MicroDigitalEd.com for additional information on using H-Bridges.

Figure 17-18 shows the connection of the L293 to an PIC18. Be aware that the L293 will generate heat during operation. For sustained operation of the motor, use a heat sink. Example 17-6 shows control of the L293.

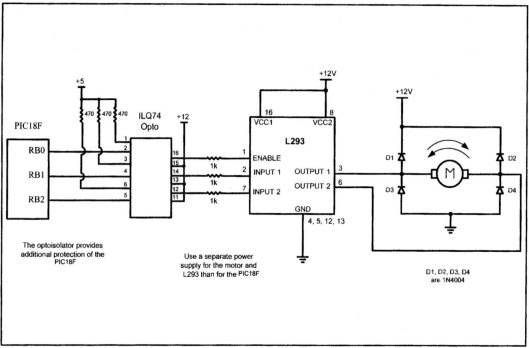

Figure 17-18. Bidirectional Motor Control Using an L293 Chip

Example 17-6

Figure 17-18 shows the connection of an L293. Add a switch to pin RD7 (PORTD.7). Write a program to monitor the status of SW and perform the following:

(a) If SW = 0, the DC motor moves clockwise.
(b) If SW = 1, the DC motor moves counterclockwise.

Solution:

```
        BCF    TRISB,0
        BCF    TRISB,1
        BCF    TRISB,2
        BSF    TRISD,7
        BSF    PORTB,0        ;enable the chip
CHK     BTFSS  PORTD,7
        BRA    CWISE
        BCF    PORTB,1        ;turn the motor counterclockwise
        BSF    PORTB,2
        BRA    CHK
CWISE   BSF    PORTB,1
        BCF    PORTB,2        ;turn motor clockwise
        BRA    CHK
```

Pulse width modulation (PWM)

The speed of the motor depends on three factors: (a) load, (b) voltage, and (c) current. For a given fixed load we can maintain a steady speed by using a method called *pulse width modulation* (PWM). By changing (modulating) the width of the pulse applied to the DC motor we can increase or decrease the amount of power provided to the motor, thereby increasing or decreasing the motor speed. Notice that, although the voltage has a fixed amplitude, it has a variable duty cycle. That means the wider the pulse, the higher the speed. PWM is so widely used in DC motor control that some microcontrollers come with the PWM circuitry embedded in the chip. In such microcontrollers all we have to do is load the proper registers with the values of the high and low portions of the desired pulse, and the rest is taken care of by the microcontroller. This allows the microcontroller to do other things. For microcontrollers without PWM circuitry, we must create the various duty cycle pulses using software, which prevents the microcontroller from doing other things. The ability to control the speed of the DC motor using PWM is one reason that DC motors are preferable over AC motors. AC motor speed is dictated by the AC frequency of the voltage applied to the motor and the frequency is generally fixed. As a result, we cannot control the speed of the AC motor when the load is increased. As was shown earlier, we can also change the DC motor's direction and torque. See Figure 17-19 for PWM comparisons.

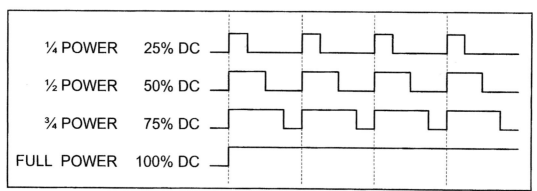

Figure 17-19. Pulse Width Modulation Comparison

DC motor control with optoisolator

As we discussed in the first section of this chapter, the optoisolator is indispensable in many motor control applications. Figures 17-20 and 17-21 show the connections to a simple DC motor using a bipolar and a MOSFET transistor. Notice that the PIC18 is protected from EMI created by motor brushes by using an optoisolator and a separate power supply.

Figures 17-20 and 17-21 show optoisolators for control of single directional motor control, and the same principle should be used for most motor applications. Separating the power supplies of the motor and logic will reduce the possibility of damage to the control circuitry.

Figure 17-20 shows the connection of a bipolar transistor to a motor. Protection of the control circuit is provided by the optoisolator. The motor and PIC18 use separate power supplies. The separation of power supplies also allows the use of high-voltage motors. Notice that we use a decoupling capacitor across the motor; this helps reduce the EMI created by the motor. The motor is switched on by clearing bit P1.0.

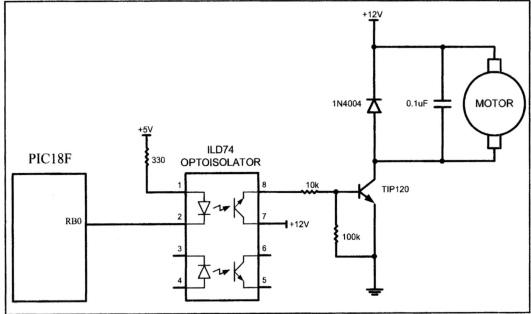

Figure 17-20. DC Motor Connection using a Darlington Transistor

Figure 17-21 shows the connection of a MOSFET transistor. The optoisolator protects the PIC18 from EMI. The zener diode is required for the transistor to reduce gate voltage below the rated maximum value. See Example 17-7.

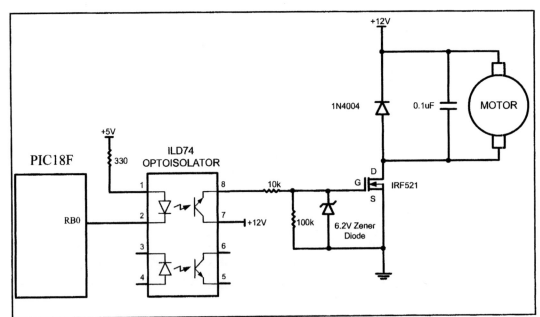

Figure 17-21. DC Motor Connection using a MOSFET Transistor

Example 17-7

Refer to the figure in this example. Write a program to monitor the status of the switch and perform the following:

(a) If PORTD.7 = 1, the DC motor moves with 25% duty cycle pulse.
(b) If PORTD.7 = 0, the DC motor moves with 50% duty cycle pulse.

Solution:

```
            BCF  TRISB,RB0        ;PORTB.0 as output
            BSF  TRISD,RD7        ;PORTD.7 as input
            BCF  PORTB,RB0        ;turn off motor
CHK
            BTFSS PORTD,RD7
            BRA  PWM_50
            BSF  PORTB,RB0        ;high portion of pulse
            CALL DELAY
            BCF  PORTB,RB0        ;low portion of pulse
            CALL DELAY
            CALL DELAY
            CALL DELAY
            BRA  CHK
PWM_50
            BSF  PORTB,RB0        ;high portion of pulse
            CALL DELAY
            CALL DELAY
            BCF  PORTB,RB0        ;low portion of pulse
            CALL DELAY
            CALL DELAY
            BRA CHK
```

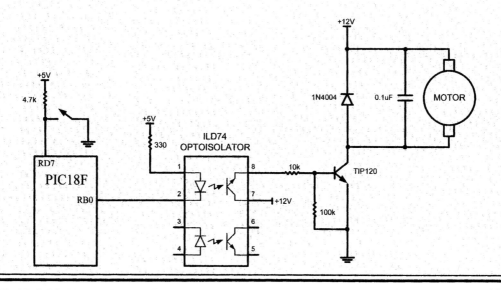

DC motor control and PWM using C

Examples 17-8 through 17-10 show the PIC18 C version of the earlier programs controlling the DC motor.

Example 17-8

Refer to Figure 17-18 for connection of the motor. A switch is connected to pin RD7. Write a C program to monitor the status of SW and perform the following:

(a) If SW = 0, the DC motor moves clockwise.
(b) If SW = 1, the DC motor moves counterclockwise.

Solution:

```c
#include <p18f458.h>

#define SW PORTDbits.RD7
#define ENABLE PORTBbits.RB0
#define MTR_1 PORTBbits.RB1
#define MTR_2 PORTBbits.RB2

void main()
   {
   TRISD=0x80;     //make RD7 input pin
   TRISB=0x0;      //make PORTB output
   SW = 1;
   ENABLE = 0;
   MTR_1 = 0;
   MTR_2 = 0;

   while(1)
      {
        ENABLE = 1;
        if(SW == 1)
           {
             MTR_1 = 1;
             MTR_2 = 0;
           }
        else
           {
             MTR_1 = 0;
             MTR_2 = 1;
           }
      }
   }
```

Example 17-9

Refer to the figure in this example. Write a C program to monitor the status of SW and perform the following:

(a) If SW = 0, the DC motor moves with 50% duty cycle pulse.
(b) If SW = 1, the DC motor moves with 25% duty cycle pulse.

Solution:
```c
#include <p18f458.h>
#define SW PORTDbits.RD7
#define MTR  PORTBbits.RB1
void MSDelay(unsigned int value);
void main()
  {
  TRISD=0x80;          //make RD7 input pin
  TRISB=0xFD;          //make RB1 output pin
  while(1)
    {
      if(SW == 1)
        {
          MTR = 1;
          MSDelay(25);
          MTR = 0;
          MSDelay(75);
        }
      else
        {
          MTR = 1;
          MSDelay(50);
          MTR = 0;
          MSDelay(50);
        }
    }
  }
void MSDelay(unsigned int value)
  {
    unsigned char x, y;
    for(x=0; x<1275; x++)
      for(y=0; y<value; y++);
  }
```

Example 17-10

Refer to Figure 17-20 for connection to the motor. Two switches are connected to pins RD0 and RD1. Write a C program to monitor the status of both switches and perform the following:

SW2 (RD1)   SW1 (RD0)

0	0	DC motor moves slowly (25% duty cycle).
0	1	DC motor moves moderately (50% duty cycle).
1	0	DC motor moves fast (75% duty cycle).
1	1	DC motor moves very fast (100% duty cycle).

Solution:

```
#include <p18f458.h>
#define MTR PORTBbits.RB1
void MSDelay(unsigned int value);

void main()
   {
   unsigned int duty;
   TRISB = 0xFD;
   TRISD = 0xFF;
   while(1)
      {
      duty = PORTD&0x03;
      duty++;
      duty *= 25;
      MTR = 1;
      MSDelay(duty);
      MTR = 0;
      MSDelay(100-duty);
      }
   }
```

Review Questions

1. True or false. The permanent magnet field DC motor has only two leads for + and – voltages.
2. True or false. Just like a stepper motor, one can control the exact angle of a DC motor's move.
3. Why do we put a driver between the microcontroller and the DC motor?
4. How do we change a DC motor's rotation direction?
5. What is stall in a DC motor?
6. True or false. PWM allows the control of a DC motor with the same phase, but different amplitude pulses.
7. The RPM rating given for the DC motor is for _____ (no-load, loaded).

SECTION 17.4: PWM MOTOR CONTROL WITH CCP

We examined the CCP (Compare Capture Pulse-Width-Modulation) part of the PIC452/458 in Chapter 15. One of the features of the CCP is the pulse width modulation (PWM) as we saw in Section 15.4 of Chapter 15. In this section we use the PWM feature of the CCP to control DC motors. Review the programming of the PWM in Section 15.4 before embarking on this section.

DC motor control with CCP

Recall from Section 15.4 that the PWM part of the CCP is programmed by using the PR2 and Timer2 registers. Program 17-2 is the rewrite of Example 17-7 using the PWM feature of the CCP1. Notice that Program 17-2 is the modified version of Program 15-5 in Chapter 15. Program 17-2C is the C version of Program 17-2. In Program 17-2 (and 17-2C), an input switch is being monitored. If the switch is low, the PIC18 creates a 50% duty cycle PWM using the CCP1 module. If the switch is high, a 25% duty cycle PWM is created. Recall from Chapter 15 that we must use PR2 and Timer2 registers for creating PWM pulses.

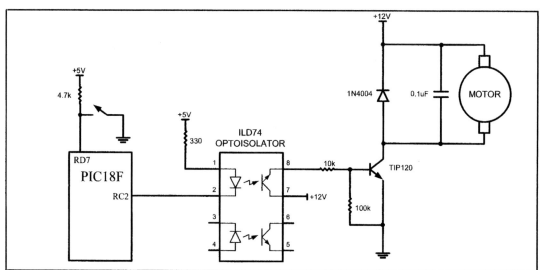

Figure 17-22: DC Motor Control Using CCP1 Pin

```
;Program 17-2
        BCF    TRISC,CCP1        ;make PWM output pin
        BSF    TRISD,RD7         ;make RD7 input pin
        MOVLW  0x3C              ;PWM MODE, 11 for DC1B1:B0
        MOVWF  CCP1CON
        MOVLW  D'100'            ;set period to 100 * Fosc/4
        MOVWF  PR2
        MOVLW  0x01              ;Timer2, 4 prescale, no postscaler
        MOVWF  T2CON
AGAIN   BTFSS  PORTD,RD7         ;Is the switch high?
        BRA    T2DUTY            ;no, then 50%
        MOVLW  D'25'             ;25% duty cycle
        BRA    LOAD
T2DUTY  MOVLW  D'50'             ;50% duty cycle
```

```
        BRA  LOAD
LOAD  MOVWF  CCPR1L            ;load duty cycle
      CLRF   TMR2              ;clear Timer2
      BSF    T2CON,TMR2ON      ;turn on Timer2
      BCF    PIR1,TMR2IF       ;clear Timer2 flag
OVER  BTFSS PIR1,TMR2IF        ;wait for end of period
      BRA    OVER
      GOTO   AGAIN             ;continue
```

The following is the C version of the above program.

```c
//Program 17-2C
#include <p18f458.h>
void main()
  {
  TRISC = 0xFB;            //make CCP1 output pin
  TRISD = 0x80;            //make RD7 input pin
  CCP1CON = 0x3C;          //PWM MODE, 11 for DC1B1:B0
  PR2=100;                 //set period to 100 * 16/Fosc
  T2CON=0x01;              //4 prescaler, no postscaler
  while(1)
    {
    if(PORTDbits.RD7==1)
      CCPR1L = 25;         //25% duty cycle
    else
      CCPR1L = 50;         //50% duty cycle
    TMR2=0x0;              //clear Timer2
    PIR1bits.TMR2IF=0;     //clear Timer2 flag
    T2CONbits.TMR2ON=1;    //start Timer2
    while(PIR1bits.TMR2IF==0);//wait for end of period
    }
  }
```

Review Questions

1. True or false. For standard CCP1, we use the RC2 pin for PWM.
2. True or false. For standard CCP1, the CCP1 pin must be configured as output.
3. In standard CCP1, we use _____ to set the period for PWM.
4. In standard CCP1, we use _____ to set the duty cycle for PWM.
5. True or false. In standard CCP1, we must use Timer1 for PWM.

SECTION 17.5: DC MOTOR CONTROL WITH ECCP

The PIC18F452/458 (or 4520/4580) comes with one standard CCP and one enhanced CCP (ECCP). Indeed, in recent years the CCP module has been deemphasized while the ECCP is becoming more prominent in the PIC18 family. The reason is that ECCP allows the implementation of the H-Bridge for bidirectional control of the DC motor in addition to the capture/compare mode present in the standard CCP. In this section, we use the ECCP feature of the PIC18 to control the DC motor. Before embarking on this section, the basic concept of ECCP programming in Chapter 15 needs to be reviewed.

Bidirectional DC motor control with ECCP

ECCP allows the implementation of the H-Bridge for bidirectional movement of the DC motor because it uses 4 pins instead of a single pin as is used in standard CCP. As we saw in Section 17.3 of this chapter, the bidirectional DC movement needs some kind of H-Bridge circuitry. The ECCP module of the PIC18 implements the entire H-Bridge circuitry internally. It uses RD7–RD4 (PORTD.7–PORTD.4) for this purpose as shown in Figures 17-23 through 17-26.

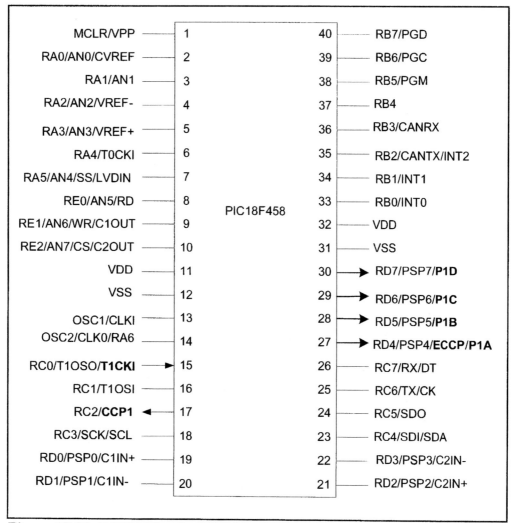

Figure 17-23. ECCP Pins for PWM in PIC18F458/4580 (452/4520)

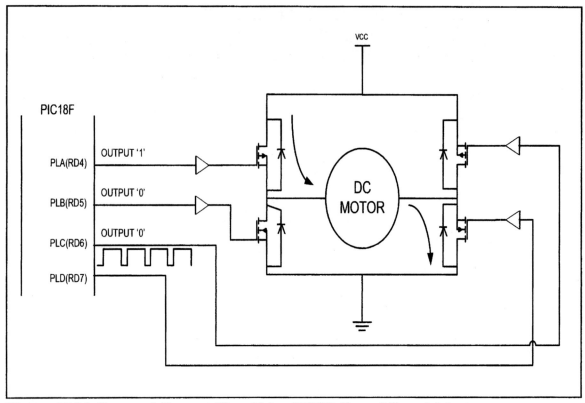

Figure 17-24. Forward Current Flow Using ECCP (from Microchip)

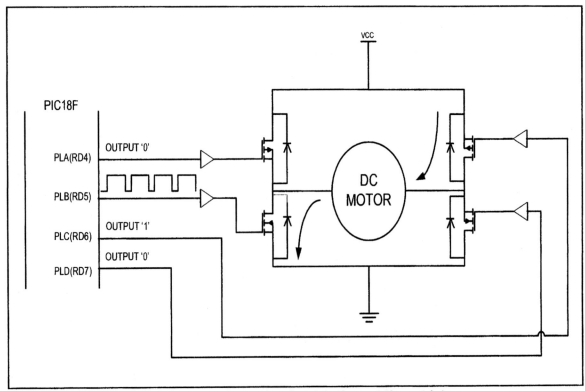

Figure 17-25. Reverse Current Flow Using ECCP (from Microchip)

EPWM1M1	EPWM1M0	EDC1B1	EDC1B0
D7			

ECCP1M3	ECCP1M2	ECCP1M1	ECCP1M0
			D0

EPWM1M1:EPWM1M0 PWM output pin configuration. It allows the use of a single pin for the capture/compare mode, or four pins for the PWM.
In compare/capture mode, only pin P1A (RD4) is used. In that case, there is no selection for these two bits.
In the PWM mode the options for these two bits are as follows:

00   P1A is used as a modulated output. P1B, P1C, and P1D are used as I/O.
01   Full-Bridge output forward. P1D modulated, P1A active. P1B and P1C inactive.

10   Half-Bridge output. P1A and P1D modulated with deadband control, P1C and P1D used as I/O.
11   Full-Bridge output reverse. P1B modulated, P1C active. P1A and P1D inactive.

EDC1B10:EDC1B1 PWM Duty Cycle least-significant bits. Used in PWM only.
The least-significant bits (Bit 1 and Bit 0) of the 10-bit duty cycle register are used in PWM. The ECCPR1L register is used as Bit 2 to Bit 9 of the 10-bit duty cycle register.

ECCP1M3–ECC1M0 ECCP1 Mode Select
        0 0 0 0   ECCP1 is off
        0 0 0 1   Reserved
        0 0 1 0   Compare Mode. Toggle ECCP1 output pin on match.
                     (ECCP1IF bit is set.)
        0 0 1 1   Reserved

        0 1 0 0   Capture mode, every falling edge
        0 1 0 1   Capture mode, every rising edge
        0 1 1 0   Capture mode, every 4th rising edge
        0 1 1 1   Capture mode, every 16th rising edge

        1 0 0 0   Compare mode. Initialize ECCP1 pin low; on compare match,
                     force CCP1 pin HIGH. (ECCP1IF is set.)
        1 0 0 1   Compare mode. Initialize CCP1 pin HIGH; on compare match,
                     force CCP1 pin LOW. (ECCP1IF is set.)
        1 0 1 0   Compare mode. Generate software interrupt on compare
                     match. (ECCP1IF bit is set, ECCP1 pin is unaffected.)
        1 0 1 1   Compare mode. Trigger special event. (ECCP1IF bit is set, and
                     Timer1 or Timer3 is reset to zero.)

        1 1 0 0   PWM Mode; P1A, P1C active-HIGH; P1B and P1D active-HIGH
        1 1 0 1   PWM Mode; P1A, P1C active-HIGH; P1B and P1D active-LOW
        1 1 1 0   PWM Mode; P1A, P1C active-LOW; P1B and P1D active-HIGH
        1 1 1 1   PWM Mode; P1A, P1C active-LOW; P1B and P1D active-LOW

Figure 17-26. ECCP1 Control Register. (This register selects one of the operation modes of Capture, Compare, or PWM of EECP1)

Program 17-3 shows Full-Bridge implementation of the PWM for ECCP module. For the implementation of Half-Bridge and other applications of PWM using the ECCP module, see the PIC18 manual.

```
;Program 17-3
      CLRF  TRISD              ;make PORTD output
      MOVLW D'100'
      MOVWF PR2                ;period = 100 * 16/Fosc
      MOVLW D'50'
      MOVWF ECCPR1L            ;duty = 50%
      MOVLW 0xCF
      MOVWF ECCP1CON           ;reverse full-bridge PWM
      MOVLW 0x24
      MOVWF T2CON              ;4 postscaler, turn on Timer2
AGAIN CLRF  TMR2               ;start pulse
      BCF   PIR1,TMR2IF        ;clear flag
WAIT  BTFSS PIR1,TMR2IF        ;wait for period
      BRA   WAIT
      BRA   AGAIN              ;do it again
```

The following is the C version of the above program.

```
//Program 17-3C
#include <p18f458.h>

void main()
  {
      TRISD=0;                 //make PORTD output
      PR2=100;                 //period = 100 * 16/Fosc
      ECCPR1L=50;              //duty = 50%
      ECCP1CON=0xCF;           //reverse full-bridge PWM
      T2CON=0x24;              //4 postscaler,turn on Timer2
while(1)
      {
      TMR2=0;                  //start pulse
      PIR1bits.TMR2IF=0;          //clear flag
      while(PIR1bits.TMR2IF==0); //wait for period
      }
  }
```

Review Questions

1. True or false. For ECCP1, we use the RD3–RD0 pins for Full-Bridge.
2. True or false. For ECCP1, the P1A to P1D pins must be configured as output.
3. In ECCP1, we use _____ to set the period for PWM.
4. In ECCP1, we use _____ to set the duty cycle for PWM.
5. True or false. In ECCP1, we must use Timer2 for PWM.

SUMMARY

This chapter continued showing how to interface the PIC18 with real-world devices. Devices covered in this chapter were the relay, optoisolator, stepper motor, and DC motor.

First, the basic operation of relays and optoisolators was defined, along with key terms used in describing and controlling their operations. Then the PIC18 was interfaced with a stepper motor. The stepper motor was then controlled via an optoisolator using PIC18 Assembly and C programming languages.

The PIC18 was interfaced with DC motors. A typical DC motor will take electronic pulses and convert them to mechanical motion. This chapter showed how to interface the PIC18 with a DC motor. Then, simple Assembly and C programs were written to show the concept of PWM.

Control systems that require motors must be evaluated for the type of motor needed. For example, you would not want to use a stepper in a high-velocity application or a DC motor for a low-speed, high-torque situation. The stepper motor is ideal in an open-loop positional system and a DC motor is better for a high-speed conveyer belt application. DC motors can be modified to operate in a closed-loop system by adding a shaft encoder, then using a microcontroller to monitor the exact position and velocity of the motor. In the last two sections, we showed how to use CCP and ECCP features of PIC18 to control DC motors.

PROBLEMS

SECTION 17.1: RELAYS AND OPTOISOLATORS

1. True or false. The minimum voltage needed to energize a relay is the same for all relays.
2. True or false. The minimum current needed to energize a relay depends on the coil resistance.
3. Give the advantages of a solid-state relay over an EM relay.
4. True or false. In relays, the energizing voltage is the same as the contact voltage.
5. Find the current needed to energize a relay if the coil resistance is 1,200 ohms and the coil voltage is 5 V.
6. Give two applications for an optoisolator.
7. Give the advantages of an optoisolator over an EM relay.
8. Of the EM relay and solid-state relay, which has the problem of back EMF?
9. True or false. The greater the coil resistance, the worse the back EMF voltage.
10. True or false. We should use the same voltage sources for both the coil voltage and contact voltage.

SECTION 17.2: STEPPER MOTOR INTERFACING

11. If a motor takes 90 steps to make one complete revolution, what is the step angle for this motor?
12. Calculate the number of steps per revolution for a step angle of 7.5 degrees.

13. Finish the normal four-step sequence clockwise if the first step is 0011 (binary).
14. Finish the normal four-step sequence clockwise if the first step is 1100 (binary).
15. Finish the normal four-step sequence counterclockwise if the first step is 1001 (binary).
16. Finish the normal four-step sequence counterclockwise if the first step is 0110 (binary).
17. What is the purpose of the ULN2003 placed between the PIC18 and the stepper motor? Can we use that for 3A motors?
18. Which of the following cannot be a sequence in the normal four-step sequence for a stepper motor?
    (a) CCH (b) DDH (c) 99H (d) 33H
19. What is the effect of a time delay between issuing each step?
20. In Question 19, how can we make a stepper motor go faster?

SECTION 17.3: DC MOTOR INTERFACING AND PWM

21. Which motor is best for moving a wheel exactly 90 degrees?
22. True or false. Current dissipation of a DC motor is proportional to the load.
23. True or false. The rpm of a DC motor is the same for no-load and loaded.
24. The rpm given in data sheets is for _____ (no-load, loaded).
25. What is the advantage of DC motors over AC motors?
26. What is the advantage of stepper motors over DC motors?
27. True or false. Higher load on a DC motor slows it down if the current and voltage supplied to the motor are fixed.
28. What is PWM, and how is it used in DC motor control?
29. A DC motor is moving a load. How do we keep the rpm constant?
30. What is the advantage of placing an optoisolator between the motor and the microcontroller?

ANSWERS TO REVIEW QUESTIONS

SECTION 17.1: RELAYS AND OPTOISOLATORS

1. With a relay we can use a 5 V digital system to control 12 V–120 V devices such as horns and appliances.
2. Because microcontroller/digital outputs lack sufficient current to energize the relay, we need a driver.
3. When the coil is not energized, the contact is closed.
4. When current flows through the coil, a magnetic field is created around the coil, which causes the armature to be attracted to the coil.
5. It is faster and needs less current to get energized.
6. It is smaller and can be connected to the microcontroller directly without a driver.

SECTION 17.2: STEPPER MOTOR INTERFACING

1. 0110, 0011, 1001, 1100 for clockwise; and 0110, 1100, 1001, 0011 for counterclockwise
2. 72
3. Because the microcontroller pins do not provide sufficient current to drive the stepper motor

SECTION 17.3: DC MOTOR INTERFACING AND PWM

1. True
2. False
3. Because microcontroller/digital outputs lack sufficient current to drive the DC motor, we need a driver.
4. By reversing the polarity of voltages connected to the leads
5. The DC motor is stalled if the load is beyond what it can handle.
6. False
7. No-load

SECTION 17.4: PWM MOTOR CONTROL WITH CCP

1. True
2. True
3. PR2
4. CCPR1L
5. False

SECTION 17.5: DC MOTOR CONTROL WITH ECCP

1. False
2. True
3. PR2
4. CCPR1L
5. True

APPENDIX A

PIC18 INSTRUCTIONS: FORMAT AND DESCRIPTION

OVERVIEW

In the first section of this appendix, we describe the instruction format of the PIC18. Special emphasis is placed on the instructions using both WREG and file registers. This section includes a list of machine cycles (clock counts) for each of the PIC18 instructions.

In the second section of this appendix, we describe each instruction of the PIC18. In many cases, a simple programming example is given to clarify the instruction.

This Appendix deals mainly with PIC18 instructions. In Section A.1, we describe the instruction formats and categories. In Section A.2, we describe each instruction of PIC18 with some examples.

SECTION A.1: PIC18 INSTRUCTION FORMATS AND CATEGORIES

As shown in Figure A-1, the PIC18 instructions fall into five categories:

1. Bit-oriented instructions
2. Intructions using a literal value
3. Byte-oriented instructions
4. Table read and write instructions
5. Control instructions using branch and call

In this section, we describe the format and syntax with special emphasis placed on byte-oriented instructions. For some of the instructions, the reader needs to review the concepts of access bank and bank registers in Chapter 6 (Section 6.3).

Bit-oriented instructions

The bit-oriented instructions perform operations on a specific bit of a file register. After the operation, the result is placed back in the same file register. For example, the "BCF f,b,a" instruction clears a specific bit of fileReg. See Table A-1. In these types of instructions, the b is the specific bit of the fileReg, which can be 0 to 7, representing the D0 to D7 bits of the register. The fileReg location can be in the bank register called access bank (if a = 0) or a location within other bank registers (if a = 1). Notice that if a = 0, the assembler assumes the access bank automatically.

Table A-1: Bit-Oriented Instructions (from Microchip datasheet)

Mnemonic, Operands	Description	Cycles
BIT-ORIENTED FILE REGISTER OPERATIONS		
BCF f, b, a	Bit Clear f	1
BSF f, b, a	Bit Set f	1
BTFSC f, b, a	Bit Test f, Skip if Clear	1 (2 or 3)
BTFSS f, b, a	Bit Test f, Skip if Set	1 (2 or 3)
BTG f, d, a	Bit Toggle f	1

Look at the examples that follow for clarification of bit-oriented instructions:

Byte-oriented File Register operations   Example Instructions

```
15              10 9  8  7                    0
|   OPCODE    | d | a |    f (FILE #)    | ADDWF  MYREG, W, B
```

    d = 0 for result destination to be WREG Register

    d = 1 for result destination to be File Register (f)

    a = 0 to force Access Bank

    a = 1 for BSR to select bank

    f = 8-bit File Register address

Byte to Byte move operations (2-word)

```
15      12 11                    0
| OPCODE |        f (Source FILE #)     | MOVFF  MYREG1, MYREG2
```

```
15      12 11                    0
|  1111    |    f (Destination FILE #)   |
```

    f = 12-bit File Register address

Bit-oriented File Register operations

```
15          12 11   9 8 7              0
|   OPCODE | b (BIT #)| a |   f (FILE #) | BSF  MYREG, bit, B
```

    b = 3-bit position of bit in File Register (f)

    a = 0 to force Access Bank

    a = 1 for BSR to select bank

    f = 8-bit File Register address

Literal operations

```
15                  8 7              0
|     OPCODE      |      k (literal)   | MOVLW  0x7F
```

    k = 8-bit immediate value

Control operations

CALL, GOTO, and Branch operations

```
15                  8 7              0
|      OPCODE      | n<7:0> (literal) | GOTO  label
```

```
15      12 11                    0
|  1111    |    n<19:8> (literal)       |
```

    n = 20-bit immediate value

Figure A-1. General Formatting of PIC18 Instructions (From MicroChip)

```
BCF   PORTB,5          ;clear bit D5 of PORTB
BCF   TRISB,4          ;clear bit D4 of TRISC reg
BTG   PORTC,7          ;toggle bit D7 of PORTC
BTG   PORTD,0          ;toggle bit D0 of PORTD
BSF   STATUS,C         ;set carry flag to one
```

The following example uses the fileReg in the access bank:

```
MyReg    SET   0x30    ;set aside loc 30H for MyReg
MOVLW 0x0              ;WREG = 0
MOVWF MyReg            ;MyReg = 0
BTG MYReg,7            ;toggle bit D7 of MyReg
BTG MYReg,5            ;toggle bit D5 of MyReg
```

The following example uses the fileReg in the access bank:

```
MyReg     SET    0x50 ;set aside loc. 50H for MyReg
MOVLW    0x0           :WREG = 0
MOVWF    MyReg         ;MyReg = 0
BTG      MYReg,2       ;toggle bit D2 of MyReg
BTG      MYReg,4       ;toggle bit D4 of MyReg
```

As we discuss in Chapter 6, when using a bank other than the access bank, we must load the BSR (bank select register) with the desired bank number, which can go from 1 to F (in hex), depending on the family member. We do that by using the MOVLB instruction. Look at the following examples.

The example below uses a location in Bank 2 (RAM locations 200–2FFH).

```
YReg   SET 0x30        ;set aside loc 30H for YReg
MOVLB 0x2         ;use Bank 2 (address loc 230H)
MOVLW 0x0             :WREG = 0
MOVWF YReg            ;YReg = 0
BTG   YReg,7,1   ;toggle bit D7 of YReg in bank 2
BTG   YReg,5,1   ;toggle bit D5 of YReg in bank 2
```

The example below uses a location in Bank 4 (RAM locations 400–4FFH).

```
ZReg   SET 0x10        ;set aside loc 10H for ZReg
MOVLB 0x4         ;use Bank 4 (address loc 410H)
MOVWL 0x0             ;WREG = 0
MOVWF ZReg            ;ZReg = 0
BSF ZReg,6,1 ;set HIGH bit D6 of ZReg in bank 4
BSF ZReg,1,1 ;set HIGH bit D1 of ZReg in bank 4
```

Notice that all the bit-oriented instructions start with letter B (bit). The branch instructions also start with letter B, like "BZ target" for branch if zero, but they are not bit-oriented.

Table A-2: Literal Instructions (from Microchip datasheet)

Mnemonic, Operands		Description	Cycles
LITERAL OPERATIONS			
ADDLW	k	Add literal and WREG	1
ANDLW	k	AND literal with WREG	1
IORLW	k	Inclusive OR literal with WREG	1
LFSR	f, k	Move literal (12-bit) 2nd word	2
		to FSRx 1st word	
MOVLB	k	Move literal to BSR<3:0>	1
MOVLW	k	Move literal to WREG	1
MULLW	k	Multiply literal with WREG	1
RETLW	k	Return with literal in WREG	2
SUBLW	k	Subtract WREG from literal	1
XORLW	k	Exclusive OR literal with WREG	1

Instructions using literal values

In this type of instruction, an operation is performed on the WREG register and a fixed value called k. See Table A-2. Because WREG is only 8-bit, the k value cannot be greater than 8-bit. Therefore, the k value is between 0–255 (00–FF in hex). After the operation, the result is placed back in WREG. Look at the following examples for clarification:

```
MOVLW     0x45 ;WREG  =  45H
ADDLW     0x24 ;WREG  =  45H  +  24H  =  69H

MOVLW     0x35 ;WREG  =  35H
ANDLW     0x0F ;WREG  =  35H ANDed with 0FH  =  05H

MOVLW     0x55 ;WREG  =  55H
XORLW     0xAA ;WREG  =  55H EX-ORed with AAH  =  FFH
```

Byte-oriented instructions

There are two groups of instructions in this category. In the first group, the operation is performed on the file register and the result is placed back in the file register. The instruction "CLRF f,a" is an example in this group. See Table A-3. In the second group, the operation involves both fileReg and WREG. As a result, we have the options of placing the result in fileReg or in WREG. As an example in this group, examine the "ADDWF f,d,a" instruction. The destination for the result can be WREG (if d = 0) or file register (if d = 1). For the fileReg location, it can be in the access bank (if a = 0) or in other bank registers (if a = 1). Also notice that if a = 0, the assembler assumes that automatically.

Table A-3: Byte-Oriented Instructions (from Microchip datasheet)

Mnemonic, Operands	Description	Cycles
BYTE-ORIENTED FILE REGISTER OPERATIONS		
ADDWF f, d, a	Add WREG and f	1
ADDWFC f, d, a	Add WREG and Carry bit to f	1
ANDWF f, d, a	Add WREG with f	1
CLRF f, a,	Clear f	1
COMF f, d, a	Complement f	1
CPFSEQ f, a,	Compare f with WREG, skip =	1
CPFSGT f, a,	Compare f with WREG, skip >	1
CPFSLT f, a,	Compare f with WREG, skip <	1
DECF f, d, a	Decrement f	1
DECFSZ f, d, a	Decrement f, Skip if 0	1
DCFSNZ f, d, a	Decrement f, Skip if Not 0	1
INCF f, d, a	Increment f	1
INCFSZ f, d, a	Increment f, Skip if 0	1
INFSNZ f, d, a	Increment f, Skip if Not 0	1
IORWF f, d, a	Inclusive OR WREG with f	1
MOVF f, d, a	Move f	1
MOVFF f_s, f_d	Move f_s(source) to 1st word f_d(destination) 2nd word	2
MOVWF f, a	Move WREG to f	1
MULWF f, a	Multiply WREG with f	1
NEGF f, a	Negate f	1
RLCF f, d, a	Rotate Left f through Carry	1
RLNCF f, d, a	Rotate Left f (No Carry)	1
RRCF f, d, a	Rotate Right f through Carry	1
RRNCF f, d, a	Rotate Right f (No Carry)	1
SETF f, a,	Set f	1
SUBFWB f, d, a	Subtract f from WREG with borrow	1
SUBWF f, d, a	Subtract WREG from f	1
SUBWFB f, d, a	Subtract WREG from f with borrow	1
SWAPF f, d, a	Swap nibbles in f	1
TSTFSZ f, a	Test f, Skip if 0	1
XORWF f, d, a	Exclusive OR WREG with f	1

Look at the following examples.

When d = 0 and a = 0:

```
MyReg   SET   0x20            ;loc 20H for MyReg
MOVLW 0x45                    ;WREG = 45H
MOVWF MyReg                   ;MyReg = 45H
MOVLW 0x23                    ;WREG = 23H
ADDWF MyReg            ;WREG = 68H (45H + 23H = 68H)
```

In the above example, the last instruction could have been coded as "ADDWF MyReg,0,0".

When d = 1 and a = 0:

```
MyReg SET   0x20             ;loc 20H for MyReg
MOVLW 0x45                    ;WREG = 45H
MOVWF MyReg                   ;MyReg = 45H
MOVLW 0x23                    ;WREG = 23H
ADDWF MyReg,F          ;MyReg = 68H (45H + 23H = 68H)
```

In the above example, the last instruction could have been coded as "ADDWF MyReg,F,0" or "ADDWF MyReg,1,0". As far as the MPLAB is concerned, they mean the same thing. Notice that the use of letter F in "ADDWF MyReg,F" is being used in place of 1.

To use banks other than the access bank, we must load the BSR register first. The following example uses a location in Bank 2 (RAM location 200–2FFH).

When d = 0 and a = 1:

```
MyReg SET 0x30 ;set aside location 30H for MyReg
MOVLB0x2            ;use Bank 2 (address loc 230H)
MOVLW 0x45                   ;WREG = 45H
MOVWF MyReg,1                ;MyReg = 45H (loc 230H)
MOVLW 0x23                   ;WREG = 23H
ADDWF MyReg,1    ;WREG = 68H (add loc 230H to W)
```

When d = 1 and a = 1:

```
MyReg   SET   0x20            ;loc 20H for MyReg
MOVLB 0x4                    ;use bank 4
MOVLW 0x45                   ;WREG = 45H
MOVWF MyReg                  ;MyReg = 45H (loc 420H)
MOVLW 0x23                   ;WREG = 23H
ADDWF MyReg,F,1             ;MyReg = 68H (loc 420)
```

Register-indirect addressing mode uses FSRx as a pointer to RAM location. We have three registers, FSR0, FSR1, and FSR2, that can be used for pointers.

Examples:

```
ADDWF POSTINC0 ;add to W data pointed to by FSR0,
               ;also increment FSR0

ADDWF POSTINC1 ;add to W data pointed to by FSR1
               ;also increment FSR1
```

See Example 6-6 in Chapter 6.

Table processing instructions

The table processing instructions allow us to read fixed data located in the program ROM of the PIC18. See Table A-4. They also allow us to write into the program ROM if it is Flash memory. Chapter 14 discusses the TBLRD and TBLWRT instructions in detail. It also shows how to use table read and write to access the EEPROM.

Table A-4: Table Processing Instructions (from Microchip datasheet)

Mnemonic, Operands	Description	Cycles
DATA ◄──► PROGRAM MEMORY OPERATIONS		
TBLRD*	Table Read	2
TBLRD*+	Table Read with post-increment	2
TBLRD*-	Table Read with post-decrement	2
TBLRD+*	Table Read with pre-increment	2
TBLWT*	Table Write	2
TBLWT*+	Table Write with post-increment	2
TBLWT*-	Table Write with post-decrement	2
TBLWT+*	Table Write with pre-increment	2

Control instructions

The control instructions such as branch and call deal mainly with flow control. See Table A-5. We must pay special attention to the target address of the control instructions. The target address for some of the branch instructions such as BZ (branch if zero) cannot be farther than 128 bytes away from the current instruction. The CALL instruction allows us to call a subroutine located anywhere in the 2M ROM space of the PIC18. See the individual instructions in the next section for further discussion on this issue.

Table A-5: Control Instructions (from Microchip datasheet)

Mnemonic, Operands		Description		Cycles
CONTROL OPERATIONS				
BC	n	Branch if Carry		1
BN	n	Branch if Negative		1
BNC	n	Branch if Not Carry		1
BNN	n	Branch if Not Negative		1
BNOV	n	Branch if Not Overflow		1
BNZ	n	Branch if Not Zero		1
BOV	n	Branch if Overflow		1
BRA	n	Branch Unconditionally		2
BZ	n	Branch if Zero		1
CALL	n, s	Call subroutine	1st word 2nd word	2
CLRWDT	—	Clear Watchdog Timer		1
DAW	—	Decimal Adjust WREG		1
GOTO	n	Go to address	1st word 2nd word	2
NOP	—	No Operation		1
NOP	—	No Operation		1
POP	—	Pop top of return stack (TOS)		1
PUSH	—	Push top of return stack (TOS)		1
RCALL	n	Relative Call		2
RESET		Software device RESET		1
RETFIE	s	Return from interrupt enable		2
RETLW	k	Return with literal in WREG		2
RETURN	s	Return from Subroutine		2
SLEEP	—	Go into standby mode		1

SECTION A.2: THE PIC18 INSTRUCTION SET

In this section we provide a brief description of each instruction with some examples.

ADDLW K	**Add Literal to WREG**

    Function: ADD literal value of k to WREG
    Syntax: ADDLW   k

This adds the literal value of k to the WREG register, and places the result back into WREG. Because register WREG is one byte in size, the operand k must also be one byte.

The ADD instruction is used for both signed and unsigned numbers. Each one is discussed separately. See Chapter 5 for discussion of signed numbers.

Unsigned addition

In the addition of unsigned numbers, the status of C, DC, Z, N, and OV may change. The most important of these flags is C. It becomes 1 when there is a carry from D7 out in 8-bit (D0–D7) operations.

Example:
```
        MOVLW 0x45          ;WREG = 45H
        ADDLW 0x4F          ;WREG = 94H (45H + 4FH = 94H)
                            ;C = 0
```

Example:
```
        MOVLW  0xFE         ;WREG = FEH
        ADDLW  0x75         ;WREG = FE + 75 = 73H
                            ;C = 1
```

Example:
```
        MOVLW  0x25         ;WREG = 25H
        ADDLW  0x42         ;WREG = 67H (25H + 42H = 67H)
                            ;C = 0
```

Notice that in all the above examples we ignored the status of the OV flag. Although ADD instructions do affect OV, it is in the context of signed numbers that the OV flag has any significance. This is discussed next.

Signed addition and negative numbers

In the addition of signed numbers, special attention should be given to the overflow flag (OV) because this indicates if there is an error in the result of the addition. There are two rules for setting OV in signed number operation. The overflow flag is set to 1:

1. If there is a carry from D6 to D7 and no carry from D7 out.
2. If there is a carry from D7 out and no carry from D6 to D7.
   Notice that if there is a carry both from D7 out and from D6 to D7, OV = 0.

Example:

```
MOVLW +D'8'      ;W = 0000 1000
ADDLW +D'4'      ;W = 0000 1100 OV = 0,
                 ;C = 0, N = 0
```

Notice that N = D7 = 0 because the result is positive, and OV = 0 because there is neither a carry from D6 to D7 nor any carry beyond D7. Because OV = 0, the result is correct [(+8) + (+4) = (+12)].

Example:

```
MOVLW +D'66'     ;W = 0100 0010
ADDLW +D'69'     ;W = 1000 0101 = -121
ADDWF            ;W = 1000 0111 = -121
      ;(INCORRECT) C = 0, N = D7 = 1, OV = 1
```

In the above example, the correct result is +135 [(+66) + (+69) = (+135)], but the result was –121. OV = 1 is an indication of this error. Notice that N = 1 because the result is negative; OV = 1 because there is a carry from D6 to D7 and C = 0.

Example:

```
MOVLW -D'12'     ;W = 1111 0100
ADDLW +D'18'     ;W = W + (+0001 0010)
                 ;W = 0000 0110 (+6) correct
                 ;N = 0, OV = 0, and C = 1
```

Notice above that the result is correct (OV = 0), because there is a carry from D6 to D7 and a carry from D7 out.

Example:

```
MOVLW -D'30'     ;W = 1110 0010
ADDLW +D'14'     ;W = W + 0000 1110
                 ;W = 1111 0000 (-16, CORRECT)
                 ;N = D7 = 1, OV = 0, C = 0
```

OV = 0 because there is no carry from D7 out nor any carry from D6 to D7.

Example:

```
MOVLW -D'126'    ;W = 1000 0010
ADDLW -D'127'    ;W = W + 1000 0001
                 ;W = 0000 0011 (+3, INCORRECT)
                 ;D7 = N = 0, OV = 1
```

C = 1 because there is a carry from D7 out but no carry from D6 to D7.

From the above discussion we conclude that while Carry is important in any addition, OV is extremely important in signed number addition because it is used to indicate whether or not the result is valid. As we will see in instruction "DAW", the DC flag is used in the addition of BCD numbers.

ADDWF   Add WREG and f

     Function: ADD WREG and fileReg
     Syntax: ADDWF f,d,a

This adds the fileReg value to the WREG register, and places the result in WREG (if d = 0) or fileReg (if d = 1).

The ADDWF instruction is used for both signed and unsigned numbers. (See ADDLW instruction.)

Example:

```
MyReg  SET  0x20    ;loc 20H for MyReg
MOVLW 0x45          ;WREG = 45H
MOVWF MyReg         ;MyReg = 45H
MOVLW 0x4F          ;WREG = 4FH
ADDWF MyReg     ;WREG = 94H (45H + 4FH = 94H)
                    ;C = 0
```

We can place the result in fileReg, as shown in the following example:

```
MyReg  SET  0x20    ;loc 20H for MyReg
MOVLW 0x45          ;WREG = 45H
MOVWF MyReg         ;MyReg = 45H
MOVLW 0x4F          ;WREG = 4FH
ADDWF MyReg,F       ;MyReg = 94H
                ;(45H + 4FH = 94H), C = 0
```

For cases of a = 0 and a = 1, see Section A.1 in this chapter.

ADDWFC   Add WREG and Carry flag to fileReg

     Function: ADD WREG and Carry bit to fileReg
     Syntax: ADDWFC f,d,a

This will add WREG and the C flag to fileReg (Destination = WREG + fileReg + C). If C = 1 prior to this instruction, 1 is also added to destination. If C = 0 prior to the instruction, source is added to destination plus 0. This instruction is used in multibyte additions. In the addition of 25F2H to 3189H, for example, we use the ADDWFC instruction as shown below.

Example when d = 0:

     Assume we have the following data in RAM locations 0x10 and 0x11

         0x10 = (F2)
         0x11 = (25)

```
Reg_L    SET  0x10 ;loc 0x10 for Reg_L
Reg_H    SET  0x11 ;loc 0x11 for Reg_H
BCF  STATUS,C      ;make carry = 0
MOVLW 89H          ;WREG = 89H
ADDWFC Reg_L,1 ;Reg_L = 89H + F2H + 0 = 7BH
```

```
                              ;and C = 1
         MOVLW 0x31           ;WREG = 31H
         ADDWFC Reg_2,1       ;Reg_H = 31H + 25H + 1 = 57H
```

Therefore the result is:

```
                    25F2H
                   +3189H
                    577BH
```

ANDLW   AND Literal byte with WREG

Function: Logical AND literal value k with WREG
Syntax: ANDLW   k

This performs a logical AND on the WREG and
the Literal byte operand, bit by bit, storing the result in
the WREG.

A	B	A AND B
0	0	0
0	1	0
1	0	0
1	1	1

Example:

```
         MOVLW    0x39    ;W = 39H
         ANDLW    0x09    ;W = 39H ANDed with 09

         39H    0011 1001
         09H    0000 1001
         09H    0000 1001
```

Example:

```
         MOVLW 32H    ;W = 32H       32H   0011 0010
         ANDLW 50H    ;AND W with    50H   0101 0000
                      ;(W = 10H)     10H   0001 0000
```

ANDWF   AND WREG with fileReg

Function: Logical AND for byte variables
Syntax: ANDWF f,d,a

This performs a logical AND on the fileReg value and the WREG register,
bit by bit, and places the result in WREG (if d = 0) or fileReg (if d = 1).

Example:

```
         MyReg    SET 0x40;set MyReg loc at 0x40
         MOVLW    0x39    ;W = 39H
         MOVWF    MyReg   ;MyReg = 39H
         MOVLW    0x09
         ANDWF    MyReg   ;39H ANDed with 09 (W = 09)

         39H    0011 1001
         09H    0000 1001
         09H    0000 1001
```

APPENDIX A: PIC18 INSTRUCTIONS: FORMAT AND DESCRIPTION   685

Example:

```
MyReg  SET 0x40;set MyReg loc at 0x40
MOVLW 0x32     ;W = 32H
MOVWF MyReg    ;MyReg = 32H
MOVLW 0x0F     ;WREG = 0FH
ANDLW MyReg    ;32H ANDed with 0FH (W = 02)
```

```
32H   0011 0010
0FH   0000 1111
02H   0000 0010
```

We can place the result in fileReg as shown in the examples below:

```
MyReg  SET 0x40;set MyReg loc at 0x40
MOVLW 0x32     ;W = 32H
MOVWF MyReg    ;MyReg = 32H
MOVLW 0x50     ;WREG = 50H
ANDLW MyReg,F  ;MyReg = 09, WREG = 50H
```

The instructions below clear (mask) certain bits of the output ports, assuming the ports are configured as output ports:

```
MOVLW 0xFE
ANDWF PORTB,F  ;mask PORTB.0 (D0 of Port B)
MOVLW 0x7F
ANDWF PORTC,F  ;mask PORTC.7 (D7 of Port C)
MOVLW 0xF7
ANDWF PORTD,F  ;mask PORTD.3 (D3 of Port D)
```

Branch Condition

Function: Conditional Branch (jump)

In this type of Branch (jump), control is transferred to a target address if certain conditions are met. The following is list of branch instructions dealing with the flags:

BC	Branch if carry	jump if C = 1
BNC	Branch if no carry	jump if C = 0
BZ	Branch if zero	jump if Z = 1
BNZ	Branch if no zero	jump if Z = 0
BN	Branch if negative	jump if N = 1
BNN	Branch if no negative	jump if N = 0
BOV	Branch if overflow	jump if OV = 1
BNOV	Branch if no overflow	jump if OV = 0

Notice that all "Branch condition" instructions are short jumps, meaning that the target address cannot be more than –128 bytes backward or +127 bytes forward of the PC of the instruction following the jump. In other words, the target address cannot be more than –128 to +127 bytes away from the current PC. What happens if a programmer needs to use a "Branch condition" to go to a target address beyond the –128 to +127 range? The solution is to use the "Branch condition" along with the unconditional GOTO instruction, as shown below.

```
             ORG   0x100
             MOVLW 0x87     ;WREG = 87H
             ADDLW 0x95     ;C = 1 after addition
             BNC   NEXT     ;branch if C = 0
             GOTO  OVER     ;target more than 128 bytes away
NEXT:        . . .
             . . .
             . . .
             ORG 0x5000
OVER:        MOVWF PORTD
```

BC	**Branch if C = 1**

Function: Branch if Carry flag bit = 1
Syntax: BC target_address

This instruction branches if C = 1.

Example:

```
        MOLW  0x0      ;WREG = 0
BACK ADDLW 0x1      ;add 1 to WREG
        BC    EXIT     ;exit if C = 1
        BRA   BACK     ;keep doing it
EXIT .....
        .....
```

Notice that this is a 2-byte instruction; therefore, the target address cannot be more than –128 to +127 bytes away from the program counter. See Branch Condition for further discussion on this issue.

BCF	**Bit Clear fileReg**

Function: Clear bit of a fileReg
Syntax: BCF f,b,a

This instruction clears a single bit of a given file register. The bit can be the directly addressable bit of a port, register, or RAM location. Here are some examples of its format:

```
BCF   STATUS,C   ;C = 0
BCF   PORTB,5    ;CLEAR PORTB.5 (PORTB.5 = 0)
BCF   PORTC,7    ;CLEAR PORTC.7 (PORTC.7 = 0)
BCF   MyReg,1    ;CLEAR D1 OF File Register MyFile
```

BN   Branch if N = 1

Function:	Jump if Negative flag bit = 1
Syntax:	BN target_address

This instruction branches if N = 1. It is used in signed number addition. See ADDLW instruction. Notice that this is a 2-byte instruction; therefore, the target address cannot be more than –128 to +127 bytes away from the program counter. See Branch Condition for further discussion on this issue.

BNC   Branch if no Carry

Function:	Branch if Carry flag is 0
Syntax:	BNC target_address

This instruction examines the C flag, and if it is zero it will jump (branch) to the target address.

Example: Find the total sum of the bytes F6H, 98H, and 8AH. Save the carries in register C_Reg.

```
C_Reg SET 0x20 ;set aside loc 0x20 for carries

        MOVLW 0x0        ;W = 0
        MOVWF C_Reg      ;C_Reg = 0
        ADDLW 0xF6
        BNC   OVER1
        INCF C_Reg,F
OVER1:  ADDLW 0x98
        BNC   OVER2
        INCF  C_Reg,F
OVER2:  ADDWF 0x8A
        BNC   OVER3
        INCF C_Reg
OVER3:
```

Notice that this is a 2-byte instruction; therefore, the target address cannot be more than –128 to +127 bytes away from the program counter. See Branch Condition for further discussion on this.

BNN   Branch if Not Negative

Function:	Branch if Negative flag bit = 0
Syntax:	BNN target_address

This instruction branches if N = 0. It is used in signed number addition. See ADDLW instruction. Notice that this is a 2-byte instruction; therefore, the target address cannot be more than –128 to +127 bytes away from the program counter. See Branch Condition for further discussion on this issue.

BNOV   **Branch if No Overflow**

Function: Jump if overflow flag bit = 0
Syntax: BNOV target_address

This instruction branches if OV = 0. It is used in signed number addition. See ADDLW instruction. Notice that this is a 2-byte instruction; therefore, the target address cannot be more than –128 to +127 bytes away from the program counter. See Branch Condition for further discussion on this issue.

BNZ   **Branch if No Zero**

Function: Jump if Zero flag is 0
Syntax: BNZ target_address

This instruction branches if Z = 0.

Example:

```
        CLRF  TRISB     ;PORTB as output
        CLRF  PORTB     ;clear PORTB
OVER    INCF  PORTB,F   ;INC PORTB
        BNZ   OVER      ;do it until it becomes zero
```

Example: Add value 7 to WREG five times.

```
        COUNTER   SET   0x20 ;loc 20H for COUNTER
        MOVLW 0x5         ;WREG = 5
        MOVWF COUNTER     ;COUNTER = 05
        MOVLW 0x0         ;WREG = 0
OVER    ADDLW 0x7         ;add 7 to WREG
        DECF  COUNTER,F   ;decrement counter
        BNZ   OVER        ;do it until counter is zero
```

Notice that this is a 2-byte instruction; therefore, the target address cannot be more than –128 to +127 bytes away from the program counter. See Branch Condition for further discussion on this issue.

BOV   **Branch if Overflow**

Function: Jump if Overflow flag = 1
Syntax: BOV target_address

This instruction jumps if OV = 1. It is used in signed number addition. See ADDLW instruction. Notice that this is a 2-byte instruction; therefore, the target address cannot be more than –128 to +127 bytes away from the program counter. See Branch Condition for further discussion on this issue.

BRA   Branch unconditional

Function:	Branch unconditionally
Syntax:	BRA target_address

BRA stands for "Branch." It transfers program execution to the target address unconditionally. The target address for this instruction must be within 1K of program memory. This is a 2-byte instruction. The first 5 bits is the opcode and the rest is the signed number displacement, which is added to the PC (program counter) of the instruction following the BRA to get the target address. Therefore, in this branch, the target address must be within –1024 to +1023 bytes of the PC (program counter) of the instruction after the BRA because the 11-bit address can take values of +1024 to –1023. This address is often referred to as a *relative address* because the target address is –1024 to +1023 bytes relative to the program counter (PC).

BSF   Bit Set fileReg

Function:	Set bit
Syntax:	BSF f, b, a

This sets HIGH the indicated bit of a file register. The bit can be any directly addressable bit of a port, register, or RAM location.

Examples:

```
BSF    PORTB,3    ;make PORTB.3 = 1
BSF    PORTC,6    ;make PORTC.6 = 1
BSF    MyReg,2    ;make bit D2 of MyReg = 1
BSF    STATUS,C   ;set Carry Flag C = 1
```

BTFSC   Bit Test fileReg, Skip if Clear

Function:	Skip the next instruction if bit is 0
Syntax:	BTFSC f, b,a

This instruction is used to test a given bit and skip the next instruction if the bit is low. The given bit can be any of the bit-addressable bits of RAM, ports, or registers of the PIC18.

Example: Monitor the PORTB.5 bit continuously and, when it becomes low, put 55H in WREG.

```
        BSF TRISB,5     ;make PORTB.5 an input bit
HERE    BTFSC PORTB,5   ;skip if PORTB.5 = 0
        BRA    HERE
        MOVLW 0x55      ;because PORTB.5 = 0,
                        ;put 55H in WREG
```

Example: See if WREG has an even number. If so, make it odd.

```
        BTFSC WREG,0    ;skip if it is odd
        BRA   NEXT
        ADDLW 0x1       ;it is even, make it odd
NEXT:   ...
```

BTFSS   Bit Test fileReg, Skip if Set

Function: Skip the next instruction if bit is 1
Syntax: BTFSS f, b, a

This instruction is used to test a given bit and skip the next instruction if the bit is HIGH. The given bit can be any of the bit-addressable bits of RAM, ports, or registers of the PIC18.

Example: Monitor the PORTB.5 bit continuously and when it becomes HIGH, put 55H in WREG.

```
        BSF  TRISB,5    ;make PORTB.5 an input bit
HERE    BTFSS PORTB,5   ;skip if PORTB.5 = 1
        BRA   HERE
        MOVLW 55H       ;because PORTB.5 = 0 WREG = 55H
```

Example: See if WREG has an odd number. If so, make it even.

```
        BTFSS WREG,0    ;skip if it is even
        BRA   NEXT
        ADDLW 0x01      ;it is even, make it odd
NEXT:   ...
```

BTG   Bit Toggle fileReg

Function: Toggle (Complement) bit
Syntax: BTG f, b, a

This instruction complements a single bit. The bit can be any bit-addressable location in the PIC18.

Example:
```
        BCF  TRISB,0    ;make PORTB.0 an output
AGAIN   BTG  PORTB,0    ;complement PORTB.0 bit
        BRA  AGAIN      ;continuously forever
```

Example: Toggle PORTB.7 a total of 150 times.

```
COUNTER    SET  0x20    ;loc 20H for COUNTER
        MOVLW 'D'150    ;WREG = 150
        MOVWF COUNTER   ;COUNTER = 150
        BCF    TRISB,7  ;make PORTB.7 an output
```

```
     OVER BTG   PORTB.7      ;toggle PORTB.7
          DECF COUNTER,F ;decrement and put it in
                            ;COUNTER
          BNZ  OVER        ;do it 150 times
```

BZ   Branch if Zero

Function: Branch if $Z = 1$
Syntax: BZ target_address

Example: Keep checking PORTB for value 99H.
```
          SETF TRISB      ;port B as input
     BACK MOVFW PORTB     ;get PORTB into WREG
          SUBLW 0x99      ;subtract 99H from it
          BZ   EXIT       ;if 0x99, exit
          BRA  BACK       ;keep checking
          . . .
EXIT:     . . .
```

Example: Toggle PORTB 150 times.
```
MyReg        SET  0x40 ;loc 40H for MyReg
          SETF TRISB      ;port B as output
          MOVLW D'150'    ;WREG = 150
          MOVWF MyReg
     BACK COMF PORTB      ;toggle PORTB
          DECF MyReg,F    ;decrement MyReg
          BZ   EXIT       ;if MyReg = 0, exit
          BRA  BACK       ;keep toggling
          . . .
EXIT:     . . .
```

Notice that this is a 2-byte instruction; therefore, the target address cannot be more than –128 to +127 bytes away from the program counter. See Branch Condition for further discussion on this.

CALL

Function: Transfers control to a subroutine
Syntax: CALL   k,s   ;s is used for fast context switching

The Call intruction is a 4-byte instruction. The first 12 bits are used for the opcode and the rest (20 bits) are set aside for the address. A 20-bit address allows us to reach the target address anywhere in the 2M ROM space of the PIC18. If calling a subroutine, the PC register (which has the address of the instruction after the CALL) is pushed onto the stack and the stack pointer (SP) is incremented by 1. Then the program counter is loaded with the new address and control is transferred to the subroutine. At the end of the procedure, when RETURN is executed, PC is popped off the stack, which returns control to the instruction after the CALL.

Notice that CALL is a 4-byte instruction, in which 12 bits are the opcode, and the other 20 bits are the 20-bit address of an even address location. Because all the PIC18 instructions are 2 bytes in size, the lowest address bit, A0, is automatically set to zero to make sure that the CALL instruction will not land at the middle of the targeted instruction. The 20-bit address of the CALL provides the A20–A1 part of the address and with the A0 = 0, we have the 21-bit address needed to go anywhere in the 2M address space of the PIC18.

We have two options for the "CALL k,s" instruction. They are s = 0, and s = 1. When s = 0, it is simply calling a subroutine. With s = 1, we are calling a subroutine and we are also asking the CPU to save the three major registers of WREG, STATUS, and BSR in internal buffers (shadow registers) for the purpose of context-switching. This fast context-switching can be used only in the main subroutine because the depth of the shadow registers is only one. That means no nested call with the s = 1. Look at the following case:

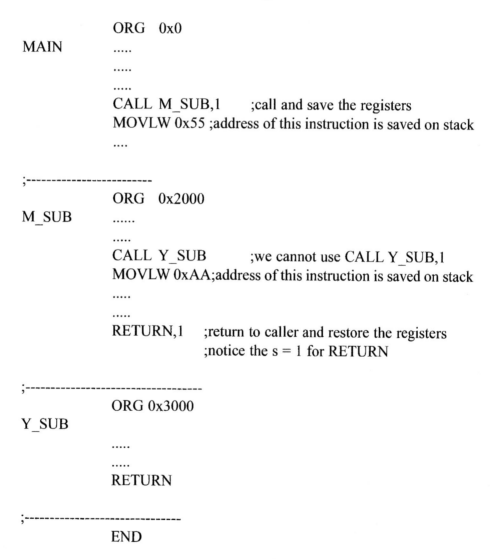

```
                ORG   0x0
MAIN            .....

                .....

                .....

                CALL M_SUB,1      ;call and save the registers
                MOVLW 0x55 ;address of this instruction is saved on stack
                ....


;------------------------
                ORG   0x2000
M_SUB           ......

                .....

                CALL Y_SUB          ;we cannot use CALL Y_SUB,1
                MOVLW 0xAA;address of this instruction is saved on stack
                .....

                .....

                RETURN,1    ;return to caller and restore the registers
                            ;notice the s = 1 for RETURN


;----------------------------------
                ORG 0x3000
Y_SUB

                .....

                .....

                RETURN


;------------------------------
                END
```

As shown in RETURN instruction, we also have two options for the RETURN: s = 0 and s = 1. If we use s = 1 for the CALL, we must also use s = 1 for the RETURN. Notice that "CALL Target" with no number after it is interpreted as s = 0 by the assembler. Likewise, the "RETURN" with no number after it is interpreted as s = 0 by the assembler.

CLRF   Clear fileReg

    Function: Clear
    Syntax: CLRF f, a

This instruction clears the entire byte in the fileReg. All bits of the register are cleared to 0.

Example:

```
MyReg     SET  0x20 ;loc 20H for MyReg
     CLRF MyReg      ;clear MyReg
     CLRF TRISB ;clear TRISB (make PORTB output)
     CLRF PORTB      ;clear PORTB
     CLRF TMR01L     ;TMR0L = 0
```

Notice that in this instruction the result can be placed in fileReg only and there is no option for the WREG to be used as the destination.

CLRWDT

    Function: Clear Watchdog Timer
    Syntax: CLRWDT

This instruction clears the Watchdog Timer.

COMF   Complement the fileReg

    Function: Complement a fileReg
    Syntax: COMF f, d, a

This complements the contents of a given fileReg. The result is the 1's complement of the register; that is, 0s become 1s and 1s become 0s. The result can be placed in WREG (if $d = 0$) or fileReg (if $d = 1$).

Example:

```
          MOVLW 0x0     ;WREG = 0
          MOVWF TRISB   ;Make PORTB an output port
          MOVLW 0x55    ;WREG = 01010101
          MOVWF PORTB
AGAIN     COMF PORTB,F  ;complement (toggle) PORTB
          CALL DELAY
          BRA  AGAIN ;continuously (notice WREG = 55H)
```

Example:

```
          MyReg  SET 0x40;set MyReg loc at 0x40
          MOVLW  0x39    ;W = 39H
          MOVWF  MyReg   ;MyReg = 39H
          COMPF  MyReg,F ;MyReg = C6H and WREG = 39H
```
Where 39H (0011 1001 bin) becomes C6H (1100 0110).

Example:

```
MyReg   SET 0x40;set MyReg loc at 0x40
MOVLW 0x55      ;W = 55H
MOVWF MyReg     ;MyReg = 55H
COMPF MyReg,F   ;MyReg AAH, WREG = 55H
```

where 55H (0101 0101) becomes AAH (1010 1010).

Example: Toggle PORTB 150 times.

```
COUNTER    SET  0x40 ;loc 40H for COUNTER
     SETF TRISB      ;port B as output
     MOVLW D'150'    ;WREG = 150
     MOVWF COUNTER   ;COUNTER = 150
     MOVLW 0x55      ;WREG = 55H
     MOVWF PORTB
BACK COMF PORTB,F    ;toggle PORTB
     DECF COUNTER,F  ;decrement COUNTER
     BNZ  BACK ;toggle until counter becomes 0
```

We can place the result in WREG as shown in the examples below:

```
MyReg   SET 0x40    ;set MyReg loc at 0x40
     MOVLW   0x39    ;W = 39H
     MOVWF   MyReg   ;MyReg = 39H
     COMPF   MyReg   ;MyReg = 39H and WREG = C6H
```

Example:
```
MyReg   SET     0x40 ;set MyReg loc at 0x40
     MOVLW 0x55      ;W = 55H
     MOVWF MyReg     ;MyReg = 55H
     COMPF MyReg     ;WREG = AA and MyReg 55H SETF
```

CPFSEQ   Compare FileReg with WREG and skip if equal (F = W)

Function: Compare fileReg and WREG and skip if they are equal
Syntax: CPFSEQ f, a

The magnitudes of the fileReg byte and WREG byte are compared. If they are equal, it skips the next instruction.

Example: Keep monitoring PORTB indefinitely for the value of 99H. Get out only when PORTB has the value 99H.

```
     SETF  TRISB    ;PORTB an input port
     MOVLW 0x99      ;WREG = 99h
BACK CPFSEQ PORTB    ;skip if PORTB has 0x99
     BRA BACK        ;keep monitoring
```

Notice that CPFSEQ skips only when fileReg and WREG have equal values.

CPFSGT   Compare FileReg with WREG and skip if greater (F > W)

Function: Compare fileReg and WREG and skip if fileReg > WREG.
Syntax: CPFSGT f, a

The magnitudes of the fileReg byte and WREG byte are compared. If fileReg is larger than the WREG, it skips the next instruction.

Example: Keep monitoring PORTB indefinitely for the value of 99H. Get out only when PORTB has a value greater than 99H.

```
        SETF  TRISB     ;PORTB an input port
        MOVLW 0x99       ;WREG = 99H
BACK    CPFSGT PORTB     ;skip if PORTB > 99H
        BRA BACK         ;keep monitoring
```

Notice that CPFSGT skips only if FileReg is greater than WREG.

CPFSLT   Compare FileReg with WREG and skip if less than (F < W)

Function: Compare fileReg and WREG and skip if fileReg < WREG.
Syntax: CPFSLT f, a

The magnitudes of the fileReg byte and WREG byte are compared. If fileReg is less than the WREG, it skips the next instruction.

Example: Keep monitoring PORTB indefinitely for the value of 99H. Get out only when PORTB has a value less than 99H.

```
        SETF  TRISB     ;PORTB an input port
        MOVLW 0x99       ;WREG = 99H
BACK:   CPFSEQ PORTB     ;skip if PORTB < 99H
        BRA BACK         ;keep monitoring
```

Notice that CPFSLT skips only if FileReg < WREG.

DAW

Function: Decimal-adjust WREG after addition
Syntax: DAW

This instruction is used after addition of BCD numbers to convert the result back to BCD. The data is adjusted in the following two possible cases:

1. It adds 6 to the lower 4 bits of WREG if it is greater than 9 or if DC = 1.
2. It also adds 6 to the upper 4 bits of WREG if it is greater than 9 or if C = 1.

Example:

```
MOVLW 0x47      ;WREG = 0100 0111
ADDLW 0x38      ;WREG = 47H + 38H = 7FH,
                ;invalid BCD
DAW             ;WREG = 1000 0101 = 85H, valid BCD
```

```
   47H
 + 38H
   7FH   (invalid BCD)
 +  6H   (after DAW)
   85H   (valid BCD)
```

In the above example, because the lower nibble was greater than 9, DAW added 6 to WREG. If the lower nibble is less than 9 but DC = 1, it also adds 6 to the lower nibble. See the following example:

```
MOVLW 0x29      ;WREG = 0010 1001
ADDLW 0x18      ;WREG = 0100 0001  INCORRECT
DAW             ;WREG = 0100 0111 = 47H VALID BCD
```

```
   29H
 + 18H
   41H   (incorrect result in BCD)
 +  6H
   47H   correct result in BCD
```

The same thing can happen for the upper nibble. See the following example:

```
MOVLW 0x52      ;WREG = 0101 0010
ADDLW 0x91      ;WREG = 1110 0011  INVALID BCD
DAW             ;WREG = 0100 0011 AND C = 1
```

```
   52H
 + 91H
   E3H          (invalid BCD)
 + 6            (after DAW, adding to upper nibble)
  143H          valid BCD
```

Similarly, if the upper nibble is less than 9 and C = 1, it must be corrected. See the following example:

```
MOVLW 0x94          ;W = 1001 0100
ADDLW 0x91          ;W = 0010 0101 INCORRECT
DAW                 ;W = 1000 0101, VALID BCD
                    ;FOR 85, C = 1
```

```
   94H
+  91H
 1 25H          (incorrect BCD)
+  6            (after DAW, adding to upper nibble)
 1 85
```

It is possible that 6 is added to both the high and low nibbles. See the following example:

```
        MOVLW   0x54     ;WREG = 0101 0100
        ADDLW   0x87     ;WREG = 1101 1011 INVALID BCD
        DAW              ;WREG = 0100 0001, C = 1 (BCD 141)


        54H
+       87H
        DBH    (invalid result in BCD)
+       66H
 1  4  1H          valid BCD
```

DECF   Decrement fileReg

Function: Decrement fileReg
Syntax: DECF f, d, a

This instruction subtracts 1 from the byte operand in fileReg. The result can be placed in WREG (if d = 0) or fileReg (if d = 1).

Example:

```
        MyReg SET 0x40 ;set aside loc 40H for MyReg
        MOVLW 0x99       ;WREG = 99H
        MOVWF MyReg      ;MyReg = 99H
        DECF MyReg,F     ;MyReg = 98H, WREG 99H
        DECF MyReg,F     ;MyReg = 97H, WREG 99H
        DECF MyReg,F     ;MyReg = 96H, WREG 99H
```

Example: Toggle PORTB 250 times.

```
COUNTER    SET  0x40 ;loc 40H for COUNTER
     SETF TRISB       ;PORTB as output
     MOVLW D'250'     ;WREG = 250
     MOVWF COUNTER    ;COUNTER = 250
     MOVLW 0x55       ;WREG = 55H
     MOVWF PORTB
BACK COMF PORTB,F     ;toggle PORTB
     DECF COUNTER,F ;decrement COUNTER
     BNZ  BACK ;toggle until counter becomes 0
```

We can place the result in WREG as shown in the examples below:

```
MyReg SET 0x40 ;set aside loc for MyReg
MOVLW 0x99      ;WREG = 99H
MOVWF MyReg     ;MyReg = 99H
DECF MyReg      ;WREG = 98H, MyReg = 99H
DECF MyReg      ;WREG = 97H, MyReg = 99H
DECF MyReg      ;WREG = 96H, MyReg = 99H
```

Example:
```
MyReg   SET     0x50 ;set MyReg loc at 0x50
MOVLW   0x39    ;W = 39H
MOVWF   MyReg   ;MyReg = 39H
DECF    MyReg   ;WREG = 38H and MyReg = 39H
DECF    MyReg   ;WREG = 37H and MyReg = 39H
DECF    MyReg   ;WREG = 36H and MyReg = 39H
DECF    MyReg   ;WREG = 35H and MyReg = 39H
```

DECFSZ   Decrement fileReg and Skip if zero

Function: Decrement fileReg and skip if fileReg has zero in it
Syntax: DECFSZ f, d, a

This instruction subtracts 1 from the byte operand of fileReg. If the result is zero, then it skips execution of the next instruction.

Example: Toggle PORTB 250 times.

```
COUNT       SET 0x40 ;loc 40H for COUNT
    CLRF TRISB       ;PORTB an output
    MOVLW D'250'     ;WREG = 250
    MOVWF COUNT      ;COUNT = 250
    MOVLW 0x55       ;WREG = 55H
    MOVWF PORTB
BACK COMF PORTB,F    ;toggle PORTB
    DECFSZ COUNT,F   ;decrement COUNT and
                     ;skip if zero
    BRA  BACK ;toggle until counter becomes 0
    . . . .
```

DECFSNZ   Decrement fileReg and skip if not zero

Function: Decrement fileReg and skip if fileReg has other than zero
Syntax: DECFSNZ f, d, a

This instruction subtracts 1 from the byte operand of fileReg. If the result is not zero, then it skips execution of the next instruction.

Example: Toggle PORTB 250 times continuously.

```
COUNT      SET  0x40  ;loc 40H for COUNT
      CLRF TRISB       ;PORTB an output
OVER MOVLW D'250'      ;WREG = 250
      MOVWF COUNT      ;COUNT = 250
      MOVLW 0x55       ;WREG = 55H
      MOVWF PORTB
BACK COMF PORTB,F      ;toggle PORTB
      DECFSNZ COUNT,F  ;decrement COUNT and
                       ;skip if zero
      BRA  OVER        ;start over
      BRA  BACK ;toggle until counter becomes 0
```

GOTO   Unconditional Branch

Function: Transfers control unconditionally to a new address.
Syntax: GOTO   k

In the PIC18 there are two unconditional branches (jumps): GOTO (long jump) and BRA (short jump). Each is described next.

1. GOTO (long jump): This is a 4-byte instruction. The first 12 bits are the opcode, and the next 20 bits are an even address of the target location. Because all the PIC18 instructions are 2 bytes in size, the lowest address bit, A0, is automatically set to zero to make sure that the GOTO instruction will not land at the middle of the targeted instruction. The 20-bit address of the GOTO provides the A20–A1 part of the address and with A0 = 0, we have the 21-bit address needed to go anywhere in the 2M address space of the PIC18.

2. BRA: This is a 2-byte instruction. The first 5 bits are the opcode and the remaining 11 bits are the signed number displacement, which is added to the PC (program counter) of the instruction following the BRA to get the target address. Therefore, for the BRA instruction the target address must be within –1023 to +1024 bytes of the PC of the instruction after the BRA because an 11-bit address can take values of +1023 to –1024.

While GOTO is used to jump to any address location within the 2M code space of the PIC18, BRA is used to jump to a location within the 1K ROM space. The advantage of BRA is the fact that it takes 2 bytes of program ROM, while GOTO takes 4 bytes. BRA is widely used in chips with a small amount of program ROM and a limited number of pins.

Notice that the difference between GOTO and CALL is that the CALL instruction will return and continue execution with the instruction following the CALL, whereas GOTO will not return.

INCF   Increment fileReg

Function: Increment

Syntax: INCF f, d, a

This instruction adds 1 to the byte operand in fileReg. The result can be placed in WREG (if d = 0) or fileReg (if d = 1).

Example:

```
MyReg SET 0x40 ;set aside loc 40H for MyReg
MOVLW 0x99     ;WREG = 99H
MOVWF MyReg
INCF MyReg,F   ;MyReg = 9AH, WREG 99H
INCF MyReg,F   ;MyReg = 9BH, WREG 99H
DECF MyReg,F   ;MyReg = 9CH, WREG 99H
```

Example: Toggle PORTB 5 times.

```
COUNTER    SET  0x40 ;loc 40H for COUNTER
     SETF TRISB    ;PORTB as output
     MOVLW D'251'  ;WREG = 251
     MOVWF COUNTER ;COUNTER = 251
     MOVLW 0x55    ;WREG = 55H
     MOVWF PORTB
BACK COMF PORTB,F   ;toggle PORTB
     INCF COUNTER,F ;INC COUNTER
     BNC  BACK ;toggle until counter becomes 0
```

We can place the result in fileReg as shown in the examples below:

```
MyReg SET 0x40 ;set aside loc for MyReg
MOVLW 0x99     ;WREG = 99H
MOVWF MyReg    ;MyReg = 99H
INCF MyReg     ;WREG = 9AH, MyReg = 99H
INCF MyReg     ;WREG = 9BH, MyReg = 99H
```

Example:

```
MyReg    SET    0x40 ;set MyReg loc at 0x40
MOVLW    0x5    ;W = 05H
MOVWF    MyReg  ;MyReg = 05H
INCF     MyReg  ;WREG = 06H and MyReg = 05H
```

INCFSZ   Increment fileReg and skip if zero

Function: Increment

Syntax: INCFSZ f, d, a

This instruction adds 1 to fileReg and if the result is zero it skips the next instruction.

Example: Toggle PORTB 156 times.

```
COUNTER     SET  0x40  ;loc 40H for COUNTER
      SETF TRISB       ;PORTB as output
      MOVLW D'156'     ;WREG = 156
      MOVWF COUNTER    ;COUNTER = 156
      MOVLW 0x55       ;WREG = 55H
      MOVWF PORTB
BACK COMF PORTB,F      ;toggle PORTB
      INCFSZ COUNTER,F ;INC COUNTER and skip if 0
      BRA  BACK ;toggle until counter becomes 0
      . . . . .
```

INCFSNZ   Increment fileReg and skip if not zero

Function: Increment
Syntax: INFSNZ f, d, a

This instruction adds 1 to the register or memory location specified by the operand. If the result is not zero, it skips the next instruction.

Example: Toggle PORTB 156 times continuously.

```
COUNTER     SET  0x40  ;loc 40H for COUNTER
      SETF TRISB       ;PORTB as output
OVER MOVLW D'156'      ;WREG = 156
      MOVWF COUNTER    ;COUNTER = 156
      MOVLW 0x55       ;WREG = 55H
      MOVWF PORTB
BACK COMF PORTB,F      ;toggle PORTB
      INCFSNZ COUNTER,F;INC COUNTER, skip if not 0
      BRA  OVER ;start over
      BRA  BACK ;toggle until counter becomes 0
```

IORLW   OR K value with WREG

Function: Logical-OR WREG with value k
Syntax: IORLW   k

This performs a logical OR on the WREG register and k value, bit by bit, and stores the result in WREG.

Example:

```
      MOVLW 0x30   ;W = 30H
      IORLW 0x09   ;now W = 39H
```

A	B	A OR B
0	0	0
0	1	1
1	0	1
1	1	1

```
39H     0011 0000
09H     0000 1001
39      0011 1001
```

Example:
```
MOVLW  0x32       ;W = 32H
IORLW  0x50       ;(W = 72H)
```

```
32H     0011 0010
50H     0101 0000
72H     0111 0010
```

IORWF   **OR FileReg with WREG**

Function: Logical-OR fileReg and WREG
Syntax: IORWF f, d, a

This performs a logical OR on the fileReg value and the WREG register,
bit by bit, and places the result in WREG (if d = 0) or fileReg (if d = 1).

Example:
```
MyReg   SET 0x40;set MyReg loc at 0x40
MOVLW   0x39      ;WREG = 39H
MOVWF   MyReg     ;MyReg = 39H
MOVLW   0x07
IORWF   MyReg     ;39H ORed with 07 (W = 3F)
```

```
39      0011 1001
07      0000 0111
3F      0011 1111
```

Example:
```
MyReg   SET 0x40;set MyReg loc at 0x40
MOVLW   0x5       ;WREG = 05H
MOVWF   MyReg     ;MyReg = 05H
MOVLW   0x30
IORWF   MyReg     ;30H ORed with 05 (W = 35H)
```

```
05      0000 0101
30      0011 0000
35      0011 0101
```

We can place the result in fileReg as shown in the examples below:

```
MOVLW 0x30         ;W = 30H
IORWF PORTB,F      ;W and PORTB are ORed and result
                   ;goes to PORTB
```

Example:

```
MyReg       SET    0x20
MOVLW 0x54        ;WREG = 54H
MOVWF MyReg
MOVLW 0x67        ;WREG = 67H
IORWF MyReg,F    ;OR WREG and MyReg
;after the operation MyReg = 77H
```

```
44H   0101 0100
67H   0110 0111
77H   0111 0111  Therefore MyReg will have 77H, WREG = 54H.
```

LFSR   Load FSR

Function: Load into FSR registers a 12-bit value of k
Syntax: LFSR f,k   ;k is between 000 and FFFH

This loads a 12-bit value into one of the FSR registers of FSR0, FSR1, or FSR2.

```
LFSR 0 , 0x200 ;FSR0 = 200H
LFSR 1 , 0x050 ;FSR1 = 050H
LFSR 2 , 0x160 ;FSR2 = 160H
```

This is widely used in register indirect addressing mode. See Chapter 6.

MOVF (or MOVFW)   Move fileReg to WREG

Function: Copy byte from fileReg to WREG
Syntax   MOVF f, d, a:

This instruction is widely used for moving data from a fileReg to WREG. Look at the following examples:

```
CLRF     TRISC      ;PORTC output
SETF     TRISB      ;PORTB as input
MOVFW    PORTB      ;copy PORTB to WREG
ANDLW    0x0F       ;mask the upper 4 bits
MOVWF    PORTC      ;put it in PORTC
```

Example:
```
CLRF     TRISD      ;PORTD as output
SETF     TRISB      ;PORTB as input
MOVFW    PORTB      ;copy PORTB to WREG
IORW     0x30       ;OR it with 30H
MOVWF    PORTD      ;put it in PORTD
```

This instruction can be used to copy the fileReg to itself in order to get the status of the N and Z flags. Look at the following example.

Example:
```
MyReg SET 0x20  ;set aside loc 0x20 to MyReg
MOVLW 0x54      ;W = 54H
MOVWF MyReg     ;MyReg = 54H
MOVFW MyReg,F   ;My Reg = 54, also N = 0 and Z = 0
```

MOVFF   Move FileReg to Filereg

Function: Copy byte from one fileReg to another fileReg
Syntax: MOVFF fs, fd

This copies a byte from the source location to the destination. The source and destination locations can be any of the file register locations, SFRs, or ports.
```
MOVFF    PORTB,MyReg
MOVFF    PORTC,PORTD
MOVFF    RCREG,PORTC
MOVFF    Reg1,REG2
```
Notice that this a 4-byte instruction because the source and destination address each take 12 bits of the instruction. That means the 24 bits of the instruction are used for the source and destination addresses. The 12-bit address allows data to be moved from any source location to any destination location within the 4K RAM space of the PIC18.

MOVLB   Move Literal 4-bit value to lower 4-bit of the BSR

Function: Move 4-bit value k to lower 4 bits of the BSR registers
Syntax: MOVLB   k   ;k is between 0 and 15 (0–F in hex)

We use this instruction to select a register bank other than the access bank. With this instruction we can load into the BSR (bank selector register) a 4-bit value representing one of 16 banks supported by the PIC18. That means the values between 0000 and 1111 (0–F in hex). For examples of the MOVLB instruction, see Chapter 6 and Section A.1 in this chapter.

MOVLW   K   Move Literal to WREG

Function: Move 8-bit value k to WREG
Syntax: MOVLW   k   ;k is between 0 and 255 (0–FF in hex)

Example:
```
MOVLW    0x55    ;WREG = 55H
MOVLW    0x0     ;clear WREG (WREG = 0)
MOVLW    0xC2    ;WREG = C2H
MOVLW    0x7F    ;WREG = 7FH
```

This instruction, along with the MOVWF, is widely used to load fixed values into any port, SFR, or fileReg location. See the next instruction to see how it is used.

MOVWF   Move WREG to a fileReg

Function:	Copy the WREG contents to a fileReg
Syntax:	MOVWF f, a

This copies a byte from WREG to fileReg. This instruction is widely used along with the MOVLW instruction to load any of the fileReg locations, SFRs, or PORTs with a fixed value. See the following examples:

Example: Toggle PORTB.
```
MOVLW      0x55        ;WREG = 55H
MOVWF      PORTB
MOVLW      0xAA        ;WREG = AAH
MOVWF      PORTB
BRA        OVER        ;keep toggling the PORTB
```

Example: Load RAM location 20H with value 50H.
```
MyReg SET 0x20 ;set aside the loc 0x20 for MyReg
MOVLW      0x50
MOVWF      MyReg       ;MyReg = 50H (loc 20H has 50H)
```

Example: Initialize the Timer0 low and high registers.
```
MOVLW      0x05        ;WREG = 05H
MOVWF      TMR0H       ;TMR0H = 0x5
MOVLW      0x30        ;WREG = 30H
MOVWF      TMR0L       ;TMR0L = 0x30
```

MULLW   Multiply Literal with WREG

Function:	Multiply k × WREG
Syntax:	MULLW k

This multiplies an unsigned byte k by an unsigned byte in register WREG and the 16-bit result is placed in registers PRODH and PRODL, where PRODL has the lower byte and PRODH has the higher byte.

Example:
```
    MOVLW 0x5         ;WREG = 5H
    MULLW 0x07        ;PRODL = 35 = 23H, PRODH = 00
```

Example:
```
    MOVLW 0x0A        ;WREG = 10
    MULLW 0x0F        ;PRODL = 10 x 15 = 150 = 96H
                      ;PRODH = 00
```

Example:
```
    MOVLW 0x25
    MULLW 0x78   ;PRODL = 58H, PRODH = 11H
    ;because 25H x 78H = 1158H
```

Example:
```
        MOVLW  D'100'   ;WREG = 100
        MULLW  D'200'   ;PRODL = 20H, PRODH = 4EH
                        ;(100 x 200 = 20,000 = 4E20H)
```

MULWF	Multiply WREG with fileReg

Function:	Multiply WREG × fileReg and place the result in PRODH:PROFDL registers
Syntax:	MULWF f, a

This multiplies an unsigned byte in WREG by an unsigned byte in the fileReg register and the result is placed in PRODL and PRODH, where PRODL has the lower byte and PRODH has the higher byte.

Example:
```
    MyReg   SET  0x20 ;MyReg has location of 0x20
        MOVLW 0x5
        MOVWF MyReg    ;MyReg has 0x5
        MOVLW 0x7      ;WREG = 0x7
        MULWF MyReg    ;PRODL = 35 = 23H, PRODH = 00
```
Example:
```
        MOVLW 0x0A
        MOVWF MyReg    ;MyReg = 10
        MOVLW 0x0F     ;WREG = 15
        MULFW MyReg    ;PRODL = 150 = 96H, PRODH = 00
```

Example:
```
        MOVLW 0x25
        MOVWF MyReg    ;MyReg = 0x25
        MOVLW 0x78     ;WREG 78H
        MULWF Myreg    ;PRODL = 58H, PRODH = 11H
                       ;(25H x 78H = 1158H)
```
Example:
```
        MOVLW  D'100'   ;WREG = 100
        MOVWF  MyReg    ;MyReg = 100
        MOVLW  D'200'   ;WREG = 200
        MULWF  MyReg    ;PRODL = 20H, PRODH = 4EH
                        ;(100 x 200 = 20,000 = 4E20H)
```

NEGF	Negate fileReg

Function:	No operation
Syntax:	NEGF f, a

This performs 2's complement on the value stored in fileReg and places it back in fileReg.

Example:

```
MyReg         SET 0x30
MOVLW 0x98      ;WREG = 0x98
MOVWF MyReg     ;MyReg = 0x98
NEGF            ;2's complement fileReg
```

```
98H  10011000
     01100111      1's complement
   +        1
     01101000      Now FileReg = 68H
```

Example:

```
MyReg         SET 0x10
MOVLW 0x75      ;WREG = 0x75
MOVWF MyReg     ;MyReg = 0x75
NEGF            ;2's complement fileReg
```

```
75H  01110101
     10001010      1's complement
   +        1
     10001011      Now FileReg = 7AH
```

Notice that in this instruction we cannot place the result in the WREG register.

NOP   **No Operation**
_____

Function: No operation
Syntax: NOP

This performs no operation and execution continues with the next instruction. It is sometimes used for timing delays to waste clock cyles. This instruction only updates the PC (program counter) to point to the next instruction following NOP. In PIC18, this a 2-byte instruction.

POP   **POP Top of Stack**
_____

Function: Pop from the stack
Syntax: POP

This takes out the top of stack (TOS) pointed to by SP (stack pointer) and discards it. It also decrements SP by 1. After the operation, the top of the stack will be the value pushed onto the stack previously.

PUSH   **PUSH Top of the Stack**
_____

Function: Push the PC onto the stack
Syntax: PUSH

This copies the program counter (PC) onto the stack and increments SP by 1, which means the previous top of the stack is pushed down.

RCALL   Relative Call

        Function: Transfers control to a subroutine within 1K space
        Syntax: RCALL target_address

There are two types of CALLs: RCALL and CALL. In RCALL, the target address is within 1K of the current PC (program counter). To reach the target address in the 2M ROM space of the PIC18, we must use CALL. In calling a subroutine, the PC register (which has the address of the instruction after the RCALL) is pushed onto the stack and the stack pointer (SP) is incremented by 1. Then the program counter is loaded with the new address and control is transferred to the subroutine. At the end of the procedure, when RETURN is executed, PC is popped off the stack, which returns control to the instruction after the RCALL.

Notice that RCALL is a 2-byte instruction, in which 5 bits are used for the opcode and the remaining 11 bits are used for the target subroutine address. An 11-bit address limits the range to –1024 to +1023. See the CALL instruction for discussion of the target address being anywhere in the 2M ROM space of the PIC18. Notice that RCALL is a 2-byte instruction while CALL is a 4-byte instruction. Also notice that the RCALL does not have the option of context saving, as CALL has.

RESET   Reset (by software)

        Function: Reset by software
        Syntax: RESET

This instruction is used to reset the PIC18 by way of software. After execution of this instruction, all the registers and flags are forced to their reset condition. The reset condition is created by activating the hardware pin MCLR. In other words, the RESET instruction is the software version of the MCLR pin.

RETFIE   Return from Interrupt Exit

        Function: Return from interrupt
        Syntax: RETFIE s

This is used at the end of an interrupt service routine (interrupt handler). The top of the stack is popped into the program counter and program execution continues at this new address. After popping the top of the stack into the program counter (PC), the stack pointer (SP) is decremented by 1.

Notice that while the RETURN instruction is used at the end of a subroutine associated with the CALL and RCALL instructions, RETFIE must be used for the interrupt service routines (ISRs).

RETLW   Return with Literal in WREG

Function:	The k value is placed in WREG and the top of the stack is the placed in PC (program counter)
Syntax:	RETLW k

After execution of this instruction, the k value is loaded into WREG and the top of the stack is popped into the program counter (PC). After popping the top of the stack into the program counter, the stack pointer (SP) is decremented by 1. This instruction is used for the implementation of a look-up table. See Section 6.3 in Chapter 6.

RETURN   Return

Function:	Return from subroutine	
Syntax:	RETURN s	;where s = 0 or s = 1

This instruction is used to return from a subroutine previously entered by instructions CALL or RCALL. The top of the stack is popped into the program counter (PC) and program execution continues at this new address. After popping the top of the stack into the program counter, the stack pointer (SP) is decremented by 1. For the case of "RETURN   s" where s = 1, the RETURN will also restore the context registers. See the CALL instruction for the case of s = 1. Notice that "RETURN 1" cannot be used for subroutines associated with RCALL.

RLCF   Rotate Left Through Carry the fileReg

Function:	Rotate fileReg left through carry
Syntax:	RLCF f, d, a

This rotates the bits of a fileReg register left. The bits rotated out of fileReg are rotated into C, and the C bit is rotated into the opposite end of the fileReg register.

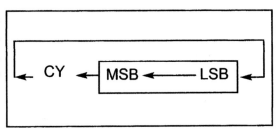

Example:
```
MyReg SET 0x30      ;set aside loc 30H for MyReg
      BCF   STATUS,C ;C = 0
      MOVLW 0x99     ;WREG = 99H
      MOVWF MyReg    ;MyReg = 99H = 10011001
      RLCF  MyReg,F  ;now MyReg = 00110010 and
                     ;C = 1
      RLCF  MyReg,F  ;now MyReg = 01100101 and
                     ;C = 0
```

RLNCF   **Rotate left not through Carry**

Function: Rotate left the fileReg

Syntax: RLNCF f, d, a

This rotates the bits of a fileReg register left. The bits rotated out of fileReg are rotated back into fileReg at the opposite end.

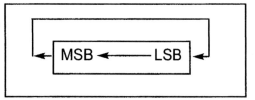

Example:

```
MyReg SET 0x20   ;set aside loc 20 for MyReg
MOVLW 0x69       ;WREG = 01101001
MOVWF MyReg      ;MyReg = 69H = 01101001
RLNCF MyReg,F    ;now MyReg = 11010010
RLNCF MyReg,F    ;now MyReg = 10100101
RLNCF MyReg,F    ;now MyReg = 01001011
RLNCF MyReg,F    ;now MyReg = 10010110
```

Notice that after four rotations, the upper and lower nibbles are swapped.

RRCF   **Rotate Right through Carry**

Function: Rotate fileReg right through carry

Syntax: RRCF f, d, a

This rotates the bits of a fileReg register right. The bits rotated out of the register are rotated into C, and the C bit is rotated into the opposite end of the register.

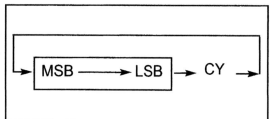

Example:

```
MyReg SET 0x20   ;set aside loc 20 for MyReg
BSF   STATUS,C   ;C = 1
MOVLW 0x99       ;WREG = 10011001
MOVWF MyReg      ;MyReg = 99H = 10011001
RRCF  MyReg,F    ;now MyReg = 11001100, C = 1
RRCF  MyReg,F    ;now MyReg = 11100110, C = 0
```

RRNCF   **Rotate Right not through Carry**

Function: Rotate fileReg right

Syntax: RRNCF f, d, a

This rotates the bits of a fileReg register right. The bits rotated out of the register are rotated back into fileReg at the opposite end.

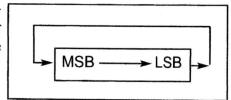

Example:
```
MyReg SET 0x20 ;set aside loc 20H for MyReg
      MOVLW 0x66    ;WREG = 66H = 01100110
      MOVWF MyReg   ;MyReg = 66H = 01100110
      RRNCF MyReg,F ;now MyReg = 00110011
      RRNCF MyReg,F ;now MyReg = 10011001
      RRNCF MyReg,F ;now MyReg = 11001100
      RRNCF MyReg,F ;now MyReg = 01100110
```

Example: We can use this instruction to swap the upper and lower nibbles.
```
MyReg SET 0x20 ;set aside loc 20H for MyReg
      MOVLW 0x36    ;WREG = 36H = 00110110
      MOVWF MyReg   ;MyReg = 36H = 00110110
      RRNCF MyReg,F ;now MyReg = 00011011
      RRNCF MyReg,F ;now MyReg = 10001101
      RRNCF MyReg,F ;now MyReg = 11000110
      RRNCF MyReg,F ;now MyReg = 01100011 = 63H
```

SETF   Set fileReg

Function: Set
Syntax: SETF f, a

This instruction sets the entire byte in fileReg to HIGH. All bits of the register are set to 1.

Examples:
```
      SETF MyReg      ;MyReg = 11111111
      SETF TRISB ;TRISB = FFH,(makes PORTB input)
      SETF PORTC      ;PORTC = 1111 1111
```

Notice that in this instruction, the result can be placed in fileReg only and there is no option for WREG to be used as the destination for the result.

SLEEP   Enter Sleep mode

Function: Put the CPU into sleep mode
Syntax: SLEEP

This instruction stops the oscillator and puts the CPU into sleep mode. It also resets the Watchdog Timer (WDT). The WDT is used mainly with the SLEEP instruction. Upon execution of the SLEEP instruction, the entire microcontroller goes into sleep mode by shutting down the main oscillator and by stopping the Program Counter from fetching the next instruction after SLEEP. There are two ways to get out of sleep mode: (a) an external event via hardware interrupt, (b) the internal WDT interrupt. Upon wake-up from a WDT interrupt, the microcontroller resumes operation by executing the next instruction after SLEEP.

Check the Microchip Corp. website for application notes on WDT.

SUBFWB　　　　　Subtract fileReg from WREG with borrow

Function:　　WREG – fileReg – #borrow ;#borrow is inverted carry
Syntax:　　　SUBFWB f, d, a

This subtracts fileReg and the Carry (borrow) flag from WREG and puts the result in WREG (d = 0) or fileReg (d = 1). The steps for subtraction performed by the internal hardware of the CPU are as follows:

1. Take the 2's complement of the fileReg byte.
2. Add this to register WREG.
3. Add the inverted Carry (borrow) flag to the result.
4. Ignore the Carry.
5. Examine the N (negative) flag for positive or negative result.

```
Example:
        MyReg SET 0x20  ;set aside loc 0x20 for MyReg
        BSF   STATUS,C  ;make Carry = 1
        MOVLW 0x45      ;WREG 45H
        MOVWF MyReg     ;MYReg = 45H
        MOVLW 0x23
        SUBWF MyReg     ;WREG = 45H - 23H - 0 = 22H
```

```
  45H       0100 0101              0100 0101
 -23H       0010 0011 2's comp  + 1101 1101
                Inverted carry +        0
 -------                          -----------
 +22H                             0010 0010
Because D7 (the N flag) is 0, the result is
positive.
```

This instruction sets the negative flag according to the following:

	N	
WREG > (fileReg + #C)	0	the result is positive
WREG = (fileReg + #C)	0	the result is 0
WREG < (fileReg + #C)	1	the result is negative and in 2's comp

SUBLW　　　　　Subtract WREG from Literal value

Function:　　Subtract WREG from literal value k (WREG = k – WREG)
Syntax:　　　SUBLW　　k

This subtracts the WREG value from the literal value k and puts the result in WREG. The steps for subtraction performed by the internal hardware of the CPU are as follows:

1. Take the 2's complement of the WREG value.
2. Add it to literal value k.
3. Ignore the Carry.
4. Examine the N (negative) flag for positive or negative result.

```
        MOVLW 0x23          ;WREG 23H
        SUBLW 0x45          ;WREG = 45H - 23H = 22H
```

```
  45H        0100 0101              0100 0101
 -23H        0010 0011 2's comp  +1101 1101
 -------                         -----------------
 +22H                              0010 0010
```
Because D7 (the N flag) is 0, the result is positive.

This instruction sets the negative flag according to the following:

	N	
Literal value k > WREG	0	the result is positive
Literal value k = WREG	0	the result is 0
Literal value < WREG	1	the result is negative and in 2's comp

Example:
```
        MOVLW 0x98          ;WREG 98H
        SUBLW 0x66          ;WREG = 66H - 98H = CEH
```

```
  66H        0110 0110              0110 0110
 -98H        1001 1000 2's comp  +0110 1000
 ------                          ----------------
  CEH                              1100 1110
```
Because D7 (the N flag) is 1, the result is negative and in 2's comp.

SUBWF   Subtract WREG from fileReg

Function: Subtract WREG from fileReg (Dest = fileReg – WREG)
Syntax: SUBWF f, d, a

This subtracts the WREG value from the fileReg value and puts the result in either WREG (d = 0) or fileReg (d = 1). The steps for subtraction performed by the internal hardware of the CPU are as follows:

1. Take the 2's complement of the WREG byte.
2. Add this to the fileReg register.
3. Ignore the carry.
4. Examine the N (negative) flag for positive or negative result.
    Example:
```
        MyReg SET 0x20 ;set aside loc 0x20 for MyReg
        MOVLW 0x45     ;WREG 45H
        MOVWF MyReg    ;MYReg = 45H
        MOVLW 0x23     ;WREG = 23H
        SUBWF MyReg,F  ;MyReg = 45H - 23H = 22H
```

```
45H          0100 0101              0100 0101
-23H         0010 0011 2's comp   +1101 1101
-------                            -----------------
+22H                               0010 0010
```
Because D7 (the N flag) is 0, the result is positive.

This instruction sets the negative flag according to the following:

	N	
fileReg > WREG	0	the result is positive
fileReg = WREG	0	the result is 0
fileReg < WREG	1	the result is negative and in 2's comp

SUBWFB   Subtract WREG from fileReg with borrow

Function: Dest = fileReg – WREG – #borrow ;#borrow is inverted carry
Syntax: SUBWFB f, d, a

This subtracts the WREG value and the inverted borrow (carry) flag from the fileReg value and puts the result in WREG (if d = 0), or fileReg (if d = 1). The steps for subtraction performed by the internal hardware of the CPU are as follows:

1. Take the 2's complement of WREG.
2. Add this to fileReg.
3. Add the inverted Carry flag to the result.
4. Ignore the carry.
5. Examine the N (negative) flag for positive or negative result.

Example:
```
        MyReg SET 0x20 ;set aside loc 0x20 for MyReg
        BSF       STATUS,C  ;C = 1
        MOVLW 0x45      ;WREG 45H
        MOVWF MyReg     ;MYReg = 45H
        MOVLW 0x23      ;WREG = 23H
        SUBWFB MyReg,F ;MyReg = 45H - 23H - 0 = 22H
```

```
45H          0100 0101              0100 0101
-23H         0010 0011 2's comp   +1101 1101
                  Inverted carry +          0
-----                            ------------
+22H                               0010 0010
```
Because D7 (the N flag) is 0, the result is positive.

This instruction sets the negative flag according to the following:

	N	
fileReg > (WREG + #C)	0	the result is positive
fileReg = (WREG + #C)	0	the result is 0
fileReg < (WREG + #C)	1	the result is negative and in 2's comp

SWAPF — Swap Nibbles in fileReg

Function: Swap nibbles within fileReg
Syntax: SAWPF f, d, a

The SWAPF instruction interchanges the lower nibble (D0–D3) with the upper nibble (D4–D7) inside fileReg. The result is placed in WREG (d = 0) or fileReg (d = 1).

Example:

```
MyReg SET 0X20 ;set aside loc 20H for MyReg
MOVLW 0x59H   ;W = 59H (0101 1001 in binary)
MOVWF MyReg    ;MyReg = 59H (0101 1001)
SWAPF MyReg,F  ;MyReg = 95H (1001 0101)
```

TBLRD — Table Read

Function: Read a byte from ROM to the TABLAT register
Syntax: TBLRD *
               TBLRD *+
               TBLRD *-
               TBLRD +*

This instruction moves (copies) a byte of data located in program (code) ROM into the TableLatch (TABLAT) register. This allows us to put strings of data, such as look-up table elements, in the code space and read them into the CPU. The address of the desired byte in the program space (on-chip ROM) is held by the TBLPTR register. Table A-6 shows the auto-increment feature of the TBLRD instruction.

Table A-6: PIC18 Table Read Instructions

Instruction	Function	
TBLRD*	Table Read	After read, TBLPTR stays the same
TBLRD*+	Table Read with post-increment (Read and increment TBLPTR)	
TBLRD*-	Table Read with post-decrement (Read and decrement TBLPTR)	
TBLRD+*	Table Read with pre-increment (increment TBLPTR and read)	

Note: A byte of data is read into the TABLAT register from code space pointed to by TBLPTR.

Example: Assume that an ASCII character string is stored in the on-chip ROM program memory starting at address 500H. Write a program to bring each character into the CPU and send it to PORTB.

```
ORG  0000H              ;burn into ROM starting at 0
```

```
        MOVLW  LOW(MESSAGE)   ;WREG = 00 low-byte addr.
        MOVWF  TBLPTRL        ;look-up table low-byte addr
        MOVLW  HIGH(MESSAGE)  ;WREG = 05 = high-byte addr
        MOVWF  TBLPTRH        ;look-up table high-byte addr
        CLRF   TBLPTRU        ;clear upper 5 bits

B8      TBLRD*+    ;read the table,then increment TBLPTR
        MOVF   TABLAT,W       ;copy to WREG (Z = 1 if null)
        BZ     EXIT           ;exit if end of string
        MOVWF  PORTB          ;copy WREG to PORTB
        BRA    B8
EXIT GOTO EXIT
;-------------------message
        ORG 0x500       ;data burned starting at 0x500
             ORG   0x500
MESSAGE      DB    "The earth is but one country and "
             DB    "mankind its citizens","Baha'u'llah",0
             END
```

In the program above, the TBLPTR holds the address of the desired byte. After the execution of the TBLRD*+ instruction, register TABLAT has the character. Notice that TBLPTR is incremented automatically to point to the next character in the MRESSAGE table.

TBLWT	Table Write
Function:	Write to Flash a block of data
Syntax:	TBLWT*
	TBLWT*+
	TBLWT*-
	TBLWT+*

This instruction writes a block of data to the program (code) space assuming that the on-chip program ROM is of Flash type. The address of the desired location in Flash ROM is held by the TBLPTR register. The process of writing to Flash ROM using the TBLWT instruction is discussed in Section 14.3 of Chapter 14.

TSTFSZ	Test fileReg, Skip if Zero
Function:	Test fileReg for zero value and skip if it is zero
Syntax:	TSTFSZ f, a

This instruction tests the entire contents of fileReg for value zero and skips the next instruction if fileReg has zero in it.

Example: Test PORTB for zero continuously.
```
        SETF   TRISB   ;make PORTB an input
        CLRF   TRISD   ;make PORTD an output
```

```
BACK TSTFSZ PORTB
     BRA    BACK
     MOVFF  PORTB,PORTD
```

Example: Toggle PORTB 250 times.

```
COUNTER   SET  0x40  ;loc 40H for COUNTER
     SETF TRISB      ;PORTB as output
     MOVLW D'250'    ;WREG = 250
     MOVWF COUNTER   ;COUNTER = 250
     MOVLW 0x55      ;WREG = 55H
     MOVWF PORTB
BACK COMF  PORTB,F   ;toggle PORTB
     DECF  COUNTER,F ;decrement COUNTER
     TSTFSZ COUNTER  ;test counter for 0
     BRA   BACK      ;keep doing it
     ......
```

XORLW   Ex-Or Literal with WREG

Function: Logical exclusive-OR Literal k and WREG
Syntax: XORLW   k

This performs a logical exclusive-OR on the Literal value and WREG operands, bit by bit, storing the result in WREG.

A	B	A XOR B
0	0	0
0	1	1
1	0	1
1	1	0

Example:
```
     MOVLW 0x39    ;WREG = 39H
     XORLW 0x09    ;WREG = 39H ORed with 09
                   ;now, WREG = 30H
     39H   0011 1001
     09H   0000 1001
     30    0011 0000
```

Example:
```
     MOVLW 0x32    ;WREG = 32H
     XORLW 0x50    ;(now, WREG = 62H)

     32H   0011 0010
     50H   0101 0000
     62H   0110 0010
```

XORWF   Ex-Or WREG with fileReg

Function: Logical exclusive-OR fileReg and WREG
Syntax: XORWF   f, d, a

This performs a logical exclusive-OR on the operands, bit by bit, storing the result in the destination. The destination can be WREG (d = 0), or fileReg (d = 1).

Example:
```
MyReg SET 0x20  ;set aside loc 20h for MyReg
MOVLW 0x39      ;WREG = 39H
MOVWF MyReg     ;MyReg = 39H
MOVLW 0x09      ;WREG = 09H
XORWF MyReg,F   ;MyReg = 39H ORed with 09
                ;MyReg = 30H


39H    0011 1001
09H    0000 1001
30     0011 0000
```

Example:
```
MyReg SET 0x15  ;set aside loc 15 for MyReg
MOVLW 0x32      ;WREG = 32H
MOVWF MyReg     ;MyReg = 32H
MOVLW 0x50      ;WREG = 50H
XORWF MyReg,F   ;now W = 62H


32H    0011 0010
50H    0101 0000
62H    0110 0010.
```

We can place the result in WREG.

Example:
```
MyReg SET 0x15  ;set aside loc 15 for MyReg
MOVLW 0x44      ;WREG = 44H
MOVWF MyReg     ;MyReg = 44H
MOVLW 0x67      ;WREG = 67H
XORWF MyReg     ;now W = 23H, and MyReg = 44H

44H    0100 0100
67H    0110 0111
23H    0010 0011
```

APPENDIX B

BASICS OF
WIRE WRAPPING

OVERVIEW

This appendix shows the basics of wire wrapping.

BASICS OF WIRE WRAPPING

Note: For this tutorial appendix, you will need the following:
Wire-wrapping tool (Radio Shack part number 276-1570)
30-gauge (30-AWG) wire for wire wrapping
(Thanks to Shannon Looper and Greg Boyle for their assistance on this section.)

The following describes the basics of wire wrapping:

1. There are several different types of wire-wrap tools available. The best one is available from Radio Shack for less than $10. The part number for the Radio Shack model is 276-1570. This tool combines the wrap and unwrap functions in the same end of the tool and includes a separate stripper. We found this to be much easier to use than the tools that combined all these features on one two-ended shaft. There are also wire-wrap guns, which are, of course, more expensive.

2. Wire-wrapping wire is available prestripped in various lengths or in bulk on a spool. The prestripped wire is usually more expensive and you are restricted to the different wire lengths you can afford to buy. Bulk wire can be cut to any length you wish, which allows each wire to be custom fit.

3. Serveral different types of wire-wrap boards are available. These are usually called perfboards or wire-wrap boards. These types of boards are sold at many electronics stores (such as Radio Shack). The best type of board has plating around the holes on the bottom of the board. These boards are better because the sockets and pins can be soldered to the board, which makes the circuit more mechanically stable.

4. Choose a board that is large enough to accommodate all the parts in your design with room to spare so that the wiring does not become too cluttered. If you wish to expand your project in the future, you should be sure to include enough room on the original board for the complete circuit. Also, if possible, the layout of the IC on the board needs to be such that signals go from left to right just like the schematics.

5. To make the wiring easier and to keep pressure off the pins, install one stand-off on each corner of the board. You may also wish to put standoffs on the top of the board to add stability when the board is on its back.

6. For power hook-up, use some type of standard binding post. Solder a few single wire-wrap pins to each power post to make circuit connections (to at least one pin for each IC in the circuit).

7. To further reduce problems with power, each IC must have its own connection to the main power of the board. If your perfboard does not have built-in power buses, run a separate power and ground wire from each IC to the main power. In other words, DO NOT daisy chain (chip-to-chip connection is called daisy chain) power connections, as each connection down the line will have more wire and more resistance to get power through. See Figure B-1. However, daisy chaining is acceptable for other connections such as data, address, and control buses.

8. You must use wire-wrap sockets. These sockets have long square pins whose edges will cut into the wire as it is wrapped around the pin.

9. Wire wrapping will not work on round legs. If you need to wrap to components, such as capacitors, that have round legs, you must also solder these connections. The best way to connect single components is to install individual wire-wrap pins into the board and then solder the components to the pins. An alternate method is to use an empty IC socket to hold small components such as resistors and wrap them to the socket.

10. The wire should be stripped about 1 inch. This will allow 7 to 10 turns for each connection. The first turn or turn-and-a-half should be insulated. This prevents stripped wire from coming in contact with other pins. This can be accomplished by inserting the wire as far as it will go into the tool before making the connection.

11. Try to keep wire lengths to a minimum. This prevents the circuit from looking like a bird nest. Be neat and use color coding as much as possible. Use only red wires for V_{CC} and black wires for ground connections. Also use different colors for data, address, and control signal connections. These suggestions will make troubleshooting much easier.

12. It is standard practice to connect all power lines first and check them for continuity. This will eliminate trouble later on.

13. It's also a good idea to mark the pin orientation on the bottom of the board. Plastic templates are available with pin numbers preprinted on them specifically for this purpose, or you can make your own from paper. Forgetting to reverse pin order when looking at the bottom of the board is a very common mistake when wire wrapping circuits.

14. To prevent damage to your circuit, place a diode (such as IN5338) in reverse bias across the power supply. If the power gets hooked up backwards, the diode will be forward biased and will act as a short, keeping the reversed voltage from your circuit.

15. In digital circuits, there can be a problem with current demand on the power supply. To filter the noise on the power supply, a 100 μF electrolytic capacitor and a 0.1 μF monolithic capacitor are connected from V_{CC} to ground, in parallel with each other, at the entry point of the power supply to the board. These two together will filter both the high- and the low-frequency noises. Instead of using two capacitors in parallel, you can use a single 20–100 μF tantalum capacitor. Remember that the long lead is the positive one.

16. To filter the transient current, use a 0.1 μF monolithic capacitor for each IC. Place the 0.1 μF monolithic capacitor between V_{CC} and ground of each IC. Make sure the leads are as short as possible.

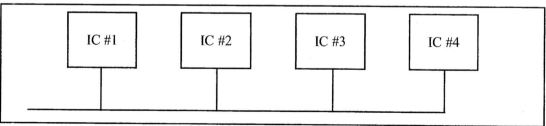

Figure B-1. Daisy Chain Connection (not recommended for power lines)

APPENDIX C

IC TECHNOLOGY AND SYSTEM DESIGN ISSUES

OVERVIEW

This appendix provides an overview of IC technology and PIC18 interfacing. In addition, we look at the microcontroller-based system as a whole and examine some general issues in system design.

First, in Section C.1, we provide an overview of IC technology. Then, in Section C.2, the internal details of PIC18 I/O ports and interfacing are discussed. Section C.3 examines system design issues.

C.1: OVERVIEW OF IC TECHNOLOGY

In this section we examine IC technology and discuss some major developments in advanced logic families. Because this is an overview, it is assumed that the reader is familiar with logic families on the level presented in basic digital electronics books.

Transistors

The transistor was invented in 1947 by three scientists at Bell Laboratory. In the 1950s, transistors replaced vacuum tubes in many electronics systems, including computers. It was not until 1959 that the first integrated circuit was successfully fabricated and tested by Jack Kilby of Texas Instruments. Prior to the invention of the IC, the use of transistors, along with other discrete components such as capacitors and resistors, was common in computer design. Early transistors were made of germanium, which was later abandoned in favor of silicon. This was because the slightest rise in temperature resulted in massive current flows in germanium-based transistors. In semiconductor terms, it is because the band gap of germanium is much smaller than that of silicon, resulting in a massive flow of electrons from the valence band to the conduction band when the temperature rises even slightly. By the late 1960s and early 1970s, the use of the silicon-based IC was widespread in mainframes and minicomputers. Transistors and ICs at first were based on P-type materials. Later on, because the speed of electrons is much higher (about two-and-a-half times) than the speed of holes, N-type devices replaced P-type devices. By the mid-1970s, NPN and NMOS transistors had replaced the slower PNP and PMOS transistors in every sector of the electronics industry, including in the design of microprocessors and computers. Since the early 1980s, CMOS (complementary MOS) has become the dominant technology of IC design. Next we provide an overview of differences between MOS and bipolar transistors. See Figure C-1.

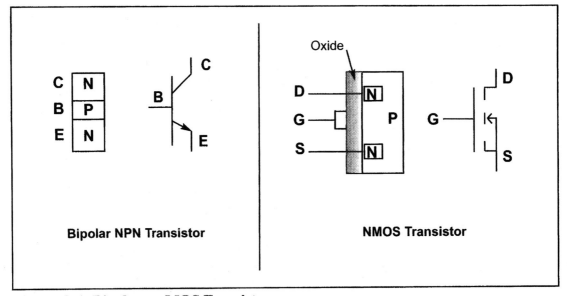

Figure C-1. Bipolar vs. MOS Transistors

MOS vs. bipolar transistors

There are two types of transistors: bipolar and MOS (metal-oxide semi-conductor). Both have three leads. In bipolar transistors, the three leads are referred to as the *emitter, base,* and *collector,* while in MOS transistors they are named *source, gate,* and *drain.* In bipolar transistors, the carrier flows from the emitter to the collector, and the base is used as a flow controller. In MOS transistors, the carrier flows from the source to the drain, and the gate is used as a flow controller. In NPN-type bipolar transistors, the electron carrier leaving the emitter must overcome two voltage barriers before it reaches the collector (see Figure C-1). One is the N-P junction of the emitter-base and the other is the P-N junction of the base-collector. The voltage barrier of the base-collector is the most difficult one for the electrons to overcome (because it is reverse-biased) and it causes the most power dissipation. This led to the design of the unipolar type transistor called MOS. In N-channel MOS transistors, the electrons leave the source and reach the drain without going through any voltage barrier. The absence of any voltage barrier in the path of the carrier is one reason why MOS dissipates much less power than bipolar transistors. The low power dissipation of MOS allows millions of transistors to fit on a single IC chip. In today's technology, putting 10 million transistors into an IC is common, and it is all because of MOS technology. Without the MOS transistor, the advent of desktop personal computers would not have been possible, at least not so soon. The bipolar transistors in both the mainframes and minicomputers of the 1960s and 1970s were bulky and required expensive cooling systems and large rooms. MOS transistors do have one major drawback: They are slower than bipolar transistors. This is due partly to the gate capacitance of the MOS transistor. For a MOS to be turned on, the input capacitor of the gate takes time to charge up to the turn-on (threshold) voltage, leading to a longer propagation delay.

Overview of logic families

Logic families are judged according to (1) speed, (2) power dissipation, (3) noise immunity, (4) input/output interface compatibility, and (5) cost. Desirable qualities are high speed, low power dissipation, and high noise immunity (because it prevents the occurrence of false logic signals during switching transition). In interfacing logic families, the more inputs that can be driven by a single output, the better. This means that high-driving-capability outputs are desired. This, plus the fact that the input and output voltage levels of MOS and bipolar transistors are not compatible mean that one must be concerned with the ability of one logic family to drive the other one. In terms of the cost of a given logic family, it is high during the early years of its introduction but it declines as production and use rise.

The case of inverters

As an example of logic gates, we look at a simple inverter. In a one-transistor inverter, the transistor plays the role of a switch, and R is the pull-up resistor. See Figure C-2. For this inverter to work most effectively in digital circuits, however, the R value must be high when the transistor is "on" to limit the current flow from V_{CC} to ground in order to have low power dissipation (P = VI, where V

= 5 V). In other words, the lower the I, the lower the power dissipation. On the other hand, when the transistor is "off", R must be a small value to limit the voltage drop across R, thereby making sure that V_{OUT} is close to V_{CC}. This is a contradictory demand on R. This is one reason that logic gate designers use active components (transistors) instead of passive components (resistors) to implement the pull-up resistor R.

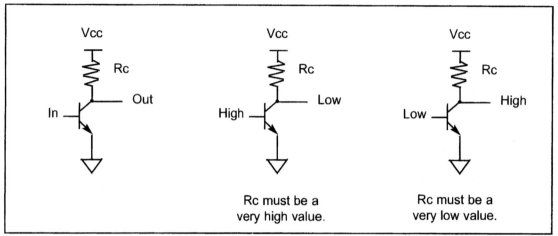

Figure C-2. One-Transistor Inverter with Pull-up Resistor

The case of a TTL inverter with totem-pole output is shown in Figure C-3. In Figure C-3, Q3 plays the role of a pull-up resistor.

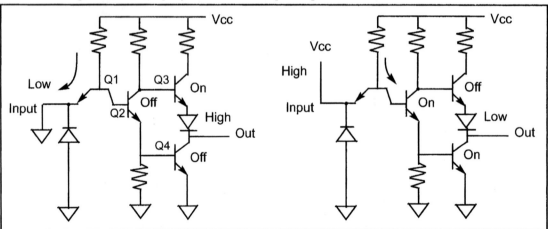

Figure C-3. TTL Inverter with Totem-Pole Output

CMOS inverter

In the case of CMOS-based logic gates, PMOS and NMOS are used to construct a CMOS (complementary MOS) inverter as shown in Figure C-4. In CMOS inverters, when the PMOS transistor is off, it provides a very high impedance path, making leakage current almost zero (about 10 nA); when the PMOS is on, it provides a low resistance on the path of V_{DD} to load. Because the speed of the hole is slower than that of the electron, the PMOS transistor is wider to compensate for this disparity; therefore, PMOS transistors take more space than NMOS transistors in the CMOS gates. At the end of this section we will see an open-collector gate in which the pull-up resistor is provided externally, thereby allowing system designers to choose the value of the pull-up resistor.

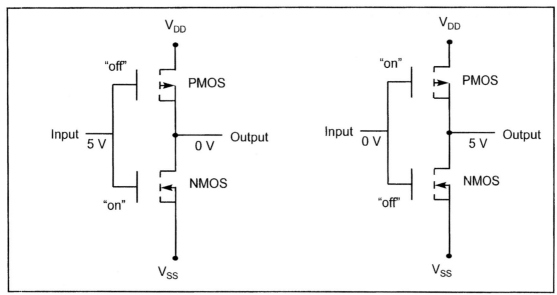

Figure C-4. CMOS Inverter

Input/output characteristics of some logic families

In 1968 the first logic family made of bipolar transistors was marketed. It was commonly referred to as the standard TTL (transistor-transistor logic) family. The first MOS-based logic family, the CD4000/74C series, was marketed in 1970. The addition of the Schottky diode to the base-collector of bipolar transistors in the early 1970s gave rise to the S family. The Schottky diode shortens the propagation delay of the TTL family by preventing the collector from going into what is called deep saturation. Table C-1 lists major characteristics of some logic families. In Table C-1, note that as the CMOS circuit's operating frequency rises, the power dissipation also increases. This is not the case for bipolar-based TTL.

Table C-1: Characteristics of Some Logic Families

Characteristic	STD TTL	LSTTL	ALSTTL	HCMOS
V_{CC}	5 V	5 V	5 V	5 V
V_{IH}	2.0 V	2.0 V	2.0 V	3.15 V
V_{IL}	0.8 V	0.8 V	0.8 V	1.1 V
V_{OH}	2.4 V	2.7 V	2.7 V	3.7 V
V_{OL}	0.4 V	0.5 V	0.4 V	0.4 V
I_{IL}	-1.6 mA	-0.36 mA	-0.2 mA	-1 μA
I_{IH}	40 μA	20 μA	20 μA	1 μA
I_{OL}	16 mA	8 mA	4 mA	4 mA
I_{OH}	-400 μA	-400 μA	-400 μA	4 mA
Propagation delay	10 ns	9.5 ns	4 ns	9 ns
Static power dissipation (f = 0)	10 mW	2 mW	1 mW	0.0025 nW
Dynamic power dissipation at f = 100 kHz	10 mW	2 mW	1 mW	0.17 mW

History of logic families

Early logic families and microprocessors required both positive and negative power voltages. In the mid-1970s, 5 V V_{CC} became standard. In the late 1970s, advances in IC technology allowed combining the speed and drive of the S family with the lower power of LS to form a new logic family called FAST (Fairchild Advanced Schottky TTL). In 1985, AC/ACT (Advanced CMOS Technology), a much higher speed version of HCMOS, was introduced. With the introduction of FCT (Fast CMOS Technology) in 1986, the speed gap between CMOS and TTL at last was closed. Because FCT is the CMOS version of FAST, it has the low power consumption of CMOS but the speed is comparable with TTL. Table C-2 provides an overview of logic families up to FCT.

Table C-2: Logic Family Overview

Product	Year Introduced	Speed (ns)	Static Supply Current (mA)	High/Low Family Drive (mA)
Std TTL	1968	40	30	–2/32
CD4K/74C	1970	70	0.3	–0.48/6.4
LS/S	1971	18	54	–15/24
HC/HCT	1977	25	0.08	–6/–6
FAST	1978	6.5	90	–15/64
AS	1980	6.2	90	–15/64
ALS	1980	10	27	–15/64
AC/ACT	1985	10	0.08	–24/24
FCT	1986	6.5	1.5	–15/64

Reprinted by permission of Electronic Design Magazine, c. 1991.

Recent advances in logic families

As the speed of high-performance microprocessors reached 25 MHz, it shortened the CPU's cycle time, leaving less time for the path delay. Designers normally allocate no more than 25% of a CPU's cycle time budget to path delay. Following this rule means that there must be a corresponding decline in the propagation delay of logic families used in the address and data path as the system frequency is increased. In recent years, many semiconductor manufacturers have responded to this need by providing logic families that have high speed, low noise, and high drive I/O. Table C-3 provides the characteristics of high-performance logic families introduced in recent years. ACQ/ACTQ are the second-generation advanced CMOS (ACMOS) with much lower noise. While ACQ has the CMOS input level, ACTQ is equipped with TTL-level input. The FCTx and FCTx-T are second-generation FCT with much higher speed. The "x" in the FCTx and FCTx-T refers to various speed grades, such as A, B, and C, where A means low speed and C means high speed. For designers who are well versed in using the FAST logic family, FASTr is an ideal choice because it is faster than FAST, has higher driving capability (I_{OL}, I_{OH}), and produces much lower noise than FAST. At the time of this writing, next to ECL and gallium arsenide logic gates, FASTr is the fastest logic family in the market (with the 5 V V_{CC}), but the power consumption is high relative to other logic families, as shown in Table C-3. The combining of high-speed bipolar TTL and the low power consumption of CMOS has given birth to what is called BICMOS. Although BICMOS seems to be the future trend in IC design, at this time it is expensive due to extra steps required in BICMOS IC fabrication, but in some cases there is no other choice. (For example, Intel's Pentium microprocessor, a BICMOS product, had to use high-speed bipolar transistors to speed up some of the internal functions.) Table C-3 provides advanced logic characteristics. The "x" is for different speeds designated as A, B, and C. A is the slowest one while C is the fastest one. The above data is for the 74244 buffer.

Table C-3: Advanced Logic General Characteristics

Family	Year	Number Suppliers	Tech Base	I/O Level	Speed (ns)	Static Current	I_{OH}/I_{OL}
ACQ	1989	2	CMOS	CMOS/CMOS	6.0	80 µA	−24/24 mA
ACTQ	1989	2	CMOS	TTL/CMOS	7.5	80 µA	−24/24 mA
FCTx	1987	3	CMOS	TTL/CMOS	4.1–4.8	1.5 mA	−15/64 mA
FCTxT	1990	2	CMOS	TTL/TTL	4.1–4.8	1.5 mA	−15/64 mA
FASTr	1990	1	Bipolar	TTL/TTL	3.9	50 mA	−15/64 mA
BCT	1987	2	BICMOS	TTL/TTL	5.5	10 mA	−15/64 mA

Reprinted by permission of Electronic Design Magazine, c. 1991.

Since the late 70s, the use of a +5 V power supply has become standard in all microprocessors and microcontrollers. To reduce power consumption, 3.3 V V_{CC} is being embraced by many designers. The lowering of V_{CC} to 3.3 V has two major advantages: (1) it lowers the power consumption, prolonging the life of the battery in systems using a battery, and (2) it allows a further reduction of line size (design rule) to submicron dimensions. This reduction results in putting more transistors in a given die size. As fabrication processes improve, the decline in the line size is reaching submicron level and transistor densities are approaching 1 billion transistors.

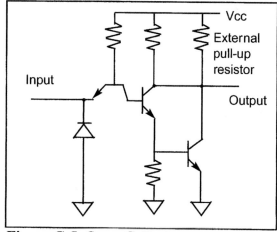

Figure C-5. Open Collector

Open-collector and open-drain gates

To allow multiple outputs to be connected together, we use open-collector logic gates. In such cases, an external resistor will serve as load. This is shown in Figures C-5 and C-6.

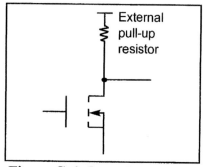

Figure C-6. Open Drain

SECTION C.2: PIC18 I/O PORT STRUCTURE AND INTERFACING

In interfacing the PIC18 microcontroller with other IC chips or devices, fan-out is the most important issue. To understand the PIC18 fan-out we must first understand the port structure of the PIC18. This section provides a detailed discussion of the PIC18 port structure and its fan-out. It is very critical that we understand the I/O port structure of the PIC18 lest we damage it while trying to interface it with an external device.

IC fan-out

When connecting IC chips together, we need to find out how many input pins can be driven by a single output pin. This is a very important issue and involves the discussion of what is called IC fan-out. The IC fan-out must be addressed for both logic "0" and logic "1" outputs. See Example C-1. Fan-out for logic LOW and fan-out for logic HIGH are defined as follows:

$$\text{fan-out (of LOW)} = \frac{I_{OL}}{I_{IL}} \qquad\qquad \text{fan-out (of HIGH)} = \frac{I_{OH}}{I_{IH}}$$

Of the above two values, the lower number is used to ensure the proper noise margin. Figure C-7 shows the sinking and sourcing of current when ICs are connected together.

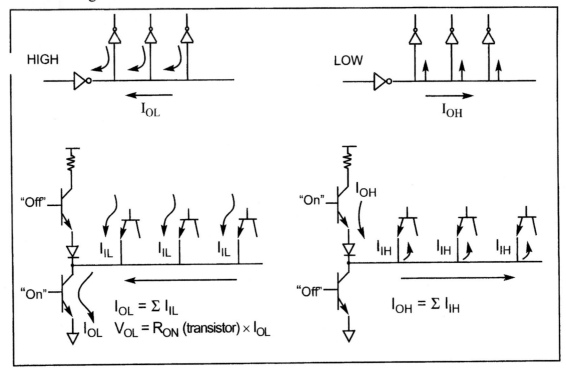

Figure C-7. Current Sinking and Sourcing in TTL

Notice that in Figure C-7, as the number of input pins connected to a single output increases, I_{OL} rises, which causes V_{OL} to rise. If this continues, the rise of V_{OL} makes the noise margin smaller, and this results in the occurrence of false logic due to the slightest noise.

Example C-1

Find how many unit loads (UL) can be driven by the output of the LS logic family.

Solution:

The unit load is defined as $I_{IL} = 1.6$ mA and $I_{IH} = 40$ μA. Table C-1 shows $I_{OH} = 400$ μA and $I_{OL} = 8$ mA for the LS family. Therefore, we have

$$\text{fan-out (LOW)} = \frac{I_{OL}}{I_{IL}} = \frac{8 \text{ mA}}{1.6 \text{ mA}} = 5$$

$$\text{fan-out (HIGH)} = \frac{I_{OH}}{I_{IH}} = \frac{400 \text{ μA}}{40 \text{ μA}} = 10$$

This means that the fan-out is 5. In other words, the LS output must not be connected to more than 5 inputs with unit load characteristics.

74LS244 and 74LS245 buffers/drivers

In cases where the receiver current requirements exceed the driver's capability, we must use buffers/drivers such as the 74LS245 and 74LS244. Figure C-8 shows the internal gates for the 74LS244 and 74LS245. The 74LS245 is used for bidirectional data buses, and the 74LS244 is used for unidirectional address buses.

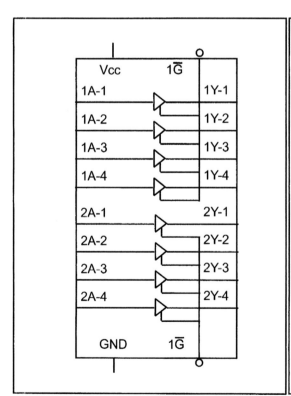

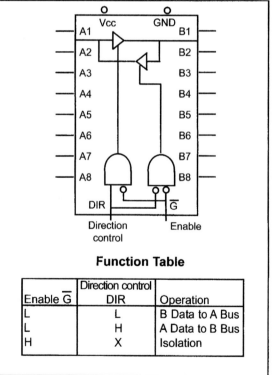

Function Table

Enable $\overline{G}$	Direction control DIR	Operation
L	L	B Data to A Bus
L	H	A Data to B Bus
H	X	Isolation

Figure C-8 (a). 74LS244 Octal Buffer

(Reprinted by permission of Texas Instruments, Copyright Texas Instruments, 1988)

Figure C-8 (b). 74LS245 Bidirectional Buffer

(Reprinted by permission of Texas Instruments, Copyright Texas Instruments, 1988)

Tri-state buffer

Notice that the 74LS244 is simply 8 tri-state buffers in a single chip. As shown in Figure C-9 a tri-state buffer has a single input, a single output, and the enable control input. By activating the enable, data at the input is transferred to the output. The enable can be an active-LOW or an active-HIGH. Notice that the enable input for the 74LS244 is an active-LOW whereas the enable input pin for Figure C-9 is active-HIGH.

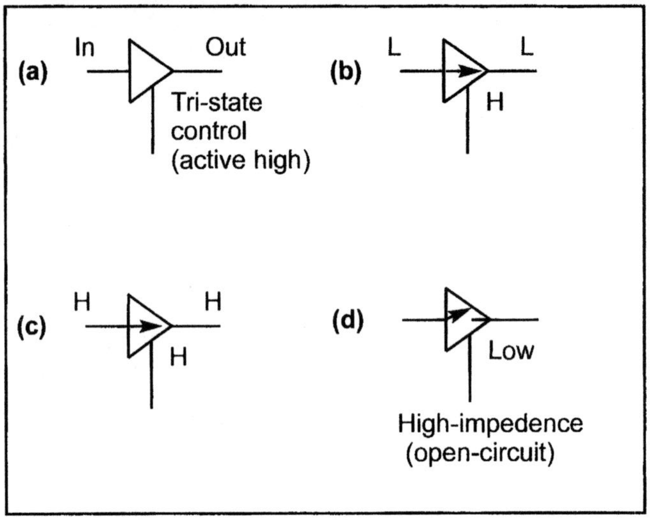

Figure C-9. Tri-State Buffer

74LS245 and 74LS244 fan-out

It must be noted that the output of the 74LS245 and 74LS244 can sink and source a much larger amount of current than that of other LS gates. See Table C-4. That is the reason we use these buffers for driver when a signal is travelling a long distance through a cable or it has to drive many inputs.

Table C-4: Electrical Specifications for Buffers/Drivers

	I_{OH} (mA)	I_{OL} (mA)
74LS244	3	12
74LS245	3	12

After this background on the fan-out, next we discuss the structure of PIC18 ports.

PIC18 port structure and operation

Because all the ports of the PIC18 are bidirectional they all have the following four components in their structure:

1. Data latch
2. Output driver
3. Input buffer
4. TRIS latch

Figure C-10 shows the structure of a port and its four components. Notice that in Figure C-10, the PIC18 ports have both the latch and buffer. Now the question is, in reading the port, are we reading the status of the input pin or are we read-

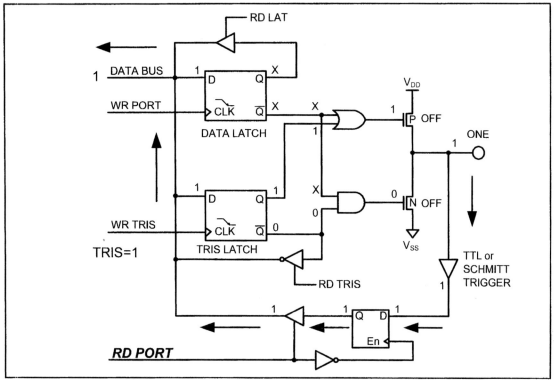

Figure C-10. Inputting (Reading) 1 from a Pin in the PIC18

ing the status of the latch? That is an extremely important question and its answer depends on which instruction we are using. Therefore, when reading the ports there are two possibilities: (1) reading the input pin, or (2) reading the latch. The above distinction is very important and must be understood lest you damage the PIC18 port. Each is described next.

Reading the pin when TRIS = 1 (Input)

As we stated in Chapter 4, to make any bits of any port of the PIC18 an input port, we first must write a 1 (logic HIGH) to the TRIS bit. Look at the following sequence of events to see why:

1. As can be seen from Figure C-10, a 1 written to the TRIS latch has "HIGH" on its Q. Therefore, Q = 1 and $\overline{Q}$ = 0. Because Q = 1, it turns off the P transistor.
2. Because $\overline{Q}$ = 0 and is connected to the gate of the N transistor, the N transistor is off.
3. When both transistors are off, they block any path to the ground or VCC for any signal connected to the input pin, and the input signal is directed to the buffer.
4. When reading the input port in instructions such as "MOVFW PORTB" we are really reading the data present at the pin. In other words, it is bringing into the CPU the status of the external pin. This instruction activates the read pin of buffer and lets data at the pins flow into the CPU's internal bus. Figures C-10 and C-11 show HIGH and LOW signals at the input, respectively.

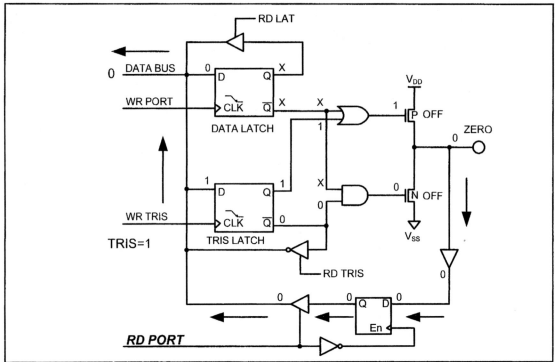

Figure C-11. Inputting (Reading) 0 from a Pin in the PIC18

Writing to pin when TRIS = 0 (Output)

The above discussion showed why we must write a "HIGH" to a port's TRIS bits in order to make it an input port. What happens if we write a "0" to TRIS that was configured as an input port? From Figure C-12 we see that when TRIS = 0, if we write a 0 to the Data latch, then Q = 0 and $\overline{Q}$ = 1. As a result of $\overline{Q}$ = 1, the N transistor is "on" and the P transistor is "off." If N is "on," it provides the path to ground for the input pin. Therefore, any attempt to read the input pin will always get the "LOW" ground signal. Figure C-13 shows what happens when we write "HIGH" to output port (Data latch) when TRIS = 0. Writing 1 to the Data latch makes $\overline{Q}$ = 0. As a result of that, the P transistor is "on" and the N transistor is "off," which allows a 1 to be provided to the output pin. Therefore, any attempt to read the input pin will always get the "HIGH" signal.

Avoid damaging the port

The following methods can be used as precautions to prevent damage to the PIC18 ports:

1. Have a 10k ohms resistor on the V_{CC} path to limit current flow.
2. Connect any input switch to a 74LS244 tri-state buffer before it is fed to the PIC18 pin.

The above points are extremely important and must be emphasized because many people damage their ports and afterwards wonder how it happened. We must also use the right instruction when we want to read the status of an input pin. Table C-5 shows the list of instructions in which reading the port reads the status of the input pin.

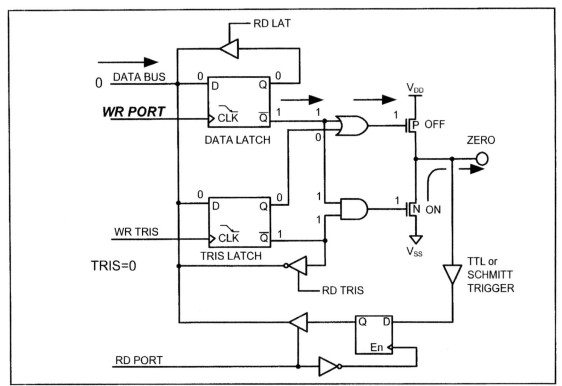

Figure C-12. Outputting (Writing) 0 to a Pin in the PIC18

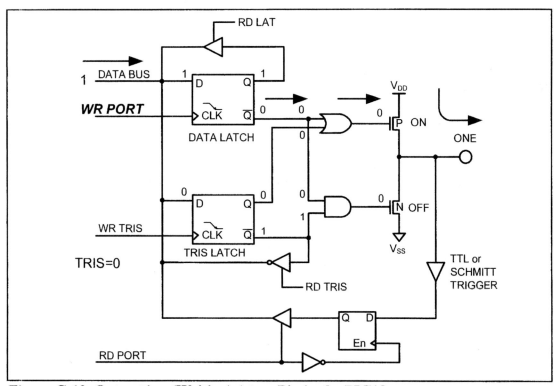

Figure C-13. Outputting (Writing) 1 to a Pin in the PIC18

Table C-5: Some of the Instructions Reading the Status of Input Port

Mnemonics	Examples
MOVFW PORTx	MOVFW PORTB
TSTFSZ f	TSTFSZ PORTC
BTFSS f,b	BTFSS PORTD,0
BTFSC f,b	BTFSC PORTB,7
CPFSEQ f	CPFSEQ PORTB

PIC18 port fan-out

Now that we are familiar with the port structure of the PIC18, we need to examine the fan-out for the PIC18 microconctroller. While the early chips were based on NMOS IC technology, today's PIC18 microcontrollers are all based on CMOS technology. Note, however, that while the core of the PIC18 microcontroller is CMOS, the circuitry driving its pins is all TTL compatible. That is, the PIC18 is a CMOS-based product with TTL-compatible pins. All the ports of the PIC18 have the same I/O structure, and therefore the same fan-out. Table C-6 provides the I/O characteristics of PIC18F458 ports.

Table C-6: PIC18 Fan-out for PORTS

Pin	Fan-out
IOL	8.5 mA
IOH	−3 mA
IIL	1 μA
IIH	1 μA

Note: Negative current is defined as current sourced by the pin.

74LS244 driving an output pin

In some cases, when an PIC18 port is driving multiple inputs, or driving a single input via a long wire or cable (e.g., printer cable), we can use the 74LS244 as a driver. When driving an off-board circuit, placing the 74LS244 buffer between your PIC18 and the circuit is essential because the PIC18 lacks sufficient current. See Figure C-14.

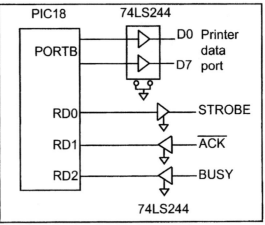

Figure C-14. PIC18 Connection to Printer Signals

738

SECTION C.3: SYSTEM DESIGN ISSUES

In addition to fan-out, the other issues related to system design are power dissipation, ground bounce, V_{CC} bounce, crosstalk, and transmission lines. In this section we provide an overview of these topics.

Power dissipation considerations

Power dissipation of a system is a major concern of system designers, especially for laptop and hand-held systems in which batteries provide the power. Power dissipation is a function of frequency and voltage as shown below:

$$Q = CV$$

$$\frac{Q}{T} = \frac{CV}{T}$$

$$\text{since} \quad F = \frac{1}{T} \qquad \text{and} \quad I = \frac{Q}{T}$$

$$I = CVF$$

$$\text{now} \quad P = VI = CV^2F$$

In the above equations, the effects of frequency and V_{CC} voltage should be noted. While the power dissipation goes up linearly with frequency, the impact of the power supply voltage is much more pronounced (squared). See Example C-2.

Example C-2

Compare the power consumption of two microcontroller-based systems. One uses 5 V and the other uses 3 V for V_{CC}.

Solution:
Because $P = VI$, by substituting $I = V/R$ we have $P = V^2/R$. Assuming that $R = 1$, we have $P = 5^2 = 25$ W and $P = 3^2 = 9$ W. This results in using 16 W less power, which means power saving of 64%. ($16/25 \times 100$) for systems using 3 V for power source.

Dynamic and static currents

Two major types of currents flow through an IC: dynamic and static. A dynamic current is $I = CVF$. It is a function of the frequency under which the component is working. This means that as the frequency goes up, the dynamic current and power dissipation go up. The static current, also called DC, is the current consumption of the component when it is inactive (not selected). The dynamic current dissipation is much higher than the static current consumption. To reduce power consumption, many microcontrollers, including the PIC18, have power-saving modes. In the PIC18, the power saving mode is called *sleep mode.* We describe the sleep mode next.

Sleep mode

In sleep mode the on-chip oscillator is frozen, which cuts off frequency to the CPU and peripheral functions, such as serial ports, interrupts, and timers. Notice that while this mode brings power consumption down to an absolute minimum, the contents of RAM and the SFR registers are saved and remain unchanged.

Ground bounce

One of the major issues that designers of high-frequency systems must grapple with is ground bounce. Before we define ground bounce, we will discuss lead inductance of IC pins. There is a certain amount of capacitance, resistance, and inductance associated with each pin of the IC. The size of these elements varies depending on many factors such as length, area, and so on.

The inductance of the pins is commonly referred to as *self-inductance* because there is also what is called *mutual inductance*, as we will show below. Of the three components of capacitor, resistor, and inductor, the property of self-inductance is the one that causes the most problems in high-frequency system design because it can result in ground bounce. Ground bounce occurs when a massive amount of current flows through the ground pin caused by many outputs changing from HIGH to LOW all at the same time. See Figure C-15(a). The voltage is related to the inductance of the ground lead as follows:

$$V = L \frac{di}{dt}$$

As we increase the system frequency, the rate of dynamic current, di/dt, is also increased, resulting in an increase in the inductance voltage L (di/dt) of the ground pin. Because the LOW state (ground) has a small noise margin, any extra voltage due to the inductance can cause a false signal. To reduce the effect of ground bounce, the following steps must be taken where possible:

1. The V_{CC} and ground pins of the chip must be located in the middle rather than at opposite ends of the IC chip (the 14-pin TTL logic IC uses pins 14 and 7 for ground and V_{CC}). This is exactly what we see in high-performance logic gates such as Texas Instruments' advanced logic AC11000 and ACT11000 families. For example, the ACT11013 is a 14-pin DIP chip in which pin numbers 4 and 11 are used for the ground and V_{CC}, instead of 7 and 14 as in the traditional TTL family. We can also use the SOIC packages instead of DIP.
2. Another solution is to use as many pins for ground and V_{CC} as possible to reduce the lead length. This is exactly why all high-performance microprocessors and logic families use many pins for V_{CC} and ground instead of the traditional single pin for V_{CC} and single pin for GND. For example, in the case of Intel's Pentium processor there are over 50 pins for ground, and another 50 pins for V_{CC}.

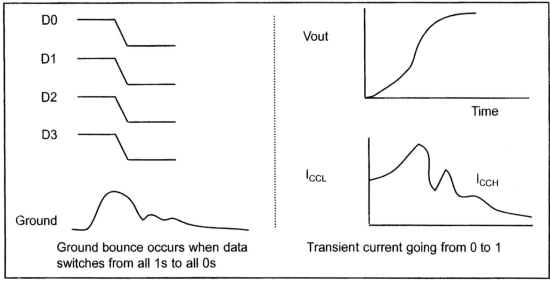

DO

D1

D2

D3

Ground

Ground bounce occurs when data
switches from all 1s to all 0s

Vout

Time

I_{CCL}

I_{CCH}

Transient current going from 0 to 1

Figure C-15. (a) Ground Bounce　　　(b) Transient Current

The above discussion of ground bounce is also applicable to V_{CC} when a large number of outputs changes from the LOW to the HIGH state; this is referred to as *V_{CC} bounce*. However, the effect of V_{CC} bounce is not as severe as ground bounce because the HIGH ("1") state has a wider noise margin than the LOW ("0") state.

Filtering the transient currents using decoupling capacitors

In the TTL family, the change of the output from LOW to HIGH can cause what is called *transient current*. In a totem-pole output in which the output is LOW, Q4 is on and saturated, whereas Q3 is off. By changing the output from the LOW to the HIGH state, Q3 turns on and Q4 turns off. This means that there is a time when both transistors are on and drawing current from V_{CC}. The amount of current depends on the R_{ON} values of the two transistors, which in turn depend on the internal parameters of the transistors. The net effect of this, however, is a large amount of current in the form of a spike for the output current, as shown in Figure C-15(b). To filter the transient current, a 0.01 μF or 0.1 μF ceramic disk capacitor can be placed between the V_{CC} and ground for each TTL IC. The lead for this capacitor, however, should be as small as possible because a long lead results in a large self-inductance, and that results in a spike on the V_{CC} line [V = L (di/dt)]. This spike is called V_{CC} bounce. The ceramic capacitor for each IC is referred to as a *decoupling capacitor*. There is also a bulk decoupling capacitor, as described next.

Bulk decoupling capacitor

If many IC chips change state at the same time, the combined currents drawn from the board's V_{CC} power supply can be massive and may cause a fluctuation of V_{CC} on the board where all the ICs are mounted. To eliminate this, a relatively large decoupling tantalum capacitor is placed between the V_{CC} and ground lines. The size and location of this tantalum capacitor varies depending on the number of ICs on the board and the amount of current drawn by each IC, but it is common to have a single 22 μF to 47 μF capacitor for each of the 16 devices, placed between the V_{CC} and ground lines.

Crosstalk

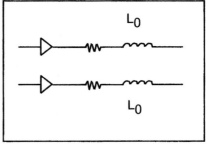

Figure C-16. Crosstalk (EMI)

Crosstalk is due to mutual inductance. See Figure C-16. Previously, we discussed self-inductance, which is inherent in a piece of conductor. *Mutual inductance* is caused by two electric lines running parallel to each other. The mutual inductance is a function of l, the length of two conductors running in parallel, d, the distance between them, and the medium material placed between them. The effect of crosstalk can be reduced by increasing the distance between the parallel or adjacent lines (in printed circuit boards, they will be traces). In many cases, such as printer and disk drive cables, there is a dedicated ground for each signal. Placing ground lines (traces) between signal lines reduces the effect of crosstalk. This method is used even in some ACT logic families where a V_{CC} and a GND pin are next to each other. Crosstalk is also called *EMI* (electromagnetic interference). This is in contrast to *ESI* (electrostatic interference), which is caused by capacitive coupling between two adjacent conductors.

Transmission line ringing

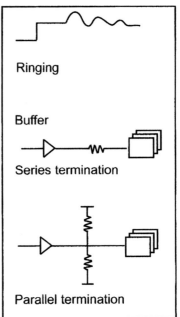

Figure C-17. Reducing Transmission Line Ringing

The square wave used in digital circuits is in reality made of a single fundamental pulse and many harmonics of various amplitudes. When this signal travels on the line, not all the harmonics respond in the same way to the capacitance, inductance, and resistance of the line. This causes what is called *ringing*, which depends on the thickness and the length of the line driver, among other factors. To reduce the effect of ringing, the line drivers are terminated by putting a resistor at the end of the line. See Figure C-17. There are three major methods of line driver termination: parallel, serial, and Thevenin.

In serial termination, resistors of 30–50 ohms are used to terminate the line. The parallel and Thevenin methods are used in cases where there is a need to match the impedance of the line with the load impedance. This requires a detailed analysis of the signal traces and load impedance, which is beyond the scope of this book. In high-frequency systems, wire traces on the printed circuit board (PCB) behave like transmission lines, causing ringing. The severity of this ringing depends on the speed and the logic family used. Table C-7 provides the length of the traces, beyond which the traces must be looked at as transmission lines.

Table C-7: Line Length Beyond Which Traces Behave Like Transmission Lines

Logic Family	Line Length (in.)
LS	25
S, AS	11
F, ACT	8
AS, ECL	6
FCT, FCTA	5

(Reprinted by permission of Integrated Device Technology, copyright IDT 1991)

APPENDIX D

FLOWCHARTS AND PSEUDOCODE

OVERVIEW

This appendix provides an introduction to writing flowcharts and pseudocode.

Flowcharts

If you have taken any previous programming courses, you are probably familiar with flowcharting. Flowcharts use graphic symbols to represent different types of program operations. These symbols are connected together into a flowchart to show the flow of execution of a program. Figure D-1 shows some of the more commonly used symbols. Flowchart templates are available to help you draw the symbols quickly and neatly.

Pseudocode

Flowcharting has been standard practice in industry for decades. However, some find limitations in using flowcharts, such as the fact that you can't write much in the little boxes, and it is hard to get the "big picture" of what the program does without getting bogged down in the details. An alternative to using flowcharts is pseudocode, which involves writing brief descriptions of the flow of the code. Figures D-2 through D-6 show flowcharts and pseudocode for commonly used control structures.

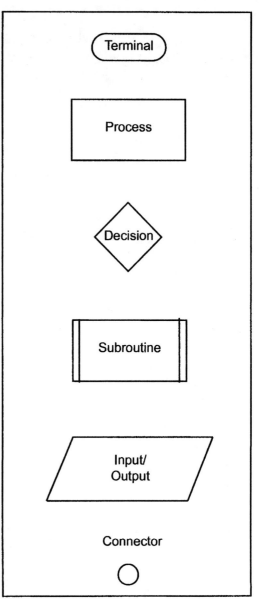

Figure D-1. Commonly Used Flowchart Symbols

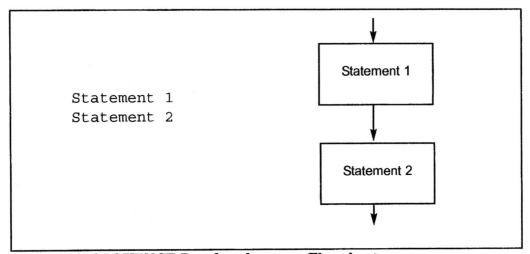

```
Statement 1
Statement 2
```

Figure D-2. SEQUENCE Pseudocode versus Flowchart

Structured programming uses three basic types of program control structures: sequence, control, and iteration. Sequence is simply executing instructions one after another. Figure D-2 shows how sequence can be represented in pseudocode and flowcharts.

Figures D-3 and D-4 show two control programming structures: IF-THEN-ELSE and IF-THEN in both pseudocode and flowcharts.

Note in Figures D-2 through D-6 that "statement" can indicate one statement or a group of statements.

Figures D-5 and D-6 show two iteration control structures: REPEAT UNTIL and WHILE DO. Both structures execute a statement or group of statements repeatedly. The difference between them is that the REPEAT UNTIL structure always executes the statement(s) at least once, and checks the condition after each iteration, whereas the WHILE DO may not execute the statement(s) at all because the condition is checked at the beginning of each iteration.

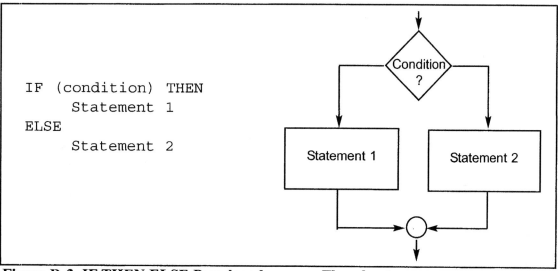

Figure D-3. IF THEN ELSE Pseudocode versus Flowchart

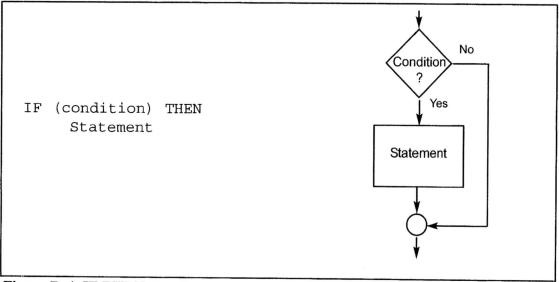

Figure D-4. IF THEN Pseudocode versus Flowchart

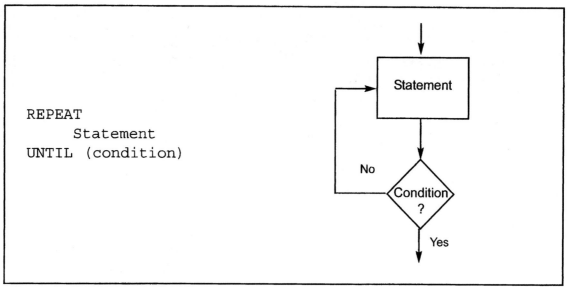

```
REPEAT
     Statement
UNTIL (condition)
```

Figure D-5. REPEAT UNTIL Pseudocode versus Flowchart

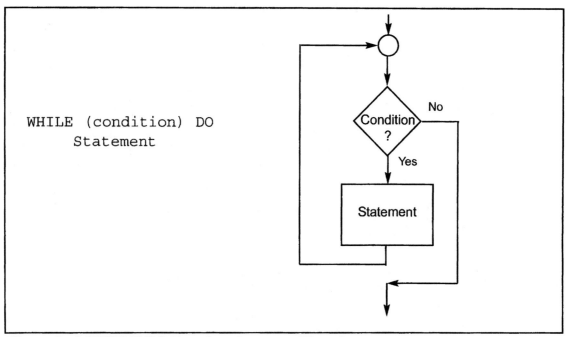

```
WHILE (condition) DO
     Statement
```

Figure D-6. WHILE DO Pseudocode versus Flowchart

Program D-1 finds the sum of a series of bytes. Compare the flowchart versus the pseudocode for Program D-1 (shown in Figure D-7). In this example, more program details are given than one usually finds. For example, this shows steps for initializing and decrementing counters. Another programmer may not include these steps in the flowchart or pseudocode. It is important to remember that the purpose of flowcharts or pseudocode is to show the flow of the program and what the program does, not the specific Assembly language instructions that accomplish the program's objectives. Notice also that the pseudocode gives the same information in a much more compact form than does the flowchart. It is important to note that sometimes pseudocode is written in layers, so that the outer level or layer shows the flow of the program and subsequent levels show more details of how the program accomplishes its assigned tasks.

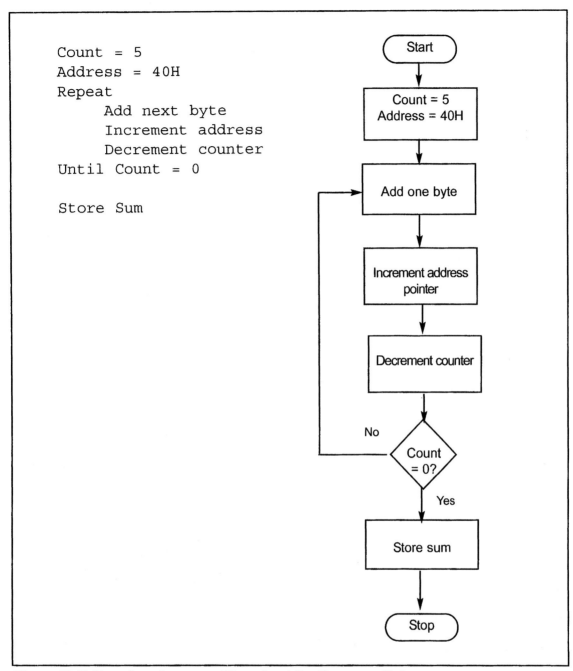

```
Count = 5
Address = 40H
Repeat
      Add next byte
      Increment address
      Decrement counter
Until Count = 0

Store Sum
```

Figure D-7. Pseudocode versus Flowchart for Program D-1

```
COUNTVAL       EQU 5               ;COUNT = 5
COUNTREG       SET 0x20            ;set aside location 20H for counter
SUM            SET 0x30            ;set aside location 30H for sum
       MOVLW   COUNTVAL            ;WREG = 5
       MOVWF   COUNTREG            ;load the counter
       LFSR    0,0x40              ;load pointer. FSR0 = 40H, RAM address
       CLRF    WREG                ;clear WREG
B5     ADDWF   POSTINC0, W         ;add RAM to WREG and increment FSR0
       DECF    COUNTREG,F          ;decrement counter
       BNZ     B5                  ;loop until counter = zero
       MOVWF   SUM                 ;store WREG in SUM
```

Program D-1

APPENDIX E.1

PIC18 PRIMER FOR x86 PROGRAMMERS

	x86	PIC18
8-bit registers:	AL, AH, BL, BH, CL, CH, DL, DH	WREG and up to 256 RAM locations in Access Bank
16-bit (data pointer):	BX, SI, DI	TBLPTR
Program Counter:	IP (16-bit)	PC (21-bit)
Input:	`MOV   DX,port addr` `IN    AL,DX`	`MOVFW PORTx ; (x = A,B,..G)`
Output:	`MOV   DX,port addr` `OUT   DX,AL`	`MOVWF PORTx  ; (x = A,B,..G)`
Loop:	`DEC CL` `JNZ TARGET`	`DECF  MyReg,F` `BNZ   TARGET`
Stack pointer:	SP (16-bit)	SP (21-bit)
	As we PUSH data onto the stack, it decrements the SP.	Push increments the SP. (Used exclusively for saving PC)
	As we POP data from the stack, it increments the SP.	Pop decrements the SP. (Used exclusively for retrieving PC)

Data movement:
  From the code segment:

	`MOV AL,CS:[SI]`	`TBLRD`

  From the data segment:

	`MOV AL,[SI]`	`MOVFW FSRx`

  From RAM:

	`MOV AL,[SI]` (Use SI, DI, or BX only.)	`MOVFW FSRx`

To RAM:	`MOV [SI],AL`	`MOVWF FSRx`

APPENDIX E.2

PIC18 PRIMER FOR
8051 PROGRAMMERS

	8051	**PIC18**
8-bit registers:	A, B, R0, R1, ...R7	WREG and up to 256 RAM locations in Access Bank

16-bit (data pointer):	DPTR	TBLPTR
Program Counter:	PC (16-bit)	PC (21-bit)
Input:		
	MOV A,Pn ; (n=0 - 3)	MOVFW PORTx ; (x = A,B,..G)

Output:		
	MOV Pn,A ; (n=0 - 3)	MOVWF PORTx ; (x = A,B,..G)

Loop:		
	DJNZ R3, TARGET (Using R0-R7)	DECF MyReg,F BNZ TARGET

Stack pointer:	SP (8-bit)	SP (21-bit)
	As we PUSH data onto the stack, it increments the SP.	Push increments the SP. (Used exclusively for saving PC)
	As we POP data from the stack, it decrements the SP.	Pop decrements the SP. (Used exclusively for retrieving PC)

Data movement:
    From the code segment:
        MOVC A,@A+PC   TBLRD
    From the data segment:
        MOVX A,@DPTR   MOVFW FSRx

    From RAM:
        MOV A, @R0   MOVFW FSRx
        (Use R0 or R1 only)
    To RAM:
        MOV @R0,A   MOVWF FSRx
        (Use R0 or R1 only)

APPENDIX F

ASCII CODES

Ctrl	Dec	Hex	Ch	Code
^@	0	00		NUL
^A	1	01	☺	SOH
^B	2	02	☻	STX
^C	3	03	♥	ETX
^D	4	04	♦	EOT
^E	5	05	♣	ENQ
^F	6	06	♠	ACK
^G	7	07	•	BEL
^H	8	08	◘	BS
^I	9	09	○	HT
^J	10	0A	◙	LF
^K	11	0B	♂	VT
^L	12	0C	♀	FF
^M	13	0D	♪	CR
^N	14	0E	♫	SO
^O	15	0F	☼	SI
^P	16	10	►	DLE
^Q	17	11	◄	DC1
^R	18	12	↕	DC2
^S	19	13	‼	DC3
^T	20	14	¶	DC4
^U	21	15	§	NAK
^V	22	16	▬	SYN
^W	23	17	↨	ETB
^X	24	18	↑	CAN
^Y	25	19	↓	EM
^Z	26	1A	→	SUB
^[	27	1B	←	ESC
^\	28	1C	∟	FS
^]	29	1D	↔	GS
^^	30	1E	▲	RS
^_	31	1F	▼	US

Dec	Hex	Ch
32	20	
33	21	!
34	22	"
35	23	#
36	24	$
37	25	%
38	26	&
39	27	'
40	28	(
41	29	)
42	2A	*
43	2B	+
44	2C	,
45	2D	–
46	2E	.
47	2F	/
48	30	0
49	31	1
50	32	2
51	33	3
52	34	4
53	35	5
54	36	6
55	37	7
56	38	8
57	39	9
58	3A	:
59	3B	;
60	3C	<
61	3D	=
62	3E	>
63	3F	?

Dec	Hex	Ch
64	40	@
65	41	A
66	42	B
67	43	C
68	44	D
69	45	E
70	46	F
71	47	G
72	48	H
73	49	I
74	4A	J
75	4B	K
76	4C	L
77	4D	M
78	4E	N
79	4F	O
80	50	P
81	51	Q
82	52	R
83	53	S
84	54	T
85	55	U
86	56	V
87	57	W
88	58	X
89	59	Y
90	5A	Z
91	5B	[
92	5C	\
93	5D	]
94	5E	^
95	5F	_

Dec	Hex	Ch
96	60	`
97	61	a
98	62	b
99	63	c
100	64	d
101	65	e
102	66	f
103	67	g
104	68	h
105	69	i
106	6A	j
107	6B	k
108	6C	l
109	6D	m
110	6E	n
111	6F	o
112	70	p
113	71	q
114	72	r
115	73	s
116	74	t
117	75	u
118	76	v
119	77	w
120	78	x
121	79	y
122	7A	z
123	7B	{
124	7C		
125	7D	}
126	7E	~
127	7F	⌂

Dec	Hex	Ch	Dec	Hex	Ch	Dec	Hex	Ch	Dec	Hex	Ch
128	80	Ç	160	A0	á	192	C0	└	224	E0	α
129	81	ü	161	A1	í	193	C1	┴	225	E1	β
130	82	é	162	A2	ó	194	C2	┬	226	E2	Γ
131	83	â	163	A3	ú	195	C3	├	227	E3	π
132	84	ä	164	A4	ñ	196	C4	─	228	E4	Σ
133	85	à	165	A5	Ñ	197	C5	┼	229	E5	σ
134	86	å	166	A6	ª	198	C6	╞	230	E6	μ
135	87	ç	167	A7	º	199	C7	╟	231	E7	τ
136	88	ê	168	A8	¿	200	C8	╚	232	E8	Φ
137	89	ë	169	A9	⌐	201	C9	╔	233	E9	θ
138	8A	è	170	AA	¬	202	CA	╩	234	EA	Ω
139	8B	ï	171	AB	½	203	CB	╦	235	EB	δ
140	8C	î	172	AC	¼	204	CC	╠	236	EC	∞
141	8D	ì	173	AD	¡	205	CD	═	237	ED	φ
142	8E	Ä	174	AE	«	206	CE	╬	238	EE	∈
143	8F	Å	175	AF	»	207	CF	╧	239	EF	∩
144	90	É	176	B0	░	208	D0	╨	240	F0	≡
145	91	æ	177	B1	▒	209	D1	╤	241	F1	±
146	92	Æ	178	B2	▓	210	D2	╥	242	F2	≥
147	93	ô	179	B3	│	211	D3	╙	243	F3	≤
148	94	ö	180	B4	┤	212	D4	?	244	F4	⌠
149	95	ò	181	B5	╡	213	D5	F	245	F5	⌡
150	96	û	182	B6	╢	214	D6	π	246	F6	÷
151	97	ù	183	B7	╖	215	D7	╫	247	F7	≈
152	98	ÿ	184	B8	╕	216	D8	╪	248	F8	≈
153	99	Ö	185	B9	╣	217	D9	┘	249	F9	·
154	9A	Ü	186	BA	║	218	DA	┌	250	FA	·
155	9B	¢	187	BB	╗	219	DB	█	251	FB	√
156	9C	£	188	BC	╝	220	DC	▄	252	FC	ⁿ
157	9D	¥	189	BD	╜	221	DD	▌	253	FD	²
158	9E	Pts	190	BE	╛	222	DE	▐	254	FE	■
159	9F	ƒ	191	BF	┐	223	DF	▀	255	FF	

APPENDIX G

ASSEMBLERS, DEVELOPMENT RESOURCES, AND SUPPLIERS

This appendix provides various sources for PIC18 assemblers and trainers. In addition, it lists some suppliers for chips and other hardware needs. While these are all established products from well-known companies, neither the authors nor the publisher assumes responsibility for any problem that may arise with any of them. You are neither encouraged nor discouraged from purchasing any of the products mentioned; you must make your own judgment in evaluating the products. This list is simply provided as a service to the reader. It also must be noted that the list of products is by no means complete or exhaustive.

PIC18 assemblers

The PIC18 assembler is provided by Microchip and other companies. Some of the companies provide shareware versions of their products, which you can download from their Web sites. However, the size of code for these shareware versions is limited to a few KB. Figure G-1 lists some suppliers of assemblers.

PIC18 trainers

There are many companies that produce and market PIC18 trainers. Figure G-2 provides a list of some of them.

Microchip Corp.
www.microchip.com

Custom Computer Services Inc
www.ccsinfo.com

Figure G-1. Suppliers of Assemblers and Compilers

Microchip Corp.
www.microchip.com

www.MicroDigitalEd.com

Custom Computer Services Inc.
www.ccsinfo.com

RSR Electronics
www.elexp.com

Figure G-2. Trainer Suppliers

Parts Suppliers

Figure G-3 provides a list of suppliers for many electronics parts.

RSR Electronics
Electronix Express
365 Blair Road
Avenel, NJ 07001
Fax: (732) 381-1572
Mail Order: 1-800-972-2225
In New Jersey: (732) 381-8020
www.elexp.com

Altex Electronics
11342 IH-35 North
San Antonio, TX 78233
Fax: (210) 637-3264
Mail Order: 1-800-531-5369
www.altex.com

Digi-Key
1-800-344-4539 (1-800-DIGI-KEY)
Fax: (218) 681-3380
www.digikey.com

Radio Shack
www.radioshack.com

JDR Microdevices
1850 South 10th St.
San Jose, CA 95112-4108
Sales 1-800-538-5000
(408) 494-1400
Fax: 1-800-538-5005
Fax: (408) 494-1420
www.jdr.com

Mouser Electronics
958 N. Main St.
Mansfield, TX 76063
1-800-346-6873
www.mouser.com

Jameco Electronic
1355 Shoreway Road
Belmont, CA 94002-4100
1-800-831-4242
(415) 592-8097
Fax: 1-800-237-6948
Fax: (415) 592-2503
www.jameco.com

B. G. Micro
P. O. Box 280298
Dallas, TX 75228
1-800-276-2206 (orders only)
(972) 271-5546
Fax: (972) 271-2462
This is an excellent source of LCDs, ICs,
keypads, etc.
www.bgmicro.com

Tanner Electronics
1100 Valwood Parkway, Suite #100
Carrollton, TX 75006
(972) 242-8702
www.tannerelectronics.com

Figure G-3. Electronics Suppliers

APPENDIX H

DATA SHEETS

25.0 INSTRUCTION SET SUMMARY

PIC18F2480/2580/4480/4580 devices incorporate the standard set of 75 PIC18 core instructions, as well as an extended set of 8 new instructions for the optimization of code that is recursive or that utilizes a software stack. The extended set is discussed later in this section.

25.1 Standard Instruction Set

The standard PIC18 instruction set adds many enhancements to the previous PICmicro® instruction sets, while maintaining an easy migration from these PICmicro instruction sets. Most instructions are a single program memory word (16 bits), but there are four instructions that require two program memory locations.

Each single-word instruction is a 16-bit word divided into an opcode, which specifies the instruction type and one or more operands, which further specify the operation of the instruction.

The instruction set is highly orthogonal and is grouped into four basic categories:

- **Byte-oriented** operations
- **Bit-oriented** operations
- **Literal** operations
- **Control** operations

The PIC18 instruction set summary in Table 25-2 lists **byte-oriented**, **bit-oriented**, **literal** and **control** operations. Table 25-1 shows the opcode field descriptions.

Most **byte-oriented** instructions have three operands:

1. The file register (specified by 'f')
2. The destination of the result (specified by 'd')
3. The accessed memory (specified by 'a')

The file register designator 'f' specifies which file register is to be used by the instruction. The destination designator 'd' specifies where the result of the operation is to be placed. If 'd' is zero, the result is placed in the WREG register. If 'd' is one, the result is placed in the file register specified in the instruction.

All **bit-oriented** instructions have three operands:

1. The file register (specified by 'f')
2. The bit in the file register (specified by 'b')
3. The accessed memory (specified by 'a')

The bit field designator 'b' selects the number of the bit affected by the operation, while the file register designator 'f' represents the number of the file in which the bit is located.

The **literal** instructions may use some of the following operands:

- A literal value to be loaded into a file register (specified by 'k')
- The desired FSR register to load the literal value into (specified by 'f')
- No operand required (specified by '—')

The **control** instructions may use some of the following operands:

- A program memory address (specified by 'n')
- The mode of the CALL or RETURN instructions (specified by 's')
- The mode of the table read and table write instructions (specified by 'm')
- No operand required (specified by '—')

All instructions are a single word, except for four double-word instructions. These instructions were made double-word to contain the required information in 32 bits. In the second word, the 4 MSbs are '1's. If this second word is executed as an instruction (by itself), it will execute as a NOP.

All single-word instructions are executed in a single instruction cycle, unless a conditional test is true or the program counter is changed as a result of the instruction. In these cases, the execution takes two instruction cycles with the additional instruction cycle(s) executed as a NOP.

The double-word instructions execute in two instruction cycles.

One instruction cycle consists of four oscillator periods. Thus, for an oscillator frequency of 4 MHz, the normal instruction execution time is 1 µs. If a conditional test is true, or the program counter is changed as a result of an instruction, the instruction execution time is 2 µs. Two-word branch instructions (if true) would take 3 µs.

Figure 25-1 shows the general formats that the instructions can have. All examples use the convention 'nnh' to represent a hexadecimal number.

The Instruction Set Summary, shown in Table 25-2, lists the standard instructions recognized by the Microchip MPASM™ Assembler.

Section 25.1.1 "Standard Instruction Set" provides a description of each instruction.

PIC18F2480/2580/4480/4580

TABLE 25-1: OPCODE FIELD DESCRIPTIONS

Field	Description
a	RAM access bit a = 0: RAM location in Access RAM (BSR register is ignored) a = 1: RAM bank is specified by BSR register
bbb	Bit address within an 8-bit file register (0 to 7).
BSR	Bank Select Register. Used to select the current RAM bank.
C, DC, Z, OV, N	ALU status bits: **Carry, Digit Carry, Zero, Overflow, Negative.**
d	Destination select bit d = 0: store result in WREG d = 1: store result in file register f
dest	Destination: either the WREG register or the specified register file location.
f	8-bit Register file address (00h to FFh), or 2-bit FSR designator (0h to 3h).
f$_s$	12-bit Register file address (000h to FFFh). This is the source address.
f$_d$	12-bit Register file address (000h to FFFh). This is the destination address.
GIE	Global Interrupt Enable bit.
k	Literal field, constant data or label (may be either an 8-bit, 12-bit or a 20-bit value)
label	Label name
mm	The mode of the TBLPTR register for the table read and table write instructions. Only used with table read and table write instructions:
*	No change to register (such as TBLPTR with table reads and writes)
*+	Post-Increment register (such as TBLPTR with table reads and writes)
*-	Post-Decrement register (such as TBLPTR with table reads and writes)
+*	Pre-Increment register (such as TBLPTR with table reads and writes)
n	The relative address (2's complement number) for relative branch instructions or the direct address for Call/Branch and Return instructions
PC	Program Counter.
PCL	Program Counter Low Byte.
PCH	Program Counter High Byte.
PCLATH	Program Counter High Byte Latch.
PCLATU	Program Counter Upper Byte Latch.
$\overline{PD}$	Power-down bit.
PRODH	Product of Multiply High Byte.
PRODL	Product of Multiply Low Byte.
s	Fast Call/Return mode select bit s = 0: do not update into/from shadow registers s = 1: certain registers loaded into/from shadow registers (Fast mode)
TBLPTR	21-bit Table Pointer (points to a Program Memory location).
TABLAT	8-bit Table Latch.
$\overline{TO}$	Time-out bit.
TOS	Top-of-Stack.
u	Unused or unchanged.
WDT	Watchdog Timer.
WREG	Working register (accumulator).
x	Don't care ('0' or '1'). The assembler will generate code with x = 0. It is the recommended form of use for compatibility with all Microchip software tools.
z$_s$	7-bit offset value for indirect addressing of register files (source).
z$_d$	7-bit offset value for indirect addressing of register files (destination).
{ }	Optional argument.
[text]	Indicates an indexed address.
(text)	The contents of text.
[expr]<n>	Specifies bit n of the register indicated by the pointer expr.
→	Assigned to.
< >	Register bit field.
∈	In the set of.
italics	User defined term (font is Courier).

Preliminary © 2004 Microchip Technology Inc.

FIGURE 25-1: **GENERAL FORMAT FOR INSTRUCTIONS**

Byte-oriented file register operations   **Example Instruction**

15	10 9	8	7	0
OPCODE	d	a	f (FILE #)	

ADDWF MYREG, W, B

d = 0 for result destination to be WREG register
d = 1 for result destination to be file register (f)
a = 0 to force Access Bank
a = 1 for BSR to select bank
f = 8-bit file register address

Byte to Byte move operations (2-word)

15	12	11	0
OPCODE		f (Source FILE #)	

15	12	11	0
1111		f (Destination FILE #)	

MOVFF MYREG1, MYREG2

f = 12-bit file register address

Bit-oriented file register operations

15	12	11	9	8	7	0
OPCODE	b (BIT #)			a	f (FILE #)	

BSF MYREG, bit, B

b = 3-bit position of bit in file register (f)
a = 0 to force Access Bank
a = 1 for BSR to select bank
f = 8-bit file register address

Literal operations

15	8	7	0
OPCODE		k (literal)	

MOVLW 7Fh

k = 8-bit immediate value

Control operations

CALL, GOTO **and Branch** operations

15	8	7	0
OPCODE		n<7:0> (literal)	

15	12	11	0
1111		n<19:8> (literal)	

GOTO Label

n = 20-bit immediate value

15	8	7	0
OPCODE		S	n<7:0> (literal)

15	12	11	0
1111		n<19:8> (literal)	

CALL MYFUNC

S = Fast bit

15	11	10	0
OPCODE		n<10:0> (literal)	

BRA MYFUNC

15	8	7	0
OPCODE		n<7:0> (literal)	

BC MYFUNC

TABLE 25-2: PIC18FXXXX INSTRUCTION SET

Mnemonic, Operands		Description	Cycles	16-Bit Instruction Word				Status Affected	Notes
				MSb			LSb		
BYTE-ORIENTED OPERATIONS									
ADDWF	f, d, a	Add WREG and f	1	0010	01da	ffff	ffff	C, DC, Z, OV, N	1, 2
ADDWFC	f, d, a	Add WREG and Carry bit to f	1	0010	00da	ffff	ffff	C, DC, Z, OV, N	1, 2
ANDWF	f, d, a	AND WREG with f	1	0001	01da	ffff	ffff	Z, N	1,2
CLRF	f, a	Clear f	1	0110	101a	ffff	ffff	Z	2
COMF	f, d, a	Complement f	1	0001	11da	ffff	ffff	Z, N	1, 2
CPFSEQ	f, a	Compare f with WREG, skip =	1 (2 or 3)	0110	001a	ffff	ffff	None	4
CPFSGT	f, a	Compare f with WREG, skip >	1 (2 or 3)	0110	010a	ffff	ffff	None	4
CPFSLT	f, a	Compare f with WREG, skip <	1 (2 or 3)	0110	000a	ffff	ffff	None	1, 2
DECF	f, d, a	Decrement f	1	0000	01da	ffff	ffff	C, DC, Z, OV, N	1, 2, 3, 4
DECFSZ	f, d, a	Decrement f, Skip if 0	1 (2 or 3)	0010	11da	ffff	ffff	None	1, 2, 3, 4
DCFSNZ	f, d, a	Decrement f, Skip if Not 0	1 (2 or 3)	0100	11da	ffff	ffff	None	1, 2
INCF	f, d, a	Increment f	1	0010	10da	ffff	ffff	C, DC, Z, OV, N	1, 2, 3, 4
INCFSZ	f, d, a	Increment f, Skip if 0	1 (2 or 3)	0011	11da	ffff	ffff	None	4
INFSNZ	f, d, a	Increment f, Skip if Not 0	1 (2 or 3)	0100	10da	ffff	ffff	None	1, 2
IORWF	f, d, a	Inclusive OR WREG with f	1	0001	00da	ffff	ffff	Z, N	1, 2
MOVF	f, d, a	Move f	1	0101	00da	ffff	ffff	Z, N	1
MOVFF	f_s, f_d	Move f_s (source) to 1st word f_d (destination) 2nd word	2	1100 1111	ffff ffff	ffff ffff	ffff	None	
MOVWF	f, a	Move WREG to f	1	0110	111a	ffff	ffff	None	
MULWF	f, a	Multiply WREG with f	1	0000	001a	ffff	ffff	None	1, 2
NEGF	f, a	Negate f	1	0110	110a	ffff	ffff	C, DC, Z, OV, N	
RLCF	f, d, a	Rotate Left f through Carry	1	0011	01da	ffff	ffff	C, Z, N	1, 2
RLNCF	f, d, a	Rotate Left f (No Carry)	1	0100	01da	ffff	ffff	Z, N	
RRCF	f, d, a	Rotate Right f through Carry	1	0011	00da	ffff	ffff	C, Z, N	
RRNCF	f, d, a	Rotate Right f (No Carry)	1	0100	00da	ffff	ffff	Z, N	
SETF	f, a	Set f	1	0110	100a	ffff	ffff	None	1, 2
SUBFWB	f, d, a	Subtract f from WREG with borrow	1	0101	01da	ffff	ffff	C, DC, Z, OV, N	
SUBWF	f, d, a	Subtract WREG from f	1	0101	11da	ffff	ffff	C, DC, Z, OV, N	1, 2
SUBWFB	f, d, a	Subtract WREG from f with borrow	1	0101	10da	ffff	ffff	C, DC, Z, OV, N	
SWAPF	f, d, a	Swap nibbles in f	1	0011	10da	ffff	ffff	None	4
TSTFSZ	f, a	Test f, skip if 0	1 (2 or 3)	0110	011a	ffff	ffff	None	1, 2
XORWF	f, d, a	Exclusive OR WREG with f	1	0001	10da	ffff	ffff	Z, N	

Note 1: When a Port register is modified as a function of itself (e.g., MOVF PORTB, 1, 0), the value used will be that value present on the pins themselves. For example, if the data latch is '1' for a pin configured as input and is driven low by an external device, the data will be written back with a '0'.

2: If this instruction is executed on the TMR0 register (and where applicable, 'd' = 1), the prescaler will be cleared if assigned.

3: If Program Counter (PC) is modified or a conditional test is true, the instruction requires two cycles. The second cycle is executed as a NOP.

4: Some instructions are two-word instructions. The second word of these instructions will be executed as a NOP unless the first word of the instruction retrieves the information embedded in these 16 bits. This ensures that all program memory locations have a valid instruction.

5: If the table write starts the write cycle to internal memory, the write will continue until terminated.

TABLE 25-2: **PIC18FXXXX INSTRUCTION SET (CONTINUED)**

Mnemonic, Operands		Description	Cycles	16-Bit Instruction Word				Status Affected	Notes
				MSb			LSb		
BIT-ORIENTED OPERATIONS									
BCF	f, b, a	Bit Clear f	1	1001	bbba	ffff	ffff	None	1, 2
BSF	f, b, a	Bit Set f	1	1000	bbba	ffff	ffff	None	1, 2
BTFSC	f, b, a	Bit Test f, Skip if Clear	1 (2 or 3)	1011	bbba	ffff	ffff	None	3, 4
BTFSS	f, b, a	Bit Test f, Skip if Set	1 (2 or 3)	1010	bbba	ffff	ffff	None	3, 4
BTG	f, d, a	Bit Toggle f	1	0111	bbba	ffff	ffff	None	1, 2
CONTROL OPERATIONS									
BC	n	Branch if Carry	1 (2)	1110	0010	nnnn	nnnn	None	
BN	n	Branch if Negative	1 (2)	1110	0110	nnnn	nnnn	None	
BNC	n	Branch if Not Carry	1 (2)	1110	0011	nnnn	nnnn	None	
BNN	n	Branch if Not Negative	1 (2)	1110	0111	nnnn	nnnn	None	
BNOV	n	Branch if Not Overflow	1 (2)	1110	0101	nnnn	nnnn	None	
BNZ	n	Branch if Not Zero	1 (2)	1110	0001	nnnn	nnnn	None	
BOV	n	Branch if Overflow	1 (2)	1110	0100	nnnn	nnnn	None	
BRA	n	Branch Unconditionally	2	1101	0nnn	nnnn	nnnn	None	
BZ	n	Branch if Zero	1 (2)	1110	0000	nnnn	nnnn	None	
CALL	n, s	Call subroutine 1st word	2	1110	110s	kkkk	kkkk	None	
		2nd word		1111	kkkk	kkkk	kkkk		
CLRWDT	—	Clear Watchdog Timer	1	0000	0000	0000	0100	$\overline{TO}, \overline{PD}$	
DAW	—	Decimal Adjust WREG	1	0000	0000	0000	0111	C	
GOTO	n	Go to address 1st word	2	1110	1111	kkkk	kkkk	None	
		2nd word		1111	kkkk	kkkk	kkkk		
NOP	—	No Operation	1	0000	0000	0000	0000	None	
NOP	—	No Operation	1	1111	xxxx	xxxx	xxxx	None	4
POP	—	Pop top of return stack (TOS)	1	0000	0000	0000	0110	None	
PUSH	—	Push top of return stack (TOS)	1	0000	0000	0000	0101	None	
RCALL	n	Relative Call	2	1101	1nnn	nnnn	nnnn	None	
RESET		Software device Reset	1	0000	0000	1111	1111	All	
RETFIE	s	Return from interrupt enable	2	0000	0000	0001	000s	GIE/GIEH, PEIE/GIEL	
RETLW	k	Return with literal in WREG	2	0000	1100	kkkk	kkkk	None	
RETURN	s	Return from Subroutine	2	0000	0000	0001	001s	None	
SLEEP	—	Go into Standby mode	1	0000	0000	0000	0011	$\overline{TO}, \overline{PD}$	

Note 1: When a Port register is modified as a function of itself (e.g., MOVF PORTB, 1, 0), the value used will be that value present on the pins themselves. For example, if the data latch is '1' for a pin configured as input and is driven low by an external device, the data will be written back with a '0'.

2: If this instruction is executed on the TMR0 register (and where applicable, 'd' = 1), the prescaler will be cleared if assigned.

3: If Program Counter (PC) is modified or a conditional test is true, the instruction requires two cycles. The second cycle is executed as a NOP.

4: Some instructions are two-word instructions. The second word of these instructions will be executed as a NOP unless the first word of the instruction retrieves the information embedded in these 16 bits. This ensures that all program memory locations have a valid instruction.

5: If the table write starts the write cycle to internal memory, the write will continue until terminated.

PIC18F2480/2580/4480/4580

TABLE 25-2:   PIC18FXXXX INSTRUCTION SET (CONTINUED)

Mnemonic, Operands		Description	Cycles	16-Bit Instruction Word				Status Affected	Notes
				MSb			LSb		
LITERAL OPERATIONS									
ADDLW	k	Add literal and WREG	1	0000	1111	kkkk	kkkk	C, DC, Z, OV, N	
ANDLW	k	AND literal with WREG	1	0000	1011	kkkk	kkkk	Z, N	
IORLW	k	Inclusive OR literal with WREG	1	0000	1001	kkkk	kkkk	Z, N	
LFSR	f, k	Move literal (12-bit) 2nd word	2	1110	1110	00ff	kkkk	None	
		to FSR(f) 1st word		1111	0000	kkkk	kkkk		
MOVLB	k	Move literal to BSR<3:0>	1	0000	0001	0000	kkkk	None	
MOVLW	k	Move literal to WREG	1	0000	1110	kkkk	kkkk	None	
MULLW	k	Multiply literal with WREG	1	0000	1101	kkkk	kkkk	None	
RETLW	k	Return with literal in WREG	2	0000	1100	kkkk	kkkk	None	
SUBLW	k	Subtract WREG from literal	1	0000	1000	kkkk	kkkk	C, DC, Z, OV, N	
XORLW	k	Exclusive OR literal with WREG	1	0000	1010	kkkk	kkkk	Z, N	
DATA MEMORY ↔ PROGRAM MEMORY OPERATIONS									
TBLRD*		Table Read	2	0000	0000	0000	1000	None	
TBLRD*+		Table Read with post-increment		0000	0000	0000	1001	None	
TBLRD*-		Table Read with post-decrement		0000	0000	0000	1010	None	
TBLRD+*		Table Read with pre-increment		0000	0000	0000	1011	None	
TBLWT*		Table Write	2	0000	0000	0000	1100	None	5
TBLWT*+		Table Write with post-increment		0000	0000	0000	1101	None	5
TBLWT*-		Table Write with post-decrement		0000	0000	0000	1110	None	5
TBLWT+*		Table Write with pre-increment		0000	0000	0000	1111	None	5

Note 1: When a Port register is modified as a function of itself (e.g., MOVF PORTB, 1, 0), the value used will be that value present on the pins themselves. For example, if the data latch is '1' for a pin configured as input and is driven low by an external device, the data will be written back with a '0'.

2: If this instruction is executed on the TMR0 register (and where applicable, 'd' = 1), the prescaler will be cleared if assigned.

3: If Program Counter (PC) is modified or a conditional test is true, the instruction requires two cycles. The second cycle is executed as a NOP.

4: Some instructions are two-word instructions. The second word of these instructions will be executed as a NOP unless the first word of the instruction retrieves the information embedded in these 16 bits. This ensures that all program memory locations have a valid instruction.

5: If the table write starts the write cycle to internal memory, the write will continue until terminated.

25.1.1   STANDARD INSTRUCTION SET

ADDLW	ADD Literal to W
Syntax:	ADDLW k
Operands:	$0 \le k \le 255$
Operation:	$(W) + k \rightarrow W$
Status Affected:	N, OV, C, DC, Z

Encoding:

0000	1111	kkkk	kkkk

Description: The contents of W are added to the 8-bit literal 'k' and the result is placed in W.

Words: 1

Cycles: 1

Q Cycle Activity:

Q1	Q2	Q3	Q4
Decode	Read literal 'k'	Process Data	Write to W

Example: ADDLW   15h

Before Instruction
  W   =   10h
After Instruction
  W   =   25h

ADDWF	ADD W to f
Syntax:	ADDWF f {,d {,a}}
Operands:	$0 \le f \le 255$ $d \in [0,1]$ $a \in [0,1]$
Operation:	$(W) + (f) \rightarrow dest$
Status Affected:	N, OV, C, DC, Z

Encoding:

0010	01da	ffff	ffff

Description: Add W to register 'f'. If 'd' is '0', the result is stored in W. If 'd' is '1', the result is stored back in register 'f' (default).

If 'a' is '0', the Access Bank is selected. If 'a' is '1', the BSR is used to select the GPR bank (default).

If 'a' is '0' and the extended instruction set is enabled, this instruction operates in Indexed Literal Offset Addressing mode whenever f ≤ 95 (5Fh). See **Section 25.2.3 "Byte-Oriented and Bit-Oriented Instructions in Indexed Literal Offset Mode"** for details.

Words: 1

Cycles: 1

Q Cycle Activity:

Q1	Q2	Q3	Q4
Decode	Read register 'f'	Process Data	Write to destination

Example: ADDWF   REG, 0, 0

Before Instruction
  W   =   17h
  REG   =   0C2h
After Instruction
  W   =   0D9h
  REG   =   0C2h

Note: All PIC18 instructions may take an optional label argument preceding the instruction mnemonic for use in symbolic addressing. If a label is used, the instruction format then becomes: {label} instruction argument(s).

ADDWFC	ADD W and Carry bit to f
Syntax:	ADDWFC f {,d {,a}}
Operands:	$0 \le f \le 255$ $d \in [0,1]$ $a \in [0,1]$
Operation:	(W) + (f) + (C) → dest
Status Affected:	N,OV, C, DC, Z

Encoding:

0010	00da	ffff	ffff

Description:
Add W, the Carry flag and data memory location 'f'. If 'd' is '0', the result is placed in W. If 'd' is '1', the result is placed in data memory location 'f'.

If 'a' is '0', the Access Bank is selected. If 'a' is '1', the BSR is used to select the GPR bank (default).

If 'a' is '0' and the extended instruction set is enabled, this instruction operates in Indexed Literal Offset Addressing mode whenever $f \le 95$ (5Fh). See **Section 25.2.3 "Byte-Oriented and Bit-Oriented Instructions in Indexed Literal Offset Mode"** for details.

Words: 1

Cycles: 1

Q Cycle Activity:

Q1	Q2	Q3	Q4
Decode	Read register 'f'	Process Data	Write to destination

Example: ADDWFC   REG, 0, 1

Before Instruction
Carry bit = 1
REG   = 02h
W   = 4Dh

After Instruction
Carry bit = 0
REG   = 02h
W   = 50h

ANDLW	AND Literal with W
Syntax:	ANDLW k
Operands:	$0 \le k \le 255$
Operation:	(W) .AND. k → W
Status Affected:	N, Z

Encoding:

0000	1011	kkkk	kkkk

Description:
The contents of W are ANDed with the 8-bit literal 'k'. The result is placed in W.

Words: 1

Cycles: 1

Q Cycle Activity:

Q1	Q2	Q3	Q4
Decode	Read literal 'k'	Process Data	Write to W

Example: ANDLW   05Fh

Before Instruction
W   =   A3h
After Instruction
W   =   03h

ANDWF	AND W with f

Syntax:	ANDWF f {,d {,a}}
Operands:	$0 \leq f \leq 255$ $d \in [0,1]$ $a \in [0,1]$
Operation:	(W) .AND. (f) → dest
Status Affected:	N, Z
Encoding:	0001 | 01da | ffff | ffff
Description:	The contents of W are AND'ed with register 'f'. If 'd' is '0', the result is stored in W. If 'd' is '1', the result is stored back in register 'f' (default). If 'a' is '0', the Access Bank is selected. If 'a' is '1', the BSR is used to select the GPR bank (default). If 'a' is '0' and the extended instruction set is enabled, this instruction operates in Indexed Literal Offset Addressing mode whenever $f \leq 95$ (5Fh). See **Section 25.2.3 "Byte-Oriented and Bit-Oriented Instructions in Indexed Literal Offset Mode"** for details.
Words:	1
Cycles:	1

Q Cycle Activity:

Q1	Q2	Q3	Q4
Decode	Read register 'f'	Process Data	Write to destination

Example: ANDWF   REG, 0, 0

Before Instruction
W	=	17h
REG	=	C2h

After Instruction
W	=	02h
REG	=	C2h

BC	Branch if Carry

Syntax:	BC n
Operands:	$-128 \leq n \leq 127$
Operation:	if Carry bit is '1' (PC) + 2 + 2n → PC
Status Affected:	None
Encoding:	1110 | 0010 | nnnn | nnnn
Description:	If the Carry bit is '1', then the program will branch. The 2's complement number '2n' is added to the PC. Since the PC will have incremented to fetch the next instruction, the new address will be PC + 2 + 2n. This instruction is then a two-cycle instruction.
Words:	1
Cycles:	1(2)

Q Cycle Activity:
If Jump:

Q1	Q2	Q3	Q4
Decode	Read literal 'n'	Process Data	Write to PC
No operation	No operation	No operation	No operation

If No Jump:

Q1	Q2	Q3	Q4
Decode	Read literal 'n'	Process Data	No operation

Example: HERE   BC   5

Before Instruction
PC	=	address (HERE)

After Instruction
If Carry	=	1;
PC	=	address (HERE + 12)
If Carry	=	0;
PC	=	address (HERE + 2)

BCF	Bit Clear f
Syntax:	BCF f, b {,a}
Operands:	$0 \le f \le 255$ $0 \le b \le 7$ $a \in [0,1]$
Operation:	$0 \rightarrow f<b>$
Status Affected:	None
Encoding:	1001 bbba ffff ffff
Description:	Bit 'b' in register 'f' is cleared. If 'a' is '0', the Access Bank is selected. If 'a' is '1', the BSR is used to select the GPR bank (default). If 'a' is '0' and the extended instruction set is enabled, this instruction operates in Indexed Literal Offset addressing mode whenever $f \le 95$ (5Fh). See **Section 25.2.3 "Byte-Oriented and Bit-Oriented Instructions in Indexed Literal Offset Mode"** for details.
Words:	1
Cycles:	1

Q Cycle Activity:

Q1	Q2	Q3	Q4
Decode	Read register 'f'	Process Data	Write register 'f'

<u>Example:</u>   BCF   FLAG_REG, 7, 0

Before Instruction
    FLAG_REG = C7h
After Instruction
    FLAG_REG = 47h

BN	Branch if Negative
Syntax:	BN n
Operands:	$-128 \le n \le 127$
Operation:	if Negative bit is '1' $(PC) + 2 + 2n \rightarrow PC$
Status Affected:	None
Encoding:	1110 0110 nnnn nnnn
Description:	If the Negative bit is '1', then the program will branch. The 2's complement number '2n' is added to the PC. Since the PC will have incremented to fetch the next instruction, the new address will be PC + 2 + 2n. This instruction is then a two-cycle instruction.
Words:	1
Cycles:	1(2)

Q Cycle Activity:
If Jump:

Q1	Q2	Q3	Q4
Decode	Read literal 'n'	Process Data	Write to PC
No operation	No operation	No operation	No operation

If No Jump:

Q1	Q2	Q3	Q4
Decode	Read literal 'n'	Process Data	No operation

<u>Example:</u>   HERE   BN   Jump

Before Instruction
    PC   =   address (HERE)
After Instruction
    If Negative   =   1;
        PC   =   address (Jump)
    If Negative   =   0;
        PC   =   address (HERE + 2)

Preliminary

BNC	Branch if Not Carry
Syntax:	BNC n
Operands:	$-128 \leq n \leq 127$
Operation:	if Carry bit is '0' (PC) + 2 + 2n → PC
Status Affected:	None

Encoding:

1110	0011	nnnn	nnnn

Description: If the Carry bit is '0', then the program will branch.

The 2's complement number '2n' is added to the PC. Since the PC will have incremented to fetch the next instruction, the new address will be PC + 2 + 2n. This instruction is then a two-cycle instruction.

Words: 1

Cycles: 1(2)

Q Cycle Activity:

If Jump:

Q1	Q2	Q3	Q4
Decode	Read literal 'n'	Process Data	Write to PC
No operation	No operation	No operation	No operation

If No Jump:

Q1	Q2	Q3	Q4
Decode	Read literal 'n'	Process Data	No operation

Example: HERE   BNC   Jump

Before Instruction
    PC   =   address (HERE)
After Instruction
    If Carry   =   0;
       PC   =   address (Jump)
    If Carry   =   1;
       PC   =   address (HERE + 2)

BNN	Branch if Not Negative
Syntax:	BNN n
Operands:	$-128 \leq n \leq 127$
Operation:	if Negative bit is '0' (PC) + 2 + 2n → PC
Status Affected:	None

Encoding:

1110	0111	nnnn	nnnn

Description: If the Negative bit is '0', then the program will branch.

The 2's complement number '2n' is added to the PC. Since the PC will have incremented to fetch the next instruction, the new address will be PC + 2 + 2n. This instruction is then a two-cycle instruction.

Words: 1

Cycles: 1(2)

Q Cycle Activity:

If Jump:

Q1	Q2	Q3	Q4
Decode	Read literal 'n'	Process Data	Write to PC
No operation	No operation	No operation	No operation

If No Jump:

Q1	Q2	Q3	Q4
Decode	Read literal 'n'	Process Data	No operation

Example: HERE   BNN   Jump

Before Instruction
    PC   =   address (HERE)
After Instruction
    If Negative   =   0;
       PC   =   address (Jump)
    If Negative   =   1;
       PC   =   address (HERE + 2)

BNOV	Branch if Not Overflow
Syntax:	BNOV n
Operands:	$-128 \le n \le 127$
Operation:	if Overflow bit is '0' $(PC) + 2 + 2n \rightarrow PC$
Status Affected:	None

Encoding:

1110	0101	nnnn	nnnn

Description: If the Overflow bit is '0', then the program will branch.

The 2's complement number '2n' is added to the PC. Since the PC will have incremented to fetch the next instruction, the new address will be PC + 2 + 2n. This instruction is then a two-cycle instruction.

Words: 1

Cycles: 1(2)

Q Cycle Activity:

If Jump:

Q1	Q2	Q3	Q4
Decode	Read literal 'n'	Process Data	Write to PC
No operation	No operation	No operation	No operation

If No Jump:

Q1	Q2	Q3	Q4
Decode	Read literal 'n'	Process Data	No operation

Example: HERE   BNOV Jump

Before Instruction
PC   =   address (HERE)
After Instruction
If Overflow =   0;
PC   =   address (Jump)
If Overflow =   1;
PC   =   address (HERE + 2)

BNZ	Branch if Not Zero
Syntax:	BNZ n
Operands:	$-128 \le n \le 127$
Operation:	if Zero bit is '0' $(PC) + 2 + 2n \rightarrow PC$
Status Affected:	None

Encoding:

1110	0001	nnnn	nnnn

Description: If the Zero bit is '0', then the program will branch.

The 2's complement number '2n' is added to the PC. Since the PC will have incremented to fetch the next instruction, the new address will be PC + 2 + 2n. This instruction is then a two-cycle instruction.

Words: 1

Cycles: 1(2)

Q Cycle Activity:

If Jump:

Q1	Q2	Q3	Q4
Decode	Read literal 'n'	Process Data	Write to PC
No operation	No operation	No operation	No operation

If No Jump:

Q1	Q2	Q3	Q4
Decode	Read literal 'n'	Process Data	No operation

Example: HERE   BNZ Jump

Before Instruction
PC   =   address (HERE)
After Instruction
If Zero   =   0;
PC   =   address (Jump)
If Zero   =   1;
PC   =   address (HERE + 2)

BRA	**Unconditional Branch**
Syntax:	BRA n
Operands:	$-1024 \leq n \leq 1023$
Operation:	$(PC) + 2 + 2n \rightarrow PC$
Status Affected:	None

Encoding:

1101	0nnn	nnnn	nnnn

Description: Add the 2's complement number '2n' to the PC. Since the PC will have incremented to fetch the next instruction, the new address will be PC + 2 + 2n. This instruction is a two-cycle instruction.

Words: 1

Cycles: 2

Q Cycle Activity:

Q1	Q2	Q3	Q4
Decode	Read literal 'n'	Process Data	Write to PC
No operation	No operation	No operation	No operation

Example: HERE   BRA   Jump

Before Instruction
PC   =   address (HERE)
After Instruction
PC   =   address (Jump)

BSF	**Bit Set f**
Syntax:	BSF f, b {,a}
Operands:	$0 \leq f \leq 255$ $0 \leq b \leq 7$ $a \in [0,1]$
Operation:	$1 \rightarrow f<b>$
Status Affected:	None

Encoding:

1000	bbba	ffff	ffff

Description: Bit 'b' in register 'f' is set.

If 'a' is '0', the Access Bank is selected. If 'a' is '1', the BSR is used to select the GPR bank (default).

If 'a' is '0' and the extended instruction set is enabled, this instruction operates in Indexed Literal Offset Addressing mode whenever $f \leq 95$ (5Fh). See **Section 25.2.3 "Byte-Oriented and Bit-Oriented Instructions in Indexed Literal Offset Mode"** for details.

Words: 1

Cycles: 1

Q Cycle Activity:

Q1	Q2	Q3	Q4
Decode	Read register 'f'	Process Data	Write register 'f'

Example: BSF   FLAG_REG, 7, 1

Before Instruction
FLAG_REG   =   0Ah
After Instruction
FLAG_REG   =   8Ah

BTFSC	Bit Test File, Skip if Clear
Syntax:	BTFSC f, b {,a}
Operands:	$0 \le f \le 255$ $0 \le b \le 7$ $a \in [0,1]$
Operation:	skip if (f<b>) = 0
Status Affected:	None

Encoding:

1011	bbba	ffff	ffff

Description:
If bit 'b' in register 'f' is '0', then the next instruction is skipped. If bit 'b' is '0', then the next instruction fetched during the current instruction execution is discarded and a NOP is executed instead, making this a two-cycle instruction.

If 'a' is '0', the Access Bank is selected. If 'a' is '1', the BSR is used to select the GPR bank (default).

If 'a' is '0' and the extended instruction set is enabled, this instruction operates in Indexed Literal Offset Addressing mode whenever f ≤ 95 (5Fh).
See **Section 25.2.3 "Byte-Oriented and Bit-Oriented Instructions in Indexed Literal Offset Mode"** for details.

Words:	1
Cycles:	1(2)
	Note: 3 cycles if skip and followed by a 2-word instruction.

Q Cycle Activity:

Q1	Q2	Q3	Q4
Decode	Read register 'f'	Process Data	No operation

If skip:

Q1	Q2	Q3	Q4
No operation	No operation	No operation	No operation

If skip and followed by 2-word instruction:

Q1	Q2	Q3	Q4
No operation	No operation	No operation	No operation
No operation	No operation	No operation	No operation

<u>Example:</u>

```
HERE    BTFSC   FLAG, 1, 0
FALSE   :
TRUE    :
```

Before Instruction
    PC   =   address (HERE)
After Instruction
    If FLAG<1> =   0;
        PC   =   address (TRUE)
    If FLAG<1> =   1;
        PC   =   address (FALSE)

BTFSS	Bit Test File, Skip if Set
Syntax:	BTFSS f, b {,a}
Operands:	$0 \le f \le 255$ $0 \le b < 7$ $a \in [0,1]$
Operation:	skip if (f<b>) = 1
Status Affected:	None

Encoding:

1010	bbba	ffff	ffff

Description:
If bit 'b' in register 'f' is '1', then the next instruction is skipped. If bit 'b' is '1', then the next instruction fetched during the current instruction execution is discarded and a NOP is executed instead, making this a two-cycle instruction.

If 'a' is '0', the Access Bank is selected. If 'a' is '1', the BSR is used to select the GPR bank (default).

If 'a' is '0' and the extended instruction set is enabled, this instruction operates in Indexed Literal Offset Addressing mode whenever f ≤ 95 (5Fh).
See **Section 25.2.3 "Byte-Oriented and Bit-Oriented Instructions in Indexed Literal Offset Mode"** for details.

Words:	1
Cycles:	1(2)
	Note: 3 cycles if skip and followed by a 2-word instruction.

Q Cycle Activity:

Q1	Q2	Q3	Q4
Decode	Read register 'f'	Process Data	No operation

If skip:

Q1	Q2	Q3	Q4
No operation	No operation	No operation	No operation

If skip and followed by 2-word instruction:

Q1	Q2	Q3	Q4
No operation	No operation	No operation	No operation
No operation	No operation	No operation	No operation

<u>Example:</u>

```
HERE    BTFSS   FLAG, 1, 0
FALSE   :
TRUE    :
```

Before Instruction
    PC   =   address (HERE)
After Instruction
    If FLAG<1> =   0;
        PC   =   address (FALSE)
    If FLAG<1> =   1;
        PC   =   address (TRUE)

BTG	Bit Toggle f

Syntax:	BTG f, b {,a}
Operands:	$0 \leq f \leq 255$ $0 \leq b < 7$ $a \in [0,1]$
Operation:	$(\overline{f<b>}) \rightarrow f<b>$
Status Affected:	None
Encoding:	0111 | bbba | ffff | ffff
Description:	Bit 'b' in data memory location 'f' is inverted. If 'a' is '0', the Access Bank is selected. If 'a' is '1', the BSR is used to select the GPR bank (default). If 'a' is '0' and the extended instruction set is enabled, this instruction operates in Indexed Literal Offset Addressing mode whenever $f \leq 95$ (5Fh). See **Section 25.2.3 "Byte-Oriented and Bit-Oriented Instructions in Indexed Literal Offset Mode"** for details.
Words:	1
Cycles:	1

Q Cycle Activity:

Q1	Q2	Q3	Q4
Decode	Read register 'f'	Process Data	Write register 'f'

Example: BTG PORTC, 4, 0

Before Instruction:
  PORTC = 0111 0101 [75h]
After Instruction:
  PORTC = 0110 0101 [65h]

BOV	Branch if Overflow

Syntax:	BOV n
Operands:	$-128 \leq n \leq 127$
Operation:	if Overflow bit is '1' $(PC) + 2 + 2n \rightarrow PC$
Status Affected:	None
Encoding:	1110 | 0100 | nnnn | nnnn
Description:	If the Overflow bit is '1', then the program will branch. The 2's complement number '2n' is added to the PC. Since the PC will have incremented to fetch the next instruction, the new address will be PC + 2 + 2n. This instruction is then a two-cycle instruction.
Words:	1
Cycles:	1(2)

Q Cycle Activity:
If Jump:

Q1	Q2	Q3	Q4
Decode	Read literal 'n'	Process Data	Write to PC
No operation	No operation	No operation	No operation

If No Jump:

Q1	Q2	Q3	Q4
Decode	Read literal 'n'	Process Data	No operation

Example: HERE BOV Jump

Before Instruction
  PC   =   address (HERE)
After Instruction
  If Overflow =   1;
    PC   =   address (Jump)
  If Overflow =   0;
    PC   =   address (HERE + 2)

BZ	Branch if Zero
Syntax:	BZ n
Operands:	$-128 \le n \le 127$
Operation:	if Zero bit is '1' $(PC) + 2 + 2n \to PC$
Status Affected:	None
Encoding:	1110 0000 nnnn nnnn
Description:	If the Zero bit is '1', then the program will branch. The 2's complement number '2n' is added to the PC. Since the PC will have incremented to fetch the next instruction, the new address will be PC + 2 + 2n. This instruction is then a two-cycle instruction.
Words:	1
Cycles:	1(2)

Q Cycle Activity:

If Jump:

Q1	Q2	Q3	Q4
Decode	Read literal 'n'	Process Data	Write to PC
No operation	No operation	No operation	No operation

If No Jump:

Q1	Q2	Q3	Q4
Decode	Read literal 'n'	Process Data	No operation

Example: HERE BZ Jump

Before Instruction
PC = address (HERE)
After Instruction
If Zero = 1;
PC = address (Jump)
If Zero = 0;
PC = address (HERE + 2)

CALL	Subroutine Call
Syntax:	CALL k {,s}
Operands:	$0 \le k \le 1048575$ $s \in [0,1]$
Operation:	$(PC) + 4 \to TOS$, $k \to PC<20:1>$, if $s = 1$ $(W) \to WS$, $(Status) \to STATUSS$, $(BSR) \to BSRS$
Status Affected:	None

Encoding:

1st word (k<7:0>)	1110	110s	$k_7$kkk	kkkk$_0$
2nd word(k<19:8>)	1111	k_{19}kkk	kkkk	kkkk$_8$

Description:	Subroutine call of entire 2-Mbyte memory range. First, return address (PC + 4) is pushed onto the return stack. If 's' = 1, the W, Status and BSR registers are also pushed into their respective shadow registers, WS, STATUSS and BSRS. If 's' = 0, no update occurs (default). Then, the 20-bit value 'k' is loaded into PC<20:1>. CALL is a two-cycle instruction.
Words:	2
Cycles:	2

Q Cycle Activity:

Q1	Q2	Q3	Q4
Decode	Read literal 'k'<7:0>,	Push PC to stack	Read literal 'k'<19:8>, Write to PC
No operation	No operation	No operation	No operation

Example: HERE CALL THERE,1

Before Instruction
PC = address (HERE)
After Instruction
PC = address (THERE)
TOS = address (HERE + 4)
WS = W
BSRS = BSR
STATUSS= Status

Preliminary

CLRF	Clear f
Syntax:	CLRF f {,a}
Operands:	$0 \le f \le 255$ $a \in [0,1]$
Operation:	000h → f 1 → Z
Status Affected:	Z

Encoding:

0110	101a	ffff	ffff

Description: Clears the contents of the specified register.

If 'a' is '0', the Access Bank is selected. If 'a' is '1', the BSR is used to select the GPR bank (default).

If 'a' is '0' and the extended instruction set is enabled, this instruction operates in Indexed Literal Offset Addressing mode whenever f ≤ 95 (5Fh). See **Section 25.2.3 "Byte-Oriented and Bit-Oriented Instructions in Indexed Literal Offset Mode"** for details.

Words: 1

Cycles: 1

Q Cycle Activity:

Q1	Q2	Q3	Q4
Decode	Read register 'f'	Process Data	Write register 'f'

Example: CLRF   FLAG_REG,1

Before Instruction
    FLAG_REG   =   5Ah
After Instruction
    FLAG_REG   =   00h

CLRWDT	Clear Watchdog Timer
Syntax:	CLRWDT
Operands:	None
Operation:	000h → WDT, 000h → WDT postscaler, 1 → $\overline{TO}$, 1 → $\overline{PD}$
Status Affected:	$\overline{TO}$, $\overline{PD}$

Encoding:

0000	0000	0000	0100

Description: CLRWDT instruction resets the Watchdog Timer. It also resets the postscaler of the WDT. Status bits $\overline{TO}$ and $\overline{PD}$ are set.

Words: 1

Cycles: 1

Q Cycle Activity:

Q1	Q2	Q3	Q4
Decode	No operation	Process Data	No operation

Example: CLRWDT

Before Instruction
    WDT Counter   =   ?
After Instruction
    WDT Counter   =   00h
    WDT Postscaler   =   0
    $\overline{TO}$ =   1
    $\overline{PD}$ =   1

PIC18F2480/2580/4480/4580

COMF	Complement f
Syntax:	COMF f {,d {,a}}
Operands:	$0 \leq f \leq 255$ $d \in [0,1]$ $a \in [0,1]$
Operation:	$(\overline{f}) \rightarrow$ dest
Status Affected:	N, Z
Encoding:	0001 | 11da | ffff | ffff
Description:	The contents of register 'f' are complemented. If 'd' is '1', the result is stored in W. If 'd' is '0', the result is stored back in register 'f' (default). If 'a' is '0', the Access Bank is selected. If 'a' is '1', the BSR is used to select the GPR bank (default). If 'a' is '0' and the extended instruction set is enabled, this instruction operates in Indexed Literal Offset Addressing mode whenever $f \leq 95$ (5Fh). See **Section 25.2.3 "Byte-Oriented and Bit-Oriented Instructions in Indexed Literal Offset Mode"** for details.
Words:	1
Cycles:	1

Q Cycle Activity:

Q1	Q2	Q3	Q4
Decode	Read register 'f'	Process Data	Write to destination

Example: COMF   REG, 0, 0

Before Instruction
  REG   =   13h
After Instruction
  REG   =   13h
  W   =   ECh

CPFSEQ	Compare f with W, Skip if f = W
Syntax:	CPFSEQ f {,a}
Operands:	$0 \leq f \leq 255$ $a \in [0,1]$
Operation:	(f) – (W), skip if (f) = (W) (unsigned comparison)
Status Affected:	None
Encoding:	0110 | 001a | ffff | ffff
Description:	Compares the contents of data memory location 'f' to the contents of W by performing an unsigned subtraction. If 'f' = W, then the fetched instruction is discarded and a NOP is executed instead, making this a two-cycle instruction. If 'a' is '0', the Access Bank is selected. If 'a' is '0', the BSR is used to select the GPR bank (default). If 'a' is '0' and the extended instruction set is enabled, this instruction operates in Indexed Literal Offset Addressing mode whenever $f \leq 95$ (5Fh). See **Section 25.2.3 "Byte-Oriented and Bit-Oriented Instructions in Indexed Literal Offset Mode"** for details.
Words:	1
Cycles:	1(2)

Note: 3 cycles if skip and followed by a 2-word instruction.

Q Cycle Activity:

Q1	Q2	Q3	Q4
Decode	Read register 'f'	Process Data	No operation

If skip:

Q1	Q2	Q3	Q4
No operation	No operation	No operation	No operation

If skip and followed by 2-word instruction:

Q1	Q2	Q3	Q4
No operation	No operation	No operation	No operation
No operation	No operation	No operation	No operation

Example: HERE   CPFSEQ REG, 0
                NEQUAL   :
                EQUAL   :

Before Instruction
  PC Address   =   HERE
  W   =   ?
  REG   =   ?
After Instruction
  If REG   =   W;
    PC   =   Address   (EQUAL)
  If REG   ≠   W;
    PC   =   Address   (NEQUAL)

Preliminary

CPFSGT	Compare f with W, Skip if f > W
Syntax:	CPFSGT f {,a}
Operands:	$0 \le f \le 255$ $a \in [0,1]$
Operation:	(f) – (W), skip if (f) > (W) (unsigned comparison)
Status Affected:	None
Encoding:	`0110` `010a` `ffff` `ffff`
Description:	Compares the contents of data memory location 'f' to the contents of the W by performing an unsigned subtraction. If the contents of 'f' are greater than the contents of WREG, then the fetched instruction is discarded and a NOP is executed instead, making this a two-cycle instruction. If 'a' is '0', the Access Bank is selected. If 'a' is '1', the BSR is used to select the GPR bank (default). If 'a' is '0' and the extended instruction set is enabled, this instruction operates in Indexed Literal Offset Addressing mode whenever f ≤ 95 (5Fh). See **Section 25.2.3 "Byte-Oriented and Bit-Oriented Instructions in Indexed Literal Offset Mode"** for details.
Words:	1
Cycles:	1(2)
Note:	3 cycles if skip and followed by a 2-word instruction.

Q Cycle Activity:

Q1	Q2	Q3	Q4
Decode	Read register 'f'	Process Data	No operation

If skip:

Q1	Q2	Q3	Q4
No operation	No operation	No operation	No operation

If skip and followed by 2-word instruction:

Q1	Q2	Q3	Q4
No operation	No operation	No operation	No operation
No operation	No operation	No operation	No operation

Example:

```
          HERE        CPFSGT REG, 0
          NGREATER  :
          GREATER   :
```

Before Instruction
```
    PC        =   Address (HERE)
    W         =   ?
```
After Instruction
```
    If REG    >   W;
        PC    =   Address (GREATER)
    If REG    ≤   W;
        PC    =   Address (NGREATER)
```

CPFSLT	Compare f with W, Skip if f < W
Syntax:	CPFSLT f {,a}
Operands:	$0 \le f \le 255$ $a \in [0,1]$
Operation:	(f) – (W), skip if (f) < (W) (unsigned comparison)
Status Affected:	None
Encoding:	`0110` `000a` `ffff` `ffff`
Description:	Compares the contents of data memory location 'f' to the contents of W by performing an unsigned subtraction. If the contents of 'f' are less than the contents of W, then the fetched instruction is discarded and a NOP is executed instead, making this a two-cycle instruction. If 'a' is '0', the Access Bank is selected. If 'a' is '1', the BSR is used to select the GPR bank (default).
Words:	1
Cycles:	1(2)
Note:	3 cycles if skip and followed by a 2-word instruction.

Q Cycle Activity:

Q1	Q2	Q3	Q4
Decode	Read register 'f'	Process Data	No operation

If skip:

Q1	Q2	Q3	Q4
No operation	No operation	No operation	No operation

If skip and followed by 2-word instruction:

Q1	Q2	Q3	Q4
No operation	No operation	No operation	No operation
No operation	No operation	No operation	No operation

Example:

```
          HERE        CPFSLT REG, 1
          NLESS   :
          LESS    :
```

Before Instruction
```
    PC        =   Address (HERE)
    W         =   ?
```
After Instruction
```
    If REG    <   W;
        PC    =   Address (LESS)
    If REG    ≥   W;
        PC    =   Address (NLESS)
```

DAW	Decimal Adjust W Register
Syntax:	DAW
Operands:	None

Operation:	If [W<3:0> >9] or [DC = 1] then
	(W<3:0>) + 6 → W<3:0>;
	else
	(W<3:0>) → W<3:0>;
	
	If [W<7:4> >9] or [C = 1] then
	(W<7:4>) + 6 → W<7:4>;
	C = 1;
	else
	(W<7:4>) → W<7:4>;

Status Affected:	C

Encoding:

0000	0000	0000	0111

Description: DAW adjusts the eight-bit value in W, resulting from the earlier addition of two variables (each in packed BCD format) and produces a correct packed BCD result.

Words: 1

Cycles: 1

Q Cycle Activity:

Q1	Q2	Q3	Q4
Decode	Read register W	Process Data	Write W

Example 1:

```
            DAW
```

Before Instruction

W	=	A5h
C	=	0
DC	=	0

After Instruction

W	=	05h
C	=	1
DC	=	0

Example 2:

Before Instruction

W	=	CEh
C	=	0
DC	=	0

After Instruction

W	=	34h
C	=	1
DC	=	0

DECF	Decrement f
Syntax:	DECF f {,d {,a}}
Operands:	$0 \le f \le 255$ $d \in [0,1]$ $a \in [0,1]$
Operation:	(f) − 1 → dest
Status Affected:	C, DC, N, OV, Z

Encoding:

0000	01da	ffff	ffff

Description: Decrement register 'f'. If 'd' is '0', the result is stored in W. If 'd' is '1', the result is stored back in register 'f' (default).

If 'a' is '0', the Access Bank is selected. If 'a' is '1', the BSR is used to select the GPR bank (default).

If 'a' is '0' and the extended instruction set is enabled, this instruction operates in Indexed Literal Offset Addressing mode whenever $f \le 95$ (5Fh). See **Section 25.2.3 "Byte-Oriented and Bit-Oriented Instructions in Indexed Literal Offset Mode"** for details.

Words: 1

Cycles: 1

Q Cycle Activity:

Q1	Q2	Q3	Q4
Decode	Read register 'f'	Process Data	Write to destination

Example: DECF   CNT,   1,   0

Before Instruction

CNT	=	01h
Z	=	0

After Instruction

CNT	=	00h
Z	=	1

DECFSZ	Decrement f, Skip if 0
Syntax:	DECFSZ f {,d {,a}}
Operands:	$0 \leq f \leq 255$ $d \in [0,1]$ $a \in [0,1]$
Operation:	$(f) - 1 \rightarrow$ dest, skip if result = 0
Status Affected:	None
Encoding:	0010	11da	ffff	ffff

Description: The contents of register 'f' are decremented. If 'd' is '0', the result is placed in W. If 'd' is '1', the result is placed back in register 'f' (default).

If the result is '0', the next instruction which is already fetched is discarded and a NOP is executed instead, making it a two-cycle instruction.

If 'a' is '0', the Access Bank is selected. If 'a' is '1', the BSR is used to select the GPR bank (default).

If 'a' is '0' and the extended instruction set is enabled, this instruction operates in Indexed Literal Offset Addressing mode whenever $f \leq 95$ (5Fh). See **Section 25.2.3 "Byte-Oriented and Bit-Oriented Instructions in Indexed Literal Offset Mode"** for details.

Words: 1

Cycles: 1(2)

Note: 3 cycles if skip and followed by a 2-word instruction.

Q Cycle Activity:

Q1	Q2	Q3	Q4
Decode	Read register 'f'	Process Data	Write to destination

If skip:

Q1	Q2	Q3	Q4
No operation	No operation	No operation	No operation

If skip and followed by 2-word instruction:

Q1	Q2	Q3	Q4
No operation	No operation	No operation	No operation
No operation	No operation	No operation	No operation

Example:
```
HERE       DECFSZ   CNT, 1, 1
           GOTO     LOOP
CONTINUE
```
Before Instruction
```
PC        =   Address (HERE)
```
After Instruction
```
CNT       =   CNT – 1
If CNT    =   0;
  PC      =   Address (CONTINUE)
If CNT    ≠   0;
  PC      =   Address (HERE + 2)
```

DCFSNZ	Decrement f, Skip if not 0
Syntax:	DCFSNZ f {,d {,a}}
Operands:	$0 \leq f \leq 255$ $d \in [0,1]$ $a \in [0,1]$
Operation:	$(f) - 1 \rightarrow$ dest, skip if result $\neq$ 0
Status Affected:	None
Encoding:	0100	11da	ffff	ffff

Description: The contents of register 'f' are decremented. If 'd' is '0', the result is placed in W. If 'd' is '1', the result is placed back in register 'f' (default).

If the result is not '0', the next instruction which is already fetched is discarded and a NOP is executed instead, making it a two-cycle instruction.

If 'a' is '0', the Access Bank is selected. If 'a' is '1', the BSR is used to select the GPR bank (default).

If 'a' is '0' and the extended instruction set is enabled, this instruction operates in Indexed Literal Offset Addressing mode whenever $f \leq 95$ (5Fh). See **Section 25.2.3 "Byte-Oriented and Bit-Oriented Instructions in Indexed Literal Offset Mode"** for details.

Words: 1

Cycles: 1(2)

Note: 3 cycles if skip and followed by a 2-word instruction.

Q Cycle Activity:

Q1	Q2	Q3	Q4
Decode	Read register 'f'	Process Data	Write to destination

If skip:

Q1	Q2	Q3	Q4
No operation	No operation	No operation	No operation

If skip and followed by 2-word instruction:

Q1	Q2	Q3	Q4
No operation	No operation	No operation	No operation
No operation	No operation	No operation	No operation

Example:
```
HERE       DCFSNZ   TEMP, 1, 0
ZERO       :
NZERO      :
```
Before Instruction
```
TEMP      =   ?
```
After Instruction
```
TEMP      =   TEMP – 1,
If TEMP   =   0;
  PC      =   Address (ZERO)
If TEMP   ≠   0;
  PC      =   Address (NZERO)
```

GOTO	Unconditional Branch
Syntax:	GOTO k
Operands:	$0 \leq k \leq 1048575$
Operation:	$k \rightarrow PC<20:1>$
Status Affected:	None

Encoding:

1st word (k<7:0>)	1110	1111	$k_7$kkk	kkkk$_0$
2nd word(k<19:8>)	1111	k_{19}kkk	kkkk	kkkk$_8$

Description:	GOTO allows an unconditional branch anywhere within entire 2-Mbyte memory range. The 20-bit value 'k' is loaded into PC<20:1>. GOTO is always a two-cycle instruction.
Words:	2
Cycles:	2

Q Cycle Activity:

Q1	Q2	Q3	Q4
Decode	Read literal 'k'<7:0>,	No operation	Read literal 'k'<19:8>, Write to PC
No operation	No operation	No operation	No operation

Example: GOTO   THERE

After Instruction
  PC =   Address (THERE)

INCF	Increment f
Syntax:	INCF f {,d {,a}}
Operands:	$0 \leq f \leq 255$ $d \in [0,1]$ $a \in [0,1]$
Operation:	$(f) + 1 \rightarrow dest$
Status Affected:	C, DC, N, OV, Z

Encoding:

0010	10da	ffff	ffff

Description:	The contents of register 'f' are incremented. If 'd' is '0', the result is placed in W. If 'd' is '1', the result is placed back in register 'f' (default). If 'a' is '0', the Access Bank is selected. If 'a' is '1', the BSR is used to select the GPR bank (default). If 'a' is '0' and the extended instruction set is enabled, this instruction operates in Indexed Literal Offset Addressing mode whenever $f \leq 95$ (5Fh). See **Section 25.2.3 "Byte-Oriented and Bit-Oriented Instructions in Indexed Literal Offset Mode"** for details.
Words:	1
Cycles:	1

Q Cycle Activity:

Q1	Q2	Q3	Q4
Decode	Read register 'f'	Process Data	Write to destination

Example: INCF   CNT, 1, 0

Before Instruction
  CNT = FFh
  Z   = 0
  C   = ?
  DC   = ?
After Instruction
  CNT = 00h
  Z   = 1
  C   = 1
  DC   = 1

PIC18F2480/2580/4480/4580

INCFSZ	Increment f, Skip if 0

Syntax:	INCFSZ f {,d {,a}}
Operands:	$0 \le f \le 255$ $d \in [0,1]$ $a \in [0,1]$
Operation:	(f) + 1 → dest, skip if result = 0
Status Affected:	None
Encoding:	`0011` `11da` `ffff` `ffff`
Description:	The contents of register 'f' are incremented. If 'd' is '0', the result is placed in W. If 'd' is '1', the result is placed back in register 'f' (default). If the result is '0', the next instruction which is already fetched is discarded and a NOP is executed instead, making it a two-cycle instruction. If 'a' is '0', the Access Bank is selected. If 'a' is '1', the BSR is used to select the GPR bank (default). If 'a' is '0' and the extended instruction set is enabled, this instruction operates in Indexed Literal Offset Addressing mode whenever f ≤ 95 (5Fh). See **Section 25.2.3 "Byte-Oriented and Bit-Oriented Instructions in Indexed Literal Offset Mode"** for details.
Words:	1
Cycles:	1(2)
	Note: 3 cycles if skip and followed by a 2-word instruction.

Q Cycle Activity:

Q1	Q2	Q3	Q4
Decode	Read register 'f'	Process Data	Write to destination

If skip:

Q1	Q2	Q3	Q4
No operation	No operation	No operation	No operation

If skip and followed by 2-word instruction:

Q1	Q2	Q3	Q4
No operation	No operation	No operation	No operation
No operation	No operation	No operation	No operation

Example:
```
HERE    INCFSZ  CNT, 1, 0
NZERO   :
ZERO    :
```
Before Instruction
```
PC      =   Address (HERE)
```
After Instruction
```
CNT     =   CNT + 1
If CNT  =   0;
PC      =   Address (ZERO)
If CNT  ≠   0;
PC      =   Address (NZERO)
```

INFSNZ	Increment f, Skip if not 0

Syntax:	INFSNZ f {,d {,a}}
Operands:	$0 \le f \le 255$ $d \in [0,1]$ $a \in [0,1]$
Operation:	(f) + 1 → dest, skip if result ≠ 0
Status Affected:	None
Encoding:	`0100` `10da` `ffff` `ffff`
Description:	The contents of register 'f' are incremented. If 'd' is '0', the result is placed in W. If 'd' is '1', the result is placed back in register 'f' (default). If the result is not '0', the next instruction which is already fetched is discarded and a NOP is executed instead, making it a two-cycle instruction. If 'a' is '0', the Access Bank is selected. If 'a' is '1', the BSR is used to select the GPR bank (default). If 'a' is '0' and the extended instruction set is enabled, this instruction operates in Indexed Literal Offset Addressing mode whenever f ≤ 95 (5Fh). See **Section 25.2.3 "Byte-Oriented and Bit-Oriented Instructions in Indexed Literal Offset Mode"** for details.
Words:	1
Cycles:	1(2)
	Note: 3 cycles if skip and followed by a 2-word instruction.

Q Cycle Activity:

Q1	Q2	Q3	Q4
Decode	Read register 'f'	Process Data	Write to destination

If skip:

Q1	Q2	Q3	Q4
No operation	No operation	No operation	No operation

If skip and followed by 2-word instruction:

Q1	Q2	Q3	Q4
No operation	No operation	No operation	No operation
No operation	No operation	No operation	No operation

Example:
```
HERE    INFSNZ  REG, 1, 0
ZERO
NZERO
```
Before Instruction
```
PC      =   Address (HERE)
```
After Instruction
```
REG     =   REG + 1
If REG  ≠   0;
PC      =   Address (NZERO)
If REG  =   0;
PC      =   Address (ZERO)
```

PIC18F2480/2580/4480/4580

IORLW	Inclusive OR Literal with W
Syntax:	IORLW k
Operands:	$0 \leq k \leq 255$
Operation:	(W) .OR. k $\rightarrow$ W
Status Affected:	N, Z
Encoding:	0000 1001 kkkk kkkk
Description:	The contents of W are ORed with the eight-bit literal 'k'. The result is placed in W.
Words:	1
Cycles:	1

Q Cycle Activity:

Q1	Q2	Q3	Q4
Decode	Read literal 'k'	Process Data	Write to W

Example: IORLW 35h

Before Instruction
W = 9Ah
After Instruction
W = BFh

IORWF	Inclusive OR W with f
Syntax:	IORWF f {,d {,a}}
Operands:	$0 \leq f \leq 255$ d $\in$ [0,1] a $\in$ [0,1]
Operation:	(W) .OR. (f) $\rightarrow$ dest
Status Affected:	N, Z
Encoding:	0001 00da ffff ffff
Description:	Inclusive OR W with register 'f'. If 'd' is '0', the result is placed in W. If 'd' is '1', the result is placed back in register 'f' (default). If 'a' is '0', the Access Bank is selected. If 'a' is '1', the BSR is used to select the GPR bank (default). If 'a' is '0' and the extended instruction set is enabled, this instruction operates in Indexed Literal Offset Addressing mode whenever f $\leq$ 95 (5Fh). See **Section 25.2.3 "Byte-Oriented and Bit-Oriented Instructions in Indexed Literal Offset Mode"** for details.
Words:	1
Cycles:	1

Q Cycle Activity:

Q1	Q2	Q3	Q4
Decode	Read register 'f'	Process Data	Write to destination

Example: IORWF RESULT, 0, 1

Before Instruction
RESULT = 13h
W = 91h
After Instruction
RESULT = 13h
W = 93h

Preliminary

© 2004 Microchip Technology Inc.

LFSR	Load FSR

Syntax:	LFSR f, k
Operands:	$0 \leq f \leq 2$ $0 \leq k \leq 4095$
Operation:	$k \rightarrow FSRf$
Status Affected:	None

Encoding:

1110	1110	00ff	k_{11}kkk
1111	0000	$k_7$kkk	kkkk

Description:	The 12-bit literal 'k' is loaded into the file select register pointed to by 'f'.
Words:	2
Cycles:	2

Q Cycle Activity:

Q1	Q2	Q3	Q4
Decode	Read literal 'k' MSB	Process Data	Write literal 'k' MSB to FSRfH
Decode	Read literal 'k' LSB	Process Data	Write literal 'k' to FSRfL

Example: LFSR 2, 3ABh

After Instruction
FSR2H   =   03h
FSR2L   =   ABh

MOVF	Move f

Syntax:	MOVF f {,d {,a}}
Operands:	$0 \leq f \leq 255$ $d \in [0,1]$ $a \in [0,1]$
Operation:	$f \rightarrow dest$
Status Affected:	N, Z

Encoding:

0101	00da	ffff	ffff

Description:	The contents of register 'f' are moved to a destination dependent upon the status of 'd'. If 'd' is '0', the result is placed in W. If 'd' is '1', the result is placed back in register 'f' (default). Location 'f' can be anywhere in the 256-byte bank. If 'a' is '0', the Access Bank is selected. If 'a' is '1', the BSR is used to select the GPR bank (default). If 'a' is '0' and the extended instruction set is enabled, this instruction operates in Indexed Literal Offset Addressing mode whenever f ≤ 95 (5Fh). See **Section 25.2.3 "Byte-Oriented and Bit-Oriented Instructions in Indexed Literal Offset Mode"** for details.
Words:	1
Cycles:	1

Q Cycle Activity:

Q1	Q2	Q3	Q4
Decode	Read register 'f'	Process Data	Write W

Example: MOVF   REG, 0, 0

Before Instruction
REG   =   22h
W   =   FFh
After Instruction
REG   =   22h
W   =   22h

MOVFF	Move f to f
Syntax:	MOVFF f$_s$,f$_d$
Operands:	$0 \le f_s \le 4095$ $0 \le f_d \le 4095$
Operation:	$(f_s) \rightarrow f_d$
Status Affected:	None

Encoding:

1st word (source)	1100 ffff	ffff	ffff$_s$
2nd word (destin.)	1111 ffff	ffff	ffff$_d$

Description: The contents of source register 'f$_s$' are moved to destination register 'f$_d$'. Location of source 'f$_s$' can be anywhere in the 4096-byte data space (000h to FFFh) and location of destination 'f$_d$' can also be anywhere from 000h to FFFh.

Either source or destination can be W (a useful special situation).

MOVFF is particularly useful for transferring a data memory location to a peripheral register (such as the transmit buffer or an I/O port).

The MOVFF instruction cannot use the PCL, TOSU, TOSH or TOSL as the destination register

Words: 2

Cycles: 2 (3)

Q Cycle Activity:

Q1	Q2	Q3	Q4
Decode	Read register 'f' (src)	Process Data	No operation
Decode	No operation No dummy read	No operation	Write register 'f' (dest)

Example: MOVFF REG1, REG2

Before Instruction
REG1 = 33h
REG2 = 11h
After Instruction
REG1 = 33h
REG2 = 33h

MOVLB	Move Literal to Low Nibble in BSR
Syntax:	MOVLW k
Operands:	$0 \le k \le 255$
Operation:	$k \rightarrow BSR$
Status Affected:	None

Encoding:

0000	0001	kkkk	kkkk

Description: The eight-bit literal 'k' is loaded into the Bank Select Register (BSR). The value of BSR<7:4> always remains '0', regardless of the value of k_7:k_4.

Words: 1

Cycles: 1

Q Cycle Activity:

Q1	Q2	Q3	Q4
Decode	Read literal 'k'	Process Data	Write literal 'k' to BSR

Example: MOVLB 5

Before Instruction
BSR Register = 02h
After Instruction
BSR Register = 05h

MOVLW	Move Literal to W
Syntax:	MOVLW k
Operands:	$0 \le k \le 255$
Operation:	$k \rightarrow W$
Status Affected:	None
Encoding:	0000 1110 kkkk kkkk
Description:	The eight-bit literal 'k' is loaded into W.
Words:	1
Cycles:	1

Q Cycle Activity:

Q1	Q2	Q3	Q4
Decode	Read literal 'k'	Process Data	Write to W

Example: MOVLW 5Ah

After Instruction
W   =   5Ah

MOVWF	Move W to f
Syntax:	MOVWF f {,a}
Operands:	$0 \le f \le 255$ $a \in [0,1]$
Operation:	$(W) \rightarrow f$
Status Affected:	None
Encoding:	0110 111a ffff ffff
Description:	Move data from W to register 'f'. Location 'f' can be anywhere in the 256-byte bank. If 'a' is '0', the Access Bank is selected. If 'a' is '1', the BSR is used to select the GPR bank (default). If 'a' is '0' and the extended instruction set is enabled, this instruction operates in Indexed Literal Offset Addressing mode whenever $f \le 95$ (5Fh). See **Section 25.2.3 "Byte-Oriented and Bit-Oriented Instructions in Indexed Literal Offset Mode"** for details.
Words:	1
Cycles:	1

Q Cycle Activity:

Q1	Q2	Q3	Q4
Decode	Read register 'f'	Process Data	Write register 'f'

Example: MOVWF REG, 0

Before Instruction
W   =   4Fh
REG   =   FFh
After Instruction
W   =   4Fh
REG   =   4Fh

MULLW	Multiply Literal with W
Syntax:	MULLW k
Operands:	$0 \le k \le 255$
Operation:	(W) x k → PRODH:PRODL
Status Affected:	None
Encoding:	`0000` `1101` `kkkk` `kkkk`
Description:	An unsigned multiplication is carried out between the contents of W and the 8-bit literal 'k'. The 16-bit result is placed in the PRODH:PRODL register pair. PRODH contains the high byte. W is unchanged.
	None of the status flags are affected.
	Note that neither overflow nor carry is possible in this operation. A zero result is possible but not detected.
Words:	1
Cycles:	1

Q Cycle Activity:

Q1	Q2	Q3	Q4
Decode	Read literal 'k'	Process Data	Write registers PRODH: PRODL

Example: MULLW   0C4h

Before Instruction
W	=	E2h
PRODH	=	?
PRODL	=	?

After Instruction
W	=	E2h
PRODH	=	ADh
PRODL	=	08h

MULWF	Multiply W with f
Syntax:	MULWF f {,a}
Operands:	$0 \le f \le 255$
	$a \in [0,1]$
Operation:	(W) x (f) → PRODH:PRODL
Status Affected:	None
Encoding:	`0000` `001a` `ffff` `ffff`
Description:	An unsigned multiplication is carried out between the contents of W and the register file location 'f'. The 16-bit result is stored in the PRODH:PRODL register pair. PRODH contains the high byte. Both W and 'f' are unchanged.
	None of the status flags are affected.
	Note that neither overflow nor carry is possible in this operation. A zero result is possible but not detected.
	If 'a' is '0', the Access Bank is selected. If 'a' is '1', the BSR is used to select the GPR bank (default).
	If 'a' is '0' and the extended instruction set is enabled, this instruction operates in Indexed Literal Offset Addressing mode whenever $f \le 95$ (5Fh). See **Section 25.2.3 "Byte-Oriented and Bit-Oriented Instructions in Indexed Literal Offset Mode"** for details.
Words:	1
Cycles:	1

Q Cycle Activity:

Q1	Q2	Q3	Q4
Decode	Read register 'f'	Process Data	Write registers PRODH: PRODL

Example: MULWF   REG, 1

Before Instruction
W	=	C4h
REG	=	B5h
PRODH	=	?
PRODL	=	?

After Instruction
W	=	C4h
REG	=	B5h
PRODH	=	8Ah
PRODL	=	94h

PIC18F2480/2580/4480/4580

NEGF	**Negate f**	

Syntax:	NEGF f {,a}
Operands:	$0 \le f \le 255$ $a \in [0,1]$
Operation:	$(\bar{f}) + 1 \rightarrow f$
Status Affected:	N, OV, C, DC, Z
Encoding:	

0110	110a	ffff	ffff

Description: Location 'f' is negated using two's complement. The result is placed in the data memory location 'f'.

If 'a' is '0', the Access Bank is selected. If 'a' is '1', the BSR is used to select the GPR bank (default).

If 'a' is '0' and the extended instruction set is enabled, this instruction operates in Indexed Literal Offset Addressing mode whenever f ≤ 95 (5Fh). See **Section 25.2.3 "Byte-Oriented and Bit-Oriented Instructions in Indexed Literal Offset Mode"** for details.

Words: 1

Cycles: 1

Q Cycle Activity:

Q1	Q2	Q3	Q4
Decode	Read register 'f'	Process Data	Write register 'f'

Example: NEGF   REG, 1

Before Instruction
REG   =   0011 1010 [3Ah]
After Instruction
REG   =   1100 0110 [C6h]

NOP	**No Operation**	

Syntax:	NOP
Operands:	None
Operation:	No operation
Status Affected:	None
Encoding:	

0000	0000	0000	0000
1111	xxxx	xxxx	xxxx

Description: No operation.

Words: 1

Cycles: 1

Q Cycle Activity:

Q1	Q2	Q3	Q4
Decode	No operation	No operation	No operation

Example:

None.

POP	**Pop Top of Return Stack**
Syntax:	POP
Operands:	None
Operation:	(TOS) → bit bucket
Status Affected:	None
Encoding:	0000 0000 0000 0110
Description:	The TOS value is pulled off the return stack and is discarded. The TOS value then becomes the previous value that was pushed onto the return stack. This instruction is provided to enable the user to properly manage the return stack to incorporate a software stack.
Words:	1
Cycles:	1

Q Cycle Activity:

Q1	Q2	Q3	Q4
Decode	No operation	POP TOS value	No operation

Example:

```
POP
GOTO    NEW
```

Before Instruction
TOS   =   0031A2h
Stack (1 level down)   =   014332h

After Instruction
TOS   =   014332h
PC   =   NEW

PUSH	**Push Top of Return Stack**
Syntax:	PUSH
Operands:	None
Operation:	(PC + 2) → TOS
Status Affected:	None
Encoding:	0000 0000 0000 0101
Description:	The PC + 2 is pushed onto the top of the return stack. The previous TOS value is pushed down on the stack. This instruction allows implementing a software stack by modifying TOS and then pushing it onto the return stack.
Words:	1
Cycles:	1

Q Cycle Activity:

Q1	Q2	Q3	Q4
Decode	PUSH PC + 2 onto return stack	No operation	No operation

Example:

```
PUSH
```

Before Instruction
TOS   =   345Ah
PC   =   0124h

After Instruction
PC   =   0126h
TOS   =   0126h
Stack (1 level down)   =   345Ah

Preliminary

PIC18F2480/2580/4480/4580

RCALL	Relative Call
Syntax:	RCALL n
Operands:	$-1024 \le n \le 1023$
Operation:	$(PC) + 2 \rightarrow TOS$, $(PC) + 2 + 2n \rightarrow PC$
Status Affected:	None

Encoding:

1101	1nnn	nnnn	nnnn

Description: Subroutine call with a jump up to 1K from the current location. First, return address (PC + 2) is pushed onto the stack. Then, add the 2's complement number '2n' to the PC. Since the PC will have incremented to fetch the next instruction, the new address will be PC + 2 + 2n. This instruction is a two-cycle instruction.

Words: 1

Cycles: 2

Q Cycle Activity:

Q1	Q2	Q3	Q4
Decode	Read literal 'n' PUSH PC to stack	Process Data	Write to PC
No operation	No operation	No operation	No operation

Example: HERE   RCALL Jump

Before Instruction
  PC = Address (HERE)
After Instruction
  PC = Address (Jump)
  TOS = Address (HERE + 2)

RESET	Reset
Syntax:	RESET
Operands:	None
Operation:	Reset all registers and flags that are affected by a MCLR Reset.
Status Affected:	All

Encoding:

0000	0000	1111	1111

Description: This instruction provides a way to execute a MCLR Reset in software.

Words: 1

Cycles: 1

Q Cycle Activity:

Q1	Q2	Q3	Q4
Decode	Start Reset	No operation	No operation

Example: RESET

After Instruction
  Registers = Reset Value
  Flags*   = Reset Value

APPENDIX H: DATA SHEETS   **787**

RETFIE	Return from Interrupt
Syntax:	RETFIE {s}
Operands:	s ∈ [0,1]
Operation:	(TOS) → PC, 1 → GIE/GIEH or PEIE/GIEL, if s = 1 (WS) → W, (STATUSS) → Status, (BSRS) → BSR, PCLATU, PCLATH are unchanged.
Status Affected:	GIE/GIEH, PEIE/GIEL.

Encoding:

0000	0000	0001	000s

Description: Return from Interrupt. Stack is popped and Top-of-Stack (TOS) is loaded into the PC. Interrupts are enabled by setting either the high or low priority global interrupt enable bit. If 's' = 1, the contents of the shadow registers, WS, STATUSS and BSRS, are loaded into their corresponding registers, W, Status and BSR. If 's' = 0, no update of these registers occurs (default).

Words: 1

Cycles: 2

Q Cycle Activity:

Q1	Q2	Q3	Q4
Decode	No operation	No operation	POP PC from stack Set GIEH or GIEL
No operation	No operation	No operation	No operation

Example: RETFIE 1

After Interrupt

PC	=	TOS
W	=	WS
BSR	=	BSRS
Status	=	STATUSS
GIE/GIEH, PEIE/GIEL	=	1

RETLW	Return Literal to W
Syntax:	RETLW k
Operands:	0 ≤ k ≤ 255
Operation:	k → W, (TOS) → PC, PCLATU, PCLATH are unchanged
Status Affected:	None

Encoding:

0000	1100	kkkk	kkkk

Description: W is loaded with the eight-bit literal 'k'. The program counter is loaded from the top of the stack (the return address). The high address latch (PCLATH) remains unchanged.

Words: 1

Cycles: 2

Q Cycle Activity:

Q1	Q2	Q3	Q4
Decode	Read literal 'k'	Process Data	POP PC from stack, Write to W
No operation	No operation	No operation	No operation

Example:

```
      CALL   TABLE  ; W contains table
                    ; offset value
                    ; W now has
                    ; table value
      :
TABLE
      ADDWF PCL    ; W = offset
      RETLW k0     ; Begin table
      RETLW k1     ;
      :
      :
      RETLW kn     ; End of table
```

Before Instruction
W = 07h
After Instruction
W = value of kn

RETURN	Return from Subroutine
Syntax:	RETURN {s}
Operands:	s ∈ [0,1]
Operation:	(TOS) → PC, if s = 1 (WS) → W, (STATUSS) → Status, (BSRS) → BSR, PCLATU, PCLATH are unchanged
Status Affected:	None
Encoding:	0000 0000 0001 001s
Description:	Return from subroutine. The stack is popped and the top of the stack (TOS) is loaded into the program counter. If 's'= 1, the contents of the shadow registers, WS, STATUSS and BSRS, are loaded into their corresponding registers, W, Status and BSR. If 's' = 0, no update of these registers occurs (default).
Words:	1
Cycles:	2

Q Cycle Activity:

Q1	Q2	Q3	Q4
Decode	No operation	Process Data	POP PC from stack
No operation	No operation	No operation	No operation

Example: RETURN

After Interrupt
    PC = TOS

RLCF	Rotate Left f through Carry
Syntax:	RLCF f {,d {,a}}
Operands:	0 ≤ f ≤ 255 d ∈ [0,1] a ∈ [0,1]
Operation:	(f<n>) → dest<n + 1>, (f<7>) → C, (C) → dest<0>
Status Affected:	C, N, Z
Encoding:	0011 01da ffff ffff
Description:	The contents of register 'f' are rotated one bit to the left through the Carry flag. If 'd' is '0', the result is placed in W. If 'd' is '1', the result is stored back in register 'f' (default). If 'a' is '0', the Access Bank is selected. If 'a' is '1', the BSR is used to select the GPR bank (default). If 'a' is '0' and the extended instruction set is enabled, this instruction operates in Indexed Literal Offset Addressing mode whenever f ≤ 95 (5Fh). See **Section 25.2.3 "Byte-Oriented and Bit-Oriented Instructions in Indexed Literal Offset Mode"** for details.

Words:	1
Cycles:	1

Q Cycle Activity:

Q1	Q2	Q3	Q4
Decode	Read register 'f'	Process Data	Write to destination

Example: RLCF   REG, 0, 0

Before Instruction
    REG   =   1110 0110
    C   =   0
After Instruction
    REG   =   1110 0110
    W   =   1100 1100
    C   =   1

RLNCF	Rotate Left f (No Carry)
Syntax:	RLNCF f {,d {,a}}
Operands:	$0 \leq f \leq 255$ $d \in [0,1]$ $a \in [0,1]$
Operation:	$(f<n>) \rightarrow dest<n + 1>$, $(f<7>) \rightarrow dest<0>$
Status Affected:	N, Z

Encoding:

0100	01da	ffff	ffff

Description: The contents of register 'f' are rotated one bit to the left. If 'd' is '0', the result is placed in W. If 'd' is '1', the result is stored back in register 'f' (default).

If 'a' is '0', the Access Bank is selected. If 'a' is '1', the BSR is used to select the GPR bank (default).

If 'a' is '0' and the extended instruction set is enabled, this instruction operates in Indexed Literal Offset Addressing mode whenever f ≤ 95 (5Fh). See **Section 25.2.3 "Byte-Oriented and Bit-Oriented Instructions in Indexed Literal Offset Mode"** for details.

Words:	1
Cycles:	1

Q Cycle Activity:

Q1	Q2	Q3	Q4
Decode	Read register 'f'	Process Data	Write to destination

Example: RLNCF   REG, 1, 0

Before Instruction
REG   =   1010 1011
After Instruction
REG   =   0101 0111

RRCF	Rotate Right f through Carry
Syntax:	RRCF f {,d {,a}}
Operands:	$0 \leq f \leq 255$ $d \in [0,1]$ $a \in [0,1]$
Operation:	$(f<n>) \rightarrow dest<n - 1>$, $(f<0>) \rightarrow C$, $(C) \rightarrow dest<7>$
Status Affected:	C, N, Z

Encoding:

0011	00da	ffff	ffff

Description: The contents of register 'f' are rotated one bit to the right through the Carry flag. If 'd' is '0', the result is placed in W. If 'd' is '1', the result is placed back in register 'f' (default).

If 'a' is '0', the Access Bank is selected. If 'a' is '1', the BSR is used to select the GPR bank (default).

If 'a' is '0' and the extended instruction set is enabled, this instruction operates in Indexed Literal Offset Addressing mode whenever f ≤ 95 (5Fh). See **Section 25.2.3 "Byte-Oriented and Bit-Oriented Instructions in Indexed Literal Offset Mode"** for details.

Words:	1
Cycles:	1

Q Cycle Activity:

Q1	Q2	Q3	Q4
Decode	Read register 'f'	Process Data	Write to destination

Example: RRCF   REG, 0, 0

Before Instruction
REG   =   1110 0110
C   =   0
After Instruction
REG   =   1110 0110
W   =   0111 0011
C   =   0

RRNCF	Rotate Right f (No Carry)
Syntax:	RRNCF f {,d {,a}}
Operands:	$0 \leq f \leq 255$ $d \in [0,1]$ $a \in [0,1]$
Operation:	$(f<n>) \rightarrow dest<n-1>,$ $(f<0>) \rightarrow dest<7>$
Status Affected:	N, Z
Encoding:	0100	00da	ffff	ffff
Description:	The contents of register 'f' are rotated one bit to the right. If 'd' is '0', the result is placed in W. If 'd' is '1', the result is placed back in register 'f' (default). If 'a' is '0', the Access Bank will be selected, overriding the BSR value. If 'a' is '1', then the bank will be selected as per the BSR value (default). If 'a' is '0' and the extended instruction set is enabled, this instruction operates in Indexed Literal Offset Addressing mode whenever $f \leq 95$ (5Fh). See **Section 25.2.3 "Byte-Oriented and Bit-Oriented Instructions in Indexed Literal Offset Mode"** for details.

Words:	1
Cycles:	1

Q Cycle Activity:

Q1	Q2	Q3	Q4
Decode	Read register 'f'	Process Data	Write to destination

Example 1:   RRNCF   REG, 1, 0

Before Instruction
REG   =   1101 0111
After Instruction
REG   =   1110 1011

Example 2:   RRNCF   REG, 0, 0

Before Instruction
W   =   ?
REG   =   1101 0111
After Instruction
W   =   1110 1011
REG   =   1101 0111

SETF	Set f
Syntax:	SETF f {,a}
Operands:	$0 \leq f \leq 255$ $a \in [0,1]$
Operation:	$FFh \rightarrow f$
Status Affected:	None
Encoding:	0110	100a	ffff	ffff
Description:	The contents of the specified register are set to FFh. If 'a' is '0', the Access Bank is selected. If 'a' is '1', the BSR is used to select the GPR bank (default). If 'a' is '0' and the extended instruction set is enabled, this instruction operates in Indexed Literal Offset Addressing mode whenever $f \leq 95$ (5Fh). See **Section 25.2.3 "Byte-Oriented and Bit-Oriented Instructions in Indexed Literal Offset Mode"** for details.
Words:	1
Cycles:	1

Q Cycle Activity:

Q1	Q2	Q3	Q4
Decode	Read register 'f'	Process Data	Write register 'f'

Example: SETF   REG, 1

Before Instruction
REG   =   5Ah
After Instruction
REG   =   FFh

APPENDIX H: DATA SHEETS   **791**

SLEEP	Enter Sleep mode
Syntax:	SLEEP
Operands:	None
Operation:	00h → WDT, 0 → WDT postscaler, 1 → $\overline{TO}$, 0 → $\overline{PD}$
Status Affected:	$\overline{TO}$, $\overline{PD}$

Encoding:

0000	0000	0000	0011

Description:	The Power-Down status bit ($\overline{PD}$) is cleared. The Time-out status bit ($\overline{TO}$) is set. Watchdog Timer and its postscaler are cleared. The processor is put into Sleep mode with the oscillator stopped.
Words:	1
Cycles:	1

Q Cycle Activity:

Q1	Q2	Q3	Q4
Decode	No operation	Process Data	Go to Sleep

Example: SLEEP

Before Instruction
$\overline{TO}$ = ?
$\overline{PD}$ = ?

After Instruction
$\overline{TO}$ = 1 †
$\overline{PD}$ = 0

† If WDT causes wake-up, this bit is cleared.

SUBFWB	Subtract f from W with Borrow
Syntax:	SUBFWB f {,d {,a}}
Operands:	$0 \le f \le 255$ $d \in [0,1]$ $a \in [0,1]$
Operation:	$(W) - (f) - (\overline{C}) \to$ dest
Status Affected:	N, OV, C, DC, Z

Encoding:

0101	01da	ffff	ffff

Description:	Subtract register 'f' and Carry flag (borrow) from W (2's complement method). If 'd' is '0', the result is stored in W. If 'd' is '1', the result is stored in register 'f' (default). If 'a' is '0', the Access Bank is selected. If 'a' is '1', the BSR is used to select the GPR bank (default). If 'a' is '0' and the extended instruction set is enabled, this instruction operates in Indexed Literal Offset Addressing mode whenever $f \le 95$ (5Fh). See **Section 25.2.3 "Byte-Oriented and Bit-Oriented Instructions in Indexed Literal Offset Mode"** for details.
Words:	1
Cycles:	1

Q Cycle Activity:

Q1	Q2	Q3	Q4
Decode	Read register 'f'	Process Data	Write to destination

Example 1: SUBFWB REG, 1, 0

Before Instruction
REG = 3
W = 2
C = 1

After Instruction
REG = FF
W = 2
C = 0
Z = 0
N = 1 ; result is negative

Example 2: SUBFWB REG, 0, 0

Before Instruction
REG = 2
W = 5
C = 1

After Instruction
REG = 2
W = 3
C = 1
Z = 0
N = 0 ; result is positive

Example 3: SUBFWB REG, 1, 0

Before Instruction
REG = 1
W = 2
C = 0

After Instruction
REG = 0
W = 2
C = 1
Z = 1 ; result is zero
N = 0

Preliminary

© 2004 Microchip Technology Inc.

SUBLW	Subtract W from Literal
Syntax:	SUBLW k
Operands:	$0 \le k \le 255$
Operation:	$k - (W) \to W$
Status Affected:	N, OV, C, DC, Z

Encoding:

0000	1000	kkkk	kkkk

Description:	W is subtracted from the eight-bit literal 'k'. The result is placed in W.
Words:	1
Cycles:	1

Q Cycle Activity:

Q1	Q2	Q3	Q4
Decode	Read literal 'k'	Process Data	Write to W

Example 1: SUBLW 02h

Before Instruction
```
W   =   01h
C   =   ?
```
After Instruction
```
W   =   01h
C   =   1     ; result is positive
Z   =   0
N   =   0
```

Example 2: SUBLW 02h

Before Instruction
```
W   =   02h
C   =   ?
```
After Instruction
```
W   =   00h
C   =   1     ; result is zero
Z   =   1
N   =   0
```

Example 3: SUBLW 02h

Before Instruction
```
W   =   03h
C   =   ?
```
After Instruction
```
W   =   FFh; (2's complement)
C   =   0   ; result is negative
Z   =   0
N   =   1
```

SUBWF	Subtract W from f
Syntax:	SUBWF f {,d {,a}}
Operands:	$0 \le f \le 255$ $d \in [0,1]$ $a \in [0,1]$
Operation:	$(f) - (W) \to dest$
Status Affected:	N, OV, C, DC, Z

Encoding:

0101	11da	ffff	ffff

Description:	Subtract W from register 'f' (2's complement method). If 'd' is '0', the result is stored in W. If 'd' is '1', the result is stored back in register 'f' (default). If 'a' is '0', the Access Bank is selected. If 'a' is '1', the BSR is used to select the GPR bank (default). If 'a' is '0' and the extended instruction set is enabled, this instruction operates in Indexed Literal Offset Addressing mode whenever $f \le 95$ (5Fh). See **Section 25.2.3 "Byte-Oriented and Bit-Oriented Instructions in Indexed Literal Offset Mode"** for details.
Words:	1
Cycles:	1

Q Cycle Activity:

Q1	Q2	Q3	Q4
Decode	Read register 'f'	Process Data	Write to destination

Example 1: SUBWF REG, 1, 0

Before Instruction
```
REG =   3
W   =   2
C   =   ?
```
After Instruction
```
REG =   1
W   =   2
C   =   1     ; result is positive
Z   =   0
N   =   0
```

Example 2: SUBWF REG, 0, 0

Before Instruction
```
REG =   2
W   =   2
C   =   ?
```
After Instruction
```
REG =   2
W   =   0
C   =   1     ; result is zero
Z   =   1
N   =   0
```

Example 3: SUBWF REG, 1, 0

Before Instruction
```
REG =   1
W   =   2
C   =   ?
```
After Instruction
```
REG =   FFh  ;(2's complement)
W   =   2
C   =   0     ; result is negative
Z   =   0
N   =   1
```

SUBWFB	Subtract W from f with Borrow
Syntax:	SUBWFB f {,d {,a}}
Operands:	0 ≤ f ≤ 255 d ∈ [0,1] a ∈ [0,1]
Operation:	(f) − (W) − ($\overline{C}$) → dest
Status Affected:	N, OV, C, DC, Z
Encoding:	0101 | 10da | ffff | ffff
Description:	Subtract W and the Carry flag (borrow) from register 'f' (2's complement method). If 'd' is '0', the result is stored in W. If 'd' is '1', the result is stored back in register 'f' (default). If 'a' is '0', the Access Bank is selected. If 'a' is '1', the BSR is used to select the GPR bank (default). If 'a' is '0' and the extended instruction set is enabled, this instruction operates in Indexed Literal Offset Addressing mode whenever f ≤ 95 (5Fh). See **Section 25.2.3 "Byte-Oriented and Bit-Oriented Instructions in Indexed Literal Offset Mode"** for details.
Words:	1
Cycles:	1

Q Cycle Activity:

Q1	Q2	Q3	Q4
Decode	Read register 'f'	Process Data	Write to destination

Example 1:   SUBWFB   REG, 1, 0

Before Instruction
```
REG   =   19h    (0001 1001)
W     =   0Dh    (0000 1101)
C     =   1
```
After Instruction
```
REG   =   0Ch    (0000 1011)
W     =   0Dh    (0000 1101)
C     =   1
Z     =   0
N     =   0      ; result is positive
```

Example 2:   SUBWFB   REG, 0, 0

Before Instruction
```
REG   =   1Bh    (0001 1011)
W     =   1Ah    (0001 1010)
C     =   0
```
After Instruction
```
REG   =   1Bh    (0001 1011)
W     =   00h
C     =   1
Z     =   1      ; result is zero
N     =   0
```

Example 3:   SUBWFB   REG, 1, 0

Before Instruction
```
REG   =   03h    (0000 0011)
W     =   0Eh    (0000 1101)
C     =   1
```
After Instruction
```
REG   =   F5h    (1111 0100)
                 ; [2's comp]
W     =   0Eh    (0000 1101)
C     =   0
Z     =   0
N     =   1      ; result is negative
```

SWAPF	Swap f
Syntax:	SWAPF f {,d {,a}}
Operands:	0 ≤ f ≤ 255 d ∈ [0,1] a ∈ [0,1]
Operation:	(f<3:0>) → dest<7:4>, (f<7:4>) → dest<3:0>
Status Affected:	None
Encoding:	0011 | 10da | ffff | ffff
Description:	The upper and lower nibbles of register 'f' are exchanged. If 'd' is '0', the result is placed in W. If 'd' is '1', the result is placed in register 'f' (default). If 'a' is '0', the Access Bank is selected. If 'a' is '1', the BSR is used to select the GPR bank (default). If 'a' is '0' and the extended instruction set is enabled, this instruction operates in Indexed Literal Offset Addressing mode whenever f ≤ 95 (5Fh). See **Section 25.2.3 "Byte-Oriented and Bit-Oriented Instructions in Indexed Literal Offset Mode"** for details.
Words:	1
Cycles:	1

Q Cycle Activity:

Q1	Q2	Q3	Q4
Decode	Read register 'f'	Process Data	Write to destination

Example: SWAPF   REG, 1, 0

Before Instruction
```
REG   =   53h
```
After Instruction
```
REG   =   35h
```

TBLRD	Table Read
Syntax:	TBLRD (*; *+; *-; +*)
Operands:	None
Operation:	if TBLRD *,

(Prog Mem (TBLPTR)) → TABLAT;
TBLPTR – No Change;
if TBLRD *+,
(Prog Mem (TBLPTR)) → TABLAT;
(TBLPTR) + 1 → TBLPTR;
if TBLRD *-,
(Prog Mem (TBLPTR)) → TABLAT;
(TBLPTR) – 1 → TBLPTR;
if TBLRD +*,
(TBLPTR) + 1 → TBLPTR;
(Prog Mem (TBLPTR)) → TABLAT;

Status Affected: None

Encoding:

0000	0000	0000	10nn
			nn=0 *
			=1 *+
			=2 *-
			=3 +*

Description:
This instruction is used to read the contents of Program Memory (P.M.). To address the program memory, a pointer, called Table Pointer (TBLPTR), is used.

The TBLPTR (a 21-bit pointer) points to each byte in the program memory. TBLPTR has a 2-Mbyte address range.

TBLPTR[0] = 0: Least Significant Byte of Program Memory Word
TBLPTR[0] = 1: Most Significant Byte of Program Memory Word

The TBLRD instruction can modify the value of TBLPTR as follows:

- no change
- post-increment
- post-decrement
- pre-increment

Words: 1

Cycles: 2

Q Cycle Activity:

Q1	Q2	Q3	Q4
Decode	No operation	No operation	No operation
No operation	No operation (Read Program Memory)	No operation	No operation (Write TABLAT)

TBLRD	Table Read (Continued)

Example 1:   TBLRD   *+ ;

Before Instruction
TABLAT	=	55h
TBLPTR	=	00A356h
MEMORY(00A356h)	=	34h

After Instruction
| TABLAT | = | 34h |
| TBLPTR | = | 00A357h |

Example 2:   TBLRD   +* ;

Before Instruction
TABLAT	=	0AAh
TBLPTR	=	01A357h
MEMORY(01A357h)	=	12h
MEMORY(01A358h)	=	34h

After Instruction
| TABLAT | = | 34h |
| TBLPTR | = | 01A358h |

TBLWT	Table Write
Syntax:	TBLWT (*; *+; *-; +*)
Operands:	None
Operation:	if TBLWT*, (TABLAT) → Holding Register; TBLPTR – No Change; if TBLWT*+, (TABLAT) → Holding Register; (TBLPTR) + 1 → TBLPTR; if TBLWT*-, (TABLAT) → Holding Register; (TBLPTR) – 1 → TBLPTR; if TBLWT+*, (TBLPTR) + 1 → TBLPTR; (TABLAT) → Holding Register;
Status Affected:	None

Encoding:

0000	0000	0000	11nn
			nn=0 *
			=1 *+
			=2 *-
			=3 +*

Description:

This instruction uses the 3 LSBs of the TBLPTR to determine which of the 8 holding registers the TABLAT is written to. The holding registers are used to program the contents of Program Memory (P.M.). (Refer to **Section 6.0 "Flash Program Memory"** for additional details on programming Flash memory.)

The TBLPTR (a 21-bit pointer) points to each byte in the program memory. TBLPTR has a 2-MBtye address range. The LSb of the TBLPTR selects which byte of the program memory location to access.

TBLPTR[0] = 0: Least Significant Byte of Program Memory Word

TBLPTR[0] = 1: Most Significant Byte of Program Memory Word

The TBLWT instruction can modify the value of TBLPTR as follows:

- no change
- post-increment
- post-decrement
- pre-increment

Words: 1

Cycles: 2

Q Cycle Activity:

Q1	Q2	Q3	Q4
Decode	No operation	No operation	No operation
No operation	No operation (Read TABLAT)	No operation	No operation (Write to Holding Register)

TBLWT	Table Write (Continued)

Example 1:   TBLWT *+;

Before Instruction

TABLAT	=	55h
TBLPTR	=	00A356h
HOLDING REGISTER (00A356h)	=	FFh

After Instructions (table write completion)

TABLAT	=	55h
TBLPTR	=	00A357h
HOLDING REGISTER (00A356h)	=	55h

Example 2:   TBLWT +*;

Before Instruction

TABLAT	=	34h
TBLPTR	=	01389Ah
HOLDING REGISTER (01389Ah)	=	FFh
HOLDING REGISTER (01389Bh)	=	FFh

After Instruction (table write completion)

TABLAT	=	34h
TBLPTR	=	01389Bh
HOLDING REGISTER (01389Ah)	=	FFh
HOLDING REGISTER (01389Bh)	=	34h

TSTFSZ	Test f, Skip if 0

Syntax:	TSTFSZ f {,a}
Operands:	$0 \le f \le 255$ $a \in [0,1]$
Operation:	skip if f = 0
Status Affected:	None

Encoding:

0110	011a	ffff	ffff

Description: If 'f' = 0, the next instruction fetched during the current instruction execution is discarded and a NOP is executed, making this a two-cycle instruction.

If 'a' is '0', the Access Bank is selected. If 'a' is '1', the BSR is used to select the GPR bank (default).

If 'a' is '0' and the extended instruction set is enabled, this instruction operates in Indexed Literal Offset Addressing mode whenever $f \le 95$ (5Fh). See **Section 25.2.3 "Byte-Oriented and Bit-Oriented Instructions in Indexed Literal Offset Mode"** for details.

Words: 1

Cycles: 1(2)

Note: 3 cycles if skip and followed by a 2-word instruction.

Q Cycle Activity:

Q1	Q2	Q3	Q4
Decode	Read register 'f'	Process Data	No operation

If skip:

Q1	Q2	Q3	Q4
No operation	No operation	No operation	No operation

If skip and followed by 2-word instruction:

Q1	Q2	Q3	Q4
No operation	No operation	No operation	No operation
No operation	No operation	No operation	No operation

Example:

```
HERE     TSTFSZ  CNT, 1
NZERO    :
ZERO     :
```

Before Instruction
```
PC          =    Address (HERE)
```
After Instruction
```
If CNT      =    00h,
PC          =    Address (ZERO)
If CNT      ≠    00h,
PC          =    Address (NZERO)
```

XORLW	Exclusive OR Literal with W

Syntax:	XORLW k
Operands:	$0 \le k \le 255$
Operation:	(W) .XOR. k $\rightarrow$ W
Status Affected:	N, Z

Encoding:

0000	1010	kkkk	kkkk

Description: The contents of W are XORed with the 8-bit literal 'k'. The result is placed in W.

Words: 1

Cycles: 1

Q Cycle Activity:

Q1	Q2	Q3	Q4
Decode	Read literal 'k'	Process Data	Write to W

Example: XORLW 0AFh

Before Instruction
```
W    =    B5h
```
After Instruction
```
W    =    1Ah
```

XORWF	Exclusive OR W with f

Syntax:	XORWF f {,d {,a}}
Operands:	0 ≤ f ≤ 255 d ∈ [0,1] a ∈ [0,1]
Operation:	(W) .XOR. (f) → dest
Status Affected:	N, Z
Encoding:	0001	10da	ffff	ffff
Description:	Exclusive OR the contents of W with register 'f'. If 'd' is '0', the result is stored in W. If 'd' is '1', the result is stored back in the register 'f' (default). If 'a' is '0', the Access Bank is selected. If 'a' is '1', the BSR is used to select the GPR bank (default). If 'a' is '0' and the extended instruction set is enabled, this instruction operates in Indexed Literal Offset Addressing mode whenever f ≤ 95 (5Fh). See **Section 25.2.3 "Byte-Oriented and Bit-Oriented Instructions in Indexed Literal Offset Mode"** for details.
Words:	1
Cycles:	1

Q Cycle Activity:

Q1	Q2	Q3	Q4
Decode	Read register 'f'	Process Data	Write to destination

Example: XORWF   REG, 1, 0

Before Instruction
REG   =   AFh
W   =   B5h
After Instruction
REG   =   1Ah
W   =   B5h

25.2 Extended Instruction Set

In addition to the standard 75 instructions of the PIC18 instruction set, PIC18F2480/2580/4480/4580 devices also provide an optional extension to the core CPU functionality. The added features include eight additional instructions that augment indirect and indexed addressing operations and the implementation of Indexed Literal Offset Addressing mode for many of the standard PIC18 instructions.

The additional features are disabled by default. To enable them, users must set the XINST configuration bit.

The instructions in the extended set can all be classified as literal operations, which either manipulate the File Select Registers or use them for indexed addressing. Two of the instructions, ADDFSR and SUBFSR, each have an additional special instantiation for using FSR2. These versions (ADDULNK and SUBULNK) allow for automatic return after execution.

The extended instructions are specifically implemented to optimize re-entrant program code (that is, code that is recursive or that uses a software stack) written in high-level languages, particularly C. Among other things, they allow users working in high-level languages to perform certain operations on data structures more efficiently. These include:

- dynamic allocation and de-allocation of software stack space when entering and leaving subroutines
- function pointer invocation
- software Stack Pointer manipulation
- manipulation of variables located in a software stack

A summary of the instructions in the extended instruction set is provided in Table 25-3. Detailed descriptions are provided in **Section 25.2.2 "Extended Instruction Set"**. The opcode field descriptions in Table 25-1 apply to both the standard and extended PIC18 instruction sets.

Note: The instruction set extension and the Indexed Literal Offset Addressing mode were designed for optimizing applications written in C; the user may likely never use these instructions directly in assembler. The syntax for these commands is provided as a reference for users who may be reviewing code that has been generated by a compiler.

25.2.1 EXTENDED INSTRUCTION SYNTAX

Most of the extended instructions use indexed arguments, using one of the File Select Registers and some offset to specify a source or destination register. When an argument for an instruction serves as part of indexed addressing, it is enclosed in square brackets ("[ ]"). This is done to indicate that the argument is used as an index or offset. MPASM™ Assembler will flag an error if it determines that an index or offset value is not bracketed.

When the extended instruction set is enabled, brackets are also used to indicate index arguments in byte-oriented and bit-oriented instructions. This is in addition to other changes in their syntax. For more details, see **Section 25.2.3.1 "Extended Instruction Syntax with Standard PIC18 Commands"**.

Note: In the past, square brackets have been used to denote optional arguments in the PIC18 and earlier instruction sets. In this text and going forward, optional arguments are denoted by braces ("{ }").

TABLE 25-3: EXTENSIONS TO THE PIC18 INSTRUCTION SET

Mnemonic, Operands		Description	Cycles	16-Bit Instruction Word				Status Affected
				MSb			LSb	
ADDFSR	f, k	Add literal to FSR	1	1110	1000	ffkk	kkkk	None
ADDULNK	k	Add literal to FSR2 and return	2	1110	1000	11kk	kkkk	None
CALLW		Call subroutine using WREG	2	0000	0000	0001	0100	None
MOVSF	z_s, f_d	Move z_s (source) to 1st word	2	1110	1011	0zzz	zzzz	None
		f_d (destination) 2nd word		1111	ffff	ffff	ffff	
MOVSS	z_s, z_d	Move z_s (source) to 1st word	2	1110	1011	1zzz	zzzz	None
		z_d (destination) 2nd word		1111	xxxx	xzzz	zzzz	
PUSHL	k	Store literal at FSR2, decrement FSR2	1	1110	1010	kkkk	kkkk	None
SUBFSR	f, k	Subtract literal from FSR	1	1110	1001	ffkk	kkkk	None
SUBULNK	k	Subtract literal from FSR2 and return	2	1110	1001	11kk	kkkk	None

25.2.2 EXTENDED INSTRUCTION SET

ADDFSR	Add Literal to FSR
Syntax:	ADDFSR f, k
Operands:	$0 \leq k \leq 63$ $f \in [0, 1, 2]$
Operation:	FSR(f) + k → FSR(f)
Status Affected:	None
Encoding:	1110 1000 ffkk kkkk
Description:	The 6-bit literal 'k' is added to the contents of the FSR specified by 'f'.
Words:	1
Cycles:	1

Q Cycle Activity:

Q1	Q2	Q3	Q4
Decode	Read literal 'k'	Process Data	Write to FSR

Example: ADDFSR 2, 23h

Before Instruction
    FSR2   =   03FFh
After Instruction
    FSR2   =   0422h

ADDULNK	Add Literal to FSR2 and Return
Syntax:	ADDULNK k
Operands:	$0 \leq k \leq 63$
Operation:	FSR2 + k → FSR2, PC = (TOS)
Status Affected:	None
Encoding:	1110 1000 11kk kkkk
Description:	The 6-bit literal 'k' is added to the contents of FSR2. A RETURN is then executed by loading the PC with the TOS. The instruction takes two cycles to execute; a NOP is performed during the second cycle. This may be thought of as a special case of the ADDFSR instruction, where f = 3 (binary '11'); it operates only on FSR2.
Words:	1
Cycles:	2

Q Cycle Activity:

Q1	Q2	Q3	Q4
Decode	Read literal 'k'	Process Data	Write to FSR
No Operation	No Operation	No Operation	No Operation

Example: ADDULNK 23h

Before Instruction
    FSR2   =   03FFh
    PC   =   0100h
    TOS   =   02AFh
After Instruction
    FSR2   =   0422h
    PC   =   02AFh
    TOS   =   TOS – 1

Note: All PIC18 instructions may take an optional label argument preceding the instruction mnemonic for use in symbolic addressing. If a label is used, the instruction syntax then becomes: {label} instruction argument(s).

CALLW	Subroutine Call Using WREG
Syntax:	CALLW
Operands:	None
Operation:	(PC + 2) → TOS, (W) → PCL, (PCLATH) → PCH, (PCLATU) → PCU
Status Affected:	None

Encoding:

0000	0000	0001	0100

Description	First, the return address (PC + 2) is pushed onto the return stack. Next, the contents of W are written to PCL; the existing value is discarded. Then, the contents of PCLATH and PCLATU are latched into PCH and PCU, respectively. The second cycle is executed as a NOP instruction while the new next instruction is fetched. Unlike CALL, there is no option to update W, Status or BSR.
Words:	1
Cycles:	2

Q Cycle Activity:

Q1	Q2	Q3	Q4
Decode	Read WREG	Push PC to stack	No operation
No operation	No operation	No operation	No operation

Example: HERE   CALLW

Before Instruction
```
PC       =   address (HERE)
PCLATH   =   10h
PCLATU   =   00h
W        =   06h
```
After Instruction
```
PC       =   001006h
TOS      =   address (HERE + 2)
PCLATH   =   10h
PCLATU   =   00h
W        =   06h
```

MOVSF	Move Indexed to f
Syntax:	MOVSF [z_s], f_d
Operands:	$0 \le z_s \le 127$ $0 \le f_d \le 4095$
Operation:	((FSR2) + z_s) → f_d
Status Affected:	None

Encoding:

1st word (source)	1110	1011	0zzz	zzzz$_s$
2nd word (destin.)	1111	ffff	ffff	ffff$_d$

Description:	The contents of the source register are moved to destination register 'f_d'. The actual address of the source register is determined by adding the 7-bit literal offset 'z_s' in the first word to the value of FSR2. The address of the destination register is specified by the 12-bit literal 'f_d' in the second word. Both addresses can be anywhere in the 4096-byte data space (000h to FFFh). The MOVSF instruction cannot use the PCL, TOSU, TOSH or TOSL as the destination register. If the resultant source address points to an indirect addressing register, the value returned will be 00h.
Words:	2
Cycles:	2

Q Cycle Activity:

Q1	Q2	Q3	Q4
Decode	Determine source addr	Determine source addr	Read source reg
Decode	No operation No dummy read	No operation	Write register 'f' (dest)

Example: MOVSF [05h], REG2

Before Instruction
```
FSR2          =   80h
Contents
of 85h        =   33h
REG2          =   11h
```
After Instruction
```
FSR2          =   80h
Contents
of 85h        =   33h
REG2          =   33h
```

MOVSS	Move Indexed to Indexed
Syntax:	MOVSS [z_s], [z_d]
Operands:	$0 \le z_s \le 127$ $0 \le z_d \le 127$
Operation:	$((FSR2) + z_s) \rightarrow ((FSR2) + z_d)$
Status Affected:	None

Encoding:

1st word (source)	1110	1011	1zzz	zzzz$_s$
2nd word (dest.)	1111	xxxx	xzzz	zzzz$_d$

Description

The contents of the source register are moved to the destination register. The addresses of the source and destination registers are determined by adding the 7-bit literal offsets 'z_s' or 'z_d', respectively, to the value of FSR2. Both registers can be located anywhere in the 4096-byte data memory space (000h to FFFh).

The MOVSS instruction cannot use the PCL, TOSU, TOSH or TOSL as the destination register.

If the resultant source address points to an indirect addressing register, the value returned will be 00h. If the resultant destination address points to an indirect addressing register, the instruction will execute as a NOP.

Words:	2
Cycles:	2

Q Cycle Activity:

Q1	Q2	Q3	Q4
Decode	Determine source addr	Determine source addr	Read source reg
Decode	Determine dest addr	Determine dest addr	Write to dest reg

Example: MOVSS [05h], [06h]

Before Instruction

FSR2	=	80h
Contents of 85h	=	33h
Contents of 86h	=	11h

After Instruction

FSR2	=	80h
Contents of 85h	=	33h
Contents of 86h	=	33h

PUSHL	Store Literal at FSR2, Decrement FSR2
Syntax:	PUSHL k
Operands:	$0 \le k \le 255$
Operation:	$k \rightarrow (FSR2)$, $FSR2 - 1 \rightarrow FSR2$
Status Affected:	None

Encoding:

1111	1010	kkkk	kkkk

Description:

The 8-bit literal 'k' is written to the data memory address specified by FSR2. FSR2 is decremented by 1 after the operation.

This instruction allows users to push values onto a software stack.

Words:	1
Cycles:	1

Q Cycle Activity:

Q1	Q2	Q3	Q4
Decode	Read 'k'	Process data	Write to destination

Example: PUSHL 08h

Before Instruction

FSR2H:FSR2L	=	01ECh
Memory (01ECh)	=	00h

After Instruction

FSR2H:FSR2L	=	01EBh
Memory (01ECh)	=	08h

Preliminary

802

SUBFSR	**Subtract Literal from FSR**
Syntax:	SUBFSR f, k
Operands:	$0 \leq k \leq 63$ $f \in [0, 1, 2]$
Operation:	FSRf – k → FSRf
Status Affected:	None

Encoding:

1110	1001	ffkk	kkkk

Description:	The 6-bit literal 'k' is subtracted from the contents of the FSR specified by 'f'.
Words:	1
Cycles:	1

Q Cycle Activity:

Q1	Q2	Q3	Q4
Decode	Read register 'f'	Process Data	Write to destination

Example: SUBFSR 2, 23h

    Before Instruction
        FSR2   =   03FFh
    After Instruction
        FSR2   =   03DCh

SUBULNK	**Subtract Literal from FSR2 and Return**
Syntax:	SUBULNK k
Operands:	$0 \leq k \leq 63$
Operation:	FSR2 – k → FSR2 (TOS) → PC
Status Affected:	None

Encoding:

1110	1001	11kk	kkkk

Description:	The 6-bit literal 'k' is subtracted from the contents of the FSR2. A RETURN is then executed by loading the PC with the TOS. The instruction takes two cycles to execute; a NOP is performed during the second cycle. This may be thought of as a special case of the SUBFSR instruction, where f = 3 (binary '11'); it operates only on FSR2.
Words:	1
Cycles:	2

Q Cycle Activity:

Q1	Q2	Q3	Q4
Decode	Read register 'f'	Process Data	Write to destination
No Operation	No Operation	No Operation	No Operation

Example: SUBULNK 23h

    Before Instruction
        FSR2   =   03FFh
        PC   =   0100h
    After Instruction
        FSR2   =   03DCh
        PC   =   (TOS)

25.2.3 BYTE-ORIENTED AND BIT-ORIENTED INSTRUCTIONS IN INDEXED LITERAL OFFSET MODE

> **Note:** Enabling the PIC18 instruction set extension may cause legacy applications to behave erratically or fail entirely.

In addition to eight new commands in the extended set, enabling the extended instruction set also enables Indexed Literal Offset Addressing mode (**Section 5.6.1 "Indexed Addressing with Literal Offset"**). This has a significant impact on the way that many commands of the standard PIC18 instruction set are interpreted.

When the extended set is disabled, addresses embedded in opcodes are treated as literal memory locations: either as a location in the Access Bank (a = 0), or in a GPR bank designated by the BSR (a = 1). When the extended instruction set is enabled and a = 0, however, a file register argument of 5Fh or less is interpreted as an offset from the pointer value in FSR2 and not as a literal address. For practical purposes, this means that all instructions that use the Access RAM bit as an argument – that is, all byte-oriented and bit-oriented instructions, or almost half of the core PIC18 instructions – may behave differently when the extended instruction set is enabled.

When the content of FSR2 is 00h, the boundaries of the Access RAM are essentially remapped to their original values. This may be useful in creating backward compatible code. If this technique is used, it may be necessary to save the value of FSR2 and restore it when moving back and forth between 'C' and assembly routines in order to preserve the Stack Pointer. Users must also keep in mind the syntax requirements of the extended instruction set (see **Section 25.2.3.1 "Extended Instruction Syntax with Standard PIC18 Commands"**).

Although the Indexed Literal Offset Addressing mode can be very useful for dynamic stack and pointer manipulation, it can also be very annoying if a simple arithmetic operation is carried out on the wrong register. Users who are accustomed to the PIC18 programming must keep in mind that, when the extended instruction set is enabled, register addresses of 5Fh or less are used for Indexed Literal Offset Addressing.

Representative examples of typical byte-oriented and bit-oriented instructions in the Indexed Literal Offset Addressing mode are provided on the following page to show how execution is affected. The operand conditions shown in the examples are applicable to all instructions of these types.

25.2.3.1 Extended Instruction Syntax with Standard PIC18 Commands

When the extended instruction set is enabled, the file register argument, 'f', in the standard byte-oriented and bit-oriented commands is replaced with the literal offset value, 'k'. As already noted, this occurs only when 'f' is less than or equal to 5Fh. When an offset value is used, it must be indicated by square brackets ("[ ]"). As with the extended instructions, the use of brackets indicates to the compiler that the value is to be interpreted as an index or an offset. Omitting the brackets, or using a value greater than 5Fh within brackets, will generate an error in the MPASM™ Assembler.

If the index argument is properly bracketed for Indexed Literal Offset Addressing, the Access RAM argument is never specified; it will automatically be assumed to be '0'. This is in contrast to standard operation (extended instruction set disabled) when 'a' is set on the basis of the target address. Declaring the Access RAM bit in this mode will also generate an error in the MPASM Assembler.

The destination argument, 'd', functions as before.

In the latest versions of the MPASM assembler, language support for the extended instruction set must be explicitly invoked. This is done with either the command line option, /y, or the PE directive in the source listing.

25.2.4 CONSIDERATIONS WHEN ENABLING THE EXTENDED INSTRUCTION SET

It is important to note that the extensions to the instruction set may not be beneficial to all users. In particular, users who are not writing code that uses a software stack may not benefit from using the extensions to the instruction set.

Additionally, the Indexed Literal Offset Addressing mode may create issues with legacy applications written to the PIC18 assembler. This is because instructions in the legacy code may attempt to address registers in the Access Bank below 5Fh. Since these addresses are interpreted as literal offsets to FSR2 when the instruction set extension is enabled, the application may read or write to the wrong data addresses.

When porting an application to the PIC18F2480/2580/4480/4580, it is very important to consider the type of code. A large, re-entrant application that is written in 'C' and would benefit from efficient compilation will do well when using the instruction set extensions. Legacy applications that heavily use the Access Bank will most likely not benefit from using the extended instruction set.

ADDWF	ADD W to Indexed (Indexed Literal Offset mode)
Syntax:	ADDWF [k] {,d}
Operands:	$0 \leq k \leq 95$ $d \in [0,1]$ $a = 0$
Operation:	(W) + ((FSR2) + k) → dest
Status Affected:	N, OV, C, DC, Z
Encoding:	0010 01d0 kkkk kkkk
Description:	The contents of W are added to the contents of the register indicated by FSR2, offset by the value 'k'. If 'd' is '0', the result is stored in W. If 'd' is '1', the result is stored back in register 'f' (default).
Words:	1
Cycles:	1

Q Cycle Activity:

Q1	Q2	Q3	Q4
Decode	Read 'k'	Process Data	Write to destination

Example: ADDWF [OFST] , 0

Before Instruction
```
W              =   17h
OFST           =   2Ch
FSR2           =   0A00h
Contents
of 0A2Ch       =   20h
```
After Instruction
```
W              =   37h
Contents
of 0A2Ch       =   20h
```

BSF	Bit Set Indexed (Indexed Literal Offset mode)
Syntax:	BSF [k], b
Operands:	$0 \leq f \leq 95$ $0 \leq b \leq 7$ $a = 0$
Operation:	1 → ((FSR2 + k))<b>
Status Affected:	None
Encoding:	1000 bbb0 kkkk kkkk
Description:	Bit 'b' of the register indicated by FSR2, offset by the value 'k', is set.
Words:	1
Cycles:	1

Q Cycle Activity:

Q1	Q2	Q3	Q4
Decode	Read register 'f'	Process Data	Write to destination

Example: BSF [FLAG_OFST], 7

Before Instruction
```
FLAG_OFST      =   0Ah
FSR2           =   0A00h
Contents
of 0A0Ah       =   55h
```
After Instruction
```
Contents
of 0A0Ah       =   D5h
```

SETF	Set Indexed (Indexed Literal Offset mode)
Syntax:	SETF [k]
Operands:	$0 \leq k \leq 95$
Operation:	FFh → ((FSR2) + k)
Status Affected:	None
Encoding:	0110 1000 kkkk kkkk
Description:	The contents of the register indicated by FSR2, offset by 'k', are set to FFh.
Words:	1
Cycles:	1

Q Cycle Activity:

Q1	Q2	Q3	Q4
Decode	Read 'k'	Process Data	Write register

Example: SETF [OFST]

Before Instruction
```
OFST           =   2Ch
FSR2           =   0A00h
Contents
of 0A2Ch       =   00h
```
After Instruction
```
Contents
of 0A2Ch       =   FFh
```

25.2.5 SPECIAL CONSIDERATIONS WITH MICROCHIP MPLAB® IDE TOOLS

The latest versions of Microchip's software tools have been designed to fully support the extended instruction set of the PIC18F2480/2580/4480/4580 family of devices. This includes the MPLAB C18 C compiler, MPASM assembly language and MPLAB Integrated Development Environment (IDE).

When selecting a target device for software development, MPLAB IDE will automatically set default configuration bits for that device. The default setting for the XINST configuration bit is '0', disabling the extended instruction set and Indexed Literal Offset Addressing mode. For proper execution of applications developed to take advantage of the extended instruction set, XINST must be set during programming.

To develop software for the extended instruction set, the user must enable support for the instructions and the Indexed Addressing mode in their language tool(s). Depending on the environment being used, this may be done in several ways:

- A menu option, or dialog box within the environment, that allows the user to configure the language tool and its settings for the project
- A command line option
- A directive in the source code

These options vary between different compilers, assemblers and development environments. Users are encouraged to review the documentation accompanying their development systems for the appropriate information.

INDEX

A

Accumulator.
　　See WREG
ADC
　ADCON0 register,　507
　ADCON1 register,　508
　ADFM bit and data formatting,　509
　block diagram,　506
　conversion time,　510
　features,　505
　interrupt programming,　513
　interrupt programming in C,　514
　polling programming in C,　513
　steps in polling programming,　511
ADC devices
　analog input, 504
　block diagram,　500
　connection,　500
　conversion signals,　504
　conversion time,　501
　data output, 502
　parallel vs. serial,　502
　reference voltage,　501
　resolution,　501
Addition in the PIC18,　156
Address bus,　14, 15
Addressing modes
　bit addressing,　214
　direct addressing mode,　195
　immediate addressing mode,　194
　register indirect addressing mode,　199
　　INDFx registers,　199, 202
　　LFSR instruction, 199
　　look-up table in RAM,　212
　　PLUSWx registers,　202
　　POSTDECx registers,　202
　　POSTINCx registers,　202
　　PREINCx registers,　202
　ROM addressing mode.
　　See Table processing
AND gate,　9
Arithmetic instructions
　ADDLW,　41, 156, 682

ADDLWC,　157
ADDWF,　49, 156, 683
ADDWFC,　684
DAW,　159, 696
DECF,　196, 698
DECFSNZ,　699
DECFSZ,　196, 699
INCF,　196, 701
INCFSNZ,　702
INCFSZ,　701
MULLW,　163, 706
MULWF,　707
NEGF,　707
SUBFWB,　162, 713
SUBLW,　161, 713
SUBWF,　714
SUBWFB,　162, 715
ASCII, 7, 8
ASCII numbers,　184
ASCII table,　752
ASCII to packed BCD conversion,　186
ASCII to packed BCD conversion in C, 272
asm file,　70, 71
Assembler directives
　DB (define byte),　205
　EXTERN,　241
　GLOBAL,　241
　INCLUDE, 237
　LIST,　313
　LOCAL,　235
　MACRO,　234
　NOEXPAND/EXPAND,　237
Assemblers,　754–755
Assembly language,　67
　assembling and linking,　70
　negative values,　350
　structure of, 68

B

Bank switching,　219
　BSR register,　219
　destination select bit, d ,　196, 222
　MOVFF instructions,　223
　RAM access bit, a,　219, 221
BCD number systems,　158, 159
　BCD addition and correction,　160